Lecture Notes in Computer Science 4117

Commenced Publication in 1973
Founding and Former Series Editors:
Gerhard Goos, Juris Hartmanis, and Jan van Leeuwen

Cynthia Dwork (Ed.)

Advances in Cryptology - CRYPTO 2006

26th Annual International Cryptology Conference
Santa Barbara, California, USA, August 20-24, 2006
Proceedings

 Springer

Volume Editor

Cynthia Dwork
Microsoft Research
1065 La Avenida, Mountain View, CA 94043, USA
E-mail: dwork@microsoft.com

Library of Congress Control Number: 2006930607

CR Subject Classification (1998): E.3, G.2.1, F.2.1-2, D.4.6, K.6.5, C.2, J.1

LNCS Sublibrary: SL 4 – Security and Cryptology

ISSN   0302-9743
ISBN-10   3-540-37432-9 Springer Berlin Heidelberg New York
ISBN-13   978-3-540-37432-9 Springer Berlin Heidelberg New York

Springer is a part of Springer Science+Business Media

springer.com

© Springer-Verlag Berlin Heidelberg 2006
Printed in Germany

Typesetting: Camera-ready by author, data conversion by Scientific Publishing Services, Chennai, India
Printed on acid-free paper   SPIN: 11818175   06/3142   5 4 3 2 1 0

Preface

These are the proceedings of Crypto 2006, the 26th Annual International Cryptology Conference. The conference was sponsored by the International Association of Cryptologic Research, in cooperation with the IEEE Computer Society Technical Committee on Security and Privacy, and the Computer Science Department of the University of California, Santa Barbara. The conference was held in Santa Barbara, California, August 20–24, 2006.

The conference received 220 submissions, out of which the Program Committee selected 34 for presentation. Submission and selection of papers was done using the IChair sofware, developed at the École Polytechnique Fédérale de Lausanne (EPFL) by Thomas Baignères and Matthieu Finiasz. Aided in part by comments from the committee and external reviewers, the authors of accepted papers had roughly six weeks in which to prepare final versions for these proceedings. These were not subject to editorial review.

The committee chose "On the Power of the Randomized Iterate," by Iftach Haitner, Danny Harnik, and Omer Reingold, to receive the Best Paper award.

The committee also invited Oded Regev and David Wagner to speak on topics of their choice. Their talks were entitled, respectively, "Lattice-Based Cryptography" and "Cryptographic Protocols for Electronic Voting."

We continued the tradition of a "Rump Session" of very brief presentations.

The cryptology community provides a collaborative and supportive environment for exciting research, and the success of previous Crypto conferences fosters enthusiasm for participation in subsequent ones. I am deeply grateful to all the authors who submitted papers, not only for their contribution to this conference but also for maintaining this tradition.

I thank Thomas Baignères and Matthieu Finiasz for kindly hosting the server – and for writing IChair in the first place. David Fuchs provided invaluable assistance in assembling the final papers into this volume. Josh Benaloh was everything one could possibly hope for in a General Chair. I thank him for his good judgement and gracious assistance at all times.

In a departure from recent tradition, submissions were not anonymous. I am grateful to Andy Clark and Kevin McCurley for their counsel regarding this course of action, and to the Program Committee for being open to change. I also warmly thank the members of the Program Committee for their energy, intelligence, wisdom, and the maturity with which they approached their task.

Finally, I thank Moni Naor, who for the past nineteen years has taught me cryptography.

June 2006   Cynthia Dwork
                                                                    Program Chair

CRYPTO 2006

August 20–24, 2006, Santa Barbara, California, USA

Sponsored by the

International Association for Cryptologic Research (IACR)

in cooperation with

IEEE Computer Society Technical Committee on Security and Privacy,

Computer Science Department, University of California, Santa Barbara

General Chair
Josh Benaloh, Microsoft, USA

Program Chair
Cynthia Dwork, Microsoft, USA

Program Committee

Boaz Barak .... Princeton University, USA
Eli Biham .... Technion, Israel
Ivan Damgård .... University of Aarhus, Denmark
Yuval Ishai .... Technion, Israel
Jonathan Katz .... University of Maryland, USA
Arjen Lenstra .... EPFL, Switzerland
Yehuda Lindell .... Bar-Ilan University, Israel
Tal Malkin .... Columbia University, USA
Mitsuru Matsui .... Mitsubishi Electric, Japan
Daniele Micciancio .... University of California, San Diego, USA
Moni Naor .... Weizmann Institute of Science, Israel
Phong Nguyen .... CNRS/École Normale Supérieure, France
Kobbi Nissim .... Ben-Gurion University, Israel
Bart Preneel .... Katholieke Universiteit Leuven, Belgium
Hovav Shacham .... Weizmann Institute of Science, Israel
Vitaly Shmatikov .... University of Texas, Austin, USA
Edlyn Teske .... University of Waterloo, Canada
Salil Vadhan .... Harvard University, USA
Yiqun Lisa Yin .... Independent Consultant, USA

Advisory Members

Victor Shoup (Crypto 2005 Program Chair) ...... New York University, USA
Alfred Menezes (Crypto 2007 Program Chair) University of Waterloo, Canada

External Reviewers

Michel Abdalla
Masayuki Abe
Adi Akavia
Elena Andreeva
Spyridon Antona-
    kopoulos
Kazumaro Aoki
Frederik Armknecht
Joonsang Baek
Elad Barkan
Lejla Batina
Peter Beelen
Amos Beimel
Mihir Bellare
Josh Benaloh
Daniel Bernstein
Alex Biryukov
Daniel Bleichenbacher
Xavier Boyen
An Braeken
Emmanuel Bresson
Justin Brickell
Jan Camenisch
Ran Canetti
Christophe De Cannière
Dario Catalano
Melissa Chase
Lily Chen
Rafi Chen
Yongxi Cheng
Seung Geol Choi
Scott Contini
Ronald Cramer
Anupam Datta
Cécile Delerablée
Anand Desai
Claus Diem
Jingtai Ding
Yan Zhong Ding
Yevgenyi Dodis
Orr Dunkelman
Phil Eisen
Ariel Elbaz

Serge Fehr
Matthias Fitzi
Lance Fortnow
Pierre-Alain Fouque
Soichi Furuya
Steven Galbraith
Juan Garay
Rosario Gennaro
Henri Gilbert
Eu-Jin Goh
Ronen Gradwohl
Louis Granboulan
Prateek Gupta
Iftach Haitner
Shai Halevi
Renen Hallak
Safuat Hamdy
Helena Handschuh
Danny Harnik
Anwar Hasan
Carmit Hazay
Alex Healy
Javier Herranz
Jonathan Herzog
Jason Hinek
Dennis Hofheinz
Nick Howgrave-Graham
Tetsu Iwata
Stas Jarecki
Ellen Jochemsz
Antoine Joux
Pascal Junod
Charanjit Jutla
Marcelo Kaihara
Yael Tauman Kalai
Alexander Kholosha
Joe Kilian
Eike Kiltz
Jongsung Kim
Vlastimil Klima
Vlad Kolesnikov
Chiu-Yuen Koo
Simon Kramer

Steve Kremer
Sebastien Kunz-Jacques
Eyal Kushilevitz
Tanja Lange
Joseph Lano
Kristin Lauter
Homin Lee
Stephane Lemieux
Matt Lepinski
Gatan Leurent
Benoit Libert
Stefan Lucks
Christoph Ludwig
Anna Lysyanskaya
Vadim Lyubashevsky
Phil MacKenzie
Mohammad Mahmoody
John Malone-Lee
Mark Manasse
Alexander May
Frank McSherry
Willi Meier
Daniele Micciancio
John Mitchell
Anton Mityagin
Peter Montgomery
Tal Moran
Ruggero Morselli
Siguna Müller
Sean Murphy
David Naccache
Arvind Narayanan
Andrew Neff
Gregory Neven
Jesper Buus Nielsen
Tatsuaki Okamoto
Michael Østergaard
Rafi Ostrovsky
Saurabh Panjwani
Souradyuti Paul
Raphael C.-W. Phan
Krzysztof Pietrzak
Benny Pinkas

David Pointcheval
Tal Rabin
Oded Regev
Omer Reingold
Leo Reyzin
Tom Ristenpart
Phil Rogaway
Alon Rosen
Amit Sahai
Yasuyuki Sakai
Louis Salvail
Christian Schaffner
Claus Schnorr
Berry Schoenmakers
Gil Segev
Jean-Pierre Seifert
Ronen Shaltiel
Taizo Shirai

Victor Shoup
Igor Shparlinski
Tom Shrimpton
Andrey Sidorenko
Alice Silverberg
Robert Silverman
Adam Smith
Martijn Stam
François-Xavier
　Standaert
Ron Steinfeld
Daisuke Suzuki
Mike Szydlo
Katsuyuki Takashima
Tamir Tassa
Tomas Toft
Eran Tromer
Toyohiro Tsurumaru

Andrew Wan
Shuhong Wang
Dai Watanabe
Brent Waters
John Watrous
Benne de Weger
Stephanie Wehner
Enav Weinreb
Susanne Wetzel
Udi Wieder
Douglas Wikström
Christopher Wolf
Duncan Wong
David Woodruff
David Xiao
Guomin Yang
Kan Yasuda
Feng Zhu

Table of Contents

Rigorous Bounds on Cryptanalytic Time/Memory Tradeoffs

Elad Barkan[1], Eli Biham[1], and Adi Shamir[2]

[1] Computer Science Department
Technion – Israel Institute of Technology
Haifa 32000, Israel
[2] Department of Computer Science and Applied Mathematics
The Weizmann Institute
Rehovot 76100, Israel

Abstract. In this paper we formalize a general model of cryptanalytic time/memory tradeoffs for the inversion of a random function $f : \{0, 1, \ldots, N - 1\} \mapsto \{0, 1, \ldots, N - 1\}$. The model contains all the known tradeoff techniques as special cases. It is based on a new notion of *stateful random graphs*. The evolution of a path in the stateful random graph depends on a *hidden state* such as the color in the Rainbow scheme or the table number in the classical Hellman scheme. We prove an upper bound on the number of images $y = f(x)$ for which f can be inverted, and derive from it a lower bound on the number of hidden states. These bounds hold for an overwhelming majority of the functions f, and their proofs are based on a rigorous combinatorial analysis. With some additional natural assumptions on the behavior of the *online phase* of the scheme, we prove a lower bound on its worst-case time complexity $T = \Omega(\frac{N^2}{M^2 \ln N})$, where M is the memory complexity. Finally, we describe new rainbowbased time/memory/data tradeoffs, and a new method for improving the time complexity of the online phase (by a small factor) by performing a deeper analysis during preprocessing.

Keywords: Time/memory tradeoff, time/memory/data tradeoff, rigorous, lower bound, hidden state, stateful random graph, Hellman, Rainbow, Cryptanalysis.

1   Introduction

In this paper we are interested in generic ("black-box") schemes for the inversion of one-way functions such as $f(x) = E_x(0)$, where E is any encryption algorithm, x is the key, and 0 is the fixed plaintext zero. For the sake of simplicity, we assume that both x and $f(x)$ are chosen from the set of N values $\{0, 1, \ldots, N - 1\}$.

The simplest example of a generic scheme is exhaustive search, in which a pre-image of $f(x)$ is found by trying all the possible pre-images x', and checking whether $f(x') = f(x)$. The worst-case time complexity T (measured by the number of applications of f) of exhaustive search is N, and the space complexity M is negligible. Another extreme scheme is holding a huge table with all the

C. Dwork (Ed.): CRYPTO 2006, LNCS 4117, pp. 1–21, 2006.
© International Association for Cryptologic Research 2006

images (in increasing order), and for each image storing one of its pre-images. This method requires a *preprocessing phase* whose time and space complexities are about N, followed by an *online inversion phase* whose running time T is negligible and space complexity M is about N. Cryptanalytic time/memory tradeoffs deal with finding a compromise between these extreme schemes, in the form of a tradeoff between the time and memory complexities of the online phase (assuming that the preprocessing phase comes for free). Cryptanalytic time/memory/data tradeoffs are a variant which accepts D inversion problems and has to be successful in at least one of them. This scenario typically arises in stream ciphers, when it suffices to invert the function that maps an internal state to the output at one point to break the cipher. However, the scenario also arises in block ciphers when the attacker needs to recover one key out of D different encryptions with different keys of the same message [4,5]. Note that for $D = 1$ the problem degenerates to the time/memory tradeoff discussed above.

1.1   Previous Work

The first and most famous cryptanalytic time/memory tradeoff was suggested by Hellman in 1980 [11]. His tradeoff requires a preprocessing phase with a time complexity of about N and allows a tradeoff curve of $M\sqrt{T} = N$. An interesting point on this curve is $M = T = N^{2/3}$. Since only values of $T \leq N$ are interesting, this curve is restricted to $M \geq \sqrt{N}$. Hellman's scheme consists of several tables, where each table covers only a small fraction of the possible values of $f(x)$ using chains of repeated applications of f. Hellman rigorously calculated a lower bound on the expected coverage of images by a single table in his scheme. However, Hellman's analysis of the coverage of images by the full scheme was highly heuristic, and in particular it made the formally unjustifiable assumption that many simple variants of f are independent of each other. Under this analysis, the success rate of Hellman's tradeoff for a random f is about 55%, which was verified using computer simulations. Shamir and Spencer proved in a rigorous way (in an unpublished manuscript from 1981) that for an overwhelming majority of the functions f, even the best Hellman table (with chains of unbounded length created from the best collection of start points, which are chosen using an unlimited preprocessing phase) has essentially the same coverage of images as a random Hellman table (up to a multiplicative logarithmic factor). However, they could not rigorously deal with the full (multi-table) Hellman scheme.

In 1982, Rivest noted that in practice, the time complexity is dominated by the number of disk accesses (random access to disk can be many orders of magnitude slower than the evaluation of f). He suggested to use distinguished points to reduce the number of disk accesses to about $\sqrt{T}$. The idea of distinguished points was described in detail and analyzed in 1998 by Borst, Preneel, and Vandewalle [8], and by Standaert, Rouvroy, Quisquater, and Legat in 2002 [15].

In 1996, Kusuda and Matsumoto [13] described how to find an optimal choice of the tradeoff parameters in order to find the optimal cost of an inversion machine. Kim and Matsumoto [12] showed in 1999 how to increase the precomputation time to allow a slightly higher success probability. In 2000, Biryukov and

Shamir [6] generalized time/memory tradeoffs to time/memory/data tradeoffs, and discussed specific applications of these tradeoffs to stream ciphers.

A new time/memory tradeoff scheme was suggested by Oechslin [14] in 2003. It claims to save a factor 2 in the worst-case time complexity compared to Hellman's original scheme (see Section 6.1 for a discussion of this point). Another interesting work on time/memory tradeoffs was performed by Fiat and Naor [9,10] in 1991. They introduce a rigorous time/memory tradeoff for inverting *any* function. Their tradeoff curve is less favorable compared to Hellman's tradeoff, but it can be used to invert any function rather than a random function.

A question which naturally arises is what is the best tradeoff curve possible for cryptanalytic time/memory tradeoffs? Yao [16] showed that $T = \Omega(\frac{N \log N}{M})$ is a lower bound on the time complexity, regardless of the structure of the algorithm, where M is measured in bits. This bound is essentially tight in case f is a single-cycle permutation.[1] However, the question remains open for functions which are not single-cycle permutations. Can there be a better cryptanalytic time/memory tradeoff than what is known today?

1.2 The Contribution of This Paper

In this paper we formalize a general model of cryptanalytic time/memory tradeoffs, which includes all the known schemes (and many new schemes). In this model, the preprocessing phase is used to create a *matrix* whose rows are long chains (where each link of a chain includes one oracle access to f), but only the *start points* and *end points* of the chains are stored in a table, which is passed to the online phase (the chains in the matrix need not be of the same length).

The main new concept in our model is that of a *hidden state*, which can affect the evolution of a chain. Typical examples of hidden states are the table number in Hellman's scheme and the color in a Rainbow scheme. The hidden state is an important ingredient of time/memory tradeoffs. Without it, the chains are paths in a single random graph, and the number of images that these chains can cover is extremely small (as shown heuristically in [11] and rigorously by Shamir and Spencer). We observe that in existing schemes, almost all of the online running time is spent on discovering the value of the hidden state (hence the name *hidden* state), which means that it is advisable to keep the number of hidden states minimal. Once the correct hidden state is found, the online phase needs to spend only about a square root of the running time to complete the inversion.

The main effect of the hidden state is that it increases the number of nodes in the graph from N to NS, where S is the number of values that the hidden state can assume. The new larger graph is called the *stateful random graph*, and the chains we create are paths in this graph. Two nodes in the stateful random graph defined by a particular f are connected by an edge:

$$\boxed{y_i} \quad \boxed{s_i} \quad \longrightarrow \quad \boxed{y_{i+1}} \quad \boxed{s_{i+1}},$$

---

[1] In [11] and in the rest of this paper, M represents the number of start points, rather than the number of bits used to represent them.

if (y_{i+1}, s_{i+1}) is the (unique) successor of (y_i, s_i) defined by a deterministic transition function, where y_i and y_{i+1} are the outputs of the f function in the two consecutive steps, and s_i, s_{i+1} are the respective values of the hidden state during the creation of y_i and y_{i+1} (see Figure 1). The evolution of the y values along a path in the stateful random graph is "somewhat random" since it is determined by the random function f applied to a possibly non-random input. However, the evolution of the hidden state (s_i and s_{i+1}) can be totally controlled by the designer of the scheme.

The larger number of nodes is what allows chains to cover a larger number of images y, by reducing the probability of collisions. We rigorously prove that for any time/memory scheme and for an overwhelming majority of the functions f, the number of images that can be covered by any collection of M chains is bounded from above by $2\sqrt{SNM \ln (SN)}$, where $M = N^\alpha$ for any $0 < \alpha < 1$. Intuitively it might seem that making S larger at the expense of N should cause the coverage to be larger (as S can be made to behave more like a permutation). Surprisingly, S and N play the same role in the bound. The product SN (which is the number of nodes in the stateful random graph) remains unchanged if we enlarge S at the expense of N or vice versa. Note that $\sqrt{SNM}$ is about the coverage that is expected with the Hellman or Rainbow schemes, and thus even for the best choice of start points and path lengths (found with unlimited preprocessing time), there is only a small factor of at most $2\sqrt{\ln SN}$ that can be gained in the coverage. We use the above upper bound to derive a lower bound on the number S of hidden states required to cover at least half of the images by the matrix.

Under some additional natural assumptions on the behavior of the online phase, we give a lower bound on the worst-case time complexity:

$$T \geq \frac{1}{1024 \ln N} \frac{N^2}{M^2},$$

where the success probability is at least $1/2$.[2] Therefore, either there are no fundamentally better schemes, or their structure will have to violate our assumptions. Finally we show a similar lower bound for time/memory/data tradeoffs:

$$T \geq \frac{1}{1024 \ln N} \frac{N^2}{D^2 M^2}.$$

1.3   Structure of the Paper

The model is formally defined in Section 2, and in Section 3 we prove the rigorous upper bound on the coverage of M chains in a stateful random graph. Section 4 uses this bound to derive a lower bound on the number of hidden states. The lower bound on the time complexity (under additional assumptions) is given in Section 5. Additional observations and notes appear in Section 6, and the paper is summarized in Section 7. Appendix A contains an extension of the bound of Section 3 to a special case needed in Section 5. We refer the reader to [11,14] for the details of previous tradeoffs schemes, or see a summary in [3, Appendix A.3].

---

[2] We use constants rather than big O notation to demonstrate that no huge constants are involved; however, we do not claim that these constants are tight.

2   The Stateful Random Graph Model

The class of time/memory tradeoffs that we consider in this paper can be seen as the following game: An adversary commits to a generic scheme with oracle accesses to a function f, which is supposed to invert f on most images y. Then, the actual choice of f is revealed to the adversary, who is allowed to perform an unbounded *precomputation* phase to construct the best collection of M chains. The chains are not necessarily of the same length, and the collection of the M chains is called the *matrix*. Then, during the *online phase*, a value y is given to the adversary, who should find x such that $f(x) = y$ using the scheme it committed to. We are interested in the time/memory complexities of schemes for which the algorithm succeeds to invert y with probability of at least $1/2$ for an overwhelming majority of random functions f.

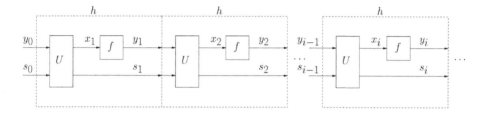

Fig. 1. A Typical Chain — A Path in the Stateful Random Graph

In the model that we consider, we are generous to the adversary by not counting the size of the memory that is needed to represent the scheme that it has committed to. Having been generous, we cannot allow the adversary to choose the scheme after f is revealed, as the adversary can use his knowledge to avoid collisions during the chain creation processes, and thus cover almost all the images using a single Hellman table.[3]

We do not impose any restrictions on the behavior of the preprocessing algorithm, but we require that it performs all oracle accesses to f through a *sub-algorithm*. When the preprocessing algorithm performs a series of oracle accesses to f, in which each oracle access can depend on the result of previous oracle accesses in the series, it is required to use the sub-algorithm. We call such a series of oracle accesses a *chain*. The *hidden state* is the internal state of the sub-algorithm (without the input/output of f).

---

[3] In the *auxiliary memory* variant of the model, we can allow the scheme to depend on an additional collection of $M \log_2 N$ bits, which the adversary chooses during the preprocessing. Thus, the adversary can customize his scheme to the specific function f by giving it a free *advice* of limited size. This variant includes schemes such as the one presented in [10]. Analysis (briefly given in Appendix A) shows that this variant is stronger than the regular model only by a small constant factor, and thus, we restrict our discussion to the regular model without loss of generality.

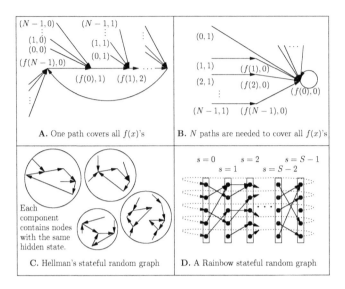

Fig. 2. Four Examples of Stateful Random Graphs

A typical chain of the sub-algorithm is depicted in Figure 1, where by U we denote the function that updates the internal state of the sub-algorithm and prepares the next input for f, and by h we denote the entire complex of U together with the oracle access to f (i.e., an application of h corresponds to a single computation step of the sub-algorithm). We denote by s_i the hidden state which accompanies the output y_i of f in the sub-algorithm. The choice of U by the adversary together with f defines the stateful random graph, and h can be seen as the function that takes us from one node (y_{i-1}, s_{i-1}) in the stateful random graph to the next node (y_i, s_i). U is assumed to be deterministic (if a non-deterministic U is desired, then the randomness can be given as part of the first hidden state s_0), and thus each node in the stateful random graph has an out-degree of 1.

Choosing U such that $s_i = s_{i-1} + 1 \pmod{N}$ and $x_i = s_{i-1}$ creates a stateful random graph that goes over all the possible images of f in a single-cycle (depicted in Figure 2.A), and thus represents exhaustive search (note that the y_{i-1} is ignored by U and thus all its N values with the same hidden state s_{i-1} converge to the same node $(f(s_{i-1}), s_{i-1}+1)$). Such a cycle is very easy to cover even with a single path, but at the heavy price of using N hidden states. At the other extreme, we can construct a stateful random graph (see Figure 2.B) that requires a full lookup table to cover all images of f by choosing U as: *if $s_{i-1} = 1$ then $x_i = y_{i-1}$ and $s_i = 0$, else $x_i = s_i = 0$.* In this function, each $(y_{i-1}, 1)$ is mapped by h to $(f(y_{i-1}), 0)$, and all these values are mapped to the same node $(f(0), 0)$.

As another example consider the mapping $x_i = y_{i-1}$ and $s_i = g(s_{i-1})$, where g is some function. This mapping creates a stateful random graph which is the direct product of the random graph induced by f, and the graph induced by g (this graph is not shown in Figure 2). We can implement Hellman's scheme by setting $x_i = y_{i-1} + s_i \pmod{N}$ and $s_i = s_{i-1}$, where s_i represents the table number to which the chain belongs. This stateful random graph (see Figure 2.C) consists of S disconnected components, where each component is defined by h and a single hidden state. Finally, we can implement a Rainbow scheme by setting $x_i = y_{i-1} + s_{i-1} \pmod{N}$ and $s_i = s_{i-1} + 1 \pmod{S}$, where S is the number of colors in the scheme. This stateful random graph (see Figure 2.D) looks like a layered graph with S columns and random connections between adjacent columns (including wrap-around links).

The preprocessing algorithm can stop the sub-algorithm at any point, using any strategy that may or may not depend on the value of the hidden states and the results of the oracle accesses, and it can use unbounded amount of additional space during its execution. For example, in Hellman's original method, the chain is stopped after t applications of f. Therefore, the internal state of the preprocessing algorithm must contain a counter that counts the length of the chain. However, the length of the chain does not affect the way the next link is computed, and therefore this counter can be part of the internal state of the preprocessing algorithm rather than the hidden state of the sub-algorithm. As a result, only the table number has to be included in the hidden state of Hellman's scheme. In the Rainbow scheme, however, the current location in the chain determines the way the next link is computed, and thus the index of the link in the chain must be part of the hidden state. The two kinds of states affect the development of chains in completely different ways: the hidden state can actually affect the values in the chain (as U depends on the hidden state), while any state which is not included in the hidden state can only stop the development of the chain, but not affect its values. As shown later, the number of hidden states strongly affects the success probability and the running time of the online phase.

The preprocessing algorithm can store in a *table* only the start points and end points of up to M chains, which are used by the *online algorithm*. Note that the requirement of passing information from the preprocessing phase to the online phase only in the form of chains does not restrict our model in any way, as the sub-algorithm that creates the chains can be designed to perform any computation. Moreover, the preprocessing algorithm can encode any information as a collection of start points, which the online algorithm can decode to receive the information. Also note that this model of a single table can accommodate multiple tables (for example, Hellman's multiple tables) by including with each start point and end point the respective value of the hidden-state.

The input of the online algorithm is y that is to be inverted, and the table generated by the preprocessing algorithm. We require that the online algorithm performs all oracle accesses to f (including chain creation) through the same sub-algorithm used during the preprocessing. In the variant of time/memory/data tradeoffs, the input of the online algorithm consists of D values $y_1, y_2, \ldots, y_D$ and the table, and it suffices that the algorithm succeeds in inverting one image. This concludes the definition of our model.

In existing time/memory tradeoffs, the online algorithm assumes that the given $y = f(x)$ is covered by the chains in the table. Therefore, y appears with some hidden state s_i, which is unfortunately unknown. The algorithm sequentially tries all the values that s_i can assume, and for each one of them it initializes the sub-algorithm on (y, s_i). The sub-algorithm is executed a certain number of steps (for example, until an end point condition has been reached). Once an end point that is stored in the table has been found, the start point is fetched, and the chain is reconstructed to reveal the x_i such that $y = f(x_i)$.[4] Existing time/memory/data tradeoffs work in a similar way, and the process is repeated for each one of the D given images.

2.1 Coverage Types and Collision of Paths in the Stateful Random Graph

A Table with M rows induces a certain coverage of the stateful random graph. Each row in the table contains a start point and an end point. For each such pair, the matrix associated with the table contains the chain of points spanned between the start point and the end point in the stateful random graph. The set of all the points (y_i, s_i) on all these chains is called the *gross coverage* of the stateful random graph that is induced by the table.

The gross coverage of the M paths is strongly affected by collisions of paths. Two paths in a graph collide once they reach a common node in the graph, i.e., two links in two different chains have the same y_i value and the same hidden state s_i. From this point on, the evolution of the paths is identical (but, the end points, which can be chosen arbitrarily on the path, might be different). As a result, the joint coverage of the two paths might be greatly reduced (compared to paths that do not collide). It is important to note that during the evolution of the paths, it is possible that the same value y_i repeats under different hidden states. However, such a repetition does not cause a collision of the paths.

To analyze the behavior of the online algorithm, we are interested in the *net coverage*, which is the number of different y_i values that appear during the evolution of the M paths, regardless of the hidden state they appear with, as this number represents the total number of images that can be inverted. Clearly, the gross coverage of the M paths is larger than or equal to the net coverage of the paths.

When we ask what is the maximum gross or net coverage that can be gained from a given start point, we can ignore the end point and allow the path to be of unbounded length, since eventually the path loops (as the graph is finite). Once the path loops, the coverage cannot grow further. An equivalent way of achieving the maximum coverage of M paths is by choosing the end point of each path to be the point (y_i, s_i) along the path whose successor is the first point seen for the second time along this path.

---

[4] Note that the fact that an end point is found does not guarantee a successful inversion of y. Such a failure in inversion is called a false alarm, and it can be caused, for example, when the chain that is recreated from y merges with a chain (of the matrix) that does not contain y.

3   A Rigorous Upper Bound on the Net Coverage of M Chains in a Stateful Random Graph

In this section we formally prove the following upper bound on the net coverage:

Theorem 1. *Let* $A = \sqrt{SNM \ln{(SN)}}$, *where* $M = N^\alpha$, *for any* $0 < \alpha < 1$. *For any* U *with* S *hidden states, with overwhelming probability over the choice of* $f : \{0, 1, \ldots, N - 1\} \mapsto \{0, 1, \ldots, N - 1\}$, *the net coverage of images* $(y = f(x))$ *values) on any collection of* M *paths of any length in the resulting stateful random graph is bounded from above by* $2A$.

This theorem shows that even though stateful random graphs can have many possible shapes, the images of f they contain can only be significantly covered by using many paths or many hidden states (or both), as defined by the implied tradeoff formula above. Without loss of generality, we can assume that $S < N$, since otherwise the claimed bound is larger than N, and clearly, the net coverage can never exceed N.

3.1   Reducing the Best Choice of Start Points to the Average Case

In the first phase of the proof, we reduce the problem of bounding the best coverage (gained by the best collection of M start points) to the problem of bounding the coverage defined by a random set of start points and a random f. We do it by constructing a huge table W (as shown in Figure 3) which contains a row for each possible function f, and a column for each possible set of M start points. In entry $W_{i,j}$ of the table we write 1 if the net coverage obtained by the set M_j of start points for function f_i (extended into paths of unbounded length) is larger than our bound $(2A)$, and we write 0 otherwise. Therefore, a row i with all zeros means that there is no set of start points for the function f_i that can achieve a net coverage larger than $2A$.

To prove the theorem, it suffices to show that the number of 1's in the table, which we denote by #1, is much smaller than the number of rows, which we denote by #r (i.e., #1 ≪ #r). From counting considerations, the vast majority of rows contain only zeros, and the correctness of the theorem follows.

We can express the number of 1's in the table by the number of entries multiplied by the probability that a random entry in the table contains 1, and require

	M_1	M_2	$\cdots$	$M_{\binom{NS}{M}}$
f_1	0	0		
f_2	1	0		
$\vdots$			$\ddots$	
f_{N^N}				

Fig. 3. A Table W denoting for each function f_i whether the net coverage obtained from the set of start points M_j is larger (1) or smaller (0) than $2A$

that the product is much smaller than $\#r$, i.e., $\#1 = \text{Prob}(W_{i,j} = 1) \cdot \#c \cdot \#r \ll \#r$, where $\#c$ is the number of columns in the table. Therefore, it suffices to show that for a random choice of the function f and a random set of start points, $\text{Prob}(W_{i,j} = 1) \cdot \#c$ is very close to zero. We have thus reduced the problem of proving that the coverage in the best case is smaller than $2A$, to bounding the number of columns multiplied by the probability that the average case is larger than $2A$. This is proven in the next few subsections.

3.2   Bounding $\text{Prob}(W_{i,j} = 1)$

We bound $\text{Prob}(W_{i,j} = 1)$ by constructing an algorithm that counts the net coverage of a given function f and a given set of M start points, and analyzing the probability that the coverage is larger than $2A$. During this analysis, we would like to consider each output of f as a new and independent coin flip, as $\text{Prob}(W_{i,j} = 1)$ is taken over a uniform choice of the function f. However, this assumption is justified only when x_i does not appear as an input to f on any previously considered point. In this case we say that x_i is *fresh*, and this freshness is a sufficient condition for f's output to be random and independent of any previous event.

---

1. For $i \in \{1, \ldots, S\}$ $Bucket_i = LowerFreshBucket_i = UpperFreshBucket_i = \phi$.
2. $NetCoverage = SeenX = \phi$.
3. Apply h to the first start point to generate the first event $\overset{x_i}{\to}(y_i, s_i)$.
4. If y_i appears in $Bucket_{s_i}$ Jump to Step 7 (Collision is detected). Otherwise:
5. Add y_i to $Bucket_{s_i}$.
6. If x_i does not appear in $SeenX$ (i.e., x_i is fresh):
   (a) If y_i does not appear in $NetCoverage$, add it to $NetCoverage$.
   (b) If $|LowerFreshBucket_{s_i}| < A/S$, add y_i to $LowerFreshBucket_{s_i}$, otherwise, add y_i to $UpperFreshBucket_{s_i}$.
7. Move to the next event:
   – Add x_i to $SeenX$ (i.e., mark that x_i is no longer fresh)
   – If a collision was detected in Step 4, apply h to the next start point (stop if there are no unprocessed start points). Otherwise: generate the next event by applying h to (y_i, s_i).
8. Jump to Step 4.

Legend:

– $SeenX$ is used to determine freshness by storing all the values of x that have been seen by now. This is the only set that stores input values of f. All the other sets store output values of f.
– $Bucket_i$ stores the all the y's that have been seen along with hidden state i (used for collision detection).
– $NetCoverage$ stores all the y's that have been seen from all chains considered so far, but without repetitions caused by different hidden states.
– For fresh values of x, $LowerFreshBucket_i$ stores the first A/S values of $y = f(x)$ seen with hidden state i (note that the x is fresh, but the y could have already appeared in other Buckets).
– For fresh values of x, $UpperFreshBucket_i$ stores the values of y after the first A/S values were seen with hidden state i (again, such a y could have already appeared in other Buckets).

Fig. 4. A Particular Algorithm for Counting the Net Coverage

Denote by $\xrightarrow{x_i}(y_i, s_i)$ the event of reaching the point (y_i, s_i), where x_i is the input of f during the application of h, i.e., $y_i = f(x_i)$. When we view the points $(y_i = f(x_i), s_i)$ as nodes in the stateful random graph, the value x_i is a property of the edge that enters (y_i, s_i), rather than a property of the node itself, since the same (y_i, s_i) might be reached from several preimages. The freshness of x_i (at a certain point in time) depends on the order in which we evolve the paths (the x_i is fresh the first time it is seen, and later occurrences of x_i are not fresh), but it should be clear that the net coverage of a set of paths is independent of the order in which the paths are considered.

The algorithm is described in Figure 4. It refers to the ratio A/S, which for the sake of simplicity we treat as an integer. Note that $A/S \geq 2\sqrt{M \ln(NS)}$ (as $S < N$), and $A/S \gg 1$ (as N grows to infinity) since $M = N^{\alpha}$. Thus, the rounding of A/S to the nearest integer causes only a negligible effect.

Lemma 1. *At the end of the algorithm* $|NetCoverage|$ *is the size of the net coverage.*

Proof. We observe that the algorithm processes all the points (y_i, s_i) that are in the coverage of the chains originating from the M start points, since it only stops a path when it encounters a collision.

A necessary condition for a $y_i = f(x_i)$ to be counted in the net coverage is that y_i appears in an event $\xrightarrow{x_i}(y_i, s_i)$ that is not a collision and in which x_i is fresh. If this condition holds, the algorithm reaches Step 6a, and adds y_i to $NetCoverage$ only if the sufficient condition $y_i \notin NetCoverage$ holds.   ∎

At the end of the algorithm $NetCoverage = \cup_{i=1}^{S}(LowerFreshBucket_i \cup UpperFreshBucket_i)$, and thus

$$|NetCoverage| \leq \sum_{i=1}^{S}(|LowerFreshBucket_i| + |UpperFreshBucket_i|),$$

since each time a y_i value is added to $NetCoverage$ (in Step 6a) it is also added to either $LowerFreshBucket$ or $UpperFreshBucket$ in Step 6b. We use this inequality to upper bound $|NetCoverage|$.

Bounding $\sum_{i=1}^{S}|LowerFreshBucket_i|$ is easy, as the condition in Step 6b assures that for each i, $|LowerFreshBucket_i| \leq A/S$, and thus their sum is at most A. Bounding $\sum_{i=1}^{S}|UpperFreshBucket_i|$ requires more effort, and we do it with a series of observations and lemmas.

Our main observation on the algorithm is that during the processing of an event $\xrightarrow{x_i}(y_i, s_i)$, the value y_i is added to $UpperFreshBucket_{s_i}$ if and only if:

1. x_i is fresh (Step 6); and
2. $LowerFreshBucket_{s_i}$ contains exactly A/S values (Step 6b); and
3. (y_i, s_i) does not collide with a previous point placed in the same bucket (Step 4).

Definition 1. *An event* $\xrightarrow{x_i}(y_i, s_i)$ *is called a coin toss if the first two conditions hold for the event.*

Therefore, a y_i is added to $UpperFreshBucket_{s_i}$ only if $\xrightarrow{x_i}(y_i, s_i)$ is a coin toss (but not vice versa), and thus the number of coin tosses serves as an upper bound on $\sum_{i=1}^{S} |UpperFreshBucket_i|$.

Our aim is to upper bound the net coverage (number of images in the coverage) by the number of different x values in the coverage (which is equal to the number of fresh x's), and to bound the number of fresh x's by A (for lower fresh buckets) plus the number of coin tosses (upper fresh buckets).

Definition 2. *A coin toss $\xrightarrow{x_i}(y_i, s_i)$ is called successful if before the coin toss $y_i \in LowerFreshBucket_{s_i}$.*

Observe that a successful coin toss causes a collision, as $LowerFreshBucket_{s_i} \subseteq Bucket_{s_i}$ at any point in time, i.e., a successful coin toss means that the node (y_i, s_i) in the graph was already visited at some previous time (the collision is detected at Step 4). Note that a collision can also be caused by events other than a successful coin toss (and these events are not interesting in the context of the proof): For example, a coin toss might cause a collision in case $y_i \in Bucket_{s_i}$ (but $y_i \notin UpperFreshBucket_{s_i} \bigcup LowerFreshBucket_{s_i}$) before the coin toss. Another example is when x_i is not fresh, and therefore, $\xrightarrow{x_i}(y_i, s_i)$ is not a coin toss, but $y_i \in Bucket_{s_i}$ before the event (x_i was marked as seen in an event of a hidden state different than s_i).

Since each chain ends with the first collision that is seen, the algorithm stops after encountering exactly M collisions, one per path. As a successful coin toss causes a collision, there can be at most M successful coin tosses in the coverage.

Note that the choice of some of the probabilistic events as coin tosses can depend on the outcome of previous events (for example, $LowerFreshBucket_s$ must contain A/S points before a coin toss can occur for hidden state s), but not on the current outcome. Therefore, once an event is designated as a coin toss we have:

Lemma 2. *A coin toss is successful with probability of exactly $A/(SN)$, and the success (or failure) is independent of any earlier probabilistic event.*

Proof. As x_i is fresh, $y_i = f(x_i)$ is truly random (i.e., chosen with uniform distribution and independently of previous probabilistic events). There are exactly A/S different values in $LowerFreshBucket_{s_i}$, and thus the probability that y_i collides with one of them is exactly $\frac{A/S}{N} = \frac{A}{SN}$. As all the other coin tosses have an x_i value different from this one, the value of $f(x_i)$ is independent of theirs.■

It is important to note that the independence of the outcomes of the coin tosses is *crucial* to the correctness of the proof.

What is the probability that the number of coin tosses in the M paths is larger than A? It is smaller than or equal to the probability that among the first A coin tosses there were fewer than M successful tosses, i.e., it is bounded by

$$\text{Prob}\,(B\,(A,q) < M),$$

where $q = A/(SN)$ and $B(A, q)$ is a random variable distributed according to the binomial distribution, namely, the number of successful coin tosses out of A independent coin tosses with success probability q for each coin toss.

Note that choosing A too large would result in a looser bound. On the other hand, choosing A too small might increase our bound for $\text{Prob}\,(W_{i,j} = 1)$ too much. We choose A such that the expected number of successes Aq in A coin tosses with probability of success q satisfies $Aq = M\ln(SN)$. This explains our choice of $A = \sqrt{SNM\ln(SN)}$.

It follows that:

$$\text{Prob}\,(W_{i,j} = 1) = \text{Prob}\,(|NetCoverage| > 2A)$$

$$\leq \text{Prob}\left(\sum_{i=1}^{S}(|LowerFreshBucket_i| + |UpperFreshBucket_i|) > 2A\right)$$

$$\leq \text{Prob}\left(A + \sum_{i=1}^{S}(|UpperFreshBucket_i|) > 2A\right)$$

$$= \text{Prob}\left(\sum_{i=1}^{S}(|UpperFreshBucket_i|) > A\right) \leq \text{Prob}\,(\text{B}\,(A,q) < M).$$

The first inequality holds due to the fact that $\sum_{i=1}^{S}(|LowerFreshBucket_i| + |UpperFreshBucket_i|) > |NetCoverage|$. The last inequality holds as the number of coin tosses upper bounds $\sum_{i=1}^{S}(|UpperFreshBucket_i|)$.

We bound $\text{Prob}\,(\text{B}\,(A,q) < M)$ by $M \cdot \text{Prob}\,(\text{B}\,(A,q) = M)$ because the binomial distribution satisfies $\text{Prob}(\text{B}(A,q) = b) \geq \text{Prob}(\text{B}(A,q) = b-1)$ as long as $b < (A+1)q$, and in our case $b \leq M$ while $(A+1)q = Aq+q = M\ln(NS)+q > M$ (as $Aq = M\ln(NS)$). Therefore, we conclude that

$$\text{Prob}\,(W_{i,j} = 1) \leq \text{Prob}\,(\text{B}\,(A,q) < M) \leq M \cdot \text{Prob}\,(\text{B}\,(A,q) = M).$$

3.3 Concluding the Proof

To complete the proof we show that $\text{Prob}(W_{i,j} = 1) \cdot \#c$ is very close to zero by bounding $\#c \cdot M \cdot \text{Prob}\,(\text{B}\,(A,q) = M)$.

In the following equations, we use the bound $\binom{x}{y} \leq x^y/y! \leq (xe/y)^y$, since from Stirling's approximation $y! \geq (y/e)^y$. We bound $(1-q)^{-M}$ by estimating that $q = \frac{A}{SN} = \sqrt{\frac{M\ln(SN)}{SN}} = \sqrt{\frac{\ln(SN)}{SN^{1-\alpha}}}$ is very close to 0, certainly lower than 0.5 (recall that $M = N^{\alpha}$, and $\alpha < 1$). Thus, $1 - q > 0.5$, and $(1-q)^{-M} < 2^M$. Moreover, as $q > 0$ is very close to 0, we approximate $(1-q)^A$ as e^{-Aq}.

Since each column in W is defined by a subset of M out of the NS start points, $\#c = \binom{NS}{M}$, and thus

$$\#c \cdot M \cdot \text{Prob}\,(\text{B}\,(A,q) = M)$$

$$= \binom{NS}{M} M \binom{A}{M} (q)^M \cdot (1-q)^{A-M} \leq Me^{-Aq}\left(\frac{2e^2 AqNS}{M^2}\right)^M$$

and substitute $Aq = M\ln(SN)$

$$=Me^{-M\ln(NS)}\left(\frac{2e^2NSM\ln(NS)}{M^2}\right)^M = M(NS)^{-M}\left(\frac{2e^2NS\ln(NS)}{M}\right)^M$$

$$=M\left(\frac{2e^2\ln(NS)}{M}\right)^M = N^\alpha\left(\frac{2e^2\ln(NS)}{N^\alpha}\right)^{N^\alpha}.$$

When $N \to \infty$ the expression converges to zero, which concludes the proof.

4   A Lower Bound for S

We now analyze the minimum S required by the scheme. By Section 3, the net coverage of even the best set of M chains contains at most $2\sqrt{SNM\ln(SN)}$ distinct y_i values. To make the success probability at least one half, we need a net coverage of at least $N/2$. Therefore (recalling that $S \le N$),

$$N/2 \le 2\sqrt{SNM\ln(SN)} \le 2\sqrt{SNM\ln(N^2)}.$$

From this, we derive a rigorous lower bound on the number of hidden states in any tradeoff scheme which covers at least half the images for almost all f:

$$S \ge \frac{N}{32M\ln N}.$$

5   A Lower Bound on the Time Complexity

So far we bounded the net coverage of the matrix produced by the preprocessing phase. In order to use this result to bound the running time of the online phase, we have to make an assumption on the behavior of the online phase.

As a motivation for the assumption consider the following simplistic "proof" for the lower bound on the time to invert an image y: As shown in the previous section, the preprocessing phase uses at least $S \ge \frac{N}{32M\ln N}$ hidden states for the overwhelming majority of the functions. How much time is spent per hidden state? The online algorithm assumes that y is covered by the table with some hidden state, but it does not know with which. Therefore, for each hidden state s_i, the algorithm tries y with s_i by repeatedly applying h on (y, s_i) until it can rule out s_i as the correct hidden state. The expected number of applications of h is at least the width of the matrix (i.e., $\frac{N}{2M}$). Multiply the number of hidden states by the expected running time per hidden state to receive the "bound":

$$T \ge \frac{N^2}{64M^2\ln N}.$$

However, it should be clear that this proof is incorrect, since there can be a correlation between the hidden state and the length of the path we have to explore. One example of such a correlation is the Rainbow scheme, in which some hidden states appear only near the end points. Moreover, there can be more hidden states close to the end points than hidden states far from the end points, which shifts the average run per hidden state towards the end points.

In the rest of the section we rigorously lower bound the running time in the worst case, based only on the following assumption:

– Given y, the online algorithm works by sequentially trying the hidden states (in any order). For each hidden state s, it applies h on (y, s) at least t_s times in case (y, s) does not appear in a chain in the matrix, where t_s is the largest distance from any point with hidden state s in the matrix to its corresponding end point. In some cases (e.g., the Rainbow scheme) each t_s is a known constant. In other cases, the t_s values can depend on the specific matrix that results from the precomputation (and thus depend on the function f). The algorithm might not know the exact value of t_s, but can use any upper bound on t_s to limit the length of the chains it traverses, and in this case, its running time will be longer than our bound.

The assumption can be seen as a combination of three smaller principles: First, all the points in the precomputed matrix must be reachable by the online algorithm (therefore, if a point (y, s) appears in the matrix in a column which is t_s steps away from the end point, the algorithm must develop the chain for image y and hidden state s for at least t_s links). Second, the algorithm cannot tell one image from another (therefore, the algorithm must develop the chain for at least t_s links not only for that y, but also for all images tried with hidden state s). Third, the algorithm cannot know if an image y is covered by the matrix of the preprocessing, and assuming that it is covered, with which hidden state (therefore, all hidden states must be tried).

As a preparation for the proof, shift the chains in the matrix until their end points are aligned in the rightmost column. Consider the $l = \frac{N}{4M}$ columns which are adjacent to the end points. The sub-matrix which constitutes these l columns contains at most $N/4$ different images $f(x)$. We call this sub-matrix the *right sub-matrix*, and the rest of the matrix the *left sub-matrix*. As $M = N^\alpha$, l is large enough so we can round it to the nearest integer (with negligible effect).

The worst case (with regards to the time complexity) is when the input y to the algorithm is not an image under f, or y is an image under f but is not covered by the matrix. Then, the time complexity is at least the sum of all the lengths t_s. We divide the hidden states into two categories: *short hidden states* for which $t_s \leq l$, and *long hidden states* for which $t_s > l$. We would like to show that the number of long hidden states S_L is large, and use the time complexity spent just on the long hidden states as a lower bound on the total time complexity.

The net coverage of $f(x)$ images in the left sub-matrix must be at least $N/4$ images which do not appear in the right sub-matrix (since the total net coverage is at least $N/2$). Note that all the $N/4$ images in the left sub-matrix must be covered only by the S_L long hidden states, as all the appearances of short hidden states are concentrated in the right sub-matrix. In other words, the left sub-matrix can be viewed as a particular coverage of at least $N/4$ images by M continuous paths that contain only the S_L long hidden states.

It is not difficult to adapt the coverage theorem to bound the coverage of the left sub-matrix (using only long hidden states). The combinatorial heart of the proof remains the same, but the definitions of the events are slightly changed. For more details see Appendix A. The adapted coverage theorem implies that in order to have a net coverage of at least $N/4$ images, the number of long hidden states must satisfy

$$S_L \geq \frac{N}{64M \ln((SN)^2)} \geq \frac{N}{256 \ln N}$$

for an overwhelming majority of the functions. Since for each long hidden state $t_s \geq l$, the total time complexity in the worst case is at least

$$T \geq l \cdot S_L \geq \frac{N}{4M} \frac{N}{256M \ln N} \geq \frac{1}{1024 \ln N} \frac{N^2}{M^2}.$$

Note that we had to restrict the length of t_s such that it includes all occurrences of the hidden state s in the matrix, as otherwise (and using the unlimited preprocessing), each chain could start with a prefix consisting of all the values of $f(x)$, and thus any image in the rest of the chain (the suffix) cannot be a fresh occurrence. The algorithm can potentially encode in the hidden state information about the x_i and $f(x_i)$ values seen in the prefix, in such a way that it can change the probability of collision (and in particular, avoid collisions). Note that the preprocessed chains in this case are very long, but the online phase can be very fast if it covers only short suffixes of each path. As a result, we cannot use the methods of our proof without making the assumption.

See [3, Appendix 5.9] for an algorithm that violates the assumption by spending less time on each one of the many wrong guesses of the hidden state compared to the correct guess of the hidden state. The key idea is to use a new variant of Hellman's method with distinguished points: Let p be the probability of a point to be distinguished. As a result, the expected chain length is about p^{-1}, and the standard deviation is also about p^{-1}. The algorithm takes advantage of the large variation in length by trying k times more start points than we need, and storing in the table only the longest chains. The resulting matrices are called *stretched* matrices. The gain in time is achieved due to two facts: First, we have to search fewer tables (as each table covers more values due to the longer chains). Second, we spend on average only p^{-1} applications of h on each wrong guess of the hidden state (which is several times shorter than the average chain length in the matrix). The marginal gain in time decreases fast as k grows larger (as we gain from the tail of the distribution). A few experimental results show that by trying four times as many start points ($k = 4$) during the preprocessing, we can save a factor of about 4 in the time complexity of the online algorithm, and with $k = 8$ we can save a factor of about 4.8. Note that these results are based only on the expected size of the coverage, and ignore other issues such as the time spent on false-alarms.

5.1 A Lower Bound on the Time Complexity of Cryptanalytic Time/Memory/Data Tradeoffs

The common approach to construct a time/memory/data tradeoff is to use an existing time/memory tradeoff, but reduce the coverage (as well as the preprocessing) of the table by a factor of D. Thus, out of the D images, one is likely to be covered by the table. The decreased coverage reduces the number of hidden states, and thus the time complexity per image is reduced by a factor of D^3. However, the online algorithm is applied D times in the worst case (for the D images), which results in an overall decrease in the time complexity by a factor of D^2 (note that the D time/memory tradeoffs can be executed in parallel, which can reduce the average time complexity in some cases). Using similar arguments and assumptions to the ones in the case of time/memory tradeoff, and assuming that the net coverage of the table is at least $N/(2D)$, it follows that the worst-case time complexity can be lower bounded by

$$T' \geq D\frac{1}{1024D^3 \ln N}\frac{N^2}{M^2} = \frac{1}{1024D^2 \ln N}\frac{N^2}{M^2}.$$

Note that this analysis works for cases where D is not too large (and thus S is larger than 1). Otherwise, a tighter bound can be reached as S cannot be lower than 1 (but the bound itself is correct for all choices of D).

6 Notes on Rainbow-Like Schemes

6.1 A Note on the Rainbow Scheme

The worst-case time complexity of the original Rainbow scheme was claimed to be half that of Hellman's scheme. However, the reasoning behind the claim estimates M by considering only the number of start points and end points, and completely disregards the actual number of bits that are needed to represent these points. What [14] ignores is that the start points and end points in Hellman's scheme can be represented by half the number of bits required in the Rainbow scheme. If we double M in Hellman's scheme to get a fair comparison, we can reduce T by a factor of four via the time/memory tradeoff, which actually outweights the claimed improvement by a factor of two in the Rainbow scheme (ignoring issues such as the number of false alarms, which are difficult to analyze). The basic idea is that the starts points need not be chosen at random. For example in Hellman's scheme (with $T = M = N^{2/3}$), the first $1/3$ of the bits in a start point can be chosen to be zero. The second $1/3$ of the bits can be chosen to be identical to the table number, and the last $1/3$ of the bits can be chosen to be an index of the row. Only the last $1/3$ of the bits have to be actually stored in memory (i.e., we need only $(\log N)/3$ bits for each start point). In Rainbow tables by contrast, we can choose the first $1/3$ of the bits to be zero, but as there is only one table, the remaining $2/3$ of the bits must be the index. Thus, we have to store twice as many bits for each start point compared to Hellman's scheme. In both methods the number of bits that are required to store end points is considerably smaller than the number of bits that are required for the start points (and we therefore ignore it): The end points in Hellman (and Rainbow) can be compressed by storing only a little more than the last $(\log N)/3$ bits (respectively, $2(\log N)/3$ bits). Moreover, as the end points are sorted (and thus the difference between subsequent end points is expected to be small), we can further compress the end points by storing only the differences between subsequent end points.

6.2   Notes on Rainbow Time/Memory/Data Tradeoffs

The original Rainbow scheme does not provide a time/memory/data tradeoff, but only a time/memory tradeoff. A possible adaptation of the Rainbow scheme to time/memory/data tradeoffs is presented in [7], but the resulting tradeoff curve of $TM^2D = N^2$ is far inferior to the $TM^2D^2 = N^2$ curve presented in [6] for the Hellman method. We suggest two new ways of implementing a Rainbow-based time/memory/data tradeoff, with a curve similar to [6]. In both ways, we reduce the number of colors in the table, but the colors are organized in different ways along the chains. This section describes the key ideas of the new methods; the full analysis (which was verified by computer simulations) can be found in [3].

The first method is to reduce the number of colors to S by repeating the series of colors t times:

$$f_0 f_1 f_2 \dots f_{S-1} f_0 f_1 f_2 \dots f_{S-1} f_0 f_1 f_2 \dots f_{S-1} \dots f_0 f_1 f_2 \dots f_{S-1},$$

we call the resulting matrix a *thin-Rainbow* matrix. The stateful random graph can be described by $x_i = y_{i-1} + s_{i-1} \pmod{N}$ and $s_i = s_{i-1} + 1 \pmod{S}$. The resulting tradeoff[5] is $TM^2D^2 = N^2$, which is similar to the tradeoff in [6], i.e., we lose the claimed improvement (by a factor of 2) of the original Rainbow time/memory tradeoff. However, like the Rainbow scheme, this method still requires twice as many bits to represent its start points, and thus it is slightly inferior to [6]. In the online phase, each color is sequentially tried by continuing the chain for at most tS links.

As a preliminary to the second method, consider a scheme in which we group the colors together in groups of t, and thus a typical row looks like:

$$\underbrace{f_0 f_0 f_0 \dots f_0}_{t \text{ times}} \underbrace{f_1 f_1 f_1 \dots f_1}_{t \text{ times}} \underbrace{f_2 f_2 f_2 \dots f_2}_{t \text{ times}} \dots \underbrace{f_{S-1} f_{S-1} f_{S-1} \dots f_{S-1}}_{t \text{ times}},$$

we call the resulting matrix a *thick-Rainbow* matrix. Note, however, that during the online phase the algorithm needs to guess not only the "flavor" i of f_i, but also the phase of f_i among the other f_i's (except for the last f_i). In fact, the hidden state is larger than S and includes the phase, as the phase affects the development of the chain. Therefore, the number of hidden states is $t(S-1)+1$ (which is

---

[5] When we write a time/memory/data tradeoff curve, the relations between the parameters relate to the expected worst-case behavior when the algorithm fails to invert y, and neglecting false-alarms.

almost identical to the number of hidden states in the original Rainbow scheme), and we get an inferior tradeoff of $TM^2D = N^2$. On the other hand, in some cases we retain the claimed savings of 2 in the time complexity. This example demonstrates the difference between "flavors" of f and the concept of a hidden state.

We propose to implement a Rainbow-based time/memory/data tradeoff by using the notion of distinguished points not only to determine the end of the chain, but also to determine the points in which we switch from one flavor of f to the next. In this case, the number of hidden states is equal to the number of flavors, and does not have to include any additional information. We can specify U as: $x_i = y_{i-1} + s_{i-1} \pmod{N}$, and if y_{i-1} is special, then $s_i = s_{i-1} + 1 \pmod{S}$ else $s_i = s_{i-1}$, where y_{i-1} is special if its $\log_2 t$ bits are zeros. We call the resulting matrix a *fuzzy-Rainbow* matrix, as each hidden state appears in slightly different locations in different rows of the matrix. In the online phase, the colors are tried in the same order as in the Rainbow scheme. Analysis shows that the tradeoff curve is $2TM^2D^2 = N^2 + ND^2M$, with $T \geq D^2$. The factor two savings is gained when $N^2 \gg ND^2M \Rightarrow D^2M \ll N$ (which happens when $T \gg D^2$). The number of disk accesses is about $\sqrt{2T}$, when $D^2M \ll N$, but is never more than in thin-Rainbow scheme for the same memory complexity.

7   Summary

In this paper, we proved that in our very general model, and under the natural assumption on the behavior of the online phase, there are no cryptanalytic time/memory tradeoffs which are better than existing time/memory tradeoffs, up to a logarithmic factor.

Acknowledgements

We would like to thank Joel Spencer for his contribution to the proof of the single table coverage bound in 1981, and Eran Tromer for his careful review and helpful comments on earlier versions of the paper. The second author was supported in part by the Israel MOD Research and Technology Unit.

References

1. Gildas Avoine, Pascal Junod, Philippe Oechslin, *Time-Memory Trade-Offs: False Alarm Detection Using Checkpoints (Extended Version)*, Available online on http://lasecwww.epfl.ch/pub/lasec/doc/AJO05a.pdf, 2005.
2. Steve Babbage, *A Space/Time Tradeoff in Exhaustive Search Attacks on Stream Ciphers*, European Convention on Security and Detection, IEE Conference Publication No. 408, 1995. Also presented at the rump session of Eurocrypt '96. Available online on http://www.iacr.org/conferences/ec96/rump/.
3. Elad Barkan, *Cryptanalysis of Ciphers and Protocols*, Ph.D. Thesis, http://www.cs.technion.ac.il/users/wwwb/cgi-bin/tr-info.cgi?2006/PHD/PHD-2006-04, 2006.

4. Eli Biham, *How to decrypt or even substitute DES-encrypted messages in 2^{28} steps*, Information Processing Letters, Volume 84, Issue 3, pp. 117–124, 2002.
5. Alex Biryukov, *Some Thoughts on Time-Memory-Data Tradeoffs*, IACR ePrint Report 2005/207, http://eprint.iacr.org/2005/207.pdf, 2005.
6. Alex Biryukov, Adi Shamir, *Cryptanalytic Time/Memory/Data Tradeoffs for Stream Ciphers*, Advances in Cryptology, proceedings of Asiacrypt 2000, Lecture Notes in Computer Science 1976, Springer-Verlag, pp. 1–13, 2000.
7. Alex Biryukov, Sourav Mukhopadhyay, Palash Sarkar, *Improved Time-Memory Trade-Offs with Multiple Data*, proceedings of SAC 2005, LNCS 3897, pp. 110–127, Springer-Verlag, 2006.
8. Johan Borst, Bart Preneel, Joos Vandewalle, *On the Time-Memory Tradeoff Between Exhaustive Key Search and Table Precomputation*, Proceedings of 19th Symposium on Information Theory in the Benelux, Veldhoven (NL), pp. 111–118, 1998.
9. Amos Fiat, Moni Naor, *Rigorous Time/Space Tradeoffs for Inverting Functions*, STOC 1991, ACM Press, pp. 534-541, 1991.
10. Amos Fiat, Moni Naor, *Rigorous Time/Space Tradeoffs for Inverting Functions*, SIAM Journal on Computing, 29(3): pp. 790-803, 1999.
11. Martin E. Hellman, *A Cryptanalytic Time-Memory Trade-Off*, IEEE Transactions on Information Theory, Vol. IT-26, No. 4, pp. 401–406, 1980.
12. Il-Jun Kim, Tsutomu Matsumoto, *Achieving Higher Success Probability in Time-Memory Trade-Off Cryptanalysis without Increasing Memory Size*, IEICE Transactions on Fundamentals, Vol. E82-A, No. 1, pp. 123–129, 1999.
13. Koji Kusuda, Tsutomu Matsumoto, *Optimization of Time-Memory Trade-Off Cryptanalysis and Its Application to DES, FEAL-32, and Skipjack*, IEICE Transactions on Fundamentals, Vol. E79-A, No. 1, pp. 35–48, 1996.
14. Philippe Oechslin, *Making a Faster Cryptanalytic Time-Memory Trade-Off*, Advances in Cryptology, proceedings of Crypto 2003, Lecture Notes in Computer Science 2729, Springer-Verlag, pp. 617–630, 2003.
15. Francois-Xavier Standaert, Gael Rouvroy, Jean-Jacques Quisquater, Jean-Didier Legat, *A Time-Memory Tradeoff Using Distinguished Points: New Analysis & FPGA Results*, proceedings of CHESS 2002, Lecture Notes in Computer Science 2523, Springer-Verlag, pp. 593–609, 2003.
16. Andrew Chi-Chih Yao, *Coherent Functions and Program Checkers (Extended Abstract)*, STOC 1990, ACM Press, pp. 84-94, 1990.

A   The Extended Coverage Theorem

We can extend the coverage theorem to bound the net coverage that can be obtained by M paths, where the paths contain only a subset of size S' out of the $S \geq S'$ hidden states that U can use. In Section 5, the S' hidden states we are interested in (long hidden states) appear in the left sub-matrix, and the rest of the hidden states (short hidden states) appear *only* in the right sub-matrix.

One way of viewing the proof of Theorem 1 is as an algorithm that constructs the best possible coverage given an *advice* chosen from a set of $\#c = \binom{NS}{M}$ possible advices. The proof shows that for an overwhelming majority of the functions f, even the best advice (which can depend on the specific choice of f) cannot lead to a coverage which is larger than the bound. In the proof we did not use any properties of the advice, except for the number of possible advices.

We can model the coverage of the left sub-matrix using a similar algorithm, only now the advice is larger. It contains not only the set of M start points, but also a termination point for each start point, and the number S' of long hidden states $(1 \leq S' \leq S)$ that the coverage includes. When the algorithm reaches a termination point, it stops the development of the chain. Clearly, there are $\#c' = S\binom{(NS)^2}{M}$ possible advices. To accommodate for the larger number of advices, we update the values of q and A to $q' = A'/(S'N)$, where $A' = \sqrt{S'NM \ln{(SN)^2}}$ (A' is chosen such that $A'q' = M \ln((SN)^2)$ to deal with the effect of the larger advice). The only remaining change compared to the original proof is that if the algorithm encounters more than S' hidden states it halts and sets its net coverage to zero, as the advice is inconsistent.

Note that we can allow $S > N$, but the model would not be fair if we allow S to be arbitrarily large, as too much information on f can be encoded by every choice of the hidden state (and we do not count the memory complexity of representing U). For example, if $S = N^N$, then with $S' = 1$ we can encode all the information on f by the specific choice of single hidden state. However, a huge amount of $N \log_2 N$ bits are required just to represent that single hidden state. Therefore, we are only interested in $S \leq N^k$ for some constant k. Then, $A' \leq \sqrt{S'NM \ln{(N^k N)^2}} = \sqrt{S'NM2(k+1) \ln N}$.

In the auxiliary memory model (described in a footnote in Section 2), the adversary is allowed to customize U to the specific f using $M \log_2 N$ bits of memory. The proof of the coverage theorem in this case is very similar to the above proof, only now there are $\#c'' = N^M \binom{(NS)}{M}$ possible advices. As a result, the coverage theorem remains correct if we replace the original A by $A'' = \sqrt{SNM \ln{(SN^2)}}$, which increase the bound by a small constant factor.

On the Power of the Randomized Iterate

Iftach Haitner[1,*], Danny Harnik[2,**], and Omer Reingold[3,*]

[1] Dept. of Computer Science and Applied Math., Weizmann Institute of Science,
Rehovot, Israel
iftach.haitner@weizmann.ac.il
[2] Dept. of Computer Science, Technion, Haifa, Israel
harnik@cs.technion.ac.il
[3] Incumbent of the Walter and Elise Haas Career Development Chair, Dept. of
Computer Science and Applied Math., Weizmann Institute of Science, Rehovot, Israel
omer.reingold@weizmann.ac.il

Abstract. We consider two of the most fundamental theorems in Cryptography. The first, due to Håstad et al. [HILL99], is that pseudorandom generators can be constructed from any one-way function. The second due to Yao [Yao82] states that the existence of weak one-way functions (i.e. functions on which every efficient algorithm fails to invert with some noticeable probability) implies the existence of full fledged one-way functions. These powerful plausibility results shape our understanding of hardness and randomness in Cryptography. Unfortunately, the reductions given in [HILL99, Yao82] are not as security preserving as one may desire. The main reason for the security deterioration is the input blow up in both of these constructions. For example, given one-way functions on n bits one obtains by [HILL99] pseudorandom generators with seed length $\Omega(n^8)$.

This paper revisits a technique that we call the *Randomized Iterate*, introduced by Goldreich, et. al. [GKL93]. This technique was used in [GKL93] to give a construction of pseudorandom generators from *regular* one-way functions. We simplify and strengthen this technique in order to obtain a similar reduction where the seed length of the resulting generators is as short as $\mathcal{O}(n \log n)$ rather than $\Omega(n^3)$ in [GKL93]. Our technique has the potential of implying seed-length $\mathcal{O}(n)$, and the only bottleneck for such a result is the parameters of current generators against space bounded computations. We give a reduction with similar parameters for security amplification of *regular* one-way functions. This improves upon the reduction of Goldreich et al. [GIL+90] in that the reduction does not need to know the regularity parameter of the functions (in terms of security, the two reductions are incomparable). Finally, we show that the randomized iterate may even be useful in the general context of [HILL99]. In Particular, we use the randomized iterate to replace the basic building block of the [HILL99] construction. Interestingly, this modification improves efficiency by an n^3 factor and reduces the seed length to $\mathcal{O}(n^7)$ (which also implies improvement in the security of the construction).

---

* Research supported in part by a grant of the Israeli Science Foundation (ISF).
** This research was conducted at the Weizmann Institute.

1   Introduction

In this paper we address two fundamental problems in cryptography: constructing pseudorandom generators from one-way functions and transforming weak one-way functions into strong one-way functions. The common thread linking the two problems in our discussion is the technique we use. This technique that we call the *Randomized Iterate* was introduced by Goldreich, Krawczyk and Luby [GKL93] in the context of constructing pseudorandom generators from regular one-way functions. We revisit this method, both simplify existing proofs and utilize our new view to achieve significantly better parameters for security and efficiency. We further expand the application of the randomized iterate to constructing pseudorandom generators from any one-way function. Specifically we revisit the seminal paper of Håstad, Impagliazzo, Levin and Luby [HILL99] and show that the randomized iterate can help improve the parameters within. Finally, we use the randomized iterate method to both simplify and strengthen previous results regarding efficient hardness amplification of regular one-way functions. We start by introducing the randomized iterate in the context of pseudorandom generators, and postpone the discussion on amplifying weak to strong one-way function to subsection 1.2.

1.1   Pseudorandom Generators and the Randomized Iterate

Pseudorandom Generators, a notion first introduced by Blum and Micali [BM82] and stated in its current, equivalent form by Yao [Yao82], are one of the cornerstones of cryptography. Informally, a pseudorandom generator is a polynomial-time computable function G that stretches a short random string x into a long string $G(x)$ that "looks" random to any efficient (i.e., polynomial-time) algorithm. Hence, there is no efficient algorithm that can distinguish between $G(x)$ and a truly random string of length $|G(x)|$ with more than a negligible probability. Originally introduced in order to convert a small amount of randomness into a much larger number of effectively random bits, pseudorandom generators have since proved to be valuable components for various cryptographic applications, such as bit commitments [Nao91], pseudorandom functions [GGM86] and pseudorandom permutations [LR88], to name a few.

The first construction of a pseudorandom generator was given in [BM82] based on a particular one-way function and was later generalized in [Yao82] into a construction of a pseudorandom generator based on any one-way permutation. We refer to the resulting construction as the BMY construction. The BMY generator works by iteratively applying the one-way permutation on its own output. More precisely, for a given function f and input x define the ith iterate recursively as $x^i = f(x^{i-1})$ where $x^0 = f(x)$. To complete the construction, one needs to take a hardcore-bit at each iteration. If we denote by $b(x)$ the hardcore-bit of x (take for instance the Goldreich-Levin [GL89] predicate), then the BMY generator on seed x outputs the hardcore-bits $b(x^0), \ldots, b(x^\ell)$.

The natural question arising from the BMY generator was whether one-way permutations are actually necessary for pseudorandom generators or can one do with a more relaxed notion. Specifically, is any one-way function sufficient for pseudorandom generators? Levin [Lev87] observed that the BMY construction works for any "one-way function on its iterates", that is, a one-way function that remains one-way when applied sequentially on its own outputs. However, a general one-way function does not have this property since the output of f may have very little randomness in it, and a second application of f may be easy to invert. A partial solution was suggested by Goldreich et al. [GKL93] that showed a construction of a pseudorandom generator based on any *regular one-way function* (referred to as the GKL generator). A regular function is a function such that every element in its image has the same number of preimages. The GKL generator uses the technique at the core of this paper, that we call the *randomized iterate*. Rather than simple iterations, an extra randomization step is added between every two applications of f. More precisely:

Definition (Informal): (The Randomized Iterate) *For function f, input x and random hash functions $h_1, \ldots, h_\ell$, recursively define the ith randomized iterate (for $i \le \ell$) by:*
$$f^i(x, h_1, \ldots, h_\ell) = x^i = f(h_i(x^{i-1}))$$
where $x^0 = f(x)$.

The rational is that $h_i(x^i)$ is now uniformly distributed, and the challenge is to show that f, when applied to $h_i(x^i)$, is hard to invert even when the randomizing hash functions $h_1, \ldots, h_\ell$ are made public. Once this is shown, the generator is similar in nature to the BMY generator (the generator outputs $b(x^0), \ldots, b(x^\ell), h_1, \ldots, h_\ell$).

Finally, Håstad et al. [HILL99], culminated this line of research by showing a construction of a pseudorandom generator using any one-way function (called here the HILL generator). This result is one of the most fundamental and influential theorems in cryptography. It introduced many new ideas that have since proved useful in other contexts, such as the notion of pseudo-entropy and the implicit use of family of pairwise-independent hash functions as randomness extractors. We note that HILL departs from GKL in its techniques, taking a significantly different approach.

The Complexity and Security of the Previous Constructions. While the HILL generator fully answers the question of the plausibility of a generator based on any one-way function, the construction is highly involved and very inefficient. Other than the evident contrast between the simplicity and elegance of the BMY generator to the complex construction and proof of the HILL generator, the parameters achieved in the construction are far worse, rendering the construction impractical.

In practice, it is not necessarily sufficient that a reduction translates polynomial security into polynomial security. In order for reductions to be of any practical use, the concrete overhead introduced by the reduction comes into play. There are various factors involved in determining the security of a reduction. In this discussion, however, we focus only on one central parameter, which is the length m of the generator's seed compared to the length n of the input to the underlying one-way function. The BMY generator takes a seed of length

$m = \mathcal{O}(n)$, the GKL generator takes a seed of length $m = \Omega(n^3)$ while the HILL construction produces a generator with seed length on the order of $m = \Omega(n^8)$.[1]

The length of the seed is of great importance to the security of the resulting generator. While it is not the only parameter, it serves as a lower bound to how good the security may be. For instance, the HILL generator on m bits has security that is at best comparable to the security of the underlying one-way function, but on only $\mathcal{O}(\sqrt[8]{m})$ bits. To illustrate the implications of this deterioration in security, consider the following example: Suppose that we only trust a one-way function when applied to inputs of at least 100 bits, then the GKL generator can only be trusted when applied to a seed of length of at least one million bits, while the HILL generator can only be trusted on seed lengths of 10^{16} and up (both being highly impractical). Thus, trying to improve the seed length towards a linear one (as it is in the BMY generator) is of great importance in making these constructions practical.

Our Results on Pseudorandom Generators

Regular One-Way Functions: We give a construction of a pseudorandom generator from any regular one-way function with seed length $\mathcal{O}(n \log n)$. We note that our approach has the potential of reaching a construction with a linear seed, the bottleneck being the efficiency of the current bounded-space generators. Our construction follows the randomized iterate method and is achieved in two steps:

- We give a significantly simpler proof that the GKL generator works, allowing the use of a family of hash functions which is pairwise-independent rather than n-wise independent (as used in [GKL93]). This gives a construction with seed length $m = \mathcal{O}(n^2)$ (see Theorem 5).
- The new proof allows for the derandomization of the choice of the randomizing hash functions via the *bounded-space generator* of Nisan [Nis92], further reducing the seed length to $m = \mathcal{O}(n \log n)$ (see Theorem 6).

The Proof Method: Following is a high-level description of our proof method. For simplicity we focus on a single randomized iteration, that is on $x^1 = f^1(x, h) = f(h(f(x)))$. In general, the main task at hand is to show that it is hard to find $x^0 = f(x)$ when given $x^1 = f^1(x, h)$ and h. This follows by showing that any procedure A for finding x^0 given (x^1, h) enables to invert the one-way function f. Specifically, we show that for a random image $z = f(x)$, if we choose a random and independent hash h' and feed the pair (z, h') to A, then A is likely to return a value $f(x')$ such that $h'(f(x')) \in f^{-1}(z)$ (and thus we obtain an inverse of z).

Ultimately, we assume that A succeeds on the distribution of (x^1, h), where h is such that $x^1 = f^1(x, h)$, and want to prove A is also successful on the distribution of (x^1, h') where h' is chosen independently. Our proof is inspired by a technique used by Rackoff in his proof of the Leftover Hash Lemma (in

---

[1] The seed length actually proved in [HILL99] is $\mathcal{O}(n^{10})$, however it is mentioned that a more careful analysis can get to $\mathcal{O}(n^8)$. A formal proof for the $\mathcal{O}(n^8)$ seed length construction is given by Holenstein [Hol06].

[IZ89]). Rackoff proves that a distribution is close to uniform by showing that it has *collision-probability*[2] that is very close to that of the uniform distribution. We would like to follow this scheme and consider the collision-probability of the two aforementioned distributions. However, in our case the two distributions could actually be very far from each other. Yet, with the analysis of the collision-probabilities, we manage to prove that the probability of any event under the first distribution is *polynomially related* to the probability of the same event under the second distribution. This proof generalizes nicely also to the case of many iterations.

The derandomization using bounded-space follows directly from the new proof. In particular, consider the procedure that takes two random inputs x_0 and x_1 and random $h_1, \ldots, h_\ell$, and compares $f^\ell(x_0, h_1, \ldots, h_\ell)$ and $f^\ell(x_1, h_1, \ldots, h_\ell)$. This procedure can be run in bounded-space since it simply needs to store the two intermediate iterates at each point. Also, this procedure accepts with probability that is exactly the collision-probability of $(f^\ell(x, h_1, \ldots, h_\ell), h_1, \ldots, h_\ell)$. Thus, replacing $h_1, \ldots, h_\ell$ with the output of a bounded-space generator cannot change the acceptance rate by much, and the collision-probability is thus unaffected. The proof of security of the derandomized pseudorandom generator now follows as in the proof when using independent randomizing hash functions.

Any One-Way Function: The HILL generator takes a totally different path than the GKL generator. We ask whether the technique of randomized-iterations can be helpful for the case of any one-way function, and give a positive answer to this question. Interestingly, this method also improves the efficiency by an n^3 factor and reduces the seed length by a factor of n (which also implies improvement in the security of the construction) over the original HILL generator. All in all, we present a pseudorandom generator from *any* one-way function with seed length $\mathcal{O}(n^7)$ (Corollary 10) which is the best known to date.

Unlike in the case of regular functions, the hardness of inverting the randomized iterate deteriorates quickly when using any one-way function. Therefore we use only the first randomized iterate of a function, that is $x^1 = f(h(f(x)))$. Denote the degeneracy of y by $D_f(y) = \lceil \log |f^{-1}(y)| \rceil$ (this is a measure that divides the images of f to n categories according to their preimage size). Let b denote a hardcore-bit (again we take the Goldreich-Levin hardcore-bit [GL89]). Loosely speaking, we consider the bit $b(x^0)$ when given the value (x^1, h) (recall that $x^0 = f(x)$) and make the following observation: When $D_f(x^0) \geq D_f(x^1)$ then $b(x^0)$ is (almost) fully determined by (x^1, h), as opposed to when $D_f(x^0) < D_f(x^1)$ where $b(x^0)$ is essentially uniform. But in addition, when $D_f(x^0) = D_f(x^1)$ then $b(x^0)$ is computationally-indistinguishable from uniform (that is, looks uniform to any efficient observer), even though it is actually fully determined. The latter stems from the fact that when $D_f(x^0) = D_f(x^1)$ the behavior is close to that of a regular function.

---

[2] The collision-probability of a distribution is the probability of getting the same element twice when taking two independent samples from the distribution.

As a corollary we get that the bit $b(x^0)$ has entropy of no more than $\frac{1}{2}$ (the probability of $D_f(x^0) < D_f(x^1)$), but has entropy of at least $\frac{1}{2} + \frac{1}{\mathcal{O}(n)}$ in the eyes of any computationally-bounded observer (the probability of $D_f(x^0) \le D_f(x^1)$). In other words, $b(x^0)$ has entropy $\frac{1}{2}$ but *pseudo-entropy* of $\frac{1}{2} + \frac{1}{\mathcal{O}(n)}$. It is this gap of $\frac{1}{\mathcal{O}(n)}$ between the entropy and pseudo-entropy that eventually allows the construction of a pseudorandom generator.

Indeed, a function with similar properties lies at the basis of the HILL construction. HILL give a different construction that has entropy p but pseudo-entropy of at least $p + \frac{1}{\mathcal{O}(n)}$. However, in the HILL construction the entropy threshold p is unknown (i.e., not efficiently computable), while with the randomized iterate the threshold is $\frac{1}{2}$. This is a real advantage since knowledge of this threshold is essential for the overall construction. To overcome this, the HILL generator enumerates all values for p (up to an accuracy of $\Omega(\frac{1}{n})$), runs the generator with every one of these values and eventually combines all generators using an XOR of their outputs. This enumeration costs an additional factor n to the seed length as well an additional factor of n^3 to the number of calls to the underlying function f.

On pseudorandomness in NC^1: For the most part, the HILL construction is "depth" preserving. In particular, given two "non-uniform" hints of $\log n$ bits each (that specify two different properties of the one-way function), the reduction gives generators in NC^1 from any one-way function in NC^1. Unfortunately, without these hints, the depth of the construction is polynomial (rather than logarithmic). Our construction eliminates the need for one of these hints, and thus can be viewed as a step towards achieving generators in NC^1 from any one-way function in NC^1 (see [AIK04] for the significance of such a construction).

Related Work: Recently, Holenstein [Hol06] gave a generalized proof to the HILL construction. His proof formally proves the best known seed length for the HILL construction $\mathcal{O}(n^8)$, and further shows that if the underlying one-way function has exponential security (e.g. 2^{-Cn} for a constant C) then the seed length can be as low as $\mathcal{O}(n^5)$, or even $\mathcal{O}(n^4 \log^2 n)$ if the security of the PRG is not required to be exponential (but rather superpolynomial). In subsequent work [HHR06], we show a construction of a PRG based on exponentially strong one-way functions with seed length of only $\mathcal{O}(n^2)$ or respectively $\mathcal{O}(n \log^2 n)$ for a PRG with just superpolynomial security. The new construction follows by further developing the techniques introduced in this paper.

1.2   One-Way Functions - Amplification from Weak to Strong

The existence of one-way functions is essential to almost any task in cryptography (see for example [IL89]) and also sufficient for numerous cryptographic primitives, such as the pseudorandom generators discussed above. In general, for constructions based on one-way functions, we use what are called *strong* one-way functions. That is, functions that can only be inverted efficiently with negligible success probability. A more relaxed definition is that of an α-weak one-way function where $\alpha(n)$ is a polynomial fraction. This is a function that any efficient algorithm fails to invert on almost an $\alpha(n)$ fraction of the inputs. This definition is significantly weaker, however, Yao [Yao82] showed how to convert any weak one-way function into a strong one. The new strong one-way function simply consists of many independent copies of the weak function concatenated to each other. The solution of Yao, however, incurs a blow-up factor of at least $\omega(1)/\alpha(n)$ to the input length of the strong function[3], which translates to a significant loss in the security (as in the case of pseudorandom generators).

With this security loss in mind, several works have tried to present an efficient method of amplification from weak to strong. Goldreich et al. [GIL+90] give a solution for one-way permutations that has just a linear blowup in the length of the input. This solution generalizes to "known-regular" one-way functions (regular functions whose image size is efficiently computable), where its input length varies according to the required security. The input length is linear when security is at most $2^{\Omega(\sqrt{n})}$, but deteriorates up to $\mathcal{O}(n^2)$ when the required security is higher (e.g., security $2^{\mathcal{O}(n)}$).[4] Their construction uses a variant of randomized iterates where the randomization is via one random step on an expander graph.

Our Contribution to Hardness Amplification: We present an alternative efficient hardness amplification for regular one-way functions. Specifically, we show that the mth randomized iterate of a weak one-way function along with the randomizing hash functions form a strong one-way function (for the right parameter m). Moreover, in Theorem 13 we show that the latter holds also for the derandomized version of the randomized iterate, giving an almost linear construction. Our construction is arguably simpler and has the following advantages:

1. While the [GIL+90] construction works only for known regular weak one-way functions, our amplification works for any regular weak one-way functions (whether its image size is efficiently computable or not).
2. The input length of the resulting strong one-way function is $\mathcal{O}(n \log n)$ regardless of the required security. Thus, for some range of the parameters our solution is better than that of [GIL+90] (although it is worse than [GIL+90] for other ranges).

Note that our method may yield an $\mathcal{O}(n)$ input construction if bounded-space generators with better parameters become available.

The Idea: At the basis of all hardness amplification lies the fact that for any inverting algorithm, a weak one-way function has a set that the algorithm fails upon, called here the *failing-set* of this algorithm. The idea is that a large enough number of randomly chosen inputs are bound to hit every such failing-set and thus fail every algorithm. Taking independent random samples works well, but when trying to generate the inputs to f sequentially this rationale fails. The

---

[3] The $\omega(1)$ factor stands for the logarithm of the required security. For example, if the security is $2^{\mathcal{O}(n)}$ then this factor of order n.

[4] Loosely speaking, one can think of the security as the probability of finding an inverse to a random image $f(x)$ simply by choosing a random element in the domain.

reason is that sequential applications of f are not likely to give random output, and hence are not guaranteed to hit a failing-set. Instead, the natural solution is to use randomized iterations. However, it might be easy for an inverter to find some choice of randomizing hash functions so that all the iterates are outside of the required failing-set. To overcome this, the randomizing hash functions are also added to the output, and thus the inverter is required to find an inverse that includes the original randomizing hash functions. In the case of permutations it is obvious that outputting the randomizing hash functions is harmless, and thus the mth randomized iterate of a weak one-way permutation is a strong one-way permutation. However, the case of regular functions requires our analysis that shows that the randomized iterate of a regular one-way function remains hard to invert when the randomizing hash functions are public. We also note that the proof for regular functions has another subtlety. For permutations the randomized iterate remains a permutation and therefore has only a single inverse. Regular functions, on the other hand, can have many inverses. This comes into play in the proof, when an inverting algorithm might not return the right inverse that is actually needed by the proof.

A major problem with the randomized iterate approach is that choosing fully independent randomizing hash functions requires an input as long as that of Yao's solution (an input of length $\mathcal{O}(n \cdot \omega(1)/\alpha(n))$). What makes this approach appealing after all, is the derandomization of the hash functions using space-bounded generators, which reduces the input length to only $\mathcal{O}(n \log n)$. Note that in this application of the derandomization, it is required that the bounded-space generator not only approximate the collision-probability well, but also maintain the high probability of hitting any failing-set.

We note that there have been several attempts to formulate such a construction, using all of the tools mentioned above. Goldreich et al. [GIL+90] did actually consider following the GKL methodology, but chose a different (though related) approach. Phillips [Phi93] gives a solution with input length $\mathcal{O}(n \log n)$ using bounded-space generators but only for the simple case of permutations (where [GIL+90] has better parameters). Di Crescenzo and Impagliazzo [DI99] give a solution for regular functions, but only in a model where public randomness is available (in the mold of [HL92]). Their solution is based on pairwise-independent hash functions that serve as the public randomness. We are able to combine all of these ingredients into one general result, perhaps due to our simplified proof.

1.3   Additional Issues

- **On Non-Length-Preserving Functions:** Throughout the paper we focus on length preserving one-way functions. In the full version we demonstrate how our proofs may be generalized to use non-length preserving functions. This generalization requires the use of a construction of a family of *almost* pairwise-independent hash functions.
- **The Results in the Public Randomness Model:** Similarly to previous works, our results also give linear reductions in the public randomness model. This model (introduced by Herzberg and Luby [HL92]) allows the use of

public random coins that are not regarded a part of the input. However, our results introduce significant savings in the amount of public randomness that is necessary.

Paper Organization: In Section 2, we present our construction of pseudorandom generators from regular one-way functions. In Section 3, we present our improvement to the HILL construction of pseudorandom generators from any one-way function. Finally, in Section 4, we present our hardness amplification of regular one-way functions. Due to space limitations, we only give the outlines of some of the proofs. For the same reasons, we omit the standard definitions and notations. Both the full proofs and the definitions can be found in the paper's full version [HHR05].

2   Pseudorandom Generators from Regular One-Way Functions

2.1   Some Motivation and the Randomized Iterate

Recall that the BMY generator simply iterates the one-way permutation f on itself, and outputs a hardcore-bit of the intermediate step at each iteration. The crucial point is that the output of the function is also uniform in $\{0,1\}^n$ since f is a permutation. Hence, when applying f to the output, it is hard to invert this last application of f, and therefore hard to predict the new hardcore-bit (Yao shows [Yao82] that the unpredictability of bits implies pseudorandomness). Since the seed is essentially just an n bit string and the output is as long as the number of iterations, the generator actually stretches the seed.

  We want to duplicate this approach for general one-way functions, but unfortunately the situation changes drastically when the function f is not a permutation. After a single application of f, the output may be very far from uniform, and in fact, may be concentrated on a very small and easy fraction of the inputs to f. Thus, reapplying f to this output gives no hardness guarantees at all. In an attempt to salvage the BMY framework, Goldreich et. al. [GKL93] suggested to add a randomization step between every two applications of f, thus making the next input to f a truly random one. This modification that we call randomized iterates lies at the core of our work and is defined next:

Definition 1 (The kth Randomized Iterate of f). *Let $f : \{0,1\}^n \to \{0,1\}^n$ and let $\mathcal{H}$ be an efficient family of pairwise-independent hash functions from $\{0,1\}^n$ to $\{0,1\}^n$. For input $x \in \{0,1\}^n$ and $h_1, \ldots, h_{k-1} \in \mathcal{H}$ define the kth* Randomized Iterate $f^k : \{0,1\}^n \times \mathcal{H}^k \to Im(f)$ *recursively as:*

$$f^k(x, h_1, \ldots, h_k) = f(h_k(f^{k-1}(x, h_1, \ldots, h_{k-1}))),$$

where $f^0(x) = f(x)$. For convenience we denote by $x^k \stackrel{def}{=} f^k(x, h_1, \ldots, h_k)$.[5]

---

[5] We make use the notation x^k only when the values of $h_1, \ldots, h_k$ and x are clear by the presentation.

Another handy notation is the kth explicit randomized iterate $\widehat{f^k} : \{0,1\}^n \times \mathcal{H}^k \to Im(f) \times \mathcal{H}^k$ defined as:

$$\widehat{f^k}(x, h_1, \ldots, h_k) = (f^k(h_1, \ldots, h_k), h_1, \ldots, h_k).$$

The application of the randomized iterate for pseudorandom generators is a bit tricky. On the one hand, such a randomization costs a large number of random bits, much larger than what can be compensated for by the hardcore-bits generated in each iteration. So in order for the output to actually be longer than the input, we also output the descriptions of the hash functions (in other words, use the explicit randomized iterate $\widehat{f^k}$). But on the other hand, handing out the randomizing hash gives information on intermediate values such as $h_i(x^i)$. Hence, f might no longer be hard to invert when applied to such an input. Somewhat surprisingly, the last randomized iterate of a regular one-way function remains hard to invert even when the hash functions are known. This fact, which is central to the whole approach, was proved in [GKL93] when using a family of n-wise independent hash functions. We give a simpler proof that extends to pairwise-independent hash functions as well.

Remark: In the definition of randomized iterate we define $f^0(x) = f(x)$. This was chosen for ease of notation and consistency with the results for general OWFs (Section 3). For the regular OWF construction it suffices to define $f^0(x) = x$, thus saving a single application of the function f.

2.2   The Last Randomized Iteration is Hard to Invert

In this section we formally state and prove the key observation mentioned above. That is, that after applying k randomized-iterations of a regular one-way function f, it is hard to invert the last iteration, even if given access to all of the hash functions leading up to this point.

Lemma 2. *Let f be a length-preserving regular one-way function, $\mathcal{H}$ be an efficient family of pairwise-independent length-preserving hash functions and x^k be the kth randomized iterates of f (Definition 1). Then for any PPT A and every $k \in poly(n)$, we have:*

$$\Pr_{(x,h_1,\ldots,h_k) \leftarrow (U_n, \mathcal{H}^k)} [A(x^k, h_1, \ldots, h_k) = x^{k-1}] \in neg(n)$$

where the probability is also taken over the random coins of A.

More precisely, if such a PPT A succeeds with probability ε, then there exists a probabilistic polynomial time oracle machine M^A that succeeds in inverting f with probability at least $\varepsilon^3/8(k+1)$ with essentially the same running time as A.

We briefly give some intuition to the proof, illustrated with regard to the first randomized iterate. Suppose that we have an algorithm A that *always* finds x^0 given $x^1 = f^1(x, h)$ and h. In order to invert the one-way function f on an element $z \in Im(f)$, we simply need to find a hash h' that is consistent with z, in the sense that there exists an x' such that $z = f^1(x', h')$. Now we simply run $y = A(z, h')$, and output $h'(y)$ (and indeed $f(h'(y)) = z$). The point is that if f is a regular function, then finding a consistent hash is easy, because a random and independent h' is likely to be consistent with z. The actual proof follows this framework, but is far more involved due to the fact that the reduction starts with an algorithm A that has only a *small* (yet polynomial) success probability.

Proof. Suppose for sake of contradiction that there exists an efficient algorithm A that given $(x^k, h_1, \ldots, h_k)$ computes x^{k-1} with probability ε for some polynomial fraction $\varepsilon(n) = 1/poly(n)$ (for simplicity we write ε). In particular, A inverts the last-iteration of $\widehat{f^k}$ with probability at least ε, that is

$$\Pr_{(x, h_1, \ldots, h_k) \leftarrow (U_n, \mathcal{H}^k)} [f(h(A(\widehat{f^k}(x, h_1, \ldots, h_k)))) = f^k(x, h_1, \ldots, h_k)] \geq \varepsilon$$

Our goal is to use this procedure A in order to break the one-way function f. Consider the procedure M^A for this task:

M^A **on input** $z \in Im(f)$:
1. Randomly (and independently) choose $h_1, \ldots, h_k \in \mathcal{H}$.
2. Apply $A(z, h_1, \ldots, h_k)$ to get an output y.
3. If $f(h_k(y)) = z$ output $h_k(y)$, otherwise abort.

The rest of the proof of Lemma 2 shows that M^A succeeds with probability at least $\varepsilon^3/8(k+1)$ on inputs $z \in Im(f)$.

We start by focusing our attention only on those inputs for which A succeeds reasonably well. Recall that the success probability of A is taken over the choice of inputs to A, as induced by the choice of $x \in \{0, 1\}^n$ and $h_1, \ldots, h_k \in \mathcal{H}$ and the internal coin-tosses of A. The following Markov argument (proof omitted) implies that the probability of getting an element in the set that A succeeds on is not very small:

Claim 3. *Let* $S_A \subseteq Im(\widehat{f^k})$ *be the subset defined as:*

$$S_A = \left\{ (y, h_1, \ldots, h_k) \in Im(\widehat{f^k}) \mid \Pr[f(h_k(A(y, h_1, \ldots, h_k))) = y] > \frac{\varepsilon}{2} \right\},$$

then

$$\Pr_{(x, h_1, \ldots, h_k) \leftarrow (U_n, \mathcal{H}^k)} [\widehat{f^k}(x, h_1, \ldots, h_k) \in S_A] \geq \frac{\varepsilon}{2}.$$

Now that we identified a subset of polynomial weight of the inputs that A succeeds upon, we want to say that M^A has a fair (polynomially large) chance to hit outputs induced by this subset. This is formally shown in the following lemma.

Lemma 4. *For every set* $T \subseteq Im(\widehat{f^k})$, *if*

$$\Pr_{(x, h_1, \ldots, h_k) \leftarrow (U_n, \mathcal{H}^k)} [\widehat{f^k}(x, h_1, \ldots, h_k) \in T] \geq \delta,$$

then

$$\Pr_{(z,h_1,\ldots,h_k) \leftarrow (f(U_n),\mathcal{H}^k)}[(z,h_1,\ldots,h_k) \in T] \geq \delta^2/(k+1).$$

We stress that the probability in the latter inequality is over z drawn from $f(U_n)$ and an independently *chosen $h_1,\ldots,h_k \in \mathcal{H}$.*

Assuming Lemma 4, we may conclude the proof of Lemma 2. By Claim 3 we have that $\Pr[(x^k,h_1,\ldots,h_k) \in S_A] \geq \frac{\varepsilon}{2}$. By Lemma 4, taking $T = S_A$ and $\delta = \varepsilon/2$, we get that $\Pr[(z,h_1,\ldots,h_k) \in S_A] \geq \varepsilon^2/4(k+1)$. Thus, M^A has a $\varepsilon^2/4(k+1)$ chance of hitting the set S_A on which it will succeed with probability at least $\varepsilon/2$. Altogether, M^A succeeds in inverting f with the polynomial probability $\varepsilon^3/8(k+1)$, contradicting the one-wayness of f. ∎

Proof. (of Lemma 4) The lemma essentially states that with respect to $\widehat{f^k}$, any large subset of inputs induces a large subset of outputs. Thus, there is a fairly high probability of hitting this output set simply by sampling independent z and $h_1,\ldots,h_k$. Intuitively, if a large set of inputs induces a small set of outputs, then there must be many collisions in this set (a collision means that two different inputs lead to the same output). However, we show that this is impossible by proving that the collision-probability of the function $\widehat{f^k}$ is small. The proof therefore follows by analyzing the collision-probability of $\widehat{f^k}$. For every two inputs $(x_0,h_1^0,\ldots,h_k^0)$ and $(x_1,h_1^1,\ldots,h_k^1)$ to $\widehat{f^k}$, in order to have a collision we must first have that $h_i^0 = h_i^1$ for every $i \in [k]$, which happens with probability $(1/|\mathcal{H}|)^k$. Now, given that $h_i^0 = h_i^1 = h_i$ for all i (with a random $h_i \in \mathcal{H}$), we require also that $x_0^k = f^k(x_0,h_1,\ldots,h_k)$ equals $x_1^k = f^k(x_1,h_1,\ldots,h_k)$. If $f(x_0) = f(x_1)$ (happens with probability $1/|Im(f)|$), then a collision is assured. Otherwise, there must be an $i \in [k]$ for which $x_0^{i-1} \neq x_1^{i-1}$ but $x_0^i = x_1^i$ (where x_0 denotes the input x). Since $x_0^{i-1} \neq x_1^{i-1}$, due to the pairwise-independence of h_i, the values $h_i(x_0^{i-1})$ and $h_i(x_1^{i-1})$ are uniformly random values in $\{0,1\}^n$, and thus $f(h_i(x_0^{i-1})) = f(h_i(x_1^{i-1}))$ happens with probability $1/|Im(f)|$. Altogether:

$$CP(\widehat{f^k}(U_n,\mathcal{H}^k)) \leq \frac{1}{|\mathcal{H}|^k} \sum_{i=0}^{k} \frac{1}{|Im(f)|} \leq \frac{k+1}{|\mathcal{H}|^k|Im(f)|} \tag{1}$$

On the other hand, we check the probability of getting a collision inside the set T, which is a lower bound on the probability of getting a collision at all. We first request that both $(x_0,h_1^0,\ldots,h_k^0) \in T$ and $(x_1,h_1^1,\ldots,h_k^1) \in T$. This happens with probability at least δ^2. Then, once inside T, we know that the probability of collision is at least $1/|T|$. Altogether:

$$CP(\widehat{f^k}(U_n,\mathcal{H}^k)) \geq \delta^2 \frac{1}{|T|} \tag{2}$$

Combining (1) and (2) we get $\frac{|T|}{|\mathcal{H}|^k|Im(f)|} \geq \frac{\delta^2}{k+1}$.
But the probability of getting a value in T when choosing a random element in $Im(f) \times \mathcal{H}^k$ is exactly $\frac{|T|}{|\mathcal{H}|^k|Im(f)|}$. Thus, $\Pr[(z,h_1,\ldots,h_k) \in T] \geq \delta^2/(k+1)$ as requested. ∎

Remark: The proof of Lemma 4 is where the regularity of the one-way function is required. In the case of general one-way functions, we cannot apply the above proof, since the collision-probability at the heart of the proof (i.e., $\widehat{CP(f^k}(U_n, \mathcal{H}^k)))$ might be much larger than the collision probability of the uniform distribution over $Im(f)$. Alternatively, we could prove the lemma in the case that the last element in a sequence of applications is at least as heavy as all the elements along the sequence. Unfortunately, for general OWFs this occurs with probability that deteriorates linearly (in the length of the sequence). Thus, using a long sequence of iterations is likely to lose the hardness of the original OWF. [6]

2.3   A Pseudorandom Generator from a Regular One-Way Function

After showing that the randomized-iterations of a regular one-way function are hard to invert, it is natural to follow the footsteps of the BMY construction to construct a pseudorandom generator. Rather than using simple iterations of the function f, randomized-iterations of f are used instead, with fresh randomness in each application. As in the BMY case, a hardcore-bit of the current input is taken at each stage. Formally:

Theorem 5. *Let $f : \{0,1\}^n \to \{0,1\}^n$ be a regular length-preserving one-way function and let $\mathcal{H}$ be an efficient family of pairwise-independent length preserving hash functions, let G be:*

$$G(x, h_1 \ldots, h_n, r) = (b_r(x^0), \ldots, b_r(x^n), h_1, \ldots, h_n, r),$$

where:
- *$x \in \{0,1\}^n$ and $h_1 \ldots, h_n \in \mathcal{H}$.*
- *Recall that $x^0 = f(x)$ and for $1 \le i \le n$ $x^i = f(h_i(x^{i-1}))$.*
- *$b_r(x^i)$ denotes the GL-hardcore bit of x^i.*

Then G is a pseudorandom generator.

Note that the above generator does not require the knowledge of the preimage size of the regular one-way function. The generator requires just $n+1$ calls to the underlying one-way function f (each call is on an n bit input). The generator's input is of length $m = \mathcal{O}(n^2)$ and it stretches the output to $m + 1$ bits. The proof of security follows by a standard hybrid argument and is given in the full version of the paper [HHR05].

2.4   An Almost-Linear-Input Construction from a Regular One-Way Function

The pseudorandom generator presented in the previous section (Theorem 5) stretches a seed of length $\mathcal{O}(n^2)$ by one bit. Although this is an improvement over the GKL generator, it still translates to a rather high loss of security. That

---

[6] In [HHR06] we use a variant of the latter idea to get an efficient pseudorandom from exponentially hard one-way functions.

is, the security of the generator on m bits relies on the security of regular one-way function on $\sqrt{m}$ bits. In this section we give a modified construction of the pseudorandom generator that takes a seed of length only $m = \mathcal{O}(n \log n)$.

Notice that the input length of the generator is dominated by the description of the n independent hash functions $h_1, \ldots, h_n$. The idea of the new construction is to give a derandomization of the choice of the n hash functions. Thus, $h_1, \ldots, h_n$ are no longer chosen independently, but are chosen in a way that is sufficient for the proof to go through. The derandomization uses generators against bounded-space distinguishers. Specifically, we can use the generator of Nisan [Nis92] (or that of Impagliazzo, Nisan and Wigderson [INW94]). An important observation is that calculating the randomized iterate of an input can be viewed as a bounded-space algorithm, alternatively presented here as a bounded-width layered branching-program. More accurately, at each step the branching program gets a random input h_i and produces $x^{i+1} = f(h_i(x^i))$. We will show that indeed when replacing $h_1, \ldots, h_n$ with the output of a generator that fools related branching programs, then the proof of security still holds (and specifically the proof of Lemma 4).

For our application we use the bounded-space generator with parameters $t = n$, $S = 2n$ and $\ell = 2n$ (or more generally, ℓ is taken to be the description length of a hash function in $\mathcal{H}$). Finally, the error is chosen to be $\varepsilon = 2^{-n}$. The generator therefore stretches $\mathcal{O}(n \log n)$ bits to $n \cdot 2n$ bits. Denote the bounded-space generator by $BSG : \{0,1\}^{cn \log n} \to \{0,1\}^{2n^2}$, where c is a universal constant. For convenience denote $\tilde{n} = cn \log n$.

The New Pseudorandom Generator

Theorem 6. *For any regular length-preserving one-way function f, let G' be:*

$$G'(x, \tilde{h}, r) = (b_r(x^0), \ldots, b_r(x^n), \tilde{h}, r),$$

where:

- *$x \in \{0,1\}^n$ and $\tilde{h} \in \{0,1\}^{\tilde{n}}$.*
- *$(h_1, \ldots, h_n) = BSG(\tilde{h})$.*
- *Recall that $x^0 = f(x)$ and for $1 \leq i \leq n$, $x^i = f(h_i(x^{i-1}))$.*
- *$b_r(x^i)$ denotes the GL-hardcore bit of x^i.*

Then G' is a pseudorandom generator.

Proof outline: The proof of the derandomized version follows in the steps of the proof of Theorem 5. We give a high-level outline of this proof, focusing only on the main technical lemma that changes slightly.

The proof first shows that given the kth randomized iterate x^k and all of the randomizing hash functions, it is hard to compute x^{k-1} (analogously to Lemma 2), only now this also holds when the hash functions are chosen as the output of the bounded-space generator. The proof is identical to the proof of 2, only replacing appearances of $(h_1, \ldots, h_k)$ with the seed $\tilde{h}$. Again, the key to the proof is the following technical lemma (slightly modified from Lemma 4, the proof of the lemma is given in the full version [HHR05]).

Lemma 7. *For every set* $T \subseteq Im(f) \times \{0,1\}^{\tilde{n}}$, *if*

$$\Pr_{(x,\tilde{h}) \leftarrow (U_n, U_{\tilde{n}})} [(x^k, \tilde{h}) \in T] \geq \delta,$$

then

$$\Pr_{(z,\tilde{h}) \leftarrow (f(U_n), U_{\tilde{n}})} [(z, \tilde{h}) \in T] \geq \delta^2/(k+2),$$

where probability is over $z \in f(U_n)$ *and an* independently *chosen* $\tilde{h} \in \{0,1\}^{\tilde{n}}$.

Once we know that x^{k-1} is hard to compute, we deduce that one cannot predict a hardcore-bit $b_r(x^{k-1})$ given x^k and the seed to the bounded-space generator. From here, the proof follow just as the proof of Theorem 5 in showing that the output of G' is an unpredictable sequence and therefore a pseudorandom sequence. ∎

Remark: It is tempting to think that one should replace Nisan/INW generator in the above proof with the generator of Nisan and Zuckerman [NZ96]. That generator may have seed of size $\mathcal{O}(n)$ (rather than $\mathcal{O}(n \log n)$) when S=2n as in our case. Unfortunately, with such a short seed, that generator will incur an error $\varepsilon = 2^{-n^{1-\gamma}}$ for some constant γ, which is too high for our proof to work. In order for the proof to go through we need that $\varepsilon < poly(n)/|Im(f)|$. Interestingly, this means that we get a linear-input construction when the image size is significantly smaller than 2^n. In order to achieve a linear-input construction in the general case, we need better generators against LBPs (that have both short seed and small error).

3   Pseudorandom Generator from Any One-Way Function

Our implementation of a pseudorandom generator from any one-way function follows the route of [HILL99] (we follow the presentation and proof of the HILL generator given in [Hol06]), but takes a totally different approach in the implementation of its initial step.

The "pseudo-entropy" of a distribution is at least k, if it is computationally-indistinguishable from some distribution that has entropy k. The basic building block of the HILL generator is a "pseudo-entropy pair".[7] Informally, the latter is a pair of a function and predicate on the same input with the following property: When given the output of the function, the pseudo-entropy of the predicate's output is noticeably larger than the real (conditional) entropy of this bit. In their construction [HILL99] exploit this gap between real and pseudo entropy to construct a pseudorandom generator. We show that the first explicit randomized iterate of a one-way function together with a standard hardcore predicate forms a pseudo-entropy pair. Moreover, this pair has better properties than the original one and hence "plugging" it as the first step of the HILL construction results in a better overall construction. Let us now turn to a more formal discussion. We define the pseudo-entropy pair as follows:

---

[7] We note that [HILL99] used this notion implicitly without giving it an explicit definition.

Definition 8. *[Pseudo-entropy pair (PEP)] Let δ and γ be some positive functions over $\mathbb{N}$ and let $g : \{0,1\}^n \to \{0,1\}^{\ell(n)}$ and $b : \{0,1\}^n \to \{0,1\}$ be polynomial-time computable functions. We say that (g,b) is a (δ, γ)-PEP if*

1. $H(b(U_n) \mid g(U_n)) \le \delta(n)$.
2. b is a $(\delta(n) + \gamma(n))$-hard predicate of g.

[HILL99] show how to construct a (δ, α)-PEP, where $\delta \in [0,1]$ is some **unknown** value and α is any fraction noticeably smaller than $\frac{1}{2n}$, using any one-way function.[8] Then they present a construction of a pseudorandom generator using a $(\delta, \frac{1}{\mathcal{O}(n)})$-PEP where δ is **known**. To overcome this gap, the HILL generator enumerates all values for δ (up to an accuracy of $\Omega(\frac{1}{n})$), runs the generator with every one of these values and eventually combines all generators using an XOR of their outputs. This enumeration costs an additional factor of n to the seed length as well as n^3 times more calls to the underlying one-way function.

We prove that the first explicit randomized iterate of a one-way function can be used to construct a $(\frac{1}{2}, \alpha)$-PEP, where α is any fraction noticeably smaller than $\frac{1}{2n}$. By combining our PEP with the second part of the [Hol06] construction, we get a pseudorandom generator that is more efficient and has better security than the original construction due to [HILL99]/[Hol06] (the efficiency improves by a factor of n^3 and the security by a factor of n). Formally, we get the following theorem.

Theorem 9. *Let $f : \{0,1\}^n \to \{0,1\}^n$ be an efficiently computable function and let $\varepsilon, \gamma : \mathbb{N} \to [0,1]$. Then there is an efficiently computable function G with the following properties:*

- *G is length expanding.*
- *G has input length $\mathcal{O}(n^6 \log(\frac{1}{\varepsilon(n)}))$.*
- *Any algorithm A that distinguishes the output of G from the uniform distribution with advantage γ, can be used to construct an algorithm that inverts f with probability $\Omega(\frac{1}{n^{13}})$ and runs in time $poly(\frac{1}{\Omega(\frac{\gamma}{n^7 \log(\frac{1}{\varepsilon})}) - \varepsilon}, n, T_A(n))$, where T_A is the running time of A.*

As a corollary we deduce the main statement of this section:

Corollary 10. *Let $f : \{0,1\}^n \to \{0,1\}^n$ be a one-way function, then there exists a pseudorandom generator with seed length $\mathcal{O}(n^7)$.*

---

[8] [HILL99] actually prove somewhat stronger result. Not only that the predicate of their PEP is $(\delta + \alpha)$-hard, but the hardness comes from the existence of a "hardcore-set" of density $\delta + \frac{1}{2n}$. Where the latter is a subset of the input such that the value of b is computationally unpredictable over it. This additional property was used by [HILL99] original proof, but it is not required by the new proof due to [Hol06]. We note that our PEP, presented next, also has such a hardcore-set.

3.1   A Pseudo-Entropy Pair Based on the Randomized Iterate

For a given one-way function f, we have defined (Definition 1) its first explicit randomized iterate as $\widehat{f^1}(x,h) = (f(h(f(x))), h)$. We present an "extended" version of the above function with the following properties: First, it maintains some hardness of the original one-way function. The hardness is maintained in a sense that with probability $\frac{1}{2} + \frac{1}{2n}$ it is hard to compute the value of $x^0 = f(x)$ given the output. Second, we show that with probability $\frac{1}{2}$ the value of x^0 can be determined w.h.p. from the output. Formally,

Definition 11 (The Extended Randomized Iterate). *Let* $f : \{0,1\}^n \to \{0,1\}^n$ *be a one-way function, let* $m = \lceil 3\log(n) + 8 \rceil$ *and let* $\mathcal{H}$ *and* $\mathcal{H}_E$ *be two families of pairwise-independent hash functions from* $\{0,1\}^n$ *to* $\{0,1\}^n$ *and* $\{0,1\}^n$ *to* $\{0,1\}^m$ *respectively. We define* g, the extended randomized iterate of f, *as:*

$$g(x, h, h_E) = (\widehat{f^1}(x,h), h_E(f(x)), h_E)$$

where $x \in \{0,1\}^n$, $h \in \mathcal{H}$ *and* $h_E \in \mathcal{H}_E$.

Lemma 12. *Let* $\mathcal{H}$, $\mathcal{H}_E$ *and* g *be as in Definition 11. For* $r \in \{0,1\}^n$, *let* $g'(x, h, h_E, r) = (g(x, h, h_E), r)$ *and let* $b(x, h, h_E, r) = b_r(f(x))$, *where* b_r *is the Goldreich-Levin predicate. Let* W *be a random variable uniformly distributed over* $Dom(g)$ *and let* α *be noticeably smaller than* $\frac{1}{2n}$, *then the following hold:*

1. $H(b(W) \mid g'(W)) \le \frac{1}{2}$.
2. b *is a* $\left(\frac{1}{2} + \alpha\right)$-*hard predicate of* g', *for any* α *that is noticeably smaller than* $\frac{1}{2n}$.

Hence (g', b) *is a* $\left(\frac{1}{2}, \alpha\right)$-PEP.

4   Hardness Amplification Of Regular One-Way Functions

In this section we present an efficient hardness amplification of any regular weak one-way function. As mentioned in the introduction (Section 1.2), the key to hardness amplification lies in the fact that every α-weak one-way function has a *failing-set* for every efficient algorithm. This is a set of density almost α that the algorithm fails to invert f upon. Sampling sufficiently many independent inputs to f is bound to hit *every* failing set and thus fail every algorithm. Indeed, the basic hardness amplification of Yao [Yao82] does exactly this. Since independent sampling requires a long input, we turn to use the randomized iterate, which together with the derandomization method, reduces the input length to $\mathcal{O}(n \log n)$.

Theorem 13. *Let* $f : \{0,1\}^n \to \{0,1\}^n$ *be a regular* $\alpha(n)$-*weak one-way function, let* $m = \lceil \frac{4n}{\alpha(n)} \rceil$ *and let* $\widehat{f^m}$ *and* $\mathcal{H}$ *be as in Definition 1. Let* BSG *be a bounded-space generator against* $(2n, n+1, 2n)$-*layered-branching-programs with* *seed length* $\tilde{n} \in \mathcal{O}(n \log n)$ *and error* 2^{-2n}. *Define* $f' : \{0,1\}^n \times \{0,1\}^{\tilde{n}} \to$ $\{0,1\}^n \times \{0,1\}^{\tilde{n}}$ *as*

$$f'(x, \tilde{h}) = (f^m(x, h_1, \ldots, h_m), \tilde{h})$$

where $x \in \{0,1\}^n$, $\tilde{h} \in \{0,1\}^{\tilde{n}}$ *and* $h_1, \ldots, h_m = BSG(\tilde{h})$. *Then* f' *is a (strong) a one-way function.*

Acknowledgments

We are grateful to Oded Goldreich, Moni Naor, Asaf Nussbaum, Eran Ofek and Ronen Shaltiel for helpful conversations. We also thank Tal Moran and Ariel Gabizon for reading a preliminary version of this manuscript.

References

[AIK04]   B. Applebaum, Y. Ishai, and E. Kushilevitz. Cryptography in NC^0. In *45th FOCS*, pages 166–175, 2004.

[BM82]   M. Blum and S. Micali. How to generate cryptographically strong sequences of pseudo random bits. In *23th FOCS*, pages 112–117, 1982.

[DI99]   G. Di Crescenzo and R. Impagliazzo. Security-preserving hardness-amplification for any regular one-way function. In *31st STOC*, pages 169–178, 1999.

[GGM86]   O. Goldreich, S. Goldwasser, and S. Micali. How to construct random functions. *Journal of the ACM*, 33(2):792–807, 1986.

[GIL+90]   O. Goldreich, R. Impagliazzo, L. Levin, R. Venkatesan, and D. Zuckerman. Security preserving amplification of hardness. In *31st FOCS*, pages 318–326, 1990.

[GKL93]   O. Goldreich, H. Krawczyk, and M. Luby. On the existence of pseudorandom generators. *SIAM Journal of Computing*, 22(6):1163–1175, 1993.

[GL89]   O. Goldreich and L.A. Levin. A hard-core predicate for all one-way functions. In *21st STOC*, pages 25–32, 1989.

[HHR05]   I. Haitner, D. Harnik, and O. Reingold. On the power of the randomized iterate. ECCC, TR05-135, 2005.

[HHR06]   I. Haitner, D. Harnik, and O. Reingold. Efficient pseudorandom generators from exponentially hard one-way functions. In ICALP 2006, 2006.

[HILL99]   J. Håstad, R. Impagliazzo, L. A. Levin, and M. Luby. A pseudorandom generator from any one-way function. *SIAM Journal of Computing*, 29(4):1364–1396, 1999.

[HL92]   A. Herzberg and M. Luby. Pubic randomness in cryptography. In *CRYPTO '92, LNCS*, volume 740, pages 421–432. Springer, 1992.

[Hol06]   T. Holenstein. Pseudorandom generators from one-way functions: A simple construction for any hardness. In *TCC '06*, LNCS. Springer, 2006.

[IL89]   R. Impagliazzo and M. Luby. One-way functions are essential for complexity based cryptography. In *30th FOCS*, pages 230–235, 1989.

[INW94]   R. Impagliazzo, N. Nisan, and A. Wigderson. Pseudorandomness for network algorithms. In *26th STOC*, pages 356–364, 1994.

[IZ89]   R. Impagliazzo and D. Zuckerman. How to recycle random bits. In *30th FOCS*, pages 248–253, 1989.

[Lev87]   L. A. Levin. One-way functions and pseudorandom generators. *Combinatorica*, 7:357–363, 1987.

[LR88]   M. Luby and C. Rackoff. How to construct pseudorandom permutations from pseudorandom functions. *SIAM Journal of Computing*, 17(2):373–386, 1988.

[Nao91]   M. Naor. Bit commitment using pseudorandomness. *Journal of Cryptology*, 4(2):151–158, 1991.

[Nis92]   N. Nisan. Pseudorandom generators for space-bounded computation. *Combinatorica*, 12(4):449–461, 1992.

[NZ96]   N. Nisan and D. Zuckerman. Randomness is linear in space. *Journal of Computer and System Sciences (JCSS)*, 52(1):43–52, 1996.

[Phi93]   S. Phillips. Security preserving hardness amplification using PRGs for bounded space. Preliminary Report, Unpublished, 1993.

[Yao82]   A. C. Yao. Theory and application of trapdoor functions. In *23rd FOCS*, pages 80–91, 1982.

Strengthening Digital Signatures Via Randomized Hashing

Shai Halevi and Hugo Krawczyk

IBM T.J. Watson Research Center, Yorktown Heights, New York 10598
shaih@alum.mit.edu, hugo@ee.technion.ac.il

Abstract. We propose randomized hashing as a mode of operation for cryptographic hash functions intended for use with standard digital signatures and without necessitating of any changes in the internals of the underlying hash function (e.g., the SHA family) or in the signature algorithms (e.g., RSA or DSA). The goal is to free practical digital signature schemes from their current reliance on strong collision resistance by basing the security of these schemes on significantly weaker properties of the underlying hash function, thus providing a *safety net* in case the (current or future) hash functions in use turn out to be less resilient to collision search than initially thought.

We design a specific mode of operation that takes into account engineering considerations (such as simplicity, efficiency and compatibility with existing implementations) as well as analytical soundness. Specifically, the scheme consists of a regular use of the hash function with randomization applied only to the message before it is input to the hash function. We formally show the sufficiency of weaker than collision-resistance assumptions for proving the security of the scheme.

1 Introduction

Recent cryptanalytical advances in the area of collision-resistant hash functions (CRHF) [8,5,15,6,16,29,30,31,32], especially the attacks against MD5 and SHA-1, have shaken our confidence in the security of existing hash functions as well as in our ability to design secure CRHF. These attacks remind us that cryptography is founded on heuristic constructions whose security may be endangered unexpectedly, and highlight the importance of designing mechanisms that require as little as possible from their basic cryptographic building blocks. In particular, they indicate that one should move away (to the extent possible) from schemes whose security fundamentally depends on full collision resistance.

The most prominent example of schemes that are endangered by these attacks are digital signatures where collisions in the hash function directly translate into forgeries. The goal of this work is to free existing (standardized) signature schemes from their dependence on full collision-resistance of the underlying hash function, while making as small as possible modifications to the signature and hashing algorithms. Specifically, we propose a *randomized mode of operation* for hash functions such that (i) no change to the underlying hash functions

C. Dwork (Ed.): CRYPTO 2006, LNCS 4117, pp. 41–59, 2006.
© International Association for Cryptologic Research 2006

(e.g., the SHA family), the signature algorithms (e.g., RSA, DSA), or their implementations is required; (ii) changes to the signing process are the minimal required by any randomization scheme, namely, the choice of a short random string by the signer and transporting this string as part of, or in addition to, the existing signature; and (iii) *security of the resultant signature scheme does not depend on the resistance of the hash function to off-line collision attacks.*

Armoring signature schemes with this mode of operation provides a *safety net* for the security of digital signatures in the case that the collision resistance of the underlying hash function is broken or weakened. At the same time, by following important engineering considerations, as in points (i) and (ii) above, one facilitates the adoption of the resultant schemes into practice. In particular, treating the hash function as a black box allows to preserve existing implementations of these functions or their future replacements. Moreover, the many other applications of hash functions (e.g., MAC, PRFs, commitments, fingerprinting, etc.) need not be changed or adapted to a new hash design; instead, these applications may use the hash function, or any "drop-in replacement", exactly as they do now.

Collision resistance and the Merkle-Damgård (M-D) construction. Roughly, a (length-decreasing) function H is collision resistant if given the description of H, no feasible attacker can find two messages M, M' such that $H(M) = H(M')$, except with insignificant probability.[1] Contemporary constructions of (allegedly) collision-resistant hash functions follow the so called Merkle-Damgård (M-D) iterated construction [17,9]. Such constructions start with a *compression function* h that maps input pairs (c, m) into an output c' where c and c' are of fixed length n (e.g., $n = 160$) and m is of fixed length b (e.g., $b = 512$). Given such a compression function h and a fixed n-bit initial value (IV), denoted c_0, a hash function H on arbitrary-length inputs is defined by iterating h as follows: On input message M, the message is broken into b-bit blocks $M = (m_1, m_2, \ldots, m_L)$, then one computes $c_i \leftarrow h(c_{i-1}, m_i)$ for $i = 1, 2, \ldots, L$ and finally the last n-bit value c_L is output as the result of $H(M)$.

Throughout this paper we typically denote the (iterated) hash function by H and its compression function by h. For an M-D function H, we also sometimes write $H^{c_0}(M)$ when we want to explicitly name the initial value c_0. To handle non-suffix-free message spaces, the M-D construction appends the length at the end of the input and uses some padding rules. We will initially ignore these issues and assume that the input consists of an integral number of blocks and that it is taken from a suffix-free set; we will return to the issue of length appending in Section 5.

Target Collision Resistance. Signature schemes that do not depend on full collision resistance were constructed in the influential work of Naor and Yung [20] who introduced the notion of *universal one-way hash functions, or UOWHF*. Later, Bellare and Rogaway [3] renamed them to the more descriptive (and

---

[1] Formalizing collision resistance requires to consider H as a family of functions rather than as a single function; we ignore this technicality for the informal discussion here.

catchy) name of *target collision resistant* (TCR) hash functions, a term that we adopt here. Roughly, a family of hash functions $\{H_r\}_{r \in R}$ (for some set R) is target collision resistant if no efficient attacker A can win the following game, except with insignificant probability: A chooses a first message M, then receives a random value $r \in_R R$, and it needs to find a second message $M' \neq M$ such that $H_r(M') = H_r(M)$. The value r is called a hashing key, or a *salt*. In [3], TCR families were used to extend signature schemes on short messages into signature schemes on arbitrary messages: To sign a long message M, the signer chooses a fresh random salt r and then uses the underlying signature scheme to sign the pair $(r, H_r(M))$.

Enhanced target collision resistance. Standard signature schemes such as RSA and DSA are defined using a deterministic hash function H and designed to only sign the hash value $H(M)$. When moving to a TCR-based scheme one replaces $H(M)$ with $H_r(M)$ but also needs to sign the salt r. Unfortunately, certain signature schemes, such as DSA, do not accommodate the signing of r in addition to $H_r(M)$. Even in cases, such as RSA, where both r and $H_r(M)$ can be included under the signed block (RSA moduli are long enough to accommodate both values) signing r requires changing the message encoding standards (such as PKCS#1) which may be impractical in some settings. To solve these difficulties, we propose a strengthened mode of operation, that provides a stronger security guarantee than TCR. The new notion, that we call *enhanced TCR* (eTCR), is sufficiently strong to ensure security of the resultant signatures even if we only apply the underlying signature to $H_r(M)$ and *do not sign the salt r*. Specifically, the TCR attack scenario is modified so that after committing to M and receiving the salt r, the attacker can supply a second message M' *and a second salt r'*, and it is considered successful if $(r, M) \neq (r', M')$ but $H_r(M) = H_{r'}(M')$.

1.1   Our Randomization Schemes

The main contribution of this work is in presenting simple and efficient randomization schemes that use any Merkle-Damgård iterated hash function as-is (i.e., in a black-box fashion) and have short salts, and which we prove to be TCR and/or eTCR under weak assumptions on the underlying compression function. Specifically, we relate the security of our schemes to "second-preimage resistance" properties of the compression function. Recall that a compression function h is called second preimage resistant (SPR) if given a uniformly random pair (c, m) it is infeasible to find a second pair $(c', m') \neq (c, m)$ such that $h(c', m') = h(c, m)$.

The first scheme that we consider, denoted H_r, simply XOR's a random b-bit block r to each message block before inputting it into the hash function (see below for a comparison of this scheme to one in [26]). That is,

$$H_r^c(m_1, \ldots, m_L) \stackrel{\text{def}}{=} H^c(m_1 \oplus r, \ldots, m_L \oplus r). \tag{1}$$

We show this scheme to be TCR under SPR-like assumption on the underlying compression function h (see below). On the other hand, the H_r construction is clearly not eTCR; in order to obtain an eTCR scheme we modify H_r by prepending the salt block r to the message (in addition to xor-ing r to every message block as before), thus computing

$$\tilde{H}_r^c(M) \overset{\text{def}}{=} H_r^c(0|M) = H^c(r, m_1 \oplus r, \ldots, m_L \oplus r) \tag{2}$$

Remarkably, for any M-D iterated hash function H we can prove that $\tilde{H}_r$ is enhanced-TCR under *the same relaxed conditions on the compression function* that we use to prove that H_r is TCR.

These precise conditions are presented in Section 2. Roughly, we show two properties of the compression function h that are sufficient for H_r to be TCR (resp. for $\tilde{H}_r$ to be eTCR). Both properties are related to the second-preimage resistance of h. One property, c-SPR, is rather natural but also rather strong, and is provided mostly as a toy problem for cryptanalysts to sink their teeth in. The other property, e-SPR, is less natural but is closer to SPR and reflects more accurately the "real hardness" of the constructions H_r and $\tilde{H}_r$. We also note that c-SPR is related to the "hierarchy of collision resistance" of Mironov [18], and that e-SPR is a weaker property than the L-order TCR property of Hong et al. [14].

We stress that ultimately, the question of whether or not a particular compression function (such as those in the SHA family) is e-SPR can only be addressed by cryptanalysts trying to attack this property. As we explain in Section 2, breaking the compression function in the sense of e-SPR is a much harder task than just finding collisions for it. Moreover, the randomized setting represents a *fundamental shift* in the attack scenario relative to the traditional use of deterministic hash functions. In the latter case, the attacker can find collisions via (heavy) *off-line* computation and later use these collisions in a signature forgery attack at any time and with any signer. In contrast, in the randomized setting, a meaningful collision is one that utilizes a random, unpredictable r *chosen by the signer anew with each signature* and revealed to the attacker only with the signature on the message M (in particular, only after the attacker committed to the message M).

Can the signer cheat? A possible objection to the randomized setting is that it may allow a signer to find two messages with the same hash value. This, however, represents no violation of signature security (c.f. the standard definition of Goldwasser, Micali, and Rivest [12]). Specifically, the potential ability of the signer to find collisions does not allow any other party to forge signatures thus protecting the signer against malicious parties. At the same time, recipients of signatures are also protected since the signer cannot disavow a signature by presenting a collision; the simple rule is, as in the case of human signatures, that the signer is liable for any message carrying his valid signature (even if the signer can show a collision between two messages). It is also worth noting that our schemes have the property that as long as the underlying hash function is collision resistant then even the signer cannot find collisions. More generally, the signature schemes resulting from our randomization modes are never weaker (or less functional) than the schemes that use a deterministic hash function, not even when the underlying hash function is truly collision resistant; and, of course, our schemes are much more secure once the hash function ceases to be collision resistant.

Practical Considerations. As seen, our randomized hashing schemes require no change to the underlying (compression and iterated) hash function. These schemes can be used, as described before, with algorithms such as RSA and DSA, to provide for digital signatures that remain secure even in the presence of collisions for the underlying hash functions. For this, the value $H(M)$ currently signed by the standard schemes needs to be replaced with $H_r(M)$ or with $\tilde{H}_r(M)$. However, while in the case of H_r one needs to also sign the salt r, in the $\tilde{H}_r$ case the latter is not needed. In this sense, $\tilde{H}_r$ provides for a more flexible and practical scheme and hence it is the one that we suggest for adoption into practice. In particular, using $\tilde{H}_r$, the salt r may be transported by an application as part of (or in addition to) a message, or r can be included under the signature itself (under the signed value in RSA or by re-using the random $r = g^k$ component in DSA). We further discuss these issues in Section 5. A full specification of the $\tilde{H}_r$ mode is included in [13].

Related Schemes. Bellare and Rogaway [3] explored the problem of constructing TCR hashing for long messages from TCR functions that work on short inputs, and showed that the M-D iteration of a TCR compression function *does not necessarily* result in a long-message TCR. Instead, they presented a few other constructions of long-message TCR hash functions from short-message TCR compression functions. Shoup [26] continued this line of research and offered the following elegant scheme that comes close to the original M-D iterated construction. The salt in Shoup's scheme consists of one b-bit message block r and a set S of $\log L$ chaining variables $s_0, s_1, \ldots$, where L is the number of message blocks. The scheme consists of a regular M-D construction except that (i) every message block is xor-ed with r before it is input to the compression function h, and (ii) the intermediate values c_i output by each iteration of h are xor-ed with one of the s_i's to form the chaining value input into the following application of h. That is, the M-D scheme is modified as follows: for $i = 1, 2, \ldots, L$, $c_i \leftarrow h(c_{i-1} \oplus s_{j_i}, m_i \oplus r)$ where j_i chooses one of the elements in the set S. Shoup proved that if the compression function is SPR then the construction yields a TCR function on messages of arbitrary length.

Note that our H_r scheme (Eqn. (1)) is similar to Shoup's in that the input blocks are xor-ed with the random r. However, the two schemes differ substantially in that: (i) H_r uses the hash function as a black box while Shoup's scheme requires the change of the internals of H to accommodate the xor-ing of the s_i's values with the partial results of the compression function; and (ii) in H_r the amount of randomness is fixed while in Shoup's it is logarithmic in the length of the message and hence non-constant; this requires significantly more random bits and results in longer signatures. In addition, Shoup's scheme (like H_r) is not eTCR. For the latter property, we need the modified $\tilde{H}_r$ scheme (Eqn. (2)).

HMAC. One interesting question is why not take as the randomization mode a well established construction such as HMAC. It is easy to see that for the purpose of TCR hashing the HMAC scheme (with the salt used as a random but known key) is not stronger than just pre-pending a random value to a message, a measure of limited effect against recent cryptanalysis. Other measures, such as appending randomness at the end, or even in the middle of a message, are even worse in that they do not help against collision attacks. In particular, this shows that the simplicity of our schemes cannot be explained by the naïve view that "any randomization of the input to the hash function will work." Yet, one analogy to the case of HMAC is illustrative here: the fact that the HMAC design (as a message authentication code) has not been endangered by the latest collision attacks is due to its deliberate non-reliance on full collision resistance. We believe that randomized hashing could have a similar significant effect on the future security of digital signatures.

Relations between various notions. In our presentation we use quite a few notions of security for hash functions, some new and others well known (see [24]). Articulating the relations between these notions is *not* the focus of this paper. In some places we comment about implication or separation between these notions, but we do not explicitly show separation examples in this extended abstract. (This will divert attention from what we see as the heart of this work, namely the analysis of the practical constructions H_r and $\tilde{H}_r$.)

Organization. In Section 2 we present and discuss the SPR-like notions that we use in proving the TCR and eTCR properties of our schemes. These proofs are presented in Section 3 and Section 4, respectively. In Section 5 we discuss some practical aspects related to the integration of our schemes with standard signature schemes. We end with some open problems in Section 6.

2   Variants of Second-Preimage Resistant Hashing

Our goal in this work is to analyze the randomized modes of operation defined in Equations 1 and 2, finding "as weak as possible SPR-like properties" of h that suffice to ensure that H_r is TCR and $\tilde{H}_r$ is eTCR. Below we describe several "games" representing different forms of second-preimage resistance (SPR) of the compression function h (recall that h represents a single, unkeyed, function with two arguments, c and m, and H^c represents the M-D iteration of h with IV$= c$). In all cases the goal of the attacker (that we denote by S) is to present pairs (c, m) and (c', m') such that $(c, m) \neq (c', m')$ and yet $h(c, m) = h(c', m')$. The difference between the games is how the values c, m, c', m' are determined and by whom.

The case where S chooses all four values is known as the "pseudo-collision" search problem for h and is at the basis of the Merkle-Damgård (M-D) construction of collision resistance. However, in the games below the adversary's task is significantly harder since some of these values are chosen randomly and are not under the control of S. More specifically, the games are variants of "second-preimage resistance" of the function h since *in all the value m is chosen independently at random* and c', m' are chosen by the attacker at will after seeing m.

The difference is in the way c is determined: at random in r-SPR, chosen (by the attacker) in c-SPR, and evaluated (as a function of m) in e-SPR:

r-SPR: S receives random m and c, and it chooses m', c'.

c-SPR: S receives random m, and it chooses c, c', m'.

e-SPR: The game is parametrized by an IV c_0. S chooses $\ell \geq 1$ values Δ_i, $i = 1, \ldots, \ell$, each of length b bits; then S receives a random $r \in \{0,1\}^b$ and c, m are set to $m = r \oplus \Delta_\ell$ and $c = H^{c_0}(r \oplus \Delta_1, \ldots, r \oplus \Delta_{\ell-1})$. Finally S chooses c', m'.[2]

Clearly, r-SPR corresponds to the standard notion of second-preimage resistance of h, while c-SPR and e-SPR are variants that we use in the analysis of our schemes, and we elaborate on them below.

The IV. The game e-SPR can be considered in either the "uniform setting" where c_0 is chosen at random and given to S as input, or the "non-uniform setting" where c_0 is a parameter of the game and S may depend on that parameter. The discussion below applies equally to both settings.

Attack parameters. When discussing the security (or hardness) of different games we denote by t and L the attacker's resources (i.e., time and length of messages used in the attack) and by ε the probability of success by the attacker. In particular, if G denotes one of the SPR or TCR games discussed in the paper then we say that a function (either h or the family H_r) is $G(\varepsilon, L, t)$ if no attacker with resources t and L can win the game with probability better than ε. (In some of the games the parameter L is irrelevant and then omitted.)

2.1   The c-SPR Game

We remark that the game c-SPR is a significantly easier variant (for the attacker) than the r-SPR and e-SPR games, since the attacker gets to choose c, after seeing m, in addition to choosing c' and m'. For example, the c-SPR game is clearly vulnerable to generic birthday attacks, whereas r-SPR and e-SPR are not.

   We observe that for a compression function $h(c, m)$, the c-SPR property is equivalent to the property that the family $h_r(m, c) = h(c, m \oplus r)$ is $CR_n(n+b, n)$ as defined by Mironov [18].[3] A family of functions $\{h_r\}_r$ belongs to $CR_\ell(n', n)$ if each h_r maps n' bits to n bits, and in addition no feasible attacker can win the following game except with insignificant probability:

1. The attacker first commits to $n' - \ell$ bits of the first message (call this $(n' - \ell)$-bit string M_1),
2. Then the attacker is given the salt r,
3. The attacker wins if it can find an ℓ-bit string M_2 and a second n'-bit message M' such that for $M = M_1 | M_2$ it holds that $h_r(M) = h_r(M')$.

---

[2] We choose to define $m = r \oplus \Delta_\ell$ for notational convenience in our proofs; however, note that the xor with Δ_ℓ does not change the fact that the value of m is determined uniformly at random (and independent of the Δ values).

[3] Here we reverse the order of the inputs to h_r in order to match Mironov's formulation.

Now, consider the following equivalent formulation of the c-SPR game: the attacker first commits to a b-bit message m, then it gets the randomness r and it needs to find c, c' and m' such that $h(c, m \oplus r) = h(c', m' \oplus r)$. Clearly, under this formulation we see that $\{h_r\}_r$ is c-SPR if and only if $\{h_r\}_r$ belongs to the class $CR_n(n+b, n)$. Mironov proved that the existence of $CR_n(n+b, n)$ families implies the existence of collision-resistant families (from $n+1$ to n bits). Hence c-SPR implies the *existence* of collision-resistant hashing families (in particular, this shows that Simon's [27] separation result applies to black-box constructions of c-SPR; it does *not* mean, however, that the M-D iteration of a c-SPR compression function is necessarily collision resistant). Still, for a particular compression function h, breaking c-SPR is also significantly harder (for the attacker) than the traditional pseudo-collision search. For example, although a pseudo-collision attack on MD5 is known for many years, breaking it in the sense of c-SPR seems beyond the state of the art in cryptanalysis, and even more so for SHA-1. (Also, assuming that c-SPR functions exist, it is easy to construct compression functions that are c-SPR but are not resistant to pseudo-collisions.) We believe that the c-SPR game provides a useful toy-example for cryptanalysts to develop tools for TCR-type attack models.

2.2   The e-SPR Game

The game e-SPR is significantly harder for the attacker than c-SPR (and in particular it is not open to generic birthday attacks). It is also closer to r-SPR. There are, however, two important differences between r-SPR and e-SPR:

- The distribution from which m is chosen is uniform, but c is determined as a function of m. In particular, the joint distribution of c and m has b bits of entropy, while the pair (c, m) is $(b + n)$-bits long. This difference can in principle make e-SPR either easier or harder than r-SPR, depending on the underlying hash function.
- The attacker S gets to influence the distribution on c via its choice of the Δ_i's (but note that S must commit to the Δ_i's before it gets the random value r !)

Due to the first point above, e-SPR and r-SPR are formally incomparable, and it is possible to concoct examples of compression functions where one is easy and the other is hard (assuming r-SPR/e-SPR functions exist).

Relation of e-SPR to L-order TCR. As noted earlier, the fact that a family of compression functions is TCR does not necessarily imply that the multi-block extension of that family obtained via M-D iterations is TCR. Therefore, it is natural to search for (stronger) properties of compression functions that suffice to ensure that the M-D extension is TCR. Hong, Preneel and Lee [14] identified such a property, that they called *L-order TCR* (we recall it below), and proved that if a compression family is L-order TCR then the M-D extension is TCR on inputs of (up to) L blocks. Thus, one way to prove that our H_r construction (Eqn. (1)) is TCR is to assume that the compression function family $\{h_r\}_r$ defined by

Eqn. (1) (i.e. $h_r(c,m) = h(c, m \oplus r)$) is L-order TCR. While this property may be a plausible assumption on practical compression functions (and hence can be used as evidence to support the TCR property of our H_r scheme), it turns out that our Theorem 1 provides for a stronger result. Indeed, that theorem only assumes h_r to be e-SPR and we show in Proposition 1 below that e-SPR is a weaker property (i.e., harder to break) of the h_r scheme than L-order TCR. (We note that this relative strength of e-SPR and L-order TCR may not hold for all families of compression functions but it does hold for our specific scheme).

Hong et al. [14] define a function family $\{h_r\}_r$ to be L-order TCR if no feasible adversary can win the following game, except with insignificant probability: As in the TCR game, the attacker commits to a message m, then it is shown the salt r and it needs to find another message m' such that $h_r(m) = h_r(m')$. The difference is that before committing to the first message m the attacker gets to learn some information about the salt r by means of L adaptive queries m_i to the hash function for which the attacker gets as a response the corresponding hash values $h_r(m_i)$.

Proposition 1. *Let $h : \{0,1\}^{b+n} \to \{0,1\}^b$ be a compression function. If the family $\{h_r\}_{r\in\{0,1\}^b}$ defined by $h_r(c,m) = h(c, m \oplus r)$ is order-L $\mathrm{TCR}(\varepsilon, t)$, then $h(\cdot)$ is e-$\mathrm{SPR}(\varepsilon, L, t - O(L))$.*

The proof is presented in the full version of the paper.

3   Achieving Target Collision Resistance

Recall the definition of the construction H_r from Eqn. (1),

$$H_r^c(m_1, \dots, m_L) \stackrel{\text{def}}{=} H^c(m_1 \oplus r, \dots, m_L \oplus r).$$

Using the games from Section 2 we can establish some relations between the corresponding SPR flavors of h and the TCR property of H_r. For this we define the TCR game for a family H_r as follows:

TCR Game: The attacker T knows the fixed IV c_0. It chooses a message M of length $L > 0$ blocks, receives a random $r \in_R \{0,1\}^b$, and outputs a second message M' of length L'. Attacker T wins if $M' \neq M$ and $H_r^{c_0}(M') = H_r^{c_0}(M)$.

For simplicity of presentation, we first state and prove our results in the case where the TCR attacker can only provide M, M' of length an integer multiple of the block length, and it is not allowed to output M, M' such that one is a suffix of the other. This restriction on the attacker T is not significant in practice since actual implementations pads the messages and append the length to the last block, thus forcing full-block suffix-freeness. (To address the case where such length appending is not necessarily done one needs an additional "one-wayness assumption", as described in Section 3.1. Also, see Section 5 for a discussion on how to handle partial blocks.) Formally, we consider a modified game (denoted TCR*) where the attacker does not win the game if it violates the condition on the messages.

TCR* Game: Same as TCR but the attacker wins only if in addition to the regular TCR conditions it also holds that M, M' consists of an integral number of b-bit blocks and neither of them is suffix of the other.

Theorem 1. *(TCR Theorem, suffix-free case)*

1. *If h is c-SPR(ε, t) then H_r is TCR*$(L\varepsilon, L, t - O(L))$*
2. *If h is e-SPR(ε, L, t) then H_r is TCR*$(L\varepsilon, L, t - O(L))$*

This theorem is proven below. We note that these are sufficient conditions for TCR-ness *but not necessary ones*. In other words, the failure of a compression function to one of the above attacks does not necessarily mean that the induced family H_r is not TCR.

It is also worth commenting on the tightness of the reductions in the above theorem. In both cases, c-SPR and e-SPR, there is a linear degradation of security when going from SPR to TCR. Note, however, that there is a significant difference between the *generic security* of the two games c-SPR and e-SPR. Against the former there is a trivial birthday-type generic attack while against e-SPR generic attacks achieve only linear advantage. Thus, the reduction from e-SPR shows a worst-case "quadratic degradation" generic TCR attack against H_r (and not "cubic degradation" as the reduction cost in the c-SPR to TCR case could indicate). Moreover, TCR attacks with quadratic degradation (i.e., with success probability in the order of $tL/2^n$) against H_r are indeed possible since SPR attacks against H can be translated into TCR attacks against H_r and we know that such birthday-type SPR attacks against H exist [11,16].

This motivates two questions: Can we have a flavor of SPR for which there is a tight reduction to TCR? Can this SPR game be defined such that it is only vulnerable to linear generic attacks? The answer to the first question is YES: in the full version of this paper we present such a game, called m-SPR (m for multiple), and its tight reduction to TCR. The answer to the second question is obviously NO, since as said SPR birthday attacks against H_r do exist. It is possible, however, to design variants of M-D hash functions (e.g., appending a sequence number to each block or using the "dithering" technique proposed by Rivest [23]) for which the best generic attacks achieve linear degradation only. Thus, the main motivation and usefulness of the game m-SPR and its tight reduction to TCR is for the analysis of such variants.

3.1   Proof of Theorem 1

In the proof below we assume that the attacker T never outputs M, M' where one is the suffix of the other. We also make the convention that when algorithm A calls another algorithm B and B aborts then A aborts too.

Lemma 1. *If h is c-SPR(ε, t) then H_r is TCR*$(L\varepsilon, L, t - O(L))$.*

Before proving this lemma, we comment that the exact assertion of this lemma (and thus the proof) depend on whether we view the IV c_0 as a random input to the TCR attacker or a parameter of the TCR game (cf. comment on page 47).

In the reduction below we suppress c_0 with the understanding that if it is a parameter then the same parameter is used in the reduction, and if it is an input then S chooses c_0 at random and gives it to T.

Proof. Given TCR attacker T we build c-SPR attacker S:

1. S gets input m, invokes T and gets from T the first message $M=(m_1,\ldots,m_L)$.
2. S chooses $\ell \in_R \{1,\ldots,L\}$, sets $r = m \oplus m_\ell$, and returns r to T.
3. T outputs the second message $M' = (m'_1,\ldots,m'_{L'})$ or $\perp$.
4. S checks that there exists $\ell' \in \{1,\ldots,L'\}$ such that (1) either $m'_{\ell'} \neq m_\ell$ or $H_r(m'_1,\ldots,m'_{\ell'-1}) \neq H_r(m_1,\ldots,m_{\ell-1})$; and (2) $H_r(m'_1,\ldots,m'_{\ell'}) = H_r(m_1,\ldots,m_\ell)$. If no such ℓ' exists then S aborts.
5. Otherwise S outputs $c = H_r(m_1,\ldots,m_{\ell-1})$, $c' = H_r(m'_1,\ldots,m'_{\ell'-1})$ and $m' = m'_{\ell'} \oplus r$. (If there is more than one index with the properties from above then S chooses ℓ' arbitrarily among all these indexes.)

Turning to the analysis of S, we first observe that if T does not abort then there must exist a pair of indexes i, i' that satisfy the properties from Step 4, namely $H_r(m'_1,\ldots,m'_{i'}) = H_r(m_1,\ldots,m_i)$ and either $m'_{i'} \neq m_i$ or $H_r(m'_1,\ldots,m'_{i'-1}) \neq H_r(m_1,\ldots,m_{i-1})$.

To see that, note that since M, M' are suffix-free there exists an integer $j < \min(L, L')$ such that $m_{L-j} \neq m'_{L'-j}$. Consider the smallest such j (i.e., the last block where M, M' differ when their ends are aligned). Also, let k be smallest non-negative integer such that $H_r(m'_1,\ldots,m'_{L'-j+k}) = H_r(m_1,\ldots,m_{L-j+k})$ (i.e., the first "collision" after these "last differing blocks" in the computation of $H_r(M'), H_r(M)$). Clearly, such k exists since T did not abort and so there must be some collision because $H_r(M') = H_r(M)$.

Now setting $i = L-j+k$ and $i' = L'-j+k$ we get the pair that we need. This is because we have $H_r(m'_1,\ldots,m'_{L'-j+k}) = H_r(m_1,\ldots,m_{L-j+k})$ by definition, and (a) either $k = 0$ in which case $m_{L-j+k} \neq m'_{L'-j+k}$, or (b) $k > 0$ in which case $H_r(m'_1,\ldots,m'_{L'-j+k-1}) \neq H_r(m_1,\ldots,m_{L-j+k-1})$.

Next, we argue that the random choice of ℓ is independent of the transcript that T sees, and in particular it is independent of the values of any such pair i, i' as above. This is easy to see, since the distributions of all the variables in the execution of S remains unchanged if instead of choosing m at random and setting $r = m_\ell \oplus m$ we choose r at random and set $m = m_\ell \oplus r$. In the new game, however, we can postpone the choice of ℓ (and m) until the end, and so it is clearly independent of M, M'.

It follows that whenever T does not abort, there exists a pair i, i' as above, and in this case the probability that S chooses $\ell = i$ is (at least) $1/L$. If that happens then there exists ℓ' as needed (i.e., $\ell' = i'$) so S does not abort. It remains to show that in this case S wins the c-SPR game. This follows since by the condition on ℓ, ℓ' we have:

either $c=H_r(m_1,\ldots,m_{\ell-1}) \neq H_r(m'_1,\ldots,m'_{\ell'-1})=c'$ or $m'=m'_{\ell'}\oplus(m_\ell\oplus m)\neq m$

and also (due to $m = r \oplus m_\ell$ and $m' = r \oplus m'_{\ell'}$):

$$h(c,m)=h(c,r\oplus m_\ell)=H_r(m_1,\ldots,m_\ell)=H_r(m'_1,\ldots,m'_{\ell'})=h(c',r\oplus m'_{\ell'})=h(c',m')$$

The attacker in the proof of Lemma 1 makes little use of the full freedom provided by the c-SPR game in (adaptively) choosing c. Indeed the value of c used by S is set as soon as S chooses ℓ, and this choice is *independent of m*. This motivates the e-SPR game where, by definition, S can influence c only through the choice of values Δ to which it commits before seeing r. Thus, e-SPR represents a better approximation of the TCR game, by giving the attacker in e-SPR less artificial freedom than in c-SPR, and making it much closer to SPR (in particular, by disallowing generic birthday attacks against it). This makes the following lemma much stronger than Lemma 1 though its proof follows the same lines. Note that we are still paying a linear (in L) degradation of security but this time the reduction is to a "non-birthday" problem. (See also the discussion about security degradation preceding Section 3.1.).

Lemma 2. *If h is e-SPR(ε, L, t) then H_r is TCR*$^*(L\varepsilon, L, t - O(\ell))$.*

Proof. This is a simple adaptation of the proof of Lemma 1. Given TCR attacker T we build e-SPR attacker S:

1. S invokes T and gets from T the first message $M = (m_1, \ldots, m_L)$.
2. S chooses $\ell \in_R \{1, \ldots, L\}$ and sets $\Delta_i = m_i$ for $i = 1, \ldots, \ell$.
3. S receives a random r and forward it to T. (The values of c, m are evaluated as $c = H_r(\Delta_1, \ldots, \Delta_{\ell-1}) = H_r(m_1, \ldots, m_{\ell-1})$ and $m = \Delta_\ell \oplus r = m_\ell \oplus r$.)
4. T outputs the second message $M' = (m'_1, \ldots, m'_{L'})$ or $\bot$.
5. S checks that there exists $\ell' \in \{1, \ldots, L'\}$ such that (1) either $m'_{\ell'} \neq m_\ell$ or $H_r(m'_1, \ldots, m'_{\ell'-1}) \neq H_r(m_1, \ldots, m_{\ell-1})$; and (2) $H_r(m'_1, \ldots, m'_{\ell'}) = H_r(m_1, \ldots, m_\ell)$. If no such ℓ' exists then S aborts.
6. Otherwise S outputs $c' = H_r(m'_1, \ldots, m'_{\ell'-1})$ and $m' = m'_{\ell'} \oplus r$. (If there is more than one index with the properties from above then S chooses ℓ' arbitrarily among all these indexes.)

The analysis of S is nearly identical to the analysis in the proof of Lemma 1. The only difference is that here it is even easier to argue that the choice of ℓ is independent of the transcript of T.

Extension to Non-suffix-free Messages. We stated Theorem 1 and its proof in terms of the TCR* game, namely, the TCR game in which the attacker is restricted to suffix-free pairs M, M'. For the sake of generality, we extend these results here to the case where inputs to H_r may not be suffix free. For this extension we need the following one-wayness assumption.

Assumption 2 (OWH). *The assumption OWH for the compression function h asserts that given a random c, it is hard to find a non-null message M such that $H^c(M) = c$.*

We use our usual notation OWH(ε, t) to denote the assumption that no t-time attacker can violate the assumption OWH with probability better than ε.

Some comments are in order here. First, note that this assumption is stated in the "uniform" setting where the IV c is chosen at random. Its "non-uniform" counterpart (with respect to a parameter c_0) does not make formal sense, for the same reason that the assertion "SHA256 as per FIPS-180-2 is collision resistant" does not make formal sense. This means that formally, Theorem 3 below only applies to this "uniform" setting.[4]

Second, the above assumption is clearly implied by the assumption that for a random c it is hard to find c', m such that $h(c', m) = c$. Furthermore, under a very mild condition on the structure of h, this last assumption is equivalent to the standard assumption that h is a one-way function (i.e., for random c, m it is hard given $h(c, m)$ to find c', m' such that $h(c', m') = h(c, m)$). Specifically, the condition that we need is that for random c, m, the distribution over $h(c, m)$ is statistically close to (or at least indistinguishable from) the uniform distribution on $\{0, 1\}^n$.

It is well known that one-wayness is implied by second-preimage resistance, and is it easy to verify that the same holds also for our notions of c-SPR and e-SPR (the latter in its "uniform" interpretation where the IV is random). It follows that under that mild structural condition, the assumption OWH is implied by the assumptions e-SPR and c-SPR and so is redundant in Theorem 3 below. Still, we prefer to state the theorem with this assumption since (a) it makes it easier to understand, and (b) it applies also to functions h for which that structural condition does not hold.

Theorem 3. *(TCR Theorem , non-suffix-free case)*
1. *If h is c-SPR(ε, t) and OWH(ε', t) then H_r is TCR$(\varepsilon' + L\varepsilon, L, t - O(L))$.*
2. *If h is e-SPR(ε, L, t) and OWH(ε', t) then H_r is TCR$(\varepsilon' + L\varepsilon, L, t - O(L))$.*

The proof is a small adaptation of the proofs of lemmas 1 and 2. One just observes that in the case where one of M, M' is a proper suffix of the other, then either the prefix of the longer message violates the one-wayness assumption (i.e., we have $H^c(\text{prefix}) = c$), or else the same analysis from lemmas 1 and 2 applies. The reason that ε' is not multiplied by L in the expressions above is that in the former case we do not care about the value of ℓ that was chosen at the beginning of the reduction. (See the proof of Theorem 4 for the formal argument.)

4   Enhanced Target Collision Resistance

Recall the definition of the construction $\tilde{H}_r$ from Eqn. (2),

$$\tilde{H}_r^c(M) \stackrel{\text{def}}{=} H_r^c(0|M) = H^c(r, m_1 \oplus r, \ldots, m_L \oplus r)$$

We show that under the same c-SPR and e-SPR assumptions, the construction $\tilde{H}_r$ is enhanced TCR (eTCR). We start by defining the eTCR game.

---

[4] However, the reduction that proves Theorem 3 is meaningful also for the case of fixed IV, showing a constructive transformation from TCR attack to either an SPR attack or to an attack that violates the one-wayness assumption.

eTCR Game: Attacker T chooses a message M of length $L \geq 0$ blocks, receives a random $r \in_R \{0,1\}^b$, and outputs another $r' \in \{0,1\}^b$ and a message M' of length L' blocks. Attacker T wins if $(M',r') \neq (M,r)$ and $H_{r'}^{c_0}(M') = H_r^{c_0}(M)$.

Note that the attacker can win this game even when $M = M'$, so long as $(M',r') \neq (M,r)$. We remark that as opposed to the case of the construction H_r and the standard TCR game, here assuming suffix-freeness does not help us, because the attacker can specify $r' \neq r$ and so even if M, M' are suffix free the messages $M \oplus r$, $M' \oplus r'$ perhaps are not. (However, we can go back to the suffix-free case by using length padding outside of the randomness, as discussed in Section 5.) Below we therefore use the OWH assumption as in Theorem 3, thus getting:

Theorem 4 (eTCR security)

1. If h is c-SPR(ε,t) and OWH(ε',t) then $\tilde{H}_r$ is eTCR$(\varepsilon'+(L+1)\varepsilon, L, t-O(L))$.
2. If h is e-SPR$(\varepsilon, L+1,t)$ and OWH(ε',t) then $\tilde{H}_r$ is eTCR$(\varepsilon' + (L+1)\varepsilon, L, t - O(L))$.

Proof. Below we only prove part 2. Part 1 can be obtained as a corollary, since an e-SPR attacker can be trivially converted to a c-SPR attacker with the same success probability. The proof is a small adaptation of the proof of Lemma 2. The idea is that S sets $\Delta_0 = 0$ (and the other Δ_i's as before) and then uses r' instead of r to compute c' and m'. In more details, given an enhanced-TCR attacker T be build e-SPR attacker S:

1. On input an IV c, S invokes $T(c)$ and gets from T the first message $M = (m_1, \ldots, m_L)$.
2. S sets $m_0 = 0$, chooses $\ell \in_R \{0, \ldots, L\}$ and sets $\Delta_i = m_i$ for $i = 0, \ldots, \ell$.
3. S receives a random r and forward it to T. (Also c, m are evaluated as $c = H_r^c(\Delta_0, \Delta_1, \ldots, \Delta_{\ell-1}) = H_r^c(0, m_1, \ldots, m_{\ell-1})$ and $m = \Delta_\ell \oplus r = m_\ell \oplus r$.)
4. T outputs either the second salt r' and message $M' = (m'_1, \ldots, m'_{L'})$ or $\perp$.
5. S sets $m'_0 = 0$, and searches for an index $\ell' \in \{0, \ldots, L'\}$ such that (1) either $m'_{\ell'} \oplus r' \neq m_\ell \oplus r$ or $H_{r'}^c(0, \ldots, m'_{\ell'-1}) \neq H_r^c(0, m_1, \ldots, m_{\ell-1})$; and (2) $H_{r'}^c(0, m'_1, \ldots, m'_{\ell'}) = H_r^c(0, m_1, \ldots, m_\ell)$. If no such ℓ' exists then S aborts.
6. Otherwise S outputs $c' = H_{r'}^c(0, m'_1, \ldots, m'_{\ell'-1})$ and $m' = m'_{\ell'} \oplus r'$. (If there is more than one index with the properties from above then S chooses ℓ' arbitrarily among all these indexes.)

To analyze S, let SUFF denote the event in which one of the messages

$$X = (r, m_1 \oplus r, \ldots, m_L \oplus r) \text{ and } X' = (r', m'_1 \oplus r', \ldots, m'_{L'} \oplus r')$$

in the game above is a suffix of the other, and in addition hashing only the prefix of the longer message yields back the IV c. In other words, the event SUFF occurs whenever we have either $X' = Y|X$ or $X = Y|X'$ with some $Y \neq \Lambda$ for which $H^c(Y) = c$.

Let $\varepsilon_{\text{suff}}$ be the probability of the event SUFF, and let ε^* be the probability of the event in which T succeeds but the event SUFF does not happen. In the latter case we can apply the same analysis as in Lemma 2, showing that there must exist a pair of indexes i, i' that satisfy the conditions in step 5, that with probability at least $1/(L+1)$ we have $\ell = i$, and if that happens then S succeeds.

The only part of the analysis that is slightly different than in the proof of Lemma 2 is proving the existence of the indexes i, i' in the case where one of X, X' is a prefix of the other but $H^c(Y) \neq c$. In this case, denote by $X|_i$ the $(i+1)$-block prefix of X (i.e., blocks 0 through i), and similarly denote $X'|_i$. Let j be the smallest integer such that $H^c(X|_{L-j}) \neq H^c(X'|_{L'-j})$ (i.e., the last place where the chaining values do not agree). Since $H^c(Y) \neq c$ then there exists such index $j \leq \min(L+1, L'+1)$, and on the other hand $j > 0$ since the success of T implies that $H^c(X_L) = H^c(X'_{L'})$. Then setting $i = L - j + 1$ and $i' = L' - j + 1$ we get the pair of indexes that we need.

We thus conclude that the success probability of S is at least $\varepsilon^*/(L + 1)$. Since (by definition) the success probability of T is at most $\varepsilon^* + \varepsilon_{\text{suff}}$, we can complete the proof by showing a OWH-attacker with success probability $\varepsilon_{\text{suff}}$. Such attacker S' is obvious: It gets c as input and runs the same execution as the e-SPR attacker S from above (choosing itself the random r), and at the end if the event SUFF occurs then it outputs the prefix Y.

5 Using Randomized Hashing in Signatures

The randomized hashing modes presented in this paper are intended to replace the deterministic hashing used by standardized signature schemes such as RSA [21] and DSS [10]. As shown throughout the paper, this replacement frees these schemes from their current essential dependency on full collision resistance by basing their security on much harder to break properties of the underlying hash functions. Of the two schemes analyzed here, the $\tilde{H}_r$ mode (Eqn. (2)) is best suited for this task as it does not require the explicit signing of the salt r and hence allows for more implementation flexibility. We present a full specification of this mode and its use in digital signature in [13]. Here we provide a short (and partial) discussion of possible approaches.

The two main approaches depend on whether an application can accommodate the sending of the salt r in addition to the message and signature or whether any increase on the size of the information is not possible. The first case is simpler and requires the least changes to standards. Most applications will be able to send the the extra salt, especially that r needs not be too long, say in the range of 128-256 bits (see below). Examples of such applications are digital certificates, code signing, XML signatures, etc. In this case, upon the need to sign a message M the modified application will: (i) choose a random salt r; (ii) transform the message $M = (m_1, \ldots, m_L)$ into the message $M' = (r, m_1 \oplus r, \ldots, m_L \oplus r)$; (iii) invoke the *existing* signature operation (including the existing hash operation) on M'; (iv) when the message M and its signature are to be transmitted, transmit r, M and the computed signature (on M'). The verification side computes M' from the received r and M and applies its *existing* signature verification operation on M'. This approach allows for preserving the existing processing order (including a one-pass over the signed message) and the possible pre-computation (ahead of signature generation) of the pair $(r, \tilde{H}_r(M))$.

Note that in this case the application can use its existing signature and verification processing (whether in software, hardware or both) *without any change*. In this case, signature standards [21,10] need no change except for adding the "front end" message randomization (to generate M'). The details of implementation of the required changes is application and implementation dependent. The extra operations may be placed at the application itself or in a "signing module" invoked by the application and which will be responsible for the signature generation as well as for the generation and transmission of r.

Applications where an increase in the size of messages or signatures is impractical will need to resort to a different approach: either re-use the existing randomness (in the case of probabilistic signatures) or encode the salt r "under the signature". The former is the case for the probabilistic schemes DSS [10] and RSA-PSS [21] where the already existing random values r (used internally by the PSS operation in the case of RSA-PSS and as the signature component $r = g^k$ in the case of DSS) can be re-used as the hashing salt and thus require no additional transmission. The latter case, i.e., encoding r "under the signature", is the case of the traditional deterministic RSA encoding of PKCS#1 v1.5 [21]. In this case, instead of padding the (deterministic) hash value to a full modulus size as done today, one will pad the concatenation of r and $\tilde{H}_r(M)$ to a full modulus size and then apply to it the private RSA operation. In this case, the salt r can be retrieved by the verifying party using the RSA public operation and hence no extra transmission is required.

Evaluating the complexity of performing the above changes depends in particular settings and applications. Clearly, no change to existing applications, signature modules or standards is "too simple" and one cannot ignore the cost of engineering the above changes. However, considering that applications that compute digital signatures will be upgraded in any case to use new hash functions, one should use this opportunity to also upgrade to randomized hashing. In particular, it is worth noting that many difficult issues for applications such as preserving backwards compatibility, the signaling/negotiation of algorithms and capabilities, version rolling, etc. are not made worse by our proposal than what is already needed to support a simple hash function upgrade [4]. Considering the simplicity and minimalistic nature of our randomization scheme, we believe that the extra changes (transmission or re-use of r) are well worth the substantial security gain they provide to existing and future signature schemes.

5.1 Specification Details

As said, a detailed specification of the $\tilde{H}_r$ mode is presented in [13]. Here we mention two elements that we omitted so far and that need to be included in a practical instantiation.

Shorter salt. The construction $\tilde{H}_r$ as defined in Eqn. (2) uses salt of length b, i.e. one message block. In practical instantiations, we propose to choose a salt string r' of length between 128 and b, and create the b-bit salt r by concatenating r' with itself as needed, possibly with the last r' repetition being truncated (e.g., in the case of $b = 512$ and a 160-bit r', one would have $r = r'||r'||r'||r''$ where r'' represents the first 32 bits of r'). The analysis in this paper applies to this modified salt, except that now the assumptions c-SPR, e-SPR are stated in terms of random messages distributed according to the distribution induced by r'. Of course, the plausibility of these assumptions needs to be evaluated; however, as long as r' is not too short, these modified SPR-type assumptions seem very reasonable (it is even possible that the added "structure" in these repetitions makes the cryptanalytical problem more difficult).

Last block padding. The Merkle-Damgård construction pads the input message to an integral number of blocks and includes an encoding of the length of the original message in the last block (thus ensuring a suffix-free message space). The analysis in this work assumes that the salt r is xor-ed to the padded message, but in practice it is likely that the xor will happen first (as part of the randomized mode-of-operation) and the padding will then be applied to the randomized message (as part of the underlying hash function). Simply switching the order and using randomize-then-pad may mean that the last block is only randomized in its first few bits. A more robust alternative is to use two levels of padding. Namely, we first use suffix-free padding of the last block to a full block minus 64 bits (i.e., 448 bits), then xor the salt to this padded message, and finally apply the original hashing scheme (which will include the length of the padded message in the remaining 64 bits). See [13].

6   Open Problems

In light of the results in this paper, we feel that more focus should be placed on evaluating current and future hash functions against the c-SPR and e-SPR attack scenarios. Specifically, we would like to offer the c-SPR scenario as a "toy example" for developing cryptanalytical tools that may prove useful in assessing randomized hashing, and then study the e-SPR scenario as a stronger foundation for our scheme. Other open problems that arise from this work are briefly discussed next.

A different construction for eTCR. Another natural proposal for obtaining enhanced-TCR is to define $\overline{H}_r^c(M) = H^c(r|H_r^c(M))$. It may be interesting to study this variant and see what advantages (and disadvantages) it offers vis-a-vis the construction $\tilde{H}$ from Eqn. (2).

The random-oracle model and weak hash functions. One inherent difficulty in formally proving the security of TCR and eTCR schemes in the context of DSS and RSA as specified in the official standards [21,10] is that these schemes do not have a proof of security (not even in the case that the underlying hash function is fully collision resistant). The "closest relatives", namely, the DSA-II variant of Brickel, Pointcheval, Vaudenay and Yung [7] or the RSA-PSS scheme of Bellare and Rogaway [2], have a proof of security in the random oracle model. Interestingly, these proofs use in an essential way the randomization of the hash function, not unlike the TCR or eTCR constructions. In some sense one can think, for example, of the use of DSS with a eTCR scheme as an "instantiation" of DSS-II. The obvious question is: how can we state anything related to random oracles when we are dealing with relatively weak (at least not CR) hash functions (clearly, random oracles are "very strong hash functions"). See some results related to this question in the full version of this paper and in [19].

We also point out that the random-oracle proofs for some of the above (randomized) signature schemes do not essentially use the collision resistance property of the random-oracle (as evidenced, for example, by the fact that some of these proofs, such as those following [22], remain meaningful even when you have a random-oracle with relatively short output, e.g., 80 bits). This brings up the interesting question of exhibiting variants of the random-oracle model where one can argue about functions that "behave randomly but are not collision-resistant" (e.g., a random oracle H with an associated oracle that outputs random H collisions upon request).

Acknowledgments. We thank Ran Canetti, Orr Dunkelman, Charanjit Jutla, Burt Kaliski, Tom Shrimpton and Lisa Yin for helpful discussions.

References

1. Mihir Bellare, Ran Canetti, Hugo Krawczyk, "Keying Hash Functions for Message Authentication". CRYPTO 1996. 1-15
2. Mihir Bellare and Phillip Rogaway, "The Exact Security of Digital Signatures – How to Sign with RSA and Rabin", EUROCRYPT 96.
3. Mihir Bellare and Phillip Rogaway, "Collision-Resistant Hashing: Towards Making UOWHFs Practical", CRYPTO 97, LNCS 1294, 1997
4. Steven M. Bellovin and Eric K. Rescorla, "Deploying a New Hash Algorithm", NDSS'06. http://www.cs.columbia.edu/~smb/papers/new-hash.pdf
5. Eli Biham and Rafi Chen, "Near-Collisions of SHA-0", CRYPTO 2004.
6. Eli Biham, Rafi Chen, Antoine Joux, Patrick Carribault, Christophe Lemuet and William Jalby, "Collisions of SHA-0 and Reduced SHA-1", EUROCRYPT 2005.
7. Ernest Brickell, David Pointcheval, Serge Vaudenay, and Moti Yung, "Design and Validations for Discrete Logarithm Based Signature Schemes", PKC'2000.
8. Florent Chabaud and Antoine Joux, "Differential Collisions in SHA-0", CRYPTO 98.
9. Ivan Damgård, "A design principle for hash functions", CRYPTO 1989.
10. Digital Signature Standard (DSS), FIPS 186, May 1994.
11. Richard Drews Dean, "Formal Aspects of Mobile Code Security", Ph.D Dissertation, Princeton University, January 1999.
12. Shafi Goldwasser, Silvio Micali and Ronald L. Rivest, "A Digital Signature Scheme Secure Against Adaptive Chosen-Message Attacks", *SIAM J. Comput.* 17(2): 281-308 (1988)
13. Shai Halevi and Hugo Krawczyk, "Randomized Hashing: Specification and Deployment", in preparation.

14. Deukjo Hong, Bart Preneel and Sangjin Lee, "Higher Order Universal One-Way Hash Functions", ASIACRYPT 2004.

15. Antoine Joux, "Multicollisions in Iterated Hash Functions. Application to Cascaded Constructions", CRYPTO 2004.

16. John Kelsey and Bruce Schneier, "Second Preimages on n-Bit Hash Functions for Much Less than 2^n Work", EUROCRYPT 2005.

17. Ralph Merkle, "One way hash functions and DES", CRYPTO 1989.

18. Ilya Mironov, "Hash Functions: From Merkle-Damgård to Shoup", EUROCRYPT 2001: 166-181.

19. Ilya Mironov, "Collision-Resistant No More: Hash-and-Sign Paradigm Revisited", Public Key Cryptography 2006: 140-156.

20. Moni Naor and Moti Yung, "Universal One-Way Hash Functions and their Cryptographic Applications", STOC 1989.

21. PKCS #1 v2.1: RSA Cryptography Standard, RSA Laboratories, June 14, 2002

22. David Pointcheval and Jacques Stern, "Security Arguments for Digital Signatures and Blind Signatures", *J.Cryptology* (2000) 13:361-396.

23. Ron Rivest, "Abelian square-free dithering for iterated hash functions", Presented at ECrypt Hash Function Workshop, June 21, 2005, Cracow.

24. Phillip Rogaway, Thomas Shrimpton, "Cryptographic Hash-Function Basics: Definitions, Implications, and Separations for Preimage Resistance, Second-Preimage Resistance, and Collision Resistance". FSE 2004, 371-388.

25. John Rompel, "One-way functions are necessary and sufficient for secure signatures", STOC 1990, pp. 387-394.

26. Victor Shoup, "A Composition Theorem for Universal One-Way Hash Functions", EUROCRYPT 2000.

27. Dan Simon, "Finding collisions on a one-way street: Can secure hash functions be based on general assumptions?", EUROCRYPT 98, pp. 334-345.

28. Michael Szydlo and Yiqun Lisa Yin, "Collision-Resistant usage of MD5 and SHA-1 via Message Preprocessing", Cryptology ePrint Archive, Report 2005/248.

29. Xiaoyun Wang, Xuejia Lai, Dengguo Feng, Hui Chen, and Xiuyuan Yu, "Cryptanalysis of the Hash Functions MD4 and RIPEMD", EUROCRYPT 2005.

30. Xiaoyun Wang and Hongbo Yu, "How to Break MD5 and Other Hash Functions", EUROCRYPT 2005.

31. Xiaoyun Wang, Hongbo Yu, and Yiqun Lisa Yin, "Efficient Collision Search Attacks on SHA-0", CRYPTO 2005.

32. Xiaoyun Wang, Yiqun Lisa Yin, and Hongbo Yu "Finding Collisions in the Full SHA-1", CRYPTO 2005.

Round-Optimal Composable Blind Signatures in the Common Reference String Model

Marc Fischlin*

Darmstadt University of Technology, Germany
marc.fischlin@gmail.com
www.fischlin.de

Abstract. We build concurrently executable blind signatures schemes in the common reference string model, based on general complexity assumptions, and with optimal round complexity. Namely, each interactive signature generation requires the requesting user and the issuing bank to transmit only one message each. We also put forward the definition of universally composable blind signature schemes, and show how to extend our concurrently executable blind signature protocol to derive such universally composable schemes in the common reference string model under general assumptions. While this protocol then guarantees very strong security properties when executed within larger protocols, it still supports signature generation in two moves.

1 Introduction

Blind signatures, introduced by Chaum [8], allow a bank to interactively issue signatures to users such that the signed message is hidden from the bank (blindness) while at the same time users cannot output more signatures than interactions with the bank took place (unforgeability). Numerous blind signature schemes have been proposed, mostly under specific number-theoretic assumptions, some relying also on the random oracle model [26,1,3,4] and some forgoing random oracles [9,18,24]. Only the work by Juels et al. [17] addresses the construction of blind signatures under general assumptions explicitly, and deploys general two-party protocols and oblivious transfer based on trapdoor permutations.

Interestingly, almost all of the aforementioned blind signature schemes require three or more moves (most of them even in the random oracle model) and concurrent executions of the signature generation protocol are often a concern (cf. [25,26,24]). For instance, making the solution by Juels et al. [17] concurrently secure would require a high round complexity. This follows from results by Lindell [19,20] showing, among others, that in the plain model the number of rounds in blind signature schemes with black-box security proofs is bounded from below by the number of concurrent executions.

A notable exception to the problems with interleaving executions are schemes with an optimal two-move signature generation protocol, solving the concurrency

---

* This work was supported by the Emmy Noether Program Fi 940/2-1 of the German Research Foundation (DFG).

C. Dwork (Ed.): CRYPTO 2006, LNCS 4117, pp. 60–77, 2006.

problem immediately. This includes Chaum's original RSA-based blind signature protocol and the pairing-based discrete-log version thereof [4]. Unfortunately, the security proofs [3,4] for these schemes need the random oracle model —by which they can bypass Lindell's lower bound for the plain model— and rely on the so-called one-more RSA or one-more discrete-log assumptions, which are less investigated than the corresponding standard problems.

Here we show that one can build secure blind signature schemes with a two-move signature generation protocol under general assumptions (namely, trapdoor permutations). Our scheme does not rely on random oracles, yet to bridge the lower bound on the number of rounds we work in the common reference string model. We note that instead of falling back on general multi-party paradigms as in [17] and inheriting the round complexity of the underlying oblivious transfer protocols for general assumptions we give a dedicated solution to the blind signature problem.

Construction Idea. The basic (and simplified) construction idea of our blind signature scheme is as follows. The user commits to the message m and sends this commitment U to the bank. The bank signs U with a signature scheme and sends the signature B back to user. The user finally derives the blind signature for m by computing a commitment[1] C of $U\|B$ and proving with a non-interactive zero-knowledge proof π (based on the common reference string) that this commitment C contains a valid signature B for U and that U itself is a commitment of the message m. The blind signature to m is given by the commitment C and the proof π.

Using standard non-interactive zero-knowledge (NIZK) proofs our protocol above provides a weaker unforgeability notion than postulated in [26,17]. That is, a malicious user can potentially generate several signatures from a single interaction with the bank by generating multiple proofs $\pi_1, \pi_2, \ldots$ for the same commitment C. All these pairs $C\|\pi_i$ would be valid blind signatures for the same message m, while the standard unforgeability definition asks that a malicious user cannot create more signatures than interactions happened. In the full version [14] we show how to thwart such attacks by switching to so-called *unique* non-interactive zero-knowledge proofs, recently introduced by Lepinski et al. [21]. In this version here we only treat the weaker notion where the malicious user cannot create a signature for a new message.

Universal Composition. As explained, our two-move signature generation protocol overcomes the concurrency problem effortlessly. More general a slight variation of our scheme yields a secure blind signature scheme in the universal composition (UC) framework [5]. Secure schemes in this UC framework enable interdependent executions with other protocols while preserving the main security characteristics. Our modified scheme now requires the same assumptions as before as well as a simulation-sound NIZK proof of knowledge, which can also be derived from trapdoor permutations [28,11].

---

[1] For the security proof we require that the commitment C is actually done through an encryption scheme.

Towards proving the universal composability result we first formalize an ideal functionality $\mathcal{F}_{\text{BlSig}}$, prescinding the basic requirements of blind signatures such as completeness, unforgeability and blindness. Since such UC blind signatures can be used to build UC commitments it follows from an impossibility result in [7] that UC blind signatures cannot be realized through two-party protocols in the plain model. But, as our solution shows, augmenting the model by common reference strings one can build UC blind signatures (against non-adaptive corruptions) under general assumptions. Compared to general feasibility results in the UC framework [10], showing that essentially any functionality can be securely realized in the common random string, our solution still needs only two-moves to generate signatures.

2   Blind Signatures in the Common Reference String Model

In this section we recall the security definition of blind signatures and present our two-move solution and prove its security.

2.1   Blind Signatures and Their Security

For the interactive signature generation protocol of a blind signature scheme we introduce the following notation. For two interactive algorithms $\mathcal{X}$, $\mathcal{Y}$ we denote by $(a, b) \leftarrow \langle \mathcal{X}(x), \mathcal{Y}(y) \rangle$ the joint execution of $\mathcal{X}$ for input x and $\mathcal{Y}$ for input y, where $\mathcal{X}$'s private output at the end of the execution equals a and $\mathcal{Y}$'s private output is b. For an algorithm $\mathcal{Y}$ we write $\mathcal{Y}^{\langle \mathcal{X}(x), \cdot \rangle^{\infty}}$ if $\mathcal{Y}$ can invoke an unbounded number of executions of the interactive protocol with $\mathcal{X}$ in arbitrarily interleaved order. Accordingly, $\mathcal{X}^{\langle \cdot, \mathcal{Y}(y_0) \rangle^1, \langle \cdot, \mathcal{Y}(y_1) \rangle^1}$ can invoke arbitrarily interleaved executions with $\mathcal{Y}(y_0)$ and $\mathcal{Y}(y_1)$ but interact with each algorithm only once.

Definition 1 (Blind Signature Scheme). *A blind signature scheme (in the common reference string model) consists of a tuple of efficient algorithms* $\mathsf{BS} = (\mathcal{C}_{BS}, \mathsf{KG}_{BS}, \langle \mathcal{B}, \mathcal{U} \rangle, \mathsf{Vf}_{BS})$ *where*

CRS Generation. $\mathcal{C}_{BS}$ *on input* 1^n *outputs a common reference (or random) string* crs_{BS}.

Key Generation. $\mathsf{KG}_{BS}(crs_{BS})$ *generates a key pair* (sk_{BS}, pk_{BS}).

Signature Issuing. *The joint execution of algorithm* $\mathcal{B}(crs_{BS}, sk_{BS})$ *and algorithm* $\mathcal{U}(crs_{BS}, pk_{BS}, m)$ *generates an output* S *of the user,* $(\perp, S) \leftarrow \langle \mathcal{B}(crs_{BS}, sk_{BS}), \mathcal{U}(crs_{BS}, pk_{BS}, m) \rangle$.

Verification. $\mathsf{Vf}_{BS}(crs_{BS}, pk_{BS}, m, S)$ *outputs a bit.*

It is assumed that the scheme is complete, *i.e., for any* $crs_{BS} \leftarrow \mathcal{C}_{BS}(1^n)$, *any* $(sk_{BS}, pk_{BS}) \leftarrow \mathsf{KG}_{BS}(crs_{BS})$, *any message* $m \in \{0, 1\}^n$ *and any* S *output by* $\mathcal{U}$ *in the joint execution of* $\mathcal{B}(crs_{BS}, sk_{BS})$ *and* $\mathcal{U}(crs_{BS}, pk_{BS}, m)$ *we have* $\mathsf{Vf}_{BS}(crs_{BS}, pk_{BS}, m, S) = 1$.

Security of blind signatures consists of two requirements [26,17]. Unforgeability says that it should be infeasible for a malicious user $\mathcal{U}^*$ to generate $k+1$ valid signatures given that k interactions with the honest bank took place (where the adversary adaptively decides on the number k of interactions). Blindness says that it should be infeasible for a malicious bank $\mathcal{B}^*$ to determine the order in which two messages m_0, m_1 have been signed in executions with an honest user. This should hold, of course, as long as the interactive signature generation produces two valid signatures and the bank $\mathcal{B}^*$ for example does not abort deliberately in one of the two executions.

Definition 2 (Secure Blind Signature Scheme). *A blind signature scheme* $\mathsf{BS} = (\mathcal{C}_{BS}, \mathsf{KG}_{BS}, \langle\mathcal{B},\mathcal{U}\rangle, \mathsf{Vf}_{BS})$ *in the common reference string model is called secure if the following holds:*

Unforgeability. *For any efficient algorithm* $\mathcal{U}^*$ *the probability that experiment* $\mathsf{Forge}^{BS}_{\mathcal{U}^*}(n)$ *evaluates to 1 is negligible (as a function of n) where*

> **Experiment** $\mathsf{Forge}^{BS}_{\mathcal{U}^*}(n)$
> $\quad crs_{BS} \leftarrow \mathcal{C}_{BS}(1^n)$
> $\quad (sk_{BS}, pk_{BS}) \leftarrow \mathsf{KG}_{BS}(crs_{BS})$
> $\quad ((m_1, S_1), \ldots, (m_{k+1}, S_{k+1})) \leftarrow \mathcal{U}^{*\langle\mathcal{B}(crs_{BS}, sk_{BS}),\cdot\rangle^\infty}(crs_{BS}, pk_{BS})$
> $\quad$ *Return 1 iff*
> $\qquad m_i \neq m_j$ *for* $1 \leq i < j \leq k+1$, *and*
> $\qquad \mathsf{Vf}_{BS}(crs_{BS}, pk_{BS}, m_i, S_i) = 1$ *for all* $i = 1, 2, \ldots, k+1$, *and*
> $\qquad$ *at most k interactions with* $\langle\mathcal{B}(crs_{BS}, sk_{BS}),\cdot\rangle^\infty$ *were initiated.*

Blindness. *For any efficient algorithm* $\mathcal{B}^*$ *(working in modes* find, issue *and* guess*) the probability that the following experiment* $\mathsf{Blind}^{BS}_{\mathcal{B}^*}(n)$ *evaluates to 1 is negligibly close to $1/2$, where*

> **Experiment** $\mathsf{Blind}^{BS}_{\mathcal{B}^*}(n)$
> $\quad crs_{BS} \leftarrow \mathcal{C}_{BS}(1^n)$
> $\quad (pk_{BS}, m_0, m_1, \beta_{\mathsf{find}}) \leftarrow \mathcal{B}^*(\mathsf{find}, crs_{BS})$
> $\quad b \leftarrow \{0, 1\}$
> $\quad \beta_{\mathsf{issue}} \leftarrow \mathcal{B}^{*\langle\cdot,\mathcal{U}(crs_{BS}, pk_{BS}, m_b)\rangle^1, \langle\cdot,\mathcal{U}(crs_{BS}, pk_{BS}, m_{1-b})\rangle^1}(\mathsf{issue}, \beta_{\mathsf{find}})$
> $\qquad$ *and let* S_b, S_{1-b} *denote the (possibly undefined) local outputs*
> $\qquad$ *of* $\mathcal{U}(crs_{BS}, pk_{BS}, m_b)$ *resp.* $\mathcal{U}(crs_{BS}, pk_{BS}, m_{1-b})$.
> $\quad b^* \leftarrow \mathcal{B}^*(\mathsf{guess}, S_0, S_1, \beta_{\mathsf{issue}})$
> $\quad$ *Return a random bit if* $S_0 = \bot$ *or* $S_1 = \bot$, *else return 1 iff* $b = b^*$.

As pointed out in the introduction, in a stronger unforgeability notion the malicious user is already deemed to be successful if it outputs $k + 1$ distinct message-signature pairs (m_i, S_i) after k interactions with the bank. In this case the user may also try to find another valid signature to a previously signed message m. Our scheme here satisfies the unforgeability notion in Definition 2 but can be turned into one achieving the higher level (see the full version [14]).

In a *partially* blind signature scheme [2] the bank and the user first agree on some information info which is attached to the blind signature. There, unforgeability demands that a malicious user cannot find $k + 1$ distinct but valid tuples

($\mathsf{info}_i, m_i, S_i$). The definition of blindness then allows a malicious bank to determine info together with the messages m_0, m_1 such that the bank still cannot decide the order in which the users execute the issuing protocol for info, m_0 and info, m_1, respectively. Jumping ahead we note that we can easily turn our blind signature scheme into a partially blind one.

2.2 Construction

The high-level idea of our blind signature protocol is as follows. To obtain a blind signature from the bank the user commits to the message m and sends this commitment U to the bank. The bank signs the commitment with a regular signature scheme and returns the signature B to the user. The user derives the blind signature for m by computing another commitment C of the signature B together with U, and a non-interactive zero-knowledge proof π showing the validity of C.

To prevent trivial "size-measuring" attacks of a malicious bank we assume that the signature scheme is *length-invariant*, i.e., that public keys pk_{Sig} as well as signatures for security parameter n are all of the same length $s(n)$. We note that this can always be achieved by standard padding techniques, and length-invariant signature schemes exist if one-way functions exist [23,27].

We furthermore assume that the commitment scheme $(\mathcal{C}_{\mathsf{Com}}, \mathsf{Com})$ in the common random string model, given by algorithms $\mathcal{C}_{\mathsf{Com}}$ generating the string crs_{Com} and algorithm $\mathsf{Com}(crs_{\mathsf{Com}}, \cdot, \cdot) : \{0,1\}^n \times \{0,1\}^n \rightarrow \{0,1\}^{c(n)}$ mapping strings from $\{0,1\}^n$ with n-bit randomness to commitments, is length-invariant, too. That is, the reference strings as well as commitments are all of length $c(n)$ for parameter n. We also need that the commitment scheme is statistically binding. Such commitment schemes can also be derived for example from one-way functions [16,22].

In order to turn the above idea into a provably secure scheme the proof π needs to allow extraction of U and B encapsulated in C. We accomplish this by using an IND-CPA secure encryption scheme $(\mathsf{KG}_{\mathsf{Enc}}, \mathsf{Enc}, \mathsf{Dec})$ to "commit" to $U\|B$ in C, where the public key pk_{Enc} of the encryption algorithm becomes part of the common reference string.[2] We presume that the encryption scheme is also length-invariant and that public keys pk_{Enc} and ciphertexts $C \leftarrow \mathsf{Enc}(pk_{\mathsf{Enc}}, U\|B; v)$ for $U\|B \in \{0,1\}^{c(n)+s(n)}$ and randomness $v \in \{0,1\}^n$ are all of length $e(n)$. This is again without loss of generality.

Finally, the non-interactive zero-knowledge (NIZK) proof $(\mathcal{C}_{\mathsf{ZK}}, \mathcal{P}, \mathcal{V})$ with the common reference string generator $\mathcal{C}_{\mathsf{ZK}}$, the prover $\mathcal{P}$ and the verifier $\mathcal{V}$ first of all obeys the two basic properties completeness (the verifier $\mathcal{V}$ accepts all honestly generated proofs of the prover $\mathcal{P}$) and soundness (no malicious prover can make the verifier accepts proofs for invalid statements). Furthermore, the system should be multiple zero-knowledge in the sense of [15], i.e., one can prove several statements in zero-knowledge for the same common reference string $crs_{\mathsf{ZK}} \leftarrow \mathcal{C}_{\mathsf{ZK}}(1^n)$. Such proofs exist if trapdoor permutations exist [15].

---

[2] We can also assume that we have dense public-key schemes [13] and that the common *random* string contains such a public key.

The underlying relation of the NIZK proof is described by (a sequence of) circuits C_n^{BS} indexed by a parameter n. Circuit C_n^{BS} takes as input a statement $x = C||pk_{Enc}||crs_{Com}||pk_{Sig}||m$ of bit length $\chi(n) = c(n) + 2e(n) + s(n) + n$ and a witness $w = u||v||B$ of length $\omega(n) = 2n + s(n)$, and returns an output bit which is determined as follows. The circuit is built from the descriptions of algorithms Com, Enc, Vf_{Sig} and checks for the signature's verification algorithm that $Vf_{Sig}(pk_{Sig}, Com(crs_{Com}, m; u), B) = 1$ and that the value C equals the ciphertext $Enc(pk_{Enc}, Com(crs_{Com}, m; u)||B; v)$. If and only if both tests evaluate to true then the circuit outputs 1. The corresponding relation (parameterized by n) is given by $R_n^{BS} = \left\{ (x, w) \in \{0,1\}^{\chi(n)} \times \{0,1\}^{\omega(n)} \mid C_n^{BS}(x, w) = 1 \right\}$.

Construction 1 (Blind Signature Scheme). *Let* $(KG_{Sig}, Sig, Vf_{Sig})$ *be a signature scheme,* (KG_{Enc}, Enc, Dec) *be an encryption scheme,* (C_{Com}, Com) *be a commitment scheme, and let* (C_{ZK}, P, V) *be a non-interactive zero-knowledge proof system for* R^{BS}. *Define the following four procedures:*

CRS Generation. *Algorithm* $C_{BS}(1^n)$ *generates* $crs_{ZK} \leftarrow C_{ZK}(1^n)$, $crs_{Com} \leftarrow C_{Com}(1^n)$ *as well as* $(pk_{Enc}, sk_{Enc}) \leftarrow KG_{Enc}(1^n)$. *It outputs the string* $crs_{BS} \leftarrow (crs_{ZK}, crs_{Com}, pk_{Enc})$.

Key Generation. *The bank's key generation algorithm* $KG_{BS}(crs_{BS})$ *generates a signature key pair* $(pk_{Sig}, sk_{Sig}) \leftarrow KG_{Sig}(1^n)$. *It returns* $(pk_{BS}, sk_{BS}) \leftarrow (pk_{Sig}, sk_{Sig})$.

Signature Issue Protocol. *The interactive signature issue protocol is described in Figure 1.*

Signature Verification. *The verification algorithm* $Vf_{BS}(crs_{BS}, m, S)$ *parses the signature* S *as* $S = C||\pi$ *and returns the output* $V(crs_{ZK}, x, \pi)$ *for the value* $x = C||pk_{Enc}||crs_{Com}||pk_{Sig}||m$.

Theorem 2. *Let* $(KG_{Sig}, Sig, Vf_{Sig})$ *be a length-invariant signature scheme which is unforgeable against adaptive chosen-message attacks,* (KG_{Enc}, Enc, Dec) *be a length-invariant IND-CPA secure encryption scheme, and let* (C_{Com}, Com) *be a length-invariant non-interactive commitment scheme in the common random string model which is statistically binding. Let* (C_{ZK}, P, V) *be a non-interactive zero-knowledge proof system for* R^{BS}. *Then the scheme defined in Construction 1 is a secure blind signature scheme.*

Proof. We first show unforgeability and then blindness.

Unforgeability. Assume that there exists an adversary U^* such that with noticeable probability the following holds. On input crs_{BS}, pk_{BS} the adversary U^* manages to output $k + 1$ valid signatures $S_i = C_i||\pi_i$ for messages m_i after at most k interactions with the honest bank B (and where the messages m_i, m_j are pairwise distinct). Given such an adversary we construct a successful adversary A against the security of the signature scheme $(KG_{Sig}, Sig, Vf_{Sig})$.

Adversary A is given as input a public key pk_{Sig} of the signature scheme and is granted oracle access to a signature oracle $Sig(sk_{Sig}, \cdot)$. This adversary first

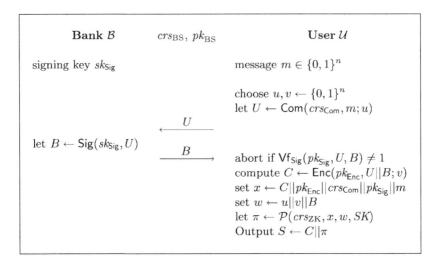

Fig. 1. Blind Signature Scheme: Issue Protocol

generates $crs_{ZK} \leftarrow \mathcal{C}_{ZK}(1^n)$ for the zero-knowledge proof, $crs_{Com} \leftarrow \mathcal{C}_{Com}(1^n)$ for the commitments, and $(pk_{Enc}, sk_{Enc}) \leftarrow \mathsf{KG}_{Enc}(1^n)$ and defines $crs_{BS} \leftarrow (crs_{ZK}, crs_{Com}, pk_{Enc})$. It next invokes a black-box simulation of $\mathcal{U}^*$ for input (crs_{BS}, pk_{Sig}). Each time the user $\mathcal{U}^*$ initiates the issue protocol with the bank, algorithm $\mathcal{A}$ uses the signature oracle to answer the request U by a signature $B \leftarrow \mathsf{Sig}(sk_{Sig}, U)$. When the adversary finally outputs the message/signature pairs (m_i, S_i) algorithm $\mathcal{A}$ parses each S_i as $S_i = C_i || \pi_i$ and decrypts each C_i to $U_i || B_i$. Algorithm $\mathcal{A}$ outputs the first one of these pairs (U_i, B_i) for which U_i has not been submitted to its signing oracle (or returns $\perp$ if no such value exists).

For the analysis assume that $\mathcal{U}^*$ succeeds with noticeable probability. Since $k + 1$ pairs (m_i, S_i) are valid, all the proofs π_i in $S_i = C_i || \pi_i$ are valid as well. Hence, with overwhelming probability over the choice of crs_{ZK} each C_i is a valid ciphertext of some $U_i || B_i$ under pk_{Enc}, and each U_i commits to the corresponding message m_i and each B_i is a valid signature under pk_{Sig} for U_i. Furthermore, by the statistically-binding property of the commitment scheme all U_i's must be distinct for different m_i's, with overwhelming probability over the choice of crs_{Com}. It follows that $\mathcal{A}$ successfully outputs a valid signature B_i for a previously not signed value U_i with noticeable probability, too, contradicting our initial assumption about the success probability of $\mathcal{U}^*$.

Blindness. To prove blindness we consider an adversarial controlled bank $\mathcal{B}^*$ in experiment $\mathsf{Blind}_{\mathcal{B}^*}^{BS}(n)$. We gradually transform the way the signatures $S_0 = C_0 || \pi_0$ and $S_1 = C_1 || \pi_1$ are computed such that, at then end, they are completely independent of bit b.

In the first step we replace all the steps involving the prover by the output of the zero-knowledge simulator $\mathcal{Z}$ (generating a string crs together with some trapdoor information σ in mode crs and faking proofs π for statements x with the help of σ in mode prove). More precisely, consider the following modified procedures of the blind signature scheme (key generation and verification remain unchanged):

CRS Generation. Algorithm $\mathcal{C}_{\mathrm{BS}}(1^n)$ generates $(crs_{\mathrm{ZK}}, \sigma) \leftarrow \mathcal{Z}(\mathrm{crs}, 1^n)$, and $crs_{\mathrm{Com}} \leftarrow \mathcal{C}_{\mathrm{Com}}(1^n)$ and $(pk_{\mathrm{Enc}}, sk_{\mathrm{Enc}}) \leftarrow \mathsf{KG}_{\mathrm{Enc}}(1^n)$. It outputs the string $crs_{\mathrm{BS}} \leftarrow (crs_{\mathrm{ZK}}, crs_{\mathrm{Com}}, pk_{\mathrm{Enc}})$.

Signature Issue Protocol. For the signature issuing the user now also picks $u, v \leftarrow \{0,1\}^n$, and again sends the commitment $U \leftarrow \mathsf{Com}(crs_{\mathrm{Com}}, m; u)$ to the bank $\mathcal{B}^*$ which replies with some B. The user aborts if verification fails, $\mathsf{Vf}_{\mathrm{Sig}}(pk_{\mathrm{Sig}}, U, B) \neq 1$, and else computes $C \leftarrow \mathsf{Enc}(pk_{\mathrm{Enc}}, U \| B; v)$ as well as $\pi \leftarrow \mathcal{Z}(\mathrm{prove}, \sigma, x)$. Output $S \leftarrow C \| \pi$.

Denote the modified scheme by BS'. It follows easily from the zero-knowledge property that experiments $\mathsf{Blind}_{\mathcal{B}^*}^{\mathsf{BS}}(n)$ and $\mathsf{Blind}_{\mathcal{B}^*}^{\mathsf{BS}'}(n)$ return 1 with the same probability (except for a negligible probability).

In the next step we further modify the signature scheme by replacing the commitments U by commitments to all-zero strings. More precisely, we change the signature issue protocol of the blind signature scheme BS' as follows (recall that this modified scheme already uses the zero-knowledge simulator to prepare the signatures):

Signature Issue Protocol. The user picks $u, v \leftarrow \{0,1\}^n$, but now sends $U \leftarrow \mathsf{Com}(crs_{\mathrm{Com}}, 0^n; u)$ to the bank to get a signature B (which is also checked). It then computes again $C \leftarrow \mathsf{Enc}(pk_{\mathrm{Enc}}, U \| B; v)$, $\pi \leftarrow \mathcal{Z}(\mathrm{prove}, \sigma, x)$ and outputs $S \leftarrow C \| \pi$.

We call this modified scheme BS''. It is easy to see that, by the secrecy of the commitment scheme, the difference in the output distributions of experiments $\mathsf{Blind}_{\mathcal{B}^*}^{\mathsf{BS}'}(n)$ and $\mathsf{Blind}_{\mathcal{B}^*}^{\mathsf{BS}''}(n)$ is negligible.

Finally, we replace the encryption of U and B by an encryption of the all-zero string, i.e., we modify the signature issuing protocol from BS'' further to obtain:

Signature Issue Protocol. The user selects $u, v \leftarrow \{0,1\}^n$ as before and again computes $U \leftarrow \mathsf{Com}(crs_{\mathrm{Com}}, 0^n; u)$. For the bank's reply B to U it checks validity and this time sets $C \leftarrow \mathsf{Enc}(pk_{\mathrm{Enc}}, 0^{c(n)+s(n)}; v)$, $\pi \leftarrow \mathcal{Z}(\mathrm{prove}, \sigma, x)$ and outputs $S \leftarrow C \| \pi$.

We call this modified scheme by BS'''. By the IND-CPA security of the encryption scheme we conclude that the difference in the output distributions of experiments $\mathsf{Blind}_{\mathcal{B}^*}^{\mathsf{BS}''}(n)$ and $\mathsf{Blind}_{\mathcal{B}^*}^{\mathsf{BS}'''}(n)$ is also negligible.

In experiment $\mathsf{Blind}_{\mathcal{B}^*}^{\mathsf{BS}'''}(n)$ the signatures $S = C \| \pi$ are independent of the data U, B in the signature generation protocol. Hence, the adversary $\mathcal{B}^*$ cannot predict b better than with probability $1/2$. Conclusively, the probability of the original experiment $\mathsf{Blind}_{\mathcal{B}^*}^{\mathsf{BS}}(n)$ to return 1 must be negligibly close to $1/2$, proving blindness. $\qquad\square$

To get a partially blind signature scheme, where the signer and the user share some public information info to be included in the blind signature, we let the bank simply sign the user's commitment U together with info, i.e., $B \leftarrow \mathsf{Sig}(sk, \mathsf{info}\|U)$ and also let the user base the correctness proof π on info. The security proof carries over straightforwardly.

3   Universally Composable Blind Signatures

As mentioned, the blind signature scheme in the previous section allows concurrent executions of the signature generation protocol, i.e., when the protocol is interleaved with itself. More generally, one would like to have a guarantee that such a scheme still supports the basic security properties, even when run as a building block within larger protocols, independently how the execution is intertwined with other steps. Such a guarantee is provided by the universal composition (UC) framework [5].

In the UC framework one defines an idealized version of the primitive in question, capturing the desired security properties in an abstract way and ensuring that the functionality is secure in interdependent settings. Given an appropriate formalization of some functionality $\mathcal{F}$ in the UC framework, one next shows that this functionality can be securely realized by an interactive protocol between the parties (without the trusted interface). Here, securely realizing means that, in any environment (modeled through an algorithm $\mathcal{Z}$) in which the protocol may be run, for this environment executions of the interactive protocol in presence of an adversary $\mathcal{A}$ are indistinguishable from executions in the ideal model with the trustworthy functionality $\mathcal{F}$ and an ideal-model adversary $\mathcal{S}$. A formal introduction to the UC framework is beyond the scope of our paper; we refer to [5] to a comprehensive definition. We remark that we consider non-adaptive adversaries here which corrupt parties at the beginning of executions only.

3.1   Definition

Our definition of an ideal blind signature functionality $\mathcal{F}_{\mathrm{BlSig}}$ follows the one $\mathcal{F}_{\mathrm{Sig}}$ of regular signature schemes given by Canetti [6]. The definition of $\mathcal{F}_{\mathrm{Sig}}$ essentially lets the adversary choose the public verification key and determine the signature value S upon a signing request ($\mathtt{Sign}, sid, m$). Verification requests for previously generated signatures are always accepted and otherwise the adversary is again allowed to specify whether a tested signature S to a message m is valid or not. See [6] for a discussion of this definition.

The formal description of the blind signature functionality is given in Figure 2. It is partly a verbatim copy of the functionality $\mathcal{F}_{\mathrm{Sig}}$ in [6]. An important difference for blind signatures is that the adversary should not learn the message of honest users and the signatures must not be linkable to the signing request. To ensure this we let the adversary (instead of the bank, analogously to the choice of the public verification key in $\mathcal{F}_{\mathrm{Sig}}$) in $\mathcal{F}_{\mathrm{BlSig}}$ provide (the description of) a stateless, possibly probabilistic algorithm BlSig to the ideal functionality $\mathcal{F}_{\mathrm{BlSig}}$. This is already done in the key generation step where the adversary chooses the public verification key, such that BlSig is used in all subsequent signature generation runs. Whenever an honest user later requests a signature for a message

---

Functionality $\mathcal{F}_{\text{BlSig}}$

Key Generation: Upon receiving a value (KeyGen, sid) from a party $\mathcal{B}$, verify that $sid = (\mathcal{B}, sid')$ for some sid'. If not, then ignore. Else, hand (KeyGen, sid) to the adversary. Upon receiving (VerificationKey, $sid, pk_{\text{BS}}, \text{BlSig}$) from the adversary, output (VerificationKey, sid, pk_{BS}) to $\mathcal{B}$ and record the pair $(\mathcal{B}, pk_{\text{BS}}, \text{BlSig})$.

Signature Generation: Upon receiving a value (Sign, sid, m, pk_{BS}) for $m \in \{0,1\}^n$ from some party $\mathcal{U}$, verify that $sid = (\mathcal{B}, sid')$ for some sid'. If not, then ignore. Else, do the following:

- If the user $\mathcal{U}$ is honest then inform $\mathcal{B}$ and the adversary through (Signature, sid) that a signature request takes place and then generate $S \leftarrow \text{BlSig}(m)$ and output (Signature, sid, m, S) to $\mathcal{U}$.
- If the user $\mathcal{U}$ is corrupt then send (Sign, sid, m) to the adversary to obtain (Signature, sid, m, S); abort if $(m, S, pk_{\text{BS}}, 0)$ has been recorded before. Send (Signature, sid) to $\mathcal{B}$.

In either case record $(m, S, pk_{\text{BS}}, 1)$.

Signature Verification: Upon receiving a value (Verify, $sid, m, S, pk'_{\text{BS}}$) from some party $\mathcal{P}$ hand (Verify, $sid, m, S, pk'_{\text{BS}}$) to the adversary. Upon receiving (Verified, sid, m, S, ϕ) from the adversary do:

1. If $pk_{\text{BS}} = pk'_{\text{BS}}$ and the entry $(m, S, pk_{\text{BS}}, 1)$ is recorded, then set $f = 1$ (completeness condition).
2. Else, if $pk_{\text{BS}} = pk'_{\text{BS}}$, the bank is not corrupt, and no entry $(m, \star, pk_{\text{BS}}, 1)$ is recorded, then set $f = 0$ and record the entry $(m, S, pk_{\text{BS}}, 0)$ (unforgeability condition).
3. Else, if there is an entry $(m, S, pk'_{\text{BS}}, f')$ recorded, then let $f = f'$ (consistency condition).
4. Else, let $f = \phi$ and record the entry $(m, S, pk'_{\text{BS}}, \phi)$.

Output (Verified, sid, m, S, f) to $\mathcal{P}$.

Fig. 2. Blind Signature Functionality $\mathcal{F}_{\text{BlSig}}$

m this algorithm $\text{BlSig}(m)$ generates a signature S but *without disclosing the message m to the adversary*, enforcing unlinkability of signatures.

If a corrupt user —that is, the adversary on behalf of the corrupt user— requests a signature, however, the ideal functionality does not run BlSig. Instead it asks the adversary about the potential signature S this user could produce from an interaction with the bank. Note that a corrupt user may not output any signature right after an interaction with the bank (or ever), the interaction merely guarantees that this user can generate this signature in principle. Hence, the functionality does not return the adversary's potential signature S to the user.

Finally, for any signature request we inform the bank $\mathcal{B}$ about the request, but without disclosing the actual message m nor the signature. This captures the fact that signature generations require active participation of the bank.

It follows quite easily that one can realize universally composable commitment schemes in the presence of functionality $\mathcal{F}_{\text{BlSig}}$. As a consequence of the hardness of constructing such commitments [7,12] we conclude (for the proof see the full version [14]):

Proposition 1. *Bilateral and terminating (i.e., only two parties are active and honest parties faithfully give output) blind signature schemes securely realizing $\mathcal{F}_{\text{BlSig}}$ in the plain model do not exist. Furthermore, blind signature schemes securely realizing $\mathcal{F}_{\text{BlSig}}$ in the common reference string model imply key agreement, and imply oblivious transfer in the common random string model. This all holds for non-adaptive corruptions.*

By the general feasibility results of Canetti et al. [5,10] functionality $\mathcal{F}_{\text{BlSig}}$ can be realized in the multi-party setting for honest majorities (in the plain model) and dishonest majorities (in the common random string model). Instead of relying on the general construction in [10] we construct a simpler two-move scheme in the common reference string model directly, based on the scheme in the previous section.

3.2   Construction

The construction in the standard model gives a good starting point for a solution in the UC framework. We augment the scheme by the following steps. First, we let the user in the first round when sending U also encrypt all the data from which the proof is later derived. This ciphertext E should contain the randomness u, v and message m. The encryption scheme itself needs to be IND-CPA secure and we simply use the same scheme as for the computation for $C \leftarrow \text{Enc}(pk_{\text{Enc}}, U\|B; v)$ but with an independent key pair $(pk'_{\text{Enc}}, sk'_{\text{Enc}})$.

In addition to the ciphertext E the user should prove with another NIZK proof that the data is valid and matches the data committed to by U. For this we require *simulation-sound* NIZK proofs [28,11] where a malicious prover is not able to find an accepted proof for an invalid statement, even if it sees proofs of the zero-knowledge simulator before (possibly for invalid but different theorems). In our case here the underlying relation R^{ss} is defined by (a sequence of) circuits C_n^{ss} evaluating to 1 if and only if for statement $x = U\|E\|pk'_{\text{Enc}}\|crs_{\text{Com}}$ of length $\chi(n) = 2c(n) + 2e'(n)$ and witness $w = m\|u\|v\|u'$ of length $\omega(n) = 4n$ it holds that $E = \text{Enc}(pk'_{\text{Enc}}, m\|u\|v; u')$ and $U = \text{Com}(crs_{\text{Com}}, m; u)$. Such simulation-sound NIZK proofs exist if trapdoor permutations exist [11].

The final step is to make the signature algorithm Sig of the bank's unforgeable signature scheme deterministic. This can be accomplished by adding a key of a pseudorandom function to the secret signing key. Each time a signature for U is requested the signing algorithm first applies the pseudorandom function to U to get the randomness s with which the signature $B \leftarrow \mathsf{Sig}(sk_{\text{Sig}}, U; s)$ is computed deterministically. The advantage of this modification, which does not require an additional complexity assumption, is that identical requests are now also answered identically. For the same consistency reason we also presume that the verification algorithm of the regular UNIZK proof is deterministic.

We note that our protocol is defined in the common reference string model. In contrast to the case of UC commitments in [7] where a fresh common reference string for each commitment through $\mathcal{F}_{\mathsf{Com}}$ is required, in our case the once generated common reference string can be used for *several* signature generations by different users; we merely need an independent common reference string for each party taking the role of a bank.

Construction 3 (Universally Composable Blind Signature Scheme).
Let $(\mathsf{KG}_{\mathsf{Sig}}, \mathsf{Sig}, \mathsf{Vf}_{\mathsf{Sig}})$ *be a signature scheme,* $(\mathsf{KG}_{\mathsf{Enc}}, \mathsf{Enc}, \mathsf{Dec})$ *be an encryption scheme, and* $(\mathcal{C}_{\mathsf{Com}}, \mathsf{Com})$ *be a commitment scheme. Let* $(\mathcal{C}_{ZK}, \mathcal{P}, \mathcal{V})$ *be a non-interactive zero-knowledge proof system for* R^{BS} *and let* $(\mathcal{C}_{ss}, \mathcal{P}_{ss}, \mathcal{V}_{ss})$ *be a non-interactive zero-knowledge proof system for* R^{ss}. *Define the following four procedures:*

CRS Generation. *Algorithm* $\mathcal{C}_{BS}$ *on input* 1^n *generates* $crs_{ZK} \leftarrow \mathcal{C}_{ZK}(1^n)$, $crs_{\mathsf{Com}} \leftarrow \mathcal{C}_{\mathsf{Com}}(1^n)$, $crs_{ss} \leftarrow \mathcal{C}_{ss}(1^n)$ *and pairs* $(pk_{\mathsf{Enc}}, sk_{\mathsf{Enc}}), (pk'_{\mathsf{Enc}}, sk'_{\mathsf{Enc}}) \leftarrow \mathsf{KG}_{\mathsf{Enc}}(1^n)$. *It outputs* $crs_{BS} \leftarrow (crs_{ZK}, crs_{\mathsf{Com}}, pk_{\mathsf{Enc}}, pk'_{\mathsf{Enc}}, crs_{ss})$.

Key Generation. *If party* $\mathcal{B}$ *with access to* crs_{BS} *receives* $(\mathtt{KeyGen}, sid)$ *it checks that* $sid = (\mathcal{B}, sid')$ *for some* sid'. *If not, it ignores. Else it generates a signature key pair* $(pk_{\mathsf{Sig}}, sk_{\mathsf{Sig}}) \leftarrow \mathsf{KG}_{\mathsf{Sig}}(1^n)$. *It sets* $(pk_{BS}, sk_{BS}) \leftarrow (pk_{\mathsf{Sig}}, sk_{\mathsf{Sig}})$, *stores* sk_{BS} *and outputs* $(\mathtt{VerificationKey}, sid, pk_{BS})$.

Signature Issue Protocol. *If party* $\mathcal{U}$ *is invoked with input* $(\mathtt{Sign}, sid, m, pk_{BS})$ *for* $sid = (\mathcal{B}, sid')$ *it initiates a run of the interactive protocol in Figure 3 with the bank* $\mathcal{B}$, *where the user gets* m *and* pk_{BS} *as input and the bank uses* sk_{BS} *as input. The user outputs* $(\mathtt{Signature}, sid, m, S)$ *for the derived signature value* S.

Signature Verification. *If a party receives* $(\mathtt{Verify}, sid, m, S, pk'_{BS})$ *it parses* S *as* $S = C||\pi$, *computes* $\phi \leftarrow \mathcal{V}(crs, x, \pi)$ *for* $x = C||pk_{\mathsf{Enc}}||crs_{\mathsf{Com}}||pk_{\mathsf{Sig}}||m$ *and outputs* $(\mathtt{Verified}, sid, m, \sigma, \phi)$.

Theorem 4. *Let* $(\mathsf{KG}_{\mathsf{Sig}}, \mathsf{Sig}, \mathsf{Vf}_{\mathsf{Sig}})$ *be a length-invariant signature scheme which is unforgeable against adaptive chosen-message attacks and for which* Sig *is deterministic. Let* $(\mathsf{KG}_{\mathsf{Enc}}, \mathsf{Enc}, \mathsf{Dec})$ *be a length-invariant IND-CPA secure encryption scheme,* $(\mathcal{C}_{\mathsf{Com}}, \mathsf{Com})$ *be a length-invariant non-interactive commitment scheme in the common reference string model which is statistically binding. Also let* $(\mathcal{C}_{ZK}, \mathcal{P}, \mathcal{V})$ *be a non-interactive zero-knowledge proof system for* R^{BS} *with deterministic verifier* $\mathcal{V}$. *Let* $(\mathcal{C}_{ss}, \mathcal{P}_{ss}, \mathcal{V}_{ss})$ *be a simulation-sound non-interactive zero-knowledge proof for* R^{ss}. *Then the scheme defined in Construction 3 securely realizes functionality* $\mathcal{F}_{BlSig}$ *for non-adaptive corruptions.*

The idea of the proof is as follows. Algorithm BlSig for functionality $\mathcal{F}_{BlSig}$, supposed to be determined by the ideal-model adversary and to create signatures for honest users, ignores its input m entirely, but instead prepares a dummy encryption $C \leftarrow \mathsf{Enc}(pk_{\mathsf{Enc}}, 0^{c(n)+s(n)}; v)$ and appends a fake correctness proof π generated by the zero-knowledge simulator. The output $C||\pi$ is thus indistinguishable from genuinely generated signatures of honest users in the actual

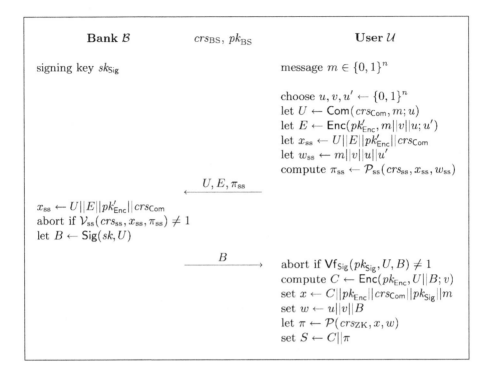

Fig. 3. UC Blind Signature Scheme: Issue Protocol

scheme. On the other hand, the additional encryption E and the simulation-sound zero-knowledge proof allows the ideal-model adversary to extract potential signatures $C||\pi$ of malicious users (in black-box simulations), and to provide them to the functionality. The completeness and consistency condition of $\mathcal{F}_{\mathsf{BISig}}$ is realized by the completeness of the underlying scheme, and unforgeability follows as for the basic scheme with concurrent security.

Proof. (of Theorem 4) We have to show that for each adversary $\mathcal{A}$ attacking the real-world protocol there exist an ideal-model adversary (aka. simulator) $\mathcal{S}$ in the ideal world with dummy parties and functionality $\mathcal{F}_{\mathsf{BISig}}$ such that no environment $\mathcal{Z}$ can distinguish whether it is facing an execution in the real world with $\mathcal{A}$ or one in the ideal world with $\mathcal{S}$.

We build the ideal-model adversary $\mathcal{S}$ by black-box simulation of $\mathcal{A}$, relaying all communication between the environment $\mathcal{Z}$ and the (simulated) adversary $\mathcal{A}$, and acting on behalf of the honest parties in this simulation. Algorithm $\mathcal{S}$ also corrupts a dummy party in the ideal model whenever $\mathcal{A}$ asks to corrupt the corresponding party in the simulation. By assumption this is only done before the execution starts.

The ideal-model simulator $\mathcal{S}$ first generates a reference string crs_{BS} for the black-box simulation by picking encryption keys $(sk_{\mathsf{Enc}}, pk_{\mathsf{Enc}}), (sk'_{\mathsf{Enc}}, pk'_{\mathsf{Enc}}) \leftarrow \mathsf{KG}_{\mathsf{Enc}}(1^n)$, generating $crs_{\mathsf{Com}} \leftarrow \mathcal{C}_{\mathsf{Com}}(1^n)$ and running the zero-knowledge simulators to generates crs_{ss} as well as crs_{ZK} for the zero-knowledge proofs. It outputs $crs_{BS} = (pk_{Enc}, pk'_{Enc}, crs_{ss}, crs_{Com}, crs_{ZK})$. We next describe the simulation of the honest parties in the black-box simulation:

- Suppose the simulator is woken up through a call (KeyGen, sid) from $\mathcal{F}_{BlSig}$ in the ideal model where $sid = (\mathcal{B}, sid')$. Then the simulator generates $(pk_{BS}, sk_{BS}) \leftarrow KG_{BS}(1^n)$ as specified by the scheme's description and lets BlSig be the algorithm that on input $m \in \{0,1\}^n$ computes the encryption $C \leftarrow Enc(pk_{Enc}, 0^{c(n)+s(n)}; v)$, and generates π through the zero-knowledge simulator for relation R^{BS} where $x \leftarrow C||pk_{Enc}||crs_{Com}||pk_{Sig}||m$, and outputs $S \leftarrow C||\pi$. Simulator $\mathcal{S}$ returns (VerificationKey, sid, pk_{BS}, BlSig) to $\mathcal{F}_{BlSig}$ in the ideal model. In the black-box simulation it sends pk_{BS} to all parties.
- Suppose that the adversary lets a corrupt user in the black-box simulation initiate a protocol run with the honest bank by sending values (U, E, π_{ss}). Then the simulator first checks the validity of the proof π_{ss}; if this check fails then it ignores this message. Else, $\mathcal{S}$ uses the secret key sk'_{Enc} to recover $m||v||u$ from E (and aborts if it fails) and submits (Sign, sid, m, pk_{BS}) on behalf of the user to the ideal functionality. It immediately receives a request (Signature, sid, m) from $\mathcal{F}_{BlSig}$. To answer, $\mathcal{S}$ computes the signature $B \leftarrow Sig(sk_{Sig}, U)$ under the unforgeable signature scheme, an encryption $C \leftarrow Enc(pk_{Enc}, U||B; v)$ and a proof $\pi \leftarrow \mathcal{P}(crs_{ZK}, x, w)$ for the extracted values and sends (Signature, sid, m, S) for $S \leftarrow C||\pi$ back to the functionality. It also returns B in the black-box simulation to the corrupt user.
- If an honest user requests a signature (Sign, sid, m, pk_{BS}) in the ideal model and waits to receive (Signature, sid, m, S), generated by the functionality through algorithm BlSig, then the ideal-model adversary generates strings $U \leftarrow Com(crs_{Com}, 0^n; u)$, $E \leftarrow Enc(pk_{Enc}, 0 \ldots 0; u')$ and a proof π_{ss} via the zero-knowledge simulator of the simulation-sound scheme and lets the user in the black-box simulation send these values (U, E, π_{ss}). If the bank is honest, then $\mathcal{S}$ uses the secret signing key sk_{Sig} to compute $B \leftarrow Sig(sk_{Sig}, U)$, else it waits to receive a value B from the adversarial controlled bank.
- If $\mathcal{S}$ in the ideal model gets a request (Verify, sid, m, S, pk'_{BS}) then it computes $\phi \leftarrow \mathcal{V}(crs, PK, x, \pi)$ and returns (Verified, sid, m, S, ϕ).

This gives a full description of the ideal-model simulator. For the analysis note that there are two main differences between the ideal-model and the black-box simulation compared to an actual attack on the protocol. First, in the black-box simulation we use fake values (commitments and encryptions of zero-strings, simulated proofs etc.). The second point to address is that the verification algorithm in the ideal model returns 0 if there is no recorded pair $(m, \star, pk_{BS}, 1)$ while in the real-life protocol Vf_{BS} may output 1; for any other verification requests the answers are identical as the verification algorithm Vf_{BS} merely runs the deterministic verification algorithm of the NIZK system (and thus guarantees completeness and, especially, consistency).

We claim that the differences are undetectable for the environment $\mathcal{Z}$. This is proven through a sequence of games transforming an execution in the ideal-model scenario into one which is equal to the one of the actual protocol. In these games we will have full control over the setting, in particular over the functionality and, in contrast to the ideal-model adversary, we will also read the inputs of $\mathcal{Z}$ to honest users. This is admissible since our goal is to emulate $\mathcal{Z}$'s environment and to use differences in the output behavior to contradict the security of the underlying cryptographic primitives.

- Experiment $\mathsf{Game}_0(n)$ describes the original attack of $\mathcal{Z}$ on the ideal-model simulation (including the black-box simulation of $\mathcal{A}$).
- In $\mathsf{Game}_1(n)$ we change the way the commitments U on behalf of honest users are computed in $\mathcal{A}$'s black-box simulation. Originally, the simulator $\mathcal{S}$ computes a fake commitment $U \leftarrow \mathsf{Com}(crs_{\mathsf{Com}}, 0^n; u)$. Now, whenever the simulator is supposed to create such a commitment, we let $U \leftarrow \mathsf{Com}(crs_{\mathsf{Com}}, m; u)$ for the right value m (given as input to the honest user by $\mathcal{Z}$), and $u, v \leftarrow \{0,1\}^n$ are picked at random. Because of the secrecy of Com it is easy to see that $\mathcal{Z}$'s output behavior will not change significantly when facing $\mathsf{Game}_1(n)$ instead of $\mathsf{Game}_0(n)$.
- Next, in $\mathsf{Game}_2(n)$, we replace every encryption $C \leftarrow \mathsf{Enc}(pk_{\mathsf{Enc}}, 0^{c(n)+s(n)}; v)$ in the computations of algorithm BlSig through an encryption of the actual values $U \| B$, i.e., $C \leftarrow \mathsf{Enc}(pk_{\mathsf{Enc}}, U \| B; v)$, where we provide the values U and B transmitted in the black-box simulation "adaptively" to algorithm BlSig. By the security of the encryption scheme this is indistinguishable from the environment's viewpoint.
- In $\mathsf{Game}_3(n)$ we replace every steps of the zero-knowledge proof in the computation of BlSig through steps of the actual proof system, i.e., generation of crs_{ZK} through $crs_{\mathsf{ZK}} \leftarrow \mathcal{C}_{\mathsf{ZK}}(1^n)$, and every generation of π through the prover (for the now genuine witness $w = u \| v \| B$). By the zero-knowledge property this substitution is indistinguishable for the environment $\mathcal{Z}$.
- Now we turn to the difference between the ideal-model verification through list comparisons and the real-life verification through $\mathsf{Vf}_{\mathsf{BS}}$. In $\mathsf{Game}_4(n)$ every time the verification algorithm in $\mathsf{Game}_3(n)$ was called by some honest user about input $(\mathtt{Verify}, sid, m, S, pk_{\mathsf{BS}})$ then we run the verification $\mathsf{Vf}_{\mathsf{BS}}(crs_{\mathsf{BS}}, pk_{\mathsf{BS}}, m, S)$ instead (the case $pk'_{\mathsf{BS}} \neq pk_{\mathsf{BS}}$ is treated as before). Consider in $\mathsf{Game}_3(n)$ the event that some user requests the functionality to verify a signature $(\mathtt{Verify}, sid, m, S, pk_{\mathsf{BS}})$ such that there is no entry $(m, \star, pk_{\mathsf{BS}}, 1)$ stored by the functionality (and the bank $\mathcal{B}$ is honest) but such that $\mathsf{Vf}_{\mathsf{BS}}$ returns 1. This would clearly allow to distinguish the two games (note that the other direction, that $\mathsf{Vf}_{\mathsf{BS}}$ yields 0 but there is an entry, cannot happen by the completeness of the blind signature scheme). We claim that if such a request should occur with noticeable probability this would contradict the unforgeability of the blind signature scheme in the previous section (there, unforgeability was proven for arbitrary, unforgeable signature scheme for the bank, and thus holds for the deterministic one we consider here as well).
  Specifically, we show how to turn a run of $\mathsf{Game}_3(n)$ with $\mathcal{A}$ and $\mathcal{Z}$ into an attack according to experiment $\mathsf{Forge}_{\mathcal{U}^*}^{\mathsf{BS}}(n)$. For this we run the same experiment as in $\mathsf{Game}_3(n)$ but this time we are use the oracle access to the bank's

signing oracle $\mathsf{Sig}(sk_{\mathsf{Sig}}, \cdot)$ instead of generating the key pair $(sk_{\mathsf{Sig}}, pk_{\mathsf{Sig}})$ ourselves (recall that the bank is assumed to be honest). Each time an honest user receives input $(\mathtt{Sign}, sid, m, pk_{\mathsf{BS}})$ we generate U, E, π_{ss} as in $\mathsf{Game}_3(n)$, including a valid commitment $U \leftarrow \mathsf{Com}(crs_{\mathsf{Com}}, m; u)$, and submit U to the bank to receive B. From this answer we honestly compute the signature $S \leftarrow C||\pi$. We memorize the pair (m, S) and the values (U, B).

If a corrupt user submits U, E, π_{ss} in the black-box simulation then we also check the proof π_{ss} (and do nothing if it is invalid). If the proof is accepted then we check if we have stored a pair (U, B) for some B. If so, we return the same B as before in the black-box simulation but without contacting the bank in the attack $\mathsf{Forge}_{\mathcal{U}^*}^{\mathsf{BS}}(n)$ (since the bank's signature would be identical we answer consistently). If there is no such pair (U, B) we use sk'_{Enc} to extract appropriate values $m||v||u$. By the simulation soundness of the NIZK this extraction works with overwhelming probability (because U has never appeared for an honest user in the execution before). We submit U to the bank's signature oracle to receive a value B, which we return to the corrupt user in the black-box simulation. We also deduce the signature S with the help of the extracted values and record (m, S) and (U, B).

Suppose that a user at some point sends a request $(\mathtt{Verify}, sid, m, S, pk_{\mathsf{BS}})$ for a message m which we have not stored (and which is thus new) but for which $\mathsf{Vf}_{\mathsf{BS}}$ accepts. Then we immediately stop and output all previously stored k message/signature pairs together with (m, S). Note that this implies that all our $k + 1$ pairs are accepted by $\mathsf{Vf}_{\mathsf{BS}}$, although we only had k interactions with the bank. Hence, if a mismatch in $\mathsf{Game}_3(n)$ happens with noticeable probability it would refute the unforgeability of the blind signature scheme of the previous section. It follows that $\mathsf{Game}_3(n)$ and $\mathsf{Game}_4(n)$ are indistinguishable.

– In $\mathsf{Game}_5(n)$ we can omit the extraction step where the ideal-model simulator decrypts E from submissions of corrupt users, in particular, we do not need to know the secret key sk'_{Enc} anymore for the simulation. This is so since the verification now only relies on $\mathsf{Vf}_{\mathsf{BS}}$ instead of lists. Furthermore every time we gave a dummy encryption $E \leftarrow \mathsf{Enc}(pk'_{\mathsf{Enc}}, 0 \ldots 0; u')$ for an honest user, we now encrypt the true values $E \leftarrow \mathsf{Enc}(pk'_{\mathsf{Enc}}, m||v||u; u')$ to prepare the correct values U and C. By the security of the encryption scheme this is indistinguishable for the environment.

– In $\mathsf{Game}_6(n)$ we replace the simulation of proofs π_{ss} for honest users through proofs computed by the prover's algorithm for witness $m||v||u$ (with respect to a truly random string crs_{ss}). The zero-knowledge property ensures that this will not significantly affect the environment's output.

All the steps in the final game now are exactly as in an attack on the real protocol with adversary $\mathcal{A}$. Therefore, the environment's output in the ideal-model simulation ($\mathsf{Game}_0(n)$) and the real-world execution ($\mathsf{Game}_6(n)$) are indistinguishable. □

We finally note that, analogously to the case of our concurrently-secure scheme, we can also extend the functionality $\mathcal{F}_{\mathsf{BlSig}}$ to handle partially blind signatures, and can realize this functionality by starting with the partially blind version of our concurrently-secure blind signature scheme and adding the encryption and the simulation-sound proof as before.

Acknowledgments

We thank the anonymous reviewers for comprehensive comments.

References

1. Masayuki Abe. *A Secure Three-Move Blind Signature Scheme for Polynomially Many Signatures.* Advances in Cryptology — Eurocrypt 2001, Volume 2045 of Lecture Notes in Computer Science, pages 136–151. Springer-Verlag, 2001.
2. Masayuki Abe and Eiichiro Fujisaki. *How to Date Blind Signatures.* Advances in Cryptology — Asiacrypt'96, Volume 1163 of Lecture Notes in Computer Science, pages 244–251. Springer-Verlag, 1996.
3. Mihir Bellare, Chanathip Namprempre, David Pointcheval, and Michael Semanko. *The One-More-RSA-Inversion Problems and the Security of Chaum's Blind Signature Scheme.* Journal of Cryptology, 16(3):185–215, 2003.
4. Alexandra Boldyreva. *Efficient Threshold Signatures, Multisignatures and Blind Signatures Based on the Gap-Diffie-Hellman-Group Signature Scheme.* Public-Key Cryptography (PKC) 2003, Volume 2567 of Lecture Notes in Computer Science, pages 31–46. Springer-Verlag, 2003.
5. Ran Canetti. *Universally Composable Security: A new Paradigm for Cryptographic Protocols.* Proceedings of the Annual Symposium on Foundations of Computer Science (FOCS) 2001. IEEE Computer Society Press, for an updated version see `eprint.iacr.org`, 2001.
6. Ran Canetti. *On Universally Composable Notions of Security for Signature, Certification and Authentication.* Proceedings of Computer Security Foundations Workshop (CSFW) 2004. IEEE Computer Society Press, for an updated version see `eprint.iacr.org`, 2004.
7. Ran Canetti and Marc Fischlin. *Universally Composable Commitments.* Advances in Cryptology — Crypto 2001, Volume 2139 of Lecture Notes in Computer Science, pages 19–40. Springer-Verlag, 2001.
8. David Chaum. *Blind Signatures for Untraceable Payments.* Advances in Cryptology — Crypto'82, pages 199–203. Plemum, New York, 1983.
9. Jan Camenisch, Maciej Koprowski, and Bogdan Warinschi. *Efficient Blind Signatures Without Random Oracles.* Security in Communication Networks, Volume 3352 of Lecture Notes in Computer Science, pages 134–148. Springer-Verlag, 2004.
10. Ran Canetti, Yehuda Lindell, Rafail Ostrovsky, and Amit Sahai. *Universally Composable Two-Party and Multi-Party Secure Computation.* Proceedings of the Annual Symposium on the Theory of Computing (STOC) 2002, pages 494–503. ACM Press, 2002.
11. Alfredo De Santis, Giovanni Di Crescenzo, Rafail Ostrovsky, Giuseppe Persiano, and Amit Sahai. *Robust Non-interactive Zero Knowledge.* Advances in Cryptology — Crypto 2001, Volume 2139 of Lecture Notes in Computer Science, pages 566–598. Springer-Verlag, 2001.

12. Ivan Damgård and Jens Groth. *Non-interactive and Reusable Non-Malleable Commitment Schemes.* Proceedings of the Annual Symposium on the Theory of Computing (STOC) 2003, pages 426–437. ACM Press, 2003.

13. Alfredo De Santis and Giuseppe Persiano. *Zero-Knowledge Proofs of Knowledge Without Interaction.* Proceedings of the Annual Symposium on Foundations of Computer Science (FOCS)'92, pages 427–436. IEEE Computer Society Press, 1992.

14. Marc Fischlin. *Round-Optimal Composable Blind Signatures in the Common Reference String Model.* full version. available at www.fischlin.de, 2006.

15. Uriel Feige, Dror Lapidot, and Adi Shamir. *Multiple NonInteractive Zero Knowledge Proofs Under General Assumption.* SIAM Journal on Computing, 29(1):1–28, 1999.

16. Johan Håstad, Russel Impagliazzo, Leonid Levin, and Michael Luby. *A Pseudorandom Generator from any One-way Function.* SIAM Journal on Computing, 28(4):1364–1396, 1999.

17. Ari Juels, Michael Luby, and Rafail Ostrovsky. *Security of Blind Digital Signatures.* Advances in Cryptology — Crypto'97, Volume 1294 of Lecture Notes in Computer Science, pages 150–164. Springer-Verlag, 1997.

18. Aggelos Kiayias and Hong-Sheng Zhou. *Two-Round Concurrent Blind Signatures without Random Oracles.* Number 2005/435 in Cryptology eprint archive. eprint.iacr.org, 2005.

19. Yehuda Lindell. *Bounded-Concurrent Secure Two-Party Computation Without Setup Assumptions.* Proceedings of the Annual Symposium on the Theory of Computing (STOC) 2003, pages 683–692. ACM Press, 2003.

20. Yehuda Lindell. *Lower Bounds for Concurrent Self Composition.* Theory of Cryptography Conference (TCC) 2004, Volume 2951 of Lecture Notes in Computer Science, pages 203–222. Springer-Verlag, 2004.

21. Matt Lepinski, Silvio Micali, and Abhi Shelat. *Fair Zero-Knowledge.* Theory of Cryptography Conference (TCC) 2005, Volume 3378 of Lecture Notes in Computer Science, pages 245–263. Springer-Verlag, 2005.

22. Moni Naor. *Bit Commitment Using Pseudo-Randomness.* Journal of Cryptology, 4(2):151–158, 1991.

23. Moni Naor and Moti Yung. *Universal One-Way Hash Functions and Their Cryptographic Applications.* Proceedings of the Annual Symposium on the Theory of Computing (STOC) 1989, pages 33–43. ACM Press, 1989.

24. Tatsuaki Okamoto. *Efficient Blind and Partially Blind Signatures Without Random Oracles.* Theory of Cryptography Conference (TCC) 2006, Volume 3876 of Lecture Notes in Computer Science, pages 80–99. Springer-Verlag, 2006.

25. David Pointcheval. *Strengthened Security for Blind Signatures.* Advances in Cryptology — Eurocrypt'98, Volume 1403 of Lecture Notes in Computer Science, pages 391–405. Springer-Verlag, 1998.

26. David Pointcheval and Jacques Stern. *Security Arguments for Digital Signatures and Blind Signatures.* Journal of Cryptology, 13(3):361–396, 2000.

27. John Rompel. *One-Way Functions are Necessary and Sufficient for Secure Signatures.* Proceedings of the Annual Symposium on the Theory of Computing (STOC) 1999, pages 387–394. ACM Press, 1990.

28. Amit Sahai. *Non-Malleable Non-Interactive Zero Knowledge and Adaptive Chosen-Ciphertext Security.* Proceedings of the Annual Symposium on Foundations of Computer Science (FOCS) 1999. IEEE Computer Society Press, 1999.

On Signatures of Knowledge

Melissa Chase and Anna Lysyanskaya

Computer Science Department
Brown University
Providence, RI 02912
{mchase, anna}@cs.brown.edu

Abstract. In a traditional signature scheme, a signature σ on a message m is issued under a public key PK, and can be interpreted as follows: "The owner of the public key PK and its corresponding secret key has signed message m." In this paper we consider schemes that allow one to issue signatures on behalf of any NP statement, that can be interpreted as follows: "A person in possession of a witness w to the statement that $x \in L$ has signed message m." We refer to such schemes as *signatures of knowledge*.

We formally define the notion of a signature of knowledge. We begin by extending the traditional definition of digital signature schemes, captured by Canetti's ideal signing functionality, to the case of signatures of knowledge. We then give an alternative definition in terms of games that also seems to capture the necessary properties one may expect from a signature of knowledge. We then gain additional confidence in our two definitions by proving them equivalent.

We construct signatures of knowledge under standard complexity assumptions in the common-random-string model.

We then extend our definition to allow signatures of knowledge to be *nested* i.e., a signature of knowledge (or another accepting input to a UC-realizable ideal functionality) can itself serve as a witness for another signature of knowledge. Thus, as a corollary, we obtain the first *delegatable* anonymous credential system, i.e., a system in which one can use one's anonymous credentials as a secret key for issuing anonymous credentials to others.

1   Introduction

Digital signature schemes constitute a cryptographic primitive of central importance. In a traditional digital signature scheme, there are three algorithms: (1) the key generation algorithm KeyGen through which a signer sets up his public and secret keys; (2) the signing algorithm Sign; and (3) the verification algorithm Verify. A signature in a traditional signature scheme can be thought of as an assertion *on behalf of a particular public key*. One way to interpret (m, σ) where Verify$(PK, m, \sigma) = Accept$, is as follows: "the person who generated public key PK and its corresponding secret key SK has signed message m."

We ask ourselves the following question: Can we have a signature scheme in which a signer can speak *on behalf of any NP statement to which he knows a*

C. Dwork (Ed.): CRYPTO 2006, LNCS 4117, pp. 78–96, 2006.

witness? For example, let ϕ be a Boolean formula. Then we want anyone who knows a satisfying assignment w to be able to issue tuples of the form (m, σ), where $\mathsf{Verify}(\phi, m, \sigma) = Accept$, that can be interpreted as follows: "a person who knows a satisfying assignment to formula ϕ has signed message m." Further, we ask whether we can have a signature that just reveals that statement but nothing else; in particular, it reveals nothing about the witness. Finally, what if we want to use a signature issued this way as a witness for issuing another signature?

Online, you are what you know, and access to data is what empowers a user to authenticate her outgoing messages. The question is: *what* data? Previously, it was believed that a user needed a public signing key associated with her identity, and knowledge of the corresponding secret key is what gave her the power to sign. Surprisingly, existence of signatures of knowledge means that if there is *any* NP statement $x \in L$ is associated with a user's identity, the knowledge of a corresponding and hard-to-find witness w for this statement is sufficient to empower the user to sign.

WHY WE NEED SIGNATURES OF KNOWLEDGE AS A NEW PRIMITIVE. Suppose that a message m is signed under some public key PK, and σ is the resulting signature. This alone is not sufficient for any application to trust the message m, unless this application has reason to trust the public key PK. Thus, in addition to (m, σ, PK), such an application will also request some proof that PK is trustworthy, e.g., a certification chain rooted at some trusted PK_0. In order to convince others to accept her signature, the owner of the public key PK has to reveal a lot of information about herself, namely, her entire certification chain. Yet, all she was trying to communicate was that the message m comes from someone trusted by the owner of PK_0. Indeed, this is all the information that the application needs to accept the message m. If instead the user could issue a *signature of knowledge* of her SK, PK, and the entire certification chain, she would accomplish the same goal without revealing all the irrelevant information.

More generally, for any polynomial-time Turing machine M_L, we want to be able to sign using knowledge of a witness w such that $M_L(x, w) = Accept$. We think of M_L as a procedure that decides whether w is a valid witness for $x \in L$ for the NP language L. We call the resulting signature *a signature of knowledge of w that is a witness to $x \in L$, on message m*, or sometimes just a signature of knowledge of w on message m, or sometimes a signature of knowledge on behalf of $x \in L$ on message m.

OTHER APPLICATIONS. Our simplest example is a ring signature [20]. In a ring signature, a signer wishes to sign a message m in such a way that the signature cannot be traced to her specifically, but instead to a group of N potential signers, chosen at signing time. A ring signature can be realized by issuing a signature of knowledge of one of the secret keys corresponding to N public keys. Moreover, following Dodis et al. [16] using cryptographic accumulators [4], the size of this ring signature need not be proportional to N: simply accumulate all public keys into one accumulator A using a public accumulation function, and then issue a signature of knowledge of a secret key corresponding to a public key in A.

Next, let us show how signatures of knowledge give rise to a simple group signature scheme [11,7,1,2,5]. In a group signature scheme, we have group members, a group manager, and an anonymity revocation manager. Each member can sign on behalf of the group, and a signature reveals no information about who signed it, unless the anonymity revocation manager gets involved. The anonymity revocation manager can trace the signature to the group member who issued it; moreover it is impossible, even if the group manager and the revocation manager collude, to create a signature that will be traced to a group member who did not issue it.

Consider the following simple construction. The group's public key consists of (PK_s, PK_E, f), where PK_s is a signature verification key for which the group manager knows the corresponding secret key; PK_E is an encryption public key for which the anonymity revocation manager knows the corresponding decryption key; and f is a one-way function. To become a group member, a user picks a secret x, gives $f(x)$ to the group manager and obtains a group membership certificate $g = \sigma_{PK}.(f(x))$. To issue a group signature, the user picks a random string R, encrypts his identity using randomness R: $c = Enc(PK_E, f(x); R)$ and produces a signature of knowledge σ of (x, g, R) such that c is an encryption of $f(x)$ using randomness R, and g is a signature on $f(x)$. The resulting group signature consists of (c, σ). To trace a group signature, the revocation manager decrypts c. It is not hard to see (only intuitively, since we haven't given any formal definitions) that this construction is a group signature scheme. Indeed, at a high level, this is how existing practical and provably secure group signatures work [1,5].

Unlike the two applications above that have already been studied and where signatures of knowledge offer just a conceptual simplification, our last application was not known to be realizable prior to this work.

Consider the problem of delegatable anonymous credentials. The problem can be explained using the following example. Suppose that, as Brown University employees, we have credentials attesting to that fact, and we can use these credentials to open doors to campus facilities. We wish to be able do so anonymously because we do not want the janitors to monitor our individual whereabouts. Now suppose that we have guests visiting us. We want to be able to issue them a guest pass using our existing credential as a secret key, and without revealing any additional information about ourselves, even to our guests. In turn, our visitors should be able to use their guest passes in order to issue credentials to their taxi drivers, so these drivers can be allowed to drive on the Brown campus. So we have a credential delegation chain, from the Brown University certification authority (CA) that issues us the employee credential, to us, to our visitors, to the visitors' taxi drivers, and each participant in the chain does not know who gave him/her the credential, but (1) knows the length of his credential chain and knows that this credential chain is rooted at the Brown CA; and (2) can extend the chain and issue a credential to the next person.

Although it may seem obvious how to solve this problem once we cast everything in terms of signatures of knowledge and show how to realize signatures of knowledge, we must stress that this fact eluded researchers for a very long time, dating back to Chaum's original vision of the world with anonymous credentials [10]. More recently this problem was raised in the anonymous credentials literature [19,6,18]. And it is still elusive when it comes to practical protocols: our solution is not efficient enough to be used in practice.

ON DEFINING SIGNATURES OF KNOWLEDGE. The first definition of any new primitive is an attempt to formalize intuition. We see from the history of cryptographic definitions (from defining security for encryption, signatures, multiparty computation) that it requires a lot of effort and care. Our approach is to give two definitions, each capturing our intuition in its own way, and then prove that they are equivalent.

One definitional approach is to give an ideal functionality that captures our intuition for a signature of knowledge. Our ideal functionality will guarantee that a signature will only be accepted if the functionality sees the witness w either when generating the signature or when verifying it; and, moreover, signatures issued by signers through this functionality will always be accepted. At the same time, the signatures that our functionality will generate will contain no information about the witness. This seems to capture the intuitive properties we require of a signature of knowledge, although there are additional subtleties we will discuss in Section 2.1. For example, this guarantees that an adversary cannot issue a signature of knowledge of w on some new message m unless he knows w, even with access to another party who does know w. This is because the signatures issued by other parties do not reveal any information about w, while in order to obtain a valid signature, the adversary must reveal w to our ideal functionality. Although this definition seems to capture the intuition, it does not necessarily give us any hints as to how a signature of knowledge can be constructed. Our second definition helps with that.

Our second definition is a game-style one [22,3]. This definition requires that a signature of knowledge scheme be in the public parameter model (where the parameters are generated by some trusted process called Setup) and consist of two algorithms, Sign and Verify. Besides the usual correctness property that requires that Verify accept all signatures issued by Sign, we also require that (1) signatures do not reveal anything about the witness; this is captured by requiring that there exist a simulator who can undetectably forge signatures of knowledge without seeing the witness using some trapdoor information about the common parameters; and (2) valid signatures can only be generated by parties who know corresponding witnesses; this is captured by requiring that there exist an extractor who can, using some trapdoor information about the common parameters, extract the witness from any signature of knowledge, even one generated by an adversary with access to the oracle producing simulated signatures. This definition is presented in Section 2.2. (We call this definition *SimExt-security*, for *sim*ulation and *ext*raction.)

We prove that the two definitions are equivalent: namely, a scheme UC-realizes our ideal functionality if and only if it is SimExt-secure.

Our ideal signature of knowledge functionality can be naturally extended to a signature of knowledge of an accepting input to another ideal functionality. For example, suppose that F_Σ is the (regular) signature functionality. Suppose w is a signature on the value x under public key PK, issued by the ideal F_Σ functionality. Then our functionality F_{SOK} can issue a signature σ on message m, whose meaning is as follows: "The message m is signed by someone who knows w, where w is a signature produced by F_Σ under public key PK on message x." In other words, a signature w on message x under public key PK that causes the verification algorithm for F_Σ to accept, can be used as a witness for a signature of knowledge. A complication in defining the signature of knowledge functionality this way is that, to be meaningful, the corresponding instance of the F_Σ functionality must also be accessible somehow, so that parties can actually obtain signatures under public key PK. Further, for F_{SOK} to be UC-realizable, we must require that the functionality that decides that w is a witness for x, also be UC-realizable. See Section 4 to see how we tackled these definitional issues. As far as we know, this is the first time that an ideal functionality is defined as a function of other ideal functionalities, which may be of independent interest to the study of the UC framework.

OUR CONSTRUCTIONS. In Section 3, we show how to construct signatures of knowledge for any polynomial-time Turing machine M_L deciding whether w is a valid witness for $x \in L$. We use the fact (proved in Section 2.3) that SimExt-security is a necessary and sufficient notion of security, and give a construction of a SimExt-secure signature of knowledge. Our construction is based on standard assumptions. In the common random string model, it requires a dense cryptosystem [15,14] and a simulation-sound non-interactive zero-knowledge proof scheme with efficient provers[21,13] (which can be realized assuming trapdoor permutations for example).

We then show in Section 4 that, given any UC-realizable functionality $\mathcal{F}$ that responds to verification queries and is willing to publish its verification algorithm, the functionality which generates signatures of knowledge of an accepting input to $\mathcal{F}$ is also UC-realizable. We then explain why this yields a delegatable anonymous credential scheme.

THE HISTORY OF THE TERMINOLOGY. The term "signature of knowledge" was introduced by Camenisch and Stadler [7], who use this term to mean a proof of knowledge (more specifically, a Σ-protocol [12]) turned into a signature using the Fiat-Shamir heuristic. Many subsequent papers on group signatures and anonymous credentials used this terminology as well. However, existing literature does not contain definitions of security for the term. Every time a particular construction uses a signature of knowledge as defined by Camenisch and Stadler, the security of the construction is analyzed from scratch, and the term "signature of knowledge" is used more for ease of exposition than as a cryptographic building block whose security properties are well-defined. This frequent informal use of signatures of knowledge indicates their importance in practical constructions and therefore serves as additional motivation of our formal study.

2   Signatures of Knowledge of a Witness for $x \in L$

A signature of knowledge scheme must have two main algorithms, Sign and Verify. The Sign algorithm takes a message and allows anyone holding a witness to a statement $x \in L$ to issue signatures on behalf of that statement. The Verify algorithm takes a message, a statement $x \in L$, and a signature σ, and verifies that the signature was generated by someone holding a witness to the statement.

Signatures of knowledge are essentially a specialized version of noninteractive zero knowledge proofs of knowledge: If a party P can generate a valid signature of knowledge on any message m for a statement $x \in L$, that should mean that, first of all, the statement is true, and secondly, P knows a witness for that statement. This intuitively corresponds to the soundness and extraction properties of a non-interactive proof of knowledge system. On the other hand, just as in a zero-knowledge proof, the signature should reveal nothing about the witness w. We know that general NIZK proof systems are impossible without some common parameters. Thus, our signatures of knowledge will require a setup procedure which outputs shared parameters for our scheme.

Thus, we can define the algorithms in a signature of knowledge schemes as follows: Let $\{Mes_k\}$ be a set of message spaces, and for any language $L \in NP$, let M_L denoted a polynomial time Turing machine which accepts input (x, w) iff w is a witness showing that $x \in L$. Let Setup be an algorithm that outputs public parameters $p \in \{0, 1\}^k$ for some parameter k. Let $\mathsf{Sign}(p, M_L, x, w, m)$ be an algorithm that takes as input some public parameters p, a TM M_L for a language L in NP, a value $x \in L$, a valid witness w for x, and $m \in Mes_k$, a message to be signed. Sign outputs a signature of knowledge for instance $x \in L$ on the message m. Let $\mathsf{Verify}(p, M_L, x, m, \sigma)$ be an algorithm that takes as input the values p, M_L, x, the message m, and a purported signature σ, and either accepts or rejects.

2.1   An Ideal Functionality for a Signature of Knowledge

Canetti's Universal Composability framework gives a simple way to specify the desired functionality of a protocol. Furthermore, the UC Theorem guarantees that protocols will work as desired, no matter what larger system they may be operating within. We begin by giving a UC definition of signatures of knowledge.

We begin by recalling Canetti's signature functionality. For a detailed discussion and justification for Canetti's modelling choices see [9].

Note that this functionality is allowed to produce an error message and halt, or quit, if things go wrong. That means that it is trivially realizable by a protocol that always halts. We will therefore only worry about protocols that realize our functionalities *non-trivially*, i.e. never output an error message.

The session id(sid) of $\mathcal{F}_{SIG}$ captures the identity P of the signer; all participants in the protocol with this session id agree that P is the signer. In a signature of knowledge, we do not have one specific signer, so P should not be included in the session id. But all participants in the protocol should agree on the language that they are talking about. Thus, we have a language $L \in \mathbf{NP}$

$\mathcal{F}_{SIG}$: Canetti's signature functionality

Key Generation Upon receiving a value (KeyGen,sid) from some party P, verify that $sid = (P, sid')$ for some sid'. If not, then ignore the request. Else, hand (KeyGen,sid) to the adversary. Upon receiving (Algorithms, sid, Verify, Sign) from the adversary, where Sign is a description of a PPT ITM, and Verify is a description of a *deterministic* polytime ITM, output (VerificationAlgorithm, sid, Verify) to P.

Signature Generation Upon receiving a value (Sign,sid,m) from P, let $\sigma \leftarrow$ Sign(m), and verify that Verify(m, σ) = 1. If so, then output (Signature, sid, m, σ) to P and record the entry (m, σ). Else, output an error message (Completeness error) to P and halt.

Signature Verification Upon receiving a value (Verify, sid, m, σ, Verify$'$) from some party V, do: If Verify$'$ = Verify, the signer is not corrupted, Verify(m, σ') = 1, and no entry (m, σ') for any σ' is recorded, then output an error message (Unforgeability error) to V and halt. Else, output (Verified,sid,m,Verify$'(m,\sigma)$) to V.

and a polynomial-time Turing machine M_L and a polynomial p, such that $x \in L$ iff there exists a witness w such that $|w| = p(|x|) \wedge M_L(x, w) = 1$. Let us capture the fact that everyone is talking about the same L by requiring that the session id begin with the description of M_L. As mentioned above, signatures of knowledge inherently require some setup. Just as in the key generation interface of $\mathcal{F}_{SIG}$ above, a signature of knowledge functionality ($\mathcal{F}_{SOK}$) setup procedure will determine the algorithm Sign that computes signatures and the algorithm Verify for verifying signatures. However, since anyone who knows a valid witness w can issue a signature of knowledge on behalf of $x \in L$, both Sign and Verify will have to be available to any party who asks for them. In addition, the setup procedure will output algorithms Simsign and Extract that we will explain later.

There are three things that the signature generation part of the $\mathcal{F}_{SOK}$ functionality must capture. The first is that in order to issue a signature, the party who calls the functionality must supply (m, x, w) where w is a valid witness to the statement that $x \in L$. This is accomplished by having the functionality check that it is supplied a valid w. The second is that a signature reveals nothing about the witness that is used. This is captured by issuing the formal signature σ via a procedure that does not take w as an input. We will call this procedure Simsign and require that the adversary provide it in the setup step. Finally, the signature generation step must ensure that the verification algorithm Verify is complete, i.e., that it will accept the resulting signature σ. If it find that Verify is incomplete, $\mathcal{F}_{SOK}$ will output and error message (Completeness error) and halt, just as $\mathcal{F}_{SIG}$ does.

The signature verification part of $\mathcal{F}_{SOK}$ should, of course, accept signatures (m, x, σ) if m was previously signed on behalf of $x \in L$, and σ is the resulting signature (or another signature such that Verify(m, x, σ) = 1). However, unlike $\mathcal{F}_{SIG}$, just because m was not signed on behalf of x through the signing interface, that does not mean that σ should be rejected, even if the signer is uncorrupted. Recall that anyone who knows a valid witness should be able to generate

---

$\mathcal{F}_{SOK}(L)$: **signature of knowledge of a witness for** $x \in L$

Setup Upon receiving a value (Setup,sid) from any party P, verify that $sid = (M_L, sid')$ for some sid'. If not, then ignore the request. Else, if this is the first time that (Setup,sid) was received, hand (Setup,sid) to the adversary; upon receiving (Algorithms, sid, Verify, Sign, Simsign, Extract) from the adversary, where Sign, Simsign, Extract are descriptions of PPT TMs, and Verify is a description of a deterministic polytime TM, store these algorithms. Output the stored (Algorithms, sid, Sign, Verify) to P.

Signature Generation Upon receiving a value (Sign,sid,m,x,w) from P, check that $M_L(x, w) = 1$. If not, ignore the request. Else, compute $\sigma \leftarrow$ Simsign(m, x), and check that Verify$(m, x, \sigma) = 1$. If so, then output (Signature,sid,m,x,σ) to P and record the entry (m, x, σ). Else, output an error message (Completeness error) to P and halt.

Signature Verification Upon receiving a value (Verify, sid,m,x,σ) from some party V, do: If (m, x, σ') is stored for some σ', then output (Verified,sid,m,x,σ, Verify(m, x, σ)) to V. Else let $w \leftarrow$ Extract(m, x, σ); if $M_L(x, w) = 1$, output (Verified,sid,m,x,σ, Verify(m, x, σ)) to V. Else if Verify$(m, x, \sigma) = 0$, output (Verified,sid,m,x,σ,0) to V. Else output an error message (Unforgeability error) to V and halt.

---

acceptable signatures! Therefore, the verification algorithm must somehow check that whoever generated σ knew the witness w. Recall that in the setup stage, the adversary provided the algorithm Extract. This algorithm is used to try to extract a witness from a signature σ that was not produced via a call to $\mathcal{F}_{SOK}$. If Extract(m, x, σ) produces a valid witness w, then $\mathcal{F}_{SOK}$ will output the outcome of Verify(m, x, σ). If Extract(m, x, σ) fails to produce a valid witness, and Verify(m, x, σ) rejects, then $\mathcal{F}_{SOK}$ will reject. What happens if Extract(m, x, σ) fails to produce a valid witness, but Verify(m, x, σ) accepts? This corresponds to the case when a signature σ on m on behalf of x was produced without a valid witness w, and yet σ is accepted by Verify. If this is ever the case, then there is an unforgeability error, and so $\mathcal{F}_{SOK}$ should output (Unforgeabilityerror) and halt. Unlike $\mathcal{F}_{SIG}$, here we need not worry about whether the requesting party supplied a correct verification algorithm, since here everyone is on the same page and is always using the same verification algorithm (determined in the setup phase).

In the UC framework, each instance of the ideal functionality is associated with a unique sid, and it ignores all queries which are not addressed to this sid. Since our $\mathcal{F}_{SOK}$ functionalities require that $sid = M_L \circ sid'$, this means that each $\mathcal{F}_{SOK}$ functionality handles queries for exactly one language.

Now consider the following language U.

Definition 1 (Universal language). *For polynomial p, define universal language U_p s.t. x would contain a description of a Turing machine M and an instance x' such that $x \in U_p$ iff there exists w s.t. $M(x', w)$ halts and accepts in time at most $p(|x|)$.*

Notice that $\mathcal{F}_{SOK}(U_p)$ allows parties to sign messages on behalf of any instance x of any language L which can be decided in non-deterministic $p(|x|)$ time.

Thus, if we have Setup, Sign, and Verify algorithms which realize $\mathcal{F}_{SOK}(U_p)$, we can use the same algorithms to generate signatures of knowledge for all such instances and languages. In particular, this means we do not need a separate setup algorithm (in implementation, a separate CRS or set of shared parameters) for each language. Readers familiar with UC composability may notice that any protocol which realizes $\mathcal{F}_{SOK}(U_p)$ for all polynomials p will also realize the multisession extension of $\mathcal{F}_{SOK}$. For more information, see full version.

2.2   A Definition in Terms of Games

We now give a second, games style definition for signatures of knowledge. We will show that this definition is equivalent to (necessary and sufficient for) the UC definition given in the previous section. Informally, a signature of knowledge is SimExt-secure if it is correct, simulatable and extractable.

The **correctness** property is similar to that of a traditional signature scheme. It requires that any signature issued by the algorithm Sign should be accepted by Verify.

The **simulatability** property requires that there exist a simulator which, given some trapdoor information on the parameters, can create valid signatures without knowing any witnesses. This captures the idea that signatures should reveal nothing about the witness used to create them. Since the trapdoor must come from somewhere, the simulator is divided into Simsetup that generates the public parameters (possibly from some different but indistinguishable distribution) together with the trapdoor, and Simsign which then signs using these public parameters. We require that no adversary can tell that he is interacting with Simsetup and Simsign rather than Setup and Sign.

The **extraction** property requires that there exist an extractor, which given a signature of knowledge for an $x \in L$, and appropriate trapdoor information, can produce a valid witness showing $x \in L$. This captures the idea that it should be impossible to create a valid signature of knowledge without knowing a witness. In defining the extraction property, we require that any adversary that interacts with the simulator Simsetup and Simsign (rather than the Setup and Sign) not be able to produce a signature from which the extractor cannot extract a witness. The reason that in the definition, the adversary interacts with Simsetup instead of Setup is because the extractor needs a trapdoor to be able to extract. Note that it also interacts with Simsign instead of Sign. The adversary could run Sign itself, so access to Simsign gives it a little bit of extra power.

Definition 2 (SimExt-security). *Let L be the language defined by a polynomial-time Turing machine M_L as explained above, such that all witnesses for $x \in L$ are of known polynomial length $p(|x|)$. Then (Setup, Sign, Verify) constitute a SimExt-secure signature of knowledge of a witness for L, for message space $\{Mes_k\}$ if the following properties hold:*

Correctness. *There exists a negligible function ν such that for all $x \in L$, valid witnesses w for x (i.e. witnesses w such that $M_L(x, w) = 1$), and $m \in Mes_k$*

$$\Pr[p \leftarrow \mathsf{Setup}(1^k); \sigma \leftarrow \mathsf{Sign}(p, M_L, x, w, m) \ : \\ \mathsf{Verify}(p, M_L, x, m, \sigma) = Accept] = 1 - \nu(k)$$

Simulatability. *There exists a polynomial time simulator consisting of algorithms* Simsetup *and* Simsign *such that for all probabilistic polynomial-time adversaries* $\mathcal{A}$ *there exists a negligible functions* ν *such that for all polynomials* f, *for all* k, *for all auxiliary input* $s \in \{0,1\}^{f(k)}$

$$\left| \begin{array}{l} \Pr[(p, \tau) \leftarrow \mathsf{Simsetup}(1^k); b \leftarrow \mathcal{A}^{\mathsf{Sim}(p, \tau, \cdot, \cdot, \cdot, \cdot)}(s, p) \ : \ b = 1] \\ - \ \Pr[p \leftarrow \mathsf{Setup}(1^k); b \leftarrow \mathcal{A}^{\mathsf{Sign}(p, \cdot, \cdot, \cdot, \cdot)}(s, p) \ : \ b = 1] \end{array} \right| = \nu(k)$$

where the oracle Sim *receives the values* (M_L, x, w, m) *as inputs, checks that the witness* w *given to it was correct and returns* $\sigma \leftarrow \mathsf{Simsign}(p, \tau, M_L, x, m)$. τ *is the additional trapdoor value that the simulator needs in order to simulate the signatures without knowing a witness.*

Extraction. *In addition to* (Simsetup, Simsign), *there exists an extractor algorithm* Extract *such that for all probabilistic polynomial time adversaries* $\mathcal{A}$ *there exists a negligible function* ν *such that for all polynomials* f, *for all* k, *for all auxiliary input* $s \in \{0,1\}^{f(k)}$

$$\Pr\left[(p, \tau) \leftarrow \mathsf{Simsetup}(1^k); (M_L, x, m, \sigma) \leftarrow \mathcal{A}^{\mathsf{Sim}(p, \tau, \cdot, \cdot, \cdot, \cdot)}(s, p); \right. \\ w \leftarrow \mathsf{Extract}(p, \tau, M_L, x, m, \sigma) \ : \\ \left. M_L(x, w) \vee (M_L, x, m, w) \in Q \vee \neg\mathsf{Verify}(p, M_L, x, m, \sigma)\right] = 1 - \nu(k)$$

where Q *denotes the query tape which lists all previous queries* (M_L, x, m, w) $\mathcal{A}$ *has sent to the oracle* Sim.

Note that the above definition captures, for example, the following intuition: suppose that Alice is the only one in the world who knows the witness w for $x \in L$, and it is infeasible to compute w. Then Alice can use x as her signing public key, and her signature σ on a message m can be formed using a signature of knowledge w. We want to make sure that the resulting signature should be existentially unforgeable against chosen message attacks [17]. Suppose it is not. Then there is a forger who can output (m, σ), such that σ is accepted by the verification algorithm without a query m to Alice. Very informally, consider the following four games:

Adversary vs. Alice: The parameters are generated by Setup. Alice chooses a random x, w pair and publishes x. The adversary sends Alice messages to be signed and Alice responds to each using x, w and Sign. Adversary outputs a purported forgery. Let p_0 be the probability that the forgery is successful.

Adversary vs. Simulator: The simulator generates parameters using Simsetup. The simulator chooses a random x, w pair and publishes x. The adversary sends the simulator messages to be signed, and he responds using x, w and Sim. The adversary outputs a purported forgery. Let p_1 be the probability that the forgery is successful.

Adversary vs. Extractor: The extractor generates parameters using Simsetup. He then chooses a random x, w pair and publishes x. The adversary sends the simulator messages to be signed, and he responds using x, w and Sim. The adversary outputs a purported forgery. The extractor runs Extract to obtain a potential witness w. Let p_2 be the probability that w is a valid witness.

Adversary vs. Reduction: The reduction is given and instance x, which it publishes. It then generates parameters using Simsetup.The adversary sends messages to be signed, and the reduction responds using x and Simsign. The adversary outputs a purported forgery. The reduction runs Extract to obtain w. Let p_3 be the probability that w is a valid witness.

By the simulatability property, the difference between p_0 and p_1 must be negligible. By the extraction property, the difference between p_1 and p_2 must be negligible. Since Sim ignores w and runs Simsign, p_2 and p_3 must be identical. Thus, generating forgeries is at least as hard as deriving a witness w for a random instance x. If the algorithm used to sample (x, w) samples hard instances with their witnesses, then we know that the probability of forgery is negligible. For a formal proof see full version.

2.3   Equivalence of the Definitions

As was mentioned above, signatures of knowledge cannot exist without some trusted setup procedure which generates shared parameters. In the UC model, shared parameters are captured by the $\mathcal{F}_{CRS}^D$ functionality [8]. This functionality generates values from a given distribution D (the desired distribution of shared parameters), and makes them available for all parties in the protocol. Thus, protocols requiring shared parameters can be defined in the $\mathcal{F}_{CRS}$-hybrid model, where real protocols are given access to the ideal shared parameter functionality.

Formally, the $\mathcal{F}_{CRS}^D$ functionality receives queries of the form (CRS,sid) from a party P. If a value v for this sid has not been stored, it chooses a random value v from distribution D and stores it. It returns (CRS, sid, v) to P and also sends (CRS,sid, v) to the adversary.

Let $\Sigma = $ (Setup, Sign, Verify) be a signature of knowledge scheme. Let k be the security parameter. We define a F_{CRS}^D-hybrid signature of knowledge protocol π_Σ, where D is the distribution of Setup(1^k).

When a party P running π_Σ receives an input (Setup,sid) from the environment, it checks that $sid = (M_L, sid')$ for some sid'. If not it ignores the request. It then queries the $\mathcal{F}_{CRS}$ functionality, receives (CRS,v), and stores v. It returns (Algorithms,sid, Sign($v, M_L, \cdot, \cdot, \cdot$), Verify($v, M_L, \cdot, \cdot$)) to the environment.

When P receives a request (Sign, sid, m, x, w) from the environment, it retrieves the stored v. It checks that $M_L(x, w) = 1$. If not, it ignores the request, otherwise it returns (Signature, sid, m, x, Sign(v, M_L, x, w, m)). When P receives a request (Verify, sid, m, x, σ) from the environment, it again retrieves the stored v, and then returns (Verified, sid, m, x, σ, Verify(v, M_L, x, m, σ)).

Theorem 1. *For any polynomial p, π_Σ UC-realizes $\mathcal{F}_{SOK}(U_p)$ in the $\mathcal{F}_{CRS}^D$-hybrid model if and only if Σ is SimExt-secure.*

Proof. (Sketch: See the full version for the full proof.) Suppose that Σ is SimExt-secure. Then let us show that π_Σ UC-realizes $\mathcal{F}_{SOK}(U_p)$. Consider the ideal adversary (simulator) S that works as follows: Upon receiving (Setup,sid) from

$\mathcal{F}_{SOK}$, S will parse $sid = (M_U, sid')$. It obtains $(p, \tau) \leftarrow \mathsf{Simsetup}(1^k)$ and sets $\mathsf{Sign} = \mathsf{Sign}(p, \cdot, \cdot, \cdot, \cdot)$ (so Sign will have four inputs: the language M_L — note that since we are realizing $\mathcal{F}_{SOK}(U_p)$, any instance will start with M_L,— the instance $x \in L$, the witness w, and the message m), $\mathsf{Verify} = \mathsf{Verify}(p, \cdot, \cdot, \cdot, \cdot)$, $\mathsf{Simsign} = \mathsf{Simsign}(p, \tau, \cdot, \cdot, \cdot, \cdot)$, and $\mathsf{Extract} = \mathsf{Extract}(p, \tau, \cdot, \cdot, \cdot, \cdot)$. Finally, it sends $(\mathtt{Algorithms}, sid, \mathsf{Sign}, \mathsf{Verify}, \mathsf{Simsign}, \mathsf{Extract})$ back to $\mathcal{F}_{SOK}$. When the adversary A queries the $\mathcal{F}^D_{CRS}$ functionality, S outputs p.

Let Z be any environment and A be an adversary. We wish to show that Z cannot distinguish interactions with A and π_Σ from interactions with S and $\mathcal{F}_{SOK}$. Let us do that in two steps. First, we show that the event E that $\mathcal{F}_{SOK}$ halts with an error message has negligible probability. Next, we will show that, conditioned on E not happening, Z's view in its interaction with S and $\mathcal{F}_{SOK}$ is indistinguishable from its view in interactions with A and π_Σ.

There are two types of errors that lead to event E: $\mathcal{F}_{SOK}$ halts with Completeness error or Unforgeability error. The only way to induce a completeness error is to cause Verify to reject a signature issued by $\mathsf{Simsign}$, which contradicts either the simulatability or the correctness requirement. The only way to induce an unforgeability error is to cause Verify to accept a signature which was not issued by $\mathsf{Simsign}$ and from which no witness can be extracted. This contradicts the extractability requirement.

We have shown that the probability of event E is negligible. Conditioned on $\bar{E}$, Z's view when interacting with $\mathcal{F}_{SOK}$ and S is indistinguishable from its view when interacting with a real adversary A and the real protocol π_Σ, because if it were distinguishable, then this would contradict the simulatability requirement.

The converse is fairly straightforward and appears in the full version.   □

3   Construction

Here we present Σ, a construction of a SimExt-secure signature of knowledge. By Theorem 1, this also implies a protocol π_Σ that UC-realizes the $\mathcal{F}_{SOK}$ functionality presented in Section 2.1.

Our construction has two main building blocks: CPA secure dense cryptosystems [15,14] and simulation-sound non-interactive zero knowledge proofs [21,13]. Let $(\mathcal{G}, \mathsf{Enc}, \mathsf{Dec})$ be a dense cryptosystem, and let $(\mathsf{NIZKProve}, \mathsf{NIZKSimsetup}, \mathsf{NIZKSim}, \mathsf{NIZKVerify})$ be a simulation-sound non-interactive zero-knowledge proof system.

Setup. Let p be a common random string. Parse p as follows: $p = PK \circ \rho$, where PK is a k-bit public key of our cryptosystem.

Signature Generation. In order to sign a message $m \in Mes_k$ using knowledge of witness w for $x \in L$, let $c = \mathsf{Enc}(PK, (m, w), R)$, where R is the randomness needed for the encryption process; let $\pi \leftarrow \mathsf{NIZKProve}(\rho, (m, M_L, x, c, PK), (\exists(w, R) : c = \mathsf{Enc}(PK, (m, w), R) \wedge M_L(x, w)), (w, R))$. Output $\sigma = (c, \pi)$.

Verification. In order to verify a signature of knowledge of witness w for $x \in L$, $\sigma = (c, \pi)$, run $\mathsf{NIZKVerify}(\rho, \pi, (m, M_L, x, c, PK), (\exists(w, R) : c = \mathsf{Enc}(PK, (m, w), R) \wedge M_L(x, w)))$.

Intuitively, the semantic security of the cryptosystem together with the zero knowledge property of the proof system ensure that the signature reveals no information about the witness. The simulation soundness property of the proof system means that the adversary cannot prove false statements. Thus any signature that verifies must include a ciphertext which is an encryption of the given message and of a valid witness. Clearly, if he is interacting only with a simulator who does not know any witnesses, this implies that the adversary should know the witness. Further, by simulatability, the adversary cannot gain any advantage by communicating with valid signers. The key is that the witness and message are encrypted together, and there is a single proof that the encryption is correct. Thus it is not possible to simply replace the message with a different one.

Theorem 2. *The construction above is a SimExt-secure signature of knowledge.*

Proof. (Sketch: See full version for full proof) First we argue simulatability. In the Simsetup phase, our simulator will choose a key pair (PK, SK) of the dense cryptosystem, and will obtain the string ρ together with trapdoor τ' by running NIZKSimsetup. In the Simsign phase, the simulator will always let c be the encryption of $0^{|m|+l\cdot}$, and will create (fake) proof π by invoking NIZKSim.

We show that the resulting simulation is successful using a two-tier hybrid argument. First, by the unbounded zero-knowledge property of the underlying NIZK proof system, signatures obtained by replacing calls to NIZKProve by calls to NIZKSim will be distributed indistinguishably from real signatures. Second, by semantic security of the dense cryptosystem used, using $c \leftarrow \mathsf{Enc}(PK, (m, w))$ versus $c \leftarrow \mathsf{Enc}(PK, (0^{|m|+l\cdot}))$ results in indistinguishable distributions.

Second, let us argue extraction. Recall that, as part of the trapdoor τ, Simsetup above retains SK, the secret key for the cryptosystem. The extractor simply decrypts the c part of the signature σ to obtain the witness w. By the simulation-soundness property of the underlying NIZK proof system, no adversary can produce a signature acceptable to the Verify algorithm without providing c that decrypts to a correct witness w. [1]   □

4 $\mathcal{F}_{SOK}$ for Generalized Languages, and Applications

Recall from the introduction that a signature of knowledge may be used in order to construct a group signature scheme. Let PK_s be the public signing key of the group manager, and suppose that the group manager can sign under this public key (using the corresponding secret key SK_s). Let PK_E be a public encryption key such that the anonymity revocation manager knows the corresponding secret key SK_E. A user must pick a secret key x and a public key $p = f(x)$ where

---

[1] Note that a CPA secure cryptosystem is sufficient: We only need the security of the encryption scheme to guarantee that the encryptions of $0^{|m|+l\cdot}$ generated by the simulator are indistinguishable from encryptions of valid witness/message pairs. This is only necessary in the proof of simulatability, in a scenario where pairs are encrypted but never decrypted.

f is some one-way function. She then obtains a group membership certificate $g = \sigma_{PK.}(p)$, the group manager's signature on her public key. In order to sign on behalf of the group, the user encrypts her public key and obtains a ciphertext $c = Enc(PK_E, p; R)$, where R is the randomness used for encryption. Finally, her group signature on message m is a signature of knowledge of (x, p, g, R) such that $c = Enc(PK_E, p, R)$, $p = f(x)$, and g is a valid signature on p under PK_s.

Now let us consider more closely the language L used in the signature of knowledge. In the example above, $c \in L$ and (x, p, g, R) is the witness. This language is determined by the parameters of the system, (f, PK_s, PK_E). This is not a general language, but instead it depends on the system parameters, which in turn depend on three other building blocks, a one-way function, an encryption scheme and a signature scheme. We want to show that even in this context, the use of a signature of knowledge has well-understood consequences for the security of the rest of the system.

To that end, we consider signatures of knowledge for languages that are defined by secure functionalities realizing particular tasks. In this example, this corresponds to the one-way function, encryption and signing functionalities. Encryption is used to incorporate the encrypted identity, c, of the signer into her group signature. A signing functionality is used to issue group membership certificates, g, to individual group members. Finally, we have a one-way function f that takes a user's secret x and maps it to her public p.

We could choose a specific realization of each of these primitives, combine these realizations, and use the resulting TM to define our language L for a signature of knowledge as described in Section 2. However, we would like to be able to define an abstract signature of knowledge functionality whose language is defined by ideal functionalities and not dependent on any specific realizations.

In this section, we wish to create a framework where, given ideal functionalities $\mathcal{F}_f$, $\mathcal{F}_{Enc}$ and $\mathcal{F}_\Sigma$ for these three primitives, we can define a signature of knowledge functionality $\mathcal{F}_{SOK}$ for the language L, where L is defined in terms of the outputs of functionalities $\mathcal{F}_f$, $\mathcal{F}_{Enc}$, and $\mathcal{F}_\Sigma$. Such $\mathcal{F}_{SOK}$ can be used to realize group signatures as above, as well as other cryptographic protocols.

First, in Section 4.1, we will characterize functionalities that define such generalized languages L. These are functionalities which, when they receive an input (x, w), verify that this is indeed an accepting input, i.e. that w constitutes a witness for $x \in L$. In Section 4.2, we will define $\mathcal{F}_{SOK}(\mathcal{F}_0)$, a signature of knowledge of an accepting input to one ideal functionality, $\mathcal{F}_0$. Then, we prove Theorem 3: given a SimExt-secure scheme, $\mathcal{F}_{SOK}(\mathcal{F}_0)$ is UC-realizable in the CRS model if and only if $\mathcal{F}_0$ is UC-realizable. In the full version we explain several generalizations of this ideal functionality which allow us to apply Theorem 3 to group signatures, delegatable credentials, and other similar scenarios.

As far as we know, prior literature on the UC framework did not address the issues of defining an ideal functionality as an extension of another ideal functionality or of a set of other functionalities. (In contrast, it addressed the case when a *real* protocol used an ideal functionality as a sub-routine.)

4.1   Explicit Verification Functionalities

Only certain functionalities make sense as a language for a signature of knowledge. In particular, they need to allow us to determine whether a given element is in the language given a potential witness: we call this "verification." We also assume that everyone knows how language membership is determined. Thus, we also require that the functionality be willing to output code which realizes its verification procedure. In this section, we formally define the functionalities which can be used to define the language for a signature of knowledge.

Consider Canetti's signature functionality $\mathcal{F}_{SIG}$. Once the key generation algorithm has been run, this functionality defines a language: namely, the language of messages that have been signed. A witness for membership in such a language is the signature σ. In a Verify_ query this functionality will receive (m, σ) and will accept if m has been signed and Verify$(m, \sigma) = Accept$, where Verify is the verification algorithm supplied to $\mathcal{F}_{SIG}$ by the ideal adversary S. Moreover, if it so happens that Verify(m, σ) accepts while m has not been signed, or if it is the case that Verify(m, σ) rejects a signature generated by $\mathcal{F}_{SIG}$, $\mathcal{F}_{SIG}$ will halt with an error. $\mathcal{F}_{SIG}$ is an example of a verification functionality, defined below:

Definition 3 ((Explicit) Verification functionality). *A functionality $\mathcal{F}$ is a verification functionality if (1) there exists some start$(\mathcal{F})$ query such that $\mathcal{F}$ ignores all queries until it receives a start query; (2) during the start query $\mathcal{F}$ obtains from the ideal adversary S a deterministic polynomial-time verification algorithm Verify; (3) in response to (Verify,sid, input, witness) queries, $\mathcal{F}$ either responds with the output of Verify(input, witness) or halts with an error. $\mathcal{F}$ is an explicit verification functionality if, once a start$(\mathcal{F})(sid)$ query has taken place, it responds to a query (VerificationAlgorithm,sid) from any party P by returning the algorithm Verify.*

Note that $start(\mathcal{F})$ is a specific command that depends on the functionality $\mathcal{F}$. For example, if $\mathcal{F}$ is a signature functionality, $start(\mathcal{F})$ =Keygen. If $\mathcal{F}$ is another signature of knowledge functionality, $start(\mathcal{F})$ =Setup.

An explicit verification functionality not only defines a language L, but also makes available a Turing machine M_L deciding whether w is a witness for $x \in L$.

4.2   Signatures of Knowledge of an Accepting Input to $\mathcal{F}_0$

Let $\mathcal{F}_0$ be any explicit verification functionality. (Our running example is Canetti's signature functionality, or our own $\mathcal{F}_{SOK}$ functionality, augmented so that it responds to VerificationAlgorithm queries with the Verify algorithm obtained from the ideal adversary.) We want to build a signature of knowledge functionality $\mathcal{F}_{SOK}(\mathcal{F}_0)$ that incorporates $\mathcal{F}_0$. It creates an instance of $\mathcal{F}_0$ and responds to all the queries directed to that instance. So, if $\mathcal{F}_0$ is a signature functionality, then $\mathcal{F}_{SOK}(\mathcal{F}_0)$ will allow some party P to run key generation and signing, and will also allow anyone to verify signatures. In addition, any party in possession of (x, w) such that $\mathcal{F}_0$'s verification interface will accept (x, w), can sign on behalf of the statement "There exists a value w such that $\mathcal{F}_0(sid_0)$ accepts (x, w)." For example, if $\mathcal{F}_0$ is a signing functionality, m is a message, and σ_0 is a signature on m created by P with session id sid_0, then through $\mathcal{F}_{SOK}(\mathcal{F}_0)$, any party knowing (m, σ_0) can issue a signature σ_1, which is a signature of knowledge of a signature σ_0 on m, where σ_0 was created by signer P. Moreover, any party can verify the validity of σ_1.

To define $\mathcal{F}_{SOK}(\mathcal{F}_0)$, we start with our definition of $\mathcal{F}_{SOK}(L)$ and modify it in a few places. In the protocol description below, these places are underlined.

The main difference in the setup, signature generation, and signature verification interfaces is that the TM M_L that decides whether w is a valid witness for $x \in L$, is no longer passed to the functionality $\mathcal{F}_{SOK}$. Instead, language membership is determined by queries to the verification procedure of $\mathcal{F}_0$, as well as by an algorithm M_L that $\mathcal{F}_0$ returns when asked to provide its verification algorithm. (Sign, Verify, Simsign, Extract) returned by the adversary now take M_L as input. M_L is supposed to be an algorithm that UC-realizes the verification procedure of $\mathcal{F}_0$. Note, however, that just because $M_L(x, w)$ accepts, does not mean that $\mathcal{F}_0$'s verification procedure necessarily accepts. Instead $\mathcal{F}_{SOK}$ expects that $M_L(x, w)$ accepts iff $\mathcal{F}_0$ accepts, and should $\mathcal{F}_{SOK}$ be given (x, w) where this is not the case, $\mathcal{F}_{SOK}$ will output an error message (Error with $\mathcal{F}_0$) and halt.

The setup procedure of $\mathcal{F}_{SOK}(\mathcal{F}_0)$ differs from that of $\mathcal{F}_{SOK}(L)$ in two places. First, it used to check that the session id contains the description M_L of the language L; instead now it checks that it contains a description of the functionality $\mathcal{F}_0$ and a session id sid_0 with which $\mathcal{F}_0$ should be invoked. Second, it must now invoke $\mathcal{F}_0$ to determine the language L and the Turing machine M_L (see below).

The signing and verification procedures of $\mathcal{F}_{SOK}(\mathcal{F}_0)$ differs from that of $\mathcal{F}_{SOK}(L)$ only in that, instead of just checking that $M_L(x, w) = 1$, they check that $\mathcal{F}_0$ accepts (x, w) and that M_L faithfully reflects what $\mathcal{F}_0$ does.

Let us explain how the language L is determined. During the first setup query, $\mathcal{F}_{SOK}$ must somehow determine the set of accepted (x, w), i.e., get the language L. To that end, it creates an instance of $\mathcal{F}_0$, and runs the start query for $\mathcal{F}_0$. It also queries $\mathcal{F}_0$ to obtain its verification algorithm M_L. We describe how this is done separately by giving a procedure we call GetLanguage($\mathcal{F}_0, sid_0$), as a subroutine of the setup phase of $\mathcal{F}_{SOK}$.

Note that this instance of $\mathcal{F}_0$ is created $inside$ of $\mathcal{F}_{SOK}$, and outside parties cannot access it directly. Instead, if they want to use $\mathcal{F}_0$ and send a query to it of the form $(query, sid_0, data)$, they should instead query $\mathcal{F}_{SOK}$ with a query of the form $(\mathcal{F}_0\text{-}query, sid, data)$, where $sid = (sid_0, sid_1)$ is the session id of $\mathcal{F}_{SOK}$. We specify this more rigorously in the actual description of $\mathcal{F}_{SOK}(\mathcal{F}_0)$. Note that $\mathcal{F}_{SOK}$ will ignore any queries until the first setup query — this is done so that one cannot query $\mathcal{F}_0$ before it is actually created.

Also note that $\mathcal{F}_0$ may require input from the adversary. Whenever this is the case, the messages that $\mathcal{F}_0$ wants to send to the adversary are forwarded to the adversary, and the adversary's responses are forwarded back to $\mathcal{F}_0$.

Finally, we want $\mathcal{F}_{SOK}(\mathcal{F}_0)$ itself to be a explicit verification functionality (as explained in Section 4.1), and so it must be able to respond to queries asking it to provide its verification algorithm.

Theorem 3. *Let $\mathcal{F}_0$ be an explicit verification functionality. Assuming SimExt-secure signatures of knowledge, $\mathcal{F}_{SOK}(\mathcal{F}_0)$ is nontrivially UC-realizable in the $\mathcal{F}_{CRS}^D$ hybrid model iff $\mathcal{F}_0$ is nontrivially UC-realizable in the $\mathcal{F}_{CRS}^D$ hybrid model, where we consider a realization to be nontrivial if it never halts with an error message.* For the proof see the full version.

$\mathcal{F}_{SOK}(\mathcal{F}_0)$: signature of knowledge of an accepting input to $\mathcal{F}_0$

For any sid, ignore any message received prior to (Setup, sid).

Setup Upon receiving a value (Setup,sid) from any party P, verify that $sid = (\mathcal{F}_0, sid_0, sid_1)$ for some sid_0, sid_1. If not, then ignore the request. Else, if this is the first time that (Setup,sid) was received, let $M_L = \mathsf{GetLanguage}(\mathcal{F}_0, sid_0)$, store M_L, and hand (Setup,sid) to the adversary; upon receiving (Algorithms, sid, Verify, Sign, Simsign, Extract) from the adversary, where Sign, Simsign, Extract are descriptions of PPT ITMs, and Verify is a description of a deterministic polytime ITM, store these algorithms. Output the (Algorithms, sid, Sign$(M_L, \cdot, \cdot, \cdot)$,Verify$(M_L, \cdot, \cdot, \cdot)$) to P.

Signature Generation Upon receiving a value (Sign,sid, m, x, w) from P, check that $\mathcal{F}_0$ accepts (Verify,sid_0, x, w) when invoked by P. If not, ignore the request. Else, if $M_L(x, w) = 0$, output an error message (Error with $\mathcal{F}_0$) to P and halt. Else, compute $\sigma \leftarrow \mathsf{Simsign}(M_L, m, x)$, and verify that Verify$(M_L, m, x, \sigma) = 1$. If so, then output (Signature,sid, m, x, σ) to P and record the entry (m, x, σ). Else, output an error message (Completeness error) to P and halt.

Signature Verification Upon receiving a value (Verify,sid, m, x, σ) from some party V, do: If (m, x, σ') is stored for some σ', then output (Verified,sid, m, x, σ, Verify(m, x, σ)) to V. Else let $w \leftarrow \mathsf{Extract}(M_L, m, x, \sigma)$. If $M_L(x, w) = 1$: if $\mathcal{F}_0$ does not accept (Verify,sid_0, x, w), output and error message (Error with $\mathcal{F}_0$) to P and halt; else output (Verified,sid, m, x, σ, Verify(M_L, m, x, σ)) to V. Else if Verify$(M_L, m, x, \sigma) = 0$, output (Verified,$sid, m, x, \sigma, 0$) to V. Else output an error message (Unforgeability error) to V and halt.

Additional routines:

GetLanguage($\mathcal{F}_0, sid_0$) Create an instance of $\mathcal{F}_0$ with session id sid_0. Send to $\mathcal{F}_0$ the message $(start(\mathcal{F}_0), sid_0)$ on behalf of P, the calling party. Send to $\mathcal{F}_0$ the message (VerificationAlgorithm,sid_0). In response, receive from $\mathcal{F}_0$ the message (VerificationAlgorithm,sid_0, M). Output M.

Queries to $\mathcal{F}_0$ Upon receiving a message ($\mathcal{F}_0$-$query$, $sid_0, sid_1, data$) from a party P, send ($query, sid_0, data$) to $\mathcal{F}_0$ on behalf of P. Upon receiving ($response, sid_0, data$) from $\mathcal{F}_0$, forward ($\mathcal{F}_0$-$response$, $sid, data$) to P.

$\mathcal{F}_0$'s interactions with the adversary When $\mathcal{F}_0$ wants to send ($command, sid_0, data$) to the adversary, give to the adversary the message ($\mathcal{F}_0$-$command$, $sid, sid_0, data$). When receive a message ($\mathcal{F}_0$-$header$, $sid, sid_0, data$) from the adversary, give ($header, sid_0, data$) to $\mathcal{F}_0$ on behalf of the adversary.

Providing the verification algorithm Upon receiving a message (VerificationAlgorithm,sid) from any party P, output (VerificationAlgorithm,sid, Verify$(M_L, \cdot, \cdot, \cdot)$) to P.

Acknowledgments. The authors wish to thank Jan Camenisch, Mira Meyerovich, and Leo Reyzin for helpful comments and discussion. Melissa Chase is supported by NSF grant CNS-0374661 and NSF Graduate Research Fellowship. Anna Lysyanskaya is supported by NSF CAREER grant CNS-0374661.

References

1. G. Ateniese, J. Camenisch, M. Joye, and G. Tsudik. A practical and provably secure coalition-resistant group signature scheme. In *CRYPTO 2000*, 255–270.
2. M. Bellare, D. Micciancio, and B. Warinschi. Foundations of group signatures: Formal definitions, simplified requirements, and a construction based on general assumptions. In *EUROCRYPT 2003*, pages 614–629.
3. M. Bellare and P. Rogaway. The game-playing technique. http://eprint.iacr.org/2004/331, 2004.
4. J. Benaloh and M. de Mare. One-way accumulators: A decentralized alternative to digital signatures. In *EUROCRYPT '93*, pages 274–285.
5. D. Boneh, X. Boyen, and H. Shacham. Short group signatures. In *CRYPTO 2004*, pages 41–55.
6. J. Camenisch and A. Lysyanskaya. Efficient non-transferable anonymous multi-show credential system with optional anonymity revocation. In *EUROCRYPT 2001*, pages 93–118.
7. J. Camenisch and M. Stadler. Efficient group signature schemes for large groups. In *CRYPTO '97*, pages 410–424.
8. R. Canetti. Universally composable security: A new paradigm for cryptographic protocols. In *FOCS*, pages 136–145, 2001.
9. R. Canetti. Universally composable security: A new paradigm for cryptographic protocol. http://eprint.iacr.org/2000/067, 2005.
10. D. Chaum. Security without identification: Transaction systems to make big brother obsolete. *Communications of the ACM*, 28(10):1030–1044, Oct. 1985.
11. D. Chaum and E. van Heyst. Group signatures. In *EUROCRYPT '91*, 257–265.
12. R. Cramer. *Modular Design of Secure yet Practical Cryptographic Protocol.* PhD thesis, University of Amsterdam, 1997.
13. A. de Santis, G. di Crescenzo, R. Ostrovsky, G. Persiano, and A. Sahai. Robust non-interactive zero knowledge. In *CRYPTO 2001*, pages 566–598.
14. A. de Santis, G. di Crescenzo, and G. Persiano. Necessary and sufficient assumptions for non-interactive zero-knowledge proofs of knowledge for all NP relations. In *ICALP 2000*, pages 451–462.
15. A. de Santis and G. Persiano. Zero-knowledge proofs of knowledge without interaction (extended abstract). In *FOCS 1992*, pages 427–436.
16. Y. Dodis, A. Kiayias, A. Nicolosi, and V. Shoup. Anonymous identification in ad hoc groups. In *EUROCRYPT 2004*, pages 609–626.
17. S. Goldwasser, S. Micali, and R. Rivest. A digital signature scheme secure against adaptive chosen-message attacks. *SIAM Journal on Computing*, 17(2):281–308.
18. A. Lysyanskaya. *Signature schemes and applications to cryptographic protocol design.* PhD thesis, Massachusetts Institute of Technology, Sept. 2002.
19. A. Lysyanskaya, R. Rivest, A. Sahai, and S. Wolf. Pseudonym systems. In *Selected Areas in Cryptography*.

20. R. L. Rivest, A. Shamir, and Y. Tauman. How to leak a secret. In *ASIACRYPT 2001*, pages 552–565.
21. A. Sahai. Non-malleable non-interactive zero knowledge and adaptive chosen-ciphertext security. In *FOCS 1999*, pages 543–553. IEEE, 1999.
22. V. Shoup. Sequences of games: a tool for taming complexity in security proofs. http://eprint.iacr.org/2004/332, 2004.

Non-interactive Zaps and New Techniques for NIZK

Jens Groth[*], Rafail Ostrovsky[**], and Amit Sahai[***]

UCLA, Computer Science Department
4732 Boelter Hall
Los Angeles, CA 90095, USA
{jg, rafail, sahai}@cs.ucla.edu

Abstract. In 2000, Dwork and Naor proved a very surprising result: that there exist "Zaps", two-round witness-indistinguishable proofs in the plain model without a common reference string, where the Verifier asks a single question and the Prover sends back a single answer. This left open the following tantalizing question: does there exist a *non-interactive* witness indistinguishable proof, where the Prover sends a single message to the Verifier for some non-trivial NP-language? In 2003, Barak, Ong and Vadhan answered this question affirmatively by derandomizing Dwork and Naor's construction under a complexity theoretic assumption, namely that Hitting Set Generators against co-nondeterministic circuits exist.

In this paper, we construct non-interactive Zaps for all NP-languages. We accomplish this by introducing new techniques for building Non-Interactive Zero Knowledge (NIZK) Proof and Argument systems, which we believe to be of independent interest, and then modifying these to yield our main result. Our construction is based on the Decisional Linear Assumption, which can be seen as a bilinear group variant of the Decisional Diffie-Hellman Assumption.

Furthermore, our single message witness-indistinguishable proof for Circuit Satisfiability is of size $O(k|C|)$ bits, where k is a security parameter, and $|C|$ is the size of the circuit. This is much more efficient than previous constructions of 1- or 2-move Zaps.

Keywords: Non-interactive zero-knowledge, witness indistinguishability, bilinear groups, Decisional Linear Assumption.

1 Introduction

In 2000, Dwork and Naor [DN00] proved a very surprising result: that there exist "Zaps", two-round Witness-Indistinguishable (WI) proofs in the plain model

---

[*] Supported by NSF grant No. 0456717, and NSF Cybertrust grant.

[**] Supported in part by a gift from Teradata,Intel equipment grant, NSF Cybertrust grant No. 0430254, OKAWA research award, B. John Garrick Foundation and Xerox Innovation group Award.

[***] Supported by grant No. 0456717 from the NSF ITR and Cybertrust programs, an equipment grant from Intel, and an Alfred P. Sloan Foundation Research Fellowship.

without a common reference string, where the Verifier asks a single question and the Prover sends back a single answer. This left open the following tantalizing question: does there exist a *non-interactive* witness indistinguishable proof, where the Prover sends a single message to the Verifier for some non-trivial NP-language? Such zaps were shown to have a number of fascinating and important applications, beyond the numerous applications of WI proofs already present in the literature.

In this paper, we introduce new techniques for constructing Non-Interactive Zero Knowledge (NIZK) Proofs and Arguments, based on the hardness of computational problems that arise in bilinear groups. Based on these new techniques, we are able to construct *non-interactive* Witness-Indistinguishable proofs for any NP relation, without any setup assumptions, based on a number-theoretic computational assumption. Furthermore, our construction is significantly more efficient than previous constructions of zaps, as we discuss below. We believe our new techniques for NIZK will have a number of other applications, as well. In the remainder of this introduction, we describe our setting and our results, and present our results in the context of previous work.

OUR SETTING. Throughout the paper we will make use of groups of prime order equipped with non-trivial bilinear maps. In other words, we let $\mathbb{G}, \mathbb{G}_T$ be abelian groups of order p, and let $e : \mathbb{G} \times \mathbb{G} \to \mathbb{G}_T$ be a non-degenerate bilinear map such that $e(u^a, v^b) = e(u, v)^{ab}$. Such groups have been widely used in cryptography in recent years.

Our underlying security assumption is the Decisional Linear Assumption: Given groups elements $(g, f = g^x, h = g^y, f^r, h^s, g^d)$ for $x, y \leftarrow \mathbb{Z}_p^*$ and $r, s \leftarrow \mathbb{Z}_p$, it is hard to distinguish between the case where $d = r + s$ or d is random. The assumption was introduced by Boneh, Boyen and Shacham in [BBS04]. The assumption gives rise to an ElGamal-like cryptosystem with public key $pk = (p, \mathbb{G}, \mathbb{G}_T, e, g, f, h)$, where $f = g^x, h = g^y$ and the secret key is $sk = (x, y)$. Encryption of $m \in \mathbb{G}$ is done by picking $r, s \leftarrow \mathbb{Z}_p$ at random and letting the ciphertext be $(f^r, h^s, g^{r+s}m)$. An encryption of 1 is called a *linear tuple* (with respect to f, h, g).

OUR TECHNIQUES AND RESULTS. The conceptual starting point for our work is our recent work [GOS06], which constructed NIZK proofs and arguments for any NP relation, based on a different computational assumption for bilinear groups of composite order, called the Subgroup Decision Assumption. In that paper, we gave a construction for NIZK proof systems, such that if the Common Reference String (CRS) was of one form, it would be perfectly sound and computational ZK; whereas if the CRS was of a different form, then the *same* construction would yield a system that is computationally sound but perfectly ZK.

Our key idea for achieving non-interactive WI proofs *without* a CRS is as follows: If we could somehow force the prover to produce a perfect soundness CRS on its own, we would be done – but this is not possible. Instead, can we somehow force a prover to produce *two* CRS's, such that at least *one* is of the perfect soundness type?

Unfortunately, in the original GOS proof system, the CRS's that force perfectly sound proofs are negligibly rare, and are computationally indistinguishable from CRS's that give only computational soundness (and indeed these CRS's have trapdoors allowing proofs of false theorems).

The main technical contribution of our paper is to construct a new NIZK system based on the Decisional Linear Assumption where perfect soundness CRS's are common, whereas computational soundness CRS's are negligibly rare. Furthermore, in our new system, we show that a simple operation – multiplication of one element in the CRS by a generator – can always transform a computational soundness CRS into a perfect soundness CRS. (Roughly speaking, we accomplish the following: if the CRS is a linear tuple, then we obtain a computationally sound proof system; whereas if the CRS is *any* non-linear tuple, then we obtain a perfectly sound proof system.) This allows us to achieve non-interactive WI proofs as follows: The prover can generate a CRS on its own, but it must provide proofs under *both* the chosen CRS *and* the transformation of that CRS. This forces perfect soundness. We show that the WI property still holds because of a hybrid argument.

We note that our constructions yield NIZK proofs and non-interactive WI proofs for Circuit Satisfiability where the proof size is $O(k|C|)$ bits, where k is the security parameter, and C is the size of the circuit. For NIZK proofs this matches the previous best bound by [GOS06], which relies on the Subgroup Decision assumption. Our NIZK proofs[1] have the advantage of being realizable in the Common *Random* String model, whereas the constructions of [GOS06] required the Common Reference String Model. For WI proofs, as far as we know, our proof size is a significant improvement over all previous constructions of zaps for NP relations.

We believe our techniques and ideas for constructing NIZK proofs using the Decisional Linear Assumption will have other applications, as well. In a companion paper, Groth [Gro06] constructs a wide variety of novel and efficient NIZK proofs under the Decisional Linear Assumption, and uses these to obtain group signatures and other important applications.

PREVIOUS WORK AND CONTEXT FOR OUR WORK. NIZK proofs were introduced by Blum, Feldman, and Micali [BFM88], following the introduction of interactive Zero-Knowledge proofs by Goldwasser, Micali, and Rackoff [GMR89]. Witness-Indistinguishable protocols were introduced by Feige and Shamir [FS90].

Dwork and Naor [DN00] constructed 2-round WI proofs, called zaps[2], for any NP relation (assuming trapdoor permutations exist), and showed a wide variety of applications for zaps. Furthermore, [DN00] showed that their constructions allowed for the first message (from Verifier to Prover) to be reused – so that between a particular pair of prover and verifier, only one message from verifier to prover is required even if many statements are to be proven. Barak, Ong, and

---

[1] These are computational zero knowledge, perfectly sound proofs.

[2] In the spirit of the name, we interpret zaps to mean WI proofs that require 2 rounds *or less*.

Vadhan [BOV03] constructed the first non-interactive zaps for any NP relation by applying derandomization techniques to the construction of Dwork and Naor, based on trapdoor permutations and the assumption that (very good) Hitting Set Generators (HSG) against co-nondeterministic circuits exist. It is known that such HSG's can be built if there is a function in E that requires exponential-size *nondeterministic* circuits – *i.e.* the assumption states that some uniform exponential deterministic computations can (only) be sped up by at most a constant power (Time 2^{cn} becomes $2^{\varepsilon n}$), when given the added power of nondeterminism and advice specific to the length of the input.

We mainly wish to emphasize that our construction is completely different and uses completely different, number-theoretic computational assumptions. Furthermore, our construction is much more efficient than both the constructions of Dwork-Naor and Barak-Ong-Vadhan (even when these constructions are instantiated with very efficient NIZK proofs such as [GOS06]).

A further point of comparison would be to look more closely at the assumptions used, for instance in the context of Naor's classification of assumptions based on falsifiability [Nao03]. While our assumption, the Decisional Linear Assumption, is an "efficiently falsifiable" assumption according to Naor's classification, it appears that the assumption about the existence of HSG's against co-nondeterministic circuits, or the assumption about functions in E with large nondeterministic circuits, are "none of the above" assumptions according to Naor's classification, since we wouldn't have time to actually "run" a suggested nondeterministic (or co-nondeterministic) circuit that claims to break the assumption.[3]

2   Definitions: Non-interactive Proofs

Let R be an efficiently computable binary relation. For pairs $(x, w) \in R$ we call x the statement and w the witness. Let L be the language consisting of statements in R.

A non-interactive proof system for a relation R consists of a CRS generation algorithm K, a prover P and a verifier V. The CRS generation algorithm produces a common reference string σ. The prover takes as input (σ, x, w) and

---

[3] We note that there is some uncertainty as to how to interpret Naor's classification with respect to these derandomization-style assumptions. We take a view that we think is consistent with the spirit of Naor's classification by asking the question – if the assumption is false, then is there necessarily a reasonably efficient (PPT) algorithmic demonstration of the falsehood of this assumption? To us, it appears that the answer is "Yes" for our assumption, but appears to be "No" for the [BOV03] assumptions; this is simply because for the latter assumptions, it is important that the breaking algorithm could be non-deterministic – and if it is, then how can we efficiently verify that it indeed does break the assumption? It would be very interesting if in fact there were a positive answer to this. Of course the question of falsifiability is less important than the question of whether an assumption is actually true; alas, we find ourselves unequipped to address this issue.

produces a proof π. The verifier takes as input (σ, x, π) and outputs 1 if the proof is acceptable and 0 if rejecting the proof. We call (K, P, V) a proof system for R if it has the completeness and soundness properties described below.

PERFECT COMPLETENESS. For all adversaries $\mathcal{A}$ we have

$$\Pr\left[\sigma \leftarrow K(1^k); (x, w) \leftarrow \mathcal{A}(\sigma); \pi \leftarrow P(\sigma, x, w) : V(\sigma, x, \pi) = 1 \text{ if } (x, w) \in R\right] = 1.$$

PERFECT SOUNDNESS. For all adversaries $\mathcal{A}$ we have

$$\Pr\left[\sigma \leftarrow K(1^k); (x, \pi) \leftarrow \mathcal{A}(\sigma) : V(\sigma, x, \pi) = 0 \text{ if } x \notin L\right] = 1.$$

COMPUTATIONAL ZERO-KNOWLEDGE [FLS99]. We call (K, P, V) an NIZK proof for R if there exists a simulator $S = (S_1, S_2)$ with the following zero-knowledge property. For all non-uniform polynomial time adversaries $\mathcal{A}$ we have

$$\Pr\left[\sigma \leftarrow K(1^k) : \mathcal{A}^{P(\sigma,\cdot,\cdot)}(\sigma) = 1\right] \approx \Pr\left[(\sigma, \tau) \leftarrow S_1(1^k) : \mathcal{A}^{S(\sigma,\tau,\cdot,\cdot)}(\sigma) = 1\right],$$

where $S(\sigma, \tau, x, w) = S_2(\sigma, \tau, x)$ for $(x, w) \in R$ and both oracles output failure if $(x, w) \notin R$.

2.1 Witness Indistinguishablity

A prerequisite for NIZK proofs is the common reference string. However, many times a witness indistinguishable proof is sufficient. Witness indistinguishability means that an adversary cannot tell which of two possible witnesses w_1, w_2 that has been used in constructing the proof. We will show how to construct a WI proof system without any setup assumptions.

COMPUTATIONAL WITNESS INDISTINGUISHABILITY. We call (K, P, V) a noninteractive zap for R or a non-interactive WI proof for R in the plain model if for all non-uniform polynomial time interactive adversaries $\mathcal{A}$ we have

$$\Pr\left[(x, w_1, w_2) \leftarrow \mathcal{A}(1^k); \pi \leftarrow P(1^k, x, w_1) : \mathcal{A}(\pi) = 1 \text{ and } (x, w_1), (x, w_2) \in R\right]$$
$$\approx \Pr\left[(x, w_1, w_2) \leftarrow \mathcal{A}(1^k); \pi \leftarrow P(1^k, x, w_2) : \mathcal{A}(\pi) = 1 \text{ and } (x, w_1), (x, w_2) \in R\right].$$

A hybrid argument shows that this definition is equivalent to one where we give the adversary access to multiple proofs using either witness w_1 or w_2. The definition of perfect WI is similar, except there we require equality of the above probabilities for all adversaries.

3   Bilinear Groups

BILINEAR GROUPS. We use two cyclic groups $\mathbb{G}, \mathbb{G}_T$ of order p, where p is a prime. We make use of a bilinear map $e : \mathbb{G} \times \mathbb{G} \to \mathbb{G}_T$. I.e., for all $u, v \in \mathbb{G}$

and $a, b \in \mathbb{Z}$ we have $e(u^a, v^b) = e(u, v)^{ab}$. We require that $e(g, g)$ is a generator of $\mathbb{G}_T$ if g is a generator of $\mathbb{G}$. We also require that group operations, group membership and the bilinear map be efficiently computable.

Throughout the paper we let $\mathcal{G}$ be a randomized algorithm that takes a security parameter as input and outputs $(p, \mathbb{G}, \mathbb{G}_T, e, g)$ such that p is prime, $\mathbb{G}, \mathbb{G}_T$ are descriptions of groups of order p, $e : \mathbb{G} \times \mathbb{G} \to \mathbb{G}_T$ is a bilinear map as described above and g is a random generator of $\mathbb{G}$.

Boneh and Franklin [BF03] give an example of a bilinear group. Let $p = 2 \bmod 3$ be a prime, and choose a small ℓ so $q = \ell p - 1$ is prime and $p^2 \nmid q + 1$. Then the elliptic curve $y^2 = x^3 + 1$ over $\mathbb{Z}_q$ has ℓp points. We can let $\mathbb{G}$ be the order p subgroup of this curve and $\mathbb{G}_T = \mathbb{F}_{q^2}^*$. The bilinear map is the modified Weil-pairing. To get a random generator g for this group, pick x at random such that $x^3 + 1$ is a square and let y be a randomly chosen squareroot. Then $g = (x, y)^\ell$ is a random generator for $\mathbb{G}$ provided $g \neq 1$.

We say the bilinear group is verifiable, if there is a verification algorithm that outputs 1 if and only if $(p, \mathbb{G}, \mathbb{G}_T, e, g)$ is a bilinear group. The bilinear group from [BF03] described above is verifiable, we just need to check that $p = 2 \bmod 3$ is a prime and g is a generator for $\mathbb{G}$.

Definition 1 (Decisional Linear Assumption). *We say the Decisional Linear Assumption holds for the bilinear group generator $\mathcal{G}$ if for all non-uniform polynomial time adversaries $\mathcal{A}$ we have*

$$\Pr\Big[(p, \mathbb{G}, \mathbb{G}_1, e, g) \leftarrow \mathcal{G}(1^k); x, y \leftarrow \mathbb{Z}_p^*; r, s \leftarrow \mathbb{Z}_p :$$

$$\mathcal{A}(p, \mathbb{G}, \mathbb{G}_T, e, g, g^x, g^y, g^{xr}, g^{ys}, g^{r+s}) = 1\Big]$$

$$\approx \Pr\Big[(p, \mathbb{G}, \mathbb{G}_1, e, g) \leftarrow \mathcal{G}(1^k); x, y \leftarrow \mathbb{Z}_p^*; r, s, d \leftarrow \mathbb{Z}_p :$$

$$\mathcal{A}(p, \mathbb{G}, \mathbb{G}_T, e, g, g^x, g^y, g^{xr}, g^{ys}, g^d) = 1\Big].$$

The Decisional Linear Assumption was first introduced by Boneh, Boyen and Shacham [BBS04] and has since been used in several cryptographic constructions. We call a tuple of the form (f^r, h^s, g^{r+s}) a *linear tuple* with respect to (f, h, g). When the basis (f, h, g) is obvious from context, we omit mention of it.

4 Homomorphic Encryption and Commitment from Bilinear Maps

4.1 A Homomorphic Cryptosystem

We recall the homomorphic cryptosystem given by [BBS04]. It uses ideas similar to ElGamal encryption, but since the Decisional Diffie-Hellman (DDH) problem is easy in bilinear groups, we have to insert an extra element in the ciphertext.

Key generation
1. $(p, \mathbb{G}, \mathbb{G}_1, e, g) \leftarrow \mathcal{G}(1^k)$
2. Let $x, y \leftarrow \mathbb{Z}_p^*$; let $f = g^x, h = g^y$

3. Let $pk = (p, \mathbb{G}, \mathbb{G}_T, e, g, f, h)$
4. Let $sk = (pk, x, y)$
5. Return (pk, sk)

Encryption: To encrypt $m \in \mathbb{G}$, let $r, s \leftarrow \mathbb{Z}_p$, and return $(u, v, w) = E(m; r, s) = (f^r, h^s, g^{r+s}m)$.

Decryption: To decrypt ciphertext $(u, v, w) \in \mathbb{G}^3$, return $m = D_{sk}(u, v, w) = u^{-1/x}v^{-1/y}w$.

The cryptosystem (K_{cpa}, E, D) has several nice properties. The Decisional Linear Assumption for $\mathcal{G}$ implies semantic security under chosen plaintext attack. All triples $(u, v, w) \in \mathbb{G}^3$ are valid ciphertexts. Also, the cryptosystem is homomorphic in the sense that

$$E(m_1; r_1, s_1)E(m_2, r_2, s_2) = E(m_1 m_2; r_1 + r_2, s_1 + s_2).$$

4.2   A Homomorphic Commitment Scheme

We will use the cryptosystem to create a homomorphic commitment scheme with the property that depending on how we generate the public key we get either a perfectly hiding trapdoor commitment scheme or a perfectly binding commitment scheme.

Perfectly hiding key generation
1. $(pk, sk) \leftarrow K_{\text{cpa}}(1^k)$
2. $r_u, s_v \leftarrow \mathbb{Z}_p$
3. $(u, v, w) = E(1; r_u, s_v) = (f^{r_u}, h^{s_v}, g^{r_u+s_v})$
4. Return $ck = (pk, u, v, w)$

Perfectly binding key generation
1. $(pk, sk) \leftarrow K_{\text{cpa}}(1^k)$
2. $r_u, s_v \leftarrow \mathbb{Z}_p$
3. $(u, v, w) = E(m; r_u, s_v) = (f^{r_u}, h^{s_v}, g^{r_u+s_v}m)$, where $m = g^{\pm 1}$ can be arbitrarily chosen
4. Return $ck = (pk, u, v, w)$

Commitment: To commit to message $m \in \mathbb{Z}_p$ do
1. $r, s \leftarrow \mathbb{Z}_p$
2. Return $c = (c_1, c_2, c_3) = \text{com}(m; r, s) = (u^m f^r, v^m h^s, w^m g^{r+s})$

Trapdoor opening: Given a commitment $c = \text{com}(m; r, s)$ under a perfectly hiding commitment key we have $c = \text{com}(m'; r - (m' - m)r_u, s - (m' - m)s_v)$. So we can create a perfectly hiding commitment and open it to any value we wish if we have the trapdoor key (r_u, s_v).

The semantic security of the cryptosystem implies that no polynomial time adversary can distinguish between perfectly hiding keys and perfectly binding keys. This implies that the perfectly binding commitment scheme is computationally hiding, and the perfectly hiding commitment scheme is computationally binding.

5   NIZK Proofs for Circuit Satisfiability

In this section, we show how to construct NIZK proofs for Circuit Satisfiability based on the Decisional Linear Assumption. To do this, we follow but somewhat change the general outline of the [GOS06] construction. We review this now:

In the GOS construction, and in ours, the overall approach is to commit[4] to the value of all the wires in the circuit (including the input wires) using an *additively homomorphic* commitment scheme, and then prove that for every gate in the circuit (W.L.O.G. a NAND gate), the 3 wires incident to the gate obey its rule. In [GOS06], we then showed how to reduce this task to just proving that a committed value is either 0 or 1. This is done by way of the homomorphic properties of the commitment scheme, together with the following simple observation: for three values $b_0, b_1, b_2 \in \{0, 1\}$, we have that $b_0 + b_1 + 2b_2 - 2 \in \{0, 1\}$ iff $b_2 = \neg(b_0 \wedge b_1)$.

In [GOS06], then all that was needed was a NIZK proof that a committed value is either 0 or 1. Here, we look a little closer at the GOS methodology, and take a slightly different route. This consists of two main observations:

1. First, we take a look at our homomorphic commitment scheme (given in the last section), and observe the following: Given a commitment $c = (c_1, c_2, c_3)$, the committed value being either 0 or 1 is equivalent to the following statement: that either c is a commitment to 0, *or* that $c' = (c_1/u, c_2/v, c_3/w)$ is a commitment to 0. Further, we note that a commitment (c_1, c_2, c_3) is a commitment to 0 iff it forms a linear tuple. Thus, we can equivalently prove that given two tuples, that either one or the other is a linear tuple, i.e., of the form (f^r, h^s, g^{r+s}).

2. Second, we take a closer look at the simulation strategy. The overall strategy is as follows: The CRS consists of the parameters for the homomorphic commitment scheme. As we have already observed, however, the Decisional Linear Assumption implies that a CRS that leads to perfectly binding commitments is indistinguishable from one that leads to perfectly hiding commitments. If we want perfect soundness for our NIZK proof system, then the "real-life" CRS should lead to perfectly binding commitments. The simulation can use a CRS of the perfectly hiding type, and the simulator can remember the trapdoor information that allows it to produce equivocal commitments that it can later open to any value.

   A key observation we make here is that the homomorphic properties of the commitment *preserves* equivocality: if one applies the homomorphic operations to multiple equivocal commitments, then the resulting commitment is still equivocal. So, we observe that the simulation can simply produce such equivocal commitments for each wire value, and then when it comes to

---

[4] In [GOS06] we called this an encryption. The fact that it was an encryption and not just a commitment is not important for the ZK property, and was used there to achieve proofs of knowledge. We can also obtain NIZK proofs of knowledge, but that is not our focus here.

proving that one of two commitments (that were generated via homomorphic operations) is a commitment to zero, the simulation will actually have the necessary information to prove this for *both* commitments. What this means is that we need the proof that one of two commitments is a commitment to zero (*i.e.* that one out of two tuples is a linear tuple) to merely be *witness-indistinguishable* rather than fully NIZK.

Before giving the NIZK proof for Circuit Satisfiability more formally, we first construct a (perfect) WI proof for one out of two tuples being a linear tuple.

5.1 Perfect WI Proof

Consider the following situation. We have two tuples (A_1, A_2, A_3) and (B_1, B_2, B_3) with discrete logarithms (a_1, a_2, a_3) and (b_1, b_2, b_3) with respect to (f, h, g), where $f = g^x$ and $h = g^y$. We want to prove that $a_1 + a_2 + a_3 = 0$ or $b_1 + b_2 + b_3 = 0$. Note, this corresponds to $(A_1^{-1}, A_2^{-1}, A_3)$ or $(B_1^{-1}, B_2^{-1}, B_3)$ being a linear tuple. We will do this by showing that

$$0 = \sum_{i=1}^{3} \sum_{j=1}^{3} a_i b_j = (a_1 + a_2 + a_3)(b_1 + b_2 + b_3).$$

We first give the intuition behind our scheme, and then the formal description and proof of correctness. Using the bilinear map, we can compute

$$
\begin{array}{ll}
e(A_1, B_1) = e(f, f)^{a_1 b_1} & e(A_1, B_2)e(A_2, B_1) = e(f, h)^{a_1 b_2 + a_2 b_1} \\
e(A_2, B_2) = e(h, h)^{a_2 b_2} & e(A_1, B_3)e(A_3, B_1) = e(f, g)^{a_1 b_3 + a_3 b_1} \\
e(A_3, B_3) = e(g, g)^{a_3 b_3} & e(A_3, B_2)e(A_2, B_3) = e(h, g)^{a_2 b_3 + a_3 b_2}
\end{array}
$$

The goal is to show that these six exponents sum to 0.

Consider the following matrix

$$
M = \begin{pmatrix}
e(A_1, B_1) & e(f, h)^t e(A_1, B_2) & e(f, g)^{-t} e(A_1, B_3) \\
e(h, f)^{-t} e(A_2, B_1) & e(A_2, B_2) & e(h, g)^t e(A_2, B_3) \\
e(g, f)^t e(A_3, B_1) & e(g, h)^{-t} e(A_3, B_2) & e(A_3, B_3)
\end{pmatrix},
$$

with $t \leftarrow \mathbb{Z}_p$ chosen at random.

If both $a_1 + a_2 + a_3 = 0$ and $b_1 + b_2 + b_3 = 0$, then this matrix is distributed identically to its transpose. To see this, we observe that since $a_1(b_1 + b_2 + b_3) = b_1(a_1 + a_2 + a_3) = 0$, we have that $a_1 b_2 - a_2 b_1 = a_3 b_1 - a_1 b_3$. Similarly, we have that $a_1 b_2 - a_2 b_1 = a_2 b_3 - a_3 b_2$. Therefore if we set $t' = t + (a_1 b_2 - a_2 b_1) = t + (a_3 b_1 - a_1 b_3) = t + (a_2 b_3 - a_3 b_2)$, but interchange the roles of a_1, a_2, a_3 and b_1, b_2, b_3, we have the same matrix. This is what will give us witness indistinguishability. If $a_1 + 2_3 + a_3 \neq 0$ or $b_1 + b_2 + b_3 \neq 0$ we only have one witness and therefore we automatically have witness indistinguishability.

So, W.L.O.G., assume that we know a_1, a_2, a_3. We can rearrange the matrix as

$$
\begin{pmatrix}
e(f, B_1^{a_1}) & e(f, h^t B_2^{a_1}) & e(f, g^{-t} B_3^{a_1}) \\
e(h, f^{-t} B_1^{a_2}) & e(h, B_2^{a_2}) & e(h, g^t B_3^{a_2}) \\
e(g, f^t B_1^{a_3}) & e(g, h^{-t} B_2^{a_3}) & e(g, B_3^{a_3})
\end{pmatrix}.
$$

In our proof system, we will reveal the 9 right-hand-side inputs to the bilinear maps for each entry of the matrix.

Observe, that we have $e(A_i, B_i) = M_{ii}$, so M_{ii} has exponent $a_i b_i$. We also have $e(A_i, B_j)e(A_j, B_i) = M_{ij}M_{ji}$, which has exponent $a_i b_j + a_j b_i$ for $i \neq j$. The verifier can check these equations, meaning he knows the sum of all the 9 exponents of M is $\sum_{i=1}^{3} \sum_{j=1}^{3} a_i b_j$. We therefore just need to show that the exponents of each of the 3 column vectors of M is 0.

Observe in the matrix above that for $j = 1, 2, 3$ we have $M_{1j}M_{2j}M_{3j} = 1$. This means we do not need to reveal M_{3j}, the verifier can compute it as $M_{3j} = M_{1j}^{-1}M_{2j}^{-1}$ himself. This also corresponds to asserting the fact that the column logarithms sum to 0: Taking discrete logarithms of these elements we have $m_{1j} + m_{2j} + m_{3j} = 0$.

These are the ideas in the WI proof, let us now write down the protocol.

Statement: A bilinear group $(p, \mathbb{G}, \mathbb{G}_T, e, g)$ and generators (f, h). The claim is that at least one of two given tuples (c_1, c_2, c_3) and (d_1, d_2, d_3) is a linear tuple with respect to f, h, g.

Witness: The witness is of the form (r, s) so $c = (f^r, h^s, g^{r+s})$ or $d = (f^r, h^s, g^{r+s})$.

Proof: Define $a_1 = -r, a_2 = -s, a_3 = r + s$. This means $a_1 + a_2 + a_3 = 0$.
If the prover has a witness for c then let $B_1 = d_1^{-1}, B_2 = d_2^{-1}, B_3 = d_3$, else let $B_1 = c_1^{-1}, B_2 = c_2^{-1}, B_3 = c_3$.
Choose $t \leftarrow \mathbb{Z}_p$ and let

$$\pi_{11} = B_1^{a_1} \qquad\qquad \pi_{12} = h^t B_2^{a_1} \qquad\qquad \pi_{13} = g^{-t} B_3^{a_1}$$

$$\pi_{21} = f^{-t} B_1^{a_2} \qquad\qquad \pi_{22} = B_2^{a_2} \qquad\qquad \pi_{23} = g^t B_3^{a_3}$$

Return the proof $\pi = (\pi_{11}, \pi_{12}, \pi_{13}, \pi_{21}, \pi_{22}, \pi_{23})$.

Verification: Compute $\pi_{3j} = (\pi_{1j}\pi_{2j})^{-1}$ for $j = 1, 2, 3$. For sake of notation consistent with the intuition above, let $\tilde{c}_1 = c_1^{-1}, \tilde{c}_2 = c_2^{-1}, \tilde{c}_3 = c_3, \tilde{d}_1 = d_1^{-1}$, $\tilde{d}_1 = d_2^{-1}$, and $\tilde{d}_1 = d_3$. Accept if and only if the bilinear group is correctly formed, and

$$e(f, \pi_{11}) = e(\tilde{c}_1, \tilde{d}_1) \qquad\qquad e(f, \pi_{12})e(h, \pi_{21}) = e(\tilde{c}_1, \tilde{d}_2)e(\tilde{c}_2, \tilde{d}_1)$$
$$e(h, \pi_{22}) = e(\tilde{c}_2, \tilde{d}_2) \qquad\qquad e(f, \pi_{13})e(g, \pi_{31}) = e(\tilde{c}_1, \tilde{d}_3)e(\tilde{c}_3, \tilde{d}_1)$$
$$e(g, \pi_{33}) = e(\tilde{c}_3, \tilde{d}_3). \qquad\qquad e(h, \pi_{23})e(g, \pi_{32}) = e(\tilde{c}_2, \tilde{d}_3)e(\tilde{c}_3, \tilde{d}_2)$$

Theorem 1. *The protocol described above is a non-interactive proof system for one of (c_1, c_2, c_3) or (d_1, d_2, d_3) being a linear tuple with respect to f, h, g. It has perfect completeness, perfect soundness and perfect witness-indistinguishability. The proof consists of 6 elements from $\mathbb{G}$.*

Proof

Perfect completeness: This follows by straightforward computation.

Perfect soundness: Define r_c, s_c, t_c and r_d, s_d, t_d so $c = (f^{r_d}, h^{s_d}, g^{t_c})$ and $d = (f^{r_d}, h^{s_d}, g^{t_d})$.
For $i = 1, 2$ let

$$m_{i1} = \log_f(\pi_{i1}) \qquad m_{i2} = \log_h(\pi_{i1}) \qquad m_{i3} = \log_g(\pi_{i3}).$$

Let

$$m_{31} = -m_{11} - m_{21} \qquad m_{32} = -m_{12} - m_{22} \qquad m_{33} = -m_{13} - m_{23}.$$

From the equalities we get

$$
\begin{aligned}
m_{11} &= r_c r_d & m_{12} + m_{21} &= r_c s_d + s_c r_d \\
m_{22} &= s_c s_d & m_{13} + m_{31} &= -r_c t_d - t_c r_d \\
m_{33} &= t_c t_d. & m_{23} + m_{32} &= -s_c t_d - t_c s_d.
\end{aligned}
$$

This means

$$
\begin{aligned}
&(r_c + s_c - t_c)(r_d + s_d - t_d) \\
&= r_c r_d + r_c s_d + s_c r_d + s_c s_d + t_c t_d - (r_c t_d + t_c r_d + s_c t_d + t_c s_d) \\
&= \sum_{i=1}^{3} \sum_{j=1}^{3} m_{ij} = 0.
\end{aligned}
$$

We conclude

$$t_c = r_c + s_c \qquad \text{or} \qquad t_d = r_d + s_d.$$

Perfect witness indistinguishability: For WI, we may assume that both tuples are linear tuples. Define $a_1 = -r, a_2 = -s, a_3 = r + s$ and B_1, B_2, B_3 as in the proof. We define b_1, b_2, b_3 so $B_1 = f^{b_1}, B_2 = h^{b_2}, B_3 = g^{b_3}$, observe that $b_1 + b_2 + b_3 = 0$. In the proof we pick $t \leftarrow \mathbb{Z}_p$. Interchanging the roles of a_1, a_2, a_3 and b_1, b_2, b_3 and using $t' = t + (a_1 b_2 - a_2 b_1) = t + (a_3 b_1 - a_1 b_3) = t + (a_2 b_3 - a_3 b_2)$ leads to exactly the same proof. □

5.2   Circuit Satisfiability NIZK Construction

Based on the intuition given earlier, we now give an NIZK proof for Circuit Satisfiability, based on the (perfect) WI proof that one out of two tuples is a linear tuple, given in the last section.

Common reference string
1. $(p, \mathbb{G}, \mathbb{G}_T, e, g) \leftarrow \mathcal{G}(1^k)$
2. f, h random generators of $\mathbb{G}$
3. $u = f^{r_0}$, $v = h^{s_0}$, and $w = g^{r_0 + s_0} m$, for random r_0, s_0 in $\mathbb{Z}_p$ and $m = g$ or $m = g^{-1}$. Note that the choice of $m = g$ or $m = g^{-1}$ is arbitrary.
4. Return $\sigma = (p, \mathbb{G}, \mathbb{G}_T, e, g, f, h, u, v, w)$.

Statement: The statement is a circuit C built from NAND-gates. The claim is that there exist input bits w so $C(w) = 1$.

Proof: The prover has a witness w consisting of input bits so $C(w) = 1$.

1. Extend w to contain the bits of all wires in the circuit.
2. Commit to each bit w_i as a tuple $(c_1 = u^{w_i} f^r, c_2 = v^{w_i} h^s, c_3 = w^{w_i} g^{r+s})$ with $r, s \leftarrow \mathbb{Z}_p$ chosen independently for each wire.
3. For the output wire, create a special commitment $c^* = (u, v, w)$ that can easily be checked to be a commitment to 1, as required.
4. For each commitment $c = (c_1, c_2, c_3)$ to each wire value w_i, generate a commitment $c' = (c_1/u, c_2/v, c_3/w)$, and give a WI proof that either c or c' is a linear tuple with respect to f, h, g. Note that if $w_i = 0$, then c is a linear tuple, and if $w_i = 1$, then c' is a linear tuple.
5. For all NAND-gates, do the following. We write the input commitments tuples as $a = (a_1, a_2, a_3), b = (b_1, b_2, b_3)$, and the output commitment tuple as $c = (c_1, c_2, c_3)$. From these commitments, create two new tuples: $C = (C_1 = a_1 b_1 c_1^2 u^{-2}, C_2 = a_2 b_2 c_2^2 v^{-2}, C_3 = a_3 b_3 c_3^2 w^{-2})$ and $C' = (C_1/u, C_2/v, C_3/w)$. Note that either C or C' is a linear tuple iff the values underlying the commitments a, b, c respect the NAND gate. Then give a WI proof that either C or C' is a linear tuple, noting that the witness for this can be derived from the wire values and randomness used to prepare the commitments a, b, and c.
6. Return π consisting of all the commitments and WI proofs.

Verification: The verifier is given a circuit C and a proof π.

1. Check that all wires have a corresponding commitment tuple and that the output wire's commitment tuple is (u, v, w).
2. Check that all WI proofs showing that each wire has a committed value in $\{0, 1\}$ are valid.
3. Check that all WI proofs corresponding to NAND-gates are valid.
4. Return 1 if all checks pass, else return 0.

Remark. We note that in the common reference string, if p is a prime number, then if we let g, f, h, u, v, w be randomly chosen elements of $\mathbb{G}$, with overwhelming probability they will form a viable CRS such that (u, v, w) are a non-linear tuple with respect to (f, h, g), and therefore the resulting commitment scheme is perfectly binding. If, for instance, the group is the one suggested by Boneh and Franklin [BF03], then all that is needed to define $\mathbb{G}$ is the prime p. Thus, we can implement our NIZK Proofs in the Common *Random* String model, where the random string is first used to obtain a k-bit prime p using standard methods (just dividing up the CRS into k-bit chunks and checking one-by-one if they are prime will do), and then the remaining randomness is used to randomly determine g, f, h, u, v, w (by picking random order p points on the curve). Such an NIZK Proof will not have perfect soundness, but statistical soundness, since the probability of (u, v, w) being a linear tuple is exponentially small in k. In the common random string model this is optimal, since for any NIZK proof system with a common random string there is a risk of accidentally selecting a simulation string.

Theorem 2. *The protocol above is an NIZK proof system for Circuit Satisfiability with perfect completeness, perfect soundness and computational zero-knowledge if the Decisional Linear Assumption holds for the bilinear group generator $\mathcal{G}$.*

Proof Sketch

Perfect completeness and soundness are clear. We now argue that our NIZK proof system is computational zero knowledge. We present this in two stages.

We first examine a hybrid in which the prover uses the witness to generate a proof, but where the CRS is simulated so that (u, v, w) form a linear tuple, instead of a non-linear tuple. We note that (by means of intermediate hybrid in which (u, v, w) are random, and a reduction to the Decisional Linear Assumption) this hybrid produces computationally indistinguishable proofs.

The simulator will now produce proofs that are *distributed identically* to the hybrid above (assuming that the underlying WI proofs are perfectly WI). It starts by choosing $u = f^{r_0}, v = h^{s_0}, w = g^{r_0+s_0}$, and remembering these values $r_0, s_0 \leftarrow \mathbb{Z}_p$.

Now, for each wire w, the simulator picks a commitment $c = (c_1 = f^r, c_2 = h^s, c_3 = g^{r+s})$ as a random linear tuple. Because (u, v, w) is a linear tuple, all commitment strings are distributed identically as random linear tuples.

For generating the WI-proofs corresponding to the commitments for wires, since the simulator directly has a witness for showing that the commitment is a linear tuple, it uses this to complete the proof.

For generating the WI-proofs corresponding to NAND gates, we note that for each NAND gate, if the 3 wire commitments are $a = (f^{r_1}, h^{s_1}, g^{r_1+s_1})$, $b = (f^{r_2}, h^{s_2}, g^{r_2+s_2})$, and $c = (f^{r_3}, h^{s_3}, g^{r_3+s_3})$, then the commitment $C = (f^{r_1+r_2+2r_3-2r_0}, h^{s_1+s_2+2s_3-2s_0}, g^{(r_1+r_2+2r_3-2r_0)+(s_1+s_2+2s_3-2s_0)})$, and therefore we have a witness to C being a linear tuple, and we can use this to complete the WI proof.

The only difference between how the simulator proceeds and how the honest prover algorithm proceeds is the choice of which witnesses to use in the WI proof. Therefore, the (perfect) indistinguishability of the simulation from the hybrid follows from the (perfect) witness-indistinguishability of the WI proofs. □

6   Non-interactive Zaps for Circuit Satisfiability

We now give our construction of non-interactive zaps for Circuit Satisfiability, following the intuition presented in the Introduction.

Statement: A circuit C.

Proof: The prover given $1^k, C$ and input values w such that $C(w) = 1$ proceeds as follows:

1. Generate a *verifiable* bilinear group $(p, \mathbb{G}, \mathbb{G}_T, e, g) \leftarrow \mathcal{G}(1^k)$.
2. Choose a *perfectly hiding* CRS, namely generators f, h, and a linear tuple (u, v, w).

3. Use the NIZK prover to obtain a proof π_1 of the statement with respect to the CRS $(p, \mathbb{G}, \mathbb{G}_1, e, g, f, h, u, v, w)$.
4. Use the NIZK prover to obtain a proof π_2 of the statement with respect to the CRS $(p, \mathbb{G}, \mathbb{G}_1, e, g, f, h, u, v, wg)$. Observe, we are using $w' = wg$.
5. The resulting proof is $\pi = (p, \mathbb{G}, \mathbb{G}_T, e, g, f, h, u, v, w, \pi_1, \pi_2)$.

Verification: On input C and a proof π as described above, accept iff the following procedure succeeds:
1. Use the verification algorithm to check that $(p, \mathbb{G}, \mathbb{G}_T, e, g)$ is a bilinear group.
2. Verify that $f \neq 1, h \neq 1$, i.e., that f and h are generators of $\mathbb{G}$.
3. Verify π_1 with respect to the CRS $(p, \mathbb{G}, \mathbb{G}_1, g, f, h, u, v, w)$.
4. Verify π_2 with respect to the CRS $(p, \mathbb{G}, \mathbb{G}_T, g, f, h, u, v, wg)$.

Theorem 3. *The protocol described above is a non-interactive proof for Circuit Satisfiability with perfect completeness, perfect soundness and computational witness indistinguishability if the Decisional Linear Assumption holds for the verifiable bilinear group generator $\mathcal{G}$.*

Proof

Perfect completeness: The protocol is perfectly complete because the NIZK proofs for Circuit Satisfiability are perfectly complete.

Perfect soundness: Perfect soundness follows from the fact that at least one of the two CRS's – $(p, \mathbb{G}, \mathbb{G}_T, e, g, f, h, u, v, w)$ and $(p, \mathbb{G}, \mathbb{G}_T, g, f, h, u, v, wg)$ 1– must have perfectly binding parameters for the commitment scheme. Perfect soundness of the corresponding NIZK proof implies that C must be satisfiable.

Computational witness indistinguishability: We now argue (computational) witness indistinguishability assuming the Decisional Linear Assumption, by means of a hybrid argument:
1. The first hybrid is simply the prover algorithm above using witness w_1. That is, it chooses a group $(p, \mathbb{G}, \mathbb{G}_T, e, g)$ and a public key (f, h) and a random linear tuple (u, v, w), then uses the NIZK prover with witness w_1 to obtain π_1, and uses the NIZK prover with witness w_1 to obtain π_2.
2. The second hybrid proceeds as in the first, except that for π_1, it uses the NIZK prover with witness w_2 to obtain π_1 instead of using witness w_1. Hybrid 1 and Hybrid 2 are identically distributed, by means of an intermediate hybrid using the NIZK simulator for π_1, and the fact that the NIZK simulator is a perfect simulator in the case where the CRS is based on a linear tuple.
3. The third hybrid proceeds as the second, except that it chooses random generators (f, h, g), and a linear tuple (u, v, w'), and sets $w = w'/g$. Note that now, (u, v, w) is a perfectly binding CRS, while (u, v, w') is a perfectly hiding CRS.
   Hybrid 2 and Hybrid 3 are computationally indistinguishable by a reduction to the Decisional Linear Assumption. This is seen by means of an intermediate hybrid in which (u, v, w) are set to a random tuple.

4. The fourth hybrid proceeds as the third, except that for π_2, it uses the NIZK prover with witness w_2 to obtain π_2 instead of using witness w_1. Hybrid 3 and Hybrid 4 are identically distributed for the same reasons Hybrids 1 and 2 were identically distributed.

5. Finally, the fifth hybrid proceeds as the fourth, except that it chooses random generators (f, h, g), and a linear tuple (u, v, w), and sets $w' = wg$. This is precisely the WI prover algorithm using witness w_2.

   Hybrid 4 and Hybrid 5 are computationally indistinguishable by a reduction to the Decisional Linear Assumption, by the same argument showing that Hybrids 2 and 3 were computationally indistinguishable. $\square$

Acknowledgment

We wish to thank Brent Waters for important collaboration at an early stage of this research, in particular for his collaboration on ideas for building NIZK proof systems based on the Decisional Linear Assumption.

References

[BBS04]   Dan Boneh, Xavier Boyen, and Hovav Shacham. Short group signatures. In *proceedings of CRYPTO '04, LNCS series, volume 3152*, pages 41–55, 2004.

[BF03]   Dan Boneh and Matthew K. Franklin. Identity-based encryption from the weil pairing. *SIAM Journal of Computing*, 32(3):586–615, 2003.

[BFM88]   Manuel Blum, Paul Feldman, and Silvio Micali. Non-interactive zero-knowledge and its applications. In *proceedings of STOC '88*, pages 103–112, 1988.

[BOV03]   Boaz Barak, Shien Jin Ong, and Salil P. Vadhan. Derandomization in cryptography. In *proceedings of CRYPTO '03, LNCS series, volume 2729*, pages 299–315, 2003.

[DN00]   Cynthia Dwork and Moni Naor. Zaps and their applications. In *proceedings of FOCS '00*, pages 283–293, 2000.

[FLS99]   Uriel Feige, Dror Lapidot, and Adi Shamir. Multiple non-interactive zero knowledge proofs under general assumptions. *SIAM Journal of Computing*, 29(1):1–28, 1999. Earlier version entitled Multiple Non-Interactive Zero Knowledge Proofs Based on a Single Random String appeared at FOCS '90.

[FS90]   Uriel Feige and Adi Shamir. Witness indistinguishable and witness hiding protocols. In *proceedings of STOC '90*, pages 416–426, 1990.

[GMR89]   Shafi Goldwasser, Silvio Micali, and Charles Rackoff. The knowledge complexity of interactive proofs. *SIAM Journal of Computing*, 18(1):186–208, 1989.

[GOS06]   Jens Groth, Rafail Ostrovsky, and Amit Sahai. Perfect non-interactive zero-knowledge for np. In *proceedings of EUROCRYPT '06, LNCS series, volume 4004*, pages 339–358, 2006.

[Gro06]   Jens Groth. Simulation-sound non-interactive zero-knowledge proofs for a practical language and constant size group signatures. Manuscript, 2006.

[Nao03]   Moni Naor. On cryptographic assumptions and challenges. In *proceedings of CRYPTO '03, LNCS series, volume 2729*, pages 96–109, 2003.

Rankin's Constant and
Blockwise Lattice Reduction

Nicolas Gama[1], Nick Howgrave-Graham[2], Henrik Koy[3], and Phong Q. Nguyen[4]

[1] École normale supérieure, DI, 45 rue d'Ulm, 75005 Paris, France
nicolas.gama@ens.fr
[2] NTRU Cryptosystems, Burlington, MA, USA
NHowgraveGraham@ntru.com
[3] Deutsche Bank AG, Postfach, 60262 Frankfurt am Main, Germany
henrik.koy@db.com
[4] CNRS/École normale supérieure, DI, 45 rue d'Ulm, 75005 Paris, France
http://www.di.ens.fr/~pnguyen

Abstract. Lattice reduction is a hard problem of interest to both public-key cryptography and cryptanalysis. Despite its importance, extremely few algorithms are known. The best algorithm known in high dimension is due to Schnorr, proposed in 1987 as a block generalization of the famous LLL algorithm. This paper deals with Schnorr's algorithm and potential improvements. We prove that Schnorr's algorithm outputs better bases than what was previously known: namely, we decrease all former bounds on Schnorr's approximation factors to their $(\ln 2)$-th power. On the other hand, we also show that the output quality may have intrinsic limitations, even if an improved reduction strategy was used for each block, thereby strengthening recent results by Ajtai. This is done by making a connection between Schnorr's algorithm and a mathematical constant introduced by Rankin more than 50 years ago as a generalization of Hermite's constant. Rankin's constant leads us to introduce the so-called smallest volume problem, a new lattice problem which generalizes the shortest vector problem, and which has applications to blockwise lattice reduction generalizing LLL and Schnorr's algorithm, possibly improving their output quality. Schnorr's algorithm is actually based on an approximation algorithm for the smallest volume problem in low dimension. We obtain a slight improvement over Schnorr's algorithm by presenting a cheaper approximation algorithm for the smallest volume problem, which we call transference reduction.

1 Introduction

Lattices are discrete subgroups of $\mathbb{R}^m$. A lattice L can be represented by a basis, that is, a set of $n \leq m$ linearly independent vectors $\mathbf{b}_1, \ldots, \mathbf{b}_n$ in $\mathbb{R}^m$ such that L is equal to the set $L(\mathbf{b}_1, \ldots, \mathbf{b}_n) = \{\sum_{i=1}^{n} x_i \mathbf{b}_i, x_i \in \mathbb{Z}\}$ of all integer linear combinations of the $\mathbf{b}_i$'s. The integer n is the dimension of the lattice L. A lattice has infinitely many bases (except in trivial dimension ≤ 1), but some are more useful than others. The goal of *lattice reduction* is to find interesting lattice

C. Dwork (Ed.): CRYPTO 2006, LNCS 4117, pp. 112–130, 2006.

bases, such as bases consisting of reasonably short and almost orthogonal vectors: it can intuitively be viewed as a vectorial generalisation of gcd computations. Finding good reduced bases has proved invaluable in many fields of computer science and mathematics (see [10,7]), particularly in cryptology (see [17,21]).

Lattice reduction is one of the few potentially hard problems currently in use in public-key cryptography (see [21,17] for surveys on lattice-based cryptosystems). But the problem is perhaps more well-known in cryptology for its major applications in public-key cryptanalysis (see [21]): knapsack cryptosystems [22], RSA in special settings [9,5,4], DSA signatures in special settings [12,19], *etc.* Nevertheless, there are very few lattice reduction algorithms, and most of the (recent) theoretical results focus on complexity aspects (see [17]).

The first lattice reduction algorithm in arbitrary dimension is due to Hermite [11], and is based on Lagrange's two-dimensional algorithm [13] (often wrongly attributed to Gauss). It was introduced to show the existence of Hermite's constant (which guarantees the existence of short lattice vectors), as well as proving the existence of lattice bases with bounded orthogonality defect. The celebrated Lenstra-Lenstra-Lovász algorithm [14] (LLL) can be viewed as a relaxed variant of Hermite's algorithm, in order to guarantee a polynomial-time complexity. There are faster variants of LLL based on floating-point arithmetic (see [20,25]), but none improves the output quality of LLL, which is tightly connected to Hermite's historical (exponential) upper bound on his constant. The only (high-dimensional) polynomial-time reduction algorithm known with better output quality than LLL is due to Schnorr [24]. From a theoretical point of view, only one improvement to Schnorr's block-reduction algorithm has been found since [24]: by plugging the probabilistic AKS sieving algorithm [2], one may increase the blocksize $k = \log n / \log \log n$ to $k = \log n$ and keep polynomial-time complexity, which leads to (slightly) better output quality. Curiously, in practice, one does not use Schnorr's algorithm when LLL turns out to be insufficient: rather, one applies the so-called BKZ variants [26,27] of Schnorr's algorithm, whose complexity is unknown.

OUR RESULTS. We focus on the best high-dimensional lattice reduction algorithm known (Schnorr's semi block-$2k$ algorithm [24]) and potential improvements. Despite its importance, Schnorr's algorithm is not described in any survey or textbook, perhaps due to the technicality of the subject. We first revisit Schnorr's algorithm by rewriting it as a natural generalization of LLL. This enables to analyze both the running time and the output quality of Schnorr's algorithm in much the same way as with LLL. It also leads us to reconsider a certain constant β_k introduced by Schnorr [24], which is tightly related to the output quality of his semi block-$2k$ algorithm. Roughly speaking, β_k plays a role similar to Hermite's constant $\gamma_2 = \sqrt{4/3}$ in LLL.

We improve the best upper bound known for β_k: we show that essentially, $\beta_k \lesssim 0.38 \times k^{2\ln 2} \approx 0.38 \times k^{1.39}$, while the former upper bound [24] was $4k^2$. This leads to better bounds on the output quality of Schnorr's algorithm: for instance, the approximation factor $(6k)^{n/k}$ given in [24] can be decreased to its $(\ln 2)$-th power (note that $\ln 2 \approx 0.69$). On the other hand, Ajtai [1] recently proved that there exists $\varepsilon > 0$ such that $\beta_k \geq k^\varepsilon$, but no explicit value of ε was known. We establish the lower bound $\beta_k \geq k/12$, and our method is completely different from Ajtai's. Indeed, we use a connection between β_k and a mathematical constant introduced by Rankin [23] more than 50 years ago as a generalization of Hermite's constant.

Besides, Rankin's constant is naturally related to a potential improvement of Schnorr's algorithm, which we call block-Rankin reduction, and which may lead to better approximation factors. Roughly speaking, the new algorithm would still follow the LLL framework like Schnorr's algorithm, but instead of using Hermite-Korkine-Zolotarev (HKZ) reduction of $2k$-blocks, it would try to solve the so-called *smallest volume problem* in $2k$-blocks, which is a novel generalization of the shortest vector problem. Here, Rankin's constant plays a role similar to β_k in Schnorr's algorithm. But our lower bound on β_k actually follows from a lower bound on Rankin's constant, which suggests that there are intrinsic limitations to the quality of block-Rankin reduction. However, while Ajtai presented in [1] "worst cases" of Schnorr's algorithm which essentially matched the bounds on the output quality, this is an open question for block-Rankin reduction: perhaps the algorithm may perform significantly better than what is proved, even in the worst case. Finally, we make a preliminary study of the smallest volume problem. In particular, we show that HKZ-reduction does not necessarily solve the problem, which suggests that block-Rankin reduction might be stronger than Schnorr's semi block reduction. We also present an exact solution of the smallest volume problem in dimension 4, as well as an approximation algorithm for the smallest volume problem in dimension $2k$, which we call *transference reduction*. Because transference reduction is cheaper than the $2k$-dimensional HKZ-reduction used by Schnorr's algorithm, we obtain a slight improvement over Schnorr's algorithm: for a similar cost, we can increase the blocksize and therefore obtain better quality.

ROAD MAP. The paper is organized as follows. In Section 2, we provide necessary background on lattice reduction. In Section 3, we revisit Schnorr's algorithm and explain its main ideas. Section 4 deals with Rankin's constant and its connection with Schnorr's algorithm. In Section 5, we study the smallest volume problem, discuss its application to the so-called block-Rankin reduction, and present transference reduction.

2   Background

Let $\|.\|$ and $\langle .,. \rangle$ be the Euclidean norm and inner product of $\mathbb{R}^m$. Vectors will be written in bold, and we will use row-representation for matrices. For a matrix M whose name is a capital letter, we will usually denote its coefficients by $m_{i,j}$: if the name is a Greek letter like μ, we will keep the same symbol for both the matrix and its coefficients. The notation $\lceil x \rfloor$ denotes a closest integer to x.

2.1   Lattices

We refer to the survey [21] for a bibliography on lattices. In this paper, by the term lattice, we mean a discrete subgroup of some $\mathbb{R}^m$. The simplest lattice is

$\mathbb{Z}^n$, and for any linearly independent vectors $\mathbf{b}_1, \ldots, \mathbf{b}_n$, the set $L(\mathbf{b}_1, \ldots, \mathbf{b}_n) = \{\sum_{i=1}^n m_i \mathbf{b}_i \mid m_i \in \mathbb{Z}\}$ is a lattice. It turns out that in any lattice L, not just $\mathbb{Z}^n$, there must exist linearly independent vectors $\mathbf{b}_1, \ldots, \mathbf{b}_n \in L$ such that $L = L(\mathbf{b}_1, \ldots, \mathbf{b}_n)$. Any such n-tuple of vectors $\mathbf{b}_1, \ldots, \mathbf{b}_n$ is called a *basis* of L: a lattice can be represented by a basis, that is, a row matrix. Two lattice bases are related to one another by some matrix in $GL_n(\mathbb{Z})$. The *dimension* of a lattice L is the dimension n of the linear span of L. The lattice is full-rank if n is the dimension of the space. Let $[\mathbf{v}_1, \ldots, \mathbf{v}_k]$ be vectors: we denote by $G(\mathbf{v}_1, \ldots, \mathbf{v}_k)$ their *Gram matrix*, that is, the $k \times k$ symmetric positive definite matrix $(\langle \mathbf{v}_i, \mathbf{v}_j \rangle)_{1 \leq i,j \leq k}$ formed by all the inner products. The *volume* of $[\mathbf{v}_1, \ldots, \mathbf{v}_k]$ is $(\det G(\mathbf{v}_1, \ldots, \mathbf{v}_k))^{1/2}$, which is zero if the vectors are linearly dependent. The *volume* vol(L) (or *determinant*) of a lattice L is the volume of any basis of L.

DIRECT SUM. Let L_1 and L_2 be two lattices such that span$(L_1) \cap$ span$(L_2) = \{\mathbf{0}\}$. Then the set $L_1 \oplus L_2$ defined as $\{\mathbf{u} + \mathbf{v}, \mathbf{u} \in L_1, \mathbf{v} \in L_2\}$ is a lattice, whose dimension is dim $L_1 +$ dim L_2. It is the smallest lattice containing L_1 and L_2.

PURE SUBLATTICE. A sublattice U of a lattice L is *pure* if there exists a sublattice V of L such that $L = U \oplus V$. A set $[\mathbf{u}_1, \ldots, \mathbf{u}_k]$ of independent lattice vectors of L is *primitive* if and only if $[\mathbf{u}_1, \ldots, \mathbf{u}_k]$ can be extended to a basis of L, which is equivalent to $L(\mathbf{u}_1, \ldots, \mathbf{u}_k)$ being a pure sublattice of L. For any sublattice U of a lattice L, there exists a pure sublattice S of L such that span$(S) =$ span(U), in which case vol$(U)/$vol$(S) = [S : U]$ is an integer.

SUCCESSIVE MINIMA. The *successive minima* of an n-dimensional lattice L are the positive quantities $\lambda_1(L), \ldots, \lambda_n(L)$ where $\lambda_r(L)$ is the smallest radius of a zero-centered ball containing r linearly independent vectors of L. The first minimum is the norm of a shortest non-zero vector of L. Note that: $\lambda_1(L) \leq \cdots \leq \lambda_n(L)$.

HERMITE'S CONSTANT. The Hermite invariant of the lattice is defined by $\gamma(L) = \left(\lambda_1(L)/\text{vol}(L)^{\frac{1}{n}} \right)^2$. Hermite's constant γ_n is the maximal value of $\gamma(L)$ over all n-dimensional lattices. Its exact value is known for $1 \leq n \leq 8$ and $n = 24$, and we have [16]: $\gamma_n \leq 1 + \frac{n}{4}$. Asymptotically, the best bounds known are: $\frac{n}{2\pi e} + \frac{\log(\pi n)}{2\pi e} \leq \gamma_n \leq \frac{1.744n}{2\pi e}(1 + o(1))$ (see [8,18]). The lower bound follows from the so-called Minkowski-Hlawka theorem.

PROJECTED LATTICE. Given a basis $[\mathbf{b}_1, \ldots, \mathbf{b}_n]$ of L, let π_i denote the orthogonal projection over span$(\mathbf{b}_1, \ldots, \mathbf{b}_{i-1})^\perp$. Then $\pi_i(L)$ is an $(n + 1 - i)$-dimensional lattice. These projections are stable by composition: if $i > j$, then $\pi_i \circ \pi_j = \pi_j \circ \pi_i = \pi_i$. Note that:

$$\pi_i(L) = \pi_i(L(\mathbf{b}_i, \ldots, \mathbf{b}_n)) = L(\pi_i(\mathbf{b}_i), \ldots, \pi_i(\mathbf{b}_n))$$

2.2   Lattice Reduction

We will consider two quantities to measure the quality of a basis $[\mathbf{b}_1, \ldots, \mathbf{b}_n]$: the first one is the usual *approximation factor* $\|\mathbf{b}_1\| / \lambda_1(L)$, and the second one is $\|\mathbf{b}_1\|/\mathrm{vol}(L)^{1/n}$, which we call the *Hermite factor*. The smaller these quantities, the shorter the first basis vector. Lovász showed in [15] that any algorithm achieving a Hermite factor $\leq q$ can be used to efficiently find a basis with approximation factor $\leq q^2$ using n calls to the algorithm.

ORTHOGONALIZATION. Given a basis $B = [\mathbf{b}_1, ..., \mathbf{b}_n]$, there exists a unique lower-triangular matrix μ with unit diagonal and an orthogonal family $B^* = [\mathbf{b}_1^*, ..., \mathbf{b}_n^*]$ such that $B = \mu B^*$. They can be computed using Gram-Schmidt orthogonalization, and will be denoted the *GSO* of B. Note that $\mathrm{vol}(B) = \prod_{i=1}^n \|\mathbf{b}_i^*\|$, which will often be used. It is well-known [14,17] that:

$$\lambda_1(L(\mathbf{b}_1, ..., \mathbf{b}_n)) \geq \min_{1 \leq i \leq n} \|\mathbf{b}_i^*\| \tag{1}$$

SIZE-REDUCTION. A basis $[\mathbf{b}_1, ..., \mathbf{b}_n]$ is *size-reduced* with factor $\eta \geq 1/2$ if its GSO μ satisfies $|\mu_{i,j}| \leq \eta$ for all $1 \leq j < i$. An individual vector $\mathbf{b}_i$ is size-reduced if $|\mu_{i,j}| \leq \eta$ for all $1 \leq j < i$. Size reduction usually refers to $\eta = 1/2$, and is typically achieved by successively size-reducing individual vectors. Size reduction was introduced by Hermite.

LLL-REDUCTION. A basis $[\mathbf{b}_1, ..., \mathbf{b}_n]$ is LLL-reduced [14] with factor (δ, η) for $1/4 < \delta \leq 1$ and $1/2 \leq \eta < \sqrt{\delta}$ if the basis is size-reduced with factor η and if its GSO family satisfies the $(n-1)$ Lovász conditions $(\delta - \mu_{i+1,i}^2)\|\mathbf{b}_i^*\|^2 \leq \|\mathbf{b}_{i+1}^*\|^2$. LLL-reduction usually refers to the factor $(3/4, 1/2)$ because this was the choice considered in the original LLL paper [14]. But the closer δ and η are respectively to 1 and $1/2$, the more reduced the basis. Reduction with a factor $(1,1/2)$ is closely related to a reduction notion introduced by Hermite [11].

When the reduction factor is close to $(1,1/2)$, Lovász conditions and size-reduction imply the Siegel conditions [6]: $\|\mathbf{b}_i^*\|^2 \lesssim \frac{4}{3}\|\mathbf{b}_{i+1}^*\|^2$ for all $1 \leq i \leq n-1$, which limit the drop of the $\|\mathbf{b}_i^*\|$. Here, the $\lesssim$ symbol means that $\frac{4}{3}$ is actually $\frac{4}{3} + \varepsilon$ for some small $\varepsilon > 0$. In particular, the first vector satisfies $\|\mathbf{b}_1\|^2 \lesssim \left(\frac{4}{3}\right)^{i-1}\|\mathbf{b}_i^*\|^2$. Hence, the Hermite factor of an LLL-reduced basis is bounded by:

$$\|\mathbf{b}_1\|/\mathrm{vol}(L)^{1/n} \lesssim \left(\frac{4}{3}\right)^{(n-1)/4} = (\sqrt{\gamma_2})^{n-1}$$

and (1) implies that the approximation factor is bounded by:

$$\|\mathbf{b}_1\|/\lambda_1(L) \lesssim \left(\frac{4}{3}\right)^{(n-1)/2} = (\gamma_2)^{n-1}$$

The LLL algorithm is an iterative algorithm. At the start of each loop iteration, the first i vectors are already LLL-reduced, then the $(i+1)$-th vector is size-reduced; if it does not satisfy Lovász condition, the consecutive vectors $\mathbf{b}_{i+1}$ and $\mathbf{b}_i$ are swapped and the counter i is decremented, otherwise i is incremented. The loop goes on until i eventually reaches the value n. If L is a full-rank integer lattice of dimension n and B is an upper bound on the $\|\mathbf{b}_i\|$'s, then the complexity of the LLL algorithm described (using integral Gram-Schmidt) without fast integer arithmetic is $O(n^6 \log^3 B)$. The main reason is that the integer $\prod_{k=1}^{n} \|\mathbf{b}_k^*\|^{2(n-k)}$ decreases by at least the geometric factor δ at every swap: thus, the number of swaps is $O(n^2 \log B)$. The recent L^2 algorithm [20] by Nguyen and Stehlé achieves a factor of (δ, ν) arbitrarily close to $(1, 1/2)$ in faster polynomial time: the complexity is $O(n^5(n + \log B) \log B)$ which is essentially $O(n^5 \log^2 B)$ for large entries. This is the fastest LLL-type reduction algorithm known for large entries.

HKZ REDUCTION. A basis $[\mathbf{b}_1, \ldots, \mathbf{b}_n]$ of a lattice L is Hermite-Korkine-Zolotarev (HKZ) reduced if it is size-reduced and if $\mathbf{b}_i^*$ is a shortest vector of the projected lattice $\pi_i(L)$ for all $1 \leq i \leq n$. In particular, the first basis vector is a shortest vector of the lattice. Schnorr introduced in [24] a constant to globally measure the drop of the $\|\mathbf{b}_i^*\|$ of $2k$-dimensional HKZ bases:

$$\beta_k = \max_{\substack{L \ 2k\text{-dim. lattice} \\ H \text{ HKZ-basis of } L}} \left(\frac{\|\mathbf{h}_1^*\| \times \cdots \times \|\mathbf{h}_k^*\|}{\|\mathbf{h}_{k+1}^*\| \times \cdots \times \|\mathbf{h}_{2k}^*\|} \right)^{\frac{2}{\cdot}}$$

which we rewrite more geometrically as,

$$\beta_k = \max_{\substack{L \ 2k\text{-dim. lattice} \\ H \text{ HKZ-basis of } L}} \left(\frac{\mathrm{vol}(\mathbf{h}_1, \ldots, \mathbf{h}_k)}{\mathrm{vol}(\pi_{k+1}(\mathbf{h}_{k+1}), \ldots, \pi_{k+1}(\mathbf{h}_{2k}))} \right)^{\frac{2}{\cdot}}$$

Schnorr proved that $\beta_k \leq 4k^2$, and Ajtai recently proved in [1] that there exists $\varepsilon > 0$ such that $\beta_k \geq k^\varepsilon$, but this is an existential lower bound: no explicit value of ε is known. The value of β_k is very important to bound the output quality of Schnorr's algorithm. One can achieve an n-dimensional HKZ reduction in essentially the same time as finding the shortest vector of an n-dimensional lattice: the deterministic algorithm [24] needs $n^{O(n)}$ polynomial operations, and the probabilistic algorithm [2] needs $2^{O(n)}$ polynomial operations.

3   Revisiting Schnorr's Algorithm

In this section, we give an intuitive description of Schnorr's semi block-$2k$-reduction algorithm and show that it is very similar to LLL. The analogy between LLL and Schnorr's algorithm is summarized in Tables 2 and 1. We explain the relationship between the constant β_k and the quality of Schnorr reduced basis, and we give the main ideas for its complexity analysis. Here, we assume that the lattice dimension n is a multiple of k.

3.1   From LLL to Schnorr

In the LLL algorithm, vectors are considered two by two. At each loop iteration, the 2-dimensional lattice $L_i = [\pi_i(\mathbf{b}_i), \pi_i(\mathbf{b}_{i+1})]$, is partially reduced (through a swap) in order to decrease $\|\mathbf{b}_i^*\|$ by at least some geometric factor. When all L_i are almost reduced, every ratio $\|\mathbf{b}_i^*\| \, / \, \|\mathbf{b}_{i+1}^*\|$ is roughly less than $\gamma_2 = \sqrt{\frac{4}{3}}$, which is Siegel's condition [6].

Schnorr's semi block-$2k$-reduction is a polynomial-time block generalization of the LLL algorithm, where the vectors $\mathbf{b}_i^*$ are "replaced" by k-dimensional blocks $S_i = [\pi_{ik-k+1}(\mathbf{b}_{ik-k+1}), \ldots, \pi_{ik-k+1}(\mathbf{b}_{ik})]$ where $1 \le i \le \frac{n}{k}$. The analogue of the 2-dimensional L_i in LLL are the $2k$-dimensional large blocks $L_i = [\pi_{ik-k+1}(\mathbf{b}_{ik-k+1}), \ldots, \pi_{ik-k+1}(\mathbf{b}_{ik+k})]$ where $1 \le i \le \frac{n}{k} - 1$. The link between the small blocks $S_1, \ldots, S_{n/k}$ and the large blocks $L_1, \ldots, L_{n/k-1}$ is that S_i consists of the first k vectors of L_i, while S_{i+1} is the projection of the last k vectors of L_i over $\mathrm{span}(S_i)^\perp$. As a result, $\mathrm{vol}(L_i) = \mathrm{vol}(S_i) \times \mathrm{vol}(S_{i+1})$.

Table 1. Analogy between LLL and Schnorr's algorithm

LLL	Schnorr's semi block-$2k$ reduction
1: **while** $i \le n$ **do**	1: **while** $i \le n/k$ **do**
2: Size-reduce $\mathbf{b}_i$	2a: HKZ-reduce S_i, do the transformations on the basis vectors, not just on the projections
	2b: Size-reduce $\mathbf{b}_{ik-k+1}, \ldots, \mathbf{b}_{ik}$.
3: $B' \leftarrow$copy of B	3: $B' \leftarrow$copy of B
4: Try to decrease $\|\mathbf{b}_i^*\|$ in B':	4: Try to decrease $\mathrm{vol}(S_i)$ in B':
4a: • by swap of $(\mathbf{b}_i, \mathbf{b}_{i+1})$	4a: • by swap of $(\mathbf{b}_{ik}, \mathbf{b}_{ik+1})$
	4b: • by HKZ reducing L_i
5: **if** $\|\mathbf{b}_i^*\|$ can lose a factor δ **then**	5: **if** $\mathrm{vol}(S_i)$ can lose a factor $\frac{1}{(1+\varepsilon)}$ **then**
6: • perform the changes on B	6: • perform the changes on B
7: • $i \leftarrow \max(i-1,1)$	7: • $i \leftarrow \max(i-1,1)$
8: **else** $i \leftarrow i+1$	8: **else** $i \leftarrow i+1$
9: **endwhile**	9: **endwhile**

Formally, a basis is semi-block-$2k$-reduced if the following three conditions hold for some small $\varepsilon > 0$:

$$B \text{ is LLL-reduced} \tag{2}$$

$$\text{For all } 1 \le i \le \frac{n}{k}, \quad S_i \text{ is HKZ-reduced} \tag{3}$$

$$\text{For all } 1 \le i < \frac{n}{k}, \quad \left(\frac{\mathrm{vol}(S_i)}{\mathrm{vol}(S_{i+1})}\right)^2 \le (1+\varepsilon)\beta_k^k \tag{4}$$

Like in LLL, the large block L_i is reduced at each loop iteration in order to decrease $\mathrm{vol}(S_i)$ by a geometric factor $1/(1 + \varepsilon)$. Note that $\mathrm{vol}(S_i)/\mathrm{vol}(S_{i+1})$ decreases by $1/(1+\varepsilon)^2$. By definition of β_k, this ratio can be made smaller than $\beta_k^{k/2}$ if L_i is HKZ-reduced. For this reason, condition (4) is a block generalization of Siegel's condition which can be fulfilled by an HKZ-reduction of L_i.

Table 2. Comparison between LLL and Schnorr's semi block-$2k$ reduction

Algorithm	LLL	Schnorr's Semi-$2k$ reduction
Upper bound on $\|\mathbf{b}_1\|/\mathrm{vol}(L)^{\frac{1}{n}}$	$\approx \left(\frac{4}{3}\right)^{\frac{n}{4}}$	$\approx \beta_k^{\frac{n}{4k}}$
Upper bound on $\|\mathbf{b}_1\|/\lambda_1(L)$	$\approx \left(\frac{4}{3}\right)^{\frac{n}{2}}$	$\approx \beta_k^{\frac{n}{2k}}$
time	Poly(size of basis)	Poly(size of basis)*HKZ($2k$)
small block S_i	$\pi_i(\mathbf{b}_i) = \mathbf{b}_i^*$	$[\pi_{ik-k+1}(\mathbf{b}_{ik-k+1}),$ $\ldots, \pi_{ik-k+1}(\mathbf{b}_{ik})]$
large block L_i	$[\pi_{i-1}(\mathbf{b}_{i-1}), \pi_{i-1}(\mathbf{b}_i)]$	$[\pi_{ik-2k+1}(\mathbf{b}_{ik-2k+1}),$ $\ldots, \pi_{ik-2k+1}(\mathbf{b}_{ik})]$
size of small block	1	k
size of large block	2	$2k$
Quantity to upper bound	$\|\mathbf{b}_i^*\|/\|\mathbf{b}_{i+1}^*\|$	$\mathrm{vol}(S_i)/\mathrm{vol}(S_{i+1})$
Method	Reduce L_i by size-reduction and swap	HKZ reduce L_i
Potential	$\prod_{i=1}^{n} \|\mathbf{b}_i^*\|^{2(n-i)}$	$\prod_{i=1}^{\frac{n}{k}} \mathrm{vol}(S_i)^{2(\frac{n}{k}-i)}$

3.2   Complexity Analysis

Each time a large block L_i is reduced, $\mathrm{vol}(S_i)$ decreases by a geometric factor $1/(1+\varepsilon)$ and since $\mathrm{vol}(L_i) = \mathrm{vol}(S_i) \times \mathrm{vol}(S_{i+1})$ remains constant, $\mathrm{vol}(S_{i+1})$ increases by the same factor. So the integer quantity $\prod_{i=1}^{n/k} \mathrm{vol}(S_i)^{2(\frac{n}{k}-i)}$ decreases by $1/(1+\varepsilon)^2$. This can occur at most a polynomial number of times: hence the complexity of the reduction is Poly(size of basis)*HKZ($2k$) where HKZ($2k$) is the complexity of a $2k$-dimensional HKZ reduction as seen in Section 2.2. In order to ensure a polynomial complexity, it is necessary to keep $k \leq \log n / \log\log n$ or $k \leq \log n$ if we use the probabilistic AKS algorithm.

3.3   The Hermite Factor of Schnorr's Reduction

The Hermite factor of a semi block-$2k$-reduced basis depends mostly on Condition (4), which implies that $\mathrm{vol}(S_1) \lesssim \beta_k^{\frac{n}{4}} \mathrm{vol}(L)^{k/n}$ because $\mathrm{vol}(L) = \prod_{i=1}^{n/k} \mathrm{vol}(S_i)$. If the first vector $\mathbf{b}_1$ is the shortest vector of S_1 (which is implied by (3)), then $\|\mathbf{b}_1\| \leq \sqrt{\gamma_k}\mathrm{vol}(S_1)^{\frac{1}{k}}$ by definition of Hermite's constant, and therefore:

$$\|\mathbf{b}_1\|/\mathrm{vol}(L)^{1/n} \lesssim \sqrt{\gamma_k}\beta_k^{\frac{n}{4k}}$$

3.4   The Approximation Factor of Schnorr's Reduction

If only condition (4) holds, even if $\mathbf{b}_1$ is the shortest vector of S_1, its norm can be arbitrarily far from the first minimum of L. Indeed, consider for instance the 6-dimensional lattice generated by $\mathrm{Diag}(1, 1, 1, 1, \varepsilon, \frac{1}{\varepsilon})$, and a blocksize $k = 3$. Then the first block S_1 is the identity and is therefore HKZ-reduced. The volume of the two blocks S_1 and S_2 is 1, thus condition (4) holds. But the norm of the first vector ($\|\mathbf{b}_1\| = 1$) is arbitrarily far from the shortest vector $\|\mathbf{b}_5\| = \varepsilon$.

Compared to Hermite's factor, we require additionally that every block S_k is reduced (which follows from condition (2)) to bound the approximation factor.

Using (1), there exists an index p such that $\|\mathbf{b}_p^*\| \leq \lambda_1(L)$. Let $a = \lfloor (p-1)/k \rfloor$, (so that the position p is inside the block S_{a+1}). Since B is LLL-reduced, $\mathrm{vol}(S_a) \lesssim \frac{4}{3}^{\frac{(3\cdot -1)\cdot}{4}} \lambda_1(L)^k$, so the approximation factor is bounded by:

$$\|\mathbf{b}_1\|/\lambda_1(L) \lesssim \sqrt{\gamma_k} \frac{4}{3}^{\frac{(3\cdot -1)}{4}} \beta_k^{\frac{\cdots -2}{2}}$$

Note however that Schnorr proved in [24] that Condition (3) allows to decrease the term $(4/3)^{(3k-1)/4}$ to $O\left(k^{2+\ln k}\right)$.

4 Rankin's Constant and Schnorr's Algorithm

4.1 Rankin's Constant

If L is a n-dimensional lattice and $1 \leq m \leq n$, the Rankin invariant $\gamma_{n,m}(L)$ is defined as (cf. [23]):

$$\gamma_{n,m}(L) = \min_{\substack{\mathbf{x}_1, \ldots, \mathbf{x}_m \in L \\ \mathrm{vol}(\mathbf{x}_1, \ldots, \mathbf{x}_m) \neq 0}} \left(\frac{\mathrm{vol}(\mathbf{x}_1, \ldots, \mathbf{x}_m)}{\mathrm{vol}(L)^{m/n}} \right)^2$$

which can be rewritten as:

$$\gamma_{n,m}(L) = \min_{\substack{S \text{ sublattice of } L \\ \dim S = m}} \left(\frac{\mathrm{vol}(S)}{\mathrm{vol}(L)^{m/n}} \right)^2$$

Rankin's constant is the maximum $\gamma_{n,m} = \max \gamma_{n,m}(L)$ over all n-dimensional lattices. Clearly, $\gamma_{n,n}(L) = 1$ and $\gamma_{n,1}(L) = \gamma_n(L)$, so $\gamma_{n,n} = 1$ and $\gamma_{n,1} = \gamma_n$. Rankin's constants satisfy the following three relations, which are proved in [16,23]:

$$\forall n \in \mathbb{N}, \ \gamma_{n,1} = \gamma_n \tag{5}$$

$$\forall n, m \text{ with } m < n \ \gamma_{n,m} = \gamma_{n,n-m} \tag{6}$$

$$\forall r \in [m+1, n-1], \ \gamma_{n,m} \leq \gamma_{r,m}(\gamma_{n,r})^{m/r} \tag{7}$$

The only known values of Rankin's constants are $\gamma_{4,2} = \frac{3}{2}$, which is reached for the $\mathbb{D}_4$ lattice, and those corresponding to the nine Hermite constants known. In the definition of $\gamma_{n,m}(L)$, the minimum is taken over sets of m linearly independent vectors of L, but we may restrict the definition to primitive sets of L or pure sublattices of L, since for any sublattice S of L, there exists a pure sublattice S_1 of L with $\mathrm{span}(S) = \mathrm{span}(S_1)$ and $\mathrm{vol}(S)/\mathrm{vol}(S_1) = [S : S_1]$. If $\mathrm{vol}(S)$ is minimal, then $[S : S_1] = 1$ so $S = S_1$ is pure.

4.2   Relation Between Rankin's Constant and Schnorr's Constant

Theorem 1. *For all $k \geq 1$, $(\gamma_{2k,k})^{2/k} \leq \beta_k$.*

Proof. Let $B = [\mathbf{b}_1, \ldots, \mathbf{b}_{2k}]$ be a HKZ-reduced basis of a lattice L, and let $h(B) = \left(\|\mathbf{b}_1^*\|^2 \times \cdots \times \|\mathbf{b}_k^*\|^2 \right) / \left(\|\mathbf{b}_{k+1}^*\|^2 \times \cdots \times \|\mathbf{b}_{2k}^*\|^2 \right)$. By definition of β_k, $h(B) \leq \beta_k^k$. On the other hand, $h(B)$ is also equal to $\left(\mathrm{vol}(\mathbf{b}_1, \ldots, \mathbf{b}_k)/\mathrm{vol}(L)^{1/2} \right)^4$ and therefore: $\gamma_{2k,k}^2(L) \leq h(B)$. Thus $(\gamma_{2k,k}(L))^{2/k} \leq \beta_k$, which completes the proof. □

4.3   Improving the Upper Bound on Schnorr's Constant

The key result of this section is:

Theorem 2. *For all $k \geq 2$, Schnorr's constant β_k satisfies: $\beta_k \leq \left(1 + \frac{k}{2}\right)^{2 \ln 2 + \frac{1}{k}}$. Asymptotically it satisfies $\beta_k \leq \frac{1}{10} k^{2 \ln 2}$.*

Without any change to Schnorr's algorithm, we deduce a much better quality for the output basis than with the former bound $\beta_k \leq 4k^2$, because both the exponent $2 \ln 2 \approx 1.386$ is much lower than 2, and the coefficient $1/2^{2 \ln 2}$ is about 10 times lower than 4. The bounds on the approximation factor and Hermite's factor of Schnorr's algorithm can be raised to the power $\ln 2 \approx 0.69$. The proof uses an easy bound mentioned by Schnorr in [24]:

$$\beta_k \leq \prod_{j=0}^{k-1} \gamma_{k+j+1}^{2/(k+j)} \tag{8}$$

Numerically, it can be verified that the product (8) is $\leq k^{1.1}$ for all $k \leq 100$ (see Figure 2). The bound $\gamma_j \leq 1 + \frac{j}{4}$ combined with an upper bound of the total exponents prove Theorem 2 for all k (see Section A in the appendix).

Surprisingly, we do not know a better upper bound on $(\gamma_{2k,k})^{2/k}$ than that of Theorem 2. The inequality (7) leads exactly to the same bound for Rankin's constant.

4.4   A Lower Bound on Rankin's Constant

In [1], Ajtai showed that $\beta_k \geq k^\varepsilon$ for small size of blocks and for some $\varepsilon > 0$, and presented worst cases for Schnorr's algorithm, which implies that the reduction power of semi block $2k$-reduction is limited. The following result proves an explicit lower bound on Rankin's constant, which suggests (but does not prove) that the approximation factor of any block-reduction algorithm (including Schnorr's semi block $2k$-reduction) based on the LLL strategy is limited.

Theorem 3. *Rankin's constant satisfies $(\gamma_{2k,k})^{\frac{2}{k}} \geq \frac{k}{12}$ for all $k \geq 1$.*

This lower bound also applies to Schnorr's constant β_k because of Theorem 1. Theorem 3 is mainly based on the following lower bound for Rankin's constant proved in [28,3] as a generalization of Minkowski-Hlawka's theorem:

$$\gamma_{n,m} \geq \left(n \frac{\prod_{j=n-m+1}^{n} Z(j)}{\prod_{j=2}^{m} Z(j)} \right)^{\frac{2}{2}}$$

where $Z(j) = \zeta(j)\Gamma(\frac{j}{2})/\pi^{\frac{j}{2}}$, $\Gamma(x) = \int_0^\infty t^{x-1}e^{-t} \cdot dt$ and ζ is Riemann's zeta function: $\zeta(j) = \sum_{p=1}^\infty p^{-j}$. As an application, for $k < 100$, it can be verified numerically that $(\gamma_{2k,k})^{\frac{2}{k}} \geq \frac{k}{9}$. More generally, we first bound $\ln Z(j)$, and we compare it to an integral to get the expected lower bound. The full proof of Theorem 3 is given in Section B of the Appendix.

5   Improving Schnorr's Algorithm

The main subroutine in Schnorr's algorithm tries to solve the following problem: given a $2k$-dimensional lattice L, find a basis $[\mathbf{b}_1, \ldots, \mathbf{b}_{2k}]$ of L such that the two k-dimensional blocks $S_1 = L(\mathbf{b}_1, \ldots, \mathbf{b}_k)$ and $S_2 = \pi_{k+1}(L)$ minimize $\mathrm{vol}(S_1)$, because $\mathrm{vol}(S_1)/\mathrm{vol}(S_2) = \mathrm{vol}(S_1)^2/\mathrm{vol}(L)$ where $\mathrm{vol}(L)$ does not change. In Schnorr's algorithm, the quality of the output basis (which was expressed as a function of β_k in Sections 3.3 and 3.4) essentially depends on the upper bound that can be achieved on the ratio $\mathrm{vol}(S_1)/\mathrm{vol}(S_2)$.

5.1   The Smallest Volume Problem

Rankin's constant and Schnorr's algorithm suggest the *smallest volume problem*: given a n-dimensional lattice L and an integer m such that $1 \leq m \leq n$, find an m-dimensional sublattice S of L such that $\mathrm{vol}(S)$ is minimal, that is, $\mathrm{vol}(S)/\mathrm{vol}(L)^{m/n} = \sqrt{\gamma_{n,m}(L)}$.

If $m = 1$, the problem is simply the shortest vector problem (SVP). If $m = n-1$, the problem is equivalent to the shortest vector problem in the dual lattice. When $(n, m) = (2k, k)$, we call this problem the *half-volume problem*. For any $m \leq n$, the minimality of the volume implies that any solution to this problem is a pure sublattice of L, so one way to solve this problem is to find a basis $[\mathbf{b}_1, \ldots, \mathbf{b}_n]$ such that $\mathrm{vol}(L(\mathbf{b}_1, \ldots, \mathbf{b}_m))$ is minimal.

We say that a basis of a n-dimensional lattice L is *m-Rankin reduced* if its first m vectors solve the smallest volume problem. Note that this is not exactly a basis reduction problem, as any notion of reduction of the basis of S is irrelevant. The only thing that matters is to minimize the volume of S. If we apply the LLL algorithm on a Rankin-reduced basis, the volume of the first m vectors can never increase: this means that LLL swaps never involve the pair $(m, m+1)$, and therefore the output basis is both LLL-reduced and Rankin-reduced. We thus have proved the following lemma:

Lemma 1. *Let L be a n-dimensional sublattice and $1 \leq m \leq n$. There exists an LLL-reduced basis of L which is m-Rankin-reduced.*

Since the number of LLL reduced bases can be bounded independently of the lattice (see [6] because LLL-reduction implies Siegel reduction), the smallest volume problem can be solved by a gigantic exhaustive search (which is constant in fixed dimension though).

5.2   Block-Rankin Reduction

A basis is $2k$-*Block-Rankin reduced* with factor $\delta \in [\frac{1}{2}; 1[$ if it is LLL-reduced
with factor $(\frac{1}{2}, \delta)$ and all the blocks S_i and L_i defined as in Section 3 satisfy:
$\mathrm{vol}(S_i)/\mathrm{vol}(S_{i+1}) \leq \frac{1}{\delta}\gamma_{2k,k}(L_i)$. Compared to Schnorr's semi block-$2k$ reduction,
this reduction notion enables to replace β_k in the bounds of the approximation
factor and Hermite's factor by $\gamma_{2k,k}^{2/k}$.

Assume that an algorithm to k-Rankin-reduce a $2k$-dimensional basis is available. Then it is easy to see that Algorithm 1, inspired from LLL and Schnorr's
semi block-$2k$ reduction, achieves block-Rankin reduction using a polynomial
number of calls to the Rankin subroutine.

Algorithm 1. 2k-block-Rankin reduction

Input: A basis $B = [\mathbf{b}_1, \ldots, \mathbf{b}_n]$ of a lattice and $\delta \in [\frac{1}{2}; 1[$
Output: A semi block $2k$-reduced basis.
  1: $i \leftarrow 1$;
  2: **while** $i \leq n/k$ **do**
  3:     LLL-reduce S_i with factor δ, do the transformations on the basis vectors, not
        just on their projections
  4:     **return** B **if** $i = n/k$.
  5:     $B_{\mathrm{tmp}} \leftarrow B$; k-Rankin reduce L_i in Btmp
  6:     **if** $\mathrm{vol}(S_i)$ in $B_{\mathrm{tmp}} \leq \delta\mathrm{vol}(S_i)$ in B **then**
  7:         $B \leftarrow B_{\mathrm{tmp}}$; $i \leftarrow i - 1$
  8:     **else**
  9:         $i \leftarrow i + 1$
 10:     **end if**
 11: **end while**

5.3   The 4-Dimensional Case

Here, we study Rankin-reduction for $(n, m) = (4, 2)$. We first notice that HKZ
reduction does not necessarily solve the half-volume problem. Consider indeed
the following HKZ-reduced row basis:

$$\begin{bmatrix} 1 & 0 & 0 & 0 \\ 0 & 1 & 0 & 0 \\ 0 & 0 & 1+\varepsilon & 0 \\ 0 & 0 & \frac{1+\varepsilon}{2} & \frac{\sqrt{3}}{2}(1+\varepsilon) \end{bmatrix}$$

The volume ratio $\frac{\|\mathbf{b}_1^*\|\|\mathbf{b}_2^*\|}{\|\mathbf{b}_3^*\|\|\mathbf{b}_4^*\|}$ is equal to $\sqrt{\frac{4}{3}}$. If we swap the two 2-blocks, the
new basis is no longer HKZ-reduced, but the ratio decreases to almost $\sqrt{\frac{3}{4}}$. This
example can easily be generalized to any even dimension, which gives an infinite
family of HKZ bases which do not reach the minimal half-volume.

However, the following lemma shows that Algorithm 2 can efficiently solve the
half-volume problem in dimension 4, given as input an HKZ basis:

Lemma 2. *Let* $(\mathbf{b}_1, ..., \mathbf{b}_4)$ *be an HKZ-reduced basis of a lattice L. To simplify notations, let* λ_1 *and* λ_2 *denote respectively* $\lambda_1(L)$ *and* $\lambda_2(L)$. *For all* $\mathbf{c}_1$ *and* $\mathbf{c}_2$ *in L such that* $\mathrm{vol}(\mathbf{c}_1, \mathbf{c}_2) \leq \mathrm{vol}(\mathbf{b}_1, \mathbf{b}_2)$ *and* $(\mathbf{c}_1, \mathbf{c}_2)$ *is reduced:* $\|\mathbf{c}_1\| \leq \|\mathbf{c}_2\|$ *and* $\|\mathbf{c}_1^*\|^2 \leq \frac{4}{3}\|\mathbf{c}_2^*\|^2$.

1. *Then* $\mathbf{c}_1$ *satisfies:* $\lambda_1^2 \leq \|\mathbf{c}_1\|^2 \leq \frac{4}{3}\lambda_1^2$.
2. *If* $\lambda_2 > \sqrt{\frac{4}{3}}\lambda_1$, *then* $\mathrm{vol}(\mathbf{c}_1, \mathbf{c}_2) = \mathrm{vol}(\mathbf{b}_1, \mathbf{b}_2)$ *given by HKZ reduction.*
3. *Otherwise* $\mathbf{c}_2$ *satisfies* $\|\mathbf{c}_2\|^2 \leq (\frac{4}{3}\lambda_1)^2$.

Proof. Because $\mathbf{c}_1$ belongs to L, $\|\mathbf{c}_1\| \geq \lambda_1$. Since $(\mathbf{b}_1, ..., \mathbf{b}_4)$ is an HKZ basis, the first vector is a shortest vector: $\|\mathbf{b}_1\| = \lambda_1$ and the second vector satisfies $\|\mathbf{b}_2^*\| \leq \lambda_2$, so $\mathrm{vol}(\mathbf{b}_1, \mathbf{b}_2) \leq \lambda_1\lambda_2$. We also know that $\mathrm{vol}(\mathbf{c}_1, \mathbf{c}_2) = \|\mathbf{c}_1\| \cdot \|\mathbf{c}_2\| \cdot \sin(\mathbf{c}_1, \mathbf{c}_2) \geq \frac{\sqrt{3}}{2}\|\mathbf{c}_1\| \cdot \|\mathbf{c}_2\|$ because $[\mathbf{c}_1, \mathbf{c}_2]$ is reduced. Since we have chosen $\|\mathbf{c}_1\| \leq \|\mathbf{c}_2\|$, then $\|\mathbf{c}_2\| \geq \lambda_2$. Thus $\lambda_1\lambda_2 \geq \frac{\sqrt{3}}{2}\|\mathbf{c}_1\| \cdot \lambda_2$, and $\|\mathbf{c}_1\|^2 \leq \frac{4}{3}\lambda_1^2$. If furthermore $\lambda_2 > \sqrt{\frac{4}{3}}\lambda_1$, then necessarily $\mathbf{c}_1 = \pm\mathbf{b}_1$, then the HKZ reduction implies the minimality of $\mathrm{vol}(\mathbf{b}_1, \mathbf{b}_2)$. If $\lambda_2 \leq \sqrt{\frac{4}{3}}\lambda_1$, then $\mathrm{vol}(\mathbf{c}_1, \mathbf{c}_2)^2 = \|\mathbf{c}_1\|^2 \cdot \|\mathbf{c}_2^*\|^2 \leq (\lambda_1\lambda_2)^2$, so $\|\mathbf{c}_2^*\|^2 \leq \lambda_2^2 \leq \frac{4}{3}\lambda_1^2$. And we also have $\|\mathbf{c}_2^*\|^2 \geq \frac{3}{4}\|\mathbf{c}_1^*\|^2$. □

Algorithm 2. 4-dimensional Rankin-reduction

Input: An HKZ reduced basis $[\mathbf{b}_1, \dots, \mathbf{b}_4]$ of a 4-dim lattice
Output: A Rankin-reduced basis $[\mathbf{c}_1, \mathbf{c}_2, \mathbf{c}_3, \mathbf{c}_4]$ minimizing $\mathrm{vol}(\mathbf{c}_1, \mathbf{c}_2)$

1: **if** $\|\mathbf{b}_2^*\| > \sqrt{\frac{4}{3}}\|\mathbf{b}_1\|$ **then**
2:    **return** $(\mathbf{b}_1, \mathbf{b}_2, \mathbf{b}_3, \mathbf{b}_4)$
3: **end if**
4: $(\mathbf{u}, \mathbf{v}) \leftarrow (\mathbf{b}_1, \mathbf{b}_2)$
5: **for** each lattice vector $\mathbf{c}_1$ shorter than $\sqrt{\frac{4}{3}}\|\mathbf{b}_1\|$ **do**
6:    find the shortest vector $\mathbf{c}_2$ in the lattice projected over $\mathbf{c}_1^{\perp}$ (We can limit the enumeration to $\|\mathbf{c}_2\|$ lower than $\frac{\mathrm{vol}(\mathbf{b}_1, \mathbf{b}_2)}{\|\mathbf{c}_1\|}$).
7:    **if** $\mathrm{vol}(\mathbf{c}_1, \mathbf{c}_2) < \mathrm{vol}(\mathbf{u}, \mathbf{v})$ **then** $(\mathbf{u}, \mathbf{v}) \leftarrow (\mathbf{c}_1, \mathbf{c}_2)$
8: **end for**
9: compute $\mathbf{c}_3$ and $\mathbf{c}_4$ a reduced basis of the lattice projected over $(\mathbf{u}, \mathbf{v})^{\perp}$
10: **return** $(\mathbf{u}, \mathbf{v}, \mathbf{c}_3, \mathbf{c}_4)$

Because the input basis is HKZ-reduced, it is easy to see that the number of vectors $\mathbf{c}_1$ enumerated in Algorithm 2 is bounded by a constant. It follows that the cost of Algorithm 2 is at most a constant times more expensive than a HKZ reduction of a 4-dimensional lattice.

If we plug Algorithm 2 into Algorithm 1, we obtain a polynomial-time reduction algorithm whose provable quality is a bit better than Schnorr's semi block-4 reduction: namely, the constant $\beta_2 \leq \gamma_4^{2/3}\gamma_3 = 2^{2/3} \approx 1.587$ in the approximation factor and the Hermite factor is replaced by the potentially smaller constant $\gamma_{4,2} = 3/2$. On the other hand, both algorithms only apply exhaustive search in dimension 4.

5.4 Higher Blocksizes

The 4-dimensional case suggests two potential improvements over Schnorr's semi block $2k$-algorithm:

- If the half-volume problem can be solved in roughly the same time (or less) than a full $2k$-HKZ reduction, then Algorithm 1 would give potentially better approximation factors at the same cost.
- If one can approximate the half-volume problem in much less time than a full $2k$-HKZ reduction, we may still obtain good approximation factors in much less time than semi block $2k$-reduction, by plugging the approximation algorithm in place of Rankin reduction in Algorithm 1.

However, we do not know how to solve the half-volume problem exactly in dimension higher than 4, without using a gigantic exhaustive search. Perhaps a good approximation can be found in reasonable blocksize, by sampling short (but not necessarily shortest) lattice vectors, and testing random combinations of such vectors.

We now present an approximation algorithm for the half volume problem, which we call *transference reduction*. Transference reduction achieves a volume ratio lower than $\frac{1}{95}k^2$ in a $2k$-dimensional lattice by making only $O(k)$ calls to a k-dimensional exhaustive search, which is thus cheaper than a full $2k$-HKZ reduction. Note that $2k$-HKZ reduction achieves a smaller volume ratio using a $2k$-dimensional exhaustive search. Let $(\mathbf{b}_1, \ldots, \mathbf{b}_{2k})$ be a basis of a $2k$-dimensional lattice. The idea of the algorithm is to perform exhaustive searches in the two halves of the basis in order to find a pair of vectors which can be highly reduced. The reduction of this pair of vectors happens in the middle of the basis so that the first half-volume decreases.

As in the previous sections, we call $S_1 = L(\mathbf{b}_1, \ldots \mathbf{b}_k)$ and $S_2 = L(\pi_{k+1}(\mathbf{b}_{k+1}), \ldots, \pi_{k+1}(\mathbf{b}_{2k}))$. Using an exhaustive search, a shortest vector of S_2 is brought on the $k+1$-th position in order to make $\left\|\mathbf{b}_{k+1}^*\right\|^2 \leq \gamma_k \mathrm{vol}(S_2)^{2/k}$. The algorithm used to perform this exhaustive search in dimension k in a projected lattice is classical. Then a second exhaustive search brings a vector of S_1 maximizing $\|\mathbf{b}_k^*\|$ on the k-th position.

Lemma 3. *Finding a basis $(\mathbf{b}_1, \ldots, \mathbf{b}_k)$ of a k-dimensional lattice S maximizing $\|\mathbf{b}_k^*\|$ reduces to finding a shortest vector of the dual lattice $S^\times$.*

Proof. The vector $\mathbf{u} = \mathbf{b}_k^* / \|\mathbf{b}_k^*\|^2$ is the last vector of the dual basis. Indeed, $\langle \mathbf{u}, \mathbf{b}_i \rangle = 0$ for $i = 1..k-1$ and $\langle \mathbf{u}, \mathbf{b}_k \rangle = 1$. If $\|\mathbf{b}_k^*\|$ is maximal, then $\mathbf{u}$ is minimal. So a simple reduction is to find a shortest vector $\mathbf{u}_k$ of the dual $S^\times$, extend it into a basis $U = (\mathbf{u}_1, \ldots, \mathbf{u}_k)$ of $S^\times$ and return the dual $U^{-t} = (\mathbf{b}_1, \ldots, \mathbf{b}_k)$. □

After maximizing $\|\mathbf{b}_k^*\|$, Hermite's inequality in the reversed dual of S_1 implies that $1/\|\mathbf{b}_k^*\|^2 \leq \gamma_k / \mathrm{vol}(S_1)^{2/k}$. At this point, the ratio $(\mathrm{vol}(S_1)/\mathrm{vol}(S_2))^{2/k}$ is lower than $\gamma_k^2 \|\mathbf{b}_k^*\|^2 / \|\mathbf{b}_{k+1}\|^2$. If the middle-vectors pair $(\pi_k(\mathbf{b}_k), \pi_k(\mathbf{b}_{k+1}))$ does not satisfy Lovász condition, then it is fully reduced and the algorithm

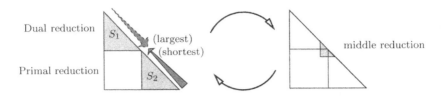

Fig. 1. Transference reduction

starts over from the beginning. The only step which can change the ratio $\text{vol}(S_1)/\text{vol}(S_2)$ is the middle-vectors reduction, in which case it drops by a geometric factor. Hence the number of swaps in the middle is at most linear in the dimension and the size of the basis. In the end, the ratio $(\text{vol}(S_1)/\text{vol}(S_2))^{2/k}$ is lower than $\frac{4}{3}\gamma_k^2$ (or $\frac{1}{12}k^2$).

The constant $4/3$ in this ratio can be reduced to almost 1 by adding further reduction conditions. Let $\hat{S}_1 = L(\mathbf{b}_1, \ldots, \mathbf{b}_{k+1})$ and $\hat{S}_2 = L(\pi_k(\mathbf{b}_k), \ldots, \pi_k(\mathbf{b}_{2k}))$ the widened blocks of S_1, S_2 and $\delta : \frac{1}{2} \leq \delta < 1$ a relaxing parameter. After minimizing $\|\mathbf{b}_{k+1}^*\|$ in S_2 and maximizing $\|\mathbf{b}_k^*\|$ in S_1 the following steps are performed: Using the third exhaustive search, a shortest vector of $\hat{S}_2$ is found. Only if the squared size of this shortest vector is smaller than $\delta \|\mathbf{b}_k^*\|^2$ this vector is brought on the k-th position and the algorithm starts over with minimizing $\|\mathbf{b}_{k+1}^*\|$ in S_2 and maximizing $\|\mathbf{b}_k^*\|$ in S_1. Otherwise the fourth exhaustive search in the dual of $\hat{S}_1$ checks if $\|\mathbf{b}_{k+1}^*\|^2$ approximates the maximized solution by the factor δ. If this condition does not hold, $\mathbf{b}_{k+1}$ is replaced by the maximized solution and the algorithm starts over from the beginning. Each of these two reduction steps decrease $\text{vol}(S_1)^2$ by the factor δ, therefore the number of steps is still bounded by $O(k)$. In case both conditions hold the algorithm stops. For these new conditions we can again apply Hermite's inequality resulting in $\delta \cdot \|\mathbf{b}_k^*\|^2 \leq \gamma_{k+1}\text{vol}(\hat{S}_2)^{2/(k+1)}$ and $\delta \cdot 1/\|\mathbf{b}_{k+1}^*\|^2 \leq \gamma_{k+1}/\text{vol}(\hat{S}_1)^{2/(k+1)}$. It follows:

$$\left(\text{vol}(\hat{S}_1)/\text{vol}(\hat{S}_2)\right)^{2/(k+1)} \leq \gamma_{k+1}^2/\delta^2 \cdot \|\mathbf{b}_{k+1}^*\|^2 / \|\mathbf{b}_k^*\|^2 .$$

Because of $\text{vol}(\hat{S}_1) = \text{vol}(S_1) \cdot \|\mathbf{b}_{k+1}^*\|$, $\text{vol}(\hat{S}_2) = \|\mathbf{b}_k^*\| \cdot \text{vol}(S_2)$ this inequality can be transformed to $(\text{vol}(S_1)/\text{vol}(S_2))^{2/k} \leq (\frac{\gamma_{k+1}}{\delta})^{2 \cdot \frac{k+1}{k}} \cdot \|\mathbf{b}_{k+1}^*\|^2 / \|\mathbf{b}_k^*\|^2$. Combining this with inequality $(\text{vol}(S_1)/\text{vol}(S_2))^{2/k} \leq \gamma_k^2 \|\mathbf{b}_k^*\|^2 / \|\mathbf{b}_{k+1}^*\|^2$ obtained after the first two exhaustive searches, the ratio $(\text{vol}(S_1)/\text{vol}(S_2))^{2/k}$ is lower than $\gamma_k(\gamma_{k+1}/\delta)^{(k+1)/k}$ (or $\gamma_k^2(1 + \varepsilon)$ with small ε if δ is near by 1). Asymptotically, Hermite's constants satisfy $\gamma_k \leq \frac{1.744k}{2\pi e}(1 + o(1))$, so this extended transference reduction provides a ratio $(\text{vol}(S_1)/\text{vol}(S_2))^{2/k}$ lower than $\frac{1}{95}k^2$.

If we use transference reduction instead of Rankin reduction in Algorithm 1, we obtain a reduction algorithm making only $(k + 1)$-dimensional exhaustive searches of a shortest vector, and providing an Hermite factor $\|\mathbf{b}_1\|/\text{vol}(L)^{1/n} \lesssim$

Table 3. Comparison between and Schnorr's semi block-$2k$ reduction and Transference reduction. (Here, SVP$(k+1)$ denotes the cost of finding the shortest lattice vector in dimension $k+1$).

Algorithm	Semi-$2k$ reduction	Transference reduction
Upper bound on $\|\mathbf{b}_1\|/\mathrm{vol}(L)^{\frac{1}{n}}$	$\approx \beta_k^{\frac{n}{4k}} \lesssim k^{n\ln 2/2k}$	$\approx \gamma_k^{\frac{n}{2k}} \lesssim k^{n/2k}$
Upper bound on $\|\mathbf{b}_1\|/\lambda_1(L)$	$\approx \beta_k^{\frac{n}{2k}} \lesssim k^{n\ln 2/k}$	$\approx \gamma_k^{\frac{n}{k}} \lesssim k^{n/k}$
Cost	Poly(size of basis) *HKZ($2k$)	Poly(size of basis) *k*SVP$(k+1)$
Reduction of large blocks	HKZ-reduction	Transference reduction

$\sqrt{\gamma_k}\gamma_k^{n/2k}$ and an approximation factor $\|\mathbf{b}_1\|/\lambda_1(L) \lesssim \gamma_k^{-\frac{3}{2}}\frac{4}{3}^{\frac{(3\cdot-1)}{4}}\gamma_k^{\frac{\cdot}{\cdot}}$. These factors are asymptotically not as good as in the semi block-$2k$ reduction, but the exhaustive searches of transference reduction are much cheaper and thus allow to use a larger k. Interestingly the Hermite factor is essentially $\gamma_k^{n/2k}$, which means that the resulting algorithm may roughly be viewed as an algorithmic version of Mordell's inequality [16]: $\gamma_n \leq \gamma_k^{\frac{\cdot-1}{\cdot-1}}$. Similarly, LLL could be viewed as the algorithmic version of Hermite's inequality $\gamma_n \leq \gamma_2^{n-1}$, which is the particular case $k=2$ of Mordell's inequality.

Acknowledgements. Part of this work, as well as a visit of the second author to the ENS, were supported by the Commission of the European Communities through the IST program under contract IST-2002-507932 ECRYPT. We would like to thank Renaud Coulangeon for useful conversations.

References

1. M. Ajtai. The worst-case behavior of Schnorr's algorithm approximating the shortest nonzero vector in a lattice. In *Proceedings of the Thirty-Fifth Annual ACM Symposium on Theory of Computing*, pages 396–406 (electronic), New York, 2003. ACM.

2. M. Ajtai, R. Kumar, and D. Sivakumar. A sieve algorithm for the shortest lattice vector problem. In *Proc. 33rd STOC*, pages 601–610. ACM, 2001.

3. M. I. Boguslavsky. Radon transforms and packings. *Discrete Appl. Math.*, 111(1-2):3–22, 2001.

4. D. Boneh. Twenty years of attacks on the RSA cryptosystem. *Notices of the AMS*, 46(2):203–213, 1999.

5. D. Boneh and G. Durfee. Cryptanalysis of RSA with private key d less than $N^{0.292}$. In *Proc. of Eurocrypt '99*, volume 1592 of *LNCS*, pages 1–11. IACR, Springer-Verlag, 1999.

6. J. W. S. Cassels. *Rational quadratic forms*, volume 13 of *London Mathematical Society Monographs*. Academic Press Inc. [Harcourt Brace Jovanovich Publishers], London, 1978.

7. H. Cohen. *A Course in Computational Algebraic Number Theory*. Springer-Verlag, 1995. Second edition.

8. J. Conway and N. Sloane. *Sphere Packings, Lattices and Groups*. Springer-Verlag, 1998. Third edition.

9. D. Coppersmith. Small solutions to polynomial equations, and low exponent RSA vulnerabilities. *J. of Cryptology*, 10(4):233–260, 1997. Revised version of two articles from Eurocrypt '96.

10. M. Grötschel, L. Lovász, and A. Schrijver. *Geometric algorithms and combinatorial optimization*, volume 2 of *Algorithms and Combinatorics: Study and Research Texts*. Springer-Verlag, Berlin, 1988.

11. C. Hermite. Extraits de lettres de M. Hermite à M. Jacobi sur différents objets de la théorie des nombres, deuxième lettre. *J. Reine Angew. Math.*, 40:279–290, 1850. Also available in the first volume of Hermite's complete works, published by Gauthier-Villars.

12. N. A. Howgrave-Graham and N. P. Smart. Lattice attacks on digital signature schemes. *Des. Codes Cryptogr.*, 23(3):283–290, 2001.

13. J. L. Lagrange. Recherches d'arithmétique. *Nouveaux Mémoires de l'Académie de Berlin*, 1773.

14. A. K. Lenstra, H. W. Lenstra, Jr., and L. Lovász. Factoring polynomials with rational coefficients. *Mathematische Ann.*, 261:513–534, 1982.

15. L. Lovász. *An Algorithmic Theory of Numbers, Graphs and Convexity*, volume 50. SIAM Publications, 1986. CBMS-NSF Regional Conference Series in Applied Mathematics.

16. J. Martinet. *Les réseaux parfaits des espaces euclidiens*. Masson, Paris, 1996.

17. D. Micciancio and S. Goldwasser. *Complexity of lattice problems*. The Kluwer International Series in Engineering and Computer Science, 671. Kluwer Academic Publishers, Boston, MA, 2002. A cryptographic perspective.

18. J. Milnor and D. Husemoller. Symmetric bilinear forms. *Math. Z*, 1973.

19. P. Q. Nguyen and I. E. Shparlinski. The insecurity of the digital signature algorithm with partially known nonces. *J. Cryptology*, 15(3):151–176, 2002.

20. P. Q. Nguyen and D. Stehlé. Floating-point LLL revisited. In *Proc. of Eurocrypt '05*, volume 3494 of *LNCS*, pages 215–233. IACR, Springer-Verlag, 2005.

21. P. Q. Nguyen and J. Stern. The two faces of lattices in cryptology. In *Proc. of CALC '01*, volume 2146 of *LNCS*. Springer-Verlag, 2001.

22. A. M. Odlyzko. The rise and fall of knapsack cryptosystems. In *Proc. of Cryptology and Computational Number Theory*, volume 42 of *Proc. of Symposia in Applied Mathematics*, pages 75–88. AMA, 1989.

23. R. A. Rankin. On positive definite quadratic forms. *J. London Math. Soc.*, 28:309–314, 1953.

24. C. P. Schnorr. A hierarchy of polynomial lattice basis reduction algorithms. *Theoretical Computer Science*, 53:201–224, 1987.

25. C. P. Schnorr. A more efficient algorithm for lattice basis reduction. *J. of algorithms*, 9(1):47–62, 1988.

26. C. P. Schnorr and M. Euchner. Lattice basis reduction: improved practical algorithms and solving subset sum problems. *Math. Programming*, 66:181–199, 1994.

27. C. P. Schnorr and H. H. Hörner. Attacking the Chor-Rivest cryptosystem by improved lattice reduction. In *Proc. of Eurocrypt '95*, volume 921 of *LNCS*, pages 1–12. IACR, Springer-Verlag, 1995.

28. J. L. Thunder. Higher-dimensional analogs of Hermite's constant. *Michigan Math. J.*, 45(2):301–314, 1998.

A   Proof of Theorem 2

The right-hand product (8) of Hermite's constants can be bounded using the absolute upper bound $\gamma_j \leq (1+j)/4$ by: $\beta_k^k \leq \prod_{j=0}^{k-1}(1+\frac{k+j+1}{4})^{\frac{2\cdot}{\cdot+\cdot}}$.

$$\beta_k^k \leq \left(1+\frac{k}{2}\right)^{\sum_{\cdot=0}^{\cdot-1}\frac{2}{1+\cdots}} \cdot \prod_{j=0}^{k-1}\left(\frac{1+\frac{k+j+1}{4}}{1+\frac{k}{2}}\right)^{\frac{\cdot}{\cdot+\cdot}}.$$

The first sum can be compared with an integral:

$$\frac{1}{k}\sum_{j=0}^{k-1}\frac{2}{1+j/k} \leq \frac{1}{k}+2\int_0^1\frac{1}{1+x}dx,$$

and the last product is always smaller than 1: more precisely, its asymptotical equivalent is $\exp\left(-k(\ln 2/2)^2\right) \approx 0.887^k$. Hence we obtain the absolute upper bound: $\beta_k^k \leq (1+\frac{k}{2})^{1+2k\cdot\ln 2}$ or $\beta_k \leq (1+\frac{k}{2})^{1/k+2\ln 2}$.

If we use the best asymptotical bound known for Hermite's constant $\gamma_j \leq \frac{1.744n}{2\pi e}(1+\circ(1))$, we obtain using the same argument, the asymptotical upper bound:

$$\beta_k \leq \exp(-k(\ln 2/2)^2)\left(\frac{1.744k}{\pi e}\right)^{1/k+2\ln 2} \leq \frac{1}{10}k^{2\ln 2}.$$

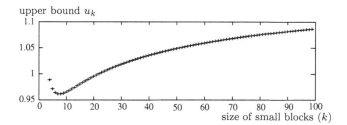

This curve shows the numerical upper bound u_k of $\ln\prod_{j=0}^{k-1}\gamma_{k+j+1}^{2/(k+j)}/\ln k$, obtained by using the exact values of Hermite's constant γ_i for $1 \leq i \leq 8$ and $i = 24$, and the bound $\gamma_i \leq 1+\frac{i}{4}$ elsewhere. Thus $u_k \leq 1.1$ for $1 \leq k \leq 100$.

Fig. 2. "Exponents" of the upper-bound on β_k

B   Proof of Theorem 3

The Stirling equivalent of Γ satisfies $0 \leq \ln(\Gamma(x+1)) - \ln\left((\frac{x}{e})^x\sqrt{2\pi x}\right) \leq \frac{K}{x}$ where $K < 0.0835$ is a constant. Since the function $k \rightarrow k^{-j}$ is decreasing for $j \geq 2$, we may compare its integral with ζ, and we deduce the following bound:

$$\zeta(j) \leq 1+\frac{1}{2^j}+\frac{1}{(j+1)2^{j+1}}$$

Combining these two relations, we obtain the following upper bound for $Z(j)$:

$$\ln(Z(j)) \le \frac{j}{2} \ln \frac{j}{2} - \frac{j}{2}(\ln \pi + 1) + \rho(j)$$

where $\rho(j) = \left(1 - \ln\left(\frac{j}{2} - 1\right) + \frac{1}{2} \ln\left(2\pi\left(\frac{j}{2} - 1\right)\right) + \frac{K}{\left(\frac{j}{2} - 1\right)}\right) + \left(\frac{1}{2^j} + \frac{1}{(j+1)2^{j+1}}\right)$.
For $j > 13$, we have $\rho(j) < 0$, therefore it can be removed from the upper bound.

$$\forall j \ge 13, \ \ln(Z(j)) \le \frac{j}{2} \ln \frac{j}{2} - \frac{j}{2}(\ln \pi + 1)$$

The lower bound is a consequence of Stirling's formula and the relation $1 \le \zeta(j)$.

$$\forall j \ge 13, \ \left(\frac{j}{2} - 1\right) \ln\left(\frac{j}{2} - 1\right) - \frac{j}{2}(\ln(\pi) + 1) \le \ln(Z(j)) \le \frac{j}{2} \ln \frac{j}{2} - \frac{j}{2}(\ln \pi + 1)$$

We now use the upper bound and an integral to bound the denominator $\prod_{j=2}^{k} Z(j)$:

$$\sum_{j=2}^{k} \ln(Z(j)) \le \sum_{j=2}^{13} \ln(Z(j)) + \int_{14}^{k+1} \left(\frac{t}{2} \ln \frac{t}{2} - \frac{t}{2}(\ln \pi + 1)\right) dt$$

$$\le \frac{(k+1)^2}{4} \left(\ln(k+1) - \ln \pi - \frac{1}{2} - \ln 2\right) + c$$

where $c = \sum_{j=2}^{13} \ln(Z(j)) - \frac{225}{4}(3 \ln 2 - \ln \pi - \frac{1}{2} - \ln 2)$. And we apply the lower bound on the numerator $\prod_{j=n-m+1}^{n} Z(j)$:

$$\sum_{j=k+1}^{2k} \ln(Z(j)) \ge \int_{k}^{2k} \left(\left(\frac{t}{2} - 1\right) \ln\left(\frac{t}{2} - 1\right) - \frac{t}{2}(\ln(\pi) + 1)\right) dt$$

$$\ge k^2 \left(\ln(k-1) - \frac{1}{4} \ln(k-2) + \frac{\ln 2}{4} - \frac{9}{8} - \frac{3}{4} \ln \pi\right) + r(k)$$

where $r(k) = k \ln(k-2) - 2 \ln(k-1) - \ln 2 + \frac{1}{2}k + \ln 2 - \ln(k-2) + \ln(k-1)$ is equivalent to $r(k) \sim -k \cdot \ln k$. Finally, we obtain a lower bound for $\gamma_{2k,k}$:

$$\ln \gamma_{2k,k}^k \ge \frac{k^2}{2} \ln(k) + \left(\frac{\ln 2}{2} - 1 - \frac{\ln \pi}{2}\right) k^2 + s(k)$$

where $s(k) = r(k) - k^2 \left(\ln\left(\frac{k-1}{k}\right) + \frac{1}{4} \ln\left(\frac{k-2}{k}\right)\right) - \left(\frac{2k+1}{4}\right)(\ln(k+1) - \ln \pi - \frac{1}{2} - \ln 2) - \frac{1}{4}k^2 \ln\left(\frac{k-1}{k}\right) + (-49 \ln 7 + \frac{195}{4} \ln \pi - \ln 2/4 + \frac{585}{8} - \sum_{j=2}^{13} \ln(Z(j)))$. We can show that this function is equivalent to $-\frac{3}{2}k \ln k$, and that for $k > 100$, $|s(k)/k^2| \le 0.06$. As a final step, we multiply the result by $2/k^2$ and apply exponentiation to obtain the bound $(\gamma_{2k,k})^{\frac{2}{k^2}} \ge \frac{k}{12}$ for all $k > 100$. Note that we already had the bound $\frac{k}{9}$ for $k \le 100$ using numerical computation of $Z(j)$. Asymptotically, we have obtained the following lower bound: $(\gamma_{2k,k})^{\frac{2}{k^2}} \ge \frac{2k}{\pi e^2}$.

Lattice-Based Cryptography

Oded Regev*

Tel Aviv University, Israel

Abstract. We describe some of the recent progress on lattice-based cryptography, starting from the seminal work of Ajtai, and ending with some recent constructions of very efficient cryptographic schemes.

1 Introduction

In this survey, we describe some of the recent progress on lattice-based cryptography. What is a lattice? It is a set of points in n-dimensional space with a periodic structure, such as the one illustrated in Figure 1. More formally, given n-linearly independent vectors $v_1, \ldots, v_n \in \mathbb{R}^n$, the lattice generated by them is the set of vectors

$$L(v_1, \ldots, v_n) := \left\{ \sum_{i=1}^{n} \alpha_i v_i \ \middle| \ \alpha_i \in \mathbb{Z} \right\}.$$

The vectors $v_1, \ldots, v_n$ are known as a *basis* of the lattice.

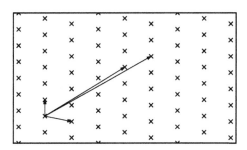

Fig. 1. A lattice in $\mathbb{R}^2$ and two of its bases

Historically, lattices were investigated since the late 18th century by mathematicians such as Lagrange, Gauss, and later Minkowski. More recently, lattices have become an active topic of research in computer science. They are used as an algorithmic tool to solve a wide variety of problems (e.g., [1,2,3]); they have found many applications in cryptanalysis (e.g., [4,5]); and they have some unique

* Supported by an Alon Fellowship, by the Binational Science Foundation, by the Israel Science Foundation, and by the EU Integrated Project QAP.

C. Dwork (Ed.): CRYPTO 2006, LNCS 4117, pp. 131–141, 2006.

properties from a computational complexity point of view (e.g., [6,7,8]). In this survey we will focus on their positive applications in cryptography, i.e., on the construction of cryptographic primitives whose security relies on the hardness of certain lattice problems.

Our starting point is Ajtai's seminal result from 1996 [9]. His surprising discovery was that lattices, which up to that point were used only as tools in cryptanalysis, can actually be used to *construct* cryptographic primitives. His work sparked a great interest in understanding the complexity of lattice problems and their relation to cryptography.

Ajtai's discovery was surprising for another reason: the security of his cryptographic primitive is based on the *worst-case hardness* of lattice problems. What this means is that if one succeeds in breaking the primitive, even with some small probability, then one can also solve *any* instance of a certain lattice problem. This remarkable property is what makes lattice-based cryptographic constructions so attractive. In contrast, virtually all other cryptographic constructions are based on some *average-case* assumption. For example, in cryptographic constructions based on factoring, the assumption is that it is hard to factor numbers chosen from a certain distribution. But how should we choose this distribution? Obviously, we should not use numbers with small factors (such as even numbers), but perhaps there are other numbers that we should avoid? In cryptographic constructions based on worst-case hardness, such questions do not even arise.

There are several other reasons for our interest in lattice-based cryptography. One is that the computations involved are very simple and often require only modular addition. This can be advantageous in certain practical scenarios when encryption is performed by a low-cost device. Another reason is that we currently do not have too many alternatives to traditional number-theoretic-based cryptography such as RSA. Such alternatives will be needed in case an efficient algorithm for factoring integers is ever found. In fact, efficient *quantum* algorithms for factoring integers and computing discrete logarithms already exist [10]. Although large-scale quantum computers are not expected to exist for at least a decade, this fact should already be regarded as a warning. There are currently no known quantum algorithms for lattice problems.

Our choice of topics for this survey is clearly biased by the author's personal taste and familiarity. One notable topic that we will not discuss here are cryptographic constructions following the design of Goldreich, Goldwasser, and Halevi [11]. In particular, this includes the highly efficient constructions developed by the company NTRU [12,13].

For other surveys on the topic, see, e.g., [14,15] and also the lecture notes [16,17]. Another useful resource is the book by Micciancio and Goldwasser [18], which also contains a wealth of information on the computational complexity aspects of lattice problems.

The rest of this survey is organized as follows. In Section 2 we define the shortest vector problem and state some known results. In Section 3 we describe the known constructions of hash functions, starting from Ajtai's work [9]. Then, in Section 4 we describe the known constructions of public key cryptosystems. The only technical part of this survey is Section 5, where we outline the construction of a lattice-based collision resistant hash function together with its security proof. We end with some open questions in Section 6.

2   Lattice Problems

The main computational problem associated with lattices is the shortest vector problem (SVP). In SVP, given a lattice basis, we are supposed to output a shortest nonzero vector in the lattice. In fact, we will be mostly interested in the approximation variant of this problem, where our goal is to output a nonzero lattice vector whose norm is greater than that of the shortest nonzero lattice vector by at most some approximation factor γ. There are other interesting lattice problems (such as SIVP), and roughly speaking, the goal in most of them is to find short vectors under some appropriate definition of 'short'. We will encounter one such problem in Section 5. The behavior of these problems is often very similar to that of SVP, so for simplicity we do not discuss them in detail here (see [18] for more details).

Part of the difficulty of SVP comes from the fact that a lattice has many different bases and that typically, the given lattice basis contains very long vectors, much longer than the shortest nonzero vector. In fact, the well-known polynomial time algorithm of Lenstra, Lenstra, and Lovász (LLL) [1] from 1982 achieves an approximation factor of $2^{O(n)}$ where n is the dimension of the lattice. As bad as this might seem, this algorithm is surprisingly useful, with applications ranging from factoring polynomials over the rational numbers, integer programming, and many applications in cryptanalysis (such as attacks on knapsack based cryptographic systems and special cases of RSA). In 1987, Schnorr presented an improved algorithm obtaining an approximation factor that is slightly subexponential, namely $2^{O(n(\log \log n)^2 / \log n)}$. This was recently improved to $2^{O(n \log \log n / \log n)}$ [19]. We should also mention that if one insists on an exact solution to SVP (or even just an approximation to within poly(n) factors), the best algorithm has a running time of $2^{O(n)}$ [19].

Given the above results, one might expect SVP to be NP-hard to approximate to within very large factors. However, the best known result only shows that approximating SVP to within factors $2^{(\log n)^{\frac{1}{2}-\epsilon}}$ is NP-hard (under randomized quasi-polynomial time reductions) [8]. Moreover, SVP is not believed to be NP-hard to approximate to within factors above $\sqrt{n/\log n}$ [20,6,7], since for such approximation factors it lies in classes such as NP ∩ coNP.

On the practical side, it is difficult to say what is the dimension n beyond which solving SVP becomes infeasible with today's computing power. A reasonable guess would be that taking n to be several hundreds makes the problem extremely difficult.

To conclude, the problem of approximating SVP to within polynomial factors n^c for $c \geq \frac{1}{2}$ seems to be very difficult (best algorithm runs in exponential time), however it is not believed to be NP-hard.

3   Hash Functions

As mentioned above, the first lattice-based cryptographic construction with
worst-case security guarantees was presented in the seminal work of Ajtai [9].
More precisely, Ajtai presented a family of one-way functions whose security
is based on the worst-case hardness of n^c-approximate SVP for some constant
$c > 0$. In other words, he showed that being able to invert a function chosen
from this family with non-negligible probability implies the ability to solve *any*
instance of n^c-approximate SVP. Shortly after, Goldreich et al. [21] improved
on Ajtai's result by constructing a stronger cryptographic primitive known as a
family of *collision resistant hash functions*. Much of the subsequent work concentrated on decreasing the constant c (thereby improving the security assumption)
[22,23,24]. In the most recent work, the constant is essentially $c = 1$ [24]. The
hash function in all these constructions is essentially the modular subset sum
function. We will see an example of such a construction in Section 5 below.

We remark that all these constructions are based on the worst-case hardness
of a problem not believed to be NP-hard. Although it seems unlikely, it is not
entirely impossible that further improvements in these constructions would lead
us to approximation factors of the form n^c for c strictly below $\frac{1}{2}$. That would
mean that we managed to base the security on the worst-case hardness of a
problem that might be NP-hard.

The constructions described above are not too efficient. For instance, $\tilde{O}(n^2)$
bits are necessary in order to specify a function in the family, where n is the
dimension of the lattice underlying the security and the $\tilde{O}$ hides poly-logarithmic
factors (in other words, the key size is $\tilde{O}(n^2)$ bits). So if, for instance, we choose
n to be several hundreds, we might need roughly a megabyte just to specify the
hash function. Recently, an improved construction was presented by Micciancio
[25]. He gives a family of one-way functions where only $\tilde{O}(n)$ bits are needed to
specify a function in the family. Its security is based on the worst-case hardness
of lattice problems on a restricted set of lattices known as cyclic lattices. Since
no better algorithms are known for this family, it is reasonable to assume that
solving lattice problems on these lattices is as hard as the general case. Finally,
in more recent work [26,27], the hash function of [25] was modified, preserving
the efficiency and achieving the stronger security property of collision resistance.

4   Public-Key Cryptography

Following Ajtai's discovery of lattice-based hash functions, Ajtai and Dwork [28]
constructed a *public-key cryptosystem* whose security is based on the worst-case
hardness of a lattice problem. Several improvements were given in subsequent
works [29,30]. Unlike the case of hash functions, the security of these cryptosystems is based on the worst-case hardness of a special case of SVP known
as unique-SVP. Here, we are given a lattice whose shortest nonzero vector is
shorter by some factor γ than all other nonparallel lattice vectors, and our goal
is to find a shortest nonzero lattice vector. The hardness of this problem is not

understood as well as that of SVP, and it is a very interesting open question whether one can base public-key cryptosystems on the (worst-case) hardness of SVP.

As is often the case in lattice-based cryptography, the cryptosystems themselves have a remarkably simple description (most of the work is in establishing their security). For example, let us describe the cryptosystem from [30]. Let N be some large integer. The private key is simply an integer h chosen randomly in the range $[\sqrt{N}, 2\sqrt{N})$. The public key consists of $m = O(\log N)$ numbers $a_1, \ldots, a_m$ in $\{0, 1, \ldots, N-1\}$ that are 'close' to integer multiples of N/h (notice that h doesn't necessarily divide N). We also include in the public key an index $i_0 \in [m]$ such that a_{i_0} is close to an *odd* multiple of N/h. We encrypt one bit at a time. An encryption of the bit 0 is the sum of a random subset of $\{a_1, \ldots, a_m\}$ reduced modulo N. An encryption of the bit 1 is done in the same way except we add $\lfloor a_{i_0}/2 \rfloor$ to the result before reducing modulo N. On receiving an encrypted word w, we consider its remainder on division by N/h. If it is small, we decrypt 0 and otherwise we decrypt 1. To establish the correctness of the decryption procedure, notice that since $a_1, \ldots, a_m$ are all close to integer multiples of N/h, any sum of a subset of them is also close to a multiple of N/h and hence encryptions of 0 are decrypted correctly. Similarly, since $\lfloor a_{i_0}/2 \rfloor$ is far from a multiple of N/h, encryptions of 1 are also far from multiples of N/h and hence we again decrypt correctly. The proof of security is more difficult and we omit it here (but see Section 5 for a related proof).

The aforementioned lattice-based cryptosystems are unfortunately quite inefficient. It turns out that when we base the security on lattices of dimension n, the size of the public key is $\tilde{O}(n^4)$ and each encrypted bit gets blown up to $\tilde{O}(n^2)$ bits. So if, for instance, we choose n to be several hundreds, the public key size is on the order of several gigabytes, which clearly makes the cryptosystem impractical.

Two recent works by Ajtai [31] and by the author [32] have tried to remedy this. Both works present cryptosystems whose public key scales like $\tilde{O}(n^2)$ (or even $\tilde{O}(n)$ if one can set up a pre-agreed random string of length $\tilde{O}(n^2)$) and each encrypted bit gets blown up to $\tilde{O}(n)$ bits. Combined with a very simple encryption process (involving only modular additions), this makes these two cryptosystems a good competitor for certain applications.

However, the security of these two cryptosystems is not as strong as that of other lattice-based cryptosystems. The security of Ajtai's cryptosystem [31] is based on a problem by Dirichlet, which is not directly related to any standard lattice problem. Moreover, his system has no worst-case hardness as the ones previously mentioned. However, his system, as well as many details in its proof of security, have the flavor of a lattice-based cryptosystem, and it might be that one day its security will be established based on the worst-case hardness of lattice problems.

The second cryptosystem [32] is based on the worst-case *quantum* hardness of the SVP. What this means is that breaking the cryptosystem implies an efficient quantum algorithm for approximating SVP. This security guarantee is incomparable to the one by Ajtai and Dwork: On one hand, it is stronger as it is based on the general SVP and not the special case of unique-SVP. On the other hand, it is weaker as it only implies a *quantum* algorithm for a lattice problem. Since no quantum algorithm is known to outperform classical algorithms for lattice problems, it is not unreasonable to conjecture that lattice problems are hard even quantumly. Moreover, it is possible that a more clever proof of security could establish the same worst-case hardness under a classical assumption. Finally, let us emphasize that the cryptosystem itself is entirely classical, and is in fact somewhat similar to the one of [30] described above.

5   An Outline of a Construction

In this section, we outline a construction of a lattice-based family of collision resistant hash functions. We will follow a simplified description of the construction in [24], without worrying too much about the exact security guarantee achieved.[1]

At the heart of the proof is the realization that by adding a sufficient amount of Gaussian noise to a lattice, one arrives at a distribution that is extremely close to uniform. An example of this effect is shown in Figure 2. This technique first appeared in [30], and is based on the work of Banaszczyk [33]. Let us denote by $\eta = \eta(L)$ the least amount of Gaussian noise required in order to obtain a distribution whose statistical distance to uniform is negligible (where by 'amount' we mean the standard deviation in each coordinate). This lattice parameter was analyzed in [24] where it was shown that it is 'relatively short' in the sense that finding nonzero lattice vectors of length at most $\mathrm{poly}(n)\eta$ is a 'hard' lattice problem as it automatically implies a solution to other, more standard, lattice problems such as an approximation to SVP to within polynomial factors.[2]

Before going on, we need to explain what exactly we mean by 'adding Gaussian noise to a lattice'. One way to rigorously define this is to consider the uniform distribution on all lattice points inside some large cube and then add Gaussian noise to this distribution. While this approach works just fine, it leads to some unnecessary technical complications due to the need to deal with the edges of the cube. Instead, we choose to take a mathematically cleaner approach (although it might be confusing at first): we work with the quotient $\mathbb{R}^n/L$. More explicitly, we define a function $h : \mathbb{R}^n \to [0,1)^n$ as follows. Given any $x \in \mathbb{R}^n$, write it as a linear combination of the lattice basis vectors $x = \sum_{i=1}^n \beta_i v_i$, and define $h(x) = (\beta_1, \ldots, \beta_n) \bmod 1$. So for instance, all points in L are mapped to $(0, \ldots, 0)$. Then the statement about the Gaussian noise above can be formally stated as follows: if we sample a point x from a Gaussian distribution in $\mathbb{R}^n$ centered around 0 of standard deviation η in each coordinate, then the statistical distance between the distribution of $h(x)$ and the uniform distribution on $[0,1)^n$ is negligible.

---

[1] A more careful analysis of the construction described below shows that its security can be based on the worst-case hardness of $\tilde{O}(n^{1.5})$-approximate SIVP, which implies a security based on $\tilde{O}(n^{2.5})$-approximate SVP using standard reductions. In order to obtain the best known factor of $\tilde{O}(n)$, one needs to use an iterative procedure.

[2] To be precise, we need slightly more than just finding vectors of length at most $\mathrm{poly}(n)\eta$; we need to be able to find n linearly independent vectors of this length. As it turns out, by repeatedly calling the procedure described below, one can obtain such vectors.

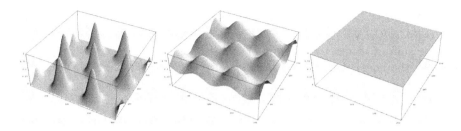

Fig. 2. A lattice with different amounts of Gaussian noise

We now turn to the construction. Our family of hash functions is the modular subset-sum function over $\mathbb{Z}_q^n$, as defined next. Fix $q = 2^{2n}$ and $m = 4n^2$. For each $a_1, \ldots, a_m \in \mathbb{Z}_q^n$, the family contains the function $f_{a_1, \ldots, a.} : \{0,1\}^m \to \{0,1\}^{n \log q}$ given by

$$f_{a_1, \ldots, a.} (b_1, \ldots, b_m) = \sum_{i=1}^m b_i a_i \bmod q.$$

Notice that with our choice of parameters, $m > n \log q$ so collisions are guaranteed to exist. Clearly, these functions are easy to compute. Our goal is therefore to show that they are collision resistant. We establish this by proving that if there exists a polynomial-time algorithm COLLISIONFIND that given $a_1, \ldots, a_m$ chosen uniformly from $\mathbb{Z}_q^n$, finds with some non-negligible probability $b_1, \ldots, b_m \in \{-1, 0, 1\}$, not all zero, such that $\sum_{i=1}^m b_i a_i = (0, \ldots, 0) \pmod{q}$, then there is a polynomial-time algorithm that finds vectors of length at most $\mathrm{poly}(n)\eta$ in any given lattice L (which, as mentioned before, implies a solution to approximate SVP).

Our first observation is that from COLLISIONFIND we can easily construct another algorithm, call it COLLISIONFIND$'$, that performs the following task: given elements $a_1, \ldots, a_m$ chosen uniformly from $[0,1)^n$, it finds with some non-negligible probability $b_1, \ldots, b_m \in \{-1, 0, 1\}$, not all zero, such that $\sum_{i=1}^m b_i a_i \in [-\frac{m}{q}, \frac{m}{q}]^n \pmod{1}$. In other words, it finds a $\{-1, 0, 1\}$ combination of $a_1, \ldots, a_m$ that is extremely close to $(0, \ldots, 0)$ modulo 1. To see this, observe that COLLISIONFIND$'$ can simply apply COLLISIONFIND to $\lfloor qa_1 \rfloor, \ldots, \lfloor qa_m \rfloor$.

Our goal now is to show that using COLLISIONFIND$'$ we can find vectors of length at most $\mathrm{poly}(n)\eta$ in any given lattice L. So let L be some lattice given by its basis $v_1, \ldots, v_n$. Our first step is to apply the LLL algorithm to $v_1, \ldots, v_n$. This makes sure that $v_1, \ldots, v_n$ are not 'unreasonably' long: namely, none of these vectors is longer than $2^n \eta$.

We now arrive at the main part of the procedure. We first choose m vectors $x_1, \ldots, x_m$ independently from the Gaussian distribution in $\mathbb{R}^n$ centered around 0 of standard deviation η in each coordinate. (To be precise, we don't know η, but we can obtain a good enough estimate by trying a polynomial number of values.) Next, we compute $a_i = h(x_i)$ for $i = 1, \ldots, m$. By the discussion above, we know that each a_i is distributed essentially uniformly on $[0,1)^n$. We can therefore apply COLLISIONFIND$'$ to $a_1, \ldots, a_m$ and obtain with non-negligible probability $b_1, \ldots, b_m \in \{-1, 0, 1\}$ such that $\sum_{i=1}^{m} b_i a_i \in [-\frac{m}{q}, \frac{m}{q}]^n \pmod 1$. Now consider the vector $y = \sum_{i=1}^{m} b_i x_i$. On one hand, this is a short vector, as it is the sum of at most m vectors of length roughly $\sqrt{n}\eta$ each. On the other hand, by the linearity of h, we have that $h(y) \in [-\frac{m}{q}, \frac{m}{q}]^n \pmod 1$. What this means is that y is extremely close to a lattice vector. Indeed, write $y = \sum \beta_i v_i$ for some reals $\beta_1, \ldots, \beta_m$. Then we have that each β_i is within $\pm\frac{m}{q}$ of an integer. Consider now the *lattice vector* $y' = \sum \lfloor \beta_i \rceil v_i$ obtained by rounding each β_i to the nearest integer. Then the distance between y and y' is

$$\|y - y'\| \le \sum_{i=1}^{n} \frac{m}{q}\|v_i\| \le \frac{mn}{q} 2^n \eta \ll \eta$$

and in particular we found a lattice vector y' of length at most $\text{poly}(n)\eta$. The procedure can now output y', which is a short lattice vector.

So are we done? Well, not completely: we still have to show that y' is *nonzero* (with some non-negligible probability). The proof of this requires some effort, so we just give the main idea. Recall that we define y' as a (rounding of a) $\{-1, 0, 1\}$ combination of $x_1, \ldots, x_m$ obtained by calling COLLISIONFIND' with $a_1, \ldots, a_m$. The difficulty in proving that $y' \ne 0$ is that we have no control over COLLISIONFIND', and in particular it might act in some 'malicious' way, trying to set the $b_1, \ldots, b_m$ so that y' ends up being the zero vector. To solve this issue, one can prove that the a_i do not contain enough information about the x_i. In other words, conditioned on any fixed values to the a_i, the x_i still have enough uncertainty in them to guarantee that no matter what COLLISIONFIND' outputs, y' is nonzero with very high probability.

To conclude, we have seen that by a single call to the collision finder, one can find in *any* given lattice, a nonzero vector of length at most $(m\sqrt{n} + 1)\eta = O(n^{2.5}\eta)$ with some non-negligible probability. Obviously, by repeating this a polynomial number of times, we can obtain such a vector with very high probability. The essence of the proof, and what makes possible the connection between the average-case collision finding problem and the worst-case lattice problem, is the realization that all lattices look the same after adding a small amount of noise — they turn into a uniform distribution.

6   Open Questions

- **Cryptanalysis:** Attacks on lattice-based cryptosystems, such as the one by Nguyen and Stern [34], seem to be limited to low dimensions (a few tens). Due to the greatly improved efficiency of the new cryptosystems in [31,32], using much higher dimensions has now become possible. It would be very interesting to see attempts to attack these new cryptographic constructions.
- **Improved cryptosystems:** As we have seen in Section 4, the situation with lattice-based cryptosystems is not entirely satisfactory: The original construction of Ajtai and Dwork, as well as some of the follow-up work, are based on the hardness of the unique-SVP and are moreover quite inefficient.

Two recent attempts [31,32] give much more efficient constructions, but with less-than-optimal security guarantees. Other constructions, such as the one by NTRU [12], are extremely efficient but have no provable security. A very interesting open question is to obtain efficient lattice-based cryptosystems based on the worst-case hardness of unique-SVP (or preferably SVP). Another interesting direction is whether specific families of lattices, such as cyclic lattices, can be used to obtain more efficient constructions.

- **Comparison with number theoretic cryptography:** Can one factor integers or compute discrete logarithms using an oracle that solves, say, $\sqrt{n}$-approximate SVP? Such a result would prove that lattice-based cryptosystems are superior to traditional number-theoretic-based ones (see [35,36] for related work).

- **Reverse reductions:** Is the security of lattice-based cryptographic constructions *equivalent* to the hardness of lattice-problems? More concretely, assuming we have an oracle that solves (say) $\sqrt{n}$-approximate SVP, can we break lattice-based cryptography? A result along these lines is known for the Ajtai-Dwork cryptosystem [34], but it is still open if the same can be shown for newer cryptosystems such as the ones in [31,32].

- **Signature schemes:** Lattices have been successfully used in constructing hash functions and public key cryptosystems. Can one also construct signature schemes with worst-case hardness guarantees and similar efficiency? See [37] for some related work.

- **Security against chosen-ciphertext attacks:** The Ajtai-Dwork cryptosystem, as well as all subsequent work, are not secure against chosen-ciphertext attacks. Indeed, it is not too difficult to see that one can extract the private key given access to the decryption oracle. In practice, there are known methods to deal with this issue. It would be interesting to find an (efficient) solution with a rigorous proof of security in the standard model (for related work, see, e.g., [38]).

- **Applications in Learning Theory:** The cryptosystems of [30,32] were recently used by Klivans and Sherstov to obtain cryptographic hardness results for problems in learning theory [39]. It would be interesting to extend this line of research.

Acknowledgements. I am grateful to Ishay Haviv, Julia Kempe, Daniele Micciancio, and Phong Nguyen for many helpful comments.

References

1. Lenstra, A.K., Lenstra, Jr., H.W., Lovász, L.: Factoring polynomials with rational coefficients. Math. Ann. **261** (1982) 515–534
2. Lenstra, Jr., H.W.: Integer programming with a fixed number of variables. Math. Oper. Res. **8** (1983) 538–548
3. Babai, L.: On Lovasz' lattice reduction and the nearest lattice point problem. Combinatorica **6** (1986) 1–13 Preliminary version in STACS 1985.

4. Lagarias, J.C., Odlyzko, A.M.: Solving low-density subset sum problems. J. Assoc. Comput. Mach. **32** (1985) 229–246

5. Coppersmith, D.: Finding small solutions to small degree polynomials. Lecture Notes in Computer Science **2146** (2001) 20–31

6. Goldreich, O., Goldwasser, S.: On the limits of nonapproximability of lattice problems. Journal of Computer and System Sciences **60** (2000) 540–563 Preliminary version in STOC 1998.

7. Aharonov, D., Regev, O.: Lattice problems in NP intersect coNP. Journal of the ACM **52** (2005) 749–765 Preliminary version in FOCS 2004.

8. Khot, S.: Hardness of approximating the shortest vector problem in lattices. In: Proc. 45th Annual IEEE Symp. on Foundations of Computer Science (FOCS). (2004) 126–135

9. Ajtai, M.: Generating hard instances of lattice problems. In: Proc. 28th ACM Symp. on Theory of Computing. (1996) 99–108 Available from ECCC at http://www.uni-trier.de/eccc/.

10. Shor, P.W.: Polynomial-time algorithms for prime factorization and discrete logarithms on a quantum computer. SIAM J. on Computing **26** (1997) 1484–1509

11. Goldreich, O., Goldwasser, S., Halevi, S.: Public-key cryptosystems from lattice reduction problems. In: Advances in cryptology. Volume 1294 of Lecture Notes in Comput. Sci. Springer (1997) 112–131

12. Hoffstein, J., Pipher, J., Silverman, J.H.: NTRU: a ring-based public key cryptosystem. In: Algorithmic number theory (ANTS). Volume 1423 of Lecture Notes in Comput. Sci. Springer (1998) 267–288

13. Hoffstein, J., Graham, N.A.H., Pipher, J., Silverman, J.H., Whyte, W.: NTRUSIGN: Digital signatures using the NTRU lattice. In: Proc. of CT-RSA. Volume 2612 of Lecture Notes in Comput. Sci., Springer-Verlag (2003) 122–140

14. Nguyen, P.Q., Stern, J.: The two faces of lattices in cryptology. In Silverman, J.H., ed.: Cryptography and Lattices, International Conference (CaLC 2001). Number 2146 in Lecture Notes in Computer Science (2001) 146–180

15. Kumar, R., Sivakumar, D.: Complexity of SVP – a reader's digest. SIGACT News **32** (2001) 40–52

16. Micciancio, D.: Lattices in cryptography and cryptanalysis (2002) Lecture notes of a course given in UC San Diego.

17. Regev, O.: Lattices in computer science (2004) Lecture notes of a course given in Tel Aviv University.

18. Micciancio, D., Goldwasser, S.: Complexity of Lattice Problems: A Cryptographic Perspective. Volume 671 of The Kluwer International Series in Engineering and Computer Science. Kluwer Academic Publishers, Boston, Massachusetts (2002)

19. Ajtai, M., Kumar, R., Sivakumar, D.: A sieve algorithm for the shortest lattice vector problem. In: Proc. 33rd ACM Symp. on Theory of Computing. (2001) 601–610

20. Lagarias, J.C., Lenstra, Jr., H.W., Schnorr, C.P.: Korkin-Zolotarev bases and successive minima of a lattice and its reciprocal lattice. Combinatorica **10** (1990) 333–348

21. Goldreich, O., Goldwasser, S., Halevi, S.: Collision-free hashing from lattice problems. Technical Report TR96-056, Electronic Colloquium on Computational Complexity (ECCC) (1996)

22. Cai, J.Y., Nerurkar, A.: An improved worst-case to average-case connection for lattice problems. In: Proc. 38th IEEE Symp. on Found. of Comp. Science. (1997) 468–477

23. Micciancio, D.: Improved cryptographic hash functions with worst-case/average-case connection. In: Proc. 34th ACM Symp. on Theory of Computing (STOC). (2002) 609–618

24. Micciancio, D., Regev, O.: Worst-case to average-case reductions based on Gaussian measures. In: Proc. 45th Annual IEEE Symp. on Foundations of Computer Science (FOCS). (2004) 372–381

25. Micciancio, D.: Generalized compact knapsacks, cyclic lattices, and efficient one-way functions from worst-case complexity assumptions. Computational Complexity (2006) To appear. Preliminary version in ECCC report TR04-095.

26. Lyubashevsky, V., Micciancio, D.: Generalized compact knapsacks are collision resistant. In: 33rd International Colloquium on Automata, Languages and Programming (ICALP). (2006)

27. Peikert, C., Rosen, A.: Efficient collision-resistant hashing from worst-case assumptions on cyclic lattices. In: 3rd Theory of Cryptography Conference (TCC). (2006) 145–166

28. Ajtai, M., Dwork, C.: A public-key cryptosystem with worst-case/average-case equivalence. In: Proc. 29th Annual IEEE Symp. on Foundations of Computer Science (FOCS). (1997) 284–293

29. Goldreich, O., Goldwasser, S., Halevi, S.: Eliminating decryption errors in the Ajtai-Dwork cryptosystem. In: Advances in cryptology. Volume 1294 of Lecture Notes in Comput. Sci. Springer (1997) 105–111

30. Regev, O.: New lattice-based cryptographic constructions. Journal of the ACM **51** (2004) 899–942 Preliminary version in STOC'03.

31. Ajtai, M.: Representing hard lattices with $O(n \log n)$ bits. In: Proc. 37th Annual ACM Symp. on Theory of Computing (STOC). (2005)

32. Regev, O.: On lattices, learning with errors, random linear codes, and cryptography. In: Proc. 37th ACM Symp. on Theory of Computing (STOC). (2005) 84–93

33. Banaszczyk, W.: New bounds in some transference theorems in the geometry of numbers. Mathematische Annalen **296** (1993) 625–635

34. Nguyen, P., Stern, J.: Cryptanalysis of the Ajtai-Dwork cryptosystem. In: Advances in cryptology (CRYPTO). Volume 1462 of Lecture Notes in Comput. Sci. Springer (1998) 223–242

35. Schnorr, C.P.: Factoring integers and computing discrete logarithms via Diophantine approximation. In Cai, J.Y., ed.: Advances in computational complexity. Volume 13 of DIMACS Series in Discrete Mathematics and Theoretical Computer Science. AMS (1993) 171–182 Preliminary version in Eurocrypt '91.

36. Adleman, L.M.: Factoring and lattice reduction (1995) Unpublished manuscript.

37. Micciancio, D., Vadhan, S.: Statistical zero-knowledge proofs with efficient provers: lattice problems and more. In: Advances in cryptology (CRYPTO). Volume 2729 of Lecture Notes in Computer Science., Springer-Verlag (2003) 282–298

38. Dwork, C., Naor, M., Reingold, O.: Immunizing encryption schemes from decryption errors. In: Advances in cryptology (EUROCRYPT). Volume 3027 of Lecture Notes in Comput. Sci. Springer (2004) 342–360

39. Klivans, A., Sherstov, A.: Cryptographic hardness results for learning intersections of halfspaces (2006) Available as ECCC report TR06-057.

A Method for Making Password-Based
Key Exchange Resilient
to Server Compromise*

Craig Gentry[1], Philip MacKenzie[2], and Zulfikar Ramzan[3]

[1] Stanford University, Palo Alto, CA, USA
cgentry@cs.stanford.edu
[2] Google, Inc., Mountain View, CA, USA
philmac@google.com
[3] Symantec, Inc., Redwood City, CA, USA
zulfikar_ramzan@symantec.com

Abstract. This paper considers the problem of password-authenticated key exchange (PAKE) in a client-server setting, where the server authenticates using a stored password file, and it is desirable to maintain some degree of security even if the server is compromised. A PAKE scheme is said to be *resilient to server compromise* if an adversary who compromises the server must at least perform an offline dictionary attack to gain any advantage in impersonating a client. (Of course, offline dictionary attacks should be infeasible in the absence of server compromise.) One can see that this is the best security possible, since by definition the password file has enough information to allow one to play the role of the server, and thus to verify passwords in an offline dictionary attack.

While some previous PAKE schemes have been proven resilient to server compromise, there was no known general technique to take an arbitrary PAKE scheme and make it provably resilient to server compromise. This paper presents a practical technique for doing so which requires essentially one extra round of communication and one signature computation/verification. We prove security in the universal composability framework by (1) defining a new functionality for PAKE with resilience to server compromise, (2) specifying a protocol combining this technique with a (basic) PAKE functionality, and (3) proving (in the random oracle model) that this protocol securely realizes the new functionality.

1 Introduction

THE BASIC PROBLEM. We start by describing the basic problem of setting up a secure channel between two parties, Alice and Bob, who only share a short secret password. Neither of them knows a public key corresponding to the other party, and neither has a certified public key (i.e., a public key whose certificate can be verified by the other party). If Alice and Bob shared a high-strength cryptographic key (i.e., a *long* secret), then this problem could be solved using

---

* This work was carried out while all the authors were at DoCoMo USA Labs.

C. Dwork (Ed.): CRYPTO 2006, LNCS 4117, pp. 142–159, 2006.

standard solutions for setting up a secure channel, such as the protocol of Bellare and Rogaway [5]. However, since Alice and Bob only share a short secret
password, they must also be concerned with *offline dictionary attacks*. An offline
dictionary attack occurs when an attacker obtains some information that can be
used to perform offline verification of password guesses. We will call this *password verification information*. For a specific example, consider the following. Say
Alice and Bob share a password π, and say an attacker somehow obtained a hash
of the password $h(\pi)$, where h is some common cryptographic hash function such
as SHA-1 [41]. Then an attacker could go offline and run through a dictionary
of possible passwords $\{\pi_1, \pi_2, \ldots\}$, testing each one against $h(\pi)$. For instance,
to test if π_i is the correct password, the attacker computes $h(\pi_i)$ and checks if
$h(\pi_i) = h(\pi)$. In general, the password verification information obtained by the
attacker may not be as simple as a hash of a password, and an attacker may not
always be able to test all possible passwords against the password verification
information, but if he can test a significant number of passwords, this is still
considered an offline dictionary attack. For some fairly recent demonstrations of
how effective an offline dictionary attack can be, see [40,44,50]. So the problem
remains: how do Alice and Bob set up a secure channel? In other words: how
do Alice and Bob bootstrap a short secret (the password) into a long secret (a
cryptographic key) that can be used to provide a secure channel?

A protocol to solve this problem is called a *password-authenticated key exchange (PAKE)* protocol. Informally, a PAKE protocol is secure if the only
feasible way to attack the protocol is to run a trivial *online dictionary attack* of
simply iteratively guessing passwords and attempting to impersonate one of the
parties. (Note that this type of attack can generally be detected and stopped by
well-known methods.) The problem of designing a secure PAKE protocol was
proposed by Bellovin and Merritt [7] and by Gong *et al.* [24], and has since
been studied extensively. Below we discuss the many techniques that have been
proposed.

RESILIENCE TO SERVER COMPROMISE. Consider a PAKE protocol run in a
client-server setting, where the client device receives a password input by a user,
but where the server stores a "password file" that contains data that can be
used to authenticate each user. In this scenario it is natural to be concerned
about the security of this password file, since an adversary that compromises
the server could obtain this password file.[1] In the case of many existing PAKE
protocols, the consequences of an adversary obtaining the server's password file
are disastrous, with the adversary obtaining enough information to impersonate
a client. That is why there has been a significant amount of work on making
PAKE schemes "as secure as possible" even if the server gets compromised.
Naturally, if an adversary obtains a server password file, he possesses password
verification information, so he can always mount an offline dictionary attack.
The goal, therefore, in improving resilience to server compromise is to make the
offline dictionary attack the best he can do.

---

[1] From the many recent reports of theft of credit cards and other personal information
from e-commerce servers, it seems that compromise of a server is a real threat.

In the remainder of the paper, a *symmetric PAKE scheme* refers to one in which the two parties use identical strings corresponding to the same password (and which, consequently is trivially insecure in the client-server setting when the server is compromised). An *asymmetric PAKE scheme* refers to one which is designed to maintain security (as discussed above) despite a server compromise. In particular, this implies that the server does not store the plaintext password.

RELATED WORK. Since the PAKE problem was introduced, it has been studied extensively, and many PAKE protocols have been proposed, e.g., [24,23,26,34,48,32]. Many of these protocols have been shown to be insecure [9,45]. More recent protocols, e.g., [3,10,1,37,51,19], have proofs of security, based on certain well-known cryptographic assumptions, in the random oracle and/or ideal cipher models. Other PAKE protocols have been proven secure in the common reference string (CRS) model, e.g., [31,18,29,14]. Finally the PAKE protocols in [20,42] were proven secure based on a general assumption (trapdoor permutations) without any setup assumptions, but with a restriction that concurrent sessions with the same password are prohibited.

The problem of PAKE with resilience to server compromise has also been studied extensively, and many protocols have been proposed, e.g., [8,27,49,33].[2] Some more recent protocols also have proofs of security based on well-known cryptographic assumptions, in the random oracle model, e.g., [10,37,35]. Although these protocols (along with the protocols of [8,27]) are based on symmetric PAKE protocols, and the techniques used to convert the symmetric PAKE protocols into asymmetric PAKE protocols seem somewhat modular, no modular versions were ever presented, and there were no attempts to prove (in a modular way) anything about the techniques themselves. Each asymmetric PAKE protocol was presented in its entirety, and was proven secure from scratch. Note that no protocols for PAKE with resilience to server compromise have yet been proven secure without relying on random oracles.

RESULTS. We were inspired by the PAK-Z protocol from MacKenzie [36], which is essentially the PAK protocol from [10] modified using the "Z-method" to be resilient to server compromise.[3] While the Z-method was claimed to be a general technique, it was only described and analyzed with respect to the PAK protocol. We first show that the general Z-method does not provide resilience to server compromise by exhibiting an attack that exploits any instantiation using discrete-log based signature schemes. Next, we present a new method, called the Ω-method, that fixes the critical flaw in the Z-method. The Ω-method is the first general and modular technique that takes any secure symmetric PAKE scheme as a building block and converts it into one that is resilient to server compromise. The Ω-method is efficient and practical, essentially adding one

---

[2] There has also been work on protecting the server password file using threshold techniques, e.g., [17,28,16,30,38].

[3] Previous to PAK-Z, there were PAK-X and PAK-Y protocols, with their own methods for modifying PAK to be resilient to server compromise.

extra round of communication and one signature generation/verification to the underlying symmetric PAKE scheme.[4]

We prove security in the universal composability (UC) framework [11] (in the random oracle model). A symmetric PAKE functionality $\mathcal{F}_{\mathsf{pwKE}}$ was recently introduced in [14]. Our original plan was to (1) extend $\mathcal{F}_{\mathsf{pwKE}}$ into an asymmetric PAKE functionality $\mathcal{F}_{\mathsf{apwKE}}$, and (2) prove that a protocol based on the Ω-method (which we call the Ω-protocol) securely realizes $\mathcal{F}_{\mathsf{apwKE}}$ in the $\mathcal{F}_{\mathsf{pwKE}}$-hybrid model. This would imply, by the universal composition theorem [11], that the Ω-protocol would be secure when instantiated with any secure symmetric PAKE scheme. For step (1) we added notions of a server setting up a password record for the client and using that password record for each session, the notion of stealing the password file, and the notion of explicitly aborting.[5] Unfortunately, step (2) was problematic, since the Ω-method relies on the notion of a protocol transcript, which does not exist in the symmetric PAKE functionality of [14]. Therefore, we added the notion of a transcript to the symmetric PAKE functionality to make a revised symmetric PAKE functionality $\mathcal{F}_{\mathsf{rpwKE}}$, and completed step (2) using the $\mathcal{F}_{\mathsf{rpwKE}}$-hybrid model. In Section 4 we discuss why adding this notion of a transcript is natural and does not have any substantial effect on whether a protocol securely realizes the functionality.

APPLICABILITY. Currently there is only one PAKE protocol that has been shown to securely realize the symmetric PAKE functionality $\mathcal{F}_{\mathsf{pwKE}}$ in the UC framework, specifically, the one of Canetti *et al.* [14]. However, we conjecture that many of the PAKE protocols cited above that were proven secure in the random oracle model, but not in the UC framework, could also be proven secure in the UC framework. Since the Ω-protocol relies on the random oracle model anyway, it would make sense to combine it with these symmetric PAKE protocols to achieve (very efficient) asymmetric PAKE protocols. Thus the results of this paper should have wide applicability.

2   Preliminaries

SYMMETRIC ENCRYPTION SCHEMES. A *symmetric encryption scheme* $\mathcal{E}$ is a pair (E, D) of algorithms, both running in polynomial time. E takes a symmetric key k and a message m as input and outputs an encryption c for m; we denote this $c \leftarrow E_k(m)$. D takes a ciphertext c and a symmetric key k as input and returns either a message m such that c is a valid encryption of m, if such an m exists, and otherwise returns $\perp$.

---

[4] As a result of this work, the PAK-Z protocol in the IEEE P1363.2 (Password-based Public-Key Cryptography) proposed standard has had the Z-method replaced with the Ω-method to provide resilience to server compromise.

[5] We do not consider the notion of explicitly aborting to be necessary for an asymmetric PAKE functionality, but it is very convenient and allows some natural protocols (including our protocol) to securely realize the functionality. There is more discussion on this notion in Section 4.

We will use specific symmetric encryption schemes based on hash functions and one-time pads.[6] The first scheme is $E_k(m) = H(k) \oplus m$, where $H()$ is a hash function with output that is the same length as m (and assumed to behave like a random oracle - see below) and where $\oplus$ is taken as a bit-wise exclusive OR operation. Note that this encryption scheme is inherently malleable. For instance, given a ciphertext c of an unknown message m under an unknown key k, one can construct a ciphertext $c' = c \oplus 00 \cdots 001$ of a related message $m' = m \oplus 00 \cdots 001$ (i.e., m with the last bit flipped), without determining m.

The second scheme is $E'_k(m) = H(k) \oplus m|H'(m)$, for a hash function $H'()$ (assumed to be one-way and to behave like a random oracle). The second hash protects against malleability since modifying the one-time pad portion requires recomputing the hash with the correct message, implying the message has been determined.

HASH FUNCTIONS. Cryptographic hash functions will be used for key generation and for producing verification values. Here we assume these functions are random oracles [4],[7] i.e., they behave like black-box perfectly random functions. In practice, one would need to verify that the actual hash function used is suitable to be used as a random oracle. See [4] for a discussion on how to instantiate random oracles, and see [25] for a discussion on key generation functions.

SIGNATURE SCHEMES. A *signature scheme* S is a triple (Gen, Sig, Verify) of algorithms, the first two being probabilistic, and all running in (probabilistic) polynomial time. Gen_S takes as input the security parameter (usually denoted as κ and represented in unary format, i.e., 1^κ) and outputs a public-key pair (pk, sk), i.e., $(pk, sk) \leftarrow \text{Gen}_S(1^\kappa)$. Sign takes a message m and a secret key sk as input and outputs a signature σ for m, i.e., $\sigma \leftarrow \text{Sig}_{sk}(m)$. Verify takes a message m, a public key pk, and a candidate signature σ' for m as input and returns the bit $b = 1$ if σ' is a valid signature for m for the corresponding private key, and otherwise returns the bit $b = 0$. That is, $b \leftarrow \text{Verify}_{pk}(m, \sigma')$. Naturally, if $\sigma \leftarrow \text{Sig}_{sk}(m)$, then $\text{Verify}_{pk}(m, \sigma) = 1$. We require signature schemes that are existentially unforgeable against chosen message attacks in the sense of [21].

3   The Ω-Method

BASIC IDEA OF THE Ω-METHOD. Similar to some previous work [10,37,35,36], the Ω-method constructs an asymmetric PAKE protocol Ω by enhancing a symmetric PAKE protocol P as follows. First, the server only stores the output of a one-way function of the password, i.e., a value $f(\pi)$ that is easy to compute from the password, but from which the password is difficult to compute. Then

---

[6] Proving security using generic encryption schemes is left as an open problem.

[7] We stress that whether schemes proven secure in the random oracle model can be instantiated securely in the real world (i.e., with polynomial-time computable functions in place of random oracles) is uncertain [13,12,43,22,6,39].

the protocol operates by first running the protocol P using $f(\pi)$ in place of π,[8] and second having the client somehow prove knowledge of a π such that $f(\pi)$ is the server's stored value.

The Ω-method uses the following specific instantiation of this basic idea.[9] The server stores a hash of the password $H(\pi)$ to be used for the protocol P. The server also stores a public/secret key pair for a secure signature scheme, with the secret key encrypted using the specific encryption scheme (E', D') defined above in which the key to the encryption scheme is the password. (Recall that this encryption scheme is a one-time pad concatenated to a cryptographic hash.) In all, the server stores $(H(\pi), pk, E'_\pi(sk))$. The protocol Ω first runs P (using $H(\pi)$). Once P is finished and has derived a cryptographically strong shared key K, the server uses a temporary session key K' derived from K to securely send $E'_\pi(sk)$ to the client, using the specific encryption scheme (E, D) defined above. (Recall that this encryption scheme is simply a one-time pad.) The client uses K and π to derive the appropriate keys, performs the necessary decryptions to obtain sk, and then creates a signature with sk on the transcript of P. In effect, this proves that the client (the one communicating with the server in protocol P) knows π. The final output of Ω is another key K'' derived from K.

The Ω-method is very similar to the Z-method in [36]. However, the Z-method specifies that both encryption schemes be simple one-time pads, which, as discussed previously, are malleable. Because of this, it can be shown that the Z-method is insecure for certain signature schemes and in particular, for certain representations of the private key generated in certain signature schemes. In fact, the Z-method is not known to be secure for any signature scheme. In contrast, the Ω-method is secure for every signature scheme.

HIGH-LEVEL DESCRIPTION. A high-level description of the Ω-method is shown in Figure 1, with some details given here.

Client Part 1: The client computes $H(\pi)$ and performs its part in the symmetric PAKE protocol P using $H(\pi)$, obtaining a shared cryptographic key K.

Server Part 1: The server performs its part in the symmetric PAKE protocol P using the value $H(\pi)$ it had stored, obtaining a shared cryptographic key K.

Server Part 2, Step 1: The server first derives key $K' = H'(K)$ and then sends $E_{K'}(E'_\pi(sk))$ to the client.

Client Part 2: The client receives the value $E_{K'}(E'_\pi(sk))$ sent by the server, computes $K' = H'(K)$, and decrypts using K' and π to obtain sk. (If the decryption fails when checking the hash, the client aborts.) Then the client signs the transcript of P, using sk, i.e., it computes $\sigma = \mathsf{Sig}_{sk}(\text{transcript})$, and sends

---

[8] We always assume a password protocol takes an arbitrary length password string, and thus would work correctly with $f(\pi)$ in place of π.

[9] Note that for security we require that the hash and encryption functions used by the Ω-method are not used by the underlying symmetric PAKE protocol P.

the result σ to the server. The client also derives a session key K'' from K (e.g., by using a cryptographic hash function $K'' = H''(K)$).

Server Part 2, Step 2: Once it receives the signature σ, the server computes $b = \mathsf{Verify}_{pk}(\text{transcript}, \sigma)$. If $b = 1$, then the server derives a session key $K'' = H''(K)$ and outputs it. Otherwise it aborts.

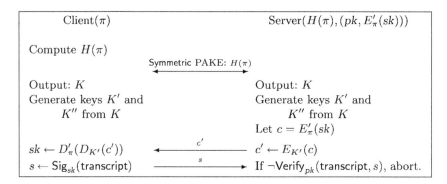

Fig. 1. The Ω-method: Augmenting a PAKE protocol to make it resilient to server compromise

The advantage of the Ω-method over the Z-method was discussed above. The advantage of the Ω-method over other asymmetric PAKE protocols is that it is modular and general. The Ω-method allows an asymmetric PAKE protocol to be constructed using any PAKE protocol and any signature scheme, and these could be chosen based on which cryptographic assumptions are used to prove their security. For instance, one could choose a PAKE protocol and a signature scheme that are based on the same cryptographic assumption. Notice also that as opposed to asymmetric PAKE protocols in which the password is used to derive the secret key of a signature scheme, e.g., [35], the Ω-method has the advantage that the secret key does not need to be computed online by the client, a potentially expensive operation.

As for efficiency, this method adds one extra round of communication along with a few hash operations, a signature calculation by the client, and a signature verification by the server, to the PAKE protocol P. The extra round of communication can often be piggybacked on actual protocol messages, as was shown for the PAK-Z protocol in [36]. But there is still the extra computation involved with the signature. Some asymmetric PAKE protocols have been designed specifically to avoid this extra computation. SRP [49] and AMP [33] are two such protocols, but neither has a proof of security, even in the random oracle model.

THE Z-METHOD AND AN ATTACK. The Z-method [36] looks exactly like the Ω-method, except that the encryption functions E and E' are simply one-time pads. That is, $E_{K'}(c) = K' \oplus c$ and $E'_\pi(sk) = H(\pi) \oplus sk$. One problem with a one-time pad encryption scheme is that in the absence of an explicit integrity-verification mechanism, the resulting ciphertext is malleable. This leads to an attack on the PAKE protocol. In particular, suppose that the Z-method is instantiated using a discrete logarithm based signature scheme (e.g., Schnorr, DSS, El-Gamal, signature schemes based on bilinear pairings, etc.). In a typical instantiation of such a scheme, we would have $(sk, pk) = (x, y)$ be the signing and verification keys respectively, where $y = g^x$ and g is a generator for a (multiplicative) group in which the discrete logarithm problem is believed to be intractable.

Now, consider an active adversary who flips the least significant bit of the server's response c'. When the client decrypts, he will either compute that $sk = x + 1$ (if the least significant bit of x were a 0) or $x - 1$ (or if the least significant bit of x were a 1). The client signs the transcript using the computed signing key. The adversary verifies the signature using yg as the public verification key. If it verifies correctly, then he deduces that the least significant bit of x was 0; otherwise he deduces that it is 1. The adversary can repeat an analogous procedure $|x|$ times to determine the remaining bits of x. Since $|x|$ is typically much smaller than the dictionary of possible passwords, we violate the security requirements of the protocol.

EVALUATION OF SECURITY. We wish to prove that the Ω-method can be combined with any (symmetric) PAKE protocol to yield an asymmetric PAKE protocol. It is often very difficult to prove such general statements. However, by using the Universal Composability (UC) framework of Canetti [11], this type of proof, while still very complicated, becomes much easier. We assume the reader is familiar with the UC framework. We remark that we focus on static adversaries that cannot corrupt parties during the execution.[10] (However, our asymmetric PAKE functionality includes a notion of an adversary stealing a password file.)

4   Password-Based Key Exchange Functionalities

THE ORIGINAL SYMMETRIC PAKE FUNCTIONALITY. We first consider the original symmetric PAKE functionality $\mathcal{F}_{pwKE}$ from Canetti $et\ al.$ [14] and presented in Figure 2.[11] The functionality is similar to that of the standard key exchange functionality $\mathcal{F}_{KE}$ given in Canetti [11]. In the $\mathcal{F}_{KE}$ functionality, the parties that start a session receive the same uniformly-distributed session key, except when one of the parties is corrupted, in which case the adversary has the power to set the session key. In the $\mathcal{F}_{pwKE}$ functionality, each party starting a session is given a password as input from the environment, and the power to set the session key for a party is also given to the adversary if it succeeds in guessing the password used by that party. When the adversary guesses the password, the party's session is marked compromised. An additional property of the definition is that a failed attempt at guessing a party's password is detected. This results

---

[10] Nevertheless, as was shown in [14], this implies the "weak corruption" model of [3] in which passwords can be adaptively obtained.

[11] Note that the variable names in the functionality have been slightly modified for consistency with protocols that we present later.

in the session being marked interrupted and the party receiving an independent uniformly-distributed session key.

A session that is neither compromised nor interrupted (and is still in progress) is considered fresh. Such sessions (between honest parties) conclude with both parties receiving the same, uniformly-distributed session key if they were given the same password as input from the environment, and otherwise with the parties receiving independent uniformly-distributed session keys. In any case, once a party receives a session key, that session is marked completed.

TWO ENHANCEMENTS TO THE ORIGINAL SYMMETRIC PAKE FUNCTIONALITY. We describe two enhancements, transcripts and explicit authentication, to the original $\mathcal{F}_{\mathsf{pwKE}}$ functionality [14].

Transcripts. The main problem with building a secure asymmetric PAKE protocol using the $\mathcal{F}_{\mathsf{pwKE}}$ functionality is that there is no way to indicate to the environment through the $\mathcal{F}_{\mathsf{pwKE}}$ functionality whether the two sessions (one for each party) involved in the key exchange are both fresh.[12] Note that extending the functionality with queries that output the state of each session (i.e., fresh or compromised) is ineffective since a real party would not know individually if its session was fresh or compromised, and thus could not output such an indication.[13] Thus allowing such a query in the ideal world makes it distinguishable from the real world.

Instead, we extend the functionality in the following way. Once a session is complete, the adversary may query the functionality with an extra value tr, which is output to the party as long as it meets the following condition: If either of the two sessions in the key exchange is not fresh, then the tr values output to each party must not be equal.

Since the query is not mandatory, any protocol that securely realizes $\mathcal{F}_{\mathsf{pwKE}}$ will securely realize the functionality extended in this way. We conjecture something stronger, in that any protocol that securely realizes $\mathcal{F}_{\mathsf{pwKE}}$ and in which each party is fresh and outputs a key if the adversary simply forwards messages between two parties in a session, and a party is not fresh if the adversary sends a message not output as the next message by the other party in a session, can be modified to securely realize the extended $\mathcal{F}_{\mathsf{pwKE}}$ functionality, in which each party outputs the tr value immediately after it outputs a key. The modification is to simply have each party output its transcript as tr once it has completed. Assuming the ideal adversary simulates the real protocol by running

---

[12] This seems critical, since after a password file compromise, it seems the adversary could compromise both client and server sessions in any symmetric PAKE functionality using the information obtained from the password file. (According to the $\mathcal{F}_{\mathsf{pwKE}}$ functionality, this would even allow the adversary to set the session keys between himself and each party to be the same.) As a man-in-the middle, he could simply forward the remaining messages between client and server sessions to complete the asymmetric PAKE protocol.

[13] A party may know if its session "succeeded" or not, but both fresh and compromised sessions may be successful.

Functionality $\mathcal{F}_{\mathsf{pwKE}}$

The functionality $\mathcal{F}_{\mathsf{pwKE}}$ is parameterized by a security parameter κ. It interacts with an adversary S and a set of parties via the following queries:

Upon receiving a query (NewSession, $sid, P_i, P_j, \pi, \text{role}$) **from party** P_i:
Send (NewSession, $sid, P_i, P_j, \text{role}$) to S. In addition, if this is the first NewSession query, or if this is the second NewSession query and there is a record (P_j, P_i, π'), then store record (P_i, P_j, π) and mark this record fresh.

Upon receiving a query (TestPwd, sid, P_i, π') **from the adversary** S:
If there is a record of the form (P_i, P_j, π) which is fresh, then do: If $\pi = \pi'$, mark the record compromised and reply to S with "correct guess". If $\pi \neq \pi'$, mark the record interrupted and reply with "wrong guess".

Upon receiving a query (NewKey, sid, P_i, k) **from** S, **where** $|k| = \kappa$:
If there is a record of the form (P_i, P_j, π), and this is the first NewKey query for P_i, then:
- If this record is compromised, or either P_i or P_j is corrupted, then output (sid, k) to player P_i.
- If this record is fresh, and there is a record (P_j, P_i, π') with $\pi' = \pi$, and a key k' was sent to P_j, and (P_j, P_i, π) was fresh at the time, then output (sid, k') to P_i.
- Any other case: pick new random key k' ($|k'| = \kappa$); send (sid, k') to P_i. Either way, mark the record (P_i, P_j, π) as completed.

Fig. 2. The password-based key-exchange functionality $\mathcal{F}_{\mathsf{pwKE}}$

---

a simulated real protocol with the real adversary, the ideal adversary could also output the transcript of a simulated session, and this obviously would preserve indistinguishability between the ideal world and real world.

Explicit Authentication. Although not critical to building an asymmetric PAKE functionality, the $\mathcal{F}_{\mathsf{pwKE}}$ functionality has another limitation in that it does not allow for some common types of explicit authentication. Specifically, a protocol that performs explicit authentication and aborts if the authentication fails, and otherwise sends one more message, will not securely realize the $\mathcal{F}_{\mathsf{pwKE}}$ functionality. Intuitively this is because an ideal adversary could not learn from the functionality whether the explicit authentication should be successful, and so could only guess whether to send the final message. (This would be easily distinguishable from the real protocol.) To allow for this type of authentication, one can add a new "test abort" query that tests whether the authentication would fail, informs the simulator, and informs the environment that the session is aborting in case of an authentication failure. A feature similar to this is mentioned in [14], in which the ideal adversary is notified if passwords for the two parties in a session do not match. This is claimed (rightly so) to weaken the definition, especially since an eavesdropper may not learn this. However, we are interested only in the case where the eavesdropper *does* learn this information, by noticing whether the protocol aborts or not. Thus in some sense we also "weaken" the definition of PAKE, but in a way that makes sense and allows some natural PAKE protocols (secure according to, say, an indistinguishability-based definition of PAKE) to securely realize the (extended) PAKE functionality. As above, since the "test abort" query is not mandatory, any protocol that securely realizes $\mathcal{F}_{pwKE}$ will securely realize the functionality extended in this way; i.e., this extended functionality does not imply a protocol must have explicit authentication, but only allows for it.

THE REVISED SYMMETRIC PAKE FUNCTIONALITY. Figure 3 describes our *revised* symmetric password-based key exchange functionality called $\mathcal{F}_{rpwKE}$ that includes a transcript query as discussed above, but not a test abort query.[14] More specifically, the revised functionality $\mathcal{F}_{rpwKE}$ is exactly $\mathcal{F}_{pwKE}$ (with a minor wording change in NewKey that has no effect on the $\mathcal{F}_{pwKE}$), but with NewTranscript queries added. We note that it is possible to prove that a protocol securely realizing $\mathcal{F}_{rpwKE}$ also is secure according to the definition of [3].

ASYMMETRIC PAKE FUNCTIONALITIES. Now we discuss our asymmetric PAKE functionality $\mathcal{F}_{apwKE}$, which is presented in Figure 4. At a high level, this functionality expands on the $\mathcal{F}_{pwKE}$ functionality to allow a server to store password data, and then use this stored password data for authentication instead of a password received as input from the environment. This functionality also accounts for the possibility that password data may be stolen by the adversary. As discussed above, this allows the adversary to perform an offline dictionary attack.[15] It also allows the adversary to impersonate the server, but not the client. In more detail, the changes are as follows.

- The $\mathcal{F}_{pwKE}$ functionality was a single-session functionality. However, asymmetric PAKE requires that a password file be used across multiple sessions, so we define the $\mathcal{F}_{apwKE}$ functionality as a multiple-session functionality. Note that this cannot be accomplished simply using "composition with joint state" [15] because the functionality itself requires shared state that needs to be maintained between sessions.

- In $\mathcal{F}_{pwKE}$, sessions are started by sending NewSession queries to two parties, including a password in each query. In $\mathcal{F}_{apwKE}$, these queries are replaced with CltSession and SvrSession queries. The CltSession queries include a password, but the SvrSession queries do not. The server password is taken from the password file, which is placed on the server using a StorePWfile query that includes the password. Note that if the server is corrupted when it receives

---

[14] The test abort query is omitted since it is not integral to our result. A test abort query is included in our asymmetric PAKE functionality because it is necessary there to handle explicit aborting in our Ω-protocol. However, the transcript query is omitted there because it is not integral to our result.

[15] Note that because this functionality accounts for each offline password guess individually, it seems to require the random oracle model (or some similar idealized model) to be securely realized.

Functionality $\mathcal{F}_{\mathsf{rpwKE}}$

The functionality $\mathcal{F}_{\mathsf{rpwKE}}$ is parameterized by a security parameter κ. It interacts with an adversary S and a set of parties via the following queries:

Upon receiving a query (NewSession, $sid, P_i, P_j, \pi,$ role) **from party** P_i:
Send (NewSession, $sid, P_i, P_j,$ role) to S. In addition, if this is the first NewSession query, or if this is the second NewSession query and there is a record (P_j, P_i, π'), then record (P_i, P_j, π) and mark this record fresh.

Upon receiving a query (TestPwd, sid, P_i, π') **from the adversary** S:
If there is a record of the form (P_i, P_j, π) which is fresh, then do: If $\pi = \pi'$, mark the record compromised and reply to S with "correct guess". If $\pi \neq \pi'$, mark the record interrupted and reply with "wrong guess".

Upon receiving a query (NewKey, sid, P_i, k) **from** S, **where** $|k| = \kappa$:
If there is a record of the form (P_i, P_j, π) that is not marked completed, then:
- If this record is compromised, or either P_i or P_j is corrupted, then output (sid, k) to player P_i.
- If this record is fresh, and there is a record (P_j, P_i, π') with $\pi' = \pi$, and a key k' was sent to P_j, and (P_j, P_i, π) was fresh at the time, then output (sid, k') to P_i.
- In any other case, pick a new random key sk' of length κ and send (sid, k') to P_i.
Either way, mark the record (P_i, P_j, π) as completed.

Upon receiving a query (NewTranscript, sid, P_i, tr) **from** S:
If there is a record of the form (P_i, P_j, π) that is marked completed, then:
- If (1) there is a record (P_j, P_i, π') for which tuple (transcript, sid, tr'') was sent to P_j, (2) either (P_i, P_j, π) or (P_j, P_i, π') was ever compromised or interrupted, and (3) $tr = tr''$, ignore the query.
- In any other case, send (transcript, sid, tr) to P_i.

Fig. 3. The revised password-based key-exchange functionality $\mathcal{F}_{\mathsf{rpwKE}}$

this query, then the adversary learns the password. However, a trusted initial setup between the client and server is generally assumed, so this would not be a problem.[16]

- In $\mathcal{F}_{\mathsf{apwKE}}$, the adversary may "steal" the server's password data using a StealPWfile query. No actual data is sent to the adversary, but after this the adversary may make queries to test passwords using OfflineTestPwd queries. These queries are "offline" as they do not correspond to any sessions.

The OfflineTestPwd queries may actually be made either before or after the StealPWfile query, but queries made before are not answered until the StealPWfile query is made. Specifically, when a StealPWfile query is made, if

---

[16] If one is concerned about this, then one could possibly change the StorePWfile to contain some data (perhaps a one-way function of the password), along with a way to verify passwords against this data. Our work focuses on the issue of password file compromise, so we did not explore these other issues.

Functionality $\mathcal{F}_{\mathsf{apwKE}}$

The functionality $\mathcal{F}_{\mathsf{apwKE}}$ is parameterized by a security parameter κ. It interacts with an adversary S and a set of parties via the following queries:

Password storage and authentication sessions

Upon receiving a query (StorePWfile, sid, P_i, π) **from party** P_j:
If this is the first StorePWfile query, store password data record (file, P_i, P_j, π) and mark it uncompromised.

Upon receiving a query (CltSession, $sid, ssid, P_j, \pi$) **from party** P_i:
Send (CltSession, $sid, ssid, P_i, P_j$) to S, and if this is the first CltSession query for $ssid$, store session record $(ssid, P_i, P_j, \pi)$ and mark it fresh.

Upon receiving a query (SvrSession, $sid, ssid$) **from party** P_j:
If there is a password data record (file, P_i, P_j, π), then send (SvrSession, $sid, ssid, P_i, P_j$) to S, and if this is the first SvrSession query for $ssid$, store session record $(ssid, P_j, P_i, \pi)$, and mark it fresh.

Stealing password data

Upon receiving a query (StealPWfile, sid) **from adversary** S:
If there is no password data record, reply to S with "no password file". Otherwise, do the following. If the password data record (file, P_i, P_j, π) is uncompromised, mark it as compromised. If there is a tuple (offline, π') stored with $\pi = \pi'$, send π to S, otherwise reply to S with "password file stolen".

Upon receiving a query (OfflineTestPwd, sid, π') **from adversary** S: If there is no password data record, or if there is a password data record (file, P_i, P_j, π) that is marked uncompromised, then store (offline, π'). Otherwise, do: If $\pi = \pi'$, reply to S with "correct guess". If $\pi \neq \pi'$, reply with "wrong guess".

Active session attacks

Upon receiving a query (TestPwd, $sid, ssid, P, \pi'$) **from adversary** S:
If there is a session record of the form $(ssid, P, P', \pi)$ which is fresh, then do: If $\pi = \pi'$, mark the record compromised and reply to S with "correct guess". Otherwise, mark the record interrupted and reply with "wrong guess".

Upon receiving a query (Impersonate, $sid, ssid$) **from adversary** S:
If there is a session record of the form $(ssid, P_i, P_j, \pi)$ which is fresh, then do: If there is a password data record (file, P_i, P_j, π) that is marked compromised, mark the session record compromised and reply to S with "correct guess", else mark the session record interrupted and reply with "wrong guess".

Key Generation and Authentication

Upon receiving a query (NewKey, $sid, ssid, P, k$) **from** S, **where** $|k| = \kappa$:
If there is a record of the form $(ssid, P, P', \pi)$ that is not marked completed, then:
• If this record is compromised, or either P or P' is corrupted, then output $(sid, ssid, k)$ to P.
• If this record is fresh, there is a session record $(ssid, P', P, \pi')$, $\pi' = \pi$, a key k' was sent to P', and $(ssid, P', P, \pi)$ was fresh at the time, then let $k'' = k'$, else pick a random key k'' of length κ. Output $(sid, ssid, k'')$ to P.
• In any other case, pick a random key k'' of length κ and output $(sid, ssid, k'')$ to P.
Finally, mark the record $(ssid, P, P', \pi)$ as completed.

Upon receiving a query (TestAbort, $sid, ssid, P$) **from** S:
If there is a record of the form $(ssid, P, P', \pi)$ that is not marked completed, then:
• If this record is fresh, there is a record $(ssid, P', P, \pi')$, and $\pi' = \pi$, let $b' = \mathsf{succ}$.
• In any other case, let $b' = \mathsf{fail}$.
Send b' to S. If $b' = \mathsf{fail}$, send (abort, $sid, ssid$) to P, and mark record $(ssid, P, P', \pi)$ completed.

Fig. 4. The Asymmetric PAKE functionality $\mathcal{F}_{\mathsf{apwKE}}$

there was a previous OfflineTestPwd query with the correct password, that password is simply returned to $\mathcal{A}$.

We also change the UC framework slightly to allow the queries StealPWfile and OfflineTestPwd to be accounted for by the environment, similar to the way Corrupt queries are accounted for.[17] Specifically, a StealPWfile query by the adversary is not allowed until the environment sends a StealPWfile message to the adversary,[18] and similarly, each OfflineTestPwd query by the adversary is not allowed until the environment sends an OfflineTestPwd message to the adversary. It is easy to see that the composition theorem holds despite these changes.[19]

- In $\mathcal{F}_{apwKE}$, in addition to TestPwd queries, an adversary can also make an Impersonate query to compromise a client session without supplying a password. This will succeed if there has already been a StealPWfile query.

- We add the TestAbort query to $\mathcal{F}_{apwKE}$. The main reason is that our asymmetric PAKE protocol specifies some verifications, where an instance will abort if a verification fails. As discussed above, the $\mathcal{F}_{pwKE}$ would need an extension to handle this type of protocol.

- In contrast to $\mathcal{F}_{rpwKE}$, we did not include a transcript query in $\mathcal{F}_{apwKE}$ – primarily because our proofs did not need it. The query could be added, and our asymmetric PAKE protocol could be modified to output a transcript.

Finally, in $\mathcal{F}_{apwKE}$, we assume that a given sid is unique to a given client/server pair for which the server stores a password. We also assume that a given $(sid, ssid)$ pair is unique to a given session between the client and server corresponding to sid. These assumptions are valid in the UC framework, as discussed in [11].

5   The Ω Protocol and Its Security

We present the Ω-protocol in the UC framework in Figure 6. It uses the revised password-based key exchange functionality $\mathcal{F}_{rpwKE}$ defined in Figure 3 and the random oracle functionality $\mathcal{F}_{RO}$ defined in Figure 5.[20] The random

---

Functionality $\mathcal{F}_{RO}$

The functionality $\mathcal{F}_{RO}$ is parameterized by an (implicit) output length ℓ.

Upon receiving query (Hash, sid, msg) **from any party** P **or adversary** S:
    If a tuple (msg, r) is stored, return r; else, generate $r \xleftarrow{R} \{0,1\}^{\ell}$. Record (msg, r) and return r.

Fig. 5. The random oracle functionality

---

[17] In fact, we could define these queries as Corrupt queries with certain parameters, which are handled by the functionality in certain specific ways, but we felt it was more clear to make them separate queries.

[18] Technically, this is enforced by the "control function" (see [11]).

[19] This is assuming that these messages are based on sid values, and the sid values used in the original and emulating protocol somehow correspond. This is the case, but we need to define it explicitly.

[20] Note in particular that $\mathcal{F}_{rpwKE}$ has no access to $\mathcal{F}_{RO}$, and thus a protocol securely realizing $\mathcal{F}_{rpwKE}$ should have no access to $\mathcal{F}_{RO}$.

UC Asymmetric PAKE Protocol Ω

Setup: This protocol uses a random oracle functionality $\mathcal{F}_{RO}$ and a revised PAKE functionality $\mathcal{F}_{rpwKE}$, as well as an existentially unforgeable signature scheme $S = (\mathsf{Gen}, \mathsf{Sig}, \mathsf{Verify})$.

Password Storage: When P_j is activated using StorePWfile(sid, P_i, w) for the first time, he does the following. He first sends (Hash, $\langle sid, d \rangle, w$) to the $\mathcal{F}_{RO}$ functionality for $d \in \{1, 2\}$, and receives responses r and k_w. He generates a signature key pair $(sk, pk) \leftarrow \mathsf{Gen}(1^\kappa)$. Next he sends (Hash, $\langle sid, 3 \rangle, sk$) to the $\mathcal{F}_{RO}$ functionality and receives response h_{sk}. He computes $c = (k_w \oplus sk, h_{sk})$ and sets file[sid] = (r, c, pk).

Protocol Steps:

1. When P_j receives input (SvrSession, $sid, ssid, P_i$), he obtains r from the tuple stored in file[sid] (aborting if this value is not properly defined), sends (NewSession, $\langle sid, ssid \rangle, P_j, P_i, r, \mathsf{server}$) to the $\mathcal{F}_{rpwKE}$ functionality, and awaits a response.

2. When P_i receives input (CltSession, $sid, ssid, P_j, w$), he sends (Hash, $\langle sid, 1 \rangle, w$) to the functionality $\mathcal{F}_{RO}$ and obtains the response r. He then sends (NewSession, $\langle sid, ssid \rangle, P_i, P_j, r, \mathsf{client}$) to the $\mathcal{F}_{rpwKE}$ functionality and awaits a response.

3. When P_j (who is a server and is awaiting a response from $\mathcal{F}_{rpwKE}$) receives responses ($\langle sid, ssid \rangle, k$) and (transcript, $\langle sid, ssid \rangle, tr$), he does the following. First he sends (Hash, $\langle sid, ssid, d \rangle, k$) for $d \in \{1, 2\}$ to receive responses k' and k'' respectively. Then he retrieves c from the tuple file[sid]. He encrypts $c' = k' \oplus c$ and sends the message (flow-zero, $sid, ssid, c'$) to P_i.

4. When P_i (who is a client and is also awaiting a response from $\mathcal{F}_{rpwKE}$) receives responses ($\langle sid, ssid \rangle, k$) and (transcript, $\langle sid, ssid \rangle, tr$), he sends (Hash, $\langle sid, ssid, d \rangle, k$) for $d \in \{1, 2\}$ to the $\mathcal{F}_{RO}$ functionality, and receives responses k' and k'' respectively.

5. When P_i receives a message (flow-zero, $sid, ssid, c'$) he computes the decryption $c = k' \oplus c'$, and parses $c = (c_1, c_2)$. He then sends (Hash, $\langle sid, 2 \rangle, w$) to the $\mathcal{F}_{RO}$ functionality, and receives the response k_w. He computes $sk = k_w \oplus c_1$, sends (Hash, $\langle sid, 3 \rangle, sk$) to $\mathcal{F}_{RO}$, receives response h_{sk}, and verifies that $h_{sk} = c_2$. If not, he outputs (abort, $\langle sid, ssid \rangle$) and terminates the session. Otherwise he computes $s = \mathsf{Sig}_{sk}(\langle sid, ssid, tr \rangle)$, sends (flow-one, $sid, ssid, s$) to P_j, outputs $(sid, ssid, k'')$, and terminates the session.

6. When P_j receives a message (flow-one, $sid, ssid, s$), he checks that $\mathsf{Verify}_{pk}(\langle sid, ssid, tr \rangle, s) = 1$. If not, he outputs (abort, $\langle sid, ssid \rangle$) and terminates the session. Otherwise he outputs $(sid, ssid, k'')$, and terminates the session.

Stealing the password file: When P_j (who is a server) receives a message (StealPWfile, sid, P_j, P_i), from the adversary $\mathcal{A}$, if file[sid] is defined, P_j sends it to $\mathcal{A}$.

Fig. 6. The UC Asymmetric PAKE Protocol Using the Ω-method

oracle functionality is parameterized by an output length ℓ, and for simplicity, this argument is implicit (albeit different depending on how the random oracle is being used). In particular if the arguments are $\langle sid, 1 \rangle, \langle sid, 3 \rangle, \langle sid, ssid, 2 \rangle$, then the output length $\ell = \kappa$. If the arguments are $\langle sid, 2 \rangle$, then $\ell = |sk|$. Finally, if the arguments are $\langle sid, ssid, 1 \rangle$, then $\ell = |c|$ (i.e., the size of the ciphertext being encrypted).

When applying the Ω-method in the UC framework to obtain the protocol Ω, there were some issues that needed to be addressed.

- In the Ω-protocol, one must use $(sid, ssid, tr)$ in place of the transcript of the symmetric PAKE protocol. This pair $(sid, ssid)$ is unique to a given pair of client/server instances, and these two instances only generate the same tr if they are fresh and use the same password. This is exactly what is needed in the proof of security, in particular, to ensure that a signature produced by a client cannot be used to impersonate a client in another session.
- We had wanted to use an ideal signature functionality instead of an explicit signature scheme. However, this does not seem possible, since the Ω-protocol explicitly encrypts and hashes the signing key, but the ideal signature functionality doesn't have any notion of a secret key.
- Similarly, it does not seem possible to use an ideal secure channel functionality in place of the symmetric encryptions, because keys are generated using a hash function. The ideal secure channel functionality does not have any notion of a secret key.

The following theorem characterizes the security of the Ω protocol and is proven in the full version of this paper.

Theorem 1. *Assume that $\mathcal{S}$ is an existentially unforgeable signature scheme. Then protocol Ω of Figure 6 securely realizes the $\mathcal{F}_{\mathsf{apwKE}}$ functionality in the $\mathcal{F}_{\mathsf{rpwKE}}, \mathcal{F}_{\mathsf{RO}}$-hybrid model, in the presence of static-corruption adversaries.*

References

1. M. Abdalla and D. Pointcheval. Simple Password-Based Encrypted Key Exchange Protocols. In *RSA Conference, Cryptographer's Track*, pp. 191–208, 2005
2. B. Barak, Y. Lindell, and T. Rabin. Protocol initialization for the framework of universal composability. In *Cryptology ePrint Archive*, Report 2004/006, http://eprint.iacr.org/, 2004.
3. M. Bellare, D. Pointcheval, and P. Rogaway. Authenticated key exchange secure against dictionary attacks. In *EUROCRYPT*, pp. 139–155, 2000.
4. M. Bellare and P. Rogaway. Random oracles are practical: A paradigm for designing efficient protocols. In *1st ACM Conference on Computer and Communications Security*, pages 62–73, 1993.
5. M. Bellare and P. Rogaway. Entity authentication and key distribution. In *CRYPTO*, pp. 232–249, 1993.
6. M. Bellare, A. Boldyreva and A. Palacio. An Uninstantiable Random-Oracle-Model Scheme for a Hybrid-Encryption Problem. In *EUROCRYPT*, pp. 171–188, 2004.

7. S. M. Bellovin and M. Merritt. Encrypted key exchange: Password-based protocols secure against dictionary attacks. In *IEEE Symp. on Research in Security and Privacy*, pp. 72–84, 1992.

8. S. M. Bellovin and M. Merritt. Augmented encrypted key exchange: A password-based protocol secure against dictionary attacks and password file compromise. In *1st ACM Conf. on Computer and Communications Security*, pp. 244–250, 1993.

9. D. Bleichenbacher. Personal communication.

10. V. Boyko, P. MacKenzie, and S. Patel. Provably secure password authentication and key exchange using Diffie-Hellman. In *EUROCRYPT*, pp. 156–171, 2000.

11. R. Canetti. Universally Composable Security: A New Paradigm for Cryptographic Protocols. In *Cryptology ePrint Archive*, Report 2000/067. http://eprint.iacr.org/, 2005.

12. R. Canetti, O. Goldreich, and S. Halevi. On the random-oracle methodology as applied to length-restricted signature schemes. In *Theory of Cryptography Conference - TCC*, pp. 40–57, 2004.

13. R. Canetti, O. Goldreich and S. Halevi. The random oracle methodology, revisited. *J. ACM*, 51(4):557–594, 2004.

14. R. Canetti, S. Halevi, J. Katz, Y. Lindell, and P. MacKenzie. Universally-composable password-based key exchange. In *EUROCRYPT*, pp. 404–421, 2005.

15. R. Canetti and T. Rabin. Universal Composition with Joint State In *CRYPTO*, pp. 265–281, 2003.

16. M. Di Raimondo and R. Gennaro. Provably Secure Threshold Password Authenticated Key Exchange. In *EUROCRYPT*, pp. 507–523, 2003.

17. W. Ford and B. S. Kaliski, Jr. Server-assisted generation of a strong secret from a password. In *5th IEEE International Workshop on Enterprise Security*, 2000.

18. R. Gennaro and Y. Lindell. A Framework for Password-Based Authenticated Key Exchange. In *EUROCRYPT*, pp. 524–543, 2003.

19. C. Gentry, P. MacKenzie, and Z. Ramzan. Password Authenticated Key Exchange Using Hidden Smooth Subgroups. In *12th ACM Conf. on Computer and Communications Security*, pp. 299–309, 2005.

20. O. Goldreich and Y. Lindell. Session-Key Generation using Human Passwords Only. In *CRYPTO*, pp. 408–432, 2001.

21. S. Goldwasser, S. Micali, and R. L. Rivest. A digital signature scheme secure against adaptive chosen-message attacks. *SIAM Journal of Computing* 17(2):281–308, April 1988.

22. S. Goldwasser and Y. Tauman Kalai. "On the (In)security of the Fiat-Shamir Paradigm." In *44th IEEE Symp. on Foundations of Computer Science (FOCS)*, pp. 102–115, 2003.

23. L. Gong. Optimal authentication protocols resistant to password guessing attacks. In *8th IEEE Computer Security Foundations Workshop*, pp. 24–29, 1995.

24. L. Gong, T. M. A. Lomas, R. M. Needham, and J. H. Saltzer. Protecting poorly chosen secrets from guessing attacks. *IEEE Journal on Selected Areas in Communications*, 11(5):648–656, June 1993.

25. IEEE Standard 1363-2000, Standard specifications for public key cryptography, 2000.

26. D. Jablon. Strong password-only authenticated key exchange. *ACM Computer Communication Review, ACM SIGCOMM*, 26(5):5–20, 1996.

27. D. Jablon. Extended password key exchange protocols immune to dictionary attack. In *WETICE'97 Workshop on Enterprise Security*, 1997.

28. D. Jablon Password authentication using multiple servers. In *Proc. RSA Conference, Cryptographer's Track, 2001*.

29. S. Jiang and G. Gong. Password based key exchange with mutual authentication. In *Workshop on Selected Areas of Cryptography (SAC)*, 2004.
30. J. Katz, P. MacKenzie, G. Taban, V. Gligor. Two-Server Password-Only Authenticated Key Exchange. In *Applied Cryptography and Network Security, 3rd Intl. Conf. (ACNS 2005)*, pp. 1–16, 2005.
31. J. Katz, R. Ostrovsky, and M. Yung. Practical password-authenticated key exchange provably secure under standard assumptions. In *EUROCRYPT*, pp. 475–494, 2001.
32. C. Kaufmann and R. Perlman. PDM: A New Strong Password-Based Protocol. In *10th Usenix Security Symposium*, 2001.
33. T. Kwon. Authentication and Key Agreement via Memorable Passwords. In *Internet Society Network and Distributed System Security Symposium (NDSS)*, 2001.
34. S. Lucks. Open key exchange: How to defeat dictionary attacks without encrypting public keys. In *Proc. of the Workshop on Security Protocols*, 1997.
35. P. MacKenzie. More Efficient Password-Authenticated Key Exchange. In *RSA Conference, Cryptographer's Track*, pp. 361–377, 2001.
36. P. MacKenzie. The PAK suite: Protocols for password-authenticated key exchange. DIMACS Technical Report 2002-46, October, 2002.
37. P. MacKenzie, S. Patel, and R. Swaminathan. Password authenticated key exchange based on RSA. In *ASIACRYPT*, pp. 599–613, 2000.
38. P. MacKenzie, T. Shrimpton, and M. Jakobsson. Threshold Password-Authenticated Key Exchange. *J. Cryptology*, 19(1):27–66, 2006.
39. U. Maurer, R. Renner, and C. Holenstein. Indifferentiability, Impossibility Results on Reductions, and Applications to the Random Oracle Methodology. In *Theory of Cryptography Conference - TCC*, pp. 21–39, 2004.
40. A. Narayanan and V. Shmatikov. Fast Dictionary Attacks on Passwords Using Time-Space Tradeoff. In *ACM Conf. on Computer and Communications Security (CCS)*, pp. 364–372, 2005.
41. National Institute of Standards and Technology (NIST). Announcing the Secure Hash Standard, FIPS 180-1, U.S. Department of Commerce, April, 1995.
42. M. Nguyen and S. Vadhan. Simpler Session-Key Generation from Short Random Passwords. In *Theory of Cryptography Conference - TCC*, pp. 428–445, 2004.
43. J. B. Nielsen. Separating Random Oracle Proofs from Complexity Theoretic Proofs: The Non-Committing Encryption Case Jesper Buus Nielsen. In *CRYPTO*, pp. 111–126, 2002
44. P. Oechslin. Making a faster cryptanalytic time-memory trade-off. In *CRYPTO*, pp. 617–630, 2003.
45. S. Patel. Number theoretic attacks on secure password schemes. In *IEEE Symposium on Research in Security and Privacy*, pages 236–247, 1997.
46. D. Pointcheval and J. Stern. Security proofs for signature schemes. In *EUROCRYPT*, pp. 387–398, 1996.
47. C. P. Schnorr. Efficient identification and signatures for smart cards. In *CRYPTO*, pp. 235–251, 1989.
48. M. Steiner, G. Tsudik, and M. Waidner. Refinement and extension of encrypted key exchange. *ACM Operating System Review*, 29:22–30, 1995.
49. T. Wu. The secure remote password protocol. In *Internet Society Network and Distributed System Security Symposium (NDSS)*, pages 97–111, 1998.
50. T. Wu. A real-world analysis of Kerberos password security. In *Internet Society Network and Distributed System Security Symposium (NDSS)*, February 1999.
51. M. Zhang. New Approaches to Password Authenticated Key Exchange Based on RSA. In *ASIACRYPT*, pp. 230–244, 2004.

Mitigating Dictionary Attacks on Password-Protected Local Storage

Ran Canetti, Shai Halevi, and Michael Steiner

IBM T.J. Watson Research Center, Hawthorne, NY, USA

Abstract. We address the issue of encrypting data in local storage using a key that is derived from the user's password. The typical solution in use today is to derive the key from the password using a cryptographic hash function. This solution provides relatively weak protection, since an attacker that gets hold of the encrypted data can mount an off-line dictionary attack on the user's password, thereby recovering the key and decrypting the stored data.

We propose an approach for limiting off-line dictionary attacks in this setting *without relying on secret storage or secure hardware*. In our proposal, the process of deriving a key from the password requires the user to solve a puzzle that is presumed to be solvable only by humans (e.g, a CAPTCHA). We describe a simple protocol using this approach: many different puzzles are stored on the disk, the user's password is used to specify which of them need to be solved, and the encryption key is derived from the password *and the solutions of the specified puzzles*. Completely specifying and analyzing this simple protocol, however, raises a host of modeling and technical issues, such as new properties of human-solvable puzzles and some seemingly hard combinatorial problems. Here we analyze this protocol in some interesting special cases.

1 Introduction

The motivation for this work is the common situation where we need to protect local storage that is physically readable by anyone, and moreover the only source of secret key material are the human users of the system. For example, think of a multi-user system where each user has an account, and where a browser lets users store personal information and site-specific passwords on the shared disk under the protection of a password. Another example is a laptop whose disk is searchable when captured and access to data is protected by a password. The common solution for this case is to derive a cryptographic key from the user-supplied password (possibly together with a public, locally stored salt), and use that key to encrypt the information (see e.g. [Kal00]).

This practice is quite problematic, however, since an attacker can perform dictionary searches for the correct password. In contrast to the case of password-based key exchange where off-line dictionary attacks can be effectively mitigated using cryptographic tools, here the lack of any secret storage seems to make such attacks inevitable. Thus, typical applications use a key-derivation-function such

C. Dwork (Ed.): CRYPTO 2006, LNCS 4117, pp. 160–179, 2006.
© International Association for Cryptologic Research 2006

as SHA1 repeated a few thousand times to derive the key from the password, in the hope of slowing down off-line dictionary attacks. Although helpful, this approach is limited, as it entails an eternal cat-and-mouse chase where the number of iterations of SHA1 continuously increases to match the increasing computing powers of potential attackers. (This approach can be thought of as a naive instantiation of the "pricing via processing" paradigm of [DN92].)

This work aims to improve the security of local storage by relying on the human user for more than just supplying the initial password. Namely, we envision an interactive key-generation process involving a human user and the automated program, at the end of which a key is generated. Specifically, we propose to make use of puzzles that are easily solvable by humans but are hard to solve for computers, as in [Naor96, vAB+03]. That is, the user would supply a password, then it would be asked to solve some puzzles, and the key would be derived from both the password and the solutions of these puzzles.

One approach for combining puzzles and passwords was proposed by Pinkas and Sandler [PS02] in the context of a client-server interaction. In their solution the server asks the client to solve some puzzles, and if the client solves these puzzles correctly then the password is used in the usual way. This approach cannot be used in our setting, however, since it requires the server to keep the correct answers in storage for the purpose of comparing them with the solutions provided by the user. In our setting there is no server that can check the user's answers, and no secret storage to keep the solutions to the challenge puzzles.

Another potential solution, proposed by Stubblefield and Simon [SS04], is to use puzzles (called Inkblots) to which each individual has its own personal and repeatable answers that are unpredictable even by other humans. Then, a (hopefully) high entropy key can be derived from the user's solutions. In essence, this method uses the answers to the puzzles as a high-entropy password. If feasible, such an enhanced password generation mechanism is indeed very attractive, but the strong unpredictability requirements severely limit the class of potentially appropriate puzzles.

We combine the traditional password mechanism with human-solvable puzzles in a different way: At system setup, a large number of puzzles are generated and stored, *without their solutions.* The user-supplied password is then used to choose a small number of puzzles out of the stored ones. The user is asked to solve these puzzles, and the key is derived from the password *and the solutions to the puzzles.* The point here is to force the adversary to solve a considerable number of puzzles per each password guess, thus limiting its power to exhaustively search for the password.

The class of puzzles that are useful for our scheme is rather broad. We only need to be able to automatically generate puzzles, have individual human users answer these puzzles in a consistent way across time, and have the answer be unpredictable to a computer. In particular, there is no need to generate puzzles together with their answers, as required in the case of CAPTCHAs [vAB+03]. In fact, there need not even exist a single "correct" answer; each individual might have its own answer, as long as it is repeatable. On the other hand, there is no need that answers by one human are unpredictable by other humans, as required for Inkblots. We call this class of puzzles by the generic term human-only solvable puzzles (HOSPs). The Appendix suggests ideas for HOSPs that are neither good CAPTCHAs nor good Inkblots but might be suitable for our scheme.

As simple as this scheme sounds, analyzing it (or even completely specifying it) takes some work. For starters, one needs to determine how many puzzles to store on the disk and how many of these should the legitimate user solve in order to get access to the encrypted data. Also, one needs to specify the function thats map user passwords to sets of puzzles and the function that maps puzzle-solutions to cryptographic keys. (We call the first function Expand and the second function Extract. These suggestive names are motivated later.) Still more difficult questions are what hardness properties we need of the puzzles in use for this scheme to be secure, or even what attack model should be considered here. We elaborate on these issues below.

1.1 Formalizing the Attack Model

The overall goal of the scheme is to generate a pseudorandom key, meaning that feàsible attackers should only have small advantage in distinguishing the key from random. The term "feasible" is typically defined as probabilistic polynomial time (PPT), so it is tempting to define security in the standard way:

Security, first attempt: *A scheme is secure if given everything that is stored on the disk and a candidate key, no PPT algorithm can distinguish between the case where the key is generated by the scheme and the case where the key is chosen at random (except perhaps with small advantage).*

However, this definition ignores the fact that the attacker may have access to humans that can solve puzzles for it. Indeed, a known attack against systems that deploy CAPTCHAs is for the attackers to ship the same CAPTCHAs to their own web sites, asking their visitors to solve them. We thus need to extend the model, allowing the attacker to recruit some humans for help.

Devising a rigorous model that captures attackers with human help raises various issues, both technical and philosophical. Indeed, we do not even have good models for describing *honest* human participants in our protocols, let alone potentially malicious human attackers. For instance, should human help be restricted to solving puzzles? If so, how to model the quality of the answers? Are they always "correct", in the sense that one human can predict the answers of another human? Does it make sense to restrict the attacker to ask for solutions of puzzles that are actually used in the scheme, or can it queries for solutions to other puzzles as well? Is it legitimate to use a human in order to determine which puzzles to ask solutions for? The problem becomes even more intricate when one wishes to somehow *quantify* the amount of human help involved.

This work takes a rather simplistic approach, assuming that humans are only accessed as puzzle-solving oracles. This assumption can be justified to some extent if we view the attacker's human helpers not as malicious but rather as basically honest users that were tricked into helping the attack. Still, one should keep in mind that this model is not completely realistic. We further simplify the model by assuming that the oracle always returns the "correct" answer provided by the legitimate user. Furthermore, we only count the overall number of puzzles that the attacker asks to solve, not differentiating between "hard" and "easy" ones. That is, our notion of security has the following flavor:

Security, better attempt: *A scheme is secure if given everything that is stored on the disk and a candidate key,* and given limited access to puzzle-solving humans, *no PPT algorithm can distinguish between the case where the key is generated by the scheme and the case where the key is chosen at random (except perhaps with small advantage).*

An additional simplifying assumption we make is that the adversary only queries its oracle on puzzles that are explicitly used in the scheme (i.e., puzzles that are written on the disk). That is, we do not consider the possibility that an adversary may be able to modify a puzzle z to obtain a puzzle z', so that the solution of the original puzzle may be easier to find given the solution for z'. (Indeed, with existing CAPTCHAs one may slightly change the puzzle without changing the solution at all.) Following [DDN00], we refer to this concern as malleability of puzzles. Intuitively, a useful puzzle systems should be "non-malleable", in the sense that any query to the human oracle can help in solving at most a single puzzle out of the puzzles used in the system.

This seems to be the first time that such malleability issues are raised in the context of human-solvable puzzles (in fact, in the context of any non-interactive puzzle system), and that formalizing reasonable non-malleability properties for puzzles is an interesting challenge. This is a topic for future research.

Computational Hardness of Puzzles. The assumption that we make on the hardness of puzzles is that given a random puzzle z (And without any help from humans), it is hard to distinguish the real solution of z from a "random solution" taken from some distribution. (Clearly, this is stronger than just assuming that computing solutions is hard.)

In a little more details, a puzzle is essentially a samplable distribution of problems with the additional property that humans can associate a solution with each problem in a repeatable way across time. This last property is captured in the model by having a puzzle-solving oracle H (for $\underline{\text{H}}$uman), which is an arbitrary deterministic function. The solution to a puzzle z is then defined as $H(z)$. The hardness assumption is formulated roughly by saying that given a random puzzle z, no PPT algorithm (that has no oracle access) can distinguish the "right solution" $H(z)$ from a "random solution" that is drawn from some distribution. The amount of hardness is measured in terms of the min-entropy of this distribution, denoted μ. See more discussion in Section 2.1.

A Generic Attack. We illustrate the capabilities of the attacker in our model by describing a generic attack against our scheme. Assume that there is an algorithm that given a puzzle generates a set of M "plausible solutions" that is guaranteed to include the right solution. (Hence this attack treats the puzzle system as if it has $\mu = \log M$ bits of pseudo-entropy.) Also, assume that the attacker has a dictionary D that is guaranteed to include the user's password.

The attacker gets an alleged key k^* and some n puzzles $z_1, \ldots, z_n$, and it needs to decide if the key is "real" or "random". It goes over the entire dictionary, expanding each password into the corresponding set of puzzles. Then it goes over all plausible solutions to each puzzle, extracts the corresponding key from each vector of plausible solutions, and keeps a list of all the solution-vectors that yield the input key k^*. These are the "consistent solutions".

The attacker then uses its human oracle in order to narrow down this list of consistent solutions. It adaptively queries the human oracle for solutions to puzzles, purging from the list all the vectors that contain a wrong solution to any puzzle. (The puzzles to be queried are chosen so as to maximize the information obtained by each query, e.g. by querying puzzle that appears in the most consistent solution vectors.) Finally, the attacker checks how many remaining consistent solutions are left for each password, and applies the maximum-likelihood decision rule to determine whether these numbers are more likely to be generated for a random key or for the "real key". The statistical advantage of this attack stems from the fact that for a "real" key k^* we expect to have one more consistent solution than for a "random" key (i.e., the actual solution).

It can be seen that to have significant advantage, the attacker must query its oracle on all but at most $m/\log M$ puzzles indexed by the actual password. Our analysis indicates that this is not just a property of this particular attack; we get essentially the same bound on the advantage of *any* attack.

1.2 Towards a Fully Specified Scheme

Below we elaborate on how to determine the functions Expand (mapping passwords to puzzles) and Extract (mapping solutions and passwords to keys). This involves fixing a number of parameters, namely the number n of puzzles stored on the disk, the number ℓ of puzzles that a user is required to solve to obtain access to the data, and the length m of the generated key.

The Expand Function. The resilience of the scheme against exhaustive password search comes from the need of the adversary to solve many puzzles (using its human oracle) in order to check each password guess. To this end, Expand must ensure that no large set of passwords are mapped a small set of puzzles.

Let q be a bound on the number of puzzle queries that the attacker can make, and let μ be the assumed hardness of the puzzle system. It is clear that to have many "missed puzzles", the attacker must only be able to query its oracle on a small fraction of the stored puzzles, so we must store on the disk $n \gg q$ puzzles. Let $I = \{i_1, i_2, \ldots, i_q\}$ be the indexes of puzzles that the attacker queries to its puzzles solver. We say that I "fully covers" a specific password pw if all the indexes in Expand(pw) are in the set I, and we say that I "almost covers" pw if at least $\ell - (m/\mu)$ of the indexes in Expand(pw) are in I. Our goal is to design the function Expand so that the attacker cannot have a small set of indexes that "almost covers" too many passwords.

This property is related to the quality of Expand as an expander graph: Consider a bipartite graph with passwords on the right and (indexes of) puzzles on the left, and where a password pw is connected to all the indexes in Expand(pw). Then a good expansion from right to left means that there is no large set of passwords that are fully covered by a small set of indexes (i.e., all their neighbors are included in a small set of indexes). The property that we need is stronger, namely that there is no large set of passwords that are even "almost covered" by a small set of indexes. It terms of expansion, we need that any subgraph of Expand (with the same set of nodes) in which the degree of nodes on the right is $\ell - \lceil m/\mu \rceil$ also has good expansion from right to left.

In Section 3.2 we analyze the cover properties of a truly random Expand function (which essentially tells us what is the best that we can expect from any function) and also briefly discuss plausible explicit constructions for Expand.

The Extract Function. Intuitively, the main property needed from function Extract is good randomness extraction, namely to compute a pseudorandom key from a set of solutions that is somewhat unpredictable to the attacker. When using the above modeling of hardness of puzzles, namely that the "right solutions" to the puzzles are indistinguishable from "random solutions" that are drawn from a distribution with sufficient min-entropy, this intuition can be formalized in a straightforward way: namely as long as the input to Extract has sufficient min-entropy, the output should be random. Thus, it is sufficient if the function Extract is a (strong) randomness extractor [NZ96].

1.3   Security Analysis

We provide two different analyses for the security of our construction. In each of these analyses we bound the advantage of an attacker as a function of the number of queries that it makes to its puzzle-solving oracle. (This is somewhat similar to the case of interactive password-based protocols where the advantage of an attacker is bounded as a function of the amount of interaction between the attacker and the honest participants.) In both analyses we reduce the security of the scheme to the assumption that an attacker cannot distinguish the "right solution" to random puzzles from "random solutions".

The first analysis (Section 4) uses a notion of security that follows the game-based approach for defining security of keys [BR93, BPR00]. The second analysis (Section 5) is done in the UC security model [Can01]. The "ideal functionality" that we present in this analysis follows the approach of [CHK+05], in that it lets ideal-world adversary make only a limited amount of password queries. This analysis requires some additional properties from the puzzles and an additional "invertibility" property from Extract; see details within.

Organization.   Section 2 defines key-generation and recovery schemes (KGR) and human-only solvable puzzles. Section 3 presents the scheme and provides some combinatorial analysis of the properties of Expand. Sections 4 and 5 contain the two analyses of the scheme as described above, Section 6 has a concrete instantiation of the scheme, and Section 7 lists some open research problems.

Appendix A suggests some potential implementations of human-solvable puzzles. Throughout, proofs are omitted for lack of space.

2 Password-Based Key Generation and Recovery Using Puzzles

Our goal is to generate a random encryption key that is recoverable by the legitimate user but still looks random to an attacker. Namely, we have two procedures: *key-generation* and *key recovery*, where the former can potentially store on the disk some information that the latter can use for recovery. In our setting, both of these procedures take as input the user's password (which should be thought of as a "weak secret"). Also, we allow them to interact with the legitimate user, modeled as a puzzle-solving oracle. The formal definition of the needed functionality proceeds as follows:

Definition 1 (Key Generation and Recovery). *A key generation and recovery scheme consists of two PPT algorithms with oracle access,* (kGen, kRec):

$(\text{key}, S) \leftarrow \text{kGen}^H(1^m, \text{pw}, \text{aux})$. *The randomized procedure* kGen *takes as input the security parameter and a password (and potentially also some other auxiliary input). It is given access to a human puzzle-solver and it outputs a key and (typically) some additional storage S.*

$\text{key} \leftarrow \text{kRec}^H(1^m, \text{pw}, S)$. *The procedure* kRec *takes as input the security parameter, a password, and the additional storage S. It is given access to the oracle H and it returns the key.*

The pair (kGen, kRec) *is* correct *with respect to a specific oracle H if for every $k \in \mathbb{N}, \text{pw} \in \{0,1\}^*$ and every* (key, S) *in the support of* $\text{kGen}^H(1^m, \text{pw})$ *it holds that* $\text{kRec}^H(1^m, \text{pw}, S) = \text{key}$. *In this case we sometime call the triple* (kGen, kRec, H) *a* Key-Generation-and-Recovery system (KGR for short).

The key generation stage can be partitioned to a system set-up stage, which would include the generation of puzzles and is common to all users, and an actual generation stage where the key is generated from the password provided by the user. We do not formally enforce this partitioning to allow for potential optimizations that involve the user in the generation of the puzzles. We also note that the auxiliary input to the key-generation procedure is meant to allow the user to provide some personal input that might help generate (or "personalize") the puzzles; see discussion in the Appendix.

2.1 Human-Only Solvable Puzzles

For our purposes, a puzzle is a randomized puzzle-generation algorithm G with the additional property that random puzzles are solvable in a mostly consistent way by most humans. That is, the same person gives the same answer to a given puzzle at different times (although this consistent answer may vary from person to person). To enable asymptotic analysis, we let the puzzle generation algorithm G take as input the security parameter and also quantify the hardness of the resulting puzzles as a function of the security parameter.

Definition 2 (Puzzles). *A puzzle-system is a pair (G, H), where G is a randomized puzzle generator that takes as input 1^m (with m the security parameter) and outputs a puzzle, $z \leftarrow G(1^m)$, and H is a solution function. That is, the value $a = H(z)$ is the* solution *of the puzzle z.*

In order to be useful for our application, we need the puzzle system to be consistently-solvable by humans but not easily solvable by computers. The specific hardness assumption that we make is that efficient algorithms cannot distinguish the right solution $H(z)$ from "random". We call this hardness assumption the *solution indistinguishability* of the puzzle system, and quantify it by the amount of randomness in the "random" solution.

Definition 3 (Solution indistinguishability). *Let $\mu = \mu(m)$ be a function of the security parameter. A puzzle system (G, H) has μ* bits of pseudo-entropy *if for every puzzle z in the support of the output of $G(1^m)$ there is a distribution $R(z)$ with min-entropy at least $\mu(m)$ such that the correct solution is indistinguishable from a solution taken randomly from $R(z)$, even when given both correct and random solution to polynomially many other puzzles. That is, for every polynomial $n = n(m)$ the following two ensembles are computationally indistinguishable:*

$$
\left\{ (z_1, a_1, (z_2, a_2, a_2'), \ldots, (z_n, a_n, a_n')) :
\begin{array}{l}
z_1, \ldots, z_n \overset{\$}{\leftarrow} G(1^m), \; a_1 \leftarrow H(z), \\
a_i \leftarrow H(z_i), \; a_i' \overset{\$}{\leftarrow} R(z_i) \; (i = 2 \ldots n)
\end{array} \right\}_m
$$

and

$$
\left\{ (z_1, a_1', (z_2, a_2, a_2'), \ldots, (z_n, a_n, a_n')) :
\begin{array}{l}
z_1, \ldots, z_n \overset{\$}{\leftarrow} G(1^m), \; a_1' \overset{\$}{\leftarrow} R(z), \\
a_i \leftarrow H(z_i), \; a_i' \overset{\$}{\leftarrow} R(z_i) \; (i = 2 \ldots n)
\end{array} \right\}_m
$$

We note that if the puzzle system is a CAPTCHA where the puzzles can be generated together with their solution, and if the distributions $R(z)$ are efficiently samplable given z, then there is no need for the tuples (z_i, a_i, a_i'). Also, a simple hybrid argument shows that t-fold repetition of (G, H) multiplies the pseudo-entropy by t:

Observation 1. *Let μ, t be functions of the security parameter where t is polynomially bounded. Let (G, H) be a puzzles system, and let (G^t, H^t) be the t-fold repetition of it, where t puzzles are generated and solved independently. If (G, H) has μ bits of pseudo-entropy then (G^t, H^t) has $t\mu$ bits of pseudo-entropy.* □

Finally, we reiterate that the security models that are described in subsequent sections give the adversary access to the same oracle H that is used by the scheme. This represents the fact that, for analyzing security, we make the worst-case assumption that all humans provide the same answer to the same question, thus the adversary can employ other humans to reproduce the answers provided by the legitimate user. One can possibly model the case where people predict the answers of each other only in a "partial", or "noisy" way. Still, it is stressed that for the correctness of the scheme we only need that the same user answers consistently with itself.

3 The Scheme

The scheme uses a puzzle system (G, H) as above, where G is used directly by the key-generation procedure and H is assumed to be available as an oracle at both key-generation and key-recovery time. The scheme depends on two internal parameters: the number of puzzles that are stored on the disk (denoted n) and the number of puzzles that the user needs to solve (denoted ℓ). We think of ℓ as a constant or a slowly increasing function, and $n = poly(k)$. In addition, we have a universe W of passwords (e.g., $W = \{0,1\}^{160}$ if using SHA1 to hash the real password before using it.) W should not be confused with the potential dictionary $D \subset W$ from which passwords are actually chosen, which is not known when the scheme is designed. The scheme uses two randomized functions, $\mathsf{Expand}_{r_1} : W \to [n]^\ell$ and $\mathsf{Extract}_{r_2} : \{0,1\}^* \to \{0,1\}^m$, that are discussed later in this section. Given all these components, denote $n = n(m)$ and $\ell = \ell(m)$, and the scheme works as follows:

$\mathsf{kGen}^H(1^m, \mathsf{pw})$. Generate n puzzles, $z_i \leftarrow G(1^m)$, $i = 1, 2, \ldots, n$, and choose the keys r_1, r_2 for Expand and $\mathsf{Extract}$ at random. (This can perhaps be carried out off-line at system setup.) Then set $\langle i_1, \ldots i_\ell \rangle \leftarrow \mathsf{Expand}_{r_1}(\mathsf{pw})$ and query the oracle H for the solutions, setting $a_{i_\cdot} \leftarrow H(z_{i_\cdot})$ for $j = 1, \ldots, \ell$. Finally, compute $\mathsf{key} \leftarrow \mathsf{Extract}_{r_2}(a_{i_1}, \ldots, a_{i_\cdot}, \mathsf{pw})$ and then all the solutions are discarded. The "additional storage" that is saved to the disk consists of the puzzles $\langle z_1, z_2, \ldots, z_n \rangle$ and the keys r_1, r_2.

$\mathsf{kRec}(1^m, \mathsf{pw}, \langle z_1, \ldots, z_n \rangle, r_1, r_2)$. Compute the indexes $\langle i_1, \ldots i_\ell \rangle \leftarrow \mathsf{Expand}_{r_1}(\mathsf{pw})$, query the oracle H for the solutions $a_{i_\cdot} \leftarrow H(z_{i_\cdot})$ for $j = 1, \ldots, \ell$, and recover the key as $\mathsf{key} \leftarrow \mathsf{Extract}_{r_2}(a_{i_1}, \ldots, a_{i_\cdot}, \mathsf{pw})$.

We remark that if the puzzle system in use is in fact a CAPTCHA system (where puzzles are generated together with their solution) then the key-generation procedure need not query the oracle H. Also, if the puzzle system is such that puzzles remain hard to solve even when the randomness of G is known (and Expand is a random oracle) then the puzzles need not be stored at all. Instead, the value $\mathsf{Expand}(\mathsf{pw})$ can be used as the randomness to G, thus generating the puzzles "on the fly".

3.1 The Function Extract

The role of $\mathsf{Extract}$ is to extract an m-bit pseudorandom key from the pseudo-entropy in the human solutions to the puzzles. Given our solution indistinguishability assumption (cf. Definition 3), this can be achieved by using a strong randomness extractor [NZ96] for the function $\mathsf{Extract}$, as long as we are willing to live with some loss of pseudo-entropy (since to get a close-to-random m-bits output from an extractor you need $m' > m$ bits of min entropy in the input). Below and throughout the analyses, we therefore assume that $\mathsf{Extract}$ is a strong extractor (e.g., a pairwise-independent hash function) and denote by m' the amount of min-entropy that is needed in the input to $\mathsf{Extract}$ in order for the output to be close to a uniform m-bit string.

3.2   The Function Expand

The role of the function Expand is to map passwords to indexes of puzzles in such a way that the attacker would have to solve many puzzles (i.e., invoke the puzzle-solving oracle H many times) to check each new password guess. Specifically, our goal is to make sure that as long as the attacker does not make too many queries to its puzzle-solving oracle, most passwords would have enough unsolved puzzles to get $m' > m$ bits of pseudo-entropy. (Recall that m' is the amount of min-entropy that is needed in the input to Extract in order for the output to be close to a uniform m-bit string.)

To make this more precise, fix the randomness r_1 and think of the function $E = \text{Expand}_{r_1}$ as a bipartite graph with the password universe W on the right and the indexes $\{1, ..., n\}$ on the left, and where each $\text{pw} \in W$ is connected to all the indexes in $E(\text{pw})$. If the puzzle system has μ bits of pseudo-entropy then we denote by $\ell^* \stackrel{\text{def}}{=} \lceil m'/\mu \rceil$ the number of puzzles that the attacker should miss in order for the key to be pseudo-random.

We say that a set I of indexes on the left almost covers a password $\text{pw} \in W$ on the right if pw has no more than ℓ^* neighbors that arc not in I. (I.e., pw has at least $\ell - \ell^*$ neighbors in I.) The almost-cover of a set I relative to mapping E, dictionary D and parameter ℓ^* is:

$$\text{acvrSet}_{\ell^*}(I, E, D) \stackrel{\text{def}}{=} \{\text{pw} \in D : |E(\text{pw}) \setminus I| \le \ell^*\}.$$

What we want is that for any set of $q \ll n$ puzzles $z_{i_1}, ..., z_{i.}$ that the attacker has solutions for, and any (large enough) potential dictionary $D \subset W$, the set $I = \{i_1, ..., i_q\}$ only covers a small fraction of the passwords in D. We denote by $\text{acvr}_{\ell^*}(q, E, D)$ the cover number of E, namely the fraction of passwords in D that can be almost-covered by q puzzles:

$$\text{acvr}_{\ell^*}(q, E, D) \stackrel{\text{def}}{=} \frac{\max\limits_{|I|=q} \left| \text{acvrSet}_{\ell^*}(I, E, D) \right|}{|D|}$$

We would like the expected cover-number of Expand_{r_1} (over the choice of r_1) to be sufficiently small . The property of having a small cover number is related to the expansion of the graph E (from right to left): We want any large enough subset of the nodes on the right to have more than q neighbors on the left, even when removing ℓ^* edges from every node. In other words, every subgraph of E where the degree of nodes on the right is $\ell - \ell^*$ should be a good expander.

To get a sense of the obtainable parameters, we first provide an analysis in the random-oracle model (i.e., assuming that Expand is a random function). Later we discuss solutions based on limited independence and speculate about other plausible constructions.

The Cover Number of a Random Function.   The following technical lemma bounds the probability of having a very large cover number in terms of various parameters of the system. This lemma depends on many parameters so as to make it applicable in many different settings. We later give an example of some specific setting.

Lemma 2. *Fix $\ell, \ell^*, n, q \in \mathbf{N}$ such that $q < n$ and $\ell^* = (1 - \alpha)\ell$ for some $\alpha > 0$, and also fix some finite set $D \subset W$. Denote $\epsilon \overset{\text{def}}{=} 2^{H_2(\alpha)}(q/n)^\alpha$ where H_2 is the binary entropy function. If ϵ^ℓ is small enough so that there exists $\rho > 0$ for which*

$$e \cdot (\epsilon^\ell)^{\rho/(1+\rho)} < \frac{(1 + \rho)H_2(q/n)}{\log(1/\epsilon^\ell)\,|D|/n}$$

then for any $\delta > 0$ we have

$$\Pr_E \left[\mathsf{acvr}_{\ell^*}(q, E, D) > (1 + \delta)\frac{(1 + \rho)H_2(q/n)}{\log(1/\epsilon^\ell)\,|D|/n} \right] < 2^{-n\delta H_2(q/n)}$$

where the probability is taken over choosing a random function $E : D \rightarrow [n]^\ell$.   □

An Example.   Assume that we have $q/n = 0.01$, $|D| = n$, $\ell = 8$ and $\ell^* = 4$. In this case we have $\alpha = (\ell - \ell^*)/\ell = 1/2$ and therefore $\epsilon = 2^{H_2(\alpha)}(q/n)^\alpha = 2\sqrt{q/n} = 0.2$, and $H_2(q/n) \approx 0.0808$. One can verify that in this case $e\log(1/\epsilon^\ell)/H_2(q/n) \approx 625 = 1/\epsilon^{\ell/2}$ which means that the requirement in the assertion of Lemma 2 is satisfied for $\rho = 1$. Plugging these values for ρ, ϵ, ℓ and $H_2(q/n)$ in the expression from Lemma 2 we get for any $\delta > 0$

$$\Pr_E \left[\mathsf{acvr}_{\ell^*}(q, E, D) > (1 + \delta) \cdot 0.0087 \right] < 2^{-n\delta/12.38}$$

It follows that the expected value of acvr over the random choice of E is at most $(1 + \delta) \cdot 0.087 + 2^{-n\delta/12.38}$. Assuming large enough value for n (e.g., $n > 4000$) and plugging a small enough value for δ, this expected value is no more than 0.9%.

Note that the trivial way of "almost covering" passwords in this case will be for the attacker to make four queries for each password, and since we assume that there are n passwords and we have $q/n = 0.01$ then the fraction of passwords that will be almost covered this way will be $0.01/4 = 0.25\%$. Hence, in this case the upper bound tells us that the attacker cannot do better than four times the obvious attack.

Limited Independence.   Storing a completely random function from the password universe W to the set of puzzle indexes $[n]$ is not realistic in most cases. A first attempt at obtaining a concrete construction is to replace a completely random mapping with an X-wise independent mapping for some X. (In terms of storage on the disk, it is acceptable to store a description of an $O(n)$-wise independent function, since we anyway need $O(n)$ storage to store the n puzzles.)

When Expand is ℓt-wise independent then one can use the t-moment inequality instead of Chernoff bound. The bound that we get for any $\delta > 0$ is

$$\Pr_E \left[\mathsf{acvr}_{\ell^*}(q, E, D) > \tau + \delta \right] \quad < \quad \binom{n}{q} \cdot \frac{\tau}{\delta^t \,|D|^{t/2}},$$

where again we have $\epsilon \overset{\text{def}}{=} 2^{H_2(\alpha)}(q/n)^{\alpha}$. Using an ℓn-wise independent mapping (so $t = n$) and setting $\delta = 2/\sqrt{|D|}$ we get

$$\Pr_{E}\left[\mathsf{acvr}_{\ell^*}(q, E, D) > \tau + \frac{2}{\sqrt{|D|}}\right] < \binom{n}{q} \cdot \frac{\tau}{2^n} < 2^{q \log n - n}$$

We thus get:

Lemma 3. *Fix* $\ell, \ell^*, n, q \in \mathbb{N}$ *such that* $\ell^* = (1 - \alpha)\ell$ *for some* $\alpha > 0$ *and* $q < n/(4 \log n)$, *and fix a finite set* D *such that* $|D| > n$.

The expected value of $\mathsf{acvr}_{\ell^*}(q, E, D)$ *over the choice of* E *as an* ℓn-*wise independent mapping from* W *to* $[n]$ *is at most* $\epsilon^{\ell} + 2/\sqrt{|D|} + 2^{q \log n - n}$ *where* $\epsilon \overset{\text{def}}{=} 2^{H_2(\alpha)}(q/n)^{\alpha}$. $\square$

More Efficient Constructions.   Although feasible, the solution of using ℓn-wise independent mapping is far from being satisfactory, as it entails very large storage and computational cost. Providing more efficient constructions that are provably good is an open problem.

One possible direction here is to extend for our purposes the result of Alon et al. [ADM+99]. In that work they considered a mapping from n balls to n buckets, and analyzed the size of the largest bucket. They proved that although "generic pairwise independent function" cannot ensure anything smaller than $n^{1/2}$, using a random linear mapping (over the binary field) has expected largest bucket of only $\tilde{O}(\log n)$. This can be thought of as a very special case of our application with $\ell = q = 1$, $\ell^* = 0$ and $|D| = n$. We thus speculate that perhaps using ℓ independent linear maps could give us a reasonable bound also for our application.

4   Game-Based Security Analysis

In this section we analyze the security of our scheme with respect to a "game-based" notion of security (as in [BR93]). In the formulation below the key generation scheme is run on a password pw that was randomly chosen from a dictionary. The adversary is then given the generated storage, plus a value that is either the real generated key or a random value of the same length. In addition, the adversary is given oracle access to H. The adversary's advantage in distinguishing between the two cases is measured as a function of the number of H-queries (q), and the size of the dictionary from which the passwords are chosen (d):

Definition 4 (Game-based security definition). *Let* $\alpha : \mathbb{N} \times \mathbb{N} \to [0, 1]$ *be a function, let* $(\mathsf{kGen}, \mathsf{kRec}, H)$ *be a key-generation and recovery system, and let* $\mathcal{C}$ *be a class of attackers with oracle access.*

We say that $(\mathsf{kGen}, \mathsf{kRec}, H)$ *is secure up to* α *with respect to* $\mathcal{C}$ *if for any attacker* $A \in \mathcal{C}$, *any polynomially related* $d, m, q \in \mathbb{N}$ *and any* $D \subset W, |D| = d$, *the following two probabilities differ by at most* $\alpha(d, q) + \mathsf{negl}(m)$, *where* negl *is a negligible function:*

$$p_{\text{real}}(A) \overset{\text{def}}{=} \Pr[\text{pw} \in_R D, \ (\text{key}, S) \leftarrow \text{kGen}^H(1^m, \text{pw}) : \ A^{H.}(\text{key}, S) = 1]$$

$$p_{\text{rand}}(A) \overset{\text{def}}{=} \Pr[\text{pw} \in_R D, \ (\text{key}, S) \leftarrow \text{kGen}^H(1^m, \text{pw}),$$
$$\text{key}' \in_R \{0, 1\}^{|\text{key}|} : \ A^{H.}(\text{key}', S) = 1]$$

where H_q answers the first q queries as H does, and answers later queries by $\perp$.

Conservative Adversaries. For our analysis to go through, we need to restrict the attacker to only query its puzzle-solving oracle H on puzzles that explicitly appear in the storage S of the scheme. This is done to prohibit "puzzle mauling attacks" as discussed in Section 2.1. Essentially, this restriction reflects the assumption that each H-query helps in answering only a single puzzle, and that getting an H-answer to a puzzle does not help solving another puzzle.

Definition 5 (Conservative adversaries). *An adversary against our KGR scheme is called* conservative *if it only queries its oracle on puzzles that are explicitly included in the "additional storage" output of the key-generation routine.*

Security Statement. We are now ready to state our main result concerning the security of our scheme. Let m be the security parameter and let $\delta, \mu, d, \ell, m', n, q$ be other parameters (that may be functions of the security parameter), and denote $\ell^* \overset{\text{def}}{=} \lceil m'/\mu \rceil$. Below we assume that that $\binom{\ell}{\ell^*}$ is polynomial in m. (Recall that ℓ is the number of puzzles that the honest user needs to solve, so we typically think of it as a constant or $\log m$, hence assuming that $\binom{\ell}{\ell^*}$ is polynomial in m is reasonable.)

Let Expand_{r_1} be a randomized mapping of passwords to indexes and let Extract_{r_2} be a strong randomness extractor [NZ96], extracting m bits that are δ away from uniform given any distribution with m' bits of min-entropy. Lemma 4 below essentially asserts that as long as there are at least m' bits of pseudo-entropy in the puzzles that are mapped to the right password but are not queried by the attacker, the attacker cannot have any significant advantage in distinguishing real from random. Specifically, we show that the advantage of the attacker is essentially the fraction of passwords in the dictionary that are almost covered by its queries to the puzzle-solving oracle.

Lemma 4. *Under the conditions from above, the scheme from Section 3 using* Expand *and* Extract *is secure up to α with respect to the class of conservative adversaries, where α is defined as*

$$\alpha(d, q) \overset{\text{def}}{=} \delta + \max_{|D|=d} E_{r_1} \left[\text{acvr}_{\ell^*}(q, \text{Expand}_{r_1}, D) \right].$$

Comment. We note that the solution indistinguishability assumption and the use of strong extractor can be replaced with the more specific assumption that the output of Extract is pseudorandom. Namely, we can assume directly that given ℓ^* puzzles without their solution and $\ell - \ell^*$ puzzles with solution and given r_2 and key, an attacker cannot tell if the key was computed as $\text{Extract}_{r_2}(a_1, a_2, \ldots)$ or was chosen at random. The proof under this (weaker) assumption is very similar to the proof of Lemma 4.

5   UC Security Analysis

In our second security analysis we use the UC security framework [Can01], and the presentation here assumes familiarity with this framework. We incorporate the human oracle H in the model by providing the adversary and the parties running the protocol with access to H. The environment is not given direct access to H; rather, it has access to H only via the adversary. This restriction represents the assumption that the puzzles used in an instance of the protocol are "local" to that instance, in the sense that they are generated within that instance, and furthermore the corresponding solutions are not affected by external events.

In addition, as in Section 4 we focus on **conservative adversaries** that ask H only on puzzles that were directly provided by the protocol. This technical restriction represents the "meta-assumption" that the puzzles are such that obtaining a solution for one puzzle does not help in solving a different puzzle. (Admittedly, this assumption may not always hold, and somewhat restricts the pool of potential implementations.)

The Ideal Password-Based Key Generation and Recovery Functionality.   We define the ideal functionality representing the security specification that we wish to obtain. The functionality, $\mathcal{F}_{\mathrm{PKGR}}$, is presented in Figure 1. $\mathcal{F}_{\mathrm{PKGR}}$ is parameterized by a dictionary D of possible passwords, the maximum number p of allowable password queries by the adversary, and the length m of the generated key. At the first activation $\mathcal{F}_{\mathrm{PKGR}}$ expects the user to provide a password pw, along with a session identifier sid. (Formally, the sid may specify the identities of the user and the computer.) $\mathcal{F}_{\mathrm{PKGR}}$ then generates a random m-bit key key and outputs it to the computer. Finally, it notifies the adversary that a key was generated. If pw is not in the dictionary then it also gives it to the adversary in full. (Formally this means that no security is guaranteed for passwords not in the dictionary. But note that the scheme itself does not depend on the actual dictionary, so we can always let the parameter D be the set of passwords that are "actually used by users".)

Next, whenever $\mathcal{F}_{\mathrm{PKGR}}$ receives a password pw' together with a request to recover the key, it outputs the key to the computer only if pw' is the same as the stored password. This request may come from anyone, not only the legitimate user. Finally, $\mathcal{F}_{\mathrm{PKGR}}$ answers up to p password guesses made by the adversary.

The main security guarantee of $\mathcal{F}_{\mathrm{PKGR}}$ is that the adversary can make only p password guesses. If none of these guesses succeeds, then the key is indistinguishable from a truly random m-bit key. For reasonable values of m, this provides strong cryptographic security. Also, $\mathcal{F}_{\mathrm{PKGR}}$ makes no mention of puzzles. Indeed, puzzles are treated as part of the implementation, geared toward limiting the number of password guesses.

Note that $\mathcal{F}_{\mathrm{PKGR}}$ obtains the password directly from the environment. This formulation (which follows the formulation in [CHK+05]) provides some strong guarantees. First, there is no a-priori assumption on the distribution from which the password is taken, as long as it is taken from D. In fact, this distribution may not even be efficiently generatable, since it may depend on the initial input of the environment. Second, we are guaranteed that in any scheme that realizes $\mathcal{F}_{\mathrm{PKGR}}$ the local storage of the computer cannot be correlated in any way with the password *and the key*.

The fact that $\mathcal{F}_{\mathrm{PKGR}}$ is parameterized by D, and furthermore provides no security for passwords not in D, is a limitation that comes from our simulation procedure. Ideally, we would like to guarantee that $\mathcal{F}_{\mathrm{PKGR}}$ does not depend on D at all and provides security for any password string. Realizing such a functionality is an interesting future challenge.

Functionality $\mathcal{F}_{\mathrm{PKGR}}(D, p, m)$

1. At the first activation, receive from the user U an input (password, sid, pw), choose a random m-bit key key, output (sid, key) to the computer V, and notify the adversary that a key has been generated. If pw $\notin D$ then also send pw to the adversary.
2. When receiving input (key-recovery, sid, pw′) from anyone, do: If pw′ = pw then output key to V. Else output an error message to V.
3. When receiving input (password-query, sid, pw′) from the adversary, do: If the adversary already made p password queries then ignore. Else, if pw′ = pw then return key to the adversary. Else return Wrong-Guess to the adversary.

Fig. 1. The ideal functionality $\mathcal{F}_{\mathrm{PKGR}}$, parameterized by a dictionary D, the maximum number p of password queries, and the length m of the generated key

Invertibility of Extract. In order to show that the scheme realizes $\mathcal{F}_{\mathrm{PKGR}}$ we need to make an additional "invertibility" assumption on function Extract with respect to solutions of puzzles: What we need is that given an m-bit key key and given $\ell - \ell^*$ solution to ℓ puzzles, one can efficiently compute ℓ^* plausible solutions to the remaining puzzles that would map the entire solution vector to the given key, Extract($\ldots$) = key.

Below let (G, H) be a puzzle system with μ bits of pseudo-entropy, where the right solution for each puzzle z is indistinguishable from a solution drawn from $R(z)$, let f_r be a randomized function, and let $\ell \geq \ell^*$ be integers.

Definition 6. *We say that f is* strongly invertible *w.r.t. the puzzle system (G, H) and the parameters ℓ, ℓ^* if there exists an efficient inversion algorithm I that given randomness r, key* key, *any string* pw, ℓ *puzzles $z_1, ..., z_\ell$ generated via G, and a set of $\ell - \ell^*$ solutions $a_{\ell^*+1}, ..., a_\ell$ such that $a_i \in R(z_i)$, outputs values $a_1, ..., a_{\ell^*}$ such that $a_i \in R(z_i)$ and $f_r(a_1, ..., a_\ell,$ pw, $) =$ key.*

Furthermore, for random r, key, $z_1, ..., z_\ell$ *and any* pw *and $a_{\ell^*+1}, ..., a_\ell$, the output of I is indistinguishable from sampling at random $a_i \in R(z_i)$ for $i = 1, ..., \ell^*$ subject to the constraint $f_r(a_1, ..., a_\ell,$ pw, $) =$ key.*

If the distributions $R(z)$ are efficiently samplable given z and m is small enough (so that 2^m is polynomial in the relevant efficiency parameters) then I can simply sample the solutions $a_1, ..., a_{\ell^*}$ from the appropriate distributions until it find a solution vector that match the given key. Alternatively, if we have a puzzle-system for which $R(z) = \{0,1\}^k$ for some k and Extract is a linear function (e.g., a linear universal hash function) then inversion is possible even for large m via linearity.

UC-Security of Our Scheme.   Consider the scheme from Section 3 with parameters n, ℓ, and m, and let D be the dictionary used by the user. Assume that (G, H) be a puzzle system with μ bits of pseudo-entropy, assume that Extract extracts m bits that are negligibly close to uniform from any distribution with $m' > m$ bits of min-entropy, and let $\ell^* = \lceil m'/\mu \rceil$. Finally assume that Extract is strongly invertible w.r.t. (G, H) with parameters ℓ, ℓ^*. Then we have

Lemma 5. *Under the conditions above, the scheme from Section 3 UC-realizes* $\mathcal{F}_{\mathrm{PKGR}}(D, p, m)$ *relative to conservative adversaries that makes at most q queries to H, where $p = $ Cover(Expand, D, q).*

6   A Concrete Example

We describe a concrete instantiation of our scheme for a "medium security" application. The example builds on the numeric example in Section 3.2. Consider trying to get an 64-bit key using our scheme, while relying on a CAPTCHA system whose assumed hardness is (say) 16 bits of pseudo-entropy per instance. In terms of our analysis, we therefore have the parameters $m = 64$ and $\mu = 16$ (so $\ell^* = 64/16 = 4$). Assuming that each puzzle takes about 4KB to encode (which is the case for common CAPTCHAs) and restricting ourselves to number of puzzles that fit on a single 4.7GB DVD-R, we would like to store about $n = 10^6$ puzzles.

We assume that the odds of the attacker guessing a weak password are one in a million (per guess) so we assume the dictionary size is $|D| = 10^6$. Also, we would like the legitimate user to solve no more than $\ell = 8$ puzzles to access the encrypted data, and we assume that the attacker cannot get a human to solve for it more than $q = 10^4$ puzzles. (This seems like a safe assumption in most settings, but maybe not all of them.) In Section 3.2 it is shown that with this setting of the parameters, and when modeling Expand as a random function, we should expect the cover-number to be no more than 0.9%. Namely, an attacker that makes up to 10,000 puzzle-solving queries has at most 0.9% chance of "almost covering" the actual password. (Note that the obvious way of "almost covering" passwords yields success probability of 0.25%.)

A simple heuristic instantiation of the scheme for this setting of the parameters would be to use a cryptographic hash function to implement both the Expand and Extract functions. More specifically, to generate the key, choose (say) two 128-bit salt values r_1 and r_2, get the user's password pw and compute $h \leftarrow$ HMAC-SHA1$_{r_1}$(pw). Parse the 160 bits of h as eight 20-bit indexes $i_1, \ldots, i_8 \in [2^{20}]$. Generate 2^{20} CAPTCHAs and store then together with r_1, r_2 on the disk (or on a DVD-R). Then obtain the solutions to the eight relevant puzzles $a_{i.} = $

$H(z_{i.})$, $j = 1 \ldots 8$, compute $h' \leftarrow \text{HMAC-SHA1}_{r_2}(a_{i_1}|\ldots|a_{i_8}|\text{pw})$, and take as many bits of h' as needed for the key. To make use of the common practice for slowing down brute-force attacks, replace SHA1 with the iterated function SHA1^k for some reasonable k, say $k = 65536$.

Although this scheme allows the key to be longer than 64 bits, the above analysis only shows that with probability more than 99.1% we get "64-bit strength". (One can similarly calculate the probability of getting "80-bit strength" which happens when the attacker misses five of the eight puzzles that are mapped to the user's password, or the probability of getting "128-bit strength" when the attacker misses all the puzzles, etc.)

7   Future Work

This work investigates a new approach for deriving cryptographic keys from human passwords, improving the resistance against off-line dictionary attacks by having the user solve some puzzles and using the solutions in the key-generation process. Still, the analysis is quite preliminary and many questions remain unanswered, regarding constructions, modeling, and analysis. Below we list a few of these questions.

Building Puzzles.   Perhaps a first challenge is to actually construct puzzle systems that are useful for schemes such as the one described in this paper. Of particular interest for our scheme would be puzzles that remain hard even when the attacker knows the randomness that was used to generate them. (As mentioned above, using such puzzle-systems we can forgo storing any puzzles on the disk, instead using the value Expand(pw) as randomness for the puzzle generation system.)

As discussed in the introduction, another important property of puzzles is "non-malleability", namely preventing the adversary from using the solution to one puzzle to solve other puzzles. This question is interesting both on the level of designing puzzles that will make it hard for the adversary to "maul" them while maintaining solvability, and on the level of mathematical formalization. On the design level, a potentially interesting approach might be to use watermarking techniques to embed in the puzzle a message to the human reader that urges to not solve the puzzle outside a certain context.

Improving the Scheme.   One obvious challenge is to find different constructions that will improve various characteristics of the current scheme (e.g., use less storage, improve security, etc.) An immediate question is to have an efficient explicit construction for Expand that provably has a small expected cover number (maybe by extending the work of Alon et al. [ADM+99]). Another direction is to split the key-generation process to two parts, where the first part generates a large amount of storage which is the same for all users and the second adds a small amount of user-specific storage that may depend on the password or on other information provided by the user. Another direction is to have a more interactive key recovery process, where the user's answer to the current question effects the questions that it is asked next.

Improving the Analysis. The analysis can be improved in a number of ways. First, the hardness assumption on puzzles may be reduced to allow for non-negligible distinguishing probability between the real solution and a random one. Furthermore, one might want to assume only that *computing* the correct answer is hard.

Another immediate set of goals is to improve the current analysis (especially the one in the UC framework), and to find ways to relate the game-based security and the UC security notions.

Also, it would be nice to be able to formally model and argue about non-conservative adversaries, namely adversaries that query the human oracle on puzzles different than the ones used in the scheme. What security properties can be guaranteed against such adversaries?

Another plausible extension of our security model is to consider an attacker that can modify the storage on the disk and then watch the output of the key-recovery procedure with this modified storage. (This seems related to the issue of puzzle malleability that we mentioned above.)

Also, it may be useful to have an algorithm to compute (or bound) the cover number of a given Expand function with respect to a given dictionary. This seems harder than computing/bounding the expansion of a graph, since here we need to compute/bound the expansion of any subgraph of a given degree.

A more speculative research direction is to try and obtain reasonable general models for human users (or even human attackers) in systems such as the ones described in this work.

References

[ADM+99] Noga Alon, Martin Dietzfelbinger, Peter Bro Miltersen, Erez Petrank, Gábor Tardos. Linear Hash Functions. J. ACM 46(5): 667–683. 1999.

[BPR00] M. Bellare, D. Pointcheval, and P. Rogaway. Authenticated Key Exchange Secure Against Dictionary Attacks. *Advances in Cryptology – Eurocrypt 2000*, LNCS vol. 1807, Springer-Verlag, pp. 139–155, 2000.

[BR93] M. Bellare and P. Rogaway. Entity Authentication and Key Distribution. *Advances in Cryptology – Crypto 1993*, LNCS vol. 773, Springer-Verlag, pp. 232–249, 1993.

[Can01] R. Canetti. Universally Composable Security: A New Paradigm for Cryptographic Protocols. *42nd IEEE Symposium on Foundations of Computer Science (FOCS)*, IEEE, pp. 136–145, 2001.

[CHK+05] R. Canetti, S. Halevi, J. Katz, Y. Lindell and P. MacKenzie. Universally composable password-based key exchange. *Advances in Cryptology – Eurocrypt 2005*, LNCS vol. 3494, Springer-Verlag, pp. 404–421, 2005.

[DDN00] Danny Dolev, Cynthia Dwork, Moni Naor. Non-malleable Cryptography. *SIAM J. Comput.* 30(2): 391-437 (2000)

[DN92] Cynthia Dwork, Moni Naor. Pricing via Processing or Combating Junk Mail. *Advances in Cryptology – Crypto 1992*, LNCS vol. 740, Springer-Verlag, pp. 139–147, 1992.

[Kal00] Burt Kaliski. PKCS #5: Password-Based Cryptography Specification Version 2.0. RFC 2898 http://www.ietf.org/rfc/rfc2898.txt. September 2000

[Naor96]   Moni Naor.   Verification of a human in the loop or identification via
           the Turing test. Manuscript, available on-line from http://www.wisdom.
           weizmann.ac.il/~naor/PAPERS/human_abs.html, September 1996.
[NP97]   Moni Naor and Benny Pinkas. Visual Authentication and Identification
           Advances in Cryptology – Crypto 1997, LNCS vol. 1294, Springer-Verlag,
           pp. 322–336, 1997
[NZ96]   Noam Nisan, David Zuckerman. Randomness is Linear in Space. *J.
           Comput. Syst. Sci.* 52(1): 43–52. 1996.
[PS02]   Benny Pinkas and Tomas Sander. Securing passwords against dictio-
           nary attacks. In *Proceedings of the 9th ACM Conference on Computer
           and Communications Security*, pages 161–170, Washington, DC, USA,
           November 2002. ACM Press.
[SS04]   Adam Stubblefield and Dan Simon. Inkblot Authentication. Microsoft
           Research Technical report MSR-TR-2004-85.
[vAB+03]   Luis von Ahn, Manuel Blum, Nicholas Hopper, and John Langford.
           CAPTCHA: Using hard AI problems for security. In *Advances in Cryp-
           tology – EUROCRYPT '2003*, volume 2656 of *Lecture Notes in Computer
           Science*. Springer-Verlag, Berlin Germany, 2003.

A   Possible Implementations of Puzzles

The notion of puzzles used in our scheme is rather relaxed, and consequently
has a large number of potential implementations. One set of implementations
are the existing and potential implementations of CAPTCHAs, e.g. the ones
described in [Naor96, vAB+03]. These implementations, however, require that
the solutions to puzzles are uniquely defined and generatable together with the
puzzles. This seems like a strong limitation on implementations. Another set
of implementations are the potential implementations of Inkblots, described in
[SS04]. These implementations, however, are required to be "private", in the
sense that the solution given by one human is unpredictable (and, in fact, pseudorandom) even to other humans. Again, this seems to be a strong limitation
on implementations.

Below we list some potential approaches for implementation that are not included in any of these classes. That is, the answers to these puzzles are somewhat
subjective and not necessarily unique. Still, answers of different individuals may
be similar. As pointed out in [NP97] in a different context, validity of these
proposals should be evaluated by testing on human individuals.

Personal ranking: Puzzles may include pictures of different persons, to be
    ranked by coolness, or age, or taste in clothing. Alternatively, puzzles may
    include different pictures or descriptions of food, to be ranked by tastiness,
    spiciness, etc.. Alternatively, a puzzle may depict several randomly generated
    drawings to be ranked by personal liking, an audio puzzle may sound several
    short melodies to be ranked by liking, etc.

Face recognition: Puzzles may include pictures of several faces, and the ques-
    tion is which if these faces most resemble people known to the user (e.g.,
    parents or other family members). Furthermore, puzzles can be "personal-
    ized" by asking the user to provide, at system set-up, pictures of close family

members. These pictures might then be mixed with random pictures to generate puzzles. (This proposal takes advantage of the fact that recognizing faces is one of the most highly specialized visual capabilities of humans.)

Personal clustering: Puzzles may include a bunch of various unrelated objects, and the question is which three objects "go together" the best, or are the most "closely related" or "look alike". The objects can be people, household items, cartoons, or a mix of all categories. (This implementation approach was proposed by Ronitt Rubinfeld.)

Imaginative inferring: Puzzles may portray a scene and ask questions about what happened a minute ago, or what will happen in a minute. Alternatively, questions can be asked regarding what happens outside the borders of the picture.

Personal association: Puzzles may depict an object (e.g., a person) and ask which familiar objects (persons) does the object in the picture remind.

Finally, we note that to prevent "mauling attacks," the puzzle generator must make sure that randomly generated puzzles are "different enough" from each other so that a solution to one randomly generated puzzle will not help in solving another randomly generated puzzle.

Rationality and Adversarial Behavior in Multi-party Computation
(Extended Abstract)

Anna Lysyanskaya and Nikos Triandopoulos

Department of Computer Science, Brown University
{anna, nikos}@cs.brown.edu

Abstract. We study multi-party computation in the model where none of n participating parties are honest: they are either rational, acting in their selfish interest to maximize their utility, or adversarial, acting arbitrarily. In this new model, which we call *the mixed-behavior model*, we define a class of functions that can be computed in the presence of an adversary using a trusted mediator. We then give a protocol that allows the rational parties to emulate the mediator and jointly compute the function such that (1) assuming that each rational party prefers that it learns the output while others do not, no rational party has an incentive to deviate from the protocol; and (2) the rational parties are protected from a malicious adversary controlling $\lceil \frac{n}{2} \rceil - 2$ of the participants: the adversary can only either cause all rational participants to abort (so no one learns the function they are trying to compute), or can only learn whatever information is implied by the output of the function.

1 Introduction

Multi-party computation (MPC) has emerged as the central problem in cryptography, of which many scenarios that arise in practice are simply a special case. In multi-party computation, we have n parties, each party P_i with its input x_i, wishing to jointly compute a function $f(x_1, \ldots, x_n)$. In the traditional secure MPC scenario (SMPC) (e.g., [5, 6]), there are two kinds of parties: honest parties that will follow a prescribed protocol, and adversarial parties that will collude and deviate from the protocol in an arbitrary way. An SMPC protocol should guarantee that no adversary controlling up to t parties will be able to learn anything about the honest parties' inputs not implied by the value $f(x_1, \ldots, x_n)$.

In this paper, we introduce the mixed-behavior model for multi-party computation (MMPC). Here, some of the parties will be *rational*, and will only follow a protocol if they have a clear incentive to do so. Others will be *adversarial*, and will collude and exhibit arbitrary, possibly irrational, behavior. The goal is to design protocols that compute the function, guarantee the same security properties as in usual SMPC and that rational parties will choose to follow.

Rational MPC vs. Mixed-Behavior Model. Halpern and Teague [8] considered rational MPC (RMPC), i.e., the setting where n *rational* parties wish

C. Dwork (Ed.): CRYPTO 2006, LNCS 4117, pp. 180–197, 2006.

to securely compute a joint function of their inputs, and there is no adversary. They showed a surprising impossibility result: suppose that each party P_i holds an additive secret share x_i of a secret $x = \sum x_i$. Suppose that each party prefers to learn x to not learning it, and prefers that as few as possible other parties learn x. In this case it was shown that there is no deterministic protocol for reconstructing x that rational parties will have an incentive to follow (or, put another way, for any deterministic strategy, there will be another one that each rational party will prefer). On the other hand, they gave a randomized protocol that did the job. The idea was to first give a protocol for $n = 3$. Then for $n > 3$, the proposed solution was to first partition the parties into three groups, have each group select a leader and have each group member send his secret share to his leader. It is easy to see that the approach above immediately fails if an adversary controls three of the parties: the adversary controlling the leaders will learn the secret, and no one else ever will. Thus, Halpern and Teague leave the problem of MPC in the mixed-behavior model totally open.

Independently of our work, Abraham et al. [1] study general function evaluation in the RMPC model, in an extended setting where rational parties can collude with each other or are allowed to have non-standard utilities, and they design protocols that are robust against this type of scenarios (e.g., resilient equilibria). Gordon and Katz [7] study rational secret sharing and MPC and provide a protocol that improves on the original protocol in [8] (e.g., it is simpler and symmetric and captures the case $n = 2$). The protocols of both works are conceptually similar to our protocol; both are for RMPC, though.

Assumptions on Communication Model vs. Assumptions on Utilities. Recently, Izmalkov et al. [9] introduced rational secure MPC (RSMPC). Their goal is to realize games without a trusted mediator, no matter what the underlying utility functions of the players are. Here, there is no adversary, but any subset of parties may be maliciously trying to collude in an attempt to maximize their utility. The main challenges of RSMPC are ensuring that players, first of all, contribute legal information into the game (e.g., if they are playing a card game, a player cannot use a card that was not dealt to him), and additionally that the players cannot collude while computing the outcome of the game, communicating to each other the information that they would not be able to communicate if the game were played using a trusted mediator. In order to address these challenges, Izmalkov et al., as well as the related works in [10, 11], place severe restrictions on the communication model (no channels of communication except those explicitly available), and utilize physical primitives such as a ballot box and a physical envelope. In contrast, we use standard communication channels, allow existence of covert channels and steganography, and make assumptions instead about the utility functions of the parties: we assume that the parties already have an incentive to participate in the protocol and to contribute their true inputs. These assumptions enable us to come up with protocols where rational parties will not be motivated to deviate in any way.

Intuition for the Construction. In a nutshell, the impossibility result of Halpern and Teague goes as follows. Suppose that this is the last round of the reconstruction protocol of a secret sharing scheme; then sending out information cannot increase utility, but may decrease it by allowing another party to learn x. How can it be, then, that a randomized protocol exists? In any given round, the players do not know whether this is supposed to be the last round (and so they would do better by keeping their information to themselves) or whether this is a test round in which no meaningful information can be revealed, but instead the parties are just being tested, and deviating from the protocol will have the consequence that others will abort the protocol. This is the key idea of in Halpern and Teague's protocol that we will use also.

Our protocol assumes that we have a *synchronous* broadcast channel. That means that, in computing message for round i, no one can take into account other parties' message for this round: waiting for those messages to arrive will mean missing your chance to speak in round i. Further, we assume that all parties, both rational and adversarial, are computationally bounded. As a subroutine, we will invoke a traditional SMPC protocol over the broadcast channel, specifically the GMW protocol [6] as analyzed in the UC model by Canetti *et al.* [3]. At each step, each party will provide a ZK proof that the data it published on the channel was computed according to the protocol as a function of the data it previously received, as well as of its input and random tape. We call a protocol where at each step each party provides such a proof, a *verifiable* MPC protocol.

Our main protocol for MMPC works roughly as follows. In a setup step, we set up the parameters of the computation, such as the common random string needed for secure multi-party computation and non-interactive zero-knowledge proofs. Next, as is usual in MPC, we run a preprocessing step as a result of which each party P_i is committed to its input x_i. Next is the key step: using a verifiable MPC protocol, the parties come up with an m-out-n secret sharing (e.g., polynomial secret sharing [12]) either of $y = 0$ (with probability $1/2$) or of the value $y = f(x_1, \ldots, x_n)$ (we assume that f never returns a 0). Here, n is the number of participants, and we will specify m later. If at any point of this protocol, any party P_j deviates (note that since we are using verifiable MPC, a deviation is detectable with high probability), the protocol tells P_i to abort.

So far, no party has an incentive to deviate from the protocol: deviation will be detected, causing everyone following the protocol to abort the computation. On the other hand, at least computationally, no information about $f(x_1, \ldots, x_n)$ has been communicated so far by the security properties of the regular SMPC. In the next step, each party P_i broadcasts its share of y and a non-interactive ZK proof of correctness. If m correct shares are broadcasted, then the parties combine them to obtain y, and if it is the case that $y \neq 0$, then they obtain $f(x_1, \ldots, x_n)$. If combining the correct shares present does not yield $f(x_1, \ldots, x_n)$, and fewer than n correct shares are present, then the protocol aborts. Otherwise, go back to the key step above.

Consider what damage an adversary controlling $m-1$ participants can inflict. It can approach $n - 2m + 2$ rational parties, give each of them $m - 1$ secret shares of the outcome, thus guaranteeing that each party will be able to *locally* (as opposed to in collaboration with other rational parties) compute the outcome and therefore will not participate in the protocol any longer. Then, the remaining $m-1$ parties, no matter what strategy they adopt between themselves, will be unable to compute anything meaningful (they do not have enough shares). This is undesirable, since we want a protocol in which, no matter what the adversary does, either everyone computes their output, or no one does. Thus, the best we can hope to do is to allow the adversary to control up to $t = m - 2$ participants, which, we will show, our protocol will achieve. A key step in the proof is to show that rational parties will have a disincentive to collude with others.

Consider what happens if only m parties are rational. Suppose that $m - 1$ of them are following the protocol, and the party P_i is considering deviating. Note that, depending on P_i's utility function, it may be a good idea for him to not broadcast his share of y. He risks causing everyone to abort (with probability $1/2$), but on the other hand, his utility might skyrocket in case he is lucky and $y \neq 0$. On the other hand, if $m + 1$ participants are rational, and we know that all of them except P_i are following the protocol, then P_i can only lose in utility if he does not broadcast his share of y_i, since in case $y = 0$, he will force everyone to abort without ever computing the function, while if $y \neq 0$, everyone will learn the outcome whether or not P_i broadcasts. Therefore, we see that the protocol is a Nash equilibrium if $m+1$ or more of the participants are rational. Accordingly, we set the value of m to be $m = \lceil \frac{n}{2} \rceil$, thus we derive a protocol that tolerates adversarial behavior for any adversary controlling up to $\lceil \frac{n}{2} \rceil - 2$ of the parties.

Note that the above observation solves an open problem posed by Halpern and Teague who asked whether it was possible to have a protocol that works independently on how much utility is attached to being the only party privy to the output of the function. Also, note that the only communication channel that our protocol needs is a synchronous broadcast channel, while Halpern and Teague assume private channels in addition to synchronous broadcast. Unlike Izmalkov *et al.*, our protocol tolerates the existence of additional communication channels between the participants.

Assumptions on Utilities and on Rationality. Halpern and Teague observe that rational parties are not going to participate in a protocol unless they have an incentive to compute the function and, moreover, they must have an incentive to submit their true individual inputs. Therefore, the function must be non-cooperatively computable (NCC) [13]. We need to extend this assumption to work in the mixed-behavior model. For example, if the function in question is computable based on inputs of a small number of parties, and covert channels between participants are present, then the parties have no incentive to participate in the protocol to begin with, since they may learn the output of the function from the adversary. Therefore, in order to tolerate t adversarial participants, we must require that it is impossible to compute the function in question based on $t + 1$ of the inputs. We define such functions, called t-NCC, in Definition 5.

Also following Halpern and Teague, we make assumptions that the more a party learns about the output, the happier he is. On the other hand, the less others learn about it, the happier he is as well. Since the adversary is not rational and is not trying to learn anything, the utility functions should ignore whether or not the adversary learns anything. We explore this in Definitions 3 and 4.

Suppose that two different strategies of a party yield indistinguishable views. Then the difference in how much this party learned is negligible, assuming polynomial-time algorithms. Our final assumption on utilities is that negligible difference in how much P_i or others have learned yield negligible differences in utilities. We call this assumption the *computational satisfaction assumption*, and state it more precisely in Definition 9.

Following Halpern and Teague, rationality is captured by postulating that a party will choose to follow a given protocol if (1) it is a Nash equilibrium, i.e., provided that other non-adversarial players follow the protocol, this party's utility is maximized by following it as well; (2) it survives the process of iterated deletion of weakly dominated strategies—i.e., those strategies for which another strategy is always at least as good and sometimes even better. However, what do these mean in the mixed-behavior model? We adapt these game-theoretic definitions to the scenario where a malicious adversary exhibiting arbitrary behavior is present. In a nutshell, a set of strategies is a Nash equilibrium in the mixed-behavior model (Definition 6) if it is a Nash equilibrium for all adversaries. A strategy σ weakly dominates another strategy τ if for all adversaries and no matter what strategy other rational parties are following, σ gives as much or better utility than τ, *and* for some adversary and some set of strategies for the remaining rational parties, σ gives strictly more utility (Definition 7).

Our protocol is not a preferred equilibrium in the strict sense because it relies on the computational intractability of certain problems that can still be solved with negligble, but positive, probability. What we show is that no deviation can increase the utility by more than a negligible amount (Definitions 6 and 7).

Paper Outline. In Section 2 we present the mixed-behavior model for multi-party computation, introduce related new concepts and formally define the notion of unconditionally or computationally secure preferred protocols. In Section 3, we describe a protocol that implements any function in the mixed-behavior model, using a special channel, and prove that it is an unconditionally secure preferred protocol. In Section 4 we describe our main protocol over a synchronous broadcast channel, and prove that it is a computationally secure preferred protocol that is ϵ-Nash and survives iterated deletion of ϵ-weakly dominated strategies, where ϵ is negligible. We refer the reader to the full version of the paper for all the details this extended abstract omits.

2   Mixed-Behavior Model for Multi-party Computation

We introduce a new model for multi-party computation, the *mixed-behavior model* (MMPC). In the standard multi-party computation setting, consider any protocol that implements a function f when executed by n parties $\mathcal{P}$; f is either unary ($f(\mathbf{x})$ is the common output) or n-ary, denoted as $\mathbf{f}$ ($\mathbf{f}(\mathbf{x}) = (y_1, \ldots, y_n)$ are the private outputs). In the mixed-behavior model, parties $\mathcal{P}$ are partitioned into *rational* parties $\mathcal{R}$ and *adversarial* parties $\mathcal{A}$. Rational parties consider the joint computation to be a game, during which they act selfishly, deviating from a prescribed protocol and exhibiting strategies aiming to increase their gained utility after the termination of the computation. At the same time, an adversary $\mathcal{A}$ controls up to t of the parties, which jointly exhibit arbitrary strategies, sharing information and acting adversarially during the computation.

We next define concepts related to computations performed in the mixed-behavior model and also new properties that protocols must satisfy, both unconditionally and computationally. We use some standard notation: $[n]$ denotes set $\{1, \ldots, n\}$; bold letters denote vectors, where $\mathbf{v} = (v_1, \ldots, v_n)$ can be written as $(v_i, \mathbf{v}_{-i})$ for any $i \in [n]$; $(v'_i, \mathbf{v}_{-i})$ denotes $\mathbf{v}' = (v_1, \ldots, v_{i-1}, v'_i, v_{i+1}, \ldots, v_n)$; $\mathbf{v}_I$ denotes vector projection to set $I \subseteq [n]$. We write $\mathcal{P} = \mathcal{R} \cup \mathcal{A}$ to denote that a set of parties $\mathcal{P}$ is partitioned into rational parties $\mathcal{R}$ and adversarial parties $\mathcal{A}$, where $\mathcal{A}$ also denotes the actual programs run by the adversary. If A is a randomized algorithm, $a \leftarrow A$ denotes that a was obtained by running A.

Definition 1 (Communication channels and structures). *Let $\mathcal{P}$ be a set of interactive TMs. A channel $C = \{\mathcal{P}_I, F_C, \text{state}, \mathcal{P}_O\}$ available to $\mathcal{P}$ is a (possibly, stateful and randomized) TM with $\mathcal{P}_I, \mathcal{P}_O \subseteq \mathcal{P}$ which operates as follows. C shares a dedicated tape t_i^I with each $P_i \in \mathcal{P}_I$ and a dedicated tape t_i^O with each $P_i \in \mathcal{P}_O$. C gets activated when a party $P_i \in \mathcal{P}_I$ writes a string I_i on t_i^I. If C is a synchronous channel, then it waits for all other parties in $\mathcal{P}_I$ to submit an input, or times out with a default value $\bot$. If activated, C computes the values $(\mathbf{O}, \text{state}') \leftarrow F_C(\mathbf{I}, \text{state})$, writing value O_i on tape t_i^O, $\forall i \in [n]$, and updating its state information. A channel structure $\mathcal{C}(\mathcal{P})$ is a set of channels available to $\mathcal{P}$, always containing channel $\{\{P_i\}, I, \bot, \{P_i\}\}$, $\forall i \in [n]$ (I: identity function).*

It is immediate how the above formulation of a communication channel as an interactive Turing machine captures the conventional notion of any channel (e.g., point-to-point or broadcast channels). Moreover, it constitutes a generalization of a communication channel: by allowing a channel to keep state and by appropriately setting its operation function F_C, we can essentially define any ideal functionality over inputs submitted by the parties in $\mathcal{P}$, thus expressing useful additional properties in computations (e.g., perfect privacy).

We consider joint computations among individual parties $\mathcal{P}$ performed over some underlying communication channel C available to $\mathcal{P}$. A computation consists of a (possibly infinite) series of rounds, each round corresponding to an activation of C. Each participating party $P_i \in \mathcal{P}$ interacts with C through its corresponding tapes: P_i runs a (possibly randomized) program that processes strings appearing on t_i^O (incoming messages) and computes a string to be written on t_i^I (outgoing messages), also updating an internal state. Also, additional channels may be used among parties in $\mathcal{P}$ in the course of a computation.

Definition 2 (Runs, views, outputs). *Let $\mathcal{P} = \mathcal{R} \cup \mathcal{A}$ be a set of parties jointly computing over channel structure $\mathcal{C}(\mathcal{P})$. A run R of the system $(\mathcal{P}, \mathcal{C}_{\mathcal{P}})$ is tuple $(\mathbf{p}, \mathbf{r}, \mathcal{A})$, where $\mathbf{p}_i$ is the (randomized) program that party $P_i \in \mathcal{R}$ runs, $\mathcal{A}$ also denotes the program that parties in $\mathcal{A}$ jointly run, and $\mathbf{r}$ is the vector of random tapes needed by parties in $\mathcal{P}$. The view of party $P_i \in \mathcal{R}$ in run R, denoted*

$VIEW_i(R)$, is tuple $(r_i, \mathbf{M}_i)$, where $\mathbf{M}_i$ is the set of all messages received by P_i; the view of the adversary $\mathcal{A}$, denoted $VIEW_\mathcal{A}(R)$, is tuple $(\mathbf{r}_\mathcal{A}, \mathbf{M}_\mathcal{A})$, where $\mathbf{M}_\mathcal{A}$ is the set of all messages received by the parties in $\mathcal{A}$. Each message in $\mathbf{M}_i$ or $\mathbf{M}_\mathcal{A}$ specifies both the channel over which it was received, and its contents (and possibly its sender). The output of a party $P_i \in \mathcal{R}$ in run R, denoted $OUT_i(R)$, is the last message received (and possibly sent) by P_i in R, if R terminates.

Definition 3 (Protocol utility and satisfaction functions). *Let* $\mathcal{P} = \mathcal{R} \cup \mathcal{A}$ *be a set of parties. We say that* $\mathbf{u}$ *is a protocol utility function with satisfaction* $\boldsymbol{\mu}$ *if for any run R: (1) for $P_i \in \mathcal{P}$, $\mu_i : \{0,1\}^* \mapsto [0,1] \cup \perp$ is any function, such that $\mu_i(\cdot) \in [0,1]$ if $P_i \in \mathcal{R}$ and $\mu_i(\cdot) = \perp$ if $P_i \in \mathcal{A}$ and (2) there exists a set of functions $\mathbf{u}'$ for parties $\mathcal{R}$, where $u_i' : [0,1]^n \mapsto [0,\infty]$ and $u_i(R) = u_i'(\boldsymbol{\mu}(R))$.*

The above definition captures our formulation of a joint computation as a game. Rational parties are each associated with a utility function, mapping (consequences of) outcomes of the computation—which is fully described by protocol runs—to positive reals as follows. A personal satisfaction function μ_i maps the view that $P_i \in \mathcal{R}$ gets out of a computation to a value in $[0,1]$ according to certain criteria for P_i about what constitutes a desired outcome; intuitively, on input $VIEW_i(R)$, μ_i measures how well P_i succeeded in computing whatever it wished to compute. Then P_i evaluates the outcome of a computation by getting (through u_i') a utility that depends on the evaluations of all parties' satisfaction functions on the protocol run. In essence, u_i is what characterizes P_i rationality, meaning that any deviation from a proposed protocol by P_i corresponds to a strategy that can be preferable or not for P_i depending solely on utility function u_i. In contrast, the adversary $\mathcal{A}$ has no specific desired outcomes; if $P_i \in \mathcal{A}$ then $\mu_i = \perp$ and there is no associated utility. This is in accordance with our intuition: $\mathcal{A}$ acts in an arbitrary, possibly irrational way. Note that it is not necessarily the case that $P_i \in \mathcal{R}$ can infer his utility u_i from his view, (since, e.g., he does not know who the adversary is); however, P_i can still act so as to maximize u_i.

We next present some minimal assumptions on the above definition, providing a concrete notion of rationality, i.e., incentives for protocol deviation, in the MMPC model. We do this by defining target functions. Intuitively, given a vector of utility functions $\mathbf{u}$, a vector of functions $\mathbf{f}$ is its corresponding target function if the utilities capture the fact that each rational P_i wishes to compute f_i, and prefers that as few as possible other rational P_j's compute f_j.

We assume that inputs $\mathbf{x}$ for computations among parties in $\mathcal{P}$ are chosen by first drawing inputs $\mathbf{x}'$ from distribution $\mathbf{X}$, then setting $\mathbf{x}_\mathcal{R} = \mathbf{x}'_\mathcal{R}$ and, finally letting $\mathbf{x}_\mathcal{A} \leftarrow \mathcal{A}(\mathbf{x}'_\mathcal{A})$. That is, $\mathbf{x}$ is formed by generating a set of values according to $\mathbf{X}$, and then replacing the entries corresponding to adversarial parties with those chosen by the adversary. We denote the above process as $\mathbf{x} \leftarrow_\mathcal{A} \mathbf{X}$.

How does μ_i captures what P_i wants to learn? The measure of how well P_i learns some function $f_i(\mathbf{x})$ is the likelihood p_i that he outputs the correct $f_i(\mathbf{x})$, where the probability is taken over $\mathbf{X}$ as well as the randomness of the particular run of the protocol. This is relatively standard in the cryptographic definitional literature, starting with semantic security. The only adjustment we want to make is to scale μ_i so that it is 0 if p_i is the same as the *a-priori* probability that P_i outputs $f_i(\mathbf{x})$, and 1 if $p_i = 1$. More formally:

Definition 4 ((Computational) target function). *For $\mathcal{P} = \mathcal{R} \cup \mathcal{A}$, let $\mathbf{u}$ be a protocol utility function with satisfaction measure $\boldsymbol{\mu}$ and let $\mathbf{x} \leftarrow_\mathcal{A} \mathbf{X}$. Then $\mathbf{f} : (\{0,1\}^*)^n \mapsto (\{0,1\}^*)^n$ is an n-ary* (computational) *target function for input distribution $\mathbf{X}$ and adversary $\mathcal{A}$ if the following conditions are met:*

$P_i \in \mathcal{R}$ **wants to learn** f_i: *μ_i is a (computational) measure of how much P_i has learnt about $f_i(\mathbf{x})$ based on the run R. Let some algorithm F_i be the best (polynomial-time) estimator of $f_i(\mathbf{x})$ both based on x_i alone (i.e., $\mu_i(x_i)$) and based on the view $VIEW_i(R)$ obtained by P_i in run R (i.e., $\mu_i(R) \triangleq \mu_i(VIEW_i(R))$). Assume that $\Pr[F_i(x_i) = f_i(\mathbf{x})] \neq 1$ (i.e., it is impossible to always correctly output $f_i(\mathbf{x})$ based on x_i alone). Then*

$$\mu_i(R) = (\Pr[F_i(R) = f_i(\mathbf{x})] - \Pr[F_i(x_i) = f_i(\mathbf{x})])/(1 - \Pr[F_i(x_i) = f_i(\mathbf{x})]).$$

The scaling and normalizing in the formula above is done so that if for a run R, $\Pr[F_i(R) = f_i(\mathbf{x})] = 1$, then $\mu_i(R) = 1$; and if for a run R, $\Pr[F_i(R) = f_i(\mathbf{x})] = \Pr[F_i(x_i) = f_i(\mathbf{x})]$ (i.e., run R did not increase P_i's chances of correctly computing $f_i(\mathbf{x})$), then $\mu_i(R) = 0$. By convention, if in run R, $\mu_i(R) = 1$, we write that $OUT_i(R) = f_i(\mathbf{x})$. Moreover, if for some $P_i \in \mathcal{R}$, and runs R and R' it is the case that $\mu_i(R) > \mu_i(R')$ and for all $j \neq i$, $\mu_j(R) = \mu_j(R')$, then $u_i(R) > u_i(R')$.

P_i **does not want others to learn:** *If for some $P_j \in \mathcal{R}$ and runs R and R' it is the case that $\mu_j(R) > \mu_j(R')$, and for all $P_i \in \mathcal{R}$, $i \neq j$, $\mu_i(R) = \mu_i(R')$, then for all $P_i \in \mathcal{R}$, $u_i(R) < u_i(R')$.*

Worst outcome: *Let R be a run such that for some $P_j \in \mathcal{R}$, $OUT_j(R) = f_j(\mathbf{x})$. Then for all $P_i \in \mathcal{R}$, if $OUT_i(R) \neq f_i(\mathbf{x})$, then $u_i(R) = 0$.*

Thus, we consider functions which for rational parties and with respect to their utility functions satisfy properties that express selfishness and antagonism in multi-party computation: $u_i(R)$ strictly increases when P_i gets closer to the target value $f_i(\mathbf{x})$ or computes $f_i(\mathbf{x})$, or when some other party P_j gets further away from its target value $f_j(\mathbf{x})$. This formulation is a generalization of the rationality in RMPC [8]. In contrast, adversarial parties behave totally arbitrarily and unpredictably: the adversary $\mathcal{A}$ acts independently of parties' rationality and its behavior may affect rational parties' utilities; e.g., $\mathcal{A}$ may consistently try to minimize or even maximize the utility of a specific rational party.

In the mixed-behavior model, we are interested in computing (target) functions for which rational parties have incentive to submit their true values as input to the joint computation. This class of functions, introduced by Shoham and Tennenholtz as *non-cooperatively computable* functions, were studied for the binary case in [13]. We next extend this concept to general n-ary functions defined on strings (rather than single bits) for computations in MMPC. In the mixed-behavior setting, we also need to capture the case where a malicious adversary changes some of the inputs: we wish to ensure that we still express honesty in submitting inputs, but capture the possible scenario where a rational party does worse by substituting his true input.

Definition 5 (t-NCC functions). *Let* $f : X \mapsto Y$ *be an n-ary function, let* $\mathcal{P} = \mathcal{R} \cup \mathcal{A}$, $|\mathcal{P}| = n$, $\mathbf{u}$ *a protocol utility function with satisfaction* $\boldsymbol{\mu}$. *Let the channel structure of* $\mathcal{P}$ *consist just of synchronous channel* $C = (\mathcal{P}, \mathbf{f}, \text{state}, \mathcal{P})$ *as well as a channel to and from the adversary* $\mathcal{A}$; C *is stateful and can only be used once. Let* $Eqv_{\mathbf{f}}(i, x_i)$ *be the maximal set of strings such that* $\forall \mathbf{x}_{-i}$ *and* $\forall x \in Eqv_{\mathbf{f}}(i, x_i)$, $\mathbf{f}(x_i, \mathbf{x}_{-i}) = \mathbf{f}(x, \mathbf{x}_{-i})$. *Let* $\Sigma_i'(x_i)$ *be the set of strategies for* P_i *that submit some* $x \in Eqv_{\mathbf{f}}(i, x_i)$ *to* C, *and output the value* y_i *received from* C. *Let* $\sigma_i^*(x_i)$ *be the strategy that submits* x_i *to* C *and outputs* y_i. *Suppose that input* $\mathbf{x}$ *is chosen with the process: (1)* $\mathbf{x}' \leftarrow \mathbf{X}$; *(2)* $\mathbf{x}_{\mathcal{R}} = \mathbf{x}_{\mathcal{R}}'$; *(3)* $\mathbf{x}_{\mathcal{A}} \leftarrow \mathcal{A}(\mathbf{x}_{\mathcal{A}}')$. *We say that* $\mathbf{f}$ *is t-non-cooperatively computable (t-NCC) under distribution* $\mathbf{X}$ *if for all* $P_i \in \mathcal{R}$, *for all adversaries* $\mathcal{A}$, $|\mathcal{A}| \leq t$: *(1)* $\mu_i(x_i, \mathbf{x}_{\mathcal{A}}) = 0$, *i.e., nothing can be learned about* $f_i(\mathbf{x})$ *based on* x_i *and* $\mathbf{x}_{\mathcal{A}}$ *alone; (2) for all* $\mathcal{A}$, i, $f_i(\mathbf{x}') = f_i(\mathbf{x})$, *i.e.,* $\mathcal{A}$ *cannot change the value of the function by changing his inputs; (3) for all* $\mathcal{A}$, P_i *gets at least as much utility from submitting his true input* x_i *to* C *as from substituting any other different input value: for all* $1 \leq i \leq n$, *all* $\sigma_i \notin \Sigma_i'$,

$$E_{\mathbf{X}}[u_i((\sigma_i(x_i), (\boldsymbol{\sigma}_{\mathcal{R}}^*)_{-i}((\mathbf{x}_{\mathcal{R}})_{-i})), \perp, \mathcal{A}(\mathbf{x}_{\mathcal{A}}))] < E_{\mathbf{X}}[u_i(\boldsymbol{\sigma}_{\mathcal{R}}^*(\mathbf{x}_{\mathcal{R}}), \perp, \mathcal{A}(\mathbf{x}_{\mathcal{A}}))].$$

For $t > 0$ the above definition is restrictive: it requires that distribution $\mathbf{X}$ outputs codewords of an error-correcting code. For instance, requiring that $\mu_i(x_i, x_{\mathcal{A}}) = 0$ is needed to exclude the following scenario: $\mathcal{A}$ gives its inputs $\mathbf{x}_{\mathcal{A}}$ to party P_i; then if $\mu_i(x_i, x_{\mathcal{A}}) > 0$, P_i and $\mathcal{A}$ abort the computation and P_i gains positive utility, whereas other rational parties gain zero utility. This demonstrates an inherent difficulty for realizing "fair" joint computations. However, the resulting class of functions is still of interest if we consider computations over values for which a secret sharing has first been computed.

We next define properties that protocols in MMPC should satisfy, either unconditionally or computationally. For brevity we do not explicitly denote that the strategies are a function of input $\mathbf{x} \leftarrow_{\mathcal{A}} \mathbf{X}$; instead we say that parties in $\mathcal{P}$ receive inputs from $\mathbf{X}$ and write run $(\boldsymbol{\sigma}_{\mathcal{R}}(\mathbf{x}_{\mathcal{R}}), \mathbf{r}, \mathcal{A}(\mathbf{x}_{\mathcal{A}}))$ as $(\boldsymbol{\sigma}_{\mathcal{R}}, \mathbf{r}, \mathcal{A})$. Also, we simplify the cumbersome $(\sigma_i, (\boldsymbol{\sigma}_{\mathcal{R}})_{-i})$ to $(\sigma_i, \boldsymbol{\sigma}_{-i})$. Finally, for the computational counterparts of the following definitions, any strategy or the adversary receives 1^k as an additional input—where k is a (security) parameter for probabilistic, polynomial-time (PPT) algorithms.

Definition 6 ((ε-)Nash equilibrium in the mixed-behavior model). *A set of (PPT) strategies* $\boldsymbol{\sigma}^*$ *is a (ε-)Nash equilibrium for* $\mathcal{P} = \mathcal{R} \cup \mathcal{A}$ *receiving inputs from* $\mathbf{X}$, *with a given channel structure* $\mathcal{C}(\mathcal{P})$ *and protocol utility function* $\mathbf{u}$ *if for all* $P_i \in \mathcal{R}$, *for all (PPT) strategies* σ_i, *and all (PPT) adversaries* $\mathcal{A}$,

$$E_{\mathbf{X}, \mathbf{r}}[u_i((\sigma_i, \boldsymbol{\sigma}_{-i}^*), \mathbf{r}, \mathcal{A})] \leq E_{\mathbf{X}, \mathbf{r}}[u_i((\sigma_i, \boldsymbol{\sigma}_{-i}^*), \mathbf{r}, \mathcal{A})] \; (+ \; \epsilon(k)).$$

The above definition extends the (canonical in game theory) concept of Nash equilibrium in the mixed-behavior and bounded computational models. In

MMPC, a strategy is a Nash equilibrium, if no deviation is strictly preferable for a rational party when all others do not deviate, and independently of what the adversary does; for an ϵ-Nash equilibrium in a computational setting it holds that, in the same environment, no deviation is preferable by more than $\epsilon(k)$.

We next consider a refinement of the Nash equilibrium, using the notion of *weakly dominated strategies* and the process of *iterated deletion* of such strategies which define a better, preferred, notion of equilibrium. We use analogues of these game-theoretic concepts in the mixed-behavior and bounded computational models to define the corresponding refined Nash equilibria. Intuitively, a strategy survives the process of iterated deletion of weakly dominated strategies if no other strategy is (always) preferable to it. This last notion of preferable strategies is expressed as a partial order over individual strategies. Strategy σ_i' of P_i is preferable to σ_i if there exists a specific environment (strategies of others and of $\mathcal{A}$) under which σ_i gives strictly less utility than σ_i', whereas at the same time, in no other environment does σ_i give strictly larger utility.

Definition 7 (Iterated deletion of (ϵ-)weakly dominated strategies). *Given $\mathcal{P} = \mathcal{R} \cup \mathcal{A}$ receiving inputs from $\mathbf{X}$, with communication structure $\mathcal{C}$ and protocol utility functions $\mathbf{u}$, let Σ' be a set of (PPT) strategies for $\mathcal{R}$. Strategy σ_i is (computationally ϵ-)weakly dominated by (PPT) strategy σ_i' restricted to Σ' (denoted $[\sigma_i < \sigma_i' \mid_{(\epsilon)} \Sigma']$) if (1) for all (PPT) strategies $\boldsymbol{\sigma}_{-i} \in \Sigma'_{-i}$, for all (PPT) adversaries $\mathcal{A}$,*

$$E_{X,\mathbf{r}}[u_i((\sigma_i, \boldsymbol{\sigma}_{-i}), \mathbf{r}, \mathcal{A})] \leq E_{X,\mathbf{r}}[u_i((\sigma_i', \boldsymbol{\sigma}_{-i}), \mathbf{r}, \mathcal{A})]$$

and (2) for some (PPT) $\boldsymbol{\sigma}'_{-i} \in \Sigma'_{-i}$, for some (PPT) adversary $\mathcal{A}$,

$$E_{\mathbf{X},\mathbf{r}}[u_i((\sigma_i, \boldsymbol{\sigma}'_{-i}), \mathbf{r}, \mathcal{A})] \ (\ +\epsilon(k)\ ) < E_{\mathbf{X},\mathbf{r}}[u_i((\sigma_i', \boldsymbol{\sigma}'_{-i}), \mathbf{r}, \mathcal{A})].$$

Let Σ^0 be the set of strategies for $\mathcal{R}$. For $\ell \geq 1$, let Δ^ℓ, Σ^ℓ be defined as follows:

$$\Delta_i^\ell = \{\sigma_i : \exists \sigma_i' \in \Sigma_i^{\ell-1} \text{ such that } [\sigma_i < \sigma_i' \mid_{(\epsilon)} \Sigma^{\ell-1}]\} \text{ and } \Sigma_i^\ell = \Sigma_i^{\ell-1} - \Delta_i^\ell.$$

We say that a strategy σ for party P_i survives iterated deletion of (computationally ϵ-)weakly dominated strategies if for all ℓ, $\sigma \notin \Delta_i^\ell$.

We finally define conditions that we wish a protocol to satisfy.

Definition 8 ((Computationally) t-secure preferred protocol for function f). *Given $\mathcal{P}$ receiving inputs from $\mathbf{X}$, with channel structure $\mathcal{C}(\mathcal{P})$ and protocol utility functions $\mathbf{u}$, a vector of (PPT) strategies $\boldsymbol{\sigma}^*$ constitutes a (computationally) t-secure preferred protocol for (polynomial-time computable) $\mathbf{f}$ under input distribution $\mathbf{X}$ if it has the following properties:*

Correctness and security: *As is standard for multi-party computation [5], (static) correctness and security is defined by requiring that for every (PPT) adversary $\mathcal{A}$, $\mathcal{A} \subset \mathcal{P}$, $|\mathcal{A}| \leq t$, there exists a (PPT) simulator S of comparable computational ability such that for all $\mathbf{x} \in \mathbf{X}$, the joint distribution of $(\boldsymbol{OUT}_{\mathcal{R}}(\boldsymbol{\sigma}_{\mathcal{R}}^*(\mathbf{x}_{\mathcal{R}}), \mathbf{r}, \mathcal{A}), VIEW_{\mathcal{A}}(\boldsymbol{\sigma}_{\mathcal{R}}^*(\mathbf{x}_{\mathcal{R}}), \mathbf{r}, \mathcal{A}))$ is indistinguishable* from the distribution sampled from as follows. S is interactive and obtains the following inputs in the following order: First, $\mathbf{x} \leftarrow \mathcal{A}$, and S is given $\mathbf{x}_{\mathcal{A}}$. Then S may output y, in which case the sample of the distribution is $(\perp_{\mathcal{R}}, y)$, or may produce query $\mathbf{x}'_{\mathcal{A}}$. In the latter case, in response to the query, S is given $\mathbf{f}_{\mathcal{A}}(\mathbf{x}')$, where $\mathbf{x}'_{\mathcal{R}} = \mathbf{x}_{\mathcal{R}}$. Finally, S outputs y. Then the sample of the distribution is $(\mathbf{f}_{\mathcal{R}}(\mathbf{x}'), y)$. Moreover, we require that there exist an adversary $\mathcal{A}$ (namely, one that follows $\sigma^*_{\mathcal{A}}$ other than for possible substitution of $\mathbf{x}'_{\mathcal{A}}$ for the inputs $\mathbf{x}_{\mathcal{A}}$), such that $\boldsymbol{OUT}_{\mathcal{R}}(\sigma^*_{\mathcal{R}}(\mathbf{x}_{\mathcal{R}}), \mathbf{r}, \mathcal{A}) \neq \perp_{\mathcal{R}}$.

Nash equilibrium: *(For all constants c,) $\sigma^*(\mathbf{x})$ is a (computational k^{-c}-)Nash equilibrium for any partitioning $\mathcal{P} = \mathcal{R} \cup \mathcal{A}$.*

Survival condition: *(For all constants c,) for any partitioning $\mathcal{P} = \mathcal{R} \cup \mathcal{A}$, for any $P_i \in \mathcal{R}$, the strategy $\sigma^*_i(x_i)$ survives iterated deletion of (computational k^{-c}-)weakly dominated strategies.*

Discussion of Definitional Framework. The definitions above are a first step in reconciling the definitional framework of SMPC with that of game theory. Our starting point in this work was our main protocol of Section 4 and the definitions presented above are meant to capture the properties of our protocol. Further definitional study of MMPC may lead to interesting insights.

3 The Ideal-World Protocol: Using a Special Channel

In this section, we assume that all participants have access to a special (ideal) communication channel C that essentially does all the computational work: the function F_C that describes the operation of C is a randomized process that computes the target functions of the parties. Using this channel, the only way that the participants influence a protocol is by contributing an input to C at the beginning, and then at every round, notifying the channel whether they wish to participate at this round or not. Additionally, the channel C notifies the parties about the other parties' participation, and enables each P_i to broadcast any string v_i.

The Channel $C_{m,\mathbf{f},c} = (\mathcal{P}_I, F_C, \text{state}, \mathcal{P}_O)$

Setting: The channel $C_{m,\mathbf{f},c}$ is a synchronous channel used by n participants $\mathcal{P} = \{P_1, \ldots, P_n\}$ (where up to t of them may be malicious). For channel $C_{m,\mathbf{f},c}, \mathcal{P}_I = \mathcal{P}_O = \mathcal{P}$, state is initially $\perp$, m and c are integer parameters and $\mathbf{f} = (f_1, \ldots, f_n)$ is a set of functions where each f_i takes as input n binary strings and outputs a binary string; operation function F_C is described below.

Initial round: In the initial round $r = 0$, the channel is activated and each participant P_i submits to $C_{m,\mathbf{f},c}$ an input x_i; if some P_i does not, then by convention, $x_i = \perp$. Define $\mathcal{P}_0 = \{P_i : x_i \neq \perp\}$, i.e., the set of participants that contributed inputs. $C_{m,\mathbf{f},c}$ announces the set $\mathcal{P}_0$ to all participants. $C_{m,\mathbf{f},c}$ stores $\text{state} = \{\mathbf{x}, r\}$. This signals the end of the initial round.

Round $r > 0$: Each round includes the following.

  – The channel is activated and each participant P_i gives $C_{m,\mathbf{f},c}$ a value $z^r_i \in \{\texttt{Compute}, \texttt{Defect}\}$; if some P_i does not, then by convention, $z^r_i =$

Defect. Let $\mathcal{P}_r = \{P_i : z_i^r = \texttt{Compute}\}$. Each participant may also contribute an additional value v_i; if P_i does not contribute, then $v_i = \bot$.

- If $c = 0$ or $r \bmod c \equiv 0$, then $C_{m,\mathbf{f},c}$ flips a coin to obtain a random bit b. Otherwise, $b = 0$.
- If $|\mathcal{P}_r| \geq m$, i.e., at least m participants wish to have the functions computed at this round, and $b = 1$, then $C_{m,\mathbf{f},c}$ sets $y_i = f_i(x_1, \ldots, x_n)$ for $1 \leq i \leq n$. Otherwise, $y_i = \bot$ for all i.
- $C_{m,\mathbf{f},c}$ sends to each participant P_i the value y_i, vector $\mathbf{v}$ and set $\mathcal{P}_r$, and stores r. This signifies the end of round r. Proceed to round $r + 1$.

Let $\mathcal{P} = \mathcal{R} \cup \mathcal{A}$, $|\mathcal{A}| = t$. We describe a protocol, a suggested strategy σ^C, for computing any NCC target function $\mathbf{f}$ over channel $C_{m,\mathbf{f},c}$ and next show that σ^C is an unconditionally t-secure preferred strategy according to Definition 8.

The Strategy $\sigma_i^C(x_i^*)$ for party $P_i \in \mathcal{R}$

Setting: For all $i \in [n]$, party P_i has input x_i^* and has access to channel $C_{m,\mathbf{f},c}$.
    Parties are also connected with each other via arbitrary communication links.

Initial round: Send $x_i = x_i^*$ to C. Receive the set $\mathcal{P}_0$ from C.

Round $r > 0$: In each round, P_i acts as follows.

- If any message has ever arrived over any communication channel other than $C_{m,\mathbf{f},c}$, never send any messages again (in any round), unless you know that the message is from the adversary $\mathcal{A}$, in which case ignore it.
- If $|\mathcal{P}_{r-1}| = n$, send $z_i^r = \texttt{Compute}$ to $C_{m,\mathbf{f},c}$, unless it was previously decided not to send any messages again. Otherwise, never send any messages again (in any round).
- Receive value y_i, vector $\mathbf{v}$ and set $\mathcal{P}_r$ from $C_{m,\mathbf{f},c}$.
- If $y_i \neq \bot$, output y_i and halt. Otherwise, proceed to round $r + 1$.

Lemma 1. *Let $\mathcal{P} = \mathcal{R} \cup \mathcal{A}$ and suppose that $\mathcal{A}$ is an entity that can be contacted. The following puppet strategy $\sigma_i^p(x_i)$ survives iterated deletion for all i: on input x_i, send x_i to $\mathcal{A}$; then follow instructions from $\mathcal{A}$ regarding messages to send to other parties and when a message from party P_j is received over any channel, forward it to $\mathcal{A}$; halt when receive $f_i(\mathbf{x})$ from $\mathcal{A}$ or when can compute it based on available information.*

Proof. (Idea) Consider the case when all P_ℓ, $\ell \neq i$, each follows $\sigma_\ell^p(x_\ell)$, and $\mathcal{A}$ acts such that either rewards or punishes P_i depending on whether or not P_i conforms with $\sigma_i^p(x_i)$, which can be detected by $\mathcal{A}$. It is within $\mathcal{A}$'s power to reward or punish P_i, because assuming that all $P_j \in \mathcal{R}$ follow $\sigma_j^p(x_j)$, $\mathcal{A}$ knows everyone's inputs. This scenario creates the situation where no other strategy can be strictly preferable to $\sigma_i^p(x_i)$ for P_i. □

We present sufficient conditions for a general class of strategies that use channel $C_{m,\mathbf{f},c}$ to survive iterated deletion even in the presence of side-channels. Let $\mathcal{P} = \mathcal{R} \cup \mathcal{A}$ be parties receiving inputs from $\mathbf{X}$ with protocol utility functions $\mathbf{u}$ and channel structure $\mathcal{C}(\mathcal{P})$ that includes $C_{m,\mathbf{f},c}$ and $\mathbf{f}$ be t-NCC target functions. We say that $\boldsymbol{\sigma}$ are *indefinite all-or-nothing strategies with visible deviations* if the following conditions are met: (1) $\boldsymbol{\sigma}(\mathbf{x})$ are strategies which, when executed jointly by $\mathcal{P}$, have the invariant that at every step of the computation, either each $P_i \in \mathcal{R}$ has already learned $f_i(\mathbf{x})$ and halted, or none of them have learned $f_i(\mathbf{x})$, and, moreover, all the messages they received so far are distributed independently of the value $f_i(\mathbf{x})$; (2) let E_s be the event that, under $\boldsymbol{\sigma}(\mathbf{x})$, after s steps of the protocol, no $P_i \in \mathcal{R}$ knows its $f_i(\mathbf{x})$; then there exists some $c(s) > 0$ such that for all views that P_i received by running $\sigma_i(x_i)$, it holds $0 < \Pr[E_{s+c(s)} \mid E_s, VIEW_i] < 1$, where the probability is over the random choices of the channels used; and (3) let the *signature* of a strategy given the inputs $\mathbf{x}$ and strategies of other parties and $\mathcal{A}$, consist of all the messages other parties received from P_i; then there exists a detection procedure that given a signature of $\sigma_i(x_i)$ can determine whether all of the actions of P_i were computed the same as if P_i was following $\sigma_i(x)$ for some x.

Lemma 2. *If $\boldsymbol{\sigma}(\mathbf{x})$ are indefinite all-or-nothing strategies with visible deviation, then any $\boldsymbol{\sigma}_i(\mathbf{x}_i)$, $i \in [n]$, survives iterated deletion of weakly dominated strategies.*

Proof. (Sketch) Consider a strategy $\sigma'(x_i)$. Suppose that there exist strategies τ_{-i} and an adversary $\mathcal{A}$ such that, if P_i follows $\sigma'(x_i)$, and $\mathcal{P}_{-i}$ follow τ_{-i}, then P_i's utility is higher when following σ_i' than σ_i. Then there also exist strategies τ_{-i}' that survive iterated deletion and an adversary $\mathcal{A}$ where P_i gets strictly more utility when following σ_i than when following σ_i'. Let strategies τ_{-i}' be just $\sigma_{-i}^p(\mathbf{x}_{-i})$, which by Lemma 1 survive iterated deletion. Suppose that $\mathcal{A}$ directs each $P_j \in \mathcal{R}_{-i}$ to send messages according to τ_j. Let d be the first deviation (from $\sigma_i(x_i)$) point in the protocol: at step d, if P_i is following σ_i' rather than σ_i, then with positive probability this will be detected. If at step d, P_i acts in a way that agrees with σ_i, then he is rewarded by $\mathcal{A}$ with $\mathbf{x}_{-i}$ and thus can compute the desired value $f_i(\mathbf{x})$. If at step d, P_i deviates, then $\mathcal{A}$ punishes P_i by directing all parties P_j not to send any more messages. Thus, in this setting and given that σ_i is an indefinite all-or-nothing strategy, following some $\sigma_i(x)$ gives better utility than $\sigma_i'(x_i)$ and no strategy σ' can weakly dominate all $\sigma_i(x)$ after any number of rounds of iterated deletion. Finally, following $\sigma_i(x_i)$ cannot be weakly dominated by following any $\sigma_i(x')$ where $x' \neq x_i$, because $\mathbf{f}$ is t-NCC.   □

Theorem 1. *Given $\mathcal{P} = \mathcal{R} \cup \mathcal{A}$ receiving inputs from $\mathbf{X}$, with channel structure $\mathcal{C}(\mathcal{P}) \supseteq C_{m,\mathbf{f},c}$ and protocol utility functions $\mathbf{u}$, strategies $\boldsymbol{\sigma}^C(\mathbf{x})$ constitute an unconditionally t-secure preferred protocol for any t-NCC target function $\mathbf{f}$.*

Proof. (Sketch) Correctness and security hold for strategy $\boldsymbol{\sigma}^C(\mathbf{x})$ in a straightforward way, since our working channel $C_{m,\mathbf{f},c}$ is an ideal one. The Nash equilibrium condition is satisfied: given that all other parties follow $\boldsymbol{\sigma}^C(\mathbf{x})$, P_i has no incentive to deviate ($\mathbf{f}$ is t-NCC). Finally, strategy $\boldsymbol{\sigma}^C(\mathbf{x})$ survives iterated deletion, by Lemma 2 and because it is a vector of indefinite all-or-nothing strategies with visible deviations: at any round, either each $P_i \in \mathcal{R}$ already knows $f_i(\mathbf{x})$ and has halted or no party knows $f_i(\mathbf{x})$; also, no party has any information related to the termination of the current run; finally, all possible deviations from $\sigma_i^C(\mathbf{x})$ are visible given some strategy τ of the other parties that appropriately uses $C_{m,\mathbf{f},c}$ or some other channel.   □

Remark. Strategy $\boldsymbol{\sigma}^C$ aborts in case of side-channel communication, so that any subset of rational parties has a disincentive to collude and thus exclude other parties. The price is that even one malicious party can cause the entire protocol to abort. In the full paper, we discuss a modified protocol and the issue of adversarial abort as it is related to covert channels and the survival condition.

4   The Real-World Protocol: Using Secure MPC

In this section we present and analyze our main protocol for computing any function in MMPC. Our protocol $\boldsymbol{\sigma}$ requires only a synchronous broadcast channel. The main idea is to implement the special channel $C_{m,\mathbf{f},c}$ using secure multiparty computation, where in each round the parties compute secret shares of either a useless value or the outputs $\mathbf{f}(\mathbf{x})$, and then they compute the target function by performing a global broadcast of such shares. At the same time, the parties conform to the protocol $\boldsymbol{\sigma}(\mathbf{x}) \triangleq \boldsymbol{\sigma}^C(\mathbf{x})$: if something goes not as expected (essentially, a Defect was ordered), this is always detected and parties abort the computation. We then use the results on $C_{m,\mathbf{f},c}$ and $\boldsymbol{\sigma}^C(\mathbf{x})$, for $t+1 < m = \lceil \frac{n}{2} \rceil$.

The Protocol $\boldsymbol{\sigma}$ – Strategy $\sigma_i(x_i)$ for party $P_i \in \mathcal{R}$

Inputs to the protocol, and general rules: $\mathbf{x} \leftarrow \mathbf{X}$, each P_i receives x_i, and security parameter 1^k. If P_i receives any message on any channel other than the broadcast channel at any point in the computation, it aborts unless it knows that the message is from the adversary, in which case it ignores it.

Setup phase: In the setup phase, all the parties jointly agree on a random string CRS of length $\ell(k)$ (to be defined later), and a public-key infrastructure with PK_i for each P_i. This is done as follows:

1. Each P_i chooses a random string CRS_i of length $\ell(k)$ and broadcasts a commitment $Com_i \leftarrow Commit(1^k, CRS_i)$ (where we use a computationally hiding, unconditionally binding cryptographic commitment) and a public key PK_i. If P_i detects that P_j failed to broadcast, then P_i aborts.
2. In turn, each P_i proves to each P_j over the broadcast channel that he knows the opening of the commitment Com_i, and a secret key corresponding to his public key PK_i, using a zero-knowledge proof of knowledge protocol with security parameter k. P_j broadcasts whether he accepts or rejects the proof. If P_l sees that some P_j does not accept the proof of some P_i (or fails to broadcast his decision), then P_l aborts.
3. Each P_i broadcasts the value CRS_i. If P_i sees that some P_j failed to broadcast, then P_i aborts.
4. In turn, each P_i proves to each P_j, using a zero-knowledge proof protocol over the broadcast channel, that the value he broadcasted corresponds to his commitment Com_i. After this, each P_j broadcasts whether he accepts or rejects the proof. If P_l sees that some P_j does not accept the proof of some P_i (or fails to broadcast his decision), then P_l aborts.
5. Obtain $CRS = \oplus_{i=1}^{n} CRS_i$, and $PKI = (PK_1, \ldots, PK_n)$.

The input phase: Parse $CRS = CRS_{COM} \circ CRS_{MPC} \circ CRS_{NIZK}$. Let $SECom$ be a simulatable and extractable commitment scheme [2] (which requires CRS_{COM} as input). Each P_i broadcasts $z_i = SECom(CRS_{COM}, x_i, r_i^{SEC})$, where r_i^{SEC} is the randomness needed to form the commitment. If P_j sees that some P_i failed to broadcast a valid commitment, P_j aborts.

The MPC phase: As a result of this phase, we want P_i to obtain m-out-n secret shares of either the strings $y_i = 0^{p_i}$ for $1 \leq i \leq n$ or the strings $y_i = out_i = f_i(\mathbf{x})$, where $p_i = |out_i|$ is known ahead of time. We want each type of output to be equally likely. This is done as follows: P_i chooses a random r_i^{MPC} (of appropriate length, to be specified later) and contributes input $(x_i, r_i^{SEC}, r_i^{MPC})$ to a secure multi-party protocol for computing an n-input function $g_{PKI,\mathbf{z}}$ over the broadcast channel [3], using CRS_{MPC} as a common random string.

Function $g_{PKI,\mathbf{z}}$ operates as follows on input $\{(x_i, r_i^{SEC}, r_i^{MPC}) : 1 \leq i \leq n\}$:

- Check that for all $1 \leq i \leq n$, $z_i = SECom(CRS_{COM}, x_i, r_i^{SEC})$. If for some i, the check fails, output the n-bit string y where $y_i = 0$ iff the check failed for z_i.
- Compute $r = \oplus_{i=1}^{n} r_i^{MPC}$. Parse $r = R_1 \circ R_2 \circ R_3$, where $|R_1| = 1$, and the lengths of R_2 and R_3 will be clear from the sequel.
- If $R_1 = 0$, then for each $1 \leq i \leq n$, come up with an m-out-n secret sharing of 0, and let $y_{i,j}$ be the j'th share. Otherwise, for each $1 \leq i \leq n$, come up with an m-out-n secret sharing of $1 \circ f_i(\mathbf{x})$, and let $y_{i,j}$ be the j'th share. Secret sharing requires randomness: use R_2 as the random tape for this step.
- Use R_3 as the random tape for forming n^2 ciphertexts as follows: $c_{i,j} = Enc(PK_j, y_{i,j})$; output these ciphertexts.

Suppose it takes $c - 1$ rounds of computation over the broadcast channel to jointly compute g. If at any round, P_j notices that P_i did not follow the protocol correctly[1] then P_j aborts. If the output of the computation indicates that some party contributed incorrect inputs, then P_j also aborts.

The possible reconstruction phase: Each P_i broadcasts a message of the form $(\{d_{j,i} : 1 \leq j \leq n\}, \pi)$, where π is a non-interactive simulation-sound zero-knowledge proof [4] that each $d_{j,i} = Enc(PK_j, y_{j,i}, r_{j,i})$ where $y_{j,i}$ is the correct decryption of $c_{j,i}$ under the public key PK_i, and $r_{j,i}$ are the cointosses for probabilistic encryption Enc. If P_i receives m messages with valid proofs, then he decrypts all ciphertexts $\{d_{i,j}\}$ addressed to him to obtain m shares either of 0 or of $1 \circ f_i(\mathbf{x})$. In the former case, if P_i received fewer than n messages with valid proofs, abort; otherwise go back to the MPC phase. In the latter case, output $f_i(\mathbf{x})$.

We analyze protocol $\boldsymbol{\sigma}$ in the mixed-behavior model and present our main result in the case where parties $\mathcal{P} = \mathcal{R} \cup \mathcal{A}$ are computationally bounded. We

---

[1] Recall that, following Goldreich, Micali and Wigderson [6], Canetti *et al.*'s protocol allows every party to detect that another party deviated from the protocol.

impose the following assumptions on protocol utilities in the computational setting: (1) if different runs produce indistinguishable views for P_i, this makes only
a negligible difference to satisfaction measure μ_i; (2) negligible differences in
satisfaction measures of all $P_j \in \mathcal{R}$ yield negligible differences in P_i's utility u_i.

Definition 9 (Computational satisfaction assumption). *Protocol utility
function* **u** *with associated satisfaction* $\boldsymbol{\mu}$ *satisfies the* computational satisfaction assumption, *if for all non-negligible functions* ϵ, *there exists some negligible
function* ν *such that* $\Pr_s[\mu_i(V_1) - \mu_i(V_2) > \epsilon(k)] = \nu(k)$, *where* $V_1 \leftarrow D_1(1^k)$ *and*
$V_2 \leftarrow D_2(1^k)$ *are views that are indistinguishable by probabilistic algorithms with
running time polynomial in* k. *Moreover, function* $u_i(R) = u'_i(\mu_1(R), ..., \mu_n(R))$
has the property that, for any negligible function $\nu_1(k)$, *there exists a negligible*
$\nu_2(k)$ *such that if* R *and* R' *are such that* $\mu_i(R) \le \mu_i(R') \le \mu_i(R) + \nu_1(k)$, *and
for all* $j \ne i$, $\mu_j(R) = \mu_j(R')$, *then* $u_i(R) \le u_i(R') \le u_i(R) + \nu_2(k)$.

Theorem 2 (Main result). *Given* $\mathcal{P} = \mathcal{R} \cup \mathcal{A}$ *receiving inputs from* **X**, *with
channel structure* $\mathcal{C}(\mathcal{P})$ *that includes a synchronous broadcast channel and protocol utility function* **u** *satisfying the computational satisfaction assumption, strategies* $\boldsymbol{\sigma}(\mathbf{x})$ *constitute a computational t-secure preferred protocol for any t-NCC
target function* **f**.

To prove our main result, we: (1) define a notion of computational reducibility
among strategies designed for different channel structures, (2) show that computational reducibility between strategies results in computational equivalence
in gained utilities, (3) show that $\boldsymbol{\sigma}^{\boldsymbol{C}}$ executed over $\mathcal{C}(\mathcal{R}) \supseteq C_{m,\mathbf{f},c}$ is reduced to
$\boldsymbol{\sigma}$ executed over any channel structure that includes the synchronous broadcast
channel, thus proving that $\boldsymbol{\sigma}$ is an ϵ-Nash equilibrium, and (4) show that $\boldsymbol{\sigma}$
survives deletion.

Definition 10 (Computational reducibility). *Suppose* $\mathcal{P} = \mathcal{R} \cup \mathcal{A}$, $P_i \in \mathcal{R}$.
Let $\mathcal{C} = \mathcal{C}_1 \cup \{C\}$, $\mathcal{C}' = \mathcal{C}_1 \cup \{C'\}$ *be channel structures available to* $\mathcal{P}$, *for
some set of channels* $\mathcal{C}_1$ *and some additional channels* C *and* C'. *Let* $(\boldsymbol{\sigma}_{\mathcal{R}})_{-i}$
and $(\boldsymbol{\sigma}'_{\mathcal{R}})_{-i}$ *be sets of strategies corresponding to* $\mathcal{C}$ *and* $\mathcal{C}'$, *respectively. Let*
$\mathcal{A}$ *be an adversary for channel structure* $\mathcal{C}$. *We say that the probabilistic poly-
time simulator* $S = (S_0, S_1, S_2, S_3)$ *computationally reduces setting* $(\mathcal{C}', (\boldsymbol{\sigma}'_{\mathcal{R}})_{-i})$
to setting $(\mathcal{C}, (\boldsymbol{\sigma}_{\mathcal{R}})_{-i})$ *if for all* **x**, *PPT strategies* τ_i, $j \in [n]$ *and PPT ad-
versaries* $\mathcal{A}$, $S_3(j, VIEW_j((S_1(\tau_i, x_i, s), (\boldsymbol{\sigma}'_{\mathcal{R}})_{-i}((\mathbf{x}_{\mathcal{R}})_{-i})), \mathbf{r}, S_2(\mathcal{A}, i, \mathbf{x}_{\mathcal{A}}, s)), s)$,
where $s \leftarrow S_0(1^k)$, *is a distribution that is computationally indistinguishable from*
$VIEW_j((\tau_i(1^k, x_i), (\boldsymbol{\sigma}_{\mathcal{R}})_{-i}((\mathbf{x}_{\mathcal{R}})_{-i})), \mathbf{r}, \mathcal{A})$. *In this setting, we say that* S *computationally translates strategy* $\tau_i(x_i)$ *and adversary* $\mathcal{A}(\mathbf{x}_{\mathcal{A}})$ *into strategy* $S_1(\tau_i,$
$x_i, s)$ *and adversary* $S_2(\mathcal{A}, \mathbf{x}_{\mathcal{A}}, s)$.

That is, S_1 transforms strategy τ_i designed for communication structure $\mathcal{C}$ into a
strategy designed for communication structure $\mathcal{C}'$, while S_2 does the same with
$\mathcal{A}$. It is possible, however, that the resulting strategy and adversary talk to each
other. Then S_3 translates the resulting view into a view that is computationally
close to the view that P_i could have gotten if it were to run τ_i in $\mathcal{C}$ with adversary

$\mathcal{A}$, or into a view that any other rational P_j would have gotten if it were to run σ_i in $\mathcal{C}$ with adversary $\mathcal{A}$ and τ_i, instead of running σ_i' in $\mathcal{C}'$.

Lemma 3. *Under the computational satisfaction assumption, if: (1)* $\mathbf{f}$ *is t-NCC, and is the computational target function for the given set of utility functions* $\mathbf{u}$; *(2) for all $i \in [n]$, for strategies $\boldsymbol{\sigma}$ and $\boldsymbol{\sigma}'$, simulator $S = (S_0, S_1, S_2, S_3)$ computationally reduces setting $(\mathcal{C}', (\sigma_{\mathcal{R}}')_{-i})$ to the setting $(\mathcal{C}, (\sigma_{\mathcal{R}})_{-i})$; (3) for all $i \in [n]$, x_i and s, $S_1(\sigma_i, x_i, s) = \sigma_i'(x_i)$; (4) for all $\mathcal{A}$, i, probabilistic poly-time τ_i, $j \neq i$, $P_j \in \mathcal{R}$, $\mu_j(S_3(j, VIEW_j((S_1(\tau_i, x_i, s), (\sigma_{\mathcal{R}}')_{-i}((\mathbf{x}_{\mathcal{R}})_{-i})), \mathbf{r}, S_2(\mathcal{A}, i, \mathbf{x}_{\mathcal{A}}, s)), s))$ is equal to $\mu_j(VIEW_j((S_1(\tau_i, x_i, s), (\sigma_{\mathcal{R}}')_{-i}((\mathbf{x}_{\mathcal{R}})_{-i})), \mathbf{r}, S_2(\mathcal{A}, i, \mathbf{x}_{\mathcal{A}}, s)))$, that is, no matter what $\mathcal{A}$ and P_i do, when transforming the view from running σ_i' to one for running σ_j, the simulator S_3 did not throw out any information relevant to μ_j. Then, if $\boldsymbol{\sigma}'$ is a Nash equilibrium, then $\boldsymbol{\sigma}$ is ϵ-Nash equilibrium.*

Proof. (Sketch) For brevity, let us denote $(\sigma_{\mathcal{R}})_{-i}((\mathbf{x}_{\mathcal{R}})_{-i})$ simply as $\boldsymbol{\sigma}_{-i}$. Consider strategy τ_i, adversary $\mathcal{A}$, inputs $\mathbf{x}$. Then we know that:

$$\mu_i(VIEW_i((\tau_i(x_i), \boldsymbol{\sigma}_{-i}), \mathbf{r}, \mathcal{A}(\mathbf{x}_{\mathcal{A}})))$$
$$\leq \mu_i(S_3(i, VIEW_i((S_1(\tau_i, x_i, s), \boldsymbol{\sigma}'_{-i}), \mathbf{r}', S_2(\mathcal{A}, i, \mathbf{x}_{\mathcal{A}}, s)), s)) + \nu_1(k)$$
$$\leq \mu_i(VIEW_i((S_1(\tau_i, x_i, s), \boldsymbol{\sigma}'_{-i}), \mathbf{r}', S_2(\mathcal{A}, i, \mathbf{x}_{\mathcal{A}}, s))) + \nu_1(k).$$

The first inequality follows by the definition of computational reducibility and the computational satisfaction assumption; the second inequality holds because S_3 may destroy some relevant information. Next, in channel structure $\mathcal{C}$, we measure μ_j for a party P_j following strategy σ_j while P_i is following τ_i. From the definition of computational reducibility, $\mu_j(VIEW_j((\tau_i(x_i), \boldsymbol{\sigma}_{-i}), \mathbf{r}, \mathcal{A}(\mathbf{x}_{\mathcal{A}})))$ $\approx \mu_j(S_3(j, VIEW_j((S_1(\tau_i, x_i, s), \boldsymbol{\sigma}'_{-i}), \mathbf{r}', S_2(\mathcal{A}, i, \mathbf{x}_{\mathcal{A}}, s)), s))$. This is approximately equal to $\mu_j(VIEW_j((S_1(\tau_i, x_i, s), \boldsymbol{\sigma}'_{-i}), \mathbf{r}', S_2(\mathcal{A}, i, \mathbf{x}_{\mathcal{A}}, s)))$, by the conditions of the lemma. Thus, for every strategy τ_i and adversary $\mathcal{A}$ in channel structure $\mathcal{C}$, there exist a strategy τ_i' (i.e., $S_1(\tau_i, x_i, s)$) and adversary $\mathcal{A}'$ (i.e., $S_2(\mathcal{A}, i, \mathbf{x}_{\mathcal{A}}, s)$), in channel structure $\mathcal{C}'$ such that:

$$u_i((\tau_i(x_i), \boldsymbol{\sigma}_{-i}), \mathbf{r}, \mathcal{A}(\mathbf{x}_{\mathcal{A}})) \leq u_i((\tau_i'(x_i), \boldsymbol{\sigma}'_{-i}), \mathbf{r}', \mathcal{A}'(\mathbf{x}_{\mathcal{A}})) + \nu_2(k)$$
$$\leq u_i(\sigma_{\mathcal{R}}'(\mathbf{x}_{\mathcal{R}}), \mathbf{r}', \mathcal{A}'(\mathbf{x}_{\mathcal{A}})) + \nu_2(k) = u_i(\sigma_{\mathcal{R}}(\mathbf{x}_{\mathcal{R}}), \mathbf{r}', \mathcal{A}(\mathbf{x}_{\mathcal{A}})) + \nu_2(k),$$

where the first inequality holds by the computational satisfaction assumption. The second inequality follows because $\boldsymbol{\sigma}'$ is a Nash equilibrium, while the last equality follows by condition (3) of the lemma and Definition 9. □

Lemma 4. *If $\mathbf{f}$ is t-NCC, then there exists a simulator $S = (S_0, S_1, S_2, S_3)$ satisfying the conditions of Lemma 3, reducing the (ideal) strategy $\boldsymbol{\sigma}'$ in channel structure $\mathcal{C}'$ that includes (ideal) channel $C_{m,\mathbf{f},c}$ to the (real) strategy $\boldsymbol{\sigma}$ in the channel structure $\mathcal{C}$ that includes a synchronous broadcast channel.*

Proof. (Sketch) The main idea is that, for rational P_i following strategy τ_i instead of σ_i, we will have simulator S_1 broadcast his input x_i and his strategy τ_i over the ideal channel $C_{m,\mathbf{f},c}$, while S_2 broadcasts the adversary's inputs and descriptions. This will ensure that the view from the resulting strategy for the ideal channel will enable S_3 to simulate what P_i and $\mathcal{A}$ would do, and create a correctly distributed view for each $P_j \in \mathcal{R}$, $j \neq i$. S_1 and S_2 will then collaboratively simulate the views for P_i and $\mathcal{A}$ using the corresponding simulators and extractors for ZK proofs of knowledge, commitments, and multi-party computation. S_3 will simply output the simulated views created by S_1 and S_2. □

Lemma 5. *Under the computational satisfaction assumption, if* f *is a t-NCC computational target function, then protocol* σ *survives iterated deletion of computational* k^{-c}*-weakly dominated strategies.*

Proof. (Sketch) We appropriately adapt the proof of Lemma 2 for strategy σ, considering all different ways in which P_i may depart from σ_i. □

Acknowledgements. We thank the anonymous reviewers and Jonathan Katz for their constructive comments, and also Silvio Micali and Moni Naor for helpful discussions. Anna Lysyanskaya is supported by NSF Career Grant CNS–0374661 and Nikos Triandopoulos by NSF Grants CCF–0311510 and IIS–0324846.

References

[1] I. Abraham, D. Dolev, R. Gonen, and J. Halpern. Distributed computing meets game theory: Robust mechanisms for rational secret sharing and multiparty computation. In *Proc. 25th ACM PODC*, 2006. (To appear.)

[2] R. Canetti and M. Fischlin. Universally composable commitments. In *Proc. CRYPTO 2001*, pages 19–40, 2001.

[3] R. Canetti, Y. Lindell, R. Ostrovsky, and A. Sahai. Universally composable two-party and multi-party secure computation. In *Proc. 34th ACM Symposium on Theory of Computing (STOC)*, pages 494–503, 2002.

[4] A. De Santis, G. Di Crescenzo, R. Ostrovsky, G. Persiano, and A. Sahai. Robust non-interactive zero knowledge. In *Proc. CRYPTO 2001*, pages 566–598, 2001.

[5] O. Goldreich. *Foundations of Cryptography*. Cambridge University Press, 2004.

[6] O. Goldreich, S. Micali, and A. Wigderson. How to play any mental game or a completeness theorem for protocols with honest majority. In *Proc. 19th ACM Symposium on Theory of Computing (STOC)*, pages 218–229, 1987.

[7] S. D. Gordon and J. Katz. Rational secret sharing, revisited, 2006. Manuscript available at http://eprint.iacr.org/2006/142. (To appear at *SCN 2006*.)

[8] J. Halpern and V. Teague. Rational secret sharing and multiparty computation: extended abstract. In *Proc. 36th ACM Symposium on Theory of Computing (STOC)*, pages 623–632, 2004.

[9] S. Izmalkov, M. Lepinski, and S. Micali. Rational secure computation and ideal mechanism design. In *Proc. 46th IEEE Symposium on Foundations of Computer Science (FOCS)*, pages 585–594, 2005.

[10] M. Lepinksi, S. Micali, and abhi shelat. Collusion-free protocols. In *Proc. 37th ACM Symposium on Theory of Computing*, pages 543–552, 2005.

[11] M. Lepinski, S. Micali, C. Peikert, and A. Shelat. Completely fair SFE and coalition-safe cheap talk. In *Proc. 23rd ACM PODC*, pages 1–10, 2004.

[12] A. Shamir. How to share a secret. *Comm. of the ACM*, 22(11):612–613, 1979.

[13] Y. Shoham and M. Tennenholtz. Non-cooperative computation: Boolean functions with correctness and exclusivity. *Theoretical Comp. Science*, 343(2):97–113, 2005.

When Random Sampling Preserves Privacy*

Kamalika Chaudhuri[1] and Nina Mishra[2]

[1] Computer Science Department, UC Berkeley, Berkeley, CA 94720
`kamalika@cs.berkeley.edu`
[2] Computer Science Department, University of Virginia, Charlottesville, VA 22904
`nmishra@cs.virginia.edu`

Abstract. Many organizations such as the U.S. Census publicly release samples of data that they collect about private citizens. These datasets are first anonymized using various techniques and then a small sample is released so as to enable "do-it-yourself" calculations. This paper investigates the privacy of the second step of this process: sampling. We observe that rare values – values that occur with low frequency in the table – can be problematic from a privacy perspective. To our knowledge, this is the first work that quantitatively examines the relationship between the number of rare values in a table and the privacy in a released random sample. If we require ϵ-privacy (where the larger ϵ is, the worse the privacy guarantee) with probability at least $1 - \delta$, we say that a value is rare if it occurs in at most $\tilde{O}(\frac{1}{\epsilon})$ rows of the table (ignoring log factors). If there are no rare values, then we establish a direct connection between sample size that is safe to release and privacy. Specifically, if we select each row of the table with probability at most ϵ then the sample is $O(\epsilon)$-private with high probability. In the case that there are t rare values, then the sample is $\tilde{O}(\epsilon\delta/t)$-private with probability at least $1 - \delta$.

1 Introduction

Private data is collected by numerous organizations for a wide variety of purposes including reporting, data mining, scientific discoveries, etc. In some circumstances, this data is in turn released in sanitized form for public consumption. The purpose of releasing such sanitized data is to enable others to discover large-scale statistical patterns, e.g., to learn averages, variances, clusters, decision trees, while hiding small-scale information, e.g., a particular individual's salary. The question we ask is: To what extent do these sanitized datasets preserve people's privacy?

While there are numerous examples of sanitization procedures, we investigate one commonly used technique: random sampling. Some organizations routinely collect data, anonymize it, and then release a sample so that others may use the data for data mining purposes. Since samples are known to preserve statistical characteristics of the data, samples can be a useful means for studying and understanding the underlying population.

---

* Research supported in part by NSF EIA-0137761.

C. Dwork (Ed.): CRYPTO 2006, LNCS 4117, pp. 198–213, 2006.

1.1 Motivating Examples

There are many examples of released samples of private data. We describe two, one from the U.S. Census and one from the Social Security Administration.

The U.S. Census Bureau releases a Public Use Microdata Sample (PUMS) [2]. This dataset contains private information in occupied housing units such as age, weight, income, and race. The Census gathers this data once every 10 years, anonymizes it and then releases a 1% or 5% sample. The purpose of releasing this microdata is to allow "do-it-yourself" calculations. Our work is motivated by such releases: Can we simply select each individual with probability 0.05 to be included in the sample? What role do rare values play in deciding what to release? What size sample is safe to release?

The Social Security Administration (SSA) also releases microdata, specifically Benefits and Earnings files [11]. Old-Age, Survivors, and Disability Insurance (OASDI) is a government-sponsored insurance program that individuals contribute to throughout their working careers. Benefits are paid to insured workers and family members when they retire or become disabled. This dataset contains annual earnings information for approximately 47 million individuals who receive OASDI benefits each month. Personal identifying information and distinguishing characteristics are removed or modified to prevent identification. Records are randomly permuted. The SSA then releases a 1% sample of this data.

Our work is a first attempt at formally understanding the privacy guarantees of just one step of the sanitization process: random sampling. In practice, these organizations employ multi-step anonymization processes prior to sampling that this paper does not analyze.

1.2 The Model

We consider the following simplified setting. The sanitizer starts with a table T consisting of k distinct private values. The k values can be anything, e.g., Boolean data over $\log k$ attributes, k real numbers, etc. The sanitizer then goes through each row of the table and includes it in the sample with probability p and does nothing with probability $1 - p$. The sample is then randomly permuted and released. We then ask the question: for what p can we guarantee privacy? In order to understand this question, we next define privacy.

1.3 Privacy

The privacy definition that we use is motivated by [5]. The authors capture the interaction between a sanitizer and a hypothetical attacker via a transcript. In our case, the transcript is a random sample S of the data. Intuitively, for any pair of tables T and T' that differ in only one position, privacy is preserved if a hypothetical attacker upon seeing the transcript S is unable to distinguish between the case when the actual table is T or T'. In other words, an attacker knowing all but one person i's private information does not gain much information about i upon seeing the sample.

We consider two definitions of privacy, one where a hypothetical attacker tries to distinguish between two tables that differ in one row, and the other, where a hypothetical attacker tries to distinguish between two tables that differ in c rows. We say that a sanitization scheme is (c, ϵ, δ)-private if for every table T, with probability at least $1 - \delta$, the scheme produces a sample S such that for any set of c rows in the table, P, and for any two sets of c assignments V and V', $\frac{\Pr(S|T_{\{P \to V\}})}{\Pr(S|T_{\{P \to V'\}})} \leq 1 + \epsilon$. The privacy definition is discussed in Section 3.

1.4 Discussion

Our results do not apply in the case that the data is a collection of distinct points, say in R^d. The reason is that if every point is different from every other point in the table then a sample of size even one violates the privacy of that individual. As a simple example, suppose the table consists of five private values $\langle 1, 2, 3, 4, 5 \rangle$ and we release the sample point 2. Then the attacker can now easily tell that the data came from the actual table versus $\langle 1, 3, 3, 4, 5 \rangle$. This violates privacy since the attacker can now distinguish between two tables that differ in one row.

Problems arise even if a value is not unique, but occurs a few times in the table. We call such a value a rare value. Observing a rare value is problematic because a rare value can be assumed by only a small group of individuals, and then observing such a value can potentially increase the hypothetical attacker's confidence about the values assumed by this small select group of people. For example, consider a table in which two individuals can have a salary of one billion dollars, and an attacker knows the salary of one of them and not the other. If we release a sample in which a row with a salary of one billion dollars appears, then the confidence of the attacker about the second individual's salary increases. This is because such a sample is much more likely to have come from a table in which two people have a salary of a billion dollars than from a table in which one person's salary is a billion dollars.

Unique values are known to be problematic in the literature. Indeed, the phrase "population unique" is used to describe those individuals that are unlike anyone else in the population, e.g., 13-year-old college graduate. Population uniques are often first removed prior to data sanitization. To the best of our knowledge, we have not seen work that quantifiably links rare values to privacy. In this paper, we find such a link. If we desire ϵ privacy with probability at least $1 - \delta$, we define a *rare value* to be one that occurs less than $O(\frac{1}{\epsilon} \log(\frac{2k}{\delta}))$ times in the table. If i rows with a certain rare value v appear in the sample, it can lead to an $O(i\epsilon/\log(2k/\delta))$ breach of privacy.

One way to deal with rare values is to suppress such rows from the table. Indeed, in practice, organizations remove population uniques. We do not consider such sanitization algorithms because then the decisions made by the sanitizer cannot be mimicked or simulated by the attacker – and as a result, information may be leaked[1]. This may seem unintuitive at first – how can private information

---

[1] The notion of simulatability is already known to be important in cryptography [8] and also in privacy research [9,3].

that we do not even release breach privacy? An example illustrates the point. Suppose the sanitizer decides to suppress all values that occur < 100 times, and rows 1 to 100 of a table take the value 0, and no other rows take the value 0. Let $p > 1/100$. Now suppose that an attacker knows the value of the first 99 rows and is trying to decide what the value of the 100th row is. In this case, not seeing any row with value 0 in the sample violates the privacy of this 100th row.

Another reason why removing rare values is problematic is that just the size of the sample can leak information. In the case where the table contains n rows and $1/10$th of the rows contain rare values, then the expected sample size is $9np/10$ instead of np. Alternatively, if the sanitization algorithm is to draw a sample and remove "sample uniques" (those individuals that are one-of-a-kind in a sample), then if every entry in the sample is unique, then nothing may be released. Thus just the size of the sample can leak information.

Because unique and rare values can lead to privacy breaches, we assume that k the number of distinct values is much smaller than n the number of individuals in the dataset. In practice, this is not true because each individual in the table has a uniquely identifying key. The assumption that $k << n$ implies that identifying information has been removed. This is admittedly a large assumption since it is an open question what information is "identifying" (see, for example, [12]). But we make this assumption so that we can focus our attention on understanding the privacy consequences of sampling.

One limitation of sampling is the probability that the sanitization algorithm fails to produce an output that preserves privacy is not negligble in n. Ideally, we would like to say that the sanitization algorithm fails with very low probability, e.g., $1/(2^n)$. With random sampling, we cannot guarantee a failure probability better than $1/n$. To see why, suppose the table has a unique value. If we sample each row of the table with probability $p = 1/n$, then the probability we pick this value is $1/n$. But once a unique value appears in the sample then we have completely breached this individual's privacy. So the probability we fail is $1/n$.

Even if there are no rare values at all, releasing a random sample of the data cannot preserve privacy with probability 1. As an example, consider a table that has $n/2$ rows with value 0 and $n/2$ rows with value 1, and a sample S from this table of size $n/2$ consisting of all 0s. The attacker knows the value of all rows in the table except for one row - that is, she is trying to decide whether the sample S came from a table with $n/2$ 0s and $n/2$ 1s or from a table with $n/2 + 1$ 0s and $n/2 - 1$ 1s. The latter event is about $n/2$ times more likely than the former. The attacker will therefore conclude that the value of the missing row is 0, and this will lead to a breach in privacy. It turns out that when we do sampling, we cannot avoid these situations entirely except to upper bound the relatively small probability that such unlikely samples occur. We also note that the probability of failure due to the occurrence of such unlikely samples is quite small compared to the probability of failure to preserve privacy because of the occurrence of one or more rows with rare values in the sample.

1.5   Contributions

Privacy. For the case where an attacker tries to distinguish between two tables that differ in one row, i.e., $(1, \epsilon, \delta)$-privacy, we define a rare value as one that occurs in at most $\frac{\log(\frac{2k}{\delta})}{\epsilon}$ rows of the table, where k is the number of distinct values in the table. We show that if there are no rare values, then a sampling frequency of at most ϵ preserves privacy. In the case where there are at most t rare values, we show that a sampling frequency of at most $\tilde{O}(\frac{\epsilon\delta}{t})$ preserves privacy. Observe that the higher the number of rare values, the lower the sampling frequency, as one would expect. We also demonstrate that the upper bound on the sampling frequency is tight up to log factors.

Furthermore, we consider the case where a hypothetical attacker already knows $n - c$ rows of the table and the goal is (c, ϵ, δ)-privacy. Now a rare value is one that occurs in at most $\frac{\log(\frac{2k}{\delta})}{\epsilon} + c$ rows of the table. We prove that when there are no rare values, a sampling frequency of at most ϵ still preserves $(c, O(c\epsilon), \delta)$-privacy. When there are at most t rare values, we show that a sampling frequency of $p < \tilde{O}(\frac{\epsilon\delta}{t})$ is $(c, O(c\epsilon), \delta)$-private.

The proof technique is similar in both cases. We partition the values in the table T into rare, infrequent and common depending on how often they occur. We then define a good sample to be one with no rare values, with infrequent values that do not occur very frequently, and common values that occur close to expectation. We prove that if we have a good sample, we have privacy. Then we prove that with high probability, random sampling produces a good sample.

Utility. For someone who is interested in discovering patterns in the released data, it is natural to ask whether sampling preserves patterns. Samples are in fact known to approximately preserve statistics about the actual table. But note again that, in practice, sampling is used in concert with other anonymization techniques. We are not making any claims about the utility of those anonymization procedures – we only discuss the utility of sampling.

When we release a random sample of the data, we are essentially releasing an estimate of the *histogram* – the frequency of each value v. Random sampling can estimate the frequency of each value v with an additive noise of $\tilde{O}(\sqrt{\frac{t}{n\epsilon\delta}})$ when there are t rare values, and we want $(1, \epsilon, \delta)$-privacy.

Note that in contrast, [5] can release the histogram of a table by adding a tiny $O(\frac{2}{n\epsilon})$ additive noise to the frequency of each possible value, a significantly smaller additive noise. We also note that unlike sampling, the error in [5] is independent of the number of rare values. Also the more privacy that is required, the better [5] does compared to sampling.

If U is the total universe of values a row in the table can take, we note that [5] needs to release $|U|$ numbers in order to release a privacy-preserving histogram. We in contrast, need to release only k numbers. When the size of U is much larger than k, releasing a random sample is a more compact way of releasing the histogram.

2   Related Work

We partition related work according to what the sanitizer does with the private data. In the input perturbation family of methods, the private data is perturbed and published as a one-time operation. The perturbed dataset must withstand an unlimited number of queries. In the output perturbation family of methods, the sanitizer receives queries about the private dataset from an attacker. The sanitizer then outputs either the true answer, a perturbed answer, or refuses to answer altogether.

2.1   Input Perturbation

While we assume that the dataset is not a collection of distinct points in R^d, another approach is to redefine privacy with respect to this higher dimensional space. Such a compelling definition is given in [3] where a point is kept private if it "blends with the crowd". That paper offers simulatable methods for perturbing the input so that privacy is preserved. Several utility results are given including learning mixtures of Gaussians and k–Center clustering.

An alternate input perturbation technique was suggested in [7]. That paper describes a method for modifying private data (adding and deleting purchase behavior) so as to enable the discovery of frequent itemsets while maintaining privacy. The privacy guarantees given in that paper are quite strong. But the notion of utility is strongly tied to frequent itemsets.

An input perturbation technique based on pseudorandom sketches was given in [10]. The idea is that each individual takes their own data over d bits, represents it as a vector of length 2^d with a 1 in the single position corresponding to their private value and a 0 everywhere else. Each bit of this vector of length 2^d is then flipped with probability p. This perturbed vector is then replaced with a slightly biased coin that forms a seed s to a pseudorandom function. The authors show that privacy is preserved in a strong sense, i.e., for all individuals x_i and for all values v, v', $\Pr(s|x_i = v) \approx \Pr(s|x_i = v')$. Furthermore, various utility results are given including estimating the fraction of individuals that satisfy any conjunction of attributes, estimating the fraction of individuals that have private values $\leq x$, etc.

2.2   Output Perturbation

A different method for preserving privacy is output perturbation [4,6,1,5,9]. One specific output perturbation result that is very relevant to this paper is due to Dwork et al [5]. The authors introduce the notion of the *sensitivity* of a function which is how much the function f can change when one row of the data changes. Privacy is then shown to be preserved if the sanitizer answers each query with additive Laplacian noise that is proportional to the function's sensitivity. Specifically, the sanitizer returns the true answer plus $\text{Lap}(\frac{sen(f)}{\epsilon})$ where $\text{Lap}(\lambda)$ is the Laplace distribution with density proportional to $p(y) \propto e^{-\frac{|y|}{\lambda}}$. The more sensitive the query, the more noise is added. The sensitivity of a sequence of queries is the extent to which the sequence can change when one row of the table changes. For example, the sensitivity of a histogram is 2 since changing one row of the table at most removes a value from one bucket and adds it to another.

3 Preliminaries

We use the term *table* to mean the original unperturbed data and denote it by T. Each entry of the table is assumed to be a tuple of the form (i, j) where i is some unique identifier, e.g., SSN, name, and j is an integer that represents the individual's private data, e.g., if the data is in binary form, then one can view j as the integer representation of the binary data.

We assume that the table has n entries, where each entry can take an integer value. We assume that the total number of distinct values taken by the rows of the table is k.

We use the term *sample* or *sanitized table* to denote the result of the sanitization process that the attacker observes and we denote it by S. Note that S is a randomized object, whereas T is a deterministic input supplied to the sanitizer.

Given a table T, the goal of the sanitizer is to release a sample S of T where the sample does not give the attacker any additional information about any row of the table beyond what the attacker already knows from looking at the rest of the table.

3.1 Privacy Definition

Our definition of privacy is closely related to $(1, \epsilon)$-privacy proposed by [5] (where it was called ϵ-indistinguishability).

Definition 1. *A sanitization mechanism is $(1, \epsilon)$-private if for every pair of tables T and T' that differ in one row*

$$\frac{\Pr[S|T]}{\Pr[S|T']} \leq 1 + \epsilon$$

Here $\Pr[S|T]$ denotes the probability that the sanitization mechanism outputs S given as input the table T and it is taken over the random choices made by the sanitizer. This definition states that the posterior probability that the sample S came from table T is almost the same as the probability it came from table T'; therefore observing S does not enable the attacker to distinguish between these two tables reliably.

As mentioned in Section 1.4, we cannot ensure privacy with probability 1 as the table may have rare values or we may simply draw an unrepresentative sample. Consequently, we allow our sanitizer a δ probability of failure.

Definition 2. *A sanitization mechanism is $(1, \epsilon, \delta)$-private if, for every table T, with probability at least $1 - \delta$, the mechanism produces S such that for all values v and v':*

$$\frac{\Pr[S|T_{\{i \rightarrow v\}}]}{\Pr[S|T_{\{i \rightarrow v'\}}]} \leq 1 + \epsilon$$

where the probability is taken over the random choices made by the sanitizer.

This definition states that regardless of the table T, with high probability, the sample S produced does not significantly help the attacker distinguish between any two values v and v' for the ith individual in the table. While this quantifies over all possible values, it includes as a special case the ith individual's actual value and any other value.

Sometimes, there may be correlations between the values of a small number of rows in a database and these correlations may be known to the attacker. For example, the HIV status of a husband and wife are probably the same. This can be thought of more generally as follows. Suppose the table is partitioned into sets of rows $\{P_i\}$ such that if the attacker knows the value of one row in a partition P_j, she knows the value of every row in the partition. In such a situation, we might want to consider an attacker who knows the value of all rows in the table except for the rows in one partition, and examine what this attacker can learn by looking at the sanitized data. This motivates the notion of (c, ϵ)-privacy proposed by [5]. (c, ϵ)-privacy ensures that the probability that the sanitized data came from two tables T and T' that differ in at most c rows is almost the same.

Typically, more noise is needed to achieve (c, ϵ)-privacy than is needed to achieve $(1, \epsilon)$-privacy. This sounds counterintuitive at first; how could it be harder to guarantee privacy for an attacker who knows the value of only $n - c$ rows of a table than it is to guarantee privacy for an attacker who knows the value of $n - 1$ rows? This happens because we say that a violation of privacy occurs when there is a deviation from what the attacker already knows. To ensure that there is no deviation from the attacker's knowledge, we need to hide more from an attacker who knows less than from an adversary who knows more.

We can think of an analogous notion of (c, ϵ, δ)-privacy as well. The definition is identical to $(1, \epsilon, \delta)$-privacy except for any set of c rows in the table T, P_i, and for any pair of states V and V' the sample does not substantially help the attacker distinguish between $T_{\{P_i \rightarrow V\}}$ and $T_{\{P_i \rightarrow V'\}}$.

It is shown in [5] that a sanitization mechanism that is $(1, \frac{\epsilon}{c})$-private is also (c, ϵ)-private. This argument can be extended to show that a mechanism which is $(1, \frac{\epsilon}{c}, \frac{\delta}{c})$-private is also (c, ϵ, δ)-private.

3.2   Some Notation

We use the following notation for the rest of the paper. Let n be the total number of items in the table, and let $n_1, n_2, \ldots, n_k$ denote the number of items in the table with value $1, 2, \ldots, k$ respectively. Let s denote the size of the sample, and $s_1, s_2, \ldots s_k$ denote the number of items in the sample with value $1, 2, \ldots, k$.

Let $V = \{v_1, v_2, \ldots, v_c\}$ be a sequence of c values. We say that a sequence of c rows has *state* V if row i in the sequence has value v_i.

We use the notation $T_{\{i \rightarrow v\}}$ to denote a table T in which row i is set to have a value v, and $T \setminus \{i\}$ to denote the set of all rows in table T except row i. Similarly, for a set of rows P_i, we use the notation $T_{\{P_i \rightarrow V\}}$ to denote a table T in which the set of rows P_i have state V, and $T \setminus \{P_i\}$ to denote the set of all rows in table T except the rows in set P_i.

4 $(1, \epsilon, \delta)$-Privacy

In this section, we show what sampling probability is $(1, \epsilon, \delta)$-private. Given ϵ, δ, a table T and k, the number of distinct values in the table, we provide p, a sampling frequency that is $(1, \epsilon, \delta)$-private.

Our guarantees can be summarized as follows.

Theorem 3. *Given a table T, let $\alpha = \frac{\delta}{2}$, k be the number of distinct values in T and t be the total number of values in T that occur less than $\frac{2 \log(\frac{k}{\alpha})}{\epsilon}$ times. Also let $\epsilon' = \max(2(p + \epsilon), 6p)$ and $p + \epsilon < \frac{1}{2}$.*

Then, a sample S of T drawn with frequency $p \leq \frac{\epsilon \log(\frac{1}{1-\alpha})}{4t \log(\frac{k}{\alpha})}$ is $(1, \epsilon', \delta)$-private when $t > 0$. When $t = 0$, a sample S of T drawn with frequency $p \leq \epsilon$ is $(1, \epsilon', \delta)$-private.

We need the assumption $p + \epsilon < \frac{1}{2}$ to make sure p is bounded away from 1 by a constant. (Any other constant than $\frac{1}{2}$ would do, but would change the constants in Theorem 3.) We want this condition because if p is too close to 1, all rows containing a certain value may appear in the sample, leading to a serious breach of privacy.

Note that for certain tables such as those consisting only of unique values, the upper bound on p according to our theorem is less than $1/n$. This should be interpreted as the fact that we cannot guarantee $(1, \epsilon, \delta)$-privacy for a sample even of size 1.

The theorem shows that for a given table T and a given failure probability δ, the lower the value of p, the better the privacy guarantee. Since $\log(\frac{1}{1-\alpha}) \approx \alpha$, p varies linearly as δ. This means that δ, the failure probability, has to be at least $\frac{1}{n}$ to ensure we draw a random sample even of size 1. This is expected, because in a table with a unique value v, the probability that any random sample selects this value is at least $\frac{1}{n}$. We see in Observation 6 that this dependence of p on δ is almost tight except for the factor of $\log(\frac{k}{\alpha})$.

Before we prove Theorem 3, we provide some intuition. For our proofs, we divide the set of values in the table T into three categories – rare values, infrequent values and common values.

Definition 4. *A value is said to be a common value if it occurs in at least $\frac{12 \log(\frac{k}{\alpha})}{p}$ rows of the table, where p is the sampling frequency. A value v is called a rare value if it occurs in at most $\frac{2 \log(\frac{k}{\alpha})}{\epsilon}$ rows of the table. A value that is neither rare nor common is called an infrequent value.*

A common value v has the property that the expected number of such values in the sample S is at least $\Omega(\log(\frac{k}{\alpha}))$, and therefore the number of such values in the sample is tightly concentrated around its mean.

If a value v is not a common value, we can only show using Chernoff Bounds that the number of occurrences of v in T is away from its expected value by at most $O(\log(\frac{k}{\alpha}))$. If about $\log(\frac{k}{\alpha})$ rows with a rare value v occur in the released sample, the posterior probability $\Pr[S|T_{\{i \to v\}}]$ can increase by more than a $(1+\epsilon)$ fraction. To deal with this, we hide all such rows. This is achieved by making p less than the inverse of the total number of such rare values.

A value that is neither common nor rare is called an *infrequent value*. Such a value may appear in a sample S, but the number of such values cannot be guaranteed to be tightly concentrated around its expectation. However, releasing about $O(\log(\frac{k}{\alpha}))$ rows with such a value does not lead to an ϵ breach in privacy.

As we showed earlier, releasing any sample drawn from a table does not ensure $(1, \epsilon, \delta)$-privacy. We show that privacy is preserved when we draw a sample with certain properties, and such a sample occurs with high probability. A sample possessing these properties is called a *good sample*.

Definition 5. *A* good sample *is one that has the following properties: (1) A rare value v does not occur. (2) An infrequent value v occurs in at most $n_v p + 2\log(\frac{k}{\alpha})$ rows. (3) A common value v occurs in at most $n_v p + \sqrt{3 n_v p \log(\frac{k}{\alpha})}$ rows.*

In Lemma 1, we show that releasing a good sample preserves privacy. In Lemma 2, we show that a good sample occurs with high probability. Combining Lemmas 1 and 2, we get a proof of Theorem 3.

Lemma 1. *Let S be a good sample drawn from table T. Then for any row i and any pair of values v and v',*

$$\frac{\Pr[S|T_{\{i \to v\}}]}{\Pr[S|T_{\{i \to v'\}}]} \leq 1 + \epsilon'$$

where $\epsilon' = \max(2(p+\epsilon), 6p)$ for $p + \epsilon < \frac{1}{2}$.

Proof. For any value u, let n_u^1 be the number of rows in table $T \setminus \{i\}$ with value u, and let s_u be as usual the number of rows with value u in the sample S. Then, $T_{\{i \to v\}}$ has $n_v^1 + 1$ rows with value v and $n_{v'}^1$ rows with value v'.

We now show that since S is a good sample, $s_v < n_v^1 + 1$. If row i of T takes any other value than v, this holds trivially; otherwise, we claim that at most n_v^1 rows of T that take value v appear in S. If v is a rare value, there are no rows with value v in a good sample. If v is an infrequent value, the maximum number of rows with value v in the sample is at most $(n_v^1 + 1)p + 2\log(\frac{k}{\alpha})$ which is at most $n_v^1(p+\epsilon) + p < n_v^1$ for $p + \epsilon < 1/2$. If v is a common value, the maximum number of rows with value v in the good sample S is at most $(n_v^1 + 1)p + \sqrt{3(n_v^1 + 1)p \log(\frac{k}{\alpha})}$,

which is at most $(n_v^1 + 1)p \left(1 + \sqrt{\frac{3\log(\frac{k}{\alpha})}{(n_v^1 + 1)p}}\right) \leq \frac{3}{2}(n_v^1 + 1)p < n_v^1 + 1$ when $p < \frac{1}{2}$.

Therefore,

$$\frac{\Pr[S|T_{\{i\to v\}}]}{\Pr[S|T_{\{i\to v'\}}]} = \frac{\binom{n_v^1+1}{s_v}\binom{n_{v'}^1}{s_{v'}}}{\binom{n_v^1}{s_v}\binom{n_{v'}^1+1}{s_{v'}}} = \frac{1 - \frac{s_{v'}}{n_{v'}^1+1}}{1 - \frac{s_v}{n_v^1+1}}$$

For a rare value v, $s_v = 0$. Therefore,

$$\frac{1 - \frac{s_{v'}}{n_{v'}^1+1}}{1 - \frac{s_v}{n_v^1+1}} = 1 - \frac{s_{v'}}{n_{v'}^1+1} \leq 1$$

For an infrequent value v, since there are at most $n_v^1 + 1$ rows with value v in the table T, $s_v \leq (n_v^1 + 1)p + 2\log(\frac{k}{\alpha})$ and $n_v^1 + 1 \geq \frac{2\log(\frac{k}{\alpha})}{\epsilon}$. This implies that, $\frac{s_v}{n_v^1+1} \leq p + \epsilon$ and assuming $p + \epsilon < 1/2$,

$$\frac{1 - \frac{s_{v'}}{n_{v'}^1+1}}{1 - \frac{s_v}{n_v^1+1}} \leq 1 + \frac{p+\epsilon}{1-(p+\epsilon)} \leq 1 + 2(p+\epsilon)$$

For a common value v, as there are either n_v^1 or $n_v^1 + 1$ rows with value v in the table, s_v is at most $(n_v^1 + 1)p + \sqrt{3(n_v^1 + 1)p\log(\frac{k}{\alpha})}$, and $(n_v^1 + 1)p \geq 12\log(\frac{k}{\alpha})$. Therefore $\frac{s_v}{n_v^1+1} \leq \frac{3}{2}p$ which implies that

$$\frac{1 - \frac{s_{v'}}{n_{v'}^1+1}}{1 - \frac{s_v}{n_v^1+1}} \leq 1 + \frac{\frac{3}{2}p}{1-\frac{3}{2}p} \leq 1 + 6p$$

for $p < 1/2$.   □

We now state a condition on p that ensures S is a good sample with high probability.

Lemma 2. *If the sampling frequency $p < \frac{\epsilon\log(\frac{1}{1-\alpha})}{4t\log(\frac{k}{\alpha})}$, the probability that a good sample is drawn is at least $(1-\alpha)^2$.*

Proof. We observe that given a fixed table T, the number of rows in S with value u is independent of the number of rows in S with any other value u'.

For a common value v, the probability that S has more than $n_v p + \sqrt{3n_v p\log(\frac{k}{\alpha})}$ rows with value v is at most

$$e^{-3\log(\frac{k}{\alpha})/3} = \frac{\alpha}{k}$$

using Chernoff Bounds.

For an infrequent value v, the probability that S has more than $n_v p + 2\log(\frac{k}{\alpha})$ rows with value v is at most

$$e^{-2\log(\frac{k}{\alpha})/2} = \frac{\alpha}{k}$$

Since there are k values altogether, the total probability that the sample has the requisite number of common and infrequent values is at least $1 - k \cdot \frac{\alpha}{k} = 1 - \alpha$.

The probability that a rare value v does not occur in S is $(1 - p)^{n_v}$. If there are at most t rare values, then the probability that none of these values occur in S is at least $(1 - p)^{2t \log(\frac{k}{\alpha})/\epsilon} \geq e^{-4pt \log(\frac{k}{\alpha})/\epsilon}$. For $p < \epsilon \frac{\log(\frac{1}{1-\alpha})}{4t \log(\frac{k}{\alpha})}$, this probability is at least $1 - \alpha$.

The total probability of seeing a good sample is therefore at least $(1 - \alpha)^2$. $\square$

Finally, we present an example to show that the upper bound on p is tight up to a $\log(\frac{k}{\alpha})$ factor.

Observation 6. *There exists a table T for which a sampling frequency $p < \frac{\epsilon \log(\frac{1}{1-\alpha})}{t}$ violates $(1, \frac{\epsilon}{2}, \alpha)$-privacy. Here $t + 1$ is the number of values with frequency at most $\frac{1}{\epsilon}$.*

We illustrate this through an example. Consider a table T with $t + 2$ distinct values; values $1, 2, \ldots, t$ each occur in $\frac{1}{\epsilon}$ rows, value $t + 1$ occurs in all the remaining rows except for one row, and value $t + 2$ occurs in row i.

Consider a sample S drawn from this table with the property that $s_u > 0$ for some $u \in [1, \ldots, t]$. Consider an attacker who knows the value of all rows of the table except for row i and is trying to find out what the value of row i is. If $s_{t+2} > 0$, $\Pr[S|T_{\{i \to u\}}] = 0$ for any u other than value $t + 2$, and $\frac{\Pr[S|T_{\{i \to t+2\}}]}{\Pr[S|T_{\{i \to u\}}]}$ is unbounded.

Otherwise, $s_{t+2} = 0$. Let u be a value in $[1, \ldots, t]$ for which $s_u > 0$. If n_v is the number of rows with value v in the table $T \setminus \{i\}$,

$$\frac{\Pr[S|T_{\{i \to t+2\}}]}{\Pr[S|T_{\{i \to u\}}]} = \frac{\binom{n_u}{s_u}\binom{n_{t+2}+1}{s_{t+2}}}{\binom{n_u+1}{s_u}\binom{n_{t+2}}{s_{t+2}}} = 1 - \frac{s_u}{n_u + 1}$$

Since $s_u \geq 1$ and $n_u = \frac{1}{\epsilon}$, this quantity is at most $1 - \frac{\epsilon}{2}$ for any $\epsilon < 1$. In other words, if we see a sample S with the property mentioned above, $(1, \epsilon/2)$-privacy is violated.

Now the probability of choosing a sample S with this property is $(1 - (1 - p)^{\frac{t}{\epsilon}}) \geq 1 - e^{-\frac{pt}{\epsilon}}$. If $p > \frac{\epsilon \log(\frac{1}{1-\alpha})}{t}$, this probability is more than α. This means that we need a sampling frequency $p < \frac{\epsilon \log(\frac{1}{1-\alpha})}{t}$ to ensure $(1, \frac{\epsilon}{2}, \alpha)$-privacy.

5 (c, ϵ, δ)-Privacy

The techniques of [5] show that a mechanism which is $(1, \frac{\epsilon}{c}, \frac{\delta}{c})$-private is also (c, ϵ, δ)-private. The proof looks at a sequence of intermediate tables, each of which differs from the previous one by one row, and shows that $(1, \frac{\epsilon}{c}, \frac{\delta}{c})$-privacy for each of these tables implies (c, ϵ, δ)-privacy for the original table. It is not apparent that the proof applies to us : we do not guarantee $(1, \epsilon, \delta)$-privacy for all tables for a uniform value of p, so a sampling frequency that is $(1, \epsilon, \delta)$-private for the starting table may not maintain the same ϵ, δ guarantees for an intermediate one. In this section, we show an upper bound on the sampling frequency p so that (c, ϵ, δ)-privacy is ensured. Our guarantees are better than the guarantees in [5] in terms of δ and slightly worse in terms of ϵ.

As in the previous section, given ϵ, δ, a table T and k, the total number of distinct values in the table and c, we provide a sampling frequency p that is (c, ϵ, δ)-private. Our main guarantees can be summarized as follows.

Theorem 7. *Given a table T, let $\alpha = \frac{1}{2}\delta$, k be the number of distinct values in T and t be the total number of values in T that occur less than $\frac{2\log(\frac{k}{\alpha})}{\epsilon} + c$ times. Also let $\epsilon' = \max\left(6c\left(1 + \frac{\epsilon c}{2\log(\frac{k}{\alpha})}\right)p, 2c\left(1 + \frac{\epsilon c}{2\log(\frac{k}{\alpha})}\right)(p + \epsilon)\right)$ and let $(1 + \frac{\epsilon c}{2\log(\frac{k}{\alpha})})(p + \epsilon) < \frac{1}{2}$. Then a sample S of T drawn with frequency $p <$*
$$\frac{\epsilon \log(\frac{1}{1-\alpha})}{4t\log(\frac{k}{\alpha})(1 + \frac{\epsilon c}{2\log(\frac{k}{\alpha})})} \text{ is } (c, \epsilon', \delta)\text{-private for } t > 0. \text{ When } t = 0, \text{ a sample } S \text{ of } T$$
drawn with frequency $p \le \epsilon$ is (c, ϵ', δ)-private.

We need the assumption $(1 + \frac{\epsilon c}{2\log(\frac{k}{\alpha})})(p + \epsilon) < \frac{1}{2}$ to make sure p is bounded away from 1 by a constant and also ϵ is small compared to c. (Any other constant than $\frac{1}{2}$ would do, but would change the constants in Theorem 7.) If p is too close to 1 or if ϵc is too big, all rows containing a rare value may appear in the sample, leading to a serious breach of privacy.

Comparing these guarantees with those in Section 4, we observe that for a table T, a sampling frequency p that is $(1, \frac{\epsilon}{c(1 + \frac{\epsilon c}{2\log(\frac{k}{\alpha})})}, \delta)$-private is also (c, ϵ, δ)-private. We therefore do better than the kind of bound given in [5] in terms of δ and a little worse in terms of ϵ.

Consider all sets of $n - c$ rows in table T and let n_v^c be the minimum number of rows with value v in any such set. In this section, we call a value v a *rare value* if $n_v^c < \frac{2\log(\frac{k}{\epsilon})}{\epsilon}$. A *common value* v has $n_v^c > \frac{12\log(\frac{k}{\alpha})}{p}$, where p is the sampling frequency. A value that is neither rare nor common is called an *infrequent value*.

Just as in the previous section, we show that privacy is preserved when we draw a sample with certain properties. We call a sample possessing these properties a *good sample*, with the same definition as in Section 4.

In Lemma 3, we show that releasing a good sample preserves privacy. Because the definition of rare and infrequent values have changed, the fact that a good sample occurs with high probability does not automatically follow from Lemma 2. Instead in Lemma 4, we show that a good sample occurs with high probability. Combining Lemmas 3 and 4, we get a proof of Theorem 7.

Lemma 3. *Let S be a good sample drawn from table T. Then for any set of c rows P_i and any pair of states V and V',*

$$\frac{\Pr[S|T_{\{P_i \to V\}}]}{\Pr[S|T_{\{P_i \to V'\}}]} \le 1 + \epsilon'$$

where $\epsilon' = \max\left(6c\left(1 + \frac{\epsilon c}{2\log(\frac{k}{\alpha})}\right)p, 2c\left(1 + \frac{\epsilon c}{2\log(\frac{k}{\alpha})}\right)(p+\epsilon)\right)$, *assuming that* $(1 + \frac{\epsilon c}{2\log(\frac{k}{\alpha})})(p+\epsilon) < \frac{1}{2}$.

Proof. Without loss of generality, we assume that the state V has c rows with value v and no rows with value V' and the state v' has no rows with value v and c rows with value v'. The proofs go through when this assumption does not hold, and so we simplify the notation accordingly.

For any value u, let n_u^c denote the number of rows in $T \setminus \{P_i\}$ with value u, and let s_u be the number of rows with value u in the sample S. Note that $T_{\{P_i \to V\}}$ has $n_v^c + c$ rows with value v and $n_{v'}^c$ rows with value v', whereas $T_{\{P_i \to V'\}}$ has $n_{v'}^c + c$ rows with value v' and n_v^c rows with value v.

We now show that since S is a good sample, $s_v < n_v^c + 1$. If the set of rows P_i in T includes no row with value v, this holds trivially; otherwise, we claim that at most n_v^c rows of T that take value v appear in S. Note that T can have at most $n_v^c + c$ rows with value v.

If v is a rare value, there are no rows with value v in a good sample. If v is an infrequent value, the maximum number of rows with value v in the sample is at most $(n_v^c + c)p + 2\log(\frac{k}{\alpha})$ which is at most $n_v^c(1 + \frac{c}{n_v^c})(p+\epsilon) < n_v^c(1 + \frac{c\epsilon}{2\log(\frac{k}{\alpha})})(p+\epsilon) < n_v^c$ for $(1 + \frac{c\epsilon}{2\log(\frac{k}{\alpha})})(p+\epsilon) < \frac{1}{2}$. If v is a common value, the maximum number of rows with value v in the good sample S is at most $(n_v^c + c)p + \sqrt{3(n_v^c + c)p\log(\frac{k}{\alpha})}$, which is at most $(n_v^c + c)p\left(1 + \sqrt{\frac{3\log(\frac{k}{\alpha})}{(n_v^c + c)p}}\right) \leq \frac{3}{2}n_v^c(1 + \frac{c\epsilon}{2\log(\frac{k}{\alpha})})p < n_v^c + 1$ when $(1 + \frac{c\epsilon}{2\log(\frac{k}{\alpha})})p < \frac{1}{2}$.

Therefore,

$$\frac{\Pr[S|T_{\{P_i \to V\}}]}{\Pr[S|T_{\{P_i \to V'\}}]} = \frac{\binom{n_v^c + c}{s_v}\binom{n_{v'}^c}{s_{v'}}}{\binom{n_v^c}{s_v}\binom{n_{v'}^c + c}{s_{v'}}} \leq \frac{1}{(1 - \frac{s_v}{n_v^c + 1})(1 - \frac{s_v}{n_v^c + 2}) \dots (1 - \frac{s_v}{n_v^c + c})}$$

For a rare value v, $s_v = 0$. Therefore,

$$\frac{1}{(1 - \frac{s_v}{n_v^c + 1})(1 - \frac{s_v}{n_v^c + 2}) \dots (1 - \frac{s_v}{n_v^c + c})} \leq 1$$

Note that for any value v, $n_v^c \leq n_v \leq n_v^c + c$. For an infrequent value v, $s_v \leq (n_v^c + c)p + 2\log(\frac{k}{\alpha})$ and $n_v \geq n_v^c \geq \frac{2\log(\frac{k}{\alpha})}{\epsilon}$. This implies that $\frac{s_v}{n_v^c + 1} \leq (1 + \frac{c}{n_v^c})(p+\epsilon) \leq (1 + \frac{c\epsilon}{2\log(\frac{k}{\alpha})})(p+\epsilon)$. Assuming $(1 + \frac{c\epsilon}{2\log(\frac{k}{\alpha})})(p+\epsilon) < 1/2$,

$$\frac{1}{(1 - \frac{s_v}{n_v^c + 1})(1 - \frac{s_v}{n_v^c + 2}) \dots (1 - \frac{s_v}{n_v^c + c})} \leq 1 + 2c\left(1 + \frac{c\epsilon}{2\log(\frac{k}{\alpha})}\right)(p+\epsilon)$$

Because v is not a rare value, this quantity is at most $1 + 2c(1 + \frac{\epsilon c}{2\log(\frac{k}{\alpha})})(p+\epsilon)$.

For a common value v, s_v is at most $(n_v^c + c)p + \sqrt{3(n_v^c + c)p\log(\frac{k}{\alpha})}$, and $n_v^c p \geq 12\log(\frac{k}{\alpha})$. Therefore $\frac{s_v}{n_v^c + 1} \leq \frac{3}{2}(1 + \frac{c}{n_v^c})p \leq \frac{3}{2}(1 + \frac{c\epsilon}{2\log(\frac{k}{\alpha})})p$.

This implies that

$$(1 - \tfrac{s_v}{n_v^c+1})(1 - \tfrac{s_v}{n_v^c+2}) \cdots (1 - \tfrac{s_v}{n_v^c+c}) \leq 1 + 6\left(1 + \frac{\epsilon c}{2\log(\frac{k}{\alpha})}\right) pc$$

for $p(1 + \frac{c\epsilon}{2\log(\frac{k}{\alpha})}) < 1/2.$ □

Lemma 4. *If the sampling frequency* $p < \dfrac{\epsilon \log\left(\frac{1}{1-\alpha}\right)}{4t\log(\frac{k}{\alpha})(1+\frac{\epsilon c}{2\log(\frac{k}{\alpha})})}$, *the probability that a good sample is drawn is at least* $(1 - \alpha)^2$.

Proof. Following exactly the same argument as in Lemma 2, the probability that the number of common and infrequent values lie within the requisite bounds is at least $1 - \alpha$.

The probability that a rare value v does not occur in S is $(1 - p)^{n_v}$. Now there are at most $c + \frac{2\log(\frac{k}{\alpha})}{\epsilon}$ rows with a rare value v. If there are at most t rare values, then the probability that none of these values occur in S is at least $(1 - p)^{2t\log(\frac{k}{\alpha})/\epsilon + tc} \geq e^{-4pt(2\log(\frac{k}{\alpha})/\epsilon + c)}$. For $p < \epsilon \dfrac{\log\left(\frac{1}{1-\alpha}\right)}{4t\log(\frac{k}{\alpha})(1+\frac{\epsilon c}{2\log(\frac{k}{\alpha})})}$, this probability is at least $1 - \alpha$.

The total probability of seeing a good sample is therefore at least $(1 - \alpha)^2$.   □

6   Future Work

There are many avenues for future work. Our work assumes that the random sample is published in unperturbed form. But it is quite possible that one can draw larger random samples if noise is added to the sample. Such a technique may be useful when k is large. One would have to also understand what impact such noise would have on utility.

Another direction for future research is the study of data streams where individual data points arrive in sequential, not necessarily random order and the question is how to maintain a random sample of the stream without breaching privacy. In this context, the attacker may know who the next person is in the stream, but not know their private values. Existing techniques for maintaining random samples over a stream such as reservoir sampling [13] violate privacy since the sample only changes to include a new person's value when that person arrives.

Finally, in practice, prior to sampling, organizations typically employ other anonymization procedures including, for example, *top-coding*, where individuals with values above a certain percentage of the distribution are placed into a single category, *geographic population thresholds*, where individuals that live in a geographic unit below a specified population level are not disclosed, *random rounding*, wherein numbers that are not multiples of say 10 are randomly rounded to one of the two nearest multiples. Analysis of the privacy/utility of these multi-step anonymization procedures that precede sampling would be an interesting direction for future work.

Acknowledgements

We are very grateful to Kobbi Nissim for extensive discussions. We thank Cynthia Dwork for her insights and Amit Agarwal for early discussion. We thank Shankar Bhamidi for useful suggestions. Finally, we thank the anonymous reviewers for their many thoughtful suggestions.

References

1. A. Blum, C. Dwork, F. McSherry, and K. Nissim. Practical privacy: the sulq framework. In *PODS*, pages 128–138, 2005.
2. U.S. Census Bureau. Public use microdata sample (pums). In *http://www.census.gov/Press-Release/www/2003/PUMS.html*, 2003.
3. S. Chawla, C. Dwork, F. McSherry, A. Smith, and H. Wee. Toward privacy in public databases. In *Theory of Cryptography Conference*, pages 363–385, 2005.
4. I. Dinur and K. Nissim. Revealing information while preserving privacy. In *PODS*, pages 202–210, 2003.
5. C. Dwork, F. McSherry, K. Nissim, and A. Smith. Calibrating noise to sensitivity in private data analysis. In *Theory of Cryptography Conference*, pages 265–284, 2006.
6. C. Dwork and K. Nissim. Privacy-preserving datamining on vertically partitioned databases. In *CRYPTO*, pages 528–544, 2004.
7. A. Evfimievski, J. Gehrke, and R. Srikant. Limiting privacy breaches in privacy preserving data mining. In *PODS*, pages 211–222, 2003.
8. O. Goldreich. *Foundations of Cryptography, Volumes I and II*. Cambridge University Press, 2004.
9. K. Kenthapadi, N. Mishra, and K. Nissim. Simulatable auditing. In *PODS*, pages 118–127, 2005.
10. N. Mishra and M. Sandler. Privacy via pseudorandom sketches. In *PODS*, 2006.
11. Social Security Administration: Office of Policy Data. Benefits and earnings public-use file. In *http://www.ssa.gov/policy/docs/microdata/earn/index.html*, 2004.
12. L. Sweeney. Guaranteeing anonymity when sharing medical data, the datafly system. In *Proceedings AMIA Annual Fall Symposium*, 1997.
13. J. Vitter. Random sampling with a reservoir. *ACM Transactions on Mathematical Software*, 11(1):37–57, March 1985.

Tight Bounds for Unconditional Authentication Protocols in the Manual Channel and Shared Key Models*

Moni Naor**, Gil Segev, and Adam Smith***

Department of Computer Science and Applied Mathematics,
Weizmann Institute of Science, Rehovot 76100, Israel
{moni.naor, gil.segev, adam.smith}@weizmann.ac.il

Abstract. We address the message authentication problem in two seemingly different communication models. In the first model, the sender and receiver are connected by an insecure channel and by a low-bandwidth auxiliary channel, that enables the sender to "manually" authenticate one short message to the receiver (for example, by typing a short string or comparing two short strings). We consider this model in a setting where no computational assumptions are made, and prove that for any $0 < \epsilon < 1$ there exists a $\log^* n$-round protocol for authenticating n-bit messages, in which only $2 \log(1/\epsilon) + O(1)$ bits are manually authenticated, and any adversary (even computationally unbounded) has probability of at most ϵ to cheat the receiver into accepting a fraudulent message. Moreover, we develop a proof technique showing that our protocol is essentially optimal by providing a lower bound of $2 \log(1/\epsilon) - 6$ on the required length of the manually authenticated string.

The second model we consider is the traditional message authentication model. In this model the sender and the receiver share a short secret key; however, they are connected only by an insecure channel. Once again, we apply our proof technique, and prove a lower bound of $2 \log(1/\epsilon) - 2$ on the required Shannon entropy of the shared key. This settles an open question posed by Gemmell and Naor (CRYPTO '93).

Finally, we prove that one-way functions are *essential* (and sufficient) for the existence of protocols breaking the above lower bounds in the computational setting.

1   Introduction

Message authentication is one of the major issues in cryptography. Protocols for message authentication provide assurance to the receiver of a message that it was sent by a specified legitimate sender, even in the presence of an adversary who

---

* A longer version, including proofs of all claims, appears as [15].
** Incumbent of the Judith Kleeman Professorial Chair. Research supported in part by a grant from the Israel Science Foundation.
*** Research supported by the Louis L. and Anita M. Perlman postdoctoral fellowship.

C. Dwork (Ed.): CRYPTO 2006, LNCS 4117, pp. 214–231, 2006.

controls the communication channel. For more than three decades, numerous authentication models have been investigated, and many authentication protocols have been suggested. The security of these protocols can be classified according to the assumed computational resources of the adversary. Security that holds when one assumes a suitable restriction on the adversary's computing capabilities is called *computational security*, while security that holds even when the adversary is computationally unbounded is called *unconditional security* or *information-theoretic security*. This paper is concerned mostly with unconditional security of a single instance of message authentication protocols. We remark that there are three main advantages to unconditional security over computational security. The first is the obvious fact that no assumptions are made about the adversary's computing capabilities or about the computational hardness of specific problems. The second, less apparent advantage, is that unconditionally secure protocols are often more efficient than computationally secure protocols. The third advantage is that unconditional security allows exact evaluation of the error probabilities.

Shared Key Authentication. The first construction of an authentication protocol in the literature was suggested by Gilbert, MacWilliams and Sloane [10] in the information-theoretic adversarial setting. They considered a communication model in which the sender and the receiver share a key, which is not known to the adversary. Gilbert et al. presented a non-interactive protocol, in which the length of the shared key is $2 \max\{n, \log(1/\epsilon)\}$; henceforth, n is the length of the input message and ϵ is the adversary's probability of cheating the receiver into accepting a fraudulent message. They also proved a lower bound of $2 \log(1/\epsilon)$ on the required entropy of the shared key in non-interactive deterministic protocols. Clearly, a trivial lower bound on this entropy is $\log(1/\epsilon)$, since an adversary can merely guess the shared key. This model, to which we refer as the *shared key model*, became the standard model for message authentication protocols. Protocols in this model should provide authenticity of messages while minimizing the length of the shared key.

Wegman and Carter [20] suggested using ϵ-*almost strongly universal$_2$* hash functions for authentication. This enabled them to construct a non-interactive protocol in which the length of the shared key is $O(\log n \log(1/\epsilon))$ bits. In 1984, Simmons [17] initiated a line of work on unconditionally secure authentication protocols (see, for example, [13,18] and more references in the full version). Gemmell and Naor [9] proposed a non-interactive protocol, in which the length of the shared key is only $\log n + 5 \log(1/\epsilon)$ bits. They also demonstrated that by introducing interaction, the length of the shared key can be made independent of the length of the input message. More specifically, they suggested a $\log^* n$-round protocol that enables the sender to authenticate n-bit messages, where the length of the shared key is only $2 \log(1/\epsilon) + O(1)$ bits. However, it was not known whether this upper bound is optimal, that is, if by introducing interaction the entropy of the shared key can be made smaller than $2 \log(1/\epsilon)$.

Manual Authentication. Recently, Vaudenay [19] formalized a realistic communication model for message authentication, in which the sender and the receiver are connected by a bidirectional insecure channel, and by a unidirectional low-bandwidth auxiliary channel, but do not share any secret information. It is assumed that the adversary has full control over the insecure channel. In particular, the adversary can read any message sent over this channel, prevent it from being delivered, and insert a new message at any point in time. The low-bandwidth auxiliary channel enables the sender to "manually" authenticate one short string to the receiver. The adversary cannot modify this short string. However, the adversary can still read it, delay it, and remove it. We refer to the auxiliary channel as the *manual channel*, and to this communication model as the *manual channel model*. Protocols in this model should provide authenticity of long messages[1] while minimizing the length of the manually authenticated string. We remark that $\log(1/\epsilon)$ is an obvious lower bound in this model as well.

The manual channel model is becoming very popular in real-world scenarios, whenever there are ad hoc networks with no trusted infrastructure. In particular, this model was found suitable for initial pairing of devices in wireless networks, such as Wireless USB [3] and Bluetooth[2] [2]. While in wired connections when a device is plugged in (i.e., when the wire is connected), the user can see that the connection is made, wireless connections may establish connection paths that are not straightforward. In fact, it may not be obvious when a device is connected or who its host is. Therefore, initial authentication in device and host connections is required so that the user will be able to validate both the device and its host.

Consider, for example, a user who wishes to connect a new DVD player to her home wireless network. Then, the user reading a short message from the display of the DVD player and typing it on a PC's keyboard constitutes a manual authentication channel from the DVD player to the PC. An equivalent channel is the user comparing two short strings displayed by the two devices, as suggested by Gehrmann et al. [8].

Constants Do Matter. The most significant constraint in the manual channel model is the length of the manually authenticated string. This quantity is determined by the environment in which the protocol is executed, and in particular by the capabilities of the user. While it is reasonable to expect a user to manually authenticate 20 or 40 bits, it is not reasonable to expect a user to manually authenticate 160 bits. Therefore, there is a considerable difference between manually authenticating $\log(1/\epsilon)$ or $2\log(1/\epsilon)$ bits, and manually authenticating a significantly longer string. This motivates the study of determining the exact lower bound on the required length of the manually authenticated string.

Our Contribution. We present an unconditionally secure authentication protocol in the manual channel model, in which the sender manually authenticates only $2\log(1/\epsilon) + O(1)$ bits. Moreover, we develop a proof technique, proving that our protocol is essentially optimal in minimizing the length of the manually authenticated string. Then, we apply this technique to the shared key model, and settle an open question posed by Gemmell and Naor [9] by deriving a similar lower bound on the required entropy of the shared key. This lower bound

---

[1] Short messages can be directly manually authenticated.

[2] However, in existing protocols for pairing of Bluetooth devices, the manual channel is assumed to provide secrecy as well.

matches the upper bound of Gemmell and Naor. Finally, we consider these two communication models in the computational setting, and prove that one-way functions are essential for the existence of protocols breaking the above lower bounds.

Paper Organization. The rest of the paper is organized as follows. We first briefly present some known definitions in Section 2. In Section 3 we describe the communication and adversarial models we deal with. Then, in Section 4 we present an overview of our results, and compare them to previous work. In Section 5 we propose an unconditionally secure message authentication protocol in the manual channel model. In Section 6 we describe the proof technique, that is then used to establish the optimality of our protocol. In Section 7 we apply the same proof technique to the shared key model, and prove a lower bound on the required entropy of the shared key. Finally, in Section 8 we prove that in the computational setting, one-way functions are essential for the existence of protocols breaking the above lower bounds.

2   Preliminaries

We first present some fundamental definitions from Information Theory. Then, we briefly present the definitions of *one-way* functions and *distributionally one-way* functions. All logarithms in this paper are to the base of 2. Let X, Y and Z denote random variables.

- The *(Shannon) entropy* of X is $H(X) = -\sum_x \Pr[X = x] \log \Pr[X = x]$.
- The *conditional entropy* of X given Y is $H(X|Y) = \sum_y \Pr[Y = y] H(X|Y = y)$.
- The *mutual information* of X and Y is $I(X;Y) = H(X) - H(X|Y)$.
- The *mutual information* of X and Y given Z is $I(X;Y|Z) = H(X|Z) - H(X|Z,Y)$.

Definition 1. *A function $f : \{0,1\}^* \to \{0,1\}^*$ is called* one-way *if it is computable in polynomial-time, and for every probabilistic polynomial-time Turing machine[3] $\mathcal{M}$, every polynomial p, and all sufficiently large n,*

$$\Pr\left[\mathcal{M}(f(x), 1^n) \in f^{-1}(f(x))\right] < \frac{1}{p(n)} \ ,$$

where the probability is taken uniformly over all the possible choices of $x \in \{0,1\}^n$ and all the possible outcomes of the internal coin tosses of $\mathcal{M}$.

Definition 2. *A function $f : \{0,1\}^* \to \{0,1\}^*$ is called* distributionally one-way *if it is computable in polynomial-time, and there exists a constant $c > 0$ such that for every probabilistic polynomial-time Turing machine $\mathcal{M}$, the distribution defined by $x \circ f(x)$ and the distribution defined by $\mathcal{M}(f(x)) \circ f(x)$ are n^{-c}-statistically far[4] when $x \in_R \{0,1\}^n$.*

---

[3] We note that uniformity is not essential to our results.

[4] The *statistical distance* between two distributions $\mathcal{D}$ and $\mathcal{F}$ is defined as $\Delta(\mathcal{D},\mathcal{F}) = \frac{1}{2}\sum_\alpha |\Pr_{x\leftarrow\mathcal{D}}[x = \alpha] - \Pr_{x\leftarrow\mathcal{F}}[x = \alpha]|$. The distributions $\mathcal{D}$ and $\mathcal{F}$ are said to be ϵ-*statistically far* if $\Delta(\mathcal{D},\mathcal{F}) \geq \epsilon$. Otherwise, $\mathcal{D}$ and $\mathcal{F}$ are ϵ-*statistically close*.

Informally, it is hard to find a random inverse of a distributionally one-way function, although finding some inverse may be easy. Clearly, any one-way function is also a distributionally one-way function, but the converse may not always be true. However, Impagliazzo and Luby [11] proved that the *existence* of both primitives is equivalent.

3   Communication and Adversarial Models

We consider the message authentication problem in a setting where the sender and the receiver are connected by a bidirectional insecure communication channel, over which an adversary has full control. In particular, the adversary can read any message sent over this channel, delay it, prevent it from being delivered, and insert a new message at any point in time.

3.1   The Manual Channel Communication Model

In addition to the insecure channel, we assume that there is a unidirectional low-bandwidth auxiliary channel, that enables the sender to "manually" authenticate one short string to the receiver. The adversary cannot modify this short string. However, the adversary can still read it, delay it, and remove it.

The input of the sender $\mathcal{S}$ in this model is a message m, which she wishes to authenticate to the receiver $\mathcal{R}$. The input message m can be determined by the adversary $\mathcal{A}$. In the first round, $\mathcal{S}$ sends the message m and an authentication tag x_1 over the insecure channel. In the following rounds only a tag x_i is sent over the insecure channel. The adversary receives each of these tags x_i and can replace them with $\widehat{x}_i$ of her choice, as well as replace the input message m with a different message $\widehat{m}$. In the last round, $\mathcal{S}$ may manually authenticate a short string s.

Notice that in the presence of a computationally unbounded adversary, additional insecure rounds (after the manually authenticated string has been sent) do not contribute to the security of the protocol. This is due to the fact that after reading the manually authenticated string, the unbounded adversary can always simulate the sender successfully (since the sender and the receiver do not share any secret information, and since the adversary has full control over the communication channel from this point on). Therefore, there is no loss of generality in assuming that the manually authenticated string is sent in the last round. This is true also in the computational setting, under the assumption that distributionally one-way functions do not exist. A generic protocol in this model is described in Figure 1.

We also allow the adversary to control the synchronization of the protocol's execution. More specifically, the adversary can carry on two separate, possibly asynchronous conversations, one with the sender and one with the receiver. However, the party that is supposed to send a message waits until it receives the adversary's message from the previous round.

When the input message m is chosen uniformly at random, the honest execution of the protocol defines a probability distribution on the message m, the tags x_i and the manually authenticated string s. We denote by M, X_i and S the random variables corresponding to m, x_i and s, respectively.

Definition 3. *An* unconditionally secure (n, ℓ, k, ϵ)-*authentication protocol in the manual channel model is a k-round protocol in the communication model described above, in which the sender wishes to authenticate an n-bit input message to the receiver, while manually authenticating at most ℓ bits. The following requirements must hold:*

1. **Completeness:** *For all input messages m, when there is no interference by the adversary in the execution, the receiver accepts m with probability at least $1/2$.*
2. **Unforgeability:** *For any computationally unbounded adversary, and for all input messages m, if the adversary replaces m with a different message $\widehat{m}$, then the receiver accepts $\widehat{m}$ with probability at most ϵ.*

In order to define the notion of a *computationally secure* authentication protocol, we actually consider a sequence of protocols by adding a security parameter t that defines the power of the adversaries against which each protocol in the sequence is secure. The completeness requirement is as in Definition 3. However, the unforgeability requirement now holds only against adversaries running in time $\text{poly}(t)$, and we allow forgery probability of $\epsilon + \text{negl}(t)$ for sufficiently large t. We refer the reader to the full version for the formal definition.

An authentication protocol in the manual channel model is said to be *perfectly complete* if for all input messages m, whenever there is no interference by the adversary in the execution, the receiver accepts m with probability 1.

3.2   The Shared Key Communication Model

In this model we assume that the sender and the receiver share a secret key s; however, they are connected only by an insecure channel. This key is not known to the adversary, but it is chosen from a probability distribution which is known to the adversary (usually the uniform distribution).

The input of the sender S in this model is a message m, which she wishes to authenticate to the receiver $\mathcal{R}$. The input message m can be determined by the adversary $\mathcal{A}$. In the first round, S sends the message m and an authentication tag x_1 over the insecure channel. In the following rounds only a tag x_i is sent over the insecure channel. The adversary receives each of these tags x_i and can replace them with $\widehat{x}_i$ of her choice, as well as replace the input message m with a different message $\widehat{m}$.

As in the manual channel model, in an honest execution we denote by S, M and X_i the random variables corresponding to s, m and x_i, respectively. A generic protocol in this model is described in Figure 1. As in the manual channel model, we allow the adversary to control the synchronization of the protocol's execution.

Definition 4. *An* unconditionally secure (n, ℓ, k, ϵ)-*authentication protocol in the shared key model is a k-round protocol in the communication model described*

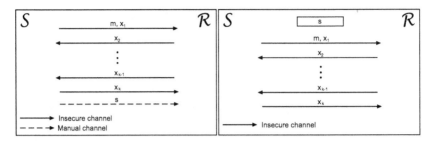

Fig. 1. Generic protocols in the manual channel model (left figure) and in the shared key model (right figure)

above, in which the sender and the receiver share an ℓ-bit secret key, and the sender wishes to authenticate an n-bit input message to the receiver. The following requirements must hold:

1. **Completeness:** *For all input messages m, when there is no interference by the adversary in the execution, the receiver accepts m with probability at least 1/2.*
2. **Unforgeability:** *For any computationally unbounded adversary, and for all input messages m, if the adversary replaces m with a different message $\widehat{m}$, then the receiver accepts $\widehat{m}$ with probability at most ϵ.*

Similarly to the definitions in the manual channel model, we refer the reader to the full version for the definitions of a *computationally secure sequence* of authentication protocols and of a *perfectly complete* protocol.

4 Overview of Our Results and Comparison with Previous Work

Vaudenay [19] formalized the manual channel model, and suggested an authentication protocol in this model. Given $0 < \epsilon < 1$, Vaudenay's protocol enables the sender to authenticate an arbitrary long message to the receiver in three rounds, by manually authenticating $\log(1/\epsilon)$ bits. This protocol guarantees that, under the assumption that a certain type of non-interactive commitment scheme exists, the forgery probability of any polynomial-time adversary is at most $\epsilon + \nu(t)$, where $\nu(\cdot)$ is a negligible function and t is a security parameter. In particular, Laur, Asokan and Nyberg [12] proved that the required assumption is the existence of a non-interactive non-malleable commitment scheme. Dolev, Dwork and Naor [6] showed how to construct an *interactive* non-malleable commitment scheme from any one-way function, and therefore we obtain the following corollary:

Corollary 5 ([6,12,19]). *If one-way functions exist, then there exists a computationally secure $(n, \ell, k, \epsilon, t)$-authentication protocol in the manual channel model, with $t = poly(n, \ell, k)$ and $\ell = \log(1/\epsilon)$.*

However, the non-malleable commitment scheme suggested by Dolev, Dwork and Naor is inefficient, as it utilizes generic zero-knowledge proofs and its number of rounds is logarithmic in its security parameter. Therefore, the protocol implied by Corollary 5 is currently not practical (this is also true if the protocols in [1,16] are used). Currently, the only known constructions of efficient non-malleable commitment schemes are in the random oracle model, or in the common random string model (see, for example, [4,5]). These are problematic for the manual channel model, since they require a trusted infrastructure. This state of affairs motivates the study of a protocol that can be proved secure under more relaxed computational assumptions or even without any computational assumptions.

In Section 5, we present an unconditionally secure perfectly complete authentication protocol in the manual channel model. For any odd integer $k \geq 3$, and any integer n and $0 < \epsilon < 1$, our k-round protocol enables the sender to authenticate an n-bit input message to the receiver, while manually authenticating at most $2\log(1/\epsilon) + 2\log^{(k-1)} n + O(1)$ bits. We prove that any adversary (even computationally unbounded) has probability of at most ϵ to cheat the receiver into accepting a fraudulent message. We note that our protocol only uses evaluations of polynomials over finite fields, for which very efficient implementations exist, and therefore it is very efficient and can be implemented on low-power devices. We prove the following theorem and corollary:

Theorem 6. *For any odd integer $k \geq 3$, and any integer n and $0 < \epsilon < 1$, there exists an unconditionally secure perfectly complete $(n, \ell = 2\log(1/\epsilon) + 2\log^{(k-1)} n + O(1), k, \epsilon)$-authentication protocol in the manual channel model.*

Corollary 7. *For any integer n and $0 < \epsilon < 1$, the following unconditionally secure perfectly complete protocols exist in the manual channel model:*

1. *A $\log^* n$-round protocol in which at most $2\log(1/\epsilon) + O(1)$ bits are manually authenticated.*
2. *A 3-round protocol in which at most $2\log(1/\epsilon) + \log\log n + O(1)$ bits are manually authenticated.*

In Section 6, we develop a proof technique for deriving lower bounds on unconditionally secure authentication protocols, which allows us to show that our $\log^* n$-round protocol is optimal with respect to the length of the manually authenticated string. Specifically, we prove the following theorem:

Theorem 8. *For any unconditionally secure (n, ℓ, k, ϵ)-authentication protocol in the manual channel model, it holds that if $n \geq 2\log(1/\epsilon) + 4$, then $\ell > 2\log(1/\epsilon) - 6$.*

In Section 7 we consider the shared key communication model. Intensive research has been devoted to proving lower bounds on the required entropy of the shared key in unconditionally secure protocols. It was proved in several papers (see, for example, [13]), that in any perfectly complete non-interactive protocol, the required entropy of the shared key is at least $2\log(1/\epsilon)$. In addition, for such protocols, Gemmell and Naor [9] proved a lower bound of $\log n + \log(1/\epsilon)$

$-\log\log(n/\epsilon) - 2$. Thus, there does not exist a perfectly complete non-interactive protocol that achieves the $2\log(1/\epsilon)$ bound. However, Gemmell and Naor also presented an *interactive* protocol that achieves the $2\log(1/\epsilon)$ bound. We remark that it was not previously known whether this bound is optimal for interactive protocols. By applying the proof technique described in Section 6, we settle this long-standing open question, proving the optimality of the protocol suggested by Gemmell and Naor.

Theorem 9. *For any unconditionally secure (n, ℓ, k, ϵ)-authentication protocol in the shared key model, it holds that $\mathrm{H}(S) \geq 2\log(1/\epsilon) - 2$, where S is the ℓ-bit shared key.*

Theorems 8 and 9 indicate that the two corresponding communication models are not equivalent: While in the manual channel model a lower bound can hold only when $n \geq \log(1/\epsilon)$, in the shared key model the lower bound holds even when authenticating only one bit. Nevertheless, the technique we develop applies to both models.

The idea underlying the lower bound proofs for the communication models under consideration can be briefly summarized as follows. First, we represent the entropies of the manually authenticated string and of the shared key by splitting them in a way that captures their reduction during the execution of the protocol. This representation allows us to prove that both the sender and the receiver must each independently reduce the entropies by at least $\log(1/\epsilon)$ bits. This is proved by considering two possible natural attacks on the given protocol. In these attacks we use the fact that the adversary is computationally unbounded in that she can sample distributions induced by the protocol. This usage of the adversary's capabilities, can alternatively be seen as randomly inverting functions given the image of a random input.

In Section 8, we take advantage of this point of view and prove that one-way functions are essential for the existence of protocols breaking the above lower bounds in the computational setting. Specifically, we show that if distributionally one-way functions do not exist, then a polynomial-time adversary can run the above mentioned attacks with almost the same success probability. The following theorem is proved (the reader is referred to the full version for a similar statement in the shared key model):

Theorem 10. *In the manual channel model, if there exists a computationally secure $(n, \ell, k, \epsilon, t)$-authentication protocol, such that $n \geq 2\log(1/\epsilon) + 4$, $\ell < 2\log(1/\epsilon) - 8$ and $t = \Omega(poly(n, k, 1/\epsilon))$, then one-way functions exist.*

A similar flavor of statement has recently been proved by Naor and Rothblum [14] in the context of memory checking, showing that one-way functions are essential for efficient on-line memory checking. Both results are based on combinatorial constructions (in our case these are the two attacks carried by an unbounded adversary), which are shown to be polynomial-time computable if one-way functions do not exist. However, we note that whereas Naor and Rothblum obtained asymptotic results (there is a multiplicative constant between the upper bound and the lower bound), we detect a sharp threshold.

5   The Message Authentication Protocol

In this section we prove Theorem 6 and Corollary 7 by constructing an authentication protocol, P_k. The protocol is based on the hashing technique suggested by Gemmell and Naor [9], in which the two parties reduce in each round the problem of authenticating the original message to that of authenticating a shorter message. In the first round the input message is sent, and then in each round the two parties *cooperatively* choose a hash function that defines a small, random "fingerprint" of the input message that the receiver should have received. If the adversary has changed the input message, then with high probability the fingerprint for the message received by the receiver will not match the fingerprint for the message that was sent by the sender. In a preliminary version of [9], this hashing technique was susceptible to synchronization attacks, as noted by Gehrmann [7]. However, in the full version of their paper, this was corrected by making *both* parties choose the random hash function used for fingerprinting the message.

We improve the hashing technique suggested by Gemmell and Naor as follows. First, we apply a different hash function, which enables us to manually authenticate a shorter string. A direct adaptation of the original hash function to the manual channel model would require the sender to manually authenticate at least $3 \log(1/\epsilon)$ bits, while our construction manages to reduce this amount to only $2 \log(1/\epsilon) + O(1)$ bits. In addition, our protocol is asymmetric in the following sense: The roles of the sender and the receiver in cooperatively choosing the hash function are switched in every round. This enables us to deal with the fact that the adversary can read and delay any manually authenticated string.

Preliminaries. Denote by $GF[Q]$ the Galois field with Q elements. For a message $m = m_1 \ldots m_k \in GF[Q]^k$ and $x \in GF[Q]$ let $C_x(m) = \sum_{i=1}^{k} m_i x^i$. In other words, m is parsed as a polynomial of degree k over $GF[Q]$ (without a constant term), and evaluated at the point x. Then, for any two different messages $m, \widehat{m} \in GF[Q]^k$ and for any $c, \widehat{c} \in GF[Q]$ the polynomials $C_x(m) + c$ and $C_x(\widehat{m}) + \widehat{c}$ are different as well, and therefore $\Pr_{x \in_R GF[Q]} [C_x(m) + c = C_x(\widehat{m}) + \widehat{c}] \leq \frac{k}{Q}$. We will use $C(\cdot)$ as a hash function to reduce the length of the message.

The Construction. In protocol P_k we apply a sequence of hash functions $C^1, \ldots, C^{k-1}$ in order to obtain a shorter and shorter message. Specifically, given the length, n, of the input message and the upper bound, ϵ, on the adversary's forgery probability, each C^j parses n_j-bit strings to polynomials over $GF[Q_j]$, where $n_1 = n$, $\frac{2^{k-j} n_j}{\epsilon} \leq Q_j < \frac{2^{k-j+1} n_j}{\epsilon}$, and $n_{j+1} = \lceil 2 \log Q_j \rceil$. The protocol is described in Figure 2. Since the adversary can replace any authentication tag sent by any one of the parties over the insecure channel, then for such a tag x we denote by $\widehat{x}$ the tag that was actually received by the other party. Note that addition and multiplication are defined by the $GF[Q_j]$ structures, and that $\langle u, v \rangle$ denotes the concatenation of the strings u and v.

Note that the two parties can combine some of their messages, and therefore the protocol requires only k rounds of communication. An alternative way to

Protocol P_k:

1. S sends $m_S^1 = m$ to R.
2. R receives m_R^1.
3. For $j = 1$ to $k - 1$:
   (a) If j is odd, then
      i. S chooses $i_S^j \in_R GF[Q_j]$ and sends it to R.
      ii. R receives $\widehat{i_S^j}$, chooses $i_R^j \in_R GF[Q_j]$, and sends it to S.
      iii. S receives $\widehat{i_R^j}$, and computes $m_S^{j+1} = \langle \widehat{i_R^j}, C_{i_R^j}^j(m_S^j) + i_S^j \rangle$.
      iv. R computes $m_R^{j+1} = \langle i_R^j, C_{i_R^j}^j(m_R^j) + \widehat{i_S^j} \rangle$.
   (b) If j is even, then
      i. R chooses $i_R^j \in_R GF[Q_j]$ and sends it to S.
      ii. S receives $\widehat{i_R^j}$, chooses $i_S^j \in_R GF[Q_j]$, and sends it to R.
      iii. R receives $\widehat{i_S^j}$, and computes $m_R^{j+1} = \langle \widehat{i_S^j}, C_{i_S^j}^j(m_R^j) + i_R^j \rangle$.
      iv. S computes $m_S^{j+1} = \langle i_S^j, C_{i_S^j}^j(m_S^j) + \widehat{i_R^j} \rangle$.
4. S manually authenticates m_S^k to R.
5. R accepts if and only if $m_S^k = m_R^k$.

Fig. 2. The k-round authentication protocol

describe the protocol is in a recursive fashion. The k-round protocol consists of S sending the message $m_1 = m$, as well as S and R exchanging i_S^1, and i_R^1. Then the two parties use protocol P_{k-1} to authenticate the message m_2, which is a computed hash value of m_1 using i_S^1, and i_R^1. Clearly, this protocol is perfectly complete.

Lemma 11. *Any computationally unbounded adversary has probability of at most ϵ to cheat the receiver into accepting a fraudulent message in protocol P_k.*

Proof. Given an execution of the protocol in which an adversary cheats the receiver into accepting a fraudulent message, it holds that $m_S^1 \neq m_R^1$ and $m_S^k = m_R^k$. Therefore, there exists an integer $1 \leq j \leq k - 1$ such that $m_S^j \neq m_R^j$ and $m_S^{j+1} = m_R^{j+1}$. Denote this event by D_j. In what follows, we bound the probability of this event, showing $\Pr[D_j] \leq \frac{\epsilon}{2^{k-j}}$. Therefore, the adversary's cheating probability is at most $\sum_{j=1}^{k-1} \Pr[D_j] \leq \sum_{j=1}^{k-1} \frac{\epsilon}{2^{k-j}} < \epsilon$.

For any variable y in the protocol and for a given execution, let $T(y)$ be the time at which the variable y is fixed, i.e., $T(i_R^j)$ denotes the time in which R sent the tag i_R^j, and $T(\widehat{i_R^j})$ denotes the time in which S received from the adversary the tag $\widehat{i_R^j}$ corresponding to i_R^j.

We assume here that j is odd, and refer the reader to the full version for the analysis in the case that j is even. There are three possible cases to consider:

1. $T(\widehat{i_R^j}) < T(i_R^j)$. In this case, the receiver chooses $i_R^j \in_R GF[Q_j]$ only after the adversary chooses $\widehat{i_R^j}$. Therefore,

$$\Pr[D_j] \le \Pr_{i_{\mathcal{R}}^j \in_{\mathrm{R}} \mathrm{GF}[Q_j]}\left[i_{\mathcal{R}}^j = \widehat{i}_{\mathcal{R}}^j\right] = \frac{1}{Q_j} \le \frac{\epsilon}{2^{k-j}} \ .$$

2. $T(\widehat{i}_{\mathcal{R}}^j) \ge T(i_{\mathcal{R}}^j)$ **and** $T(\widehat{i}_{\mathcal{S}}^j) \ge T(i_{\mathcal{S}}^j)$**.** In this case, the adversary chooses $\widehat{i}_{\mathcal{R}}^j$ not before the receiver chooses $i_{\mathcal{R}}^j$. If the adversary chooses $\widehat{i}_{\mathcal{R}}^j \ne i_{\mathcal{R}}^j$, then $m_{\mathcal{S}}^{j+1} \ne m_{\mathcal{R}}^{j+1}$, i.e., $\Pr[D_j] = 0$. Now suppose that the adversary chooses $\widehat{i}_{\mathcal{R}}^j = i_{\mathcal{R}}^j$. Since j is odd, the receiver chooses $i_{\mathcal{R}}^j$ only after he receives $\widehat{i}_{\mathcal{S}}^j$, therefore $T(i_{\mathcal{R}}^j) > T(\widehat{i}_{\mathcal{S}}^j) \ge T(i_{\mathcal{S}}^j) > T(m_{\mathcal{S}}^j)$, and also $T(i_{\mathcal{R}}^j) > T(m_{\mathcal{R}}^j)$. This means that $i_{\mathcal{R}}^j$ is chosen when $m_{\mathcal{R}}^j, \widehat{i}_{\mathcal{S}}^j, m_{\mathcal{S}}^j$ and $i_{\mathcal{S}}^j$ are fixed. Since $m_{\mathcal{S}}^j \ne m_{\mathcal{R}}^j$ and by the fact that for any choice of $i_{\mathcal{S}}^j$ and $\widehat{i}_{\mathcal{S}}^j$ the polynomials $C_{i_{\mathcal{R}}^j}^j(m_{\mathcal{R}}^j) + \widehat{i}_{\mathcal{S}}^j$ and $C_{i_{\mathcal{R}}^j}^j(m_{\mathcal{S}}^j) + i_{\mathcal{S}}^j$ are different as functions of $i_{\mathcal{R}}^j$, it follows that

$$\Pr[D_j] \le \Pr_{i_{\mathcal{R}}^j \in_{\mathrm{R}} \mathrm{GF}[Q_j]}\left[C_{i_{\mathcal{R}}^j}^j(m_{\mathcal{S}}^j) + i_{\mathcal{S}}^j = C_{i_{\mathcal{R}}^j}^j(m_{\mathcal{R}}^j) + \widehat{i}_{\mathcal{S}}^j\right]$$
$$\le \frac{1}{Q_j}\left\lceil \frac{n_j}{\log Q_j}\right\rceil \le \frac{\epsilon}{2^{k-j}} \ .$$

3. $T(\widehat{i}_{\mathcal{R}}^j) \ge T(i_{\mathcal{R}}^j)$ **and** $T(\widehat{i}_{\mathcal{S}}^j) < T(i_{\mathcal{S}}^j)$**.** As in the previous case, we can assume that the adversary chooses $\widehat{i}_{\mathcal{R}}^j = i_{\mathcal{R}}^j$. It always holds that $T(i_{\mathcal{S}}^j) > T(m_{\mathcal{S}}^j)$ and $T(i_{\mathcal{R}}^j) > T(m_{\mathcal{R}}^j)$. Since j is odd, the receiver sends $i_{\mathcal{R}}^j$ only after he receives $\widehat{i}_{\mathcal{S}}^j$, and therefore we can assume $T(\widehat{i}_{\mathcal{S}}^j) < T(i_{\mathcal{R}}^j) < T(i_{\mathcal{S}}^j)$. This implies that the sender chooses $i_{\mathcal{S}}^j \in_{\mathrm{R}} \mathrm{GF}[Q_j]$ when $m_{\mathcal{S}}^j, \widehat{i}_{\mathcal{S}}^j, m_{\mathcal{R}}^j$ and $i_{\mathcal{R}}^j$ are fixed. Hence,

$$\Pr[D_j] \le \Pr_{i_{\mathcal{S}}^j \in_{\mathrm{R}} \mathrm{GF}[Q_j]}\left[i_{\mathcal{S}}^j = C_{i_{\mathcal{R}}^j}^j(m_{\mathcal{R}}^j) + \widehat{i}_{\mathcal{S}}^j - C_{i_{\mathcal{R}}^j}^j(m_{\mathcal{S}}^j)\right] = \frac{1}{Q_j} \le \frac{\epsilon}{2^{k-j}} \ .$$

$\square$

The following claims conclude this section by showing that our choice of parameters guarantees that in protocol P_k the sender manually authenticates at most $2\log(1/\epsilon) + 2\log^{(k-1)} n + O(1)$ bits. We first show that the length n_{j+1} of the fingerprint computed in round j is roughly logarithmic in the length n_j of the fingerprint computed in round $j-1$, and then we use this fact to upper bound the length n_k of the manually authenticated fingerprint. The reader is referred to the full version of the paper for more details.

Claim 12. *If for every* $1 \le j \le k-2$ *it holds that* $n_j > \frac{2^{k-j}}{\epsilon}$*, then* $n_{k-1} \le \max\{4\log^{(k-2)} n_1 + 4\log 5 + 3,\ 27\}$.

Claim 13. *The sender manually authenticates at most* $2\log(1/\epsilon) + 2\log^{(k-1)} n + O(1)$ *bits in protocol* P_k.

6   Lower Bound in the Manual Channel Model

In this section we prove a lower bound on the length of the manually authenticated string. We present here the proof for the simplified case of a perfectly complete 3-round protocol where $n \geq 3 \log(1/\epsilon)$. The general proof is based on the same analysis, and is described in the full version of the paper. Moreover, note that by adding two more rounds, we can also assume for simplicity that in the last round the sender does not send an authentication tag x_i over the insecure channel (i.e., in the last round the sender *only* manually authenticates some string s). We prove the following theorem:

Theorem 14. *For any perfectly complete $(n, \ell, 3, \epsilon)$-authentication protocol in the manual channel model, where no authentication tag is sent in the last round, if $n \geq 3 \log(1/\epsilon)$, then $\ell \geq 2 \log(1/\epsilon) - 2$.*

As mentioned in Section 3, when the input message m is chosen uniformly at random, the honest execution of the protocol defines a probability distribution on the message m, the authentication tag x_1 (sent by the sender in the first round together with m), the authentication tag x_2 (sent by the receiver in the second round), and the manually authenticated string s (sent by the sender in the third round). We denote by M, X_1, X_2 and S the corresponding random variables.

The main idea of this proof is representing the entropy of the manually authenticated string S by splitting it as follows:

$$
\begin{aligned}
\mathrm{H}(S) &= (\mathrm{H}(S) - \mathrm{H}(S|M, X_1)) + (\mathrm{H}(S|M, X_1) - \mathrm{H}(S|M, X_1, X_2)) \\
&\quad + \mathrm{H}(S|M, X_1, X_2) \\
&= \mathrm{I}(S; M, X_1) + \mathrm{I}(S; X_2|M, X_1) + \mathrm{H}(S|M, X_1, X_2) \ .
\end{aligned}
$$

This representation captures the reduction of $\mathrm{H}(S)$ during the execution of the protocol, and allows us to prove that both the sender and the receiver must each independently reduce this entropy by at least $\log(1/\epsilon) - 1$ bits. We prove this by considering two possible man-in-the-middle attacks on the given protocol. In these attacks we use the fact that the adversary is computationally unbounded in that she can sample distributions induced by the protocol. For example, in the first attack, the adversary samples the distribution of X_2 given M, X_1 and S. While the distribution of X_2 given only M and X_1 can be sampled by merely following the protocol, this is not the case when sampling the distribution of X_2 given M, X_1 and S.

Lemma 15. *If $n \geq 2 \log \frac{1}{\epsilon}$, then $\mathrm{I}(S; M, X_1) + \mathrm{H}(S|M, X_1, X_2) > \log \frac{1}{\epsilon} - 1$.*

Proof. Consider the following attack:

1. The adversary $\mathcal{A}$ chooses $\widehat{m} \in_{\mathrm{R}} \{0,1\}^n$ and runs an honest execution with the receiver. Denote by s the manually authenticated string fixed by this execution. Now, $\mathcal{A}$'s goal is to cause the sender to manually authenticate this string.
2. $\mathcal{A}$ chooses $m \in_{\mathrm{R}} \{0,1\}^n$ as the sender's input, and receives x_1 from the sender.

3. If $\Pr[m, x_1, s] = 0$ in an honest execution, then $\mathcal{A}$ quits (in this case $\mathcal{A}$ has zero probability in convincing the sender to manually authenticate s). Otherwise, $\mathcal{A}$ samples $\widehat{x}_2$ from the distribution of X_2 given (m, x_1, s), and sends it to the sender. The sender manually authenticates some string.

4. If the sender did not authenticate s, $\mathcal{A}$ quits. Otherwise, $\mathcal{A}$ forwards s to the receiver.

By the unforgeability requirement of the protocol we obtain:

$$\epsilon \geq \Pr[\mathcal{R} \text{ accepts and } m \neq \widehat{m}] \geq \Pr[\mathcal{R} \text{ accepts}] - 2^{-n} .$$

Therefore, the assumption $n \geq 2 \log \frac{1}{\epsilon}$ implies that $\Pr[\mathcal{R} \text{ accepts}] < 2\epsilon$. Now we analyze the probability that the receiver accepts. Notice that:

- m and x_1 are chosen independently of s.
- $\widehat{x}_2$ is chosen conditioned on m, x_1 and s.
- The manually authenticated string sent by the sender is chosen conditioned on m, x_1 and x_2.

Therefore[5],

$$\Pr[\mathcal{R} \text{ accepts}] = \sum_{\substack{m, x_1, \widehat{x}_2, s: \\ \Pr[m, x_1, \widehat{x}_2, s] > 0}} \Pr[s] \Pr[m, x_1] \Pr[\widehat{x}_2 | m, x_1, s] \Pr[s | m, x_1, \widehat{x}_2]$$

$$= \sum_{\substack{m, x_1, \widehat{x}_2, s: \\ \Pr[m, x_1, \widehat{x}_2, s] > 0}} \Pr[m, x_1, s] \frac{\Pr[s]}{\Pr[s | m, x_1]} \Pr[\widehat{x}_2 | m, x_1, s] \Pr[s | m, x_1, \widehat{x}_2] \quad (6.1)$$

$$= \sum_{\substack{m, x_1, \widehat{x}_2, s: \\ \Pr[m, x_1, \widehat{x}_2, s] > 0}} \Pr[m, x_1, \widehat{x}_2, s] \, 2^{-\left\{ \log \frac{\Pr[s | m, x_1]}{\Pr[s]} + \log \frac{1}{\Pr[s | m, x_1, \widehat{x}_2]} \right\}} ,$$

where Equation (6.1) follows from Bayes' rule. By Jensen's inequality,

$$\Pr[\mathcal{R} \text{ accepts}] \geq 2^{-\sum_{\substack{m, x_1, \widehat{x}_2, s: \\ \Pr[m, x_1, \widehat{x}_2, s] > 0}} \Pr[m, x_1, \widehat{x}_2, s] \left\{ \log \frac{\Pr[s | m, x_1]}{\Pr[s]} + \log \frac{1}{\Pr[s | m, x_1, \widehat{x}_2]} \right\}}$$

$$= 2^{-\{I(S; M, X_1) + H(S | M, X_1, X_2)\}} ,$$

and therefore $I(S; M, X_1) + H(S | M, X_1, X_2) > \log \frac{1}{\epsilon} - 1$. $\qquad \square$

Lemma 16. *If $n \geq 3 \log \frac{1}{\epsilon}$ and $\ell < 2 \log(1/\epsilon) - 2$, then $I(S; X_2 | M, X_1) > \log \frac{1}{\epsilon} - 1$.*

Proof. Consider the following attack:

1. $\mathcal{A}$ chooses $m \in_R \{0, 1\}^n$, as the sender's input, and runs an honest execution with the sender. At the end of this execution, the sender manually authenticates a string s. $\mathcal{A}$ reads s, and delays it. Now, $\mathcal{A}$'s goal is to cause the receiver to accept this string together with a different input message $\widehat{m}$.

---

[5] For any random variable Z we write $\Pr[z]$ instead of $\Pr[Z = z]$.

2. $\mathcal{A}$ samples $(\widehat{m}, \widehat{x}_1)$ from the joint distribution of (M, X_1) given s, and sends them to the receiver, who answers x_2.
3. If $\Pr[\widehat{m}, \widehat{x}_1, x_2, s] = 0$, then $\mathcal{A}$ quits. Otherwise, $\mathcal{A}$ forwards s to the receiver.

As in Lemma 15, $\epsilon \geq \Pr[\mathcal{R} \text{ accepts}] - \Pr[\widehat{m} = m]$. However, in this attack, unlike the previous attack, the messages m and $\widehat{m}$ are not chosen uniformly at random and independently. First m is chosen uniformly at random, then s is picked from the distribution of S given m, and then $\widehat{m}$ is chosen from the distribution of M given s. Therefore,

$$\Pr[\widehat{m} = m] = \sum_s \Pr[s] \sum_m (\Pr[m|s])^2 \leq \sum_s \Pr[s] \max_m \Pr[m|s] \sum_m \Pr[m|s]$$

$$= \sum_s \Pr[s] \max_m \Pr[m|s] = \sum_s \max_m \Pr[m, s] \leq \sum_s \max_m \Pr[m] \quad .$$

Since the distribution of messages is uniform, and the authenticated string takes at most 2^ℓ values, we obtain $\Pr[\widehat{m} = m] \leq 2^{-n+\ell}$. From the assumptions that $\ell < 2\log(1/\epsilon) - 2$ and $n \geq 3\log(1/\epsilon)$ we get that $\Pr[\widehat{m} = m] < \epsilon$, and therefore $\Pr[\mathcal{R} \text{ accepts}] < 2\epsilon$. Now we analyze the probability that the receiver accepts. Notice that,

- $\widehat{m}$ and $\widehat{x}_1$ are chosen conditioned on s.
- x_2 is chosen conditioned only on $\widehat{m}$ and $\widehat{x}_1$.

Therefore,

$$\Pr[\mathcal{R} \text{ accepts}] = \sum_{\widehat{m}, \widehat{x}_1, s} \Pr[\widehat{m}, \widehat{x}_1, s] \sum_{\substack{x_2: \\ \Pr[\widehat{m}, \widehat{x}_1, x_2, s] > 0}} \Pr[x_2|\widehat{m}, \widehat{x}_1]$$

$$= \sum_{\substack{\widehat{m}, \widehat{x}_1, x_2, s: \\ \Pr[\widehat{m}, \widehat{x}_1, x_2, s] > 0}} \Pr[\widehat{m}, \widehat{x}_1, x_2, s] \frac{\Pr[x_2|\widehat{m}, \widehat{x}_1]}{\Pr[x_2|\widehat{m}, \widehat{x}_1, s]} \qquad (6.2)$$

$$= \sum_{\substack{\widehat{m}, \widehat{x}_1, x_2, s: \\ \Pr[\widehat{m}, \widehat{x}_1, x_2, s] > 0}} \Pr[\widehat{m}, \widehat{x}_1, x_2, s] \, 2^{-\log \frac{\Pr[x_2|\widehat{m}, \widehat{x}_1, s]}{\Pr[x_2|\widehat{m}, \widehat{x}_1]}} \quad ,$$

where Equation (6.2) follows from Bayes' rule. By Jensen's inequality,

$$\Pr[\mathcal{R} \text{ accepts}] \geq 2^{-\sum_{\substack{\widehat{m}, \widehat{x}_1, x_2, s: \\ \Pr[\widehat{m}, \widehat{x}_1, x_2, s] > 0}} \Pr[\widehat{m}, \widehat{x}_1, x_2, s] \log \frac{\Pr[x_2|\widehat{m}, \widehat{x}_1, s]}{\Pr[x_2|\widehat{m}, \widehat{x}_1]}} = 2^{-I(S; X_2|M, X_1)} \quad ,$$

and therefore $I(S; X_2|M, X_1) > \log \frac{1}{\epsilon} - 1$. $\qquad \square$

Now, Theorem 14 can be derived as follows. Suppose for contradiction that there exists a perfectly complete $(n, \ell, 3, \epsilon)$-authentication protocol, where no authentication tag is sent in the last round, and $n \geq 3\log(1/\epsilon)$ but $\ell < 2\log(1/\epsilon) - 2$. By using the fact that $\ell \geq H(S)$, we can easily derive a contradiction: The above mentioned representation of $H(S)$ and Lemmata 15 and 16 imply that $H(S) > 2\log(1/\epsilon) - 2$. Therefore $\ell \geq 2\log(1/\epsilon) - 2$ in any such protocol. This concludes the proof of Theorem 14.

7 Lower Bound in the Shared Key Model

In this section we prove a lower bound on the entropy of the shared key. This lower bound settles an open question posed by Gemmell and Naor [9], and shows that the authentication protocol proposed by Gemmell and Naor is essentially optimal with respect to the entropy of the shared key.

We present here the result for the simplified case of a perfectly complete 3-round protocol. The general proof is based on the same analysis, and is described in the full version. We prove the following theorem:

Theorem 17. *For any perfectly complete $(n, \ell, 3, \epsilon)$-authentication protocol in the shared key model, it holds that $\mathrm{H}(S) \geq 2\log(1/\epsilon)$, where S is the ℓ-bit shared key.*

As mentioned in Section 3, when the shared key s is chosen from its specified distribution, and the input message m is chosen uniformly at random, the honest execution of the protocol defines a probability distribution on the shared key s, the message m, the authentication tag x_1 (sent by the sender in the first round together with m), the authentication tag x_2 (sent by the receiver in the second round), and the authentication tag x_3 (sent by the sender in the third round). We denote by S, M, X_1, X_2 and X_3 the corresponding random variables.

We apply again the proof technique described in Section 6, and represent the entropy of the shared key S by splitting it as follows (we refer the reader to the full version for more details):

$$\mathrm{H}(S) = \mathrm{I}(S; M, X_1) + \mathrm{I}(S; X_2|M, X_1) + \mathrm{I}(S; X_3|M, X_1, X_2)$$
$$+ \mathrm{H}(S|M, X_1, X_2, X_3) \ .$$

Lemma 18. $\mathrm{I}(S; M, X_1) + \mathrm{I}(S; X_3|M, X_1, X_2) \geq \log \frac{1}{\epsilon}$.

Lemma 19. $\mathrm{I}(S; X_2|M, X_1) + \mathrm{H}(S|M, X_1, X_2, X_3) \geq \log \frac{1}{\epsilon}$.

8 Breaking the Lower Bounds Implies One-Way Functions

In this section we prove Theorem 10, namely, we show that in the computational setting one-way functions are essential for the existence of protocols breaking the lower bound stated in Theorem 8. As in Section 6 we prove here the result only for 3-round protocols, where in the last round no authentication tag x_i is sent over the insecure channel. Moreover, for simplicity we also assume that $n \geq 1/\epsilon$, and refer the reader to the full version for the proof of the general statement.

Theorem 20. *In the manual channel model, if there exists a computationally secure perfectly complete $(n, \ell, k, \epsilon, t)$-authentication protocol where no authentication tag is sent in the last round, such that $n \geq 1/\epsilon$, $\ell < 2\log(1/\epsilon) - 4$ and $t = \Omega(poly(n, k))$, then one-way functions exist.*

Proof. We show that if one-way functions do not exist, then the attacks described in Section 6 can be carried out by a polynomial-time adversary with almost the same success probability. We first focus on the attack described in Lemma 15.

Let f be a function defined as follows: f takes as input three strings r_S, r_R and m, and outputs (m, x_1, x_2, s) – the transcript of the protocol, where r_S, r_R and m are the random coins of the sender, the random coins of the receiver, and the input message, respectively. Let f' denote the function that is obtained from f by eliminating its third output, i.e., $f'(r_S, r_R, m) = (m, x_1, s)$. If one-way functions do not exist, then also distributionally one-way functions do not exist. Therefore, for any constant $c > 0$ there exists a probabilistic polynomial-time Turing machine $\mathcal{M}$ that on input (m, x_1, s) produces a distribution that is n^{-c}-statistically close to the uniform distribution on all the pre-images of (m, x_1, s) under f'. The polynomial-time adversary will use this $\mathcal{M}$ in the attack.

Let $\mathcal{A}$ denote the unbounded adversary that carried the attack described in Lemma 15, and let $\mathcal{A}^{\mathrm{PPT}}$ denote a polynomial-time adversary that carries the following attack (our goal is that the receiver will not be able to distinguish between $\mathcal{A}$ and $\mathcal{A}^{\mathrm{PPT}}$):

1. $\mathcal{A}^{\mathrm{PPT}}$ chooses $\widehat{m} \in_R \{0,1\}^n$ and runs an honest execution with the receiver. Denote by s the manually authenticated string fixed by this execution.
2. $\mathcal{A}^{\mathrm{PPT}}$ chooses $m \in_R \{0,1\}^n$ as the sender's input, and receives x_1 from the sender.
3. $\mathcal{A}^{\mathrm{PPT}}$ executes $\mathcal{M}$ on input (m, x_1, s), and then applies f to $\mathcal{M}$'s answer to compute $\widehat{x}_2$ and send it to the sender. The sender manually authenticates some string s^*.
4. $\mathcal{A}^{\mathrm{PPT}}$ forwards s^* to the receiver (who must accept $\widehat{m}$ if $s^* = s$ by the perfect completeness).

Let $\mathrm{Prob}^{\mathcal{R}}$ and $\mathrm{Prob}^{\mathrm{PPT},\mathcal{R}}$ denote the probabilities that the receiver $\mathcal{R}$ accepts $\widehat{m}$ when interacting with $\mathcal{A}$ and when interacting with $\mathcal{A}^{\mathrm{PPT}}$, respectively. Then, from the point of view of the receiver, the only difference in the these two executions is in the distribution of s^*. Therefore, $|\mathrm{Prob}^{\mathcal{R}} - \mathrm{Prob}^{\mathrm{PPT},\mathcal{R}}|$ is at most twice the statistical distance between s^* in the interaction with $\mathcal{A}$ and s^* in the interaction with $\mathcal{A}^{\mathrm{PPT}}$. By the above mentioned property of $\mathcal{M}$, this statistical difference is at most n^{-c}. Therefore, for sufficiently large t, we obtain as in Lemma 15

$$2\epsilon > \mathrm{Prob}^{\mathrm{PPT},\mathcal{R}} \geq \mathrm{Prob}^{\mathcal{R}} - 2n^{-c} \geq 2^{-\{\mathrm{I}(S;M,X_1)+\mathrm{H}(S|M,X_1,X_2)\}} - 2n^{-c} .$$

In particular, and since $n \geq 1/\epsilon$, we can choose the constant c such that $2n^{-c} < \epsilon$, and obtain $\mathrm{I}(S; M, X_1) + \mathrm{H}(S|M, X_1, X_2) > \log \frac{1}{\epsilon} - 2$.

A similar argument applied to the attack described in Lemma 16 yields $\mathrm{I}(S; X_2|M, X_1) > \log \frac{1}{\epsilon} - 2$, and therefore $\mathrm{H}(S) > 2\log \frac{1}{\epsilon} - 4$.   □

Acknowledgments. We thank Benny Pinkas, Danny Segev, Serge Vaudenay and the anonymous referees for their remarks and suggestions.

References

1. B. Barak. Constant-round coin-tossing with a man in the middle or realizing the shared random string model. In *43rd FOCS*, pages 345–355, 2002.
2. Bluetooth. http://www.bluetooth.com/bluetooth/.
3. Certified Wireless USB. http://www.usb.org/developers/wusb/.
4. G. Di Crescenzo, Y. Ishai, and R. Ostrovsky. Non-interactive and non-malleable commitment. In *30th STOC*, pages 141–150, 1998.
5. G. Di Crescenzo, J. Katz, R. Ostrovsky, and A. Smith. Efficient and non-interactive non-malleable commitment. In *EUROCRYPT '01*, pages 40–59, 2001.
6. D. Dolev, C. Dwork, and M. Naor. Non-malleable cryptography. *SIAM Journal on Computing*, 30(2):391–437, 2000.
7. C. Gehrmann. Cryptanalysis of the Gemmell and Naor multiround authentication protocol. In *CRYPTO '94*, pages 121–128, 1994.
8. C. Gehrmann, C. J. Mitchell, and K. Nyberg. Manual authentication for wireless devices. *RSA Cryptobytes*, 7:29–37, 2004.
9. P. Gemmell and M. Naor. Codes for interactive authentication. In *CRYPTO '93*, pages 355–367, 1993.
10. E. Gilbert, F. MacWilliams, and N. Sloane. Codes which detect deception. *Bell System Technical Journal*, 53(3):405–424, 1974.
11. R. Impagliazzo and M. Luby. One-way functions are essential for complexity based cryptography. In *30th FOCS*, pages 230–235, 1989.
12. S. Laur, N. Asokan, and K. Nyberg. Efficient mutual data authentication using manually authenticated strings. Cryptology ePrint Archive, Report 2005/424, 2005.
13. U. M. Maurer. Authentication theory and hypothesis testing. *IEEE Transactions on Information Theory*, 46(4):1350–1356, 2000.
14. M. Naor and G. N. Rothblum. The complexity of online memory checking. In *46th FOCS*, pages 573–584, 2005.
15. M. Naor, G. Segev, and A. Smith. Tight bounds for unconditional authentication protocols in the manual channel and shared key models. Cryptology ePrint Archive, Report 2006/175, 2006.
16. R. Pass and A. Rosen. New and improved constructions of non-malleable cryptographic protocols. In *37th STOC*, pages 533–542, 2005.
17. G. J. Simmons. Authentication theory/coding theory. In *CRYPTO '84*, pages 411–431, 1984.
18. G. J. Simmons. The practice of authentication. In *EUROCRYPT '85*, pages 261–272, 1985.
19. S. Vaudenay. Secure communications over insecure channels based on short authenticated strings. In *CRYPTO '05*, pages 309–326, 2005.
20. M. N. Wegman and L. Carter. New hash functions and their use in authentication and set equality. *Journal of Computer and System Sciences*, 22(3):265–279, 1981.

Robust Fuzzy Extractors and Authenticated Key Agreement from Close Secrets

Yevgeniy Dodis[1,*], Jonathan Katz[2,**],
Leonid Reyzin[3,***], and Adam Smith[4,†]

[1] New York University
dodis@cs.nyu.edu
[2] University of Maryland
jkatz@cs.umd.edu
[3] Boston University
reyzin@cs.bu.edu
[4] Weizmann Institute of Science
adam.smith@weizmann.ac.il

Abstract. Consider two parties holding correlated random variables W and W', respectively, that are within distance t of each other in some metric space. These parties wish to agree on a uniformly distributed secret key R by sending a single message over an insecure channel controlled by an all-powerful adversary. We consider both the *keyless* case, where the parties share no additional secret information, and the *keyed* case, where the parties share a long-term secret SK that they can use to generate a sequence of session keys $\{R_j\}$ using multiple pairs $\{(W_j, W'_j)\}$. The former has applications to, e.g., biometric authentication, while the latter arises in, e.g., the bounded storage model with errors.

Our results improve upon previous work in several respects:

- The best previous solution for the keyless case with no errors (i.e., $t = 0$) requires the min-entropy of W to exceed $2|W|/3$. We show a solution when the min-entropy of W exceeds the *minimal* threshold $|W|/2$.
- Previous solutions for the keyless case in the presence of errors (i.e., $t > 0$) required random oracles. We give the first constructions (for certain metrics) in the standard model.
- Previous solutions for the keyed case were stateful. We give the first stateless solution.

1 Introduction

A number of works have explored the problem of *secret key agreement based on correlated information* by which two parties holding instances of correlated

* Supported by NSF grants #0133806, #0311095, and #0515121.
** Supported by NSF grants #0310499, #0310751, and #0447075.
*** Supported by NSF grants #0311485 and #0515100.
† Supported by the Louis L. and Anita M. Perlman Fellowship.

C. Dwork (Ed.): CRYPTO 2006, LNCS 4117, pp. 232–250, 2006.

random variables W and W' communicate and thereby generate a shared, secret (uniformly-random) key SK. Early work [Wyn75, BBR88, Mau93, BBCM95] assumed that the parties could communicate over a *public* but *authenticated* channel or, equivalently, assumed a passive adversary. This assumption was relaxed in later work [Mau97, MW97, Wol98, MW03, RW03], which considered an active adversary who could modify all messages sent between the two parties.

The motivation of the above works was primarily to explore the possibility of *information-theoretic* security; however, this is not the only motivation. The problem also arises in the context of using noisy data (such as biometric information) for cryptographic purposes, even if computational security suffices. The problem also arises in the context of the *bounded storage model* (BSM) [Mau92] in the presence of errors [Din05, DS05]. We discuss each of these in turn.

AUTHENTICATION USING NOISY DATA. In the case of authentication using noisy data, the random variables W, W' are *close* (with respect to some metric) but not *identical*. For simplicity, we assume the noisy data represents biometric information though the same techniques apply to more general settings. In this context, two different scenarios have been considered:

"Secure authentication:" Here, a trusted server stores the "actual" biometric data W of a user; periodically, the user obtains a fresh biometric scan W' which is close, but not identical, to W. The server and user then wish to mutually authenticate and agree on a key R.

"Key recovery:" Here, a user (on his own) uses his "actual" biometric data W to generate a random key R along with some public information P, and then stores P on a (possibly untrusted) server. The key R might then be used, say, to encrypt some data for long-term storage. At a later point in time, the user obtains a fresh biometric scan W' along with the value P from the server; together, these values enable recovery of R (and hence enable decryption of the data).

In the second setting the user is, in effect, running a key agreement protocol with *himself* at two points in time, with the (untrusted) server acting as the "communication channel" between these two instances of the user. This second scenario inherently requires a *non-interactive* (i.e., one-message) key agreement protocol since W is no longer available at the later point in time. Note also that any solution for the second scenario is also a solution for the first.

Solutions for achieving secret key agreement using noisy data and an *authenticated* channel are known [BBR88, BBCM95, DORS06, JW99, FJ01, LT03]. However, existing work such as [Mau97, MW97, Wol98, MW03, RW03] does *not* solve the above problem when the parties communicate over an *unauthenticated* channel. Positive results in the unauthenticated setting were known only for two special cases: (1) when $W = W'$ and (2) when W and W' consist of (arbitrarily-many) independent realizations of the same random experiment; i.e., $W = (W^{(1)}, W^{(2)}, \ldots)$ and $W' = (W'^{(1)}, W'^{(2)}, \ldots)$. In the case of biometric data, however, W, W' are not likely to be equal and we cannot in general obtain an unbounded number of samples.

Recently, some partial solutions to the problems considered above have been obtained in the unauthenticated setting. Boyen [Boy04] shows, in the random oracle model, how to achieve *unidirectional* authentication with noisy data, as well as a weak form of security in the "key recovery" setting (essentially, R remains secret but the user can be fooled into using an incorrect key R'). Subsequent work of Boyen, et al. [BDK+05] shows two solutions: the first is non-interactive but relies on random oracles; the second solution can be used for secure authentication but does not apply to the "key recovery" scenario because it requires interaction. This second solution has some other limitations as well: since it relies on a underlying password-based key-exchange protocol, it inherently provides *computational* rather than *information-theoretic* security; furthermore, given the current state-of-the-art for password-based key exchange [BPR00, BMP00, KOY01, GL01, GL03], the resulting protocol is either impractical or else requires additional assumptions such as random oracles/ideal ciphers or public parameters.

THE BOUNDED STORAGE MODEL AND THE "KEYED" CASE. Key agreement using correlated information arises also in the context of the *bounded storage model* (BSM) [Mau92] in the presence of errors [Din05, DS05]. In the BSM, two parties share a long-term secret key SK. In each of an unlimited number of time periods $j = 1, \ldots,$ a long random string Z_j is broadcast to the parties (and observed by an adversary); the assumption is that the length of Z_j is more than what the adversary can store. The parties use SK and Z_j to generate a secret session key R_j, with $|R_j| \gg |SK|$, in each period. This process should achieve "everlasting security" [ADR02], meaning that even if SK is revealed to the adversary in some time period n, all session keys $\{R_j\}_{j<n}$ remain independently and uniformly distributed from the perspective of the adversary.

A typical paradigm for achieving the above is for the parties to sample (using SK) shorter strings W_j and W'_j, respectively, from the random string Z_j in each period. Next, the parties use W_j (resp., W'_j) and SK to generate R_j. In standard treatments of the BSM (e.g., [Mau92, ADR02]), it is assumed that both parties receive identical copies of Z_j and hence $W_j = W'_j$. In the presence of transmission errors in Z_j, however, it is possible for W_j and W'_j to be close but not identical [Din05, DS05]. The parallels to the case of biometric authentication, as discussed earlier, should now be clear. Nevertheless, the problems are incomparable: in the case of the BSM with errors there is a stronger setup assumption (the parties share a long-term key SK), but the security requirements are more stringent.

OUR CONTRIBUTIONS. We focus on the abstract problem of secret key agreement between two parties holding instances w, w' of correlated random variables W, W' that are guaranteed to be close but not necessarily identical. Specifically, we assume that w and w' are within distance t with respect to some underlying metric. Some of our results hold for arbitrary metric spaces, while others assume the Hamming metric in particular.

We consider only *non-interactive* protocols defined by procedures (Gen, Rep) that operate as follows: the first party, holding w, computes $(R, P) \leftarrow \mathsf{Gen}(w)$ and sends P to the second party; this second party computes $R' \leftarrow \mathsf{Rep}(w', P)$. (If the parties share a long-term key SK then Gen, Rep take this as additional input.) The basic requirements, informally, are

Correctness: $R = R'$ whenever w' is within distance t of w.
Security: If the entropy of W is high, R is uniformly distributed even given P.

So far, this gives exactly a *fuzzy extractor* as defined by Dodis et al. [DORS06] (although we additionally allow the possibility of a long-term key). Since we are interested in the case when the parties communicate over an *unauthenticated* channel, however, we actually want to construct *robust* fuzzy extractors [BDK+05] that additionally protect against malicious modification of P. Robustness requires that if the adversary sends any modified value $P' \neq P$, then with high probability the second player will reject (i.e., $\mathsf{Rep}(w', P) = \bot$). *Strong robustness* requires this to hold even if the adversary learns the R (held by the first player). This property is essential in settings where the first party may begin using R before the second party computes R', and is also needed for the "key recovery" scenario discussed earlier (since previous usages of R may leak information about it). *Weak robustness*, still sufficient for some applications, only requires robustness when R is not learned by the adversary.

Letting $H_\infty(X)$ denote the min-entropy of a random variable X, we now describe our results.

The keyless case with no errors. Although our focus is on the case when W, W' are *close*, we obtain improvements also in the case when they are equal (i.e., $t = 0$). Specifically, the best previous non-interactive solution in this setting is due to Maurer and Wolf[1] [MW03] who show that when $H_\infty(W) > 2|W|/3$ it is possible to achieve weak robustness and generate a shared key R of length $H_\infty(W) - 2|W|/3$. On the other hand, results of [DS02] imply that a non-interactive solution is impossible when $H_\infty(W) \leq |W|/2$.

We bridge the gap between known upper- and lower-bounds and show that whenever $H_\infty(W) > |W|/2$ it is possible to achieve weak robustness and generate a shared key R of length $2H_\infty(W) - |W|$. This improves both the required min-entropy of W as well as the length of the resulting key. Moreover, we give the first solution satisfying *strong* robustness which still works as long as $H_\infty(W) > |W|/2$ (but extracts a slightly shorter key).

The keyless case with errors. The only previously-known construction of robust fuzzy extractors [BDK+05] relies on the random oracle model. (This solution is generic and applies to any metric admitting a good error-correcting code.) We (partially) resolve the main open question of [BDK+05] by showing a construction of strongly robust fuzzy extractors *in the standard model*, for the case of the Hamming, set difference, or edit metrics. In fact, our solution

---

[1] The journal version fixes some incorrect claims made in [MW97].

is also better than the second (interactive) solution of [BDK+05] in the "secure authentication" scenario: their solution achieves computational (rather than information-theoretic) security, and is impractical unless additional assumptions (such as the existence of public parameters) are made.

The keyed case with errors. Recent work focusing on the BSM with errors [Din05, DS05] shows that a constant relative Hamming distance between W_j and W_j' (recall, these are the samples recorded by the two parties) can be tolerated using a non-interactive protocol. The solution of [Din05] is stateful (i.e., SK is updated between each time period) while the second solution [DS05] requires the parties to communicate over an authenticated channel. We construct a robust *keyed* fuzzy extractor (for generic metrics), and show that this enables a *stateless* BSM solution (for the Hamming metric) using an *unauthenticated* channel. In doing so, we retain essentially all other parameters of the previous BSM solutions.

2   Definitions

If S is a set, $x \leftarrow S$ means that x is chosen uniformly from S. If X is probability distribution, then $x \leftarrow X$ means that x is chosen according to distribution X. The notation $\Pr_X[x]$ denotes the probability assigned by X to the value x. (We often omit the subscript when the probability distribution is clear from context.) If A is a probabilistic algorithm and x an input, $A(x)$ denotes the random variable $A(x; \omega)$ where random coins ω are sampled uniformly. If X is a random variable, then $A(X)$ is defined in the analogous manner. All logarithms are base 2.

The *min-entropy* of a random variable X is $\mathbf{H}_\infty(X) = -\log(\max_x \Pr_X[x])$. We define the (average) conditional min-entropy of X given Y as $\tilde{\mathbf{H}}_\infty(X \mid Y) = -\log(\mathbf{E}_{y \leftarrow Y}(2^{-\mathbf{H}_\infty(X|Y=y)}))$. This (non-standard) definition is convenient for cryptographic purposes [DORS06, RW05].

Definition 1. *A family of efficient functions* $\mathcal{H} = \{h_i : \{0,1\}^n \to \{0,1\}^\ell\}_{i \in I}$ *is* δ-almost universal *if for all distinct* x, x' *we have* $\Pr_{i \leftarrow I}[h_i(x) = h_i(x')] \leq \delta$. *Families with* $\delta = 2^{-\ell}$ *are called* universal. $\Diamond$

Let X_1, X_2 be two probability distributions over S. Their *statistical distance* is $\mathbf{SD}(X_1, X_2) \stackrel{\text{def}}{=} \frac{1}{2} \sum_{s \in S} |\Pr_{X_1}[s] - \Pr_{X_2}[s]|$. If two distributions have statistical distance at most ε, we say they are ε-close, and write $X_1 \approx_\varepsilon X_2$. Note that ε-close distributions cannot be distinguished with advantage better than ε even by a computationally unbounded adversary.

Definition 2. *An efficient probabilistic function* Ext : $\{0,1\}^n \to \{0,1\}^\ell$ *is a* strong (m, ε)-extractor *if for all distributions* X *over* $\{0,1\}^n$ *with* $\mathbf{H}_\infty(X) \geq m$ *we have* $\mathbf{SD}((I, \mathsf{Ext}(X; I)), (I, U_\ell)) \leq \varepsilon$. *The randomness* I *is called the seed.* $\Diamond$

Lemma 1 ([BBR88, HILL99]). *Fix* $m, \varepsilon > 0$, *and* $\ell \leq m - 2\log\left(\frac{1}{\varepsilon}\right)$. *If* $\mathcal{H} = \{h_i : \{0,1\}^n \to \{0,1\}^\ell\}_{i \in I}$ *is a* $2^{-\ell}(1 + \varepsilon^2)$-almost universal family then $\mathcal{H}$ *is a strong* (m, ε)-extractor.

ONE-TIME MESSAGE AUTHENTICATION CODES (MACs). One-time MACs allow information-theoretic authentication of a message using a key shared in advance.

Definition 3. *A function family* $\{\mathsf{MAC}_\mu : \{0,1\}^{\tilde{n}} \to \{0,1\}^v\}$ *is a strongly δ-secure (one-time) MAC if: (a) for any x and σ, $\mathrm{Pr}_\mu\left[\mathsf{MAC}_\mu(x) = \sigma\right] = 2^{-v}$; and (b) for any $x \neq x'$ and any σ, σ', $\mathrm{Pr}_\mu\left[\mathsf{MAC}_\mu(x') = \sigma' \mid \mathsf{MAC}_\mu(x) = \sigma\right] \leq \delta$.* ◇

The definition above is stronger than usual, since part (b) requires security conditioned on a worst-case choice of σ, rather than taking an average over μ. However, it is convenient because it is satisfied by standard constructions, and also composes nicely with universal hash families:

Lemma 2. *If $\{\mathsf{MAC}_\mu : \{0,1\}^u \to \{0,1\}^v\}$ is a strongly δ-secure MAC and $\{h_i : \{0,1\}^{\tilde{n}} \to \{0,1\}^u\}$ is δ'-almost universal, then $\mathsf{MAC}'_{\mu,i}(x) = \mathsf{MAC}_\mu(h_i(x))$ is a strongly $(\delta + \delta')$-secure MAC for $\tilde{n}$-bit messages.*

SECURE SKETCHES AND FUZZY EXTRACTORS. We start by reviewing the definitions of (ordinary) secure sketches and fuzzy extractors from [DORS06]. Let $\mathcal{M}$ be a metric space with distance function dis.

Definition 4. *An $(m, \tilde{m}, t)$-secure sketch is a pair of efficient randomized procedures (SS, SRec) s.t.:*

1. *The sketching procedure SS on input $w \in \mathcal{M}$ returns a bit string $s \in \{0,1\}^*$. The recovery procedure SRec takes an element $w' \in \mathcal{M}$ and $s \in \{0,1\}^*$.*
2. *Correctness: If $\mathsf{dis}(w, w') \leq t$ then $\mathsf{SRec}(w', \mathsf{SS}(w)) = w$.*
3. *Security: For any distribution W over $\mathcal{M}$ with min-entropy m, the (average) min-entropy of W conditioned on s does not decrease very much. Specifically, if $\mathbf{H}_\infty(W) \geq m$ then $\widetilde{\mathbf{H}}_\infty(W \mid \mathsf{SS}(W)) \geq \tilde{m}$.*

The quantity $m - \tilde{m}$ is called the entropy loss *of the secure sketch.* ◇

For the case of the Hamming metric on $\mathcal{M} = \{0,1\}^n$, the following "code-offset" construction [BBR88, Cré97] is well known. The sketch $s = \mathsf{SS}(w)$ consists of the syndrome[2] of w with respect to some (efficiently decodable) $[n, k, 2t + 1]$-error-correcting code C. We do not need any details of this construction other than the facts that s is a (deterministic) *linear function* of w and that the entropy loss is at most $|s| = n - k$.

Definition 5. *An $(m, \ell, t, \varepsilon)$-fuzzy extractor is a pair of efficient randomized procedures (Gen, Rep) with the following properties:*

1. *The generation procedure Gen, on input $w \in \mathcal{M}$, outputs an extracted string $R \in \{0,1\}^\ell$ and a helper string $P \in \{0,1\}^*$. The reproduction procedure Rep takes an element $w' \in \mathcal{M}$ and a string $P \in \{0,1\}^*$ as inputs.*
2. *Correctness: If $\mathsf{dis}(w, w') \leq t$ and $(R, P) \leftarrow \mathsf{Gen}(w)$, then $\mathsf{Rep}(w', P) = R$.*

---

[2] For a linear code with parity check matrix H, the syndrome of w is $wH^\top$.

3. Security: *For any distribution W over $\mathcal{M}$ with min-entropy m, the string R is close to uniform even conditioned on the value of P. Formally, if $\mathbf{H}_\infty(W) \geq m$ and $(R, P) \leftarrow \mathsf{Gen}(W)$, then we have $\mathbf{SD}\left((R, P), (U_\ell, P)\right) \leq \varepsilon$.* $\Diamond$

Composing an $(m, \tilde{m}, t)$-secure sketch with a strong $(\tilde{m} - \log\left(\frac{1}{\varepsilon}\right), \varepsilon)$-extractor $\{h_i \colon \mathcal{M} \to \{0, 1\}^\ell\}$ yields a $(m, \ell, t, 2\varepsilon)$-fuzzy extractor [DORS06]. In that case $P = (\mathsf{SS}(w), i)$ and $R = h_i(w)$.

2.1 Robust Fuzzy Extractors

Fuzzy extractors, defined above, protect against a *passive* attack in which an adversary observes P, and tries to learn something about the extracted key R. However, the definition says nothing about what happens if an adversary can modify P as it is sent to the user holding w'. That is, there are no guarantees about the output of $\mathsf{Rep}(w', \tilde{P})$ for $\tilde{P} \neq P$.

Boyen et al. [BDK+05] propose the notion of *robust* fuzzy extractors which provides strong guarantees against such an attack. Specifically, Rep can output either a key or a special value $\perp$ ("fail"). We require that any value $\tilde{P} \neq P$ produced by the adversary given P causes $\mathsf{Rep}(w', \tilde{P})$ to output $\perp$. Modified versions of the correct public information P can therefore be detected.

We consider two variants of this idea, depending on whether Gen and Rep additionally share a (short) long-term key SK. Boyen et al. considered the keyless primitive; this is what we define first. Further below, we adjust the definitions to the case of a shared key.

If W, W' are two (correlated) random variables over a metric space $\mathcal{M}$, we say $\mathsf{dis}(W, W') \leq t$ if the distance between W and W' is at most t with probability one. We call (W, W') a (t, m)-*pair* if $\mathsf{dis}(W, W') \leq t$ and $\mathbf{H}_\infty(W) \geq m$.

Definition 6. *Given algorithms $(\mathsf{Gen}, \mathsf{Rep})$ and values $w, w' \in \mathcal{M}$, consider the following game involving an adversary $\mathcal{A}$: Compute $(R, P) \leftarrow \mathsf{Gen}(w)$ and $\tilde{P} = \mathcal{A}(R, P)$. The adversary succeeds if $\tilde{P} \neq P$ and $\mathsf{Rep}(w', \tilde{P}) \neq\perp$.*

$(\mathsf{Gen}, \mathsf{Rep})$ is a (strong) $(m, \ell, t, \varepsilon, \delta)$-robust fuzzy extractor if it is an $(m, \ell, t, \varepsilon)$-fuzzy extractor, and for all (t, m)-pairs (W, W') and all adversaries $\mathcal{A}$, the probability of success is at most δ. The notion of a $(m, \ell, t, \varepsilon, \delta)$-weakly-robust fuzzy extractor is defined similarly, except that $\mathcal{A}$ is given only P and not R.

See Fig. 1 for an illustration. $\Diamond$

RE-USING ROBUST EXTRACTORS. The definition of robust extractors composes with itself in some situations. For example, a generalization of the above (used in [BDK+05]) allows the adversary to output $(\tilde{P}_1, \ldots, \tilde{P}_j)$; the adversary succeeds if there exists an i with $\mathsf{Rep}(w', \tilde{P}_i) \neq\perp$. A simple union bound shows that the success probability of an adversary in this case increases at most linearly in j.

Similarly, suppose that two players (Alice and Bob) receive a sequence of correlated pairs of random variables $(W_1, W'_1), (W_2, W'_2) \ldots$, such that each pair is at distance at most t and the entropy of W_i conditioned on information from other time periods $\left\{(W_j, W'_j)\right\}_{j \neq i}$ is at least m (we call such a sequence (t, m)-*correlated*). Once again, a simple hybrid argument shows that Alice and Bob can agree on (essentially) random and uncorrelated keys $R_1, R_2, \ldots$, by having Alice apply Gen to each W_i and send P_i to Bob. Namely, after j periods the attacker's advantage at distinguishing the vector of unknown keys from random is at most $j\varepsilon$, and her probability of forging a valid message $\tilde{P}_i$ is at most δ in each period.

KEYED ROBUST FUZZY EXTRACTORS. In some scenarios, such as the bounded storage model, the parties running Gen and Rep can additionally share a short, truly random key to help them extract a (long) session key R from close variables W and W'. Syntactically, this simply means that Gen and Rep now also take an extra input SK: namely, we have $(R, P) \leftarrow \mathsf{Gen}_{\mathsf{SK}}(w)$, $R' = \mathsf{Rep}_{\mathsf{SK}}(w', P)$, and require that for any SK we have $R = R'$ whenever $\mathsf{dis}(w, w') \leq t$.

The robustness property of keyed fuzzy extractors (Def. 6) does not change with the addition of SK: in particular, the attacker does not get the secret key SK in the unforgeability game of Def. 6. At first glance, this appears to trivialize the problem of constructing keyed robust fuzzy extractors. For example, one might attempt the following transformation. Given an output (R, P) or a regular fuzzy extractor, let SK be a key to an

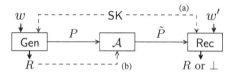

Fig. 1. Robustness of extractors (Def. 6). Dotted lines indicate variations in the definition. (a) *Keyed extractors* take an additional input SK shared by Gen and Rep. (b) For *weak robustness*, the adversary does not have access to the extracted key R.

information-theoretic MAC, and simply append to P its own tag $\mathsf{MAC}_{\mathsf{SK}}(P)$ computed using SK. This transformation is not sufficient, however, because keyed fuzzy extractors must satisfy a very strong security (i.e. extraction) condition which we define next.

Definition 7. *A keyed keyed $(m, \ell, t, \varepsilon)$-fuzzy extractor (Gen, Rep) is secure if for any distribution W over $\mathcal{M}$ with min-entropy m, the string (SK, R) is close to a pair of fresh uniform random strings, even conditioned on the value of P: if $\mathbf{H}_\infty(W) \geq m$ and $(R, P) \leftarrow \mathsf{Gen}_{\mathsf{SK}}(W)$, then $(\mathsf{SK}, R, P) \approx_\varepsilon (U_{|\mathsf{SK}|}, U_\ell, P)$.*

We say the extractor is strongly secure if $(\mathsf{SK}, R, P) \approx_\varepsilon (U_{|\mathsf{SK}|}, U_\ell, U_{|P|})$. $\diamondsuit$

The security condition for keyed extractors ensures that value of SK is essentially independent from the attacker's view. For example, the simplistic transformation above from regular fuzzy extractors leaks the tag of P (which is a known deterministic function of SK) to the attacker $\mathcal{A}$, implying that SK no longer looks random to $\mathcal{A}$ — it therefore does not satisfy Def. 7. This security condition is important for two reasons: first, it means that the session key R remains secure even if the long-term key SK is revealed in the future; second, the long-term key can be re-used (e.g., for future authentication). If Alice and Bob are given a sequence of j correlated pairs (as discussed above), then $\mathcal{A}$ has advantage at most $j\varepsilon$ in distinguishing the vector of unknown session keys from random. Similarly, her probability of forging a valid $\tilde{P}_j$ in the j-th execution of the robustness game (Def. 6) is at most $j\varepsilon + \delta$. The repeated game and its reduction to the one-time definitions presented here are given in detail in the full version of this paper.

Finally, we note that some settings require a more stringent condition called *strong* security in Def. 7. In this case the adversary's view hides both the long-term key SK and the exact distribution of W (since P is distributed identically regardless of W). The bounded storage model, discussed in Section 4, is an example of such a setting.

3 Constructing Robust Fuzzy Extractors

In this section, we describe new constructions of robust fuzzy extractors. In particular, these solve the problem of secret key generation over a completely insecure channel. We begin by analyzing the case of no errors (i.e., $t = 0$) and then consider the more challenging case where errors may occur.

3.1 The Errorless Case ($w = w'$)

Recall the "standard" solution using strong extractors, which works when the adversary is *passive*. In this case, $\mathsf{Gen}(w)$ chooses a random seed i for a strong extractor and sets $R = h_i(w)$ and $P = i$; the recovery procedure $\mathsf{Rep}(w, i)$ simply outputs $R = h_i(w)$. Unfortunately, this solution does not work if the adversary is *active*. In particular, if $i' \neq i$ there is no longer any guarantee on the output $\mathsf{Rep}(w, i')$ (and it is easy to show counterexamples where a malicious i' completely determines $\mathsf{Rep}(w, i')$ even if w is uniform). One idea is to somehow authenticate i using the extracted key R; in general, this does not work either. It turns out, however, that w itself can be used to authenticate i, at least for a particular choice of MAC and a particular strong extractor. Details follow.

CONSTRUCTION. We define the procedures $(\mathsf{Gen}, \mathsf{Rep})$. To compute $\mathsf{Gen}(w)$, parse the n-bit string w as two strings a and b of lengths $n - v$ and v, respectively (the value of v will be determined later). View a as an element of $\mathbb{F}_{2^{n-v}}$ and b as an element of $\mathbb{F}_{2^v}$ so that addition in the field corresponds to exclusive-or of bit strings. Choose a random $i \in \mathbb{F}_{2^{n-v}}$, compute $ia \in \mathbb{F}_{2^{n-v}}$, and let $[ia]_{v+1}^{n-v}$ denote the most significant $n - 2v$ bits of ia and $[ia]_1^v$ denote the remaining v bits. View $[ia]_1^v$ as an element of $\mathbb{F}_{2^v}$. Then compute $\sigma = [ia]_1^v + b$, set $P = (i, \sigma)$, and let the extracted key be $R = [ia]_{v+1}^{n-v}$.

$\mathsf{Rep}(w, P')$, where $P' = (i', \sigma')$, proceeds as follows. Parse w as two strings a and b as above. Then verify that $\sigma' = [i'a]_1^v + b$ and output $\perp$ if this is not the case. Otherwise, compute the extracted key $R = [i'a]_{v+1}^{n-v}$.

Theorem 1. *Let $|w| = n$. For an appropriate setting of v, the above construction is an $(m, \ell, 0, \varepsilon, \delta)$-weakly-robust extractor for any $m, \ell, \varepsilon, \delta$ satisfying $\ell \leq 2m - n - 2\max\left\{\log\left(\frac{1}{\delta}\right), 2\log\left(\frac{1}{\varepsilon}\right)\right\}$. It is an $(m, \ell, 0, \varepsilon, \delta)$-robust extractor when $\ell \leq \min\left(\frac{2m-n}{3} - \frac{2}{3}\log\left(\frac{1}{\delta}\right), 2m - n - 4\log\left(\frac{1}{\varepsilon}\right)\right)$.*

The proof is given in the full version. Observe that extraction is possible as long as $\mathbf{H}_\infty(W) \stackrel{\text{def}}{=} m > |W|/2$. Furthermore, in the case of weakly robust extraction (which is the notion of security considered by Maurer and Wolf [MW03]) we extract a key of length roughly $2 \cdot \mathbf{H}_\infty(W) - |W|$.

3.2 Authenticating a Message While Extracting

The above construction uses the input w to authenticate the extractor seed i. It can be extended to additionally authenticate a (bounded-length) message M; i.e., to be simultaneously a robust fuzzy extractor and an information-theoretic one-time MAC. In this case, both Gen and Rep will take an additional input M, and it should be difficult for an adversary to cause Rep to accept a different M. (We are being informal here since this is merely a stepping stone to the results of the following section.) Naturally, this could be done easily by using (a part of) R as a key for a MAC, but this would correspondingly reduce the final number of extracted bits. In contrast, the approach presented here (almost) does not reduce the length of R at all. We adapt a standard technique [WC81] for authenticating long messages using almost-universal hash functions.

CONSTRUCTION. Assume $|M| \leq L \cdot (n - v)$, where L is known to all parties in advance. Split M into L chunks $M_0, \ldots, M_{L-1}$, each $n - v$ bits long, and view these as coefficients of a polynomial $M(x) \in \mathbb{F}_{2^{n-v}}[x]$ of degree $L - 1$. Modify the construction of the previous section as follows: to compute $\mathsf{Gen}(w, M)$, parse w as $a\|b$, choose random $i \in \mathbb{F}_{2^{n-v}}$, compute $\sigma = [a^2 M(a) + ia]_1^v + b$, and set $P = (i, \sigma)$. As before, the extracted key is $R = [ia]_{v+1}^{n-v}$.

Given w, M', and $P' = (i', \sigma')$, verify that $|M'| \leq L(n - v)$ and that $\sigma' = [a^2 M'(a) + i'a]_1^v + b$. If so, compute $R = [i'a]_{v+1}^{n-v}$.

The property we need is that every distinct pair of tuples $(M, i) \neq (M', i')$ the "difference" polynomial $f(x) = x^2(M(x) - M'(x)) + (i - i')x$ is non-constant and has degree at most $L + 1$. The analysis is deferred to the full version.

3.3 Adding Error-Tolerance

We can now consider settings when the input w' held by the receiver is close, but not identical to, the value w used by the sender. An obvious first attempt, given the scheme just discussed, is to include a secure sketch $s = \mathsf{SS}(w)$ along with (i, σ), and to authenticate s using the message authentication technique discussed previously; s would allow recovery of w from w', and then verification could proceed as before. Unfortunately, this does not quite work: if the adversary modifies s, then a different value $w^* \neq w$ may be recovered; however, the results of the previous section apply only when the receiver uses the same w as used by the sender (and so in particular we can no longer claim that the adversary could not have modified s without being detected). In effect, we have a circularity: the receiver uses w to verify that s was not modified, but the receiver computes w (from w') using a possibly modified s.

We show how to break this circularity using a modification of the message authentication technique used earlier. One key idea is to use the part of w that is "independent of s" (in a way made clear below) to authenticate s.

The second key idea is to exploit algebraic structure in the metric space, and to change the message authentication code so that it remains secure *even when the adversary can influence the key* (sometimes called security against related-key attacks). Specifically, we will assume that the errors are small in the Hamming metric, and that we are given a deterministic, linear secure sketch (for example, a syndrome-based construction). In Section 3.3 we will extend the approach to related metrics such as *set difference* and *edit distance*.

Construction for Hamming Errors. Suppose the input w is an n bit string. Our starting point is a *deterministic, linear* secure sketch $s = \mathsf{SS}(w)$ that is k bits long; let $n' = n - k$. We assume that SS is a surjective, linear function (this is the case for the code-offset construction for the Hamming metric), and so there exists an $k \times n$ matrix S of rank k such that $\mathsf{SS}(w) = Sw$ (see footnote 2. Let $S^\perp$ be an $n' \times n$ matrix such that the $n \times n$ matrix $\left(\frac{S}{S^\perp}\right)$ has full rank. We let $\mathsf{SS}^\perp(w) \overset{\text{def}}{=} S^\perp w$. One can view $\mathsf{SS}^\perp(w)$ as the information remaining in w once $\mathsf{SS}(w)$ has been learned by the adversary.

$\mathsf{Gen}(w)$:

1. Set $s = \mathsf{SS}(w)$, $c = \mathsf{SS}^\perp(w)$
   - Parse c as $a\|b$ with $|a| = n' - v$ and $|b| = v$
   - Let $L = 2\lceil k/(2(n' - v)) \rceil$. Pad s with 0s to length $L(n' - v)$. Parse s as $s_{L-1}\|s_{L-2}\| \ldots \|s_0$, for $s_i \in GF(2^{n'-v})$.
2. Select $i \leftarrow GF(2^{n'-v})$
   - Define $f_{s,i}(x) = x^{L+3} + x^2(s_{L-1}x^{L-1} + s_{L-2}x^{L-2} + \cdots + s_0) + ix$
3. Set $\sigma = [f_{s,i}(a)]_1^v + b$ and output $R = [ia]_{v+1}^{n'-v}$ and $P = (s, i, \sigma)$.

To reproduce R given w' and $P' = (s', i', \sigma')$, first compute $w^* = \mathsf{SRec}(w', s')$; make sure $\mathsf{SS}(w^*) = s'$ and $\mathsf{dis}(w^*, w') \leq t$ (if not, output $\perp$). Let $c' = \mathsf{SS}^\perp(w^*)$; parse c' as $a'\|b'$. Compute σ^* as above using s', a', b', i', and check that this matches the value σ' received. If so, output $R = [i'a']_{v+1}^{n'-v}$, else output $\perp$.

The polynomial $f_{s,i}$ defined above differs from the message authentication technique in the previous section only in the leading term x^{L+3} (and the forcing of L to be even). It has the property that for any pair $(s', i') \neq (s, i)$, and for any fixed offset Δ_a, the polynomial $f_{s,i}(x) - f_{s',i'}(x + \Delta_a)$ is a non-constant polynomial of degree at most $L + 2$ (this is easy to see for $\Delta_a = 0$; if $\Delta_a \neq 0$, then the leading term is $((L + 3) \bmod 2)\Delta_a x^{L+2})$. In our analysis, we use the linearity of the scheme to understand the offset $\Delta_a = a' - a$, and conclude that the adversary succeeds only if she can guess the last v bits of $f_{s,i}(x) - f_{s',i'}(x + \Delta_a)$, which happens with low probability. Note that this definition of $f_{s,i}$ amounts to a message authentication code (MAC) provably secure against a class of related key attacks where the adversary Eve can force the receiver to use a key shifted by an offset of Eve's choice. We obtain:

Theorem 2. *Let $|w| = n$. Assume SS is an $(m, m - k, t)$-secure sketch for the Hamming metric. Then for an appropriate setting of v, the above construction* *is an $(m, \ell, t, \varepsilon, \delta)$-weakly robust fuzzy extractor for any $m, \varepsilon, \delta, \ell \geq 0$ satisfying $\ell \leq 2m - n - k - 2t \log(\frac{en}{t}) - 2 \log(\frac{n}{\varepsilon^2 \delta}) - O(1)$. It is an $(m, \ell, t, \varepsilon, \delta)$-strongly robust extractor when $\ell \leq \frac{1}{3}\left(2m - n - k - 2t \log(\frac{en}{t}) - 2 \log(\frac{n}{\varepsilon^2 \delta})\right) - O(1)$.*

The proof of this theorem is deferred to the full version. We briefly discuss the parameters in the statement. In the weakly robust case, the bound on ℓ differs in two large terms from the errorless bound $2m - n$ (assuming, say, that $\varepsilon, \delta = 2^{-o(n)}$). First, we lose the length of the sketch k. This is not surprising, since we need to publish the sketch in order to correct errors.[3] The second term $2t(\log \frac{en}{t})$ is also easy to explain, although it appears to be a technicality arising from our analysis. Our analysis essentially starts by giving the attacker the error pattern $\Delta = w' \oplus w$ "for free", which in the worst case can reduce the min-entropy w by the logarithm of volume of the Hamming ball of radius t. This logarithm is at most $t \log \frac{en}{t}$. Our analysis can, in fact, yield a more general result: if $\hat{m} = \widetilde{\mathbf{H}}_\infty(W \mid \Delta)$, then $2(m - t \log \frac{en}{t})$ in the above bounds on ℓ simply gets replaced with $2\hat{m}$. For instance, when knowing the error pattern $w' \oplus w$ does not reduce the entropy of w (say, the errors are independent of w, as in the work of Boyen [Boy04]), then the term $2t \log \frac{en}{t}$ disappears from the bounds.

The analysis gives away Δ since we can then use the linearity of the sketch to conclude that the adversary knows the difference between the original input w and the value w^* that $\mathsf{Rep}(w', \tilde{P})$ reconstructs. This means she knows $\Delta_a = a' - a$, and we can use the properties of $f_{s,i}$ to bound the forgery probability.

Extensions to Other Metrics. The analysis of the previous section relies heavily on the linearity of the secure sketch used in the protocol and on the structure of the Hamming space. We briefly indicate how it can be extended to two seemingly different metric spaces.

In the *set difference* metric, inputs in W are sets of at most r elements in a large universe $[N] = \{1, ..., N\}$, and the distance between two sets is the size of their symmetric difference. This is geometrically identical to the Hamming metric, since one can represent sets as characteristic vectors in $\{0, 1\}^N$. However, the efficiency requirement is much stricter: for set difference, we require that operations take time polynomial in the description length of the inputs, which is only $r \log N$, not N.

In order to extend the analysis of the previous section to handle this different representation of the input, we need a pair of functions $\mathsf{SS}(), \mathsf{SS}^\perp()$ that take sets and output bit strings of length k and $(r \log N) - k$, respectively. A set w of size up to r should be unique given the pair $(\mathsf{SS}(w), \mathsf{SS}^\perp(w))$, and the functions should possess the following linearity property: the addition or removal of a particular element to/from the set should correspond to adding a particular bit vector to the output. The $\mathsf{SS}()$ function of the BCH secure sketch of Dodis et al. [DORS06]

---

[3] In fact, a more naive construction would lose $2k$, since the sketch reduces the min-entropy m by k, and blindly applying the errorless bound $2m - n$ would double this loss. The use of $\mathsf{SS}^\perp$ is precisely what allows us not to lose the value k twice.

(called "PinSketch") is, in fact, linear; it outputs t values of $\log N$ bits each in order to correct up to t errors, thus producing sketches of length $k = t \log N$. For $\mathsf{SS}^{\perp}()$, we can use the last $r - t$ values computed by PinSketch with error parameter r. Since N is large, $t \log N$ is a good upper bound on the logarithm of the volume of the ball of radius t. We obtain the following statement, proved in the full version:

Corollary 1. *For any $r, m, t, \varepsilon, \delta$, there exists a weakly robust fuzzy extractor for set difference over sets of size up to r in $[N]$ with extracted key length $\ell = 2m - (r + 3t) \log N - O(\log(\frac{r \log N}{\varepsilon \delta}))$, and a strongly robust extractor with $\ell = \frac{1}{3}(2m - (r + 3t) \log N) - O(\log(\frac{r \log N}{\varepsilon \delta}))$.*

In the *edit metric*, inputs are strings over a small alphabet and distance is measured by the number of insertions and deletions required to move between strings. Dodis *et al.* defined a weak notion of a metric embedding[4] sufficient for key agreement and showed that the edit metric can be embedded into set difference with relatively little loss of entropy [DORS06, Lem. 7.3]. In our protocols, we can use embeddings along the lines of [DORS06] provided they are *deterministic*. The analysis then works as before, except it is applied to the embedded string $\psi(w)$ (the same idea may not work for randomized embeddings since $\psi(w)$ may then depend on $\tilde{P}$). The embedding of [DORS06] is indeed deterministic, and we obtain the following (for exact constants, see [DORS06, Thm 7.4]):

Corollary 2. *For any $m > n/2$, there exists a robust fuzzy extractor tolerating $t = \Omega(n \log^2 F / \log^2 n)$ edit errors in $[F]^n$ which extracts a key of length $\ell = \Omega(n \log F)$ with parameters $\varepsilon, \delta = 2^{-\Omega(n \log F)}$.*

4   Keyed Robust Fuzzy Extractors and Their Applications

In this section we show that the addition of a very short secret key SK allows us to achieve considerably better parameters when constructing *keyed* robust fuzzy extractors. The parameters are optimal up to constant factors.

To motivate our construction, let us recall the naive transformation from regular fuzzy extractors to keyed robust fuzzy extractors discussed in Section 2. Suppose we start from the generic construction of a fuzzy extractor: $P = (s, i)$, where $s \leftarrow \mathsf{SS}(w)$, and $R = \mathsf{Ext}(w; i)$ where Ext is a strong extractor. In an attempt to make this construction robust, we set $\mathsf{SK} = \mu$ and $\sigma = \mathsf{MAC}_\mu(s, i)$, and redefine P to also include the tag σ. The problem is that the value σ allows the attacker to distinguish the real key μ from a random key $U_{|\mu|}$, since the attacker knows the authenticated message (s, i). Thus this scheme fails to meet the extraction requirement (Def. 7).

We can change the scheme to avoid this. First, note that Rep must recover the input $w = \mathsf{Rec}(w', s)$ before computing R. Thus, we can add w to the authenticated message without sacrificing the correctness of the scheme: that is, set

---

[4] Roughly, a embedding of a metric space $\mathcal{M}_1$ into another space $\mathcal{M}_2$ is an efficiently computable function $\psi : \mathcal{M}_1 \to \mathcal{M}_2$ which preserves distances approximately.

$\sigma = \mathsf{MAC}_\mu(w, s, i)$. This does not strengthen the robustness property (Def. 6), which was already satisfied by the naive scheme. However, it does help satisfy extraction (Def. 7). In the naive scheme the attacker $\mathcal{A}$ *knows the message* (s, i) *we are authenticating*. In contrast, W has high entropy from $\mathcal{A}$'s point of view, even given $\mathsf{SS}(W)$ and R (for appropriate parameters). Thus, to make the pair (P, R) independent of $\mathsf{SK} = \mu$, it suffices to construct *information-theoretic* MACs whose key μ looks independent from the tag, as long as the authenticated message has high min-entropy. In other words, if we can ensure that the MAC is simultaneously a strong randomness extractor, we can solve our problem.

4.1   Extractor-MACs

Definition 8. *A family of functions* $\{\mathsf{MAC}_\mu : \{0,1\}^{\tilde{n}} \to \{0,1\}^v\}$ *is a strong* $(\tilde{m}, \varepsilon, \delta)$*-extractor-MAC if it is simultaneously a strongly δ-secure one-time MAC (Def. 3) and a $(\tilde{m}, \varepsilon)$-strong extractor (Def. 2, where the key μ is the seed).* ◇

We can construct extractor-MACS with (essentially optimal) key length $O(\log \tilde{n} + \log\left(\frac{1}{\varepsilon}\right) + \log\left(\frac{1}{\delta}\right))$. The idea is to modify the "AU" extractor construction of Srinivasan and Zuckerman [SZ99] so that it is also MAC.

Before giving an optimal construction, note that pairwise independent hash functions are simultaneously (strong) one-time MACs and strong extractors. For example, consider the function family $f_{a,b}(x) = [ax]_1^v + b$, where $a \in \mathbb{F}_{2^v}$, $b \in \mathbb{F}_{2^v}$, and $[ax]_1^v$ denotes the truncation of ax to the first v bits. This is pairwise independent [CW79], and gives an extractor-MAC with key length $u + v = \tilde{n} + \log\left(\frac{1}{\delta}\right)$. The key length needed to authenticate a u-bit message is $\kappa = u + v = u + \log\left(\frac{1}{\delta}\right)$, which is rather large. However, we can reduce the key length by first reducing the size of the input using almost-universal hashing.

Specifically, let $\{p_\beta\}$ be a $(\delta\varepsilon^2/2)$-almost-universal hash family mapping $\tilde{n}$ bits to u bits, and compose it with a pair-wise independent family $\{f_\alpha\}$ from u to v bits, where $v = \log\left(\frac{1}{\delta}\right) + 1$. That is, set $\mathsf{MAC}_{\alpha,\beta}(w) = p_\beta(f_\alpha(w))$. By Lemma 2, $\{MAC_{\alpha,\beta}\}$ is a strong δ-secure MAC, since $\delta\varepsilon^2/2 + 2^{-v} \leq \delta$. Furthermore, composing δ_1- and δ_2-almost universal families yields a $(\delta_1 + \delta_2)$-almost-universal family. Thus $\{\mathsf{MAC}_{\alpha,\beta}\}$ is $(1 + \varepsilon^2)2^{-v}$-almost-universal. By the left-over hash lemma (Lemma 1), it is a (m, ε)-extractor with $m = v + 2\log\left(\frac{1}{\varepsilon}\right)$.

It remains to set u so that we can construct a convenient almost universal hash family $\{p_\beta\}$. We can use a standard polynomial-based construction, also used in previous sections. The key β is a point in $\mathbb{F}_{2^u}$, and the message x is split into $c = \tilde{n}/u$ pieces $(x_1, \ldots, x_c)$, each of which is viewed as an element of $\mathbb{F}_{2^u}$. Now, set $p_\beta(x_1 \ldots x_c) = x_c\beta^{c-1} + \ldots + x_2\beta + x_1$. This family is $c/2^u$-almost universal with key length u. We can set $u = v + \log(\frac{\tilde{n}}{2\varepsilon^2}) = 2\log\left(\frac{1}{\varepsilon}\right) + \log\left(\frac{1}{\delta}\right) + \log \tilde{n} - 1$ to make $c/2^u < \delta\varepsilon^2/2$. This gives key length $u + 2v$, and we obtain:

Theorem 3. *For any δ, ε and $\tilde{m} \geq \log\left(\frac{1}{\delta}\right) + 2\log\left(\frac{1}{\varepsilon}\right) + 1$, there exists a $(\tilde{m}, \varepsilon, \delta)$-extractor-MAC with key length $\kappa = 2\log\tilde{n} + 3\log\left(\frac{1}{\delta}\right) + 4\log\left(\frac{1}{\varepsilon}\right)$ and tag length $v = \log\left(\frac{1}{\delta}\right) + 1$.*

4.2   Building Keyed Robust Fuzzy Extractors

We now apply the extractor-MACs to build keyed robust fuzzy extractors for $\mathcal{M}$ (which we assumed for simplicity is $\{0,1\}^n$). We start with a generic construction and set the parameters below.

Assume $(\mathsf{SS}, \mathsf{SRec})$ is a $(m, \tilde{m}, t)$-secure sketch with sketch length k, Ext is a strong $(\tilde{m} - \log\left(\frac{1}{\varepsilon}\right), \varepsilon)$-extractor having a seed i of length d and an output of length ℓ, and MAC is a $(\tilde{m} - \ell - \log\left(\frac{1}{\varepsilon}\right), \varepsilon, \delta)$-extractor-MAC from $\tilde{n} = n + k + d$ bits to v bits having a key μ of length κ. We now define a keyed robust fuzzy extractor with secret key $\mathsf{SK} = \mu$:

- $\mathsf{Gen}_\mu(w)$: compute sketch $s \leftarrow \mathsf{SS}(w)$, sample i at random, set key $R = \mathsf{Ext}(w; i)$, tag $\sigma = \mathsf{MAC}_\mu(w, s, i)$, $P = (s, i, \sigma)$ and output (R, P).
- $\mathsf{Rep}_\mu(w', (s', i', \sigma'))$: Let $\bar{w} = \mathsf{SRec}(w', s')$. If $\mathsf{MAC}_\mu(\bar{w}, s', i') = \sigma'$, then $R = \mathsf{Ext}(\bar{w}; i)$; else $R = \bot$.

Theorem 4. *The above construction is a $(m, \ell, t, 4\varepsilon, \delta)$-robust keyed fuzzy extractor, which uses a secret key SK of length κ and outputs public information P of length $k + d + v$.*

The proof of this Theorem is given in the full version. We remark that if Lemma 1 (resp. Theorem 3) is used to instantiate the extractor Ext (resp. extractor-MAC MAC) above, then a $(\tilde{m}, \varepsilon)$-extractor (resp. $(\tilde{m} - \ell, \varepsilon, \delta)$-extractor-MAC) is sufficient for Theorem 4 to hold. We also note that a variant of this construction (whose security is proven analogously) would move the extractor seed i into the secret key SK. Namely, set $\mathsf{SK} = (\mu, i)$, $\sigma = \mathsf{MAC}_\mu(w, S)$ and $P = (S, \sigma)$. The main advantage of this variant is that the scheme becomes non-interactive in the case of no errors (i.e., $t = 0$). However, in order to keep the length of SK low one must use considerably more complicated strong extractors than those given in Lemma 1.

THE PRICE OF AUTHENTICATION. We compare the parameters of Theorem 4 to the original (non-robust, non-keyed) constructions of [DORS06]. First, note that the choice of a sketch and strong extractor can be done in the same manner as for non-robust fuzzy extractors. For concreteness, assume we use almost-universal hash functions as extractors, and let us now apply Theorem 3 to choose the extractor-MAC. Then the secret key SK is the just the MAC key μ, which has length $\kappa = 2 \log \tilde{n} + 3 \log\left(\frac{1}{\delta}\right) + 4 \log\left(\frac{1}{\varepsilon}\right)$. This is $2 \log n + 3 \log\left(\frac{1}{\delta}\right) + 4 \log\left(\frac{1}{\varepsilon}\right) + O(1)$ when $d, k = O(n)$. Second, recall from Theorem 3 that the extractor-MAC is a good extractor as long as its min-entropy threshold $\tilde{m} - \ell$ is at least $v + 2 \log\left(\frac{1}{\varepsilon}\right) = 1 + \log\left(\frac{1}{\delta}\right) + 2 \log\left(\frac{1}{\varepsilon}\right)$. We get security as long as $\ell \leq \tilde{m} - 2 \log\left(\frac{1}{\varepsilon}\right) - (\log\left(\frac{1}{\delta}\right) + 1)$. Compared with non-robust extractors, which required $\ell \leq \tilde{m} - 2 \log\left(\frac{1}{\varepsilon}\right)$ [DORS06], the keyed, robust construction loses at most $\log\left(\frac{1}{\delta}\right) + 1$ bits in the possible length of the extracted key. Finally, the length of the public information P increases by the (short) tag length $v = \log\left(\frac{1}{\delta}\right) + 1$. Overall, the parameters remain similar to the corresponding "non-robust" case.

4.3   Application to the Bounded Storage Model with Errors

We briefly recall the key elements of the bounded storage model (BSM) with errors [Din05, DS05], concentrating only on the *stateless variant* of [DS05]. Our discussion will be specific to *Hamming errors*.

In the bounded storage model, the parties (say, Alice and Bob) have a long-term secret key sk, and at each period j have access to two noisy versions X_j and X'_j of a very long random string Z_j (of length N). Both the honest parties and the attacker $\mathcal{A}$ are limited in storage to fewer than N bits. More specifically, $\mathcal{A}$ can look at Z_j and store any γN bits of information about Z_j, for $\gamma < 1$ (so that on average Z_j has entropy about $(1 - \gamma)N$ from her point of view). The honest parties are also limited in their storage, but they can use their shared key to gain an advantage. For example, in "sample-and-extract" protocols [Vad04], one part of the shared key consists of a key sam for an *oblivious sampler* [BR94, Vad04]. Roughly, sam specifies n secret physical bits of X_j/X'_j which Alice and Bob will read, obtaining n-bit substrings W_j and W'_j. The properties of the sampler ensure that (a) with high probability W_j and W'_j are still close (i.e., within Hamming distance t from each other); and (b) with high probability, $\mathcal{A}$ still has some uncertainty (min-entropy $m \approx (1 - \gamma)n$) about W_j and W'_j.

This setup is quite similar to the setting of keyed robust fuzzy extractors, and provides a natural application for them. Alice and Bob can use part of the shared secret as a key SK for the robust fuzzy extractor; she can then run $\mathsf{Gen}_{\mathsf{SK}}(W_j)$ to obtain (P_j, R_j) and send P_j to Bob. Bob can run $\mathsf{Rep}_{\mathsf{SK}}$ to (hopefully) get either the key R_j or $\perp$, an indication that Alice's message was modified in transmission.

However, there is a subtle difference between the setting of the keyed robust extractors induced by the BSM and the setting we considered so far, which already causes difficulties even in the case of authenticated channels [DS05]. In our model, discussed in Section 2, the (t, m)-correlated pairs (W_j, W'_j) were arbitrary but fixed *a priori*. In contrast, in the BSM $\mathcal{A}$ can adaptively choose her storage function at each period, based on what was seen so far, and therefore affect the specific (still high-entropy) conditional distribution of each sampled W_j. In particular, if the public values P_j seen by $\mathcal{A}$ could reveal something about the long-term key $sk = (sam, \mathsf{SK})$, then Eve can affect the conditional distribution of the subsequent W_j in a manner dependent on sk, making it unsound to reuse sk in the future.

On a positive side, our definition of keyed robust extractors (Def. 7) was strong enough to ensure that the public value P is statistically independent from the key SK (meaning it is safe to reuse SK). On a negative side, it still allowed the value P to depend on the *distribution of W* (which, in turn, depends on the sampling key sam), making it insufficient for reusing the sampling key sam. And this is precisely why for this application of keyed robust extractors we will need the enhanced notion of *strongly secure* keyed robust extractors mentioned in Def. 7. Namely, the public value P not only hides the secret key SK, but even *the distribution* of W: $(\mathsf{SK}, R, P) \approx_\varepsilon (U_{|\mathsf{SK}|}, U_\ell, U_{|P|})$. A similar argument to the authenticated case of [DS05] shows that strongly secure keyed robust extractors are sufficient to solve the unauthenticated setting. Thus, we turn to constructions of such robust extractors.

STRONGLY SECURE KEYED ROBUST EXTRACTORS. Examining the keyed construction in Theorem 4, we see that the only place where the value $P = (S, I, \sigma)$ depends on the distribution of W is when computing the sketch $S \leftarrow \mathsf{SS}(W)$. Indeed, the seed I is chosen at random, and the value $\sigma = \mathsf{MAC}_\mu(W, S, I)$ looks random by the properties of the extractor-MAC. Thus, to solve our problem we only need to build an $(m, \tilde{m}, t)$-secure sketch SS such that $\mathsf{SS}(W)$ is statistically close to uniform whenever W has enough min-entropy: $\mathsf{SS}(W) \approx_\varepsilon U_{|\mathsf{SS}(W)|}$ (notice, such sketches can no longer be deterministic). Luckily, such (probabilistic) sketches, called $(m, \tilde{m}, t, \varepsilon)$-extractor-sketches (or "entropically-secure" sketches) were studied by Dodis and Smith [DS05], since they were already needed to solve the noisy BSM problem even in the authenticated channel case. In particular, [DS05] built extractor-sketches for the binary Hamming metric whose parameters were only a constant factor worse than those of regular sketches.

Theorem 5 ([DS05]). *For any min-entropy $m = \Omega(n)$, there exists efficient $(m, \tilde{m}, t, \varepsilon)$-extractor-sketches for the Hamming metric over $\{0,1\}^n$, where $\tilde{m}$, t and $\log\left(\frac{1}{\varepsilon}\right)$ are all $\Omega(n)$, and the length of the sketch is $O(n)$.*

Returning to the construction of strongly secure, keyed robust extractors, we observe that Theorem 4 indeed yields such extractors (with error 5ε instead of 4ε) if one uses $(m, \tilde{m}, t, \varepsilon)$-extractor-sketches in place of regular $(m, \tilde{m}, t)$-sketches. Combining this observation with Theorem 5, we obtain:

Theorem 6. *For any min-entropy $m = \Omega(n)$, there exists efficient $(m, \ell, t, \varepsilon, \delta)$-strongly secure, robust, keyed fuzzy extractor for the Hamming metric over $\{0,1\}^n$, which uses a secret key of length $O(\log n + \log\left(\frac{1}{\varepsilon}\right) + \log\left(\frac{1}{\delta}\right))$, tolerates $t = \Omega(n)$ errors, extracts $\ell = \Omega(n)$ bits, and has public information P of length $O(n)$.*

APPLICATION TO THE BSM. As we stated, after Alice and Bob use the shared sampling key to obtain close n-bit strings W and W', respectively, they will use a strongly secure, keyed, robust fuzzy extractor $(\mathsf{Gen}_{\mathsf{SK}}, \mathsf{Rep}_{\mathsf{SK}})$ to agree on a session key R over an unauthenticated channel. To get a specific construction, we can use Theorem 6 above. In doing so, we see that the only difference between the resulting scheme and the solution of Dodis and Smith [DS05] (for the authenticated channel) is that Alice and Bob additionally share a (short) extractor-MAC key SK, and also append a (short) extractor-MAC of (W, S, I) to the public information (S, I) that Alice sends to Bob. Therefore, our construction retains the nearly optimal parameters of [DS05], while also adding authentication.

More specifically, assume N, ε, δ are given. Since the number of read bits n can be chosen by Alice and Bob, it is convenient to specify the required number of extracted bits ℓ, and choose n afterwards. Then we obtain a stateless protocol in the BSM model with Hamming errors satisfying: (1) key reuse (stateless) and everlasting security; (2) having shared secret key sk of size $O(\log N + \log\left(\frac{1}{\varepsilon}\right) + \log\left(\frac{1}{\delta}\right))$; (3) having forgery probability at most δ against active attacker; (4)

having Alice and Bob read $n = O(\ell)$ random bits W from the source and extract ℓ bits R which are ε-close to uniform; (5) having Alice and Bob tolerate linear fraction of errors (i.e., $t = \Omega(n)$); and (6) having Alice send a single $O(\ell)$-bit message to Bob. All these parameters are optimal up to a constant factor.

Acknowledgments. We would like to thank Hoeteck Wee for his collaboration at the early stages of this work.

References

[ADR02]   Y. Aumann, Y. Ding, and M. Rabin. Everlasting security in the bounded storage model. *IEEE Trans. on Information Theory*, 48(6):1668–1680, 2002.

[BBCM95]   C. H. Bennett, G. Brassard, C. Crépeau, and U. M. Maurer. Generalized privacy amplification. *IEEE Trans. on Information Theory*, 41(6), 1995.

[BBR88]   C. Bennett, G. Brassard, and J. Robert. Privacy amplification by public discussion. *SIAM Journal on Computing*, 17(2):210–229, 1988.

[BDK$^+$05]   X. Boyen, Y. Dodis, J. Katz, R. Ostrovsky, and A. Smith. Secure remote authentication using biometric data. In *EUROCRYPT 2005*, Springer.

[BMP00]   V. Boyko, P. MacKenzie, and S. Patel. Provably-secure password-authenticated key exchange using Diffie-Hellman. In *EUROCRYPT 2000*.

[Boy04]   Xavier Boyen. Reusable cryptographic fuzzy extractors. In *11th ACM Conference on Computer and Communication Security*. ACM, 2004.

[BPR00]   Mihir Bellare, David Pointcheval, and Phillip Rogaway. Authenticated key exchange secure against dictionary attacks. In *EUROCRYPT 2000*.

[BR94]   M. Bellare and J. Rompel. Randomness-efficient oblivious sampling. In *35th Annual Symposium on Foundations of Computer Science*. IEEE, 1994.

[Cré97]   Claude Crépeau. Efficient cryptographic protocols based on noisy channels. In *Advances in Cryptology—EUROCRYPT 97*, volume 1233 of *LNCS*.

[CW79]   J.L. Carter and M.N. Wegman. Universal classes of hash functions. *Journal of Computer and System Sciences*, 18:143–154, 1979.

[Din05]   Yan Zong Ding. Error correction in the bounded storage model. In *2nd Theory of Cryptography Conference — TCC 2005*, volume 3378 of *LNCS*.

[DORS06]   Y. Dodis, R. Ostrovsky, L. Reyzin, and A. Smith. Fuzzy extractors: How to generate strong keys from biometrics and other noisy data. Technical Report 2003/235, Cryptology ePrint archive, http://eprint.iacr.org, 2006. Previous version appears in *EUROCRYPT 2004*.

[DS02]   Y. Dodis and J. Spencer. On the (non-)universality of the one-time pad. In *43rd Annual Symposium on Foundations of Computer Science*. IEEE, 2002.

[DS05]   Y. Dodis and A. Smith. Correcting errors without leaking partial information. In 37th Annual ACM Symposium on Theory of Computing, 2005.

[FJ01]   N. Frykholm and A. Juels. Error-tolerant password recovery. In *Eighth ACM Conference on Computer and Communication Security*. ACM, 2001.

[GL01]   O. Goldreich and Y. Lindell. Session-key generation using human passwords only. In *CRYPTO 2001*, volume 2139 of *LNCS*, pages 408–432, Springer.

[GL03]   R. Gennaro and Y. Lindell. A framework for password-based authenticated key exchange. In *EUROCRYPT 2003*, volume 2656 of *LNCS*.

[HILL99]   J. Håstad, R. Impagliazzo, L.A. Levin, and M. Luby. Construction of pseudorandom generator from any one-way function. *SIAM Journal on Computing*, 28(4):1364–1396, 1999.

[JW99]   A. Juels and M. Wattenberg. A fuzzy commitment scheme. In *Sixth ACM Conference on Computer and Communication Security*, pages 28–36, 1999.

[KOY01]   J. Katz, R. Ostrovsky, and M. Yung. Efficient password-authenticated key exchange using human-memorable passwords. In *EUROCRYPT 2001*.

[LT03]   J.-P. M. G. Linnartz and P. Tuyls. New shielding functions to enhance privacy and prevent misuse of biometric templates. In *AVBPA*, 2003.

[Mau92]   Ueli Maurer. Conditionally-perfect secrecy and a provably-secure randomized cipher. *Journal of Cryptology*, 5(1):53–66, 1992.

[Mau93]   Ueli Maurer. Secret key agreement by public discussion from common information. *IEEE Transactions on Information Theory*, 39(3):733–742, 1993.

[Mau97]   Ueli Maurer. Information-theoretically secure secret-key agreement by NOT authenticated public discussion. In *EUROCRYPT 97*, pp. 209–225.

[MW97]   U. Maurer and S. Wolf. Privacy amplification secure against active adversaries. In *Advances in Cryptology—CRYPTO '97*, pages 307–321.

[MW03]   U. Maurer and S. Wolf. Secret-key agreement over unauthenticated public channels — Part III: Privacy amplification. *IEEE Transactions on Information Theory*, 49(4):839–851, 2003.

[RW03]   R. Renner and S. Wolf. Unconditional authenticity and privacy from an arbitrarily weak secret. In *Advances in Cryptology—CRYPTO 2003*.

[RW05]   R. Renner and S. Wolf. Simple and tight bounds for information reconciliation and privacy amplification. In *ASIACRYPT 2005*, Springer.

[SZ99]   A. Srinivasan and D. Zuckerman. Computing with very weak random sources. *SIAM Journal on Computing*, 28(4):1433–1459, 1999.

[Vad04]   S. Vadhan. Constructing locally computable extractors and cryptosystems in the bounded-storage model. *Journal of Cryptology*, 17(1), 2004.

[WC81]   M.N. Wegman and J.L. Carter. New hash functions and their use in authentication and set equality. *J. Computer and System Sciences*, 22, 1981.

[Wol98]   S. Wolf. Strong security against active attacks in information-theoretic secret-key agreement. In *ASIACRYPT '98*, volume 1514 of *LNCS*.

[Wyn75]   A.D. Wyner. The wire-tap channel. *Bell System Technical Journal*, 54(8):1355–1387, 1975.

On Forward-Secure Storage[*]

Extended Abstract

Stefan Dziembowski

Institute of Informatics, Warsaw University
and Institute for Informatics and Telematics, CNR Pisa

Abstract. We study a problem of secure data storage in a recently
introduced *Limited Communication Model*. We propose a new cryptographic primitive that we call a *Forward-Secure Storage (FSS)*. This
primitive is a special kind of an encryption scheme, which produces huge
(5 GB, say) ciphertexts, even from small plaintexts, and has the following non-standard security property. Suppose an adversary gets access to
a ciphertext $C = E(K, M)$ and he is allowed to compute any function h
of C, with the restriction that $|h(C)| \ll |C|$ (say: $|h(C)| = 1$ GB). We
require that $h(C)$ should give the adversary no information about M,
even if he later learns K.

A practical application of this concept is as follows. Suppose a ciphertext C is stored on a machine on which an adversary can install a virus.
In many cases it is completely infeasible for the virus to retrieve 1 GB
of data from the infected machine. So if the adversary (at some point
later) learns K, then M remains secret.

We provide a formal definition of the FSS, propose some FSS schemes,
and show that FSS can be composed sequentially in a secure way. We
also show connections of the FSS to the theory of compressibility of
NP-instances (recently developed by Harnik and Naor).

1 Introduction

One of the main problems in the practical data security are the attacks of malware like Internet worms, Trojan horses or viruses. For an average user it is
quite common that his PC gets from time to time infected by some malicious
program. Once installed, such a program can take full control over the infected
machine, and it may steal some confidential message M that is stored on the
machine. A natural solution for this problem is to use encryption and to store
only a ciphertext $C = E(K, M)$, where K is some secret key. Clearly, for the
security of this method we need to assume that the key K is stored outside of
the machine: in user's memory or on some other safe device. If the key K leaks

---

[*] Part of this work was carried out during the tenure of an ERCIM fellowship. This
work was partly supported by the IST-3-016004-IP-09 *Sensoria* project and the
KBN grant 4 T11C 042 25. Part of this work was done at the Institute of Mathematics, Polish Academy of Sciences. Current address of the author: Dipartimento di
Informatica - Università La Sapienza, Via Salaria 113, 00198 Roma, Italy.

to the adversary, then it may seem that all the security is inevitably lost, since the adversary that knows C and K can decrypt M.

In this paper we show how to limit the bad consequences of the key leakage. We will use the methods of a newly-introduced [12] *Limited Communication Model (LCM)*[1]. The main idea is as follows. Of course, if the intruder knows K at the moment *when* he got access to C then he can easily decrypt M. But what if he learned K *after* he got access to C? Clearly, if C is small then the intruder can simply retrieve C and wait until he learns K. So what if C is large (e.g. 5 GB)? In this case retrieving the entire C from the infected machine may be much harder. So, suppose that the adversary got access to the machine on which C is stored (and for the moment he has no information about K). Assume that he can perform some arbitrary computation on C and that he may retrieve the result $h(C)$. Clearly, if $|h(C)| = |C|$ then he may simply set $h(C) = C$, so let us assume that $|h(C)| \ll |C|$ (e.g. $|h(C)| = 1$ GB and $|C| = 5$ GB). Suppose that later the adversary learns K. Clearly, if E is some standard encryption scheme, then the knowledge of $h(C)$ and K allows the adversary to obtain some partial information on M.

We propose a special type of encryption schemes that guarantee that the adversary that knows $h(C)$ and K does not get any substantial information on M. We call this new primitive a *Forward-Secure Storage (FSS)*. We define its security, propose implementations, and show connections to the theory of the compressibility of NP-instances [16] that was recently developed by Harnik and Naor (see Sect. 2), the Bounded Storage Model and the Hybrid Bounded Storage Model (see Sect. 3).

Some proofs and details are omitted because of the page limit and appear in an extended version of this paper [13].

1.1 How Realistic Is Our Scenario?

We believe that FSS is an interesting notion just from the theoretical point of view. Still, it may also find some practical applications. Let us now consider the question how realistic is the assumption that the key K leaks *after* the adversary lost access to C. Here we list the cases when this can happen.

1. If the key K is stored on some safe device (a floppy-disc, say) then this device may be physically stolen. The key K may also leak to the adversary if the floppy-disc is inserted in some compromised device (this can happen if the user uses the same key to encrypt data on several machines).
2. If the key K is a human-memorized password, then the adversary can crack the password by trying all possible passwords in the dictionary. While this operation may be hard to perform in an unnoticeable way directly on the victim's machine, it is usually doable if the attacker downloads $C = E(K, M)$, and then performs the dictionary attack on his own machine. This scenario was already considered in [9] in a weaker model (see Sect. 2). Note that it is

---

[1] This name was proposed in [5].

unclear how to formally model the fact that the adversary *can* crack passwords on his own machine, but he *cannot* do it on the victim's machine. We discuss it is Sect. 8.

3. Even if the key K is cryptographically strong and it does not leak, the adversary may still hope that at some point in the future he can break it, when new cryptanalytic methods, or more computing power are available.

Practicality of the Assumptions. Note that the current storage prices are extremely low. Today's price of one blank DVD (which has almost 5 GB of storage) is around 50 cents, and new HD-DVD and Blu-Ray technologies (which allow to store up to 50 GB on one disc) are entering the market right now. At the same time, downloading a 1 GB from an average PC connected to the Internet can still take considerable time. Also, observe that in many cases the adversary will not perform the attack if he can be traced down, and it may be quite hard to download 1 GB in an untraceable way (unlike downloading 1 KB of data, which can be posted e.g. on a Usenet newsgroup). One can also consider limiting the possible amount of retrieved data by constructing devices with artificially slow memory access (this was considered in [20]).

1.2 Our Contribution

We define the notion of the Forward-Secure Storage (Sect. 4), distinguishing between three levels of security: information-theoretic, computational, and hybrid (which is a mix of the former two). We show constructions of FSS schemes (Sect. 6). We prove that FSS is secure if composed sequentially (Sect. 7) and show how FSS can be used to protect encrypted data with a human-memorized key (Sect. 8). We show the connections with the theory of compressibility of NP-instances (Sect. 5). Note that (except of Sect. 8) we do not use the Random Oracle Assumption [2].

2 Related Work

The Work of [9]. Consider the scenario in Point 2 in Sect. 1.1 (i.e. K is a human-memorized password). This was studied already in [9]. The difference between our model and theirs is that they do not allow the intruder to perform arbitrary computation on the victim's machine. The only thing that the adversary can do is to retrieve some *individual bits* $(C_{i_1}, \ldots, C_{i_s})$ of the ciphertext $C = (C_1, \ldots, C_t)$. This may seem unrealistic, since the malicious programs can easily perform computations on the victim's data. In this model they propose an efficient solution (a system called *VAST*) for the problem of secure storage. Their solutions assumes the Random Oracle Model. We will later refer to the model of [9] as a *Bounded-Retrieval Model*. This term was introduced in [7].

Intrusion Resilience of [12]. The model that we consider in this paper was proposed in [12][2]. It is shown in [12] that in this model one can implement

---

[2] Some of the ideas of [12] were also independently discovered in [5].

protocols for the entity authentication and the session-key generation. The security of the protocols in [12] is proven in the Random Oracle Model. This assumption was later removed in [5]. We will refer to the model of [12] (and of our paper) as the *Limited Communication Model.*

The Theory of [16]. One of our main inspirations is the theory of the *compressibility of NP-instances* recently developed by Harnik and Naor [16] (see also [11]). Suppose that we are given an NP-language $L \subseteq \{0,1\}^*$. We say that L is π-*compressible (to another NP-language L')* if there exists a polynomial-time π-*compressing algorithm* $\mathcal{C} : \{0,1\}^* \to \{0,1\}^*$ such that (1) $x \in L$ if and only if $\mathcal{C}(x) \in L'$, and (2) for every $x \in \{0,1\}^*$, we have $|\mathcal{C}(x)| \leq \pi(|x|)$. We say that L is *compressible (to L')* if π is polynomial in $\log|x|$ and in $w(x)$, where $w(x)$ is the maximal size of the witness for the instances of length $|x|$ (otherwise we say that it is *incompressible*).

A compressing algorithm is *witness-retrievable* if there exists a polynomial-time algorithm W such that if v is an NP-witness for $x \in L$, then $W(v)$ is an NP-witness for $\mathcal{C}(x)$.

Showing that there exists an incompressible NP-language (under the standard cryptographic assumptions) is currently an open problem. In an updated version of [16][3] the authors show, however, that if one-way functions exist then there does not exist a witness-retrievable compressing algorithm for SAT.

All-Or-Nothing Transform. A scenario similar to ours was consider in [27] (and in several subsequent papers), where the notion of the *All-Or-Nothing Transform* was introduced. Namely, [27] proposes an encryption method where it is hard to get any information on M, given K and most (but not all) of the bits of $C = E(K, M)$. Note that the fundamental difference between [27] and our work is that [27] assumes that the adversary knows some *individual bits* of C (like in the Bounded Retrieval Model, described above), while in our model the adversary can compute an arbitrary function h of C (with $|h(C)| \ll |C|$). See also Sect. 9.

3   Tools

Bounded-Storage Model. We will use the results from the Bounded-Storage Model (BSM), introduced by Maurer in [24]. This model was studied in the context of *information-theoretically secure* encryption [1,15,23,28], key-agreement [4,14], oblivious transfer [3,10] and time-stamping [25]. In this model one assumes that a random string $R \in \{0,1\}^{t_{\mathrm{BSM}}}$ (called a *randomizer*) is either temporarily available to the public or broadcast by one of the legitimate parties. The honest users, Alice and Bob share a short secret *initial key* $K \in \{0,1\}^{m_{\mathrm{BSM}}}$, selected uniformly at random, and they apply a known *key-expansion function* $f : \{0,1\}^{m_{\mathrm{BSM}}} \times \{0,1\}^{t_{\mathrm{BSM}}} \to \{0,1\}^{n_{\mathrm{BSM}}}$ to compute the *derived string* $X = f(K, R) \in \{0,1\}^{n_{\mathrm{BSM}}}$ (usually $n_{\mathrm{BSM}} \gg m_{\mathrm{BSM}}$). Later X can be used,

---

[3] Available at http://www.cs.technion.ac.il/~harnik.

e.g., as a key for the one-time pad encryption. The function f must be efficiently computable and based on only a very small portion of the bits of R, so that Alice and Bob need not read the entire string R.

We assume that the adversary $\mathcal{A}_{\text{BSM}}$ (that is a computationally-unbounded Turing Machine) can compute an arbitrary function h of R, with the sole restriction that the output $U = h(R)$ of this computation has a limited size: $U \in \{0,1\}^{s_{\text{BSM}}}$, where $s_{\text{BSM}} \ll t_{\text{BSM}}$. The adversary is allowed to store in his memory only U. After R disappears, the adversary should have essentially no information about X, even if he learns K. To define the security more formally, we consider the following game:

| BSM - distinguishing game |

Phase 1: R is generated randomly and passed to the adversary. The adversary can perform arbitrary computations and he can store some value $U = h(R) \in \{0,1\}^{s_{\text{BSM}}}$. He is not allowed to store any other information. Then, the randomizer disappears.

Phase 2: The adversary learns K. Let $b \in \{0,1\}$ be chosen uniformly at random. Define

$$\hat{X} := \begin{cases} f(K, R) & \text{if } b = 0 \\ \text{a random element of } \{0,1\}^{n_{\text{BSM}}} & \text{otherwise.} \end{cases}$$

and send $\hat{X}$ to the adversary. The adversary has to guess b.

We say that *the adversary $\mathcal{A}_{\text{BSM}}$ breaks the BSM scheme f with an advantage ϵ* if his chances of guessing b correctly are $0.5 + \epsilon$. To reason about the security in the asymptotic way, let us introduce a security parameter k which is an additional input of f and of the adversary. Let us assume that the parameters $m_{\text{BSM}}, n_{\text{BSM}}$ and t_{BSM} are functions of k. For a function $\sigma : \mathbb{N} \to \mathbb{N}$ we say that function f is *σ-secure in the bounded-storage model* if any adversary with memory at most $s = \sigma(k)$ breaks the scheme f with a negligible[4] advantage. Several key expansion functions [1,15,23,28] were proven secure in the past couple of years.[5]

Observe that in some sense the assumptions in the BSM are opposite to the assumptions in the Limited Communication Model, since in the BSM we assume that transmission of large amount of data is cheaper than storing it. Nevertheless (as observed already in [12]), it turns out that the theorems proven in BSM are useful in the LCM.

***Hybrid* Bounded-Storage Model.** The *Hybrid Bounded-Storage Model* [14] is defined as follows. Suppose that K is generated by a computationally-secure key-agreement protocol. Clearly, an infinitely-powerful adversary can break such a

---

[4] A function $\epsilon : \mathbb{N} \to \mathbb{R}$ is *negligible (in k)* if for every $c \geq 1$ there exists k_0 such that for every $k \geq k_0$ we have $|\epsilon(k)| \leq k^{-c}$.

[5] In these papers security of a BSM function is defined in a slightly different way. Namely, it is required that the distribution of X is statistically close to uniform from the point of view of the adversary (who knows K and U). It is easy to see that these definitions are in fact equivalent.

key-agreement. Therefore, assume that the computational power of the adversary is restricted (to polynomial time) until the randomizer disappears (at the end of Phase 1). Of course, when the adversary later gains the unlimited computing power, he can compute K (if he recorded the transcript of the key-agreement), but this should not be dangerous, since in the BSM the security holds even when the initial key K is given to the adversary (in Phase 2). In [14] it was shown that this reasoning is not correct. Namely, they show an example of a (very artificial, but computationally-secure under the standard assumptions) key agreement that cannot be safely used in this context. For more details see [14] or [17], where this model was recently formalized and generalized.

Private Information Retrieval. A PIR [6,21] scheme $(\mathcal{U}, \mathcal{D})$ is a protocol for two parties, a user $\mathcal{U}$ and a database $\mathcal{D}$, allowing the user to access database entries in a way that $\mathcal{D}$ cannot learn which information $\mathcal{U}$ requested. More precisely, the database content can be modeled as a string $x = (x_1, \ldots, x_l) \in \{0,1\}^l$, and $\mathcal{U}$ wants to access some bits $x_{i_1}, \ldots, x_{i_q}$ of x, such that $\mathcal{D}$ does not learn $I := i_1, \ldots, i_q$. It is not relevant whether $\mathcal{U}$ learns more than $x_{i_1}, \ldots, x_{i_q}$. A typical PIR protocol proceeds in three stages. First, $\mathcal{U}$ sends a query, depending on I. Let $\mathcal{Q}(I)$ denote the query for indices in I. Second, $\mathcal{D}$ computes the reply $\mathcal{R}(\mathcal{Q}(I), x)$ and sends it to $\mathcal{U}$. Third, $\mathcal{U}$ extracts $x_{i_1}, \ldots, x_{i_q}$ from $\mathcal{R}(\mathcal{Q}(I), x)$. The scheme is computationally private if no efficient distinguisher can distinguish $\mathcal{Q}(I)$ from $\mathcal{Q}(I')$, for any I and I' such that $|I| = |I'|$. To avoid trivial solutions, we require that $|\mathcal{R}(\mathcal{Q}(I), x)| < l$.

Several PIR schemes were proven secure under different intractability assumptions. For example the scheme proposed in [21] is based on the computational difficulty of the quadratic residuosity problem, and in [22] it was shown how to construct a PIR protocol from any one-way trapdoor permutation. In [8] it was shown that the assumption that PIR exists implies the existence of the Oblivious Transfer.

Symmetric Encryption. A *symmetric encryption scheme* is a pair (E, D) of polynomial-time algorithms. Algorithm E takes as input a security parameter 1^k, a key $K \in \{0,1\}^{m_{\mathrm{sym}}}$, and a message $M \in \{0,1\}^{n_{\mathrm{sym}}}$ and outputs a *ciphertext* $C = E(K, M) \in \{0,1\}^{l_{\mathrm{sym}}}$ (we will assume that $m_{\mathrm{sym}}, n_{\mathrm{sym}}$ and l_{sym} are functions of k). Algorithm D takes as input 1^k, K and C and outputs M'. We require that always $D(K, E(K, M)) = M$. The security of an encryption scheme is defined as follows. Consider an adversary $\mathcal{A}_{\mathrm{enc}}$ that is a probabilistic polynomial-time machine that can specify two messages M^0 and M^1 (of the same length). Later, he receives $C = E(K, M^c)$ for a random key K and a random bit $c \in \{0,1\}$, and he has to guess c. If $\mathcal{A}_{\mathrm{enc}}$ guesses c correctly with probability $0.5 + \epsilon$ we will say that he *breaks* (E, D) *with advantage* ϵ. We say that (E, D) is *computationally secure* if any $\mathcal{A}_{\mathrm{enc}}$ breaks (E, D) with advantage that is negligible in k.

Pseudorandom Generators. A *pseudorandom generator (PRG)* is a polynomial-time algorithm G that takes as input a security parameter 1^k, and a *seed* $K \in \{0,1\}^{m_{\mathrm{PRG}}(k)}$ and outputs a much longer string $G(K)$. A PRG G

is *computationally-secure* if any polynomial-time adversary is not able to distinguish $G(K)$ from a truly random string (with a non-negligible advantage). This can be formalized in a similar way as the symmetric encryption in the definition above. It was shown in [18] that a pseudorandom generator can be built from any one-way function.

Oblivious Transfer. An *Oblivious Transfer (OT)* is a protocol between a Sender S (that takes as an input $(b_0, b_1) \in \{0,1\}^2$ and a security parameter 1^k) and a Receiver R (with an input $c \in \{0,1\}$ and the security parameter 1^k). After the execution of the protocol, the Receiver should know b_c (we allow a negligible probability of error). This property is called *correctness* of the protocol. Moreover: (1) the Receiver should have essentially no information on b_{1-c} (this is called *privacy of the Sender*) and (2) the Sender should have essentially no information on c (this is called *privacy of the Receiver*). In this paper we assume that the security holds in the honest-but-curious setting (i.e. even the corrupted parties follow the protocol). More formally, the security is defined as follows, let $OT(c; b_0, b_1; 1^k)$ denote the execution of the OT protocol (with the inputs c, b_0, b_1 and 1^k). To define the privacy of the Sender we require that any polynomial-time Receiver should not be able to distinguish between $OT(0; b, 0; 1^k)$ and $OT(0; b, 1; 1^k)$, and between $OT(1; 0, b; 1^k)$ and $OT(1; 1, b; 1^k)$ (for $b \in \{0,1\}$). Similarly, to define the privacy of the Receiver we require that any polynomial-time Sender should not be able to distinguish (with a non-negligible advantage) between $OT(0; b_0, b_1; 1^k)$ and $OT(1; b_0, b_1; 1^k)$ (for $b_0, b_1 \in \{0,1\}$).

An *infinitely-often Oblivious Transfer (ioOT)* is an Oblivious-Transfer protocol for whose correctness holds only for infinitely many values $k_0, k_1, \ldots$ of the security parameter k. For all values not in $\{k_0, k_1, \ldots\}$ the Receiver instead of learning b_c gets $\perp$, except with negligible probability.

An *Oblivious-Transfer with an inefficient algorithm for the Receiver* is an OT protocol where the running time of the algorithm for the Receiver is not limited, and where the privacy of the Sender holds when the Receiver is computationally-unbounded. This notion is non-standard (we are not aware of any previous use of it) and we need it for purely theoretical purposes (in Sect 6.3). Clearly, such a protocol itself has no practical significance, as the security of the Receiver is still protected only computationally, and thus the honest Receiver is assumed to have more commutating power than the dishonest Sender.

4   Definition of the Forward-Secure Storage

The main idea of the Forward-Secure Storage is as follows. It can be viewed as a randomized symmetric encryption scheme, where the encryption algorithm Encr produces (for a given secret key K and a message $M \in \{0,1\}^{s_{\mathrm{FSS}}}$) a huge ciphertext $C = \mathrm{Encr}(K, M) \in \{0,1\}^{t_{\mathrm{FSS}}}$. One can imagine that C is stored on a machine which can be attacked by an adversary. Once the adversary gets access to the infected machine, he can perform an arbitrary computation on C, with the restriction that the output U has to be of size at most $s \ll t$. Once the adversary learned U, he looses access to C. Then, we assume that the key leaks, i.e. the adversary is given K. We require that (U, K) should not give him any significant information about M. We model it in a standard indistinguishability-style, i.e. we assume that the adversary knows in advance (i.e. before he got access to C) that the message M is a random member of a (chosen by him) two-element set $\{M^0, M^1\}$ and his task is to find out whether $M = M^0$ or $M = M^1$. This reflects the fact that the adversary may have already some a priori information about M.

Formally, a *Forward-Secure Storage (FSS)* scheme is a pair of polynomial-time randomized Turing Machines $\Phi = (\text{Encr}, \text{Decr})$. The machine Encr takes as input a *security parameter* 1^k, a *key* $K \in \{0,1\}^{m_{\text{FSS}}}$ and a *plaintext* $M \in \{0,1\}^{n_{\text{FSS}}}$ and outputs a *ciphertext* $C \in \{0,1\}^{t_{\text{FSS}}}$. The algorithm Decr takes as input a security parameter 1^k, a key $K \in \{0,1\}^{m_{\text{FSS}}}$, and a ciphertext $C \in \{0,1\}^{t_{\text{FSS}}}$, and it outputs a string $M' \in \{0,1\}^{n_{\text{FSS}}}$.[6] We will assume that $m_{\text{FSS}}, n_{\text{FSS}}$ and t_{FSS} are some functions of k. The following correctness property has to be satisfied with probability 1: $\text{Decr}(1^k, K, \text{Encr}(1^k, K, M)) = M$. We will sometimes drop the parameter 1^k. To define the security of an FSS scheme take a function $\sigma : \mathbb{N} \to \mathbb{N}$ and consider a σ-*adversary* $\mathcal{A}_{\text{FSS}}$ (that we model as a randomized Turing Machine), that plays the following game against an oracle $\Omega_{\text{FSS}}^\Phi(c)$ (where the challenge bit $c \in \{0,1\}$ is random), for some fixed value of the security parameter k.

> **FSS - distinguishing game**

1. The adversary gets 1^k, produces two messages $M^0, M^1 \in \{0,1\}^{n_{\text{FSS}}(k)}$ and sends them to $\Omega_{\text{FSS}}^\Phi(c)$.
2. $\Omega_{\text{FSS}}^\Phi(c)$ selects a random key $K \in \{0,1\}^{m_{\text{FSS}}(k)}$ and computes $C = \text{Encr}(1^k, K, M^c)$.
3. The adversary gets access to C and can compute an arbitrary value $U \in \{0,1\}^{\sigma(k)}$. The adversary can store U, but he is not allowed to store any other information (i.e. the entire description of the internal state of $\mathcal{A}_{\text{FSS}}$ has to be included in U).
4. The adversary learns K and has to guess c.

We say that an adversary $\mathcal{A}_{\text{FSS}}$ *breaks the scheme Φ with an advantage ϵ* if his probability of winning the game is $0.5 + \epsilon$. We say that an FSS scheme Φ is σ-*computationally-secure* if every probabilistic polynomial-time $\mathcal{A}_{\text{FSS}}$ breaks Φ with advantage that is negligible in k. We say that an FSS scheme is σ-*information-theoretically (IT)-secure* if is secure even against an adversary that is computationally unbounded. Recall that one of the possible applications of FSS is the protection against the situation when the key K is broken because more computing power is available (see Sect. 1.1, Point 3). One can model it by

---

[6] For some applications (e.g. when we expect that the ciphertext will be decrypted often) we may prefer schemes where the algorithm Decr needs to look only on a small number of the bits in C (if the plaintext is small). Actually, the schemes that we construct in this paper have this property.

assuming that the adversary gets infinite computing power, *after* he retrieved U.[7] Such an adversary will be called a *hybrid adversary*. If an FSS scheme is secure against a hybrid σ-adversary, we will say that it is σ-*secure in the hybrid model*. Observe that this model is very closely related to the Hybrid BSM (see Section 3).

5 Connections with the Theory of [16]

In this section we discuss the connections of the theory of FSS to the theory of compressibility of NP-instances [16] (see Sect. 2). The lemmas that we show in this section have a cryptanalytic nature, i.e. we show implications of the form: if a certain language is compressible then a certain FSS scheme can be broken. This is because the theory of [16] concerns the worst-case complexity. To prove implications in the opposite direction, one would need to develop a theory of problems which are incompressible *on average* (this notion was recently defined in [11]). Let $\Phi = (\text{Encr}, \text{Decr})$ be an FSS scheme with the parameters $m_{\text{FSS}}, n_{\text{FSS}}$, and t_{FSS} as before. Define an NP-language $L_{\Phi} := \{(M, C) : \exists K.M = \text{Decr}(K, C)\}$.

Lemma 1. *Suppose Φ is such that the length n_{FSS} of the encrypted message is longer than the length m_{FSS} of the key. For a function $\sigma : \mathbb{N} \to \mathbb{N}$ let π be such that $\pi(n_{\text{FSS}}(k) + t_{\text{FSS}}(k)) = \sigma(k)$ (for every $k \in \mathbb{N}$). We have the following:*

1. *if L_{Φ} is π-compressible then Φ can be broken by a hybrid σ-adversary, and*
2. *if L_{Φ} is π-compressible with witness retrievability, then Φ can be broken by a computational σ-adversary.*

Proof. Let us first consider Point 1. Let $\mathcal{C}$ be the algorithm that compresses L_{Φ} to some language L'. We construct an adversary $\mathcal{A}_{\text{FSS}}$ that breaks Φ. First, the adversary $\mathcal{A}_{\text{FSS}}$ produces two random messages M^0 and M^1. When he gets access to C (which is equal to $\text{Encr}(K, M^c)$, for random K and c), he retrieves $U = \mathcal{C}(M^0, C)$ (clearly $|U| \leq \sigma(|C|)$). Later, when he gets infinite computing power, he checks if $(M^0, C) \in L_{\Phi}$ (by the properties of the compression algorithm, he can do it by checking if $U \in L'$). If $(M^0, C) \in L_{\Phi}$ then the adversary guesses "$c = 0$", otherwise he guesses "$c = 1$". Clearly, if $c = 0$ (which happens with probability 0.5) then his guess is correct with probability 1. If $c = 1$ then his guess may be incorrect (i.e. he may guess "$c = 0$"). But this happens only if there exist K' such that $M^0 = \text{Decr}(K', C)$ (for $C = \text{Encr}(K, M^1)$). Clearly the number of messages M^0 such that $M^0 = \text{Decr}(K', C)$ is at most equal to the number $2^{m_{\text{FSS}}}$ of the keys K'. Thus (since the total number of messages is $2^{n_{\text{FSS}}} \geq 2 \cdot 2^{m_{\text{FSS}}}$), the probability that for a random M^0 there exists K' such that $M^0 = \text{Decr}(K', C)$ is at most 0.5. Therefore, the total probability that $\mathcal{A}_{\text{FSS}}$ guesses c correctly is at least $0.5 \cdot (1 + 0.5) = 0.75$.

---

[7] Note that in this case it would make sense to consider a slightly weaker model, where the adversary does not receive K in Step 4 (and has to recover it using his unlimited computing power). For simplicity we assume that the adversary receives K anyway.

Essentially the same adversary can be used for the proof of Point 2. The only difference is that in order the determine if $U \in L'$ the (computationally-bounded) adversary obtains K and checks if $W(K)$ is a witness for U (where W is the algorithm for retrieving the witness). □

Lemma 2. *Let Φ be an FSS scheme whose length n_{FSS} of the ciphertext is polynomial in the security parameter. Moreover, suppose that the output of the algorithm* Decr *depends only on a polynomial number of the bits of the ciphertext. If SAT is compressible, then Φ can be broken by a hybrid σ-adversary, where σ is a polynomial.*

Clearly, the statement of the lemma has any practical meaning only if we can show candidates for FSS schemes whose length t_{FSS} of the ciphertext is super-polynomial in k (and the other parameters are polynomial in k). Such a candidate can, e.g., be the scheme Φ_{c2}^G that we will construct Sect. 6.3. The scheme Φ_{c2}^G will be constructed from a BSM-secure function f, and the length of the ciphertext will be equal to $t_{BSM} + n_{BSM}$ (where t_{BSM} is the length of the randomizer, and n_{BSM} is the length of the plaintext). Therefore, it is enough to show a BSM-secure function with a superpolynomial length of the randomizer. An example of such a function is the function of [15] (for an appropriate choice of the parameters). Observe that Lemma 2 implies that if Φ_{c2}^G is secure against an adversary with memory of a polynomial size (assuming that one-way functions exits, say), then SAT is incompressible.

Proof (of Lemma 2, sketch). We will show that if SAT is compressible, then L_Φ is compressible. By Lemma 1 this will suffice. It is enough to show a polynomial-time algorithm, that given an input (M, C) produces a Boolean formula $\phi_{M,C}(x_1, \ldots, x_q)$ (where $x_1, \ldots, x_q$ are free variables) such that: (1) $\phi_{M,C}$ is satisfiable if and only if $(M, C) \in L_\Phi$, and (2) the number q of the free variables is polynomial in K and $\log |(M, C)|$.[8] Such a formula can be constructed in essentially the same way as the formula in the proof of Lemma 2.14 of [16]. The details appear in [13]. □

Algorithm Encr Has to be Randomized. Further exploring the analogies between the theory of [16] and FSS, we show how the method developed in [16] can be used to prove that the algorithm Encr has to be randomized. This in particular means that algorithm Encr cannot use part of its own key K as a seed for a PRG and obtain the necessary randomness this way. Intuitively, it is quite obvious. Here we give a formal proof.

Observation 1. *Any FSS scheme with deterministic encryption can be broken by a computational σ-adversary where $\sigma(k) = k$.*

Proof (sketch). This method is taken from [16] (Claim 2.3). The adversary selects a random hash function $H : \{0,1\}^{t_{FSS}(k)} \to \{0,1\}^k$ from a family of pairwise-independent hash functions. Then, he chooses two arbitrary distinct messages

---

[8] In [16] this is called a *W-reduction*.

M^0 and M^1. When he obtains $C = \text{Encr}(K, M^i)$ he computes $U = H(C)$ and retrieves it. Later (when he receives K), he can test if $i = 0$ or $i = 1$ by finding i for which

$$H(\text{Encr}(K, M^i) = U). \tag{1}$$

By the pairwise independence of H the probability that 1 holds for both $i = 0$ and $i = 1$ is equal to 2^{-k}, and hence it is negligible.   □

Clearly, the statement of the lemma holds also for the FSS schemes that are allowed to have a random input r of a logarithmic length (i.e. $|r| = \log k$). This is because in this case a polynomial-time adversary can simply test if (1) holds by examining all possible random inputs $r \in \{0, 1\}^{\log(k)}$.

6   Construction of FSS Schemes

In this section we construct an IT-secure FSS scheme Φ_{IT} (Sect. 6.2) and an FSS scheme Φ_{c1} (Sect. 6.1), that is computationally secure, assuming that one-way functions exist. The advantage of Φ_{c1} is that it allows to encrypt messages that are much longer than the key (which, by Shannon Theorem, cannot be the case if the IT security is required).

The main drawback of the scheme Φ_{c1} is that it can be trivially broken in the hybrid model. A natural question to ask is whether there exists an FSS scheme that is secure in the hybrid model. We propose a construction of an FSS scheme Φ_{c2}^G (from a pseudorandom generator G) about which we conjecture that it is secure in the hybrid model. We are able to prove neither the hybrid nor even the computational security of Φ_{c2}^G from the assumption that G is a computationally-secure PRG. We argue about the security of Φ_{c2}^G in a non-standard way (see Sect. 6.3). Our results are summarized in the following table:

scheme	IT-security	hybrid security	computational security
Φ_{IT}	secure	secure	secure
Φ_{c2}^G	not secure	conjectured secure	conjectured secure
Φ_{c1}	not secure	not secure	secure if one-way functions exist

In our constructions we will use a function f that is σ-secure in the BSM. Let us now fix such a function (one could take e.g. the function of [15], or any other function that was proven secure in the BSM), and let $m_{\text{BSM}}, n_{\text{BSM}}$ and t_{BSM} be the parameters defined in Sect. 3.

6.1   Computationally-Secure FSS

Let (E, D) be a symmetric encryption scheme with the parameters $m_{\text{sym}}, n_{\text{sym}}$ and l_{sym} defined as in Sect. 3. In this section we construct a computationally-secure FSS scheme $\Phi_{\text{c1}} = (\text{Encr}_{\text{c1}}, \text{Decr}_{\text{c1}})$. Fix some security parameter k. The key of Φ_{c1} is interpreted as a pair (K, W), where $K \in \{0, 1\}^{m_{\text{BSM}}}$ is the initial key of f and $W \in \{0,1\}^{n_{\mathrm{BSM}}}$. Hence, the length m_{c1} of the key of Φ_{c1} is equal to $m_{\mathrm{BSM}} + n_{\mathrm{BSM}}$. We also assume that the length n_{BSM} of the string derived by the BSM function f is equal to the length of the key for the encryption scheme (E, D). The length n_{c1} of the plaintext is equal to the length n_{sym} of the plaintext of (E, D). The length t_{c1} of the ciphertext is equal to $t_{\mathrm{BSM}} + l_{\mathrm{sym}}$.

$\mathrm{Encr}_{c1}((K, W), M)$	$\mathrm{Decr}_{c1}((K, W), C)$
1. Generate a random string $R \in \{0,1\}^{t_{\mathrm{BSM}}}$.	1. Parse C as (R, X).
2. Let $K' := f(K, R) \oplus W$.	2. Let $K' := f(K, R) \oplus W$.
3. Output $C = (R, E(K', M))$.	3. Output $D(K', X)$.

Lemma 3. *If f is σ-secure in the BSM and (E, D) is a computationally-secure symmetric encryption scheme, then Φ_{c1} is a σ-computationally-secure FSS scheme.*

Proof. Let $\mathcal{A}_{\mathrm{comp}}$ be an adversary that for a security parameter k breaks Φ_{c1} with an advantage $\epsilon = \epsilon(k)$. We are going to construct adversaries: $\mathcal{A}_{\mathrm{BSM}}$ and $\mathcal{A}_{\mathrm{enc}}$, such that either $\mathcal{A}_{\mathrm{BSM}}$ breaks the BSM scheme f (using memory of size $\sigma(k)$), or $\mathcal{A}_{\mathrm{enc}}$ breaks the encryption scheme (E, D). The running time of these adversaries will be polynomial in the running time of $\mathcal{A}_{\mathrm{comp}}$. Let ϵ_{BSM} denote the advantage of $\mathcal{A}_{\mathrm{BSM}}$ in breaking f, and let ϵ_{enc} denote the advantage of $\mathcal{A}_{\mathrm{enc}}$ in breaking (E, D). The adversary $\mathcal{A}_{\mathrm{BSM}}$ simulates the adversary $\mathcal{A}_{\mathrm{comp}}$ as follows.

$\mathcal{A}_{\mathrm{BSM}}$

1. Start $\mathcal{A}_{\mathrm{comp}}$ and pass the security parameter 1^k to $\mathcal{A}_{\mathrm{comp}}$. Obtain from $\mathcal{A}_{\mathrm{comp}}$ the messages M^0 and M^1.
2. During Phase 1 of the distinguishing game (see Sect. 3) do the following. Input the randomizer $R \in \{0,1\}^{t_{\mathrm{BSM}}}$. Select a random bit $c \in \{0,1\}$ and a random string $K' \in \{0,1\}^{n_{\mathrm{BSM}}}$. Send $(R, E(K', M^c))$ to $\mathcal{A}_{\mathrm{comp}}$. Store in your memory the value $U \in \{0,1\}^{\sigma(k)}$ that $\mathcal{A}_{\mathrm{comp}}$ stores.
3. In Phase 2 receive K and $\hat{X}$, compute $W = \hat{X} \oplus K'$ and pass (K, W) to $\mathcal{A}_{\mathrm{comp}}$. If $\mathcal{A}_{\mathrm{comp}}$ guesses c correctly, then output "$b = 0$" (i.e. guess that $\hat{X} = f(K, R)$). Otherwise output "$b = 1$" (i.e. guess that $\hat{X}$ is random).

The adversary $\mathcal{A}_{\mathrm{enc}}$ simulates the adversary $\mathcal{A}_{\mathrm{comp}}$ in the following way.

$\mathcal{A}_{\mathrm{enc}}$

1. Start $\mathcal{A}_{\mathrm{comp}}$ and pass the security parameter 1^k to $\mathcal{A}_{\mathrm{comp}}$. Obtain from $\mathcal{A}_{\mathrm{comp}}$ the messages M^0 and M^1. Output them.
2. Select randomly $R \in \{0,1\}^{t_{\mathrm{BSM}}}, K \in \{0,1\}^{m_{\mathrm{BSM}}}$ and $W \in \{0,1\}^{n_{\mathrm{BSM}}}$. When you receive C (recall that $C = E(K', M^c)$, for a random K' and c, and the goal of $\mathcal{A}_{\mathrm{enc}}$ is to guess c), pass (R, C) to $\mathcal{A}_{\mathrm{comp}}$.
3. Pass (K, W) to $\mathcal{A}_{\mathrm{comp}}$. Output the bit c that $\mathcal{A}_{\mathrm{comp}}$ outputs.

Let us now look at the adversary $\mathcal{A}_{\text{BSM}}$ defined above. If $b = 0$ then $\hat{X} = f(K, R)$, and hence $K' = f(K, R) \oplus W$. Therefore in this case $\mathcal{A}_{\text{BSM}}$ simply simulated the normal execution of $\mathcal{A}_{\text{comp}}$ against the FSS scheme Φ_{c1}^{f} with the challenge bit c. Thus, in this case $\mathcal{A}_{\text{comp}}$ guesses c correctly with probability $0.5 + \epsilon$.

If $b = 1$ then K' is independent of the variable (R, K, W). Hence, $\mathcal{A}_{\text{BSM}}$ simulated $\mathcal{A}_{\text{comp}}$ in exactly the same way as $\mathcal{A}_{\text{enc}}$. Thus, the probability that $\mathcal{A}_{\text{comp}}$ guesses c correctly is $0.5 + \epsilon_{\text{enc}}$. Therefore, the probability $0.5 + \epsilon_{\text{BSM}}$, that $\mathcal{A}_{\text{BSM}}$ guesses b correctly is equal to

$$\tfrac{1}{2} \cdot P\left(\mathcal{A}_{\text{BSM}} \text{ outputs } ``b = 0" | b = 0\right) + \tfrac{1}{2} \cdot P\left(\mathcal{A}_{\text{BSM}} \text{ outputs } ``b = 1" | b = 1\right)$$
$$= (0.5 + \epsilon)/2 + (0.5 - \epsilon_{\text{enc}})/2 = (1 + \epsilon - \epsilon_{\text{enc}})/2$$

Therefore, we get that $\epsilon = \epsilon_{\text{BSM}} + \epsilon_{\text{enc}}$. Thus, if ϵ is non-negligible, then one of ϵ_{BSM} and ϵ_{enc} has to be non-negligible. Therefore, if Φ_{c1} is not secure, then either f of (E, D) is not secure either.

$\square$

Since the existence of one-way functions implies the existence of symmetric encryption schemes [18] (where the length of the plaintext is an arbitrary polynomial of the security parameter), we get the following.

Theorem 1. *If one way-functions exist, then there exists a σ-computationally-secure FSS scheme with parameters σ, m_{c1} and t_{c1} as above, where the length n_{c1} of the encrypted message is an arbitrary polynomial of the security parameter.*

6.2   IT-Secure FSS

The IT-secure FSS scheme Φ_{IT} can be constructed by substituting (in the construction from Sect. 6.1) the computationally-secure encryption scheme (E, D) with the one-time pad encryption. Clearly, in this case the length n_{IT} of the plaintext is equal to n_{BSM}, the length m_{IT} of the key is equal to $m_{\text{BSM}} + n_{\text{BSM}}$ (and hence it is greater than the length of the plaintext) and the length t_{FSS} of the ciphertext is equal to $t_{\text{BSM}} + n_{\text{BSM}}$. The security of this construction can be proven by essentially repeating the proof of Lemma 3. Therefore, get the following.

Theorem 2. *There exists a σ-IT-secure FSS scheme with the parameters σ, $m_{\text{IT}}, n_{\text{IT}}$ and t_{IT} as above.*

6.3   FSS Scheme with a Conjectured Hybrid Security

It is easy to see that the scheme Φ_{c1} can be broken in the hybrid model: the adversary can simply retrieve the second component $(E(K', M))$ of the ciphertext, and then break it by the exhaustive key-search. In this section we show the construction of an FSS scheme $\Phi_{c2}^{G} = (\text{Encr}_{c2}, \text{Decr}_{c2})$, about which we conjecture that it is secure in the hybrid model. The components for our construction are: a pseudorandom generator G (see Sect. 3), and the σ-IT-secure scheme from Sect. 6.2. The problem with this construction is that we are not able to base (even the computational) security of Φ_{c2}^G on any standard assumption. We argue about the security of Φ_{c2}^G in the following way. Informally speaking, we show that if we are given a computational adversary that breaks Φ_{c2}^G (for some PRG G) then one can construct an ioOT protocol (from this G). Although this does not guarantee that Φ_{c2}^G is secure for any choice of G, it is an indication that for a Φ_{c2}^G is secure if G is some "standard" PRG (see discussion at the end of this section). In case of the hybrid security we have a weaker security argument: we show that if one can break Φ_{c2}^G in the computational model, then one can construct from G an ioOT protocol with an inefficient algorithm for the Receiver (see Sect. 3).

Fix some security parameter k. Let G be a PRG with some short seed of length m_{PRG}. The length m_{c2} of the key of Φ_{c2}^G is equal to m_{PRG}. The length n_{c2} of the encrypted message is equal to n_{IT} and the length t_{c2} of the ciphertext is equal to t_{IT}. The Encr_{c2} and Decr_{c2} (for a key $K \in \{0,1\}^{m_{c2}}$, and a plaintext $M \in \{0,1\}^{n_{c2}}$) are defined as: $\mathrm{Encr}_{c2}(K,M) := \mathrm{Encr}_{\mathrm{IT}}(G(K),M)$ and $\mathrm{Decr}_{c2}(K,C) := \mathrm{Decr}_{\mathrm{IT}}(G(K),C)$. Clearly, $\mathrm{Decr}_{c2}(K,\mathrm{Encr}_{c2}(K,M)) = M$, so it remains to argue about the security.

Computational Security of Φ_{c2}^G. We start with considering the computational security of Φ_{c2}^G. Suppose a polynomial-time σ-adversary $\mathcal{A}_{\mathrm{comp}}$ breaks $(\mathrm{Encr}_{c2}, \mathrm{Decr}_{c2})$ with a non-negligible advantage $\epsilon(k)$ and G is a pseudorandom generator. We are going to construct an ioOT protocol (with an unconditional privacy of the Sender and a computational privacy of the Receiver). Before going to the construction let us present the intuition behind it. In [16] (Theorem 3.1) it is shown that if there exists a witness-retrievable compression algorithm for SAT then the existence of one-way functions implies existence of Oblivious Transfer. This is proven by (1) constructing a PIR protocol from a compression algorithm for SAT and then (2) using the result of [8] that PIR implies Oblivious Transfer. We proceed in a similar way, in some sense combining the proofs (1) and (2) into one. This is possible since, as we remarked before, an adversary that breaks an FSS scheme can be viewed as an compression algorithm (see Sect. 5). Note that in our proof we do not construct PIR from $\mathcal{A}_{\mathrm{comp}}$ (which may be quite hard in general). Note also that our method has some similarities with the one used recently in [11].

The details of the construction are as follows. We are going to construct an Oblivious Transfer protocol OT whose correctness holds only with probability non-negligibly greater than 0.5 (i.e. the probability that the Receiver outputs b_c after the execution of $\mathrm{OT}(b; c_0, c_1; 1^k)$ will be non-negligibly greater than 0.5, for any choice of c_0, c_1 and b). This suffices to construct an ioOT protocol, since by repeating the protocol polynomial number of times, the parties can reduce the error probability to negligible, for infinitely many values of the security parameter (the details can be found in [13]).

The main idea is that the Sender and the Receiver will simultaneously simulate the execution of $\mathcal{A}_{\mathrm{comp}}$, with the Sender holding the ciphertext C and the Receiver receiving just U. The protocol goes as follows.

$$\boxed{\mathrm{OT}(c; b_0, b_1; 1^k)}$$

1. The Receiver selects a random seed $K \in \{0,1\}^{m_{\mathrm{PRG}}}$. Let $K_c := G(K)$ and let $K_{1-c} \in \{0,1\}^{m_{\mathrm{FSS}}}$ be random. The Receiver sends (K_0, K_1) to the Sender.
2. The Sender starts the adversary $\mathcal{A}_{\mathrm{comp}}$ and passes the security parameter 1^k to him. Let M^0 and M^1 be the messages produced by $\mathcal{A}_{\mathrm{comp}}$.

   The Sender selects random bits $d, x \in \{0,1\}$ and computes $C = \mathrm{Encr}_{c2}(K_d, M^{b_d \oplus x})$. [9] He passes this value to (his simulated copy of) $\mathcal{A}_{\mathrm{comp}}$. Let U be the value that $\mathcal{A}_{\mathrm{comp}}$ stores in his memory.

   The Sender sends (U, d, x) and the random input of $\mathcal{A}_{\mathrm{comp}}$ to the Receiver.
3. If $d \neq c$ then the Receiver outputs a random bit.[10] Otherwise the Receiver simulates the execution of $\mathcal{A}_{\mathrm{comp}}$ with the random input that he got from the Sender. The Receiver sends U and K to (his copy of) the adversary $\mathcal{A}_{\mathrm{comp}}$. Let B be the output of $\mathcal{A}_{\mathrm{comp}}$. The Receiver outputs $B \oplus x$.

Lemma 4. *The protocol* OT *protects the privacy of the Receiver.*

Proof (sketch). This is quite straightforward, as the Sender that distinguishes c from $1 - c$ should also be able to distinguish a uniformly random string from a string produced by a PRG G. More formally, one can construct an adversary that breaks the PRG G by simulating the algorithm that distinguishes $\mathrm{OT}(c; b_0, b_1, 1^k)$ from $\mathrm{OT}(1 - c; b_0, b_1; 1^k)$. The details appear in [13]. □

Lemma 5. *The protocol* OT *protects the privacy of the Sender.*

Proof (sketch). We need to show that the Receiver has negligible advantage in guessing b_{1-c}. Clearly, the only value from which the Receiver could learn anything about b_{1-c} is U. If $d = c$ then this value is independent of b_{1-c}, so we are done. Otherwise (when $d \neq c$), U is a function of the ciphertext $C = \mathrm{Encr}_{\mathrm{IT}}(K_{1-c}, M^{b_d \oplus x})$ (where K_{1-c} is uniformly random). To get some non-negligible information about b_{1-c}, the Receiver would need to distinguish with non-negligible advantage between M^0 and M^1 (knowing U and K_{1-c}). This, however, would contradict the security of Φ_{IT}. □

It is easy to see that the privacy of the Sender is actually protected unconditionally.

Lemma 6. *The* OT *protocol is an Oblivious Transfer protocol whose correctness holds with probability non-negligibly greater than* 0.5.

Proof. The privacy was proven in Lemmas 4 and 5. It suffices to show the correctness. Fix some security parameter 1^k. Let $\epsilon = \epsilon(k)$ be the advantage of $\mathcal{A}_{\mathrm{comp}}$. If $d \neq c$ (which happens with probability 0.5) the probability that R outputs a correct value is clearly 0.5.

---

[9] The role of x is to guarantee that $M^{b_d \oplus x}$ is chosen *uniformly* at random from the set $\{M^0, M^1\}$, what is needed in the proof of Lemma 6.

[10] This is because in this case the Receiver has no significant information about b_c. Alternatively, we could instruct him to output some default value $\bot$.

Otherwise (if $d = c$) observe that the Receiver and the Sender simply simulated the execution of $\mathcal{A}_{\text{comp}}$ against an oracle $\Omega_{\text{FSS}}^{\Phi_{c2}^G}(b_d \oplus x)$ (see Sect. 4), where $b_d \oplus x \in \{0, 1\}$ is random. Therefore, the probability that $\mathcal{A}_{\text{comp}}$ (and hence the Receiver) outputs $B = b_c$ is $0.5 + \epsilon(k)$. Thus, the total probability that $\mathcal{A}_{\text{comp}}$ outputs b_c is equal to $\frac{1}{2} \cdot \frac{1}{2} + \frac{0.5+\epsilon}{2} = \frac{1}{2} + \frac{\epsilon}{2}$, which is clearly non-negligible. □

Hybrid Security of Φ_{c2}^G. One can essentially repeat the same construction for the hybrid adversary $\mathcal{A}_{\text{hybr}}$ (instead of the computational adversary $\mathcal{A}_{\text{comp}}$). The only difference is that, since we allow the adversary $\mathcal{A}_{\text{hybr}}$ (that attacks Φ_{c2}^G) to have infinite computing power (in the final stage), then we need to assume that also the Receiver has infinite computing power (as otherwise he would not be able to simulate $\mathcal{A}_{\text{hybr}}$ in Step 3). Thus, the resulting protocol has an inefficient algorithm for the Receiver (see Sect. 3).

Discussion. Let us start with discussing the meaning of the construction of OT from a computational adversary $\mathcal{A}_{\text{comp}}$. First, recall that constructing an OT protocol from a PRG (or, equivalently, from a one-way function) is a major open problem in cryptography. It was shown in [19] that such a construction cannot be done in black-box way (observe that our construction is non-black-box, as the adversary $\mathcal{A}_{\text{comp}}$ is allowed to use the internal structure of G) and even constructing an ioOT from a one-way function would be a breakthrough.

A natural question is: which pseudorandom generators G are safe for the use in the construction of $\mathcal{A}_{\text{hybr}}$. A highly informal answer is: "those whose security does not imply Oblivious Transfer". We leave formalizing this property as an open problem. Note that in principle there can exist (probably very artificial) pseudorandom generators which cannot be safely used in Φ_{c2}. In the next section we present an example that shows that something like this can happen if one replaces a PRG in Φ_{c2}^G with a slightly more general primitive.

Observe that one can also view our result as a win-win situation: either we win because we have FSS or we win because we have an ioOT from a one-way function. (This is how a result of a similar nature is interpreted in [11].) Interestingly, a similar reasoning was recently used also in [26] (using the language of [26] we could say that Φ_{c2}^G is *secure in Minicrypt*).

The argument for the security of Φ_{c2}^G in the hybrid model is much weaker, since we used a non-standard version on the OT protocol. Observe that (as argued in Sect. 5) showing the security of Φ_{c2}^G implies that SAT is incompressible (which may indicate that such a proof is difficult to find). In general, since there are no known techniques of basing incompressibility of NP-instances on standard assumptions, we conjecture that basing hybrid security of FSS schemes on such assumptions may also be hard.

Intuitively, the scheme Φ_{c2}^G should be secure because a computationally-limited adversary that has no (computational) information on $G(K)$ (when he has access to R) should not be stronger than a computationally-unlimited adversary that attacks the scheme Φ_{IT} (where the only difference is that instead of $G(K)$ we use a truly random key). This is a similar reasoning to the one that was used to argue about the security of the Hybrid BSM (see Sect. 3). In the next section we argue that in certain cases this intuition may be misleading (by showing an example that is similar to the one shown for the Hybrid BSM in [14]).

A Scheme $\Phi_{c2}^{G,\alpha}$. A natural question to ask is whether there exist pseudorandom generators that cannot be safely used in Φ_{c2}^{G}. We leave this as an open problem. What we can show, however, is that there exists a modification of the scheme Φ_{c2}^{G} (denote it $\Phi_{c2}^{G,\alpha} = (\text{Encr}_{c2}^{G,\alpha}, \text{Decr}_{c2}^{G,\alpha})$, where α is defined below) that has similar properties to Encr_{c2} (i.e. if one can break it, then one can construct an ioOT protocol), but for some particular choice of G and α it can be broken. This example shows that basing the security of Φ_{c2}^{G} on some standard assumptions about G may be hard. Because of the lack of space we present just a sketch of the construction. We concentrate here on the computational security; however, the same example works for the hybrid security. In the construction of α we will use a PIR scheme $(\mathcal{U}, \mathcal{D})$ (see Sect. 3). The nature of this construction is similar to the one of [14] (see also [17]).

Let $\Phi_{\text{IT}} = (\text{Encr}_{\text{IT}}, \text{Decr}_{\text{IT}})$ be the $\sigma(k)$-IT-secure scheme from Sect. 6.2. The key of the scheme $\Phi_{c2}^{G,\alpha}$ has the form $K' = (K, \rho)$, where K is a seed for a PRG G, and ρ is a random input of the PIR user $\mathcal{U}$. Define Φ_{c2}' as $\text{Encr}_{c2}^{G,\alpha}((K, \rho), M) := (\text{Encr}_{\text{IT}}(G(K), M), \alpha(K, \rho))$, where α is defined below. Recall that the first component of $\text{Encr}_{\text{IT}}(G(K), M)$ is a randomizer R, and the knowledge of a small number of bits of R (namely those that are used to compute the value of f) suffices to decrypt M (if the key $G(K)$ and the rest of the ciphertext are known). Let $I := i_1, \ldots, i_q$ be the indices of those bits. We can assume that q is much smaller than $\sigma(k)$.

Suppose that we run the user $\mathcal{U}$ (with the input I and the random input ρ). Let $\mathcal{Q}(I)$ be the query that $\mathcal{U}$ generates. We set $\alpha(K') := \mathcal{Q}(I)$. Clearly, (by the security of PIR) the knowledge of $\alpha(K')$ does not help a computationally-bounded adversary in distinguishing $G(K)$ from a random string. Hence, if there exists a computational σ-adversary $\mathcal{A}_{\text{comp}}$ that breaks $\Phi_{c2}^{G,\alpha}$, then one can construct an OT protocol (in essentially the same way as we constructed the OT protocol).

Now, a computational σ-adversary can perform the following attack. He treats R as a database and retrieves the reply $U = \mathcal{R}(\mathcal{Q}(I), R)$ (by the properties of PIR we can assume that the length of this reply is much shorter than the length of the ciphertext). Later, when the adversary learns ρ, he can simulate the user (because K' includes the random input of the user), and use $\mathcal{U} = \mathcal{R}(\mathcal{Q}(I), x)$ to compute $R_{i_1}, \ldots, R_{i_q}$. Hence, the adversary can compute M.

7 Sequential Composability

Here we argue that FSS is *sequentially composable* in the following sense. Take some σ-computationally-secure scheme FSS $\Phi = (\text{Encr}, \text{Decr})$, and let $s = \sigma(k)$ (for some fixed k). Consider the following procedure. Suppose we start with a key K_0 (that is stored on some machine $\mathcal{M}_0$) and we encrypt (using FSS) some other key, K_1, with K_0 (and we store the ciphertext $C_1 = \text{Encr}(K_0, K_1)$ on a machine $\mathcal{M}_1$, say). Then, we generate another key K_2 and encrypt if with K_1 (we store the ciphertext $C_2 = \text{Encr}(K_1, K_2)$ on another machine $\mathcal{M}_2$), and so on (a ciphertext C_i is stored on a machine $\mathcal{M}_i$). At the end we encrypt some message M with the key K_{w-1} (let $C_w = \mathrm{Encr}(K_w, M)$ be the ciphertext). We show that this procedure is safe in the following sense. Suppose an adversary can break in, and take a temporary control over each of these machines (to simplify the model, suppose that the adversary can control one machine at a time, and he cannot take control over the same machine twice). Once the adversary broke into a machine $\mathcal{M}_i$, he can perform any computation on C_i and retrieve some value $U_i \in \{0,1\}^s$. Let $(\mathcal{M}_{i_1}, \ldots, \mathcal{M}_{i_{w+1}})$ be the order in which the adversary was breaking into the machines. We assume that the adversary is allowed to break into the machines in an arbitrary order, except that the order $(\mathcal{M}_0, \ldots, \mathcal{M}_w)$ is not allowed. We say that such a scheme is computationally secure if such an adversary has negligible advantage in determining whether $M = M^0$ or $M = M^1$ (for some chosen by him messages M^0 and M^1). It can be shown that this scheme is computationally secure, assuming that the scheme Φ is computationally secure. (Observe that if $w = 1$ then the security of this scheme is equivalent to the security of Φ.) The same holds for the IT and the hybrid security. Because of the lack of space, the formal definition of the sequential composition and the security proofs are given in [13].

8   Human-Memorized Passwords

In this section we show how FSS can be used if the data is protected by a password that is memorized by a human (see Sect. 1.1, Point 2). The main problem here is how to model the fact that the adversary is not able to perform a dictionary attack on the machine of the victim, but he can do it on his own machine. Note that when defining FSS we did not differentiate between the computational power of the adversary in different stages of the attack (except of the hybrid model where we simply assumed that at a certain moment the adversary gets an unlimited computing power). We will make use of a Random Oracle Assumption [2]. Such an oracle Ω^H can be viewed as a black-box which contains a function H chosen uniformly at random from a set of functions of a type $\{0,1\}^p \to \{0,1\}^{m_{\mathrm{FSS}}}$. Normally one assumes that every party in the protocol (including the adversary) has access to Ω^H. Here (to model the fact that the adversary cannot perform the dictionary attack on the victim's machine) we assume that the adversary can get access to Ω^H only *after* he lost access to C.

We construct an FSS scheme $(\mathrm{Encr}_{pass}, \mathrm{Decr}_{pass})$ that is secure if the secret key is a human-memorized password. Let $(\mathrm{Encr}, \mathrm{Decr})$ be an computationally secure FSS scheme with a key of length m_{FSS}. Let $\pi \in \{0,1\}^p$ be a password of the user. Define $\mathrm{Encr}_{pass}(\pi, M) := \mathrm{Encr}(H(\pi), M)$ and $\mathrm{Decr}_{pass}(\pi, M) := \mathrm{Decr}(H(\pi), M)$. (Observe that in this case our protocol has actually many similarities with the protocol of [9].) One can easily prove that if $(\mathrm{Encr}, \mathrm{Decr})$ is σ-computationally-secure then also $(\mathrm{Encr}_{pass}, \mathrm{Decr}_{pass})$ is secure against an adversary that can retrieve $\sigma(k)$ bits (in the model described above).

9   Open Problems

An interesting open problem is to determine if all computationally-secure pseudo-random generators G can be safely used in the construction of Φ_{c2}^G. If it turns out that the answer is "no" (e.g. there exists examples similar to those that exists for the scheme $\Phi_{c2}^{G,\alpha}$), then another open problem is to propose new definitions of computational indistinguishability, that would capture the fact that G can be safely used in the scheme Φ_{c2}. Clearly, this questions may have different answers for the computational and for the hybrid security.

The security argument for the hybrid security of Φ_{c2}^{G} is rather weak (since we use the non-standard notion of OT with inefficient algorithm for the Receiver). A natural open problem is to strengthen this argument, or to find alternative constructions.

It would also be interesting to examine the possibility of using the theory of All-or-Nothing transforms. Recall (see also Sect. 3) that the main difference between our model and the model considered in that theory is that they do not allow the adversary to perform arbitrary computations on the ciphertext. It may well be, however, that some of their constructions are actually secure also in our model (at least in the computational case). It is an open problem to prove it (if proving it using the standard assumptions is hard, then maybe one could at least use an argument similar to the one used in Sect. 6.3).

Note also that we did not provide any concrete examples of the parameters in our schemes. Doing it remains an open task.

Acknowledgments. I would like to thank Danny Harnik for pointing to me an error in the previous version of this paper, and suggesting to me how to repair it. I am also grateful to Krzysztof Pietrzak and Bartosz Przydatek for helpful discussions, and to the anonymous CRYPTO reviewers, for lots of useful comments.

References

1. Y. Aumann, Y. Z. Ding, and M. O. Rabin. Everlasting security in the Bounded Storage Model. *IEEE Transactions on Information Theory*, 48(6):1668–1680, 2002.

2. M. Bellare and P. Rogaway. Random oracles are practical: A paradigm for designing efficient protocols. In *ACM Conference on Computer and Communications Security*, pages 62–73, 1993.

3. C. Cachin, C. Crepeau, and J. Marcil. Oblivious transfer with a memory-bounded receiver. In *39th Annual Symposium on Foundations of Computer Science*, pages 493–502, 1998.

4. C. Cachin and U. Maurer. Unconditional security against memory-bounded adversaries. In *CRYPTO'97*, volume 1294 of *LNCS*, pages 292–306. Springer, 1997.

5. D. Cash, Y. Z. Ding, Y. Dodis, W. Lee, R. Lipton, and S. Walfish. Intrusion-resilient authentication in the Limited Communication Model. Cryptology ePrint Archive, Report 2005/409, 2005. http://eprint.iacr.org/.

6. B. Chor, O. Goldreich, E. Kushilevitz, and M. Sudan. Private Information Retrieval. *Journal of the ACM*, 45(6):965–981, 1998.

7. G. Di Crescenzo, R. J. Lipton, and S. Walfish. Perfectly secure password protocols in the Bounded Retrieval Model. In *Theory of Cryptography Conference'06*, volume 3876 of *LNCS*, pages 225–244. Springer, 2006.

8. G. Di Crescenzo, T. Malkin, and R. Ostrovsky. Single Database Private Information Retrieval implies Oblivious Transfer. In *EUROCRYPT 2000*, pages 122–138, 2000.
9. D. Dagon, W. Lee, and R. J. Lipton. Protecting secret data from insider attacks. In *Financial Cryptography and Data Security*, pages 16–30, 2005.
10. Y. Z. Ding, D. Harnik, A. Rosen, and R. Shaltiel. Constant-Round Oblivious Transfer in the Bounded Storage Model. In *Theory of Cryptography Conference*, volume 2951 of *LNCS*, pages 446–472. Springer, 2004.
11. B. Dubrov and Y. Ishai. On the randomness complexity of efficient sampling. In *ACM Symposium on Theory of Computing*, pages 711–720, 2006.
12. S. Dziembowski. Intrusion-resilience via the Bounded-Storage Model. In *Theory of Cryptography Conference*, volume 3876 of *LNCS*, pages 207–224. Springer, 2006.
13. S. Dziembowski. On Forward-Secure Storage. Cryptology ePrint Archive, 2006. http://eprint.iacr.org.
14. S. Dziembowski and U. Maurer. On generating the initial key in the Bounded-Storage Model. In *EUROCRYPT '04*, volume 3027 of *LNCS*, pages 126–137. Springer, 2004.
15. S. Dziembowski and U. Maurer. Optimal randomizer efficiency in the Bounded-Storage Model. *Journal of Cryptology*, 17(1):5–26, January 2004.
16. D. Harnik and M. Naor. On the compressibility of NP instances and cryptographic applications. Electronic Colloquium on Computational Complexity, Report TR06-022, 2006.
17. D. Harnik and M. Naor. On everlasting security in the Hybrid Bounded Storage Model, July 2006. accepted to the 33rd International Colloquium on Automata, Languages and Programming.
18. J. Håstad, R. Impagliazzo, L. A. Levin, and M. Luby. A pseudorandom generator from any one-way function. *SIAM J. Comput.*, 28(4):1364–1396, 1999.
19. R. Impagliazzo and S. Rudich. Limits on the provable consequences of one-way permutations. In *ACM Symposium on Theory of Computing*, pages 44–61, 1989.
20. J. Kelsey and B. Schneier. Authenticating secure tokens using slow memory access. In *USENIX Workshop on Smart Card Technology*, pages 101–106, 1999.
21. E. Kushilevitz and R. Ostrovsky. Replication is not needed: Single database, Computationally-Private Information Retrieval. In *38th Annual Symposium on Foundations of Computer Science*, pages 364–373, 1997.
22. E. Kushilevitz and R. Ostrovsky. One-way trapdoor permutations are sufficient for non-trivial single-server Private Information Retrieval. In *EUROCRYPT*, pages 104–121, 2000.
23. C.-J. Lu. Encryption against storage-bounded adversaries from on-line strong extractors. *Journal of Cryptology*, 17(1):27–42, January 2004.
24. U. Maurer. Conditionally-perfect secrecy and a provably-secure randomized cipher. *Journal of Cryptology*, 5(1):53–66, 1992.
25. T. Moran, R. Shaltiel, and A. Ta-Shma. Non-interactive timestamping in the Bounded Storage Model. In *CRYPTO 2004*, volume 3152 of *LNCS*, pages 460–476. Springer, 2004.
26. K. Pietrzak. Composition implies adaptive security in Minicrypt, May 2006. accepted to *EUROCRYPT 2006*.
27. R. L. Rivest. All-or-nothing encryption and the package transform. In *Fast Software Encryption*, volume 1267 of *LNCS*, pages 210–218. Springer, 1997.
28. S. P. Vadhan. Constructing locally computable extractors and cryptosystems in the Bounded-Storage Model. *Journal of Cryptology*, 17(1):43–77, January 2004.

Construction of a Non-malleable Encryption Scheme from Any Semantically Secure One

Rafael Pass[1], abhi shelat[2], and Vinod Vaikuntanathan[3]

[1] Cornell University
[2] IBM ZRL
[3] MIT

Abstract. There are several candidate semantically secure encryption schemes, yet in many applications *non-malleability* of encryptions is crucial. We show how to transform *any* semantically secure encryption scheme into one that is non-malleable for arbitrarily many messages.

Keywords: Public-key Encryption, Semantic Security, Non-malleability, Non-interactive Zero-knowledge Proofs.

1 Introduction

The most basic goal of an encryption scheme is to guarantee the privacy of data. In the case of public-key encryption, the most universally accepted formalization of such privacy is the notion of *semantic security* as defined by Goldwasser and Micali [GM84]. Intuitively, semantic security guarantees that whatever a polynomial-time machine can learn about a message given its encryption, it can learn even without the encryption.

Non-malleability, as defined by Dolev, Dwork and Naor [DDN00], is a stronger notion of security for encryption schemes. In addition to the privacy guarantee, non-malleability of an encryption scheme guarantees that it is infeasible to *modify* ciphertexts $\alpha_1, \ldots, \alpha_n$ into one, or many, other ciphertexts of messages related to the decryption of $\alpha_1, \ldots, \alpha_n$. At first one might wonder whether it is possible to violate non-malleability, i.e., modify ciphertexts into ciphertexts of related messages, without violating semantic security, i.e. without learning anything about the original message.

It turns out, however, that many semantically secure encryption schemes, including the original one proposed in [GM84], are easily malleable. Thus, non-malleability is a strictly stronger requirement than semantic security. Moreover, non-malleability is often times indispensable in practical applications. For example, no one would consider secure an electronic "sealed-bid" auction in which an adversary can consistently bid exactly one more dollar than the previous bidders. The importance of non-malleability raises an important question:

> Is it possible to *immunize* any semantically secure encryption scheme against malleability attacks?

C. Dwork (Ed.): CRYPTO 2006, LNCS 4117, pp. 271–289, 2006.

The work of Dolev, Dwork and Naor partially answers this question affirmatively. They show how to perform such an immunization (for the even stronger chosen-ciphertext attack) *assuming the existence of enhanced trapdoor permutations*. Subsequently, several other constructions of non-malleable encryption schemes have been presented under various number-theoretic assumptions such as decisional Diffie-Hellman [CS98] and quadratic-residuosity [CS02]. Nevertheless, there exist some notable computational assumptions, such as computational Diffie-Hellman, and the worst-case hardness of various lattice-problems [AD97, Reg05], under which semantically secure encryption schemes exist, yet no nonmalleable encryption schemes are known.

Another partial answer comes from a *single-message* non-malleable encryption scheme based on any semantically secure scheme called DDNLite [Nao04, Dwo99]. By single-message, we mean that the DDNLite scheme is secure only when the adversary receives *one* ciphertext.[1] As was stated in [DDN00], many-message security is a *sine qua non* of non-malleable encryption, and indeed the classic motivating examples for non-malleability (such as auctions) require it.

In this paper, our main result is to fully address the posed question. We show how to immunize *any* semantically secure encryption scheme into one that is non-malleable for any number of messages *without any further computational assumptions*.

Main Theorem 1 (Informal). *Assume the existence of an encryption scheme that is semantically secure against chosen-plaintext attacks (CPA). Then there exists an encryption scheme that is many-message non-malleable against CPA.*

As an additional contribution, we show that even in the case of single-message security, the previous definitions of non-malleability and also the DDN-Lite scheme allow a subtle class of malleability attacks. We address this issue by presenting a stronger definition of non-malleability and emphasize that our construction meets this stronger notion.

1.1 Definitional Contributions

Our main conceptual contribution is to *strengthen* the definition of non-malleable encryption. Our definition has the advantage of being both technically simple and providing a *natural* explanation of non-malleability. In one sentence, our definition technically captures the following requirement:

> "No matter what encryptions the adversary receives, the *decryption* of his output will be indistinguishable."

Recall the indistinguishability-based definition of secrecy: an adversary cannot *distinguish* between the ciphertexts of any two messages of his choosing. Another way of phrasing this definition is to say that no matter which encryption an adversary receives, his *output* will be indistinguishable.

---

[1] As pointed out by Genaro and Lindell [GL03], the DDNLite scheme is in fact *malleable* when the adversary receives multiple ciphertexts.

By requiring the even the *decryption* of his output be indistinguishable, we capture the property that also the plaintexts corresponding to the ciphertexts output by the adversary are (computationally) *independent* of the messages he receives encryptions of.

Notice how in the context of auctions this requirement directly implies that an adversary's bid is independent of the previous one. In general, it naturally captures that an adversary cannot *maul* a given encryption into another one that is related—otherwise, the decryption of his output could be distinguished. Except for some simplifications, our definition is very similar to the formal description of IND-PA0 from [BS99]. However, there is some ambiguity regarding whether or not the IND-PA0 definition allows the adversary to succeed by outputting an invalid ciphertext.[2] As we discuss in the next section, this issue of invalid ciphertext can become a serious one.

The Problem of Invalid Ciphertexts. Our definition highlights a subtle technical weakness in previous definitions of non-malleability relating to how one treats an adversary who produces *invalid* ciphertexts as part of its output.

In the original definition from [DDN00], and in subsequent equivalent definitions [BDPR98, BS99], an adversary who produces any invalid ciphertexts as part of its output is considered to have lost the non-malleability game.

THE DEVIL IS IN THE DETAILS. While seemingly harmless, this small detail unleashes a subtle but devastating class of attacks against encryption schemes which only meet the [DDN00] definition.

For example, consider a "quorum voting application" in which participants encrypt their vote, either YES or NO, under the pollmaster's public key. The pollmaster decrypts all votes, counts those which decrypt to either YES or NO as *valid* votes, ignores all others (e.g., discards the "hanging chads"), and announces whether (a) there was as a *quorum* of valid votes, and if so (b) whether the "yeahs" have a majority of the valid votes. (This voting process roughly resembles how voting works in shareholder meetings of a corporation or condo association meetings.)

Now suppose a "swing voter" Alice submits an encryption of her vote x, and Bob is able to produce a ciphertext for the vote NO if x =NO and $\perp$ otherwise (in Appendix A, we show exactly such an attack against the DDNLite cryptosystem). In a tight election, such an attack enhances Bob's ability to cause the measure to fail since his vote is either NO when the swing vote is NO, or his vote undermines the legitimacy of the election by preventing the establishment of the necessary quorum. This attack is better than simply voting NO, because a NO vote by itself might be the one which establishes quorum and allows the measure to pass.

---

[2] Specifically, the formal IND-PA0 definition states one thing but the motivating discussion states another. It also appears that the main result in [BS99] is sensitive to this issue. We further discuss these issues in a forthcoming note.

Under the original definition, one can verify that Bob does not formally win the non-malleability game, and thus an encryption scheme which allows such an attack would still be considered non-malleable under CPA.[3]

MANY MESSAGES. More interestingly, we also show that this small detail is the *key factor* which allows one to prove composability of the non-malleable encryption scheme, i.e., that a single-message secure scheme will also be non-malleable when the adversary receives many messages. Indeed, Gennaro and Lindell [GL03, p.17] point out that one-message non-malleability *does not* imply many-message non-malleability under the [DDN00] definition. As mentioned above, this feature is very important in practice, since an adversary may receive *several* encryptions and we still want to guarantee non-malleability.

To understand the problem, many-message non-malleability is usually proven through a reduction of the following form: an adversary that can break the many-message non-malleability can be used to break the single-message non-malleability. Such a proof requires a *hybrid argument* in which one feeds the adversary one of two specially-constructed distribution of ciphertexts, on which it is possible to distinguish the adversary's output. However, since the input to the adversary is *hybrid*, the adversary may produce invalid ciphertexts—and such may be the basis upon which the adversary's output is distinguishable. If, in the single-message definition, the experiment unconditionally outputs 0 if the adversary produces *any* invalid ciphertexts, then we do not even have the *chance* to use the distinguisher which breaks the many-message case to break the single-message one.

MANY KEYS. There is one final issue which arises in practice. The adversary may receive encryptions under different public keys. Here too, we define a notion of non-malleability. Although this scenario is also usually handled by a standard hybrid argument, there are a few definitional issues to consider.

1.2   Overview of Our Construction

The *immunization* paradigm introduced by Dolev, Dwork and Naor is elegant and conceptually simple. To encrypt a message, one (a) generates several encryptions of the same message under independent public keys, (b) gives a non-interactive zero-knowledge proof that *all* resulting ciphertexts are encryptions of the same message, and (c) signs the entire bundle with a one-time signature.

Because one-time signatures can be constructed from one-way functions, only step (b) in the DDN construction requires an extra computational assumption—namely the existence of trapdoor permutations. Indeed, non-interactive zero-knowledge proofs have been well-studied, and it remains a pressing open

---

[3] A similar attack applies to the classic auction example presented in [DDN00]. Here, if Bob knows Alice's bid is in the range $[x, y]$ and Bob can submit $y - x$ bids, the same attack allows Bob to generate a set of bids such that exactly one bid matches Alice's bid and all the rest decrypt to invalid ciphertexts.

question whether they can be constructed without using trapdoor permutations or assumptions of comparable magnitude.[4]

In the context of non-malleable encryption, however, our main observation is that standard non-interactive zero-knowledge proofs are not necessary. Instead, we use *designated verifier* proofs in which only a verifier with some special secret information related to the common random string can verify the proof. (In contrast, standard non-interactive proofs can be verified by any third party with access to the proof and the common random string.) For example, the non-malleable encryption constructions of Cramer and Shoup [CS98, CS02] can be interpreted as using some form of designated verifier proofs based on specific number-theoretic assumptions. This was further generalized by Elkind and Sahai [ES02] through the notion of *simulation-sound* designated verifier NIZK proofs. Our main technical result is to show that

 † *Plain* designated verifier NIZK proofs are sufficient to obtain non-malleable encryption schemes, and

 †† *Plain* designated-verifier NIZK proofs can be constructed from *any* semantically secure encryption scheme.

The first of these results follows by an argument essentially similar to the one employed in [DDN00].

At a high-level, our approach to constructing a *plain* designated verifier NIZK proof is to crush a Σ-protocol for $\mathcal{NP}$ (such as Blum's Hamiltonicity protocol [Blu86]) into a non-interactive proof with only the help of a semantically-secure encryption scheme.

Towards CCA2 Security. The main results in this paper relate to non-malleability under a chosen plaintext attack. In the stronger chosen ciphertext attack (of which there are two varieties, CCA1 and CCA2), the adversary also has access to a decryption oracle at various stages of the attack. A natural question is whether our same techniques can be used to immunize against a malleability attack in these stronger CCA attack models.

Unfortunately, when designated-verifier NIZK proofs are sequentially composed, it is possible for the verifier's secret information to *leak*. In particular, a malicious prover who can interactively submit proofs and learn whether they are accepted or not might be able to *learn* about the verifier's secret information. Because a CCA1 or CCA2 attack provides precisely such an opportunity, the soundness of the designated verifier proof system will no longer hold. Thus, our technique does not provide a method to achieve full CCA security.

Organization. Our paper is organized into three sections. In §2, we present our new definition of non-malleable encryption. In §3, we define and construct designated verifier NIZK proof systems from any semantically-secure encryption scheme. Finally, in §4, we show that the DDN scheme when implemented with

---

[4] The work of [PS05] begins to address this question by presenting certain models of NIZK proofs that are either unconditional or only require one-way functions.

designated verifier NIZK proofs is non-malleable with respect to our stronger definition from §2.

2   Definitions

Preliminaries. If A is a probabilistic polynomial time (p.p.t) algorithm that runs on input x, $A(x)$ denotes the random variable corresponding to the output of A on input x and uniformly random coins. Sometimes, we want to make the randomness used by A explicit, in which case, we let $A(x; r)$ denote the output of A on input x and random coins r. We denote computational indistinguishability [GM84] of ensembles A and B by $A \overset{c}{\approx} B$.

2.1   Semantically Secure Encryption

Definition 1 (Encryption Scheme). *A triple* $(\mathsf{Gen}, \mathsf{Enc}, \mathsf{Dec})$ *is an encryption scheme, if* Gen *and* Enc *are p.p.t. algorithms and* Dec *is a deterministic polynomial-time algorithm which satisfies the following property:*

Perfect Correctness. *There exists a polynomial* $p(k)$ *and a negligible function* $\mu(k)$ *such that for every message* m*, and every random tape* r_e*,*

$$\Pr[r_g \overset{\cdot}{\leftarrow} \{0,1\}^{p(k)} \; ; \; (\text{PK}, \text{SK}) \leftarrow \mathsf{Gen}(1^k; r_g); \; \mathsf{Dec}_{\text{SK}}(\mathsf{Enc}_{\text{PK}}(m; r_e)) \neq m] \leq \mu(k).$$

Definition 2 (Indistinguishability of Encryptions). *Let* $\Pi = (\mathsf{Gen}, \mathsf{Enc}, \mathsf{Dec})$ *be an encryption scheme and let the random variable* $\mathsf{IND}_b(\Pi, A, k)$*, where* $b \in \{0, 1\}$*,* A *is a p.p.t. algorithm and* $k \in \mathbb{N}$*, denote the result of the following probabilistic experiment:*

$\mathsf{IND}_b(\Pi, A, k)$:
$\quad (\text{PK}, \text{SK}) \leftarrow \mathsf{Gen}(1^k)$
$\quad (m_0, m_1, \text{STATE}_A) \leftarrow A(\text{PK}) \; s.t. \; |m_0| = |m_1|$
$\quad y \leftarrow \mathsf{Enc}_{\text{PK}}(m_b)$
$\quad D \leftarrow A_2(y, \text{STATE}_A)$
$\quad Output \; D$

$(\mathsf{Gen}, \mathsf{Enc}, \mathsf{Dec})$ *is indistinguishable under a chosen-plaintext attack if* $\forall$ *p.p.t. algorithms* A *the following two ensembles are computationally indistinguishable:*

$$\left\{ \mathsf{IND}_0(\Pi, A, k) \right\}_{k \in \mathbb{N}} \overset{c}{\approx} \left\{ \mathsf{IND}_1(\Pi, A, k) \right\}_{k \in \mathbb{N}}$$

2.2   Our Definition of Non-malleable Encryption

Definition 3 (Non-malleable Encryption). *Let* $\Pi = (\mathsf{Gen}, \mathsf{Enc}, \mathsf{Dec})$ *be an encryption scheme and let the random variable* $\mathsf{NME}_b(\Pi, A, k, \ell)$ *where* $b \in$

$\{0,1\}$, $A = (A_1, A_2)$ and $k, \ell \in \mathbb{N}$ denote the result of the following probabilistic experiment:

$\mathsf{NME}_b(\Pi, A, k, \ell)$:

$\quad (\mathrm{PK}, \mathrm{SK}) \leftarrow \mathsf{Gen}(1^k)$

$\quad (m_0, m_1, \mathrm{STATE}_A) \leftarrow A_1(\mathrm{PK})$ s.t. $|m_0| = |m_1|$

$\quad y \leftarrow \mathsf{Enc}_{\mathrm{PK}}(m_b)$

$\quad (c_1, \ldots, c_\ell) \leftarrow A_2(y, \mathrm{STATE}_A)$

$\quad Output\ (d_1, \ldots, d_\ell)$ where $d_i = \begin{cases} \bot & \text{if } c_i = y \\ \mathsf{Dec}_{\mathrm{SK}}(c_i) & \text{otherwise} \end{cases}$

$(\mathsf{Gen}, \mathsf{Enc}, \mathsf{Dec})$ is non-malleable under a chosen-plaintext attack if $\forall$ p.p.t. algorithms $A = (A_1, A_2)$ and for any polynomial $p(k)$, the following two ensembles are computationally indistinguishable:

$$\left\{\mathsf{NME}_0(\Pi, A, k, p(k))\right\}_{k \in \mathbb{N}} \overset{c}{\approx} \left\{\mathsf{NME}_1(\Pi, A, k, p(k))\right\}_{k \in \mathbb{N}}$$

Let us remark on the natural similarity of the above definition with the definition of indistinguishable security of an encryption scheme. Indeed, the first few lines of the experiment are exactly the same. In the last step, we add the requirement that the *decryptions* (modulo copying) of the output of A_2 are indistinguishable in the two experiments. This captures the requirement that even the decryption of the adversary's output must be *computationally independent* of the values (even when if they are encrypted) received as inputs.

This definition both highlights the essence of non-malleability for encryption schemes and, similar to the original notion of indistinguishable security, provides the most technically convenient formalization to use in larger proofs.

Syntactic Differences with the IND-PA0 Definition. As we discussed in the Introduction, there is ambiguity as to how the IND-PA0 [BS99] definition treats invalid ciphertexts. Aside from this semantic difference, one can see that our definition is a simplification of IND-PA0. We simply eliminate the guess stage of the experiment.

2.3   Many Message Non-malleability

Notice that Definition 3 only applies when the adversary A_2 receives one encryption as an input. In practice, an adversary may receive *several* encryptions under many different public keys, and we would still like to guarantee non-malleability.

Definition 4 (Many Message Non-malleability). *Let $\Pi = (\mathsf{Gen}, \mathsf{Enc}, \mathsf{Dec})$ be an encryption scheme and let the random variable $\mathsf{mNM}_b(\Pi, A, k, \ell)$ where $b \in \{0, 1\}$, $A = (A_1, A_2)$ and $k, \ell, J \in \mathbb{N}$ denote the result of the following probabilistic experiment:*

$\mathsf{mNM}_b(\Pi, A, k, \ell, J)$:

   For $i = 0, \ldots, J$, $(\text{PK}_i, \text{SK}_i) \leftarrow \mathsf{Gen}(1^k)$

   $((m_1^0, m_1^1, t_1), \ldots, (m_\ell^0, m_\ell^1, t_\ell), \text{STATE}) \leftarrow A_1(\text{PK}_1, \ldots, \text{PK}_J)$ *s.t.* $|m_i^0| = |m_i^1|$

   For $j = 0, \ldots, \ell$ $y_j \leftarrow \mathsf{Enc}_{\text{PK}_\cdot}(m_j^b)$

   $((c_1, s_1), \ldots, (c_\ell, s_\ell)) \leftarrow A_2(y_1, \ldots, y_\ell, \text{STATE})$

   Output $(d_1, \ldots, d_\ell)$ *where* $d_i = \begin{cases} \perp & \text{if } s_i \notin [1, J] \\ \perp & \text{if } c_i = y_j \text{ and } s_i = t_j, \ j \in [1, J] \\ \mathsf{Dec}_{\text{SK}_\cdot}(c_i) & \text{otherwise} \end{cases}$

We say that $(\mathsf{Gen}, \mathsf{Enc}, \mathsf{Dec})$ *is non-malleable under a chosen-plaintext attack if for all p.p.t. algorithms* $A = (A_1, A_2)$ *and for all polynomials* $\ell(\cdot), J(\cdot)$, *the following two ensembles are computationally indistinguishable:*

$$\left\{ \mathsf{mNM}_0(\Pi, A, k, \ell(k), J(k)) \right\}_{k \in \mathbb{N}} \overset{c}{\approx} \left\{ \mathsf{mNM}_1(\Pi, A, k, \ell(k), J(k)) \right\}_{k \in \mathbb{N}}$$

Theorem 1. *An encryption scheme is non-malleable iff it is many message non-malleable.*

Sketch: To prove the forward implication, we use an adversary (A_1, A_2) and a distinguisher D which breaks the many-message non-malleability of Π with advantage η to to break the non-malleability of Π with advantage η/ℓ^2.

Define a new experiment $\mathsf{mNM}_{(b_1, \ldots, b_\cdot)}(\Pi, A, k, \ell, J)$ indexed by an ℓ-bit string $(b_1, \ldots, b_\ell)$ which is the same as $\mathsf{mNM}_0(\Pi, A, k, \ell, J)$ except in the fifth line (change is underlined): $y_j \leftarrow \mathsf{Enc}_{\text{PK}_\cdot}(m_j^{\underline{b_\cdot}})$. Define $B(i) = (\overbrace{0, \ldots, 0}^{i}, \overbrace{1, \ldots, 1}^{\ell - i})$ and note that $\mathsf{mNM}_0 = \mathsf{mNM}_{B(0)}$ and $\overline{\mathsf{mNM}_1} = \mathsf{mNM}_{B(\ell)}$. Because D distinguishes mNM_0 from mNM_1, there exists some $g^* \in [1, \ell]$ such that D distinguishes $\mathsf{mNM}_{B(g^*)}$ from $\mathsf{mNM}_{B(g^*+1)}$ with advantage η/ℓ. This suggests the following adversary: $A_1'(\text{PK})$ guesses value $g \in [1, \ell]$, generates $J - 1$ public keys $(\text{PK}_i, \text{SK}_i)$, and feeds $(\text{PK}_1, \ldots, \text{PK}_{g-1}, \text{PK}, \text{PK}_{g+1}, \ldots, \text{PK}_J)$ to A_1 to get a vector of triples $((m_1^0, m_1^1, t_1), \ldots, (m_\ell^0, m_\ell^1, t_\ell))$. Finally, A' outputs (m_g^0, m_g^1) as its challenge pair and outputs state information with g, all public keys, etc.

Adversary $A_2'(y, \text{STATE}')$, on input an encryption y, simulates the replaced line 5 of experiment $\mathsf{mNM}_{B(g)}$ with the exception that it uses y for the gth encryption: $y_g \leftarrow y$. It then feeds vector $\boldsymbol{y}$ to A_2 to produce $((c_1, s_1), \ldots, (c_\ell, s_\ell))$. A_2' must produce ciphertexts only for PK, and can do so by simply decrypting and re-encrypting[5] appropriately. The rest of the argument concludes by conditioning on the event that $g = g^*$. □

One can see here the importance of removing the invalid ciphertext restriction for the hybrid argument to work. Because the reduction feeds a hybrid distribution to A_2, A_2 may produce invalid ciphertexts, and these $\perp$ values may form the basis for distinguishability in the NME experiments. If one simply forces the

---

[5] There is one technicality concerning how A_2' can form invalid ciphertexts; we defer it to the full version.

single-message experiment to return 0 when invalid ciphertexts are produced, then the value of both experiments ($b = 0, 1$) will be 0 and the weaker definition will thus be met *even though* there might still be a distinguisher which could have distinguished the output of A_2'.

3   Designated Verifier NIZK

In this section, we define and construct a designated verifier NIZK proof system. The overall approach to our construction is to *crush* a 3-round Σ protocol into a one-round proof by having the prover *encrypt* all possible third-round responses to the verifier's challenge. Because we use a Σ-protocol in which the verifier's challenge is a single bit, this approach is feasible and results in short proofs. The notable benefit of this approach is that our only complexity assumption is the existence of a semantically-secure encryption scheme.

3.1   Defining Designated Verifier NIZK Proof Systems

In the designated verifier model, a non-interactive proof system has an associated polynomial-time sampleable distribution $\mathcal{D}$ over binary strings of the form (PP, SP). During a setup phase, a trusted party samples from $\mathcal{D}$, publishes PP and privately hands the Verifier SP. The Prover and Verifier then use their respective values during the proof phase.

This definition is very similar to the definition of NIZK proofs in the secret parameter model (as in [PS05]). The difference between the secret parameter model and the designated verifier model is that in the former case, the prover might be given some *secret information*, whereas this is strictly not the case in the latter. Also note that in the standard notion of NIZK proofs, the common random (resp. reference) string model can be derived as special cases of this definition by setting SP to the empty string.

Definition 5 (Designated Verifier NIZK Proof System). *A triple of algorithms, $(\mathcal{D}, P, V)$, is called a **designated verifier non-interactive zero-knowledge proof system** for an $\mathcal{NP}$-language L with witness relation R_L, if the algorithms $\mathcal{D}$ and P are probabilistic polynomial-time, the algorithm V is deterministic polynomial-time and there exists a negligible function μ such that the following three conditions hold:*

- COMPLETENESS: *For every* $(x, w) \in R_L$

$$\Pr\left[(\text{PP}, \text{SP}) \leftarrow \mathcal{D}(1^{|x|}); \; \pi \leftarrow P(\text{PP}, x, w) \; : \; V(\text{PP}, \text{SP}, x, \pi) = 1 \right] \geq 1 - \mu(|x|)$$

- SOUNDNESS: *For every prover algorithm B*

$$\Pr\left[(\text{PP}, \text{SP}) \leftarrow \mathcal{D}(1^{|x|}) \, ; \; (x', \pi') \leftarrow B(\text{PP}) \; : \; \begin{array}{l} x' \notin L \quad and \\ V(\text{PP}, \text{SP}, x', \pi') = 1 \end{array} \right] \leq \mu(|x|)$$

- ADAPTIVE ZERO-KNOWLEDGE: *For every p.p.t. theorem chooser A, there exists a p.p.t. simulator $S = (S_1, S_2)$ such that the outputs of the following experiments are indistinguishable.*

$\mathsf{ZK}_A(k)$

   $(\text{PP}, \text{SP}) \leftarrow \mathcal{D}(1^k)$

   $(x, w, \text{STATE}_A) \leftarrow A(\text{PP})$

   $\pi \leftarrow P(\text{PP}, x, w)$

   If $(x, w) \notin R_L$, output $\bot$

   Else output $(\text{PP}, \text{SP}, x, \pi, \text{STATE}_A)$

$\mathsf{ZK}_A^S(k)$

   $(\text{PP}', \text{SP}', \text{STATE}) \leftarrow S_1(1^k)$

   $(x, w, \text{STATE}_A) \leftarrow A(\text{PP}')$

   $\pi' \leftarrow S_2(\text{PP}', \text{SP}', x, \text{STATE})$

   If $(x, w) \notin R_L$, output $\bot$

   Else output $(\text{PP}', \text{SP}', x, \pi', \text{STATE}_A)$

Note that the Verifier V is a deterministic machine. This extra restriction is only used to simplify the exposition of our constructions.

3.2 Constructions

Before giving a high-level view of our protocol, let us briefly recall the structure of a 3-round honest-verifier zero-knowledge proof of knowledge for NP, also referred to as a Σ-protocol [CDS94].

Σ-Protocols. A Σ-protocol consists of four algorithms P_1, V_1, P_2, V_2. On input, a statement x and a witness w, the first algorithm $P_1(x, w)$ produces a pair (a, s) which represent a "commitment" and related state information. The Prover then sends a to the Verifier. The verifier runs $V_1(x, a)$ to produce a random challenge b and sends b to the Prover. The prover runs $P_2(s, b)$ to produce a response c_b and sends it to the Verifier. Finally, the verifier runs $V_2(x, a, b, c_b)$ to determine whether to accept the proof. An example of such a protocol is Blum's Hamiltonicity protocol. The properties of Σ-protocols, namely completeness, special soundness, and special honest-verifier zero-knowledge, are presented in [CDS94].

Construction Summary. In our construction, the prover receives k pairs of public encryption keys as the public parameter. The verifier receives the same k pairs, and also receives a secret parameter consisting of the secret key for *exactly* one key in each pair. A proof consists of k triples. To generate the ith triple of a proof, the prover runs the Σ-protocol prover algorithm on (x, w) using both 0 and 1 as the verifier's challenge to produce a triple $(a_i, c_{0,i}, c_{1,i})$. The prover encrypts this triple as $(a_i, \mathsf{Enc}_{\text{PK}_0}..(c_{0,i}), \mathsf{Enc}_{\text{PK}_1}..(c_{1,i}))$. (We note that Camenisch and Damgård use a similar idea in [CD00] to construct an interactive verifiable encryption scheme. The roots of this idea begin to appear much earlier in [KMO89].)

   To verify the proof, V considers each triple $(a_i, \alpha_{0,i}, \alpha_{1,i})$ and decrypts either the second or third component using the secret key given in SP. He then runs the Σ-protocol verifier on $(a_i, f_i, \mathsf{Dec}(\alpha_{f.,i}))$, and accepts if all k triples are accepting proofs.

   There is one additional detail to consider. The Σ-protocol for Blum Hamiltonicity requires a commitment scheme. Normally, a one-round commitment scheme requires trapdoor permutations. To get around this problem, we use a two-round commitment scheme (such as Naor [Nao91]) which can be constructed from one-way functions, which are in turn implied by the semantically-secure encryption scheme. The first round message of the commitment scheme is placed in the public parameter.

Theorem 2. *Assume there exists a semantically secure encryption scheme. Then there exists a designated verifier NIZK proof system for any language* $L \in \mathcal{NP}$.

Sampling Algorithm $\mathcal{D}(1^k)$. For $i = 1, \ldots, k$ and $b = 0, 1$, run $\mathsf{Gen}(1^k)$ $2k$ times with independent random coins, to get k key-pairs $(\mathrm{PK}_i^b, \mathrm{SK}_i^b)$. For $i = 1, \ldots, k$, flip coin $f_i \xleftarrow{\cdot} \{0, 1\}$. Generate the receiver message σ for a two-round commitment scheme.
Let $\mathrm{PP}_{dv} \overset{\text{def}}{=} [(\mathrm{PK}_i^0, \mathrm{PK}_i^1, \sigma)]_{i=1}^k$ and $\mathrm{SP}_{dv} \overset{\text{def}}{=} [f_i, \mathrm{SK}_i^{f_{\cdot}}]_{i=1}^k$. Output $(\mathrm{PP}_{dv}, \mathrm{SP}_{dv})$.
Prover $P(\mathrm{PP}_{dv}, x, w)$. For $i = 0, \ldots, k$, generate triples as follows:

$$(a_i, s_i) \leftarrow P_1(x, w)$$
$$c_{b,i} \leftarrow P_2(s, b) \text{ for both } b = 0, 1$$
$$\alpha_{b,i} \leftarrow \mathsf{Enc}_{\mathrm{PK}_{\ldots}}(c_{b,i}) \text{ for } b = 0, 1.$$

and output $\pi \overset{\text{def}}{=} [(a_i, \alpha_{0,i}, \alpha_{1,i})]_{i=1}^k$.
Verifier $V(\mathrm{PP}_{dv}, \mathrm{SP}_{dv}, x, \pi)$. Parse π into k triples of the form $(a_i, \alpha_{0,i}, \alpha_{1,i})$. For $i = 1, \ldots, k$, compute $m_i \overset{\text{def}}{=} \mathsf{Dec}_{\mathrm{SK}_{\cdot}}(\alpha_{f_{\cdot},i})$ and run the verifier $V_2(a_i, f_i, m_i)$. If all k proofs are accepted, output ACCEPT, else output REJECT.

Fig. 1. Designated Verifier NIZK Protocol

The proof that the scheme in Fig. 1 is a designated verifier NIZK proof system is standard and omitted for lack of space. □

4 Constructing a Non-malleable Encryption Scheme

In this section, we construct an encryption scheme that is non-malleable under CPA attacks. Our construction is exactly the DDN construction [DDN00] in which the standard NIZK proof is replaced with a designated-verifier NIZK proof. By the results from the previous section, our construction only relies on the assumption of the existence of a semantically secure encryption scheme.

Theorem 3 (Main Theorem, restated). *Assume there exists an encryption scheme that is semantically-secure under a CPA attack. Then, there exists an encryption scheme that is non-malleable for many messages under a CPA attack.*

Proof. (of Theorem 3) Let (Gen, Enc, Dec) be any semantically secure encryption scheme. Let (GenSig, Sign, Ver) be any existentially unforgeable *strong* one-time signature scheme.[6] Without loss of generality, assume that GenSig produces verification keys of length k.[7] Define the $\mathcal{NP}$-language L as follows:

$$\big[(c_1, \ldots, c_k), (p_1, \ldots, p_k)\big] \in L \text{ if and only if}$$
$$\exists \big[m, (r_1, \ldots, r_n)\big] \text{ such that } c_i = \mathsf{Enc}_{p.}(m; r_i) \text{ for } i = 1, \ldots, n.$$

In words, the language L contains pairs consisting of a k-tuple of ciphertexts and a k-tuple of public keys such that the ciphertexts are encryptions of the *same message m* under the k public keys.

Let $(\mathcal{D}, P, V)$ be a designated verifier NIZK proof system for L. We show that the encryption scheme $\Pi = (\mathsf{NMGen}, \mathsf{NMEnc}, \mathsf{NMDec})$ defined in Figure 4 is a non-malleable encryption scheme. The proof has two parts.

Just as in [DDN00], we define an encryption scheme $E' = (\mathsf{Gen}', \mathsf{Enc}', \mathsf{Dec}')$ in which one simply encrypts a message k times with k independently chosen public keys, and we show that E' is a semantically secure encryption scheme under the assumption that (Gen, Enc, Dec) is one. This is done in Lemma 1.

Then in Lemma 2, we show that Π is a non-malleable encryption scheme if E' is a semantically secure encryption scheme. The proof is concluded by noting that both designated verifier NIZK proofs and strong one-time signatures can be constructed given any semantically secure encryption scheme (The former is true by virtue of Theorem 2. The latter follows by combining the observation that encryption implies one-way functions, Rompel's result showing that one-way functions imply universal one-way hash functions [Rom90], and the result that universal one-way hash functions imply strong one-time signature schemes [Gol04, Lam79]).   □

The definition of the encryption scheme $E' = (\mathsf{Gen}', \mathsf{Enc}', \mathsf{Dec}')$ below and the proof of Lemma 1 are exactly as in DDN, reproduced below for the sake of completeness.

- $\mathsf{Gen}'(1^k)$: For $i = 1, \ldots, k$, run $(\mathrm{PK}_i, \mathrm{SK}_i) \leftarrow \mathsf{Gen}(1^k)$ with independent random coins. Set $\mathrm{PK} \overset{\mathrm{def}}{=} (\mathrm{PK}_1, \ldots, \mathrm{PK}_k)$ and $\mathrm{SK} \overset{\mathrm{def}}{=} (\mathrm{SK}_1, \ldots, \mathrm{SK}_k)$.
- $\mathsf{Enc}'_{\mathrm{PK}}(m)$: Output $[\mathsf{Enc}_{\mathrm{PK}_1}(m; r_1), \ldots, \mathsf{Enc}_{\mathrm{PK}.}(m; r_k)]$.
- $\mathsf{Dec}'_{\mathrm{SK}}([c_1, c_2, \ldots, c_k])$: Compute $m'_i = \mathsf{Dec}_{\mathrm{SK}.}(c_i)$. If all the m'_i are not equal, output $\perp$, else output m'_1.

Lemma 1. *If* (Gen, Enc, Dec) *is semantically secure, then* $(\mathsf{Gen}', \mathsf{Enc}', \mathsf{Dec}')$ *is semantically secure.*

Proof. Via a standard hybrid argument. Omitted for space.   □

---

[6] A strong signature is one in which, given a signature σ of a message m, it is infeasible to produce a message m' and a valid signature σ' of m', such that $(\sigma, m) \neq (\sigma', m')$. i.e, it is infeasible also to produce a different signature for the *same message*.

[7] This is without loss of generality since we can set k to be an upperbound on the length of verification keys that GenSig produces.

$\mathsf{NMGen}(1^k)$:
    1. For $i \in [1,k], b \in \{0,1\}$, run $\mathsf{Gen}(1^k)$ to generate key-pairs $(\mathrm{PK}_i^b, \mathrm{SK}_i^b)$.
    2. Run $\mathcal{D}(1^k)$ to generate $(\mathrm{PP}, \mathrm{SP})$.
    Set $\mathrm{PK} \stackrel{\text{def}}{=} \left\{ (\langle \mathrm{PK}_i^0, \mathrm{PK}_i^1 \rangle)_{i=1}^k, \mathrm{PP} \right\}$ and $\mathrm{SK} \stackrel{\text{def}}{=} \left\{ (\langle \mathrm{SK}_i^0, \mathrm{SK}_i^1 \rangle)_{i=1}^k, \mathrm{SP} \right\}$.
$\mathsf{NMEnc}_{\mathrm{PK}}(m)$:
    1. Run the signature algorithm $\mathsf{GenSig}(1^k)$ to generate $(\mathrm{SKSIG}, \mathrm{VKSIG})$.
       Let $(v_1, \ldots, v_k)$ be the binary representation of VKSIG.
    2. Compute the ciphertexts $c_i \leftarrow \mathsf{Enc}_{\mathrm{PK}_\bullet^{\cdot}}(m)$. Let $\boldsymbol{c} \stackrel{\text{def}}{=} (c_1, c_2, \ldots, c_k)$.
    3. Run the designated verifier NIZK Prover to generate a proof π that
       $[(c_1, \ldots, c_k), (\mathrm{PK}_1^{v_1}, \ldots, \mathrm{PK}_k^{v_\bullet})] \in L$.
    4. Compute the signature $\sigma \leftarrow \mathsf{Sign}_{\mathrm{SKSIG}}(\langle \boldsymbol{c}, \pi \rangle)$.
    Output the tuple $[\boldsymbol{c}, \pi, \mathrm{VKSIG}, \sigma]$.
$\mathsf{NMDec}_{\mathrm{SK}}(c)$:
    1. Verify the signature with $\mathsf{Ver}_{\mathrm{VKSIG}}[\langle \boldsymbol{c}, \pi \rangle, \sigma]$; output $\bot$ upon failure.
    2. Verify the proof with $V(\mathrm{PP}, \mathrm{SP}, (\boldsymbol{c}, \mathrm{PK}), \pi)$; output $\bot$ upon failure.
    3. Let $\mathrm{VKSIG} = (v_1, \ldots, v_k)$. Compute $m_1 = \mathsf{Dec}_{\mathrm{SK}_1^{\cdot}}(c_1)$ and output the
       result.

Fig. 2. The Non-malleable Encryption Scheme Π

Lemma 2. *If $E' = (\mathsf{Gen}', \mathsf{Enc}', \mathsf{Dec}')$ is a semantically secure encryption scheme, then Π is a non-malleable encryption scheme.*

Proof. To prove that Π is a non-malleable encryption scheme, we need to show that for any p.p.t. adversary A and for all polynomials $p(k)$,

$$\left\{ \mathsf{NME}_0(\Pi, A, k, p(k)) \right\}_{k \in \mathbb{N}} \stackrel{c}{\approx} \left\{ \mathsf{NME}_1(\Pi, A, k, p(k)) \right\}_{k \in \mathbb{N}}$$

We show this by a hybrid argument. Consider the following experiments:

Experiment. $\mathsf{NME}_b^{(1)}(\Pi, A, k, p(k))$ – *Using a Simulated NIZK Proof:* proceeds exactly like NME_b except that the simulator for the designated verifier NIZK proof system is used to generate the public parameters and to compute the challenge ciphertext (as opposed to generating an honest proof by running the prover algorithm P). Let $S = (S_1, S_2)$ denote the simulator guaranteed by the adaptive zero-knowledge of $(\mathcal{D}, P, V)$. More formally, $\mathsf{NME}_b^{(1)}$ proceeds exactly like NME_b except for the following differences:

1. The encryption key $(\mathrm{PK}, \mathrm{SK})$ is generated by (1) honestly running the key-generation algorithm Gen to generate the $2k$ encryption keys $(\mathrm{PK}_i^b, \mathrm{SK}_i^b)$, *but* (2) running the simulator $S_1(1^k)$ to generate the key-pair $(\mathrm{PP}, \mathrm{SP})$ for the designated verifier NIZK (instead of running $\mathcal{D}(1^k)$ as in NMGen).
2. Generate k encryptions of m_b (just as in Steps 1 and 2 of NMEnc). *But,* instead of using the designated verifier prover, generate a "simulated proof"

by running S_2. (Note that S_2 does not use the witness—namely, m_b and the randomness used for encryption—in order to generate the simulated proof).

Experiment. $\mathsf{NME}_b^{(2)}(\Pi, A, k, p(k))$ – *Semantic Security of E':* proceeds exactly like $\mathsf{NME}_b^{(1)}$ except for the following differences:

1. Run Gen' to get two sets of public keys $PK = \{\mathrm{PK}_i\}_{i=1}^k$ and $PK' = \{\mathrm{PK}_i'\}_{i=1}^k$, along with the corresponding secret-keys $SK = \{\mathrm{SK}_i\}_{i=1}^k$ and $SK' = \{\mathrm{SK}_i'\}_{i=1}^k$. Generate a verification key and signing key for the signature scheme $(\mathrm{VKSIG}^*, \mathrm{SKSIG}^*)$. Construct a public-key for Π as follows: Let v_i be the ith bit of VKSIG^*. Set $\mathrm{PK}_i^{v_i} = \mathrm{PK}_i$, $\mathrm{SK}_i^{v_i} = \bot$, $\mathrm{PK}_i^{1-v_i} = \mathrm{PK}_i'$ and $\mathrm{SK}_i^{1-v_i} = \mathrm{SK}_i'$. ($\mathsf{NME}_b^{(2)}$ will use the secret-keys corresponding to each PK_i', but not PK_i, later in the experiment).
2. After receiving the tuple $(\psi_1, \ldots, \psi_\ell)$ of ciphertexts from A_2, decrypt each $\psi_j = [\boldsymbol{c}_j, \pi_j, \mathrm{VKSIG}_j, \sigma_j]$ as follows: If the signature σ_j in ψ_j does not verify, output $\bot$. If $\mathrm{VKSIG}_j = \mathrm{VKSIG}^*$, output $\bot$. If the NIZK proof π_j fails verification, output $\bot$. Else, decrypt one of the components of ψ_j, for which the secret-key is known (such a component is guaranteed to exist, since $\mathrm{VKSIG}_j \neq \mathrm{VKSIG}^*$) and output the result.

We now show that these experiments are indistinguishable. The following claim follows from the adaptive zero-knowledge property of the NIZK system.

Claim 1. $\left\{ \mathsf{NME}_b(\Pi, A, k, p(k)) \right\}_{k \in \mathbb{N}} \stackrel{c}{\approx} \left\{ \mathsf{NME}_b^{(1)}(\Pi, A, k, p(k)) \right\}_{k \in \mathbb{N}}$

Proof. Assume, for contradiction, that there exists a p.p.t. algorithm D which distinguishes $\mathsf{NME}_b(\Pi, A, k, p(k))$ from $\mathsf{NME}_b^{(1)}(\Pi, A, k, p(k))$. Then, we construct a theorem-chooser A_{zk} and a ZK distinguisher D_{zk} that violate the adaptive zero-knowledge of the proof system $(\mathcal{D}, P, V)$ for the language L. That is, D_{zk} distinguishes between the experiments $\mathsf{ZK}_{A_{\mathsf{zk}}}$ and $\mathsf{ZK}_{A_{\mathsf{zk}}}^S$, where S is the zero-knowledge simulator.

On input PP, the theorem-chooser A_{zk} works as follows:

1. Run $\mathsf{Gen}(1^k)$ $2k$ times, to generate $2k$ key-pairs $(\mathrm{PK}_i^b, \mathrm{SK}_i^b)_{i \in [k], b \in \{0,1\}}$. Run the adversary A_1 on input $\left[(\mathrm{PK}_i^b)_{i \in [k], b \in \{0,1\}}, \mathrm{PP} \right]$. A_1 returns a pair of plaintexts m_0 and m_1 and a string STATE.
2. Produce the challenge ciphertext $\boldsymbol{c}$ as follows:
   – Generate a key-pair $(\mathrm{SKSIG}, \mathrm{VKSIG})$ for the signature scheme.
   – Pick a random $b \in \{0,1\}$, and for $1 \leq i \leq k$, let $c_i \leftarrow \mathsf{Enc}_{\mathrm{PK}_i^{v_i}}(m_b; r_i)$, where r_i is the randomness used for encryption.
   Let $\boldsymbol{c}$ denote $(c_1, c_2, \ldots, c_k)$ and PK denote $(\mathrm{PK}_1^{v_1}, \ldots, \mathrm{PK}_k^{v_k})$, and $\boldsymbol{r}$ denote $(r_1, r_2, \ldots, r_k)$.
3. Let $\mathbf{x} = (\boldsymbol{c}, \mathrm{PK})$ and $\mathbf{w} = (m_b, \boldsymbol{r})$. Output the theorem-witness pair $(\mathbf{x}, \mathbf{w})$. Also output the contents of the work-tape as STATE_A.

The ZK distinguisher D_{zk}, on input $(\mathrm{PP}, \mathrm{SP})$, the theorem $(\boldsymbol{c}, \mathrm{PK})$, the proof π and the state STATE_A, does the following:

1. Run A_2 on input the ciphertext $[c, \pi, \text{VKSIG}, \text{Sign}_{\text{SKSIG}}(\langle c, \pi \rangle)]$ to produce a sequence of ciphertexts $(\psi_1, \psi_2, \ldots, \psi_{p(k)})$. Run the decryption algorithm $\text{Dec}_{\text{SK}}(\psi_i)$ on each of these ciphertexts to get plaintexts $(\mu_1, \mu_2, \ldots, \mu_{p(k)})$.
2. Run distinguisher D on the sequence of plaintexts $(\mu_1, \mu_2, \ldots, \mu_{p(k)})$ and output whatever D outputs.

The experiment $\text{ZK}_{A_{zk}}$ (that is, when D_{zk} is given as input the real proof), perfectly simulates the experiment $\text{NME}_b(\Pi, A, k, p(k))$, whereas the experiment $\text{ZK}^S_{A_{zk}}$ (that is, when D_{zk} is run with a simulated proof) perfectly simulates $\text{NME}_b^{(1)}(\Pi, A, k, p(k))$. If the outputs of D in the experiments are different, then D_{zk} distinguishes between a real proof and a simulated proof, contradicting the adaptive zero-knowledge of the NIZK proof system (D, P, V). □

Next, we show that experiments $\text{NME}_0^{(1)}(\cdots)$ and $\text{NME}_0^{(2)}(\cdots)$ are statistically indistinguishable. To this end, we define three events, $\text{badNIZK}(\text{Expt})$, $\text{badSig}(\text{Expt})$ and $\text{badKey}(\text{Expt})$, corresponding to the experiment Expt. We show that the experiments $\text{NME}_b^{(1)}$ and $\text{NME}_b^{(2)}$ are *identical*, under the assumption that the events badNIZK, badSig and badKey *never* happen in these experiments. Then, we show that the bad events happen with negligible probability in both the experiments. Taken together, these two statements let us conclude that $\text{NME}_b^{(1)}$ and $\text{NME}_b^{(2)}$ are statistically indistinguishable. Details follow.

Claim 2. $\left\{ \text{NME}_0^{(1)}(\Pi, A, k, p(k)) \right\}_{k \in \mathbb{N}} \overset{s}{\approx} \left\{ \text{NME}_0^{(2)}(\Pi, A, k, p(k)) \right\}_{k \in \mathbb{N}}$

Proof. Define the event $\text{badNIZK}(\text{Expt})$, to capture the event that the adversary A violates the soundness of the NIZK proof system in experiment Expt (i.e, the adversary produces an accepting proof of a false statement). More precisely, let $(\psi_1, \psi_2, \ldots, \psi_{p(k)})$ denote the tuple of ciphertexts that A_2 returns in the experiment Expt. Let $\text{badNIZK}(\text{Expt})$ denote the following event: In experiment Expt, there exists an index j such that the NIZK proof in ψ_j is accepted by the verifier V, but all the k ciphertexts that are part of ψ_j do not decrypt to the same value (in other words, ψ_j contains an *accepting* proof of a *false* statement).

In the subclaims below, we show that $\text{badNIZK}(\text{NME}_b^{(j)})$ happens only with negligible probability.

SubClaim. For $b \in \{0, 1\}$, $\Pr[\text{badNIZK}(\text{NME}_b)] = \text{negl}(k)$

Proof. Suppose, for contradiction, that this is not true. That is, there is a polynomial $q(k)$ such that $\Pr[\text{badNIZK}(\text{NME}_b)] \geq \frac{1}{q(k)}$. Then, we construct a machine A_s that violates the soundness of the proof system (D, P, V) with probability at least $\frac{1}{p(k)q(k)}$. On input a public parameter PP, A_s works as follows:

1. Simulate the experiment NME_b using PP, until A_2 outputs $p(k)$ ciphertexts. Note that A_s does not need to know the secret parameter SP to perform these steps. This is because by the definition of NME_b, SP is used only after A_2 outputs the ciphertexts.
2. A_s picks one of the ciphertexts at random, say $[c, \pi, \text{VKSIG}, \sigma]$ and outputs the pair (c, π).

The probability that A_s outputs a false statement and an accepting proof pair is, by our assumption, at least $\frac{1}{p(k)q(k)}$, which is a contradiction to the soundness of $(\mathcal{D}, P, V)$. $\qquad\square$

SubClaim. For $b \in \{0, 1\}$, $\Pr[\texttt{badNIZK}(\mathsf{NME}_b^{(1)})] = \Pr[\texttt{badNIZK}(\mathsf{NME}_b^{(2)})] = \mathsf{negl}(k)$.

Proof. We start by noting that $\Pr[\texttt{badNIZK}(\mathsf{NME}_b^{(1)})] = \Pr[\texttt{badNIZK}(\mathsf{NME}_b^{(2)})]$. This follows because the adversary's view in experiments $\mathsf{NME}_b^{(1)}$ and $\mathsf{NME}_b^{(2)}$ are identical until the point when the adversary A_2 outputs the ciphertexts. We proceed to show that for $b \in \{0, 1\}$, $\Pr[\texttt{badNIZK}(\mathsf{NME}_b^{(1)})]$ is negligible in k. This is shown by an argument similar to the one used in the proof of Claim 1. Assume, for contradiction, that $\Pr[\texttt{badNIZK}(\mathsf{NME}_b^{(1)})]$ is non-negligible. Then, we construct a pair of machines $(A_{\mathsf{zk}}, D_{\mathsf{zk}})$ that violate the adaptive zero-knowledge of the proof system $(\mathcal{D}, P, V)$.

On input a public parameter PP for the NIZK proof system, A_{zk} and D_{zk} work exactly as in the proof of Claim 1, except that in Step 3, when A_2 returns a sequence of ciphertexts $(\psi_1, \ldots, \psi_{p(k)})$, D_{zk} looks for a ciphertext ψ_i such that not all the components of ψ_i decrypt to the same message, and the NIZK proof in ψ_i is accepting. If there exists such an i, then D_{zk} returns "Fail" and otherwise returns "OK".

Note that by definition, when D_{zk} receives a real proof, it outputs "Fail" with probability $\Pr[\texttt{badNIZK}(\mathsf{NME}_b)]$. On the other hand, when run on a simulated proof, it outputs "Fail" with probability $\Pr[\texttt{badNIZK}(\mathsf{NME}_b^{(1)})]$. However, in Step 1a, we showed that the former probability is negligible. If the latter probability is non-negligible, then D_{zk} distinguishes between a simulated proof and a real proof, contradicting the adaptive zero-knowledge property of the proof system $(\mathcal{D}, P, V)$. $\qquad\square$

Let $\psi_i = [c_i, \pi_i, \mathrm{VKSIG}_i, \sigma_i]$ denote the ith ciphertext returned by A_2. Define $\texttt{badSig}(\mathsf{NME}_b^{(j)})$ to be the event that, in experiment $\mathsf{NME}_b^{(j)}(\Pi, A, k, p(k))$, there exists an index i such that $\mathrm{VKSIG}_i = \mathrm{VKSIG}$ and $\mathsf{Ver}(\mathrm{VKSIG}_i, c_i, \pi_i) = \mathrm{ACCEPT}$. Since the signature scheme is (strongly) existentially unforgeable, it follows that, for $b \in \{0, 1\}$ and $j \in \{1, 2\}$, $\Pr[\texttt{badSig}(\mathsf{NME}_b^{(j)})] = \mathsf{negl}(k)$.

Let $\texttt{badKey}(\mathsf{NME}_b^{(j)})$ denote the event that for one of the public keys, say $\hat{\mathrm{PK}}$, generated in the experiment $\mathsf{NME}_b^{(j)}$, there exists a pair of messages m, m' and random coins r, r' such that $m \neq m'$ and $\mathsf{Enc}(\hat{pk}, m, r) = \mathsf{Enc}(\hat{pk}, m', r')$. Since the encryption scheme used is perfectly correct, by the union bound, we have $\Pr[\texttt{badKey}(\mathsf{NME}_b^{(j)})] = \mathsf{negl}(k)$.

Let $\texttt{fail}_b(\cdot)$ denote the event $\texttt{badNIZK}(\cdot) \vee \texttt{badSig}(\cdot) \vee \texttt{badKey}(\cdot)$. It follows, by a union bound, that $\Pr[\texttt{fail}_b(\mathsf{NME}_b^{(j)})] = \mathsf{negl}(k)$, for $j \in \{1, 2\}$.

We show that conditioned on the event $\texttt{fail}_b(\mathsf{NME}_b^{(j)})$ (for $j \in \{1, 2\}$) not happening, $\mathsf{NME}_b^{(1)}$ and $\mathsf{NME}_b^{(2)}$ are identical. Note that the view of A in both the experiments is (syntactically) the same. Since $\texttt{badSig}(\mathsf{NME}_b^{(j)})$ does not happen, A uses a different verification key in all the ciphertexts ψ_i it returns. This means that $\mathsf{NME}_b^{(j)}$ can decrypt at least *one* of the components of each ψ_i, using a secret-key it knows, to get a message m_i. Since $\mathtt{badNIZK}(\mathsf{NME}_b^{(j)})$ does not happen, m_i must be the message that is encrypted in all the other components of ψ_i too. Thus, ψ_i is a *valid* encryption of m_i. Also, since $\mathtt{badKey}(\mathsf{NME}_b^{(j)})$ does not happen, m_i is the *unique* such message. Thus the tuple of messages returned in both $\mathsf{NME}_b^{(1)}$ and $\mathsf{NME}_b^{(2)}$ are exactly the same, and thus the outputs of $\mathsf{NME}_b^{(1)}$ and $\mathsf{NME}_b^{(2)}$ are identical.

Combining the above with the fact that the events $\mathtt{fail}_b(\cdot)$ occur with a negligible probability, we have $\mathsf{NME}_b^{(1)}(\Pi, A, k, p(k)) \overset{s}{\approx} \mathsf{NME}_b^{(2)}(\Pi, A, k, p(k))$. $\square$

Claim 3. *For every p.p.t. machine A, there exists a p.p.t. machine B such that for $b \in \{0, 1\}$,*

$$\left\{ \mathsf{NME}_b^{(2)}(\Pi, A, k, p(k)) \right\}_{k \in \mathbb{N}} \equiv \left\{ \mathsf{IND}_b(E', B, k) \right\}_{k \in \mathbb{N}}$$

Proof. The machine B is constructed as follows. B simply simulates the experiment $\mathsf{NME}_b^{(2)}$, except that instead of generating PK by itself, it uses PK $= \{\text{PK}_i\}_{i=1}^k$ received from the outside. Let (m_0, m_1) be the pair of messages the adversary A_1 returns. B then outputs (m_0, m_1) and receives a challenge ciphertext c_b from the outside. B performs the same operations as the experiment $\mathsf{NME}_b^{(2)}$ to generate the challenge ciphertext C_b for A_2. Finally, A_2 returns a sequence of ciphertexts $(\psi_1, \psi_2, \ldots, \psi_{p(k)})$. B decrypts these ciphertexts just as in $\mathsf{NME}_b^{(2)}$ and outputs the plaintexts. (Note that $\mathsf{NME}_b^{(2)}$ uses only SK' and not SK in order to decrypt the messages).

It is easy to see that B simulates the experiment $\mathsf{NME}_b^{(2)}$ perfectly using the public-keys and ciphertexts received from the outside, and thus

$$\left\{ \mathsf{NME}_b^{(2)}(\Pi, A, k, p(k)) \right\}_{k \in \mathbb{N}} \equiv \left\{ \mathsf{IND}_b(E', B, k) \right\}_{k \in \mathbb{N}} \qquad \square$$

To conclude the proof, we combine the last three claims to conclude that for every p.p.t. adversary A, there is a p.p.t. adversary B such that

$$\mathsf{NME}_b(\Pi, A, k, p(k)) \overset{c}{\approx} \mathsf{NME}_b^{(1)}(\Pi, A, k, p(k)) \overset{s}{\approx} \mathsf{NME}_b^{(2)}(\Pi, A, k, p(k))$$
$$\equiv \mathsf{IND}_b(E', B, k)$$

Since by the semantic security of E', $\mathsf{IND}_0(E', B, k) \overset{c}{\approx} \mathsf{IND}_1(E', B, k)$, it holds that $\mathsf{NME}_0(\Pi, A, k, p(k)) \overset{c}{\approx} \mathsf{NME}_1(\Pi, A, k, p(k))$. $\square$

Acknowledgments. We thank Cynthia Dwork, Shafi Goldwasser, Yehuda Lindell, and Silvio Micali for delicious comments. We are especially grateful to Yehuda Lindell for his diligent help.

References

[AD97]   Miklós Ajtai and Cynthia Dwork. A public-key cryptosystem with worst-case/average-case equivalence. In *STOC*, pages 284–293, 1997.

[BDPR98]   M. Bellare, A. Desai, D. Pointcheval, and P. Rogaway. Relations among notions of security for public-key encryption schemes. In *CRYPTO*, 1998.

[Blu86]   Manuel Blum. How to prove a theorem so no one can claim it. In *Proc. of The International Congress of Mathematicians*, pages 1444–1451, 1986.

[BS99]   Mihir Bellare and Amit Sahai. Non-malleable encryption: Equivalence between two notions, and an indistinguishability-based characterization. In *CRYPTO*, pages 519–536, 1999.

[CD00]   Jan Camenisch and Ivan B. Damgård. Verifiable encryption, group encryption, and their applications to group signatures and signature sharing schemes. In *ASIACRYPT*, pages 331–345, 2000.

[CDS94]   Ronald Cramer, Ivan Damgard, and Berry Schoenmakers. Proofs of partial knowledge and simplified design of witness hiding protocols. In *CRYPTO*, pages 174–187, 1994.

[CS98]   Ronald Cramer and Victor Shoup. A practical public key cryptosystem provably secure against adaptive chosen ciphertext attack. In *CRYPTO*, pages 13–25, 1998.

[CS02]   Ronald Cramer and Victor Shoup. Universal hash proofs and a paradigm for adaptive chosen ciphertext secure public-key encryption. In *EUROCRYPT*, pages 45–64, 2002.

[DDN00]   Danny Dolev, Cynthia Dwork, and Moni Naor. Nonmalleable cryptography. *SIAM J. Comput.*, 30(2):391–437, 2000.

[Dwo99]   Cynthia Dwork. The non-malleability lectures. Course notes for Stanford CS 359, 1999. http://theory.stanford.edu/g̃durf/cs359-s99/.

[ES02]   Edith Elkind and Amit Sahai. A unified methodology for constructing public-key encryption schemes secure against adaptive chosen-ciphertext attack. ePrint Archive 2002/042, 2002.

[GL03]   Rosario Gennaro and Yehuda Lindell. A framework for password-based authenticated key exchange. In *EUROCRYPT*, pages 524–543, 2003.

[GM84]   Shafi Goldwasser and Silvio Micali. Probabilistic encryption. *J. Comput. Syst. Sci.*, 28(2):270–299, 1984.

[Gol04]   Oded Goldreich. *Foundations of Cryptography, Volume 2*. Cambridge University Press, 2004.

[KMO89]   Joe Kilian, Silvio Micali, and Rafail Ostrovsky. Minimum resource zero-knowledge proofs. In *FOCS*, pages 474–479, 1989.

[Lam79]   Leslie Lamport. Constructing digital signatures from a one-way function. Technical Report CSL-98, SRI International, October 1979.

[Nao91]   Naor. Bit commitment using pseudorandomness. *J. of Cryptology*, 4, 1991.

[Nao04]   Moni Naor. A taxonomy of encryption scheme security. 2004.

[PS05]   Rafael Pass and Abhi Shelat. Unconditional characterizations of non-interactive zero-knowledge. In *CRYPTO*, pages 118–134, 2005.

[Reg05]   Oded Regev. On lattices, learning with errors, random linear codes, and cryptography. In *STOC*, pages 84–93, 2005.

[Rom90]   J. Rompel. One-way functions are necessary and sufficient for secure signatures. In *STOC*, pages 387–394, 1990.

A   DDN-Lite Does Not Achieve Definition 3

DDN-Lite is a candidate single-message non-malleable encryption scheme based on any semantically secure scheme which has been informally discussed in [Nao04, Dwo99]. In this section, we show that DDNLite does not meet our stronger definition of non-malleability (Definition 3). We remark that DDNLite can, however, be proven secure under the DDN and equivalent [BDPR98, BS99]) definitions of non-malleability because in weaker notions, the adversary is considered to have lost the game if she produces invalid ciphertexts in the experiment.

Let us briefly summarize the DDN-Lite scheme. The public key consists of k pairs of encryption keys (just as in our scheme) and a universal one-way hash function h. To encrypt a message m, generate a key pair (SKSIG, VKSIG) for a one-time signature scheme, hash the verification key VKSIG to get $(b_1, \ldots, b_k) \leftarrow h(\text{VKSIG})$, compute the ciphertexts $c_i = E_{PK_i^{b_i}}(m)$, compute the signature $\sigma = \text{Sign}_{\text{SKSIG}}(c_1, \ldots, c_k)$ and output the tuple $[(c_1, c_2, \ldots, c_k), \text{VKSIG}, \sigma]$. To decrypt a ciphertext c parsed as $[(c_1, c_2, \ldots, c_k), \text{VKSIG}, \sigma]$, first verify the signature σ and output $\bot$ if Ver rejects. Otherwise, decrypt the c_i's with the corresponding secret-keys to get corresponding messages m_i. If all the m_i's are equal, then output one of the m_i's, else output $\bot$.

Claim 4. *The DDN-lite encryption scheme does not satisfy Definition 3.*

Proof. We specify an adversary $A = (A_1, A_2)$ such that the two experiments $\{\text{NME}_0(\Pi, A, k, p(k))\}_{k \in \mathbb{N}}$ and $\{\text{NME}_1(\Pi, A, k, p(k))\}_{k \in \mathbb{N}}$ are distinguishable. A works as follows:

1. A_1 outputs two arbitrary messages (m_0, m_1) and no state information.
2. On input ciphertext $c = [(e_1, \ldots, e_k), \text{VKSIG}, \sigma]$, let $(b_1, \ldots, b_k) \leftarrow h(\text{VKSIG})$. A_2 produces a new ciphertext c' as follows. Generate a new signing key (SKSIG', VKSIG'). Compute $(b'_1, \ldots, b'_k) \leftarrow h(\text{VKSIG}')$. Output ciphertexts $c' = ((x_1, \ldots, x_k), \text{VKSIG}', \sigma')$ where

$$
x_i = \begin{cases} e_i & \text{if } b'_i = b_i \\ E_{PK_i^{b'_i}}(m_0) & \text{otherwise} \end{cases}
$$

and σ' is the signature of $(x_1, \ldots, x_k)$ under the signing key SKSIG'.

Now, notice that $\text{NME}_0(\Pi, A, k, \ell) = m_0$ and $\text{NME}_1(\Pi, A, k, \ell) = \bot$ which can be easily distinguished.   □

Anonymous Hierarchical Identity-Based Encryption
(Without Random Oracles)

Xavier Boyen[1] and Brent Waters[2]

[1] Voltage Inc., Palo Alto
xb@boyen.org
[2] SRI International
bwaters@csl.sri.com

Abstract. We present an identity-based cryptosystem that features fully anonymous ciphertexts and hierarchical key delegation. We give a proof of security in the standard model, based on the mild Decision Linear complexity assumption in bilinear groups. The system is efficient and practical, with small ciphertexts of size linear in the depth of the hierarchy. Applications include search on encrypted data, fully private communication, *etc.*

Our results resolve two open problems pertaining to anonymous identity-based encryption, our scheme being the first to offer provable anonymity in the standard model, in addition to being the first to realize fully anonymous HIBE at all levels in the hierarchy.

1 Introduction

The cryptographic primitive of Identity-Based Encryption (IBE) allows a sender to encrypt a message for a receiver using only the receiver's identity as a public key. Recently, there has been interest in "anonymous" identity-based encryption systems, where the ciphertext does not leak the identity of the recipient. In addition to their obvious privacy benefits, anonymous IBE systems can be leveraged to construct Public key Encryption with Keyword Search (PEKS) schemes, as was first observed by Boneh *et al.* [9] and later formalized by Abdalla *et al.* [1]. Roughly speaking, PEKS is a form of public key encryption that allows an encryptor to make a document serarchable by keywords, and where the capabilities to search on particular keywords are delegated by a central authority. Anonymous HIBE further enables sophisticated access policies for PEKS and ID-based PEKS.

Prior to this paper, the only IBE system known to be inherently anonymous was that of Boneh and Franklin [10]. Although they did not state it explicitly, the anonymity of their scheme followed readily from their proof of semantic security. This was noticed by Boyen [12], who gave an ID-based signcryption with a formalization of sender and recipient anonymity. One drawback of the Boneh-Franklin IBE paradigm is that its security proofs are set in the random oracle model. More recently, a number of IBE schemes [14,5,6,31,16,26] have been proven secure outside of the random oracle model, but none of these schemes is anonymous. In particular, in the efficient schemes of Boneh and Boyen [5] and Waters [31], the identity is deterministically encoded in a simple manner within the exponent of an element of the bilinear group $\mathbb{G}$. When these schemes are implemented using a "symmetric" bilinear pairing $e : \mathbb{G} \times \mathbb{G} \to \mathbb{G}_T$, it becomes trivial to test whether a given ciphertext was encrypted for a candidate identity.

A tempting workaround to this problem is to use an "asymmetric" pairing $e : \mathbb{G} \times \hat{\mathbb{G}} \to \mathbb{G}_T$ in the schemes that allow it, such as Boneh and Boyen's "BB_1" and "BB_2", and Waters' by extension. In those schemes, and under the additional assumption that Decision Diffie-Hellman is hard in $\mathbb{G}$, one may prevent the use of the pairing as a direct test of whether a ciphertext is for a particular identity. Unfortunately, turning this observation into a formal security reduction would at the very least require making a strong assumption that is patently false in bilinear groups with symmetric pairings, and the approach would still fail to generalize to hierarchical IBE for fundamental reasons that are discussed later. Ideally, one would like a scheme that works indifferently with symmetric and asymmetric pairings, and generalizes to hierarchical identities.

The first anonymous IBE without random oracles was unveiled at the Rump Session of CRYPTO'05 by one of the authors, and is now described in Section 4. In a nutshell, the identity is split randomly into two blind components to prevent its recognition by the bilinear map, without making unduly strong assumptions. A second anonymous IBE without random oracles was recently proposed by Gentry [19], based on a different principle. In Gentry's scheme, the identity of a ciphertext cannot be tested because a crucial element of the ciphertext lives in the target group $\mathbb{G}_T$ rather than the bilinear group $\mathbb{G}$. Gentry's scheme is very efficient and has a number of advantages, but unfortunately relies on a strong complexity assumption and does not generalize to hierarchical IBE.

In spite of these recent achievements, creating an Anonymous Hierarchical IBE (A-HIBE) scheme has remained a challenge. Even if we avail ourselves of random oracles, there simply does not exist any known hierarchical identity-based encryption scheme which is also anonymous. In particular, the Gentry-Silverberg [20] HIBE scheme is not anonymous, despite the fact that it derives from the Boneh-Franklin IBE scheme, which is anonymous. The numerous applications to searching on encrypted data motivated Abdalla et al. [1], in their CRYPTO'05 paper, to ask for the creation of an Anonymous HIBE system, preferably without random oracles, as an important open research problem.

1.1 Our Results

Our contribution is twofold. First, we build a simple and efficient Anonymous IBE scheme, and give a proof of security without random oracles. Second, we generalize our construction into a fully Anonymous HIBE scheme (i.e., anonymous at all levels in the hierarchy), again with a proof without random oracles. Our approach gives a very efficient system in the non-hierarchical case, and remains practical for the shallow hierarchies that are likely to be encountered in most applications. The security of our systems is based on Boneh's et al. [8] Decision Linear assumption, which is arguably one of the mildest useful complexity assumptions in the realm of bilinear groups.

At first sight, our construction bears a superficial resemblance to Boneh and Boyen's "BB$_1$" HIBE scheme [5, §4] — but with at least two big differences. First, we perform "linear splittings" on various portions of the ciphertext, to thwart the trial-and-error identity guessing to which other schemes fell prey. This idea gives us provable anonymity, even under symmetric pairings. Second, we use multiple parallel HIBE systems and re-randomize the keys between them upon each delegation. This is what lets us use the linear splitting technique at all levels of the hierarchy, but also poses a technical challenge in the security reduction which must now simulate multiple interacting HIBE systems *at once*. Solving this problem was the crucial step that gave us a hierarchy without destroying anonymity.

1.2 Related Work

The concept of identity-based encryption was first proposed by Shamir [28] two decades ago. However, it was not until much later that Boneh and Franklin [10] and Cocks [17] presented the first practical solutions. The Boneh-Franklin IBE scheme was based on groups with efficiently computable bilinear maps, while the Cocks scheme was proven secure under the quadratic residuosity problem, which relies on the hardness of factoring. The security of either scheme was only proven in the random oracle model.

Canetti, Halevi, and Katz [14] suggested a weaker security notion for IBE, known as selective identity or selective-ID, relative to which they were able to build an inefficient but secure IBE scheme without using random oracles. Boneh and Boyen [5] presented two very efficient IBE systems ("BB$_1$" and "BB$_2$") with selective-ID security proofs, also without random oracles. The same authors [6] then proposed a coding-theoretic extension to their "BB$_1$" scheme that allowed them to prove security for the full notion of adaptive identity or adaptive-ID security without random oracles, but the construction was impractical. Waters [31] then proposed a much simpler extension to "BB$_1$" also with an adaptive-ID security proof without random oracles; its efficiency was further improved in two recent independent papers, [16] and [26].

The notion of hierarchical identity-based encryption was first defined by Horwitz and Lynn [22], and a construction in the random oracle model given by Gentry and Silverberg [20]. Canetti, Halevi, and Katz [14] give the first HIBE with a (selective-ID) security proof without random oracles, but that is not efficient. The first efficient HIBE scheme to be provably secure without random oracles is the "BB$_1$" system of Boneh and Boyen; further improvements include the HIBE scheme by Boneh, Boyen, and Goh [7], which features shorter ciphertexts and private keys.

Nominally adaptive-ID secure HIBE schemes have been proposed, although all constructions known to date [20,31,16,26] are depth-limited because they suffer from an exponential security degradation with the depth of the hierarchy. Qualitatively, this is no different than taking an HIBE scheme with tight selective-ID security, such as BB$_1$ or BBG, and using one of the generic transformations from [5, §7] to make it adaptive-ID secure. Quantitatively, the rate of decay will differ between those approaches, which means that the number of useful hierarchy levels will evolve similarly but not identically in function of the chosen group size and the desired security bit strength. Accordingly, it remains an important open problem in identity-based cryptography to devise an adaptive-ID secure HIBE scheme whose security degrades at most polynomially with the depth of the hierarchy, under reasonable assumptions. (In this paper, we mostly leave aside this issue of adaptive-ID security for HIBE.)

Encrypted search was studied by Song, Wagner, and Perrig [30], who presented the first scheme for searching on encrypted data. Their scheme is in the symmetric-key setting where the same party that encrypted the data would generate the keyword search capabilities. Boneh *et al.* [9] introduced Public Key Encryption with Keyword Search (PEKS), where any party with access to a public key could make an encrypted document that was searchable by keyword; they realized their construction by applying the Boneh-Franklin IBE scheme. Abdalla *et al.* [1] recently formalized the notion of Anonymous IBE and its relationship to PEKS. Additionally, they formalized the notion of Anonymous HIBE and mentioned different applications for it. Using the GS system as a starting point, they also gave an HIBE scheme that was anonymous at the first level, in the random oracle model. Another view of Anonymous IBE is as a combination of identity-based encryption with the property of key privacy, which was introduced by Bellare *et al.* [4].

1.3   Applications

In this section we discuss various applications of our fully anonymous HIBE system. The main applications can be split into several broad categories.

Fully Private Communication. The first compelling application of anonymous IBE is for fully private communication. Bellare *et al.* [4] argue that public key encryption systems that have the "key privacy" property can be used for anonymous communication: for example, if one wishes to hide the identity of a recipient one can encrypt a ciphertext with an anonymous IBE system and post it on a public bulletin board. By the anonymity property, the ciphertext will betray neither sender nor recipient identity, and since the bulletin board is public, this method will also be resistant to traffic analysis. To compound this notion of key privacy, identity-based encryption is particularly suited for untraceable anonymous communication, since, contrarily to public-key infrastructures, the sender does not even need to query a directory for the public key of the recipient. For this reason, anonymous IBE provides a very convincing solution to the problem of secure anonymous communication, as it makes it harder to conduct traffic analysis attack on directory lookups.

Search on Encrypted Data. The second main application of anonymous (H)IBE is for encrypted search. As mentioned earlier, anonymous IBE and HIBE give several application in the Public-key Encryption with Keyword Search (PEKS) domain, proposed by Boneh *et al.* [9], and further discussed by Abdalla *et al.* [1].

As a simple example of real-world application of our scheme, PEKS is a useful primitive for building secure audit logs [32,18]. Furthermore, one can leverage the hierarchical identities in our anonymous HIBE in several interesting ways. For example, we can use a two-level anonymous HIBE scheme where the first level is an identity and the second level is a keyword. This gives us the first implementation of the Identity-Based Encryption with Keyword Search (IBEKS) primitive asked for in [1]. With this primitive, someone with the private key for an identity can delegate out search capabilities for encryptions to their identity, without requiring a central authority to act as the delegator. Conversely, by using certain keywords such as "Top Secret" at the first level of the hierarchy, it is possible to broadcast innocent-looking ciphertexts that require a certain clearance to decrypt, without even hinting at the fact that their payload might be valuable. We can create more refined search capabilities with a deeper hierarchy.

As the last applications we mention, forward-secure public-key encryption [14] and forward-secure HIBE [33] are not hard to construct from HIBE systems with certain algebraic properties [7]. Without going into details, we mention that we can implement Anonymous fs-HIBE with our scheme by embedding a time component within the hierarchy, while preserving the anonymity property.

2   Background

Recall that a pairing is an efficiently computable [25], non-degenerate function, $\mathbf{e} : \mathbb{G} \times \hat{\mathbb{G}} \to \mathbb{G}_T$, with the bilinearity property that $\mathbf{e}(g^r, \hat{g}^s) = \mathbf{e}(g, \hat{g})^{rs}$. Here, $\mathbb{G}$, $\hat{\mathbb{G}}$, and $\mathbb{G}_T$ are all multiplicative groups of prime order p, respectively generated by g, $\hat{g}$, and $\mathbf{e}(g, \hat{g})$. It is *asymmetric* if $\mathbb{G} \neq \hat{\mathbb{G}}$.

We call *bilinear instance* a tuple $\mathbf{G} = [p, \mathbb{G}, \hat{\mathbb{G}}, \mathbb{G}_T, g, \hat{g}, \mathbf{e}]$. We assume an efficient generation procedure that on input a security parameter $\Sigma \in \mathbb{N}$ outputs $\mathbf{G} \xleftarrow{\$} \mathrm{Gen}(1^\Sigma)$ where $\log_2(p) = \Theta(\Sigma)$. We write $\mathbb{Z}_p = \mathbb{Z}/p\mathbb{Z}$ for the set of residues mod p and $\mathbb{Z}_p^\times = \mathbb{Z}_p \setminus \{0\}$ for its multiplicative group.

2.1   Assumptions

Since bilinear groups first appeared in cryptography half a decade ago [23], several years after their first use in cryptanalysis [24], bilinear maps or pairings have been used in a large variety of ways under many different complexity assumptions. Some of them are very strong; others are weaker. Informally, we say that an assumption is *mild* if it is tautological in the generic group model [29], and also "efficiently falsifiable" [27] in the sense that its problem instances are stated non-interactively and concisely (*e.g.*, independently of the number of adversarial queries or such large quantity). Most IBE and HIBE schemes mentioned in Introduction (except "BB₂" and the Factoring-based system by Cocks) are based on *mild* bilinear complexity assumptions, such as BDH [23,10] and Linear [8]. In this paper, our goal is to rely only on mild assumptions.

Decision BDH: The Bilinear DH assumption was first used by Joux [23], and gained popularity for its role in the Boneh-Franklin IBE system [10]. The decisional assumption posits the hardness of the D-BDH problem, which we state in asymmetric bilinear groups as:

> Given a tuple $[g, g^{z_1}, g^{z_3}, \hat{g}, \hat{g}^{z_1}, \hat{g}^{z_2}, Z] \in \mathbb{G}^3 \times \hat{\mathbb{G}}^3 \times \mathbb{G}_T$ for random exponents $[z_1, z_2, z_3] \in (\mathbb{Z}_p)^3$, decide whether $Z = \mathbf{e}(g, \hat{g})^{z_1 z_2 z_3}$.

Decision Linear: The Linear assumption was first proposed by Boneh, Boyen, and Shacham for group signatures [8]. Its decisional form posits the hardness of the D-Linear problem, which can be stated in asymmetric bilinear groups as follows:

> Given a tuple $[g, g^{z_1}, g^{z_2}, g^{z_1 z_3}, g^{z_2 z_4}, \hat{g}, \hat{g}^{z_1}, \hat{g}^{z_2}, Z] \in \mathbb{G}^5 \times \hat{\mathbb{G}}^3 \times \mathbb{G}$ for random $[z_1, z_2, z_3, z_4] \in (\mathbb{Z}_p)^4$, decide whether $Z = g^{z_3 + z_4}$.

We remark that the elements $\hat{g}, \hat{g}^{z_1}, \hat{g}^{z_2} \in \hat{\mathbb{G}}^3$ were not explicitly included in Boneh's *et al.* original formulation.

"Hard" means algorithmically non-solvable with probability $1/2 + \Omega(\text{poly}(\Sigma)^{-1})$ in time $\mathcal{O}(\text{poly}(\Sigma))$ for "bilinear instances" $[p, \mathbb{G}, \hat{\mathbb{G}}, \mathbb{G}_T, g, \hat{g}, \mathbf{e}] \xleftarrow{\$} \text{Gen}(1^{\Sigma})$ that are generated at random using an efficient algorithm, as $\Sigma \to +\infty$.

These assumptions allow but not require the groups $\mathbb{G}$ and $\hat{\mathbb{G}}$ to be distinct, and similarly we make no representation one way or the other regarding the existence of computable homomorphisms between $\mathbb{G}$ and $\hat{\mathbb{G}}$, in either direction. This is the most general formulation. It has two main benefits: (1) since it comes with fewer restrictions, it is potentially more robust and increases our confidence in the assumptions we make; and (2) it gives us the flexibility to implement the bilinear pairing on a broad variety of algebraic curves with attractive computational characteristics [2], whereas symmetric pairings tend to be confined to supersingular curves, to name this one distinction.

Note that if we let $\mathbb{G} = \hat{\mathbb{G}}$ and $g = \hat{g}$, our assumptions regain their familiar "symmetric" forms:

Given $[g, g^{z_1}, g^{z_2}, g^{z_3}, Z] \in \mathbb{G}^4 \times \mathbb{G}_T$ for random $[z_1, z_2, z_3] \in (\mathbb{Z}_p)^3$, decide whether $Z = \mathbf{e}(g, g)^{z_1 z_2 z_3}$.
Given $[g, g^{z_1}, g^{z_2}, g^{z_1 z_3}, g^{z_2 z_4}, Z] \in \mathbb{G}^5 \times \mathbb{G}$ for random $[z_1, z_2, z_3, z_4] \in (\mathbb{Z}_p)^4$, decide if $Z = g^{z_3 + z_4}$.

As a rule of thumb, the remainder of this paper may be read in the context of symmetric pairings, simply by dropping all "hats" (ˆ) in the notation. Also note that D-Linear trivially implies D-BDH.

2.2   Models

We briefly precise the security notions that are implied by the concept of Anonymous IBE or HIBE. We omit the formal definitions, which may be found in the literature [10,1].

Confidentiality: This is the usual security notion of semantic security for encryption. It means that no non-trivial information about the message can be feasibly gleaned from the ciphertext.

Anonymity: Recipient anonymity is the property that the adversary be unable
to distinguish the encryption of a chosen message for a first chosen identity from the encryption of the same message for a second chosen identity.
Equivalently, the adversary must be unable to decide whether a ciphertext
was encrypted for a chosen identity, or for a random identity.

3   Intuition

Before we present our scheme we first explain why it is difficult to implement
anonymous IBE without random oracles, as well as any form of anonymous
HIBE even in the random oracle model. We then give some intuition behind our
solution.

3.1   The Difficulty

Recall that in the basic Boneh-Franklin IBE system [10], an encryption of a
message Msg to some identity Id, takes the following form,

$$\mathsf{CT} = [\, C_1, \, C_2 \,] = [\, g^r, \, \mathbf{e}(\mathcal{H}(\mathsf{Id}), Q)^r \, \mathsf{Msg} \,] \in \mathbb{G} \times \mathbb{G}_T \,,$$

where $\mathcal{H}$ is a random oracle, r is a random exponent, and g and Q are public
system parameters. A crucial observation is that the one element of the ciphertext in the bilinear group $\mathbb{G}$, namely, g^r, is just a random element that gives no
information about the identity of the recipient. The reason why only one element
in $\mathbb{G}$ is needed is because private keys in the Boneh-Franklin scheme are deterministic — there will be no randomness in the private key to cancel out. Since
the proof of semantic security is based on the fact that C_2 is indistinguishable
from random without the private key for ID, it follows that the scheme is also
anonymous since C_2 is the only part of the ciphertext on which the recipient
identity has any bearing.

More recently, there have been a number of IBE schemes proven secure with-
out random oracles, such as BTE from [14], BB_1 and BB_2 from [5], and Waters' [31]. However, in all these schemes the proof of security requires that randomness be injected into the private key generation. Since the private keys are
randomized, some extra information is needed in the ciphertext in order to cancel out the randomness upon decryption. To illustrate, consider the encryption
of a message Msg to an identity Id in the BB_1 Boneh-Boyen system,

$$\mathsf{CT} = [\, C_1, \, C_2, \, C_3 \,] = [\, g^r, \, (g_1^{\mathsf{Id}} \, g_3)^r, \, \mathbf{e}(g_1, \hat{g}_2)^r \, \mathsf{Msg} \,] \in \mathbb{G}^2 \times \mathbb{G}_T \,,$$

where r is chosen by the encryptor and g, g_1, g_3, and $\mathbf{e}(g_1, \hat{g}_2)$ are public system
parameters. Notice, there are now two elements in $\mathbb{G}$, and between them there
is enough redundancy to determine whether a ciphertext was intended for a
given identity Id, simply by testing whether the tuple $[g, g_1^{\mathsf{Id}} \, g_3, C_1, C_2]$ is Diffie-
Hellman, using the bilinear map,

$$\mathbf{e}(C_1, \hat{g}_1^{\mathsf{Id}} \, \hat{g}_3) \overset{?}{=} \mathbf{e}(C_2, \hat{g}) \,.$$

We see that the extra ciphertext components which are seemingly necessary in IBE schemes without random oracles, in fact contribute to leaking the identity of the intended recipient of a ciphertext.

A similar argument can be made for why none of the existing HIBE schemes is anonymous, even though some of them use random oracles. Indeed, all known HIBE schemes, including the Gentry-Silverberg system in the random oracle model, rely on randomization in order to properly delegate private keys down the hierarchy in a collusion-resistant manner. Since the randomization is performed not just by the master authority, but by anyone who has the power to delegate a key, the elements needed for it are distributed as part of the public parameters. Because of this, we end up in the same situation as above, where the extra components needed to either perform or cancel the randomization will also provide a test for the addressee's identity.

Since having randomized keys seems to be fundamental to designing (H)IBE systems without random oracles, we aim to design a system where the necessary extra information will be hidden to a computationally bounded adversary. Thus, even though we cannot prevent the ciphertext from containing information about the recipient, we can design our system such that this information cannot be easily tested from the public parameters and ciphertext alone.

3.2   Our Approach

As mentioned in the introduction, we can prevent a single-level identity to be testable by performing some sort of blinding, by splitting the identity into two randomized complementary components. Indeed, building a "flat" anonymous IBE system turns out to be reasonably straightforward using our linear splitting technique to hide the recipient identity behind some randomization.

Complications arise when one tries to support hierarchical key generation. In a nutshell, to prevent collusion attacks in HIBE, "parents" must independently re-randomize the private keys they give to their "children". In all known HIBE schemes, re-randomization is enabled by listing a number of supplemental components in the public system parameters. Why this breaks anonymity is because the same mechanism that allows private keys to be publicly re-randomized, also allows ciphertexts to be publicly tested for recipient identities. Random oracles offer no protection against this.

To circumvent this obstable, we need to make the re-randomization elements non-public, and tie them to each individual private key. In practical terms, this means that private keys must convey extra components (although not too many). The real difficulty is that each set of re-randomization components constitutes a full-fledged HIBE in its own right, which must be simulated together with its peers in the security proof (their number grows linearly with the maximal depth). Because these systems are not independent but interact with each other, we are left with the task of simulating multiple HIBE subsystems that are globally constrained by a set of linear relations. A novelty of our proof technique is a method to endow the simulator with enough degrees of freedom to reduce a system of unknown keys to a single instance of the presumed hard problem.

A notable feature of our construction is that it can be implemented using all known instantiations of the bilinear pairing (whether symmetric or asymmetric, with our without a computable or invertible homomorphism, *etc.*). To cover all grounds, we first describe a "flat" anonymous IBE using the symmetric notation, for ease of exposition, and then move to the full HIBE using the general asymmetric notation without assuming any homomorphism, for maximum generality.

4   A Primer: Anonymous IBE

We start by describing an Anonymous IBE scheme that is semantically secure against selective-ID chosen plaintext attacks. This construction will illustrate our basic technique of "splitting" the bilinear group elements into two pieces to protect against the attacks described in the previous section.

For simplicity, and also to show that we get anonymity even when using symmetric pairings, we describe the IBE system (and the IBE system only) in the special case where $\mathbb{G} = \hat{\mathbb{G}}$:

Setup. The setup algorithm chooses a random generator $g \in \mathbb{G}$, random group elements $g_0, g_1 \in \mathbb{G}$, and random exponents $\omega, t_1, t_2, t_3, t_4 \in \mathbb{Z}_p$. It keeps these exponents as the master key, Msk. The corresponding system parameters are published as:

$$\mathsf{Pub} \leftarrow \left[\; \Omega = \mathbf{e}(g,g)^{t_1 t_2 \omega}, \; g, \; g_0, \; g_1, \; v_1 = g^{t_1}, \; v_2 = g^{t_2}, \; v_3 = g^{t_3}, \; v_4 = g^{t_4} \; \right]$$

Extract(Msk, Id). To issue a private key for identity Id, the key extraction authority chooses two random exponents $r_1, r_2 \in \mathbb{Z}_p$, and computes the private key as: $\mathsf{Pvk}_\mathsf{Id} = [d_0, d_1, d_2, d_3, d_4] \leftarrow$

$$\left[g^{r_1 t_1 t_2 + r_2 t_3 t_4}, \; g^{-\omega t_2}(g_0 g_1^{\mathsf{Id}})^{-r_1 t_2}, \; g^{-\omega t_1}(g_0 g_1^{\mathsf{Id}})^{-r_1 t_1}, \; (g_0 g_1^{\mathsf{Id}})^{-r_2 t_4}, (g_0 g_1^{\mathsf{Id}})^{-r_2 t_3} \; \right]$$

Encrypt(Pub, Id, M). Encrypting a message Msg $\in \mathbb{G}_T$ for an identity Id $\in \mathbb{Z}_p^\times$ works as follows. The algorithm chooses random exponents $s, s_1, s_2 \in \mathbb{Z}_p$, and creates the ciphertext as:

$$\mathsf{CT} = [\, C', C_0, C_1, C_2, C_3, C_4 \,] \leftarrow \left[\; \Omega^s M, \; (g_0 g_1^{\mathsf{Id}})^s, \; v_1^{s-s_1}, v_2^{s_1}, v_3^{s-s_2}, v_4^{s_2} \; \right]$$

Decrypt(Pvk$_\mathsf{Id}$, C). The decryption algorithm attempts to decrypt a ciphertext CT by computing:

$$C' \, \mathbf{e}(C_0, d_0) \, \mathbf{e}(C_1, d_1) \, \mathbf{e}(C_2, d_2) \, \mathbf{e}(C_3, d_3) \, \mathbf{e}(C_4, d_4) = \mathsf{Msg} \; .$$

Proving Security. We prove security using a hybrid experiment.

Let $[C', C_0, C_1, C_2, C_3, C_4]$ denote the challenge ciphertext given to the adversary during a real attack. Additionally, let R be a random element of $\mathbb{G}_T$, and R', R'' be random elements of $\mathbb{G}$. We define the following hybrid games which differ on what challenge ciphertext is given by the simulator to the adversary:

Γ_0 : The challenge ciphertext is $\mathsf{CT}_0 = [C', C_0, C_1, C_2, C_3, C_4]$.
Γ_1 : The challenge ciphertext is $\mathsf{CT}_1 = [R, C_0, C_1, C_2, C_3, C_4]$.
Γ_2 : The challenge ciphertext is $\mathsf{CT}_2 = [R, C_0, R', C_2, C_3, C_4]$.
Γ_3 : The challenge ciphertext is $\mathsf{CT}_3 = [R, C_0, R', C_2, R'', C_4]$.

We remark that the challenge ciphertext in Γ_3 leaks no information about the identity since it is composed of six random group elements, whereas in Γ_0 the challenge is well formed. We show that the transitions from Γ_0 to Γ_1 to Γ_2 to Γ_3 are all computationally indistinguishable.

Lemma 1 (semantic security). *Under the (t, ϵ)-Decision BDH assumption, there is no adversary running in time t that distinguishes between the games Γ_0 and Γ_1 with advantage greater than ϵ.*

Proof. The proof from this lemma essentially follows from the security of the Boneh-Boyen selective-ID scheme. Suppose there is an adversary that can distinguish between game Γ_0 and Γ_1 with advantage ϵ. Then we build a simulator that plays the Decision BDH game with advantage ϵ.

The simulator receives a D-BDH challenge $[g, g^{z_1}, g^{z_2}, g^{z_3}, Z]$ where Z is either $\mathbf{e}(g, g)^{z_1 z_2 z_3}$ or a random element of $\mathbb{G}_T$ with equal probability. The game proceeds as follows:

$\diamond$ *Init:* The adversary announces the identity Id^* it wants to be challenged upon.

$\diamond$ *Setup:* The simulator chooses random exponents $t_1, t_2, t_3, t_4, y \in \mathbb{Z}_p$. It retains the generator g, and sets $g_0 = (g^{z_1})^{-\mathsf{Id}^*} g^y$ and $g_1 = g^{z_1}$. The public parameters are published as:

$$\mathsf{Pub} \leftarrow \left[\; \Omega = \mathbf{e}(g^{z_1}, g^{z_2})^{t_1 t_2}, \; g, \; g_0, \; g_1, \; v_1 = g^{t_1}, v_2 = g^{t_2}, v_3 = g^{t_3}, v_4 = g^{t_4} \;\right].$$

Note that this implies that $\omega = z_1 z_2$.

$\diamond$ *Phase 1:* Suppose the adversary requests a key for identity $\mathsf{Id} \neq \mathsf{Id}^*$. The simulator picks random exponents $r_1, r_2 \in \mathbb{Z}_p$, and issues a private key as: $\mathsf{Pvk}_{\mathsf{Id}} = [d_0, d_1, d_2, d_3, d_4] \leftarrow$

$$\left[\begin{array}{c} (g^{z_2})^{\frac{-1}{\mathsf{Id}-\mathsf{Id}^*}} g^{r_1} g^{r_2 t_3 t_4}, \; ((g^{z_2})^{\frac{\cdot}{\mathsf{Id}-\mathsf{Id}^*}} (g_0 g_1^{\mathsf{Id}})^{r_1})^{-t_2}, \; ((g^{z_2})^{\frac{\cdot}{\mathsf{Id}-\mathsf{Id}^*}} (g_0 g_1^{\mathsf{Id}})^{r_1})^{-t_1}, \\ (g_0 g_1^{\mathsf{Id}})^{-r_2 t_4}, \; (g_0 g_1^{\mathsf{Id}})^{-r_2 t_3} \end{array}\right].$$

This is a well formed secret key for random exponents $\tilde{r}_1 = r_1 - z_2/(\mathsf{Id} - \mathsf{Id}^*)$ and $\tilde{r}_2 = r_2$.

$\diamond$ *Challenge:* Upon receiving a message Msg from the adversary, the simulator chooses $s_1, s_2 \in \mathbb{Z}_p$, and outputs the challenge ciphertext as:

$$\mathsf{CT} = [\, C', C_0, C_1, C_2, C_3, C_4 \,] \leftarrow \left[\begin{array}{c} Z^{-t_1 t_2} M, \; (g^{z_3})^y, (g^{z_3})^{t_1} g^{-s_1 t_1}, g^{s_1 t_2}, \\ (g^{z_3})^{t_3} g^{-s_2 t_3}, g^{s_2 t_4} \end{array}\right].$$

We can let $s = z_3$ and see that if $Z = \mathbf{e}(g, g)^{z_1 z_2 z_3}$ the simulator is playing game Γ_0 with the adversary, otherwise the simulator is playing game Γ_1 with the adversary.

⋄ *Phase 2:* The simulator answers the queries in the same way as Phase 1.

⋄ *Guess:* The simulator outputs a guess γ, which the simulator forwards as its own guess for the D-BDH game.

Since the simulator plays game Γ_0 if and only the given D-BDH instance was well formed, the simulator's advantage in the D-BDH game is exactly ϵ.

Lemma 2 (anonymity, part 1). *Under the (t, ϵ)-Decision linear assumption, no adversary that runs in time t can distinguish between the games Γ_1 and Γ_2 with advantage greater than ϵ.*

Proof. Suppose the existence of an adversary $\mathcal{A}$ that distinguishes between the two games with advantage ϵ. Then we construct a simulator that wins the Decision Linear game as follows.

The simulator takes in a D-Linear instance $[g, g^{z_1}, g^{z_2}, g^{z_1 z_3}, g^{z_2 z_4}, Z]$, where Z is either $g^{z_3 + z_4}$ or random in $\mathbb{G}$ with equal probability. For convenience, we rewrite this as $[g, g^{z_1}, g^{z_2}, g^{z_1 z_3}, Y, g^s]$ for s such that $g^s = Z$, and consider the task of deciding whether $Y = g^{z_2(s - z_3)}$ which is equivalent. The simulator plays the game in the following stages.

⋄ *Init:* The adversary $\mathcal{A}$ gives the simulator the challenge identity Id^*.

⋄ *Setup:* The simulator first chooses random exponents $\alpha, y, t_3, t_4, \omega$. It lets g in the simulation be as in the instance, and sets $v_1 = g^{z_2}$ and $v_2 = g^{z_1}$. The public key is published as:

$$
\mathsf{Pub} \leftarrow \begin{bmatrix} \Omega = \mathbf{e}(g^{z_1}, g^{z_2})^\omega, \ g, \ g_0 = (g^{z_2})^{-\mathsf{Id}^* \alpha} g^y, \ g_1 = (g^{z_2})^\alpha, \\ v_1 = (g^{z_2}), \ v_2 = (g^{z_1}), \ v_3 = g^{t_3}, \ v_4 = g^{t_4} \end{bmatrix}.
$$

If we pose $t_1 = z_2$ and $t_2 = z_1$, we note that the public key is distributed as in the real scheme.

⋄ *Phase 1:* To answer a private key extraction query for an identity $\mathsf{Id} \neq \mathsf{Id}^*$, the simulator chooses random exponents $r_1, r_2 \in \mathbb{Z}_p$, and outputs:

$$
\mathsf{Pvk}_{\mathsf{Id}} \leftarrow \begin{bmatrix} (g^{z_1})^{r_1} g^{r_2 t_3 t_4}, \ (g^{z_1})^{-\omega - \alpha(\mathsf{Id} - \mathsf{Id}^*) r_1}, \ (g^{z_2})^{-\omega - \alpha(\mathsf{Id} - \mathsf{Id}^*) r_1}, \\ (g^{z_1})^{\frac{-1}{t_3}} (g_0 g_1^{\mathsf{Id}})^{-r_2 t_4}, \ (g^{z_1})^{\frac{-1}{t_4}} (g_0 g_1^{\mathsf{Id}})^{-r_2 t_3} \end{bmatrix}.
$$

If, instead of r_1 and r_2, we consider this pair of uniform random exponents,

$$
\tilde{r}_1 = \frac{r_1 \alpha(\mathsf{Id} - \mathsf{Id}^*)}{\alpha(\mathsf{Id} - \mathsf{Id}^*) z_2 + y}, \qquad \tilde{r}_2 = r_2 + \frac{y z_1 r_1}{(t_3 t_4)(\alpha(\mathsf{Id} - \mathsf{Id}^*) z_2 + y)},
$$

then we see that the private key is well formed, since it can be rewritten as:

$$
\begin{bmatrix} g^{\tilde{r}_1 t_1 t_2 + \tilde{r}_2 t_3 t_4}, \ g^{-\omega t_2} (g_0 g_1^{\mathsf{Id}})^{-\tilde{r}_1 t_2}, g^{-\omega t_1} (g_0 g_1^{\mathsf{Id}})^{-\tilde{r}_1 t_1}, (g_0 g_1^{\mathsf{Id}})^{-\tilde{r}_2 t_4}, (g_0 g_1^{\mathsf{Id}})^{-\tilde{r}_2 t_3} \end{bmatrix}.
$$

⋄ *Challenge:* The simulator gets from the adversary a message M which it can discard, and responds with a challenge ciphertext for the identity Id^*. Pose

$s_1 = z_3$. To proceed, the simulator picks a random exponent $s_2 \in \mathbb{Z}_p$ and a random element $R \in \mathbb{G}_T$, and outputs the ciphertext as:

$$\mathsf{CT} = [\, C', C_0, C_1, C_2, C_3, C_4 \,] \leftarrow [\, R, \; (g^s)^y, Y, (g^{z_1 z_3}), (g^s)^{t_3} g^{-s_2 t_3}, g^{s_2 t_4} \,].$$

If $Y = g^{z_2(s-z_3)}$, i.e., $g^s = Z = g^{z_3 + z_4}$, then $C_1 = v_1^{s-s_1}$ and $C_2 = v_2^{s_1}$; all parts of the challenge but C' are thus well formed, and the simulator behaved as in game Γ_1. If instead Y is independent of z_1, z_2, s, s_1, s_2, which happens when Z is random, then the simulator responded as in game Γ_2.

$\diamond$ *Phase 2:* The simulator answer the query in the same way as Phase 1.

$\diamond$ *Output:* The adversary outputs a bit γ to guess which hybrid game the simulator has been playing. To conclude, the simulator forwards γ as its own answer in the Decision-Linear game.

By the simulation setup the advantage of the simulator will be exactly that of the adversary.

Lemma 3 (anonymity, part 2). *Under the (t, ϵ)-Decision linear assumption, no adversary that runs in time t can distinguish between the games Γ_2 and Γ_3 with advantage greater than ϵ.*

Proof. This argument follows almost identically to that of Lemma 2, except where the simulation is done over the parameters v_3 and v_4 in place of v_1 and v_2. The other difference is that the g^ω term that appeared in d_1, d_2 without interfering with the simulation, does not even appear in d_3, d_4.

5   The Scheme: Anonymous HIBE

We now describe our full Anonymous HIBE scheme without random oracles. Anonymity is provided by the splitting technique and hybrid proof introduced in the previous section. In addition, to thwart the multiple avenues for user collusion enabled by the hierarchy, the keys are re-randomized between all siblings and all children. Roughly speaking, this is done by using several parallel HIBE systems, which are recombined at random every time a new private key is issued. In the proof of security, this extra complication is handled by a "multi-secret simulator", that is able to simulate multiple interacting HIBE systems under a set of constraints. This is an information theoretic proof that sits on top of the hybrid argument, which is computational.

For the most part, we focus on security against selective-identity, chosen plaintext attacks, though we will briefly mention how to secure the scheme against adaptive-ID and CCA2 adversaries. Our (selective-ID) Anonymous HIBE scheme consists of the following algorithms:

Setup$(1^\Sigma, D)$. To generate the public system parameters and the corresponding master secret key, given a security parameter $\Sigma \in \mathbb{N}$ in unary, and the hierarchy's maximum depth $D \in \mathbb{N}$, the setup algorithm first generates a bilinear instance $\mathbf{G} = [p, \mathbb{G}, \hat{\mathbb{G}}, \mathbb{G}_T, g, \hat{g}, \mathbf{e}] \xleftarrow{\$} \mathrm{Gen}(1^\Sigma)$. Then:

1. Select $7 + 5\,D + D^2$ random integers modulo p (some of them non-zero):

$$\omega, [\alpha_n, \beta_n, [\theta_{n,\ell}]_{\ell=0,\ldots,D}]_{n=0,\ldots,1+D} \quad \in_\$ \ \mathbb{Z}_p^\times \times ((\mathbb{Z}_p^\times)^2 \times (\mathbb{Z}_p)^{1+D})^{2+D} \ ;$$

2. Publish $\mathbf{G}$ and the system parameters $\mathsf{Pub} \in \mathbb{G}_T \times \mathbb{G}^{2\,(1+D)\,(2+D)}$ as:

$$\Omega \leftarrow \mathbf{e}(g, \hat{g})^\omega \ ,$$

$$[[a_{n,\ell}, b_{n,\ell}]_{\ell=0,\ldots,D}]_{n=0,\ldots,1+D} \ \leftarrow \ [\,[\,g^{\alpha\cdot\,\theta\cdots}, \ g^{\beta\cdot\,\theta\cdots}\,]_{\ell=0,\ldots,D}\,]_{n=0,\ldots,1+D}$$

3. Retain the master secret key $\mathsf{Msk} \in \hat{\mathbb{G}}^{1+(3+D)\,(2+D)}$ as the elements:

$$\hat{w} \leftarrow \hat{g}^\omega \ ,$$

$$[\hat{a}_n, \hat{b}_n, [\hat{y}_{n,\ell}]_{\ell=0,\ldots,D}]_{n=0,\ldots,1+D} \leftarrow \begin{bmatrix} \hat{g}^{\alpha\cdot}, \ \hat{g}^{\beta\cdot}, \\ [\,\hat{g}^{\alpha\cdot\,\beta\cdot\,\theta\cdots}\,]_{\ell=0,\ldots,D} \end{bmatrix}_{n=0,\ldots,1+D}$$

Extract$(\mathsf{Pub}, \mathsf{Msk}, \mathsf{Id}.)$ To extract from the master key Msk a private key for an identity $\mathsf{Id} = [I_0, I_1, \ldots, I_L] \in (\mathbb{Z}_p^\times)^{1+L}$ where $L \in \{1, \ldots, D\}$ and $I_0 = 1$:

1. Pick $6 + 5\,D + D^2$ random integers:

$$[\,\rho_n, \ [\,\rho_{n,m}\,]_{m=0,\ldots,1+D}\,]_{n=0,\ldots,1+D} \quad \in_\$ \ (\mathbb{Z}_p)^{(3+D)\,(2+D)} \ .$$

2. Output the private key $\mathsf{Pvk}_{\mathsf{Id}} \in \hat{\mathbb{G}}^{(5+3\,D-L)\,(3+D)}$ constituted of the following three subkeys (for decryption, re-randomization, and delegation):

   (a) $\mathsf{Pvk}_{\mathsf{Id}}^{\mathrm{decrypt}} = k_0, [k_{n,(a)}, k_{n,(b)}]_{n=0,\ldots,1+D}$

$$\leftarrow \ \hat{w} \prod_{n=0}^{1+D} \prod_{\ell=0}^{L} (\hat{y}_{n,\ell}^{I\cdot})^{\rho\cdot} \ , \ \left[\, \hat{a}_n^{-\rho\cdot}, \ \hat{b}_n^{-\rho\cdot} \,\right]_{n=0,\ldots,1+D}$$

   (b) $\mathsf{Pvk}_{\mathsf{Id}}^{\mathrm{rerand}} = [f_{m,0}, [f_{m,n,(a)}, f_{m,n,(b)}]_{n=0,\ldots,1+D}]_{m=0,\ldots,1+D}$

$$\leftarrow \ \left[\prod_{n=0}^{1+D} \prod_{\ell=0}^{L} (\hat{y}_{n,\ell}^{I\cdot})^{\rho\cdots} \ , \ \left[\, \hat{a}_n^{-\rho\cdots}, \ \hat{b}_n^{-\rho\cdots} \,\right]_{n=0,\ldots,1+D} \right]_{m=0,\ldots,1+D}$$

   (c) $\mathsf{Pvk}_{\mathsf{Id}}^{\mathrm{deleg}} = [h_\ell, [h_{m,\ell}]_{m=0,\ldots,1+D}]_{\ell=1+L,\ldots,D}$

$$\leftarrow \ \left[\prod_{n=0}^{1+D} (\hat{y}_{n,\ell})^{\rho\cdot} \ , \ \left[\prod_{n=0}^{1+D} (\hat{y}_{n,\ell})^{\rho\cdots} \right]_{m=0,\ldots,1+D} \right]_{\ell=1+L,\ldots,D}$$

A more visual way to represent the private key is as a $(3 + D) \times (5 + 3\,D - L)$ array of elements in $\hat{\mathbb{G}}$, with $\mathsf{Pvk}_{\mathsf{Id}}^{\mathrm{decrypt}}$ as the upper left partial row, $\mathsf{Pvk}_{\mathsf{Id}}^{\mathrm{rerand}}$ as the lower left rectangle, and $\mathsf{Pvk}_{\mathsf{Id}}^{\mathrm{deleg}}$ as the entire right side block:

$$\begin{bmatrix} k_0 & k_{1,(a)} & k_{1,(b)} & \cdots & k_{1+\cdot,(a)} & k_{1+\cdot,(b)} \\ f_{0,0} & f_{0,0,(a)} & f_{0,0,(a)} & \cdots & f_{0,1+\cdot,(a)} & f_{0,1+\cdot,(a)} \\ f_{1,0} & f_{1,0,(a)} & f_{1,0,(a)} & \cdots & f_{1,1+\cdot,(a)} & f_{1,1+\cdot,(a)} \\ \vdots & & & \ddots & & \\ f_{1+\cdot,0} & f_{1+\cdot,0,(a)} & f_{1+\cdot,0,(a)} & \cdots & f_{1+\cdot,1+\cdot,(a)} & f_{1+\cdot,1+\cdot,(a)} \end{bmatrix} \begin{bmatrix} h_{1+\cdot} & \cdots & h_D \\ h_{0,1+\cdot} & \cdots & h_{0,D} \\ h_{1,1+\cdot} & \cdots & h_{1,D} \\ & \ddots & \\ h_{1+\cdot,1+\cdot} & \cdots & h_{1+\cdot,D} \end{bmatrix} .$$

Each row on the left can be viewed as a private key in an independent HIBE system (with generalized linear splitting as in Section 4). The main difference is that only $\mathsf{Pvk}_{\mathsf{Id}}^{\mathrm{decrypt}}$ contains the secret $\hat{w}$. The rows of $\mathsf{Pvk}_{\mathsf{Id}}^{\mathrm{rerand}}$ are independent HIBE keys for the same Id that do not permit decryption. The elements on the right side provide the delegation functionality: each column in $\mathsf{Pvk}_{\mathsf{Id}}^{\mathrm{deleg}}$ extends the hierarchy down one level. Delegation works as follows:

Derive$(\mathsf{Pub}, \mathsf{Pvk}_{\mathsf{Id}|L-1}, \mathsf{Id}_L)$. To derive a private key for $\mathsf{Id} = [I_0, I_1, \ldots, I_L] \in (\mathbb{Z}_p^{\times})^{1+L}$ where $L \in \{2, \ldots, D\}$ and $I_0 = 1$, given the parent's private key, $\mathsf{Pvk}_{\mathsf{Id}|L-1} = [k_0, [k_{n,(a)}, k_{n,(b)}], [f_{m,0}, [f_{m,n,(a)}, f_{m,n,(b)}]], [h_\ell, [h_{m,\ell}]]_{\ell=L,\ldots,D}]$, (where n, m range over $\{0, \ldots, 1+D\}$), do the following:

1. Pick $6 + 5D + D^2$ random integers:

$$\left[\, \pi_m,\ \left[\, \pi_{m,m'} \,\right]_{m'=0,\ldots,1+D} \,\right]_{m=0,\ldots,1+D} \ \in_{\$} \ (\mathbb{Z}_p)^{(3+D)(2+D)} \ .$$

2. Output the subordinate private key $\mathsf{Pvk}_{\mathsf{Id}} \in \hat{\mathbb{G}}^{(5+3D-L)(3+D)}$ comprised of $\mathsf{Pvk}_{\mathsf{Id}}^{\mathrm{decrypt}}$, $\mathsf{Pvk}_{\mathsf{Id}}^{\mathrm{rerand}}$, and $\mathsf{Pvk}_{\mathsf{Id}}^{\mathrm{deleg}}$, where:

   (a) To build $\mathsf{Pvk}_{\mathsf{Id}}^{\mathrm{decrypt}} = k_0', [k_{n,(a)}', k_{n,(b)}']_{n=0,\ldots,1+D}$, we set, for all n:

$$k_0' \leftarrow (k_0 \prod_{m=0}^{1+D} (f_{m,0})^{\pi_m})(h_\ell \prod_{m=0}^{1+D} (h_{m,\ell})^{\pi_m})^{I_\ell}$$

$$k_{n,(a)}', k_{n,(b)}' \leftarrow k_{n,(a)} \prod_{m=0}^{1+D} (f_{m,n,(a)})^{\pi_m}, k_{n,(b)} \prod_{m=0}^{1+D} (f_{m,n,(b)})^{\pi_m}$$

   (b) For $\mathsf{Pvk}_{\mathsf{Id}}^{\mathrm{rerand}} = [f_{m',0}', [f_{m',n,(a)}', f_{m',n,(b)}']_{n=0,\ldots,1+D}]_{m'=0,\ldots,1+D}$:

$$f_{m',0}' \leftarrow (\prod_{m=0}^{1+D} (f_{m,0})^{\pi_{m,m'}})(\prod_{m=0}^{1+D} (h_{m,\ell})^{\pi_{m,m'}})^{I_\ell}$$

$$f_{m',n,(a)}', f_{m',n,(b)}' \leftarrow \prod_{m=0}^{1+D} (f_{m,n,(a)})^{\pi_{m,m'}}, \prod_{m=0}^{1+D} (f_{m,n,(b)})^{\pi_{m,m'}}$$

   (c) And for $\mathsf{Pvk}_{\mathsf{Id}}^{\mathrm{deleg}} = [h_\ell', [h_{m',\ell}']_{m'=0,\ldots,1+D}]_{\ell=1+L,\ldots,D}$, we assign:

$$h_\ell' \leftarrow h_\ell \prod_{m=0}^{1+D} (h_{m,\ell})^{\pi_m}$$

$$h_{m',\ell}' \leftarrow \prod_{m=0}^{1+D} (h_{m,\ell})^{\pi_{m,m'}}$$

We note that *Derive* and *Extract* create private keys with the same structure and distribution. Notice that the *Derive* algorithm can be interpreted as the combination of two distinct operations: hierarchical delegation and key re-randomization.

– We start by re-randomizing the parent key, conceptually speaking, by performing random linear combinations of all its rows (in the array representation shown earlier). The first row enjoys special treatment: its coefficient into other rows' re-randomization is 0, and its own coefficient is 1.

– We then delegate by transforming the leftmost elements of $\mathsf{Pvk}_{\mathsf{Id}}^{\mathrm{decrypt}}$ and $\mathsf{Pvk}_{\mathsf{Id}}^{\mathrm{rerand}}$, in which identities are encoded. Suppose that re-randomization has already occurred, and imagine the resulting $\mathsf{Pvk}_{\mathsf{Id}}^{\mathrm{decrypt}}$, $\mathsf{Pvk}_{\mathsf{Id}}^{\mathrm{rerand}}$, $\mathsf{Pvk}_{\mathsf{Id}}^{\mathrm{deleg}}$. Delegation to a child identity I_L will "consume" the first column of $\mathsf{Pvk}_{\mathsf{Id}}^{\mathrm{deleg}}$: each element is raised to the power of I_L, and the result is aggregated into the leftmost element of $\mathsf{Pvk}_{\mathsf{Id}}^{\mathrm{decrypt}}$ or $\mathsf{Pvk}_{\mathsf{Id}}^{\mathrm{rerand}}$ on the same row, as follows:

$$
\begin{bmatrix} k_0 & \cdots & k_{n,(a)} & k_{n,(b)} & \cdots \\ f_{0,0} & \cdots & f_{0,n,(a)} & f_{0,n,(b)} & \cdots \\ \vdots & \ddots & & & \\ f_{m,0} & \cdots & f_{m,n,(a)} & f_{m,n,(b)} & \cdots \\ \vdots & & & \ddots & \end{bmatrix}
\begin{bmatrix} h. & \cdots & h_D \\ h_{0,\cdot} & \cdots & h_{0,D} \\ \vdots & \ddots & \\ h_{m,\cdot} & \cdots & h_{m,D} \\ \vdots & & \ddots \end{bmatrix}
\rightarrow
\begin{bmatrix} k'_0 & \cdots \\ f'_{0,0} & \cdots \\ \vdots & \ddots \\ f'_{m,0} & \cdots \\ \vdots & \end{bmatrix}
\begin{bmatrix} \bullet & h'_{1+\cdot} & \cdots & h'_D \\ \bullet & h'_{0,1+\cdot} & \cdots & h'_{0,D} \\ & \vdots & \ddots & \\ \bullet & h'_{m,1+\cdot} & \cdots & h'_{m,D} \\ & \vdots & & \ddots \end{bmatrix}
$$

We now turn to the encryption and decryption methods.

Encrypt(Pub, Id, Msg.) To encrypt a given message encoded as a group element $\mathsf{Msg} \in \mathbb{G}_T$ for a given identity $\mathsf{Id} = [I_0(=1), I_1, \ldots, I_L]$ at level L:

1. Select $3+D$ random integers: $r, [r_n]_{n=0,\ldots,1+D} \in_{\$} (\mathbb{Z}_p)^{3+D}$.
2. Output the ciphertext $\mathsf{CT} = E, c_0, [c_{n,(a)}, c_{n,(b)}]_{n=0,\ldots,1+D} \in \mathbb{G}_T \times \mathbb{G}^{5+2D}$, computed as:

$$
\mathsf{CT} \leftarrow \mathsf{Msg} \cdot \Omega^{-r}, g^r, \left[\left(\prod_{\ell=0}^{L} b_{n,\ell}^{I_\cdot} \right)^{r_\cdot}, \left(\prod_{\ell=0}^{L} a_{n,\ell}^{I_\cdot} \right)^{r-r_\cdot} \right]_{n=0,\ldots,1+D} .
$$

Decrypt(Pub, $\mathsf{Pvk}_{\mathsf{Id}}$, CT.) To decrypt a ciphertext CT, using the decryption subkey a private key, $\mathsf{Pvk}_{\mathsf{Id}}^{\mathrm{decrypt}} = [k_0, [k_{n,(a)}, k_{n,(b)}]_{n=0,\ldots,1+D}]$, compute:

$$
\hat{\mathsf{Msg}} \leftarrow E \cdot \mathbf{e}(c_0, k_0) \prod_{n=0}^{1+D} \mathbf{e}(c_{n,(a)}, k_{n,(a)}) \, \mathbf{e}(c_{n,(b)}, k_{n,(b)}) \in \mathbb{G}_T .
$$

Encryption can be made very cheap with a bit of caching since the exponentiation bases never change. Decryption is also fairly efficient since all the pairings in the product can be computed at once using a "multi-pairing" approach [21], which is similar to multi-exponentiation. One can also exploit the fact that all the $k_{\ldots}$ are fixed for a given recipient to perform advantageous pre-computations [3].

6   Consistency and Security

The following theorems state that extracted and delegated private keys are identically distributed, and that extraction, encryption, and decryption, are consistent. We remark that Theorem 1 is not essential for the security model, but it is nice to have and it is also useful to prove Theorem 2.

Theorem 1. *Private keys calculated by Derive and Extract have the same distribution.*

Theorem 2. *The Anonymous HIBE scheme is internally consistent.*

We now state the basic security theorems for the A-HIBE scheme. The selective-ID security reductions are almost tight and hold in the standard model. We only consider recipient anonymity, since sender anonymity is trivially attainable in an unauthenticated encryption scheme.

Theorem 3 (Confidentiality). *Suppose that* $\mathbf{G}$ *upholds the* (τ, ϵ)-*Decision BDH assumption. Then, against a selective-ID adversary that makes at most* q *private key extraction queries, the HIBE scheme of Section 5 is* $(q, \tilde{\tau}, \tilde{\epsilon})$-*IND-sID-CPA secure in* $\mathbf{G}$ *with* $\tilde{\tau} \approx \tau$ *and* $\tilde{\epsilon} = \epsilon - (3 + D)\, q/p$.

Theorem 4 (Anonymity). *Suppose that* $\mathbf{G}$ *upholds the* (τ, ϵ)-*Decision Linear assumption. Then, against a selective-ID adversary that makes* q *private key extraction queries, the HIBE scheme of Section 5 is* $(q, \tilde{\tau}, \tilde{\epsilon})$-*ANON-sID-CPA secure in* $\mathbf{G}$ *with* $\tilde{\tau} \approx \tau$ *and* $\tilde{\epsilon} = \epsilon - (2 + D)(7 + 3\,D)\, q/p$.

For completeness, we mention that based on the above theorems it is easy to secure the scheme against active adversaries, *i.e.*, adaptive-ID and CCA2. Adaptive-identity security can be obtained using the Waters [31] technique, or by using random oracles [5, §7], although these methods only work for shallow hierarchies. Adaptive chosen-ciphertext security can be achieved very effectively using one of several techniques [15,11,13], all of which are applicable here.

7   Conclusion

We presented a provably anonymous IBE and HIBE scheme without random oracles, which resolves an open question from CRYPTO 2005 regarding the existence of anonymous HIBE systems.

Our constructions make use of a novel "linear-splitting" technique which prevents an attacker from testing the intended recipient of ciphertexts, yet allows for the use of randomized private IBE keys. In the hierarchical case, we add to this a new "multi-simulation" proof device that permits multiple HIBE subsystems to concurrently re-randomize each other. Security is based solely on the Linear assumption in bilinear groups.

Our basic scheme is very efficient, a factor two slower than (non-anonymous) Boneh-Boyen BB_1 and BB_2 encryption, and quite faster than Boneh-Franklin. The full hierarchical scheme remains practical with its quadratic private key size, and its linear ciphertext size, encryption time, and decryption time, as functions of the depth of the hierarchy.

Acknowledgements

The authors would like to thank Mihir Bellare, Dan Boneh, and Hovav Shacham for helpful discussions, as well as the anonymous referees for useful comments.

References

1. Michel Abdalla, Mihir Bellare, Dario Catalano, Eike Kiltz, Tadayoshi Kohno, Tanja Lange, John Malone-Lee, Gregory Neven, Pascal Paillier, and Haixia Shi. Searchable encryption revisited: Consistency properties, relation to anonymous IBE, and extensions. In *Advances in Cryptology—CRYPTO 2005*, Lecture Notes in Computer Science, pages 205–22. Springer-Verlag, 2005.

2. Paulo S. L. M. Barreto and Michael Naehrig. Pairing-friendly elliptic curves of prime order. Cryptology ePrint Archive, Report 2005/133, 2005. http://eprint.iacr.org/.

3. Paulo S.L.M. Barreto, Hae Y. Kim, Ben Lynn, and Michael Scott. Efficient algorithms for pairing-based cryptosystems. Cryptology ePrint Archive, Report 2002/008, 2002. http://eprint.iacr.org/.

4. Mihir Bellare, Alexandra Boldyreva, Anand Desai, and David Pointcheval. Key-privacy in public-key encryption. In *Proceedings of ASIACRYPT 2001*, Lecture Notes in Computer Science, pages 566–82. Springer-Verlag, 2001.

5. Dan Boneh and Xavier Boyen. Efficient selective-ID secure identity based encryption without random oracles. In *Advances in Cryptology—EUROCRYPT 2004*, volume 3027 of *Lecture Notes in Computer Science*, pages 223–38. Springer-Verlag, 2004.

6. Dan Boneh and Xavier Boyen. Secure identity based encryption without random oracles. In *Advances in Cryptology—CRYPTO 2004*, volume 3152 of *Lecture Notes in Computer Science*, pages 443–59. Springer-Verlag, 2004.

7. Dan Boneh, Xavier Boyen, and Eu-Jin Goh. Hierarchical identity based encryption with constant size ciphertext. In *Advances in Cryptology—EUROCRYPT 2005*, volume 3494 of *Lecture Notes in Computer Science*, pages 440–56. Springer-Verlag, 2005.

8. Dan Boneh, Xavier Boyen, and Hovav Shacham. Short group signatures. In *Advances in Cryptology—CRYPTO 2004*, volume 3152 of *Lecture Notes in Computer Science*, pages 41–55. Springer-Verlag, 2004.

9. Dan Boneh, Giovanni Di Crescenzo, Rafail Ostrovsky, and Giuseppe Persiano. Public key encryption with keyword search. In *Advances in Cryptology—EUROCRYPT 2004*, volume 3027 of *Lecture Notes in Computer Science*, pages 506–22. Springer-Verlag, 2004.

10. Dan Boneh and Matthew Franklin. Identity-based encryption from the Weil pairing. *SIAM Journal of Computing*, 32(3):586–615, 2003. Extended abstract in *Advances in Cryptology—CRYPTO 2001*.

11. Dan Boneh and Jonathan Katz. Improved efficiency for CCA-secure cryptosystems built using identity based encryption. In *Proceedings of CT-RSA 2005*, volume 3376 of *Lecture Notes in Computer Science*. Springer-Verlag, 2005.

12. Xavier Boyen. Multipurpose identity-based signcryption: A Swiss Army knife for identity-based cryptography. In *Advances in Cryptology—CRYPTO 2003*, volume 2729 of *Lecture Notes in Computer Science*, pages 383–99. Springer-Verlag, 2003.

13. Xavier Boyen, Qixiang Mei, and Brent Waters. Direct chosen ciphertext security from identity-based techniques. In *ACM Conference on Computer and Communications Security—CCS 2005*. ACM Press, 2005.

14. Ran Canetti, Shai Halevi, and Jonathan Katz. A forward-secure public-key encryption scheme. In *Advances in Cryptology—EUROCRYPT 2003*, volume 2656 of *Lecture Notes in Computer Science*. Springer-Verlag, 2003.

15. Ran Canetti, Shai Halevi, and Jonathan Katz. Chosen-ciphertext security from identity-based encryption. In *Advances in Cryptology—EUROCRYPT 2004*, volume 3027 of *Lecture Notes in Computer Science*. Springer-Verlag, 2004.

16. Sanjit Chatterjee and Palash Sarkar. Trading time for space: Towards an efficient IBE scheme with short(er) public parameters in the standard model. In *Proceedings of ICISC 2005*, 2005.

17. Clifford Cocks. An identity based encryption scheme based on quadratic residues. In *Proceedings of the 8th IMA International Conference on Cryptography and Coding*, 2001.

18. Darren Davis, Fabian Monrose, and Michael K. Reiter. Time-scoped searching of encrypted audit logs. In *Proceedings of ICICS 2004*, pages 532–45, 2004.

19. Craig Gentry. Practical identity-based encryption without random oracles. In *Advances in Cryptology—EUROCRYPT 2006*, Lecture Notes in Computer Science. Springer-Verlag, 2006.

20. Craig Gentry and Alice Silverberg. Hierarchical ID-based cryptography. In *Proceedings of ASIACRYPT 2002*. Springer-Verlag, 2002.

21. Robert Granger and Nigel P. Smart. On computing products of pairings. Cryptology ePrint Archive, Report 2006/172, 2006. http://eprint.iacr.org/.

22. Jeremy Horwitz and Ben Lynn. Towards hierarchical identity-based encryption. In *Advances in Cryptology—EUROCRYPT 2002*, Lecture Notes in Computer Science, pages 466–81. Springer-Verlag, 2002.

23. Antoine Joux. A one round protocol for tripartite Diffie-Hellman. *Journal of Cryptology*, 17(4):263–76, 2004. Extended abstract in *Proceedings of ANTS IV, 2000*.

24. Alfred Menezes, Tatsuaki Okamoto, and Scott Vanstone. Reducing elliptic curve logarithms in a finite field. *IEEE Transactions on Information Theory*, 39(5):1639–46, 1993.

25. Victor Miller. The Weil pairing, and its efficient calculation. *Journal of Cryptology*, 17(4), 2004.

26. David Naccache. Secure and practical identity-based encryption. Cryptology ePrint Archive, Report 2005/369, 2005. http://eprint.iacr.org/.

27. Moni Naor. On cryptographic assumptions and challenges. In *Advances in Cryptology—CRYPTO 2003*. Springer-Verlag, 2003.

28. Adi Shamir. Identity-based cryptosystems and signature schemes. In *Advances in Cryptology—CRYPTO 1984*, volume 196 of *Lecture Notes in Computer Science*, pages 47–53. Springer-Verlag, 1984.

29. Victor Shoup. Lower bounds for discrete logarithms and related problems. In *Advances in Cryptology—EUROCRYPT 1997*, volume 1233 of *Lecture Notes in Computer Science*. Springer-Verlag, 1997.

30. Dawn Xiaodong Song, David Wagner, and Adrian Perrig. Practical techniques for searches on encrypted data. In *Proceedings of the IEEE Symposium on Security and Privacy—SP 2000*. IEEE Computer Society, 2000.

31. Brent Waters. Efficient identity-based encryption without random oracles. In *Advances in Cryptology—EUROCRYPT 2005*, volume 3494 of *Lecture Notes in Computer Science*. Springer-Verlag, 2005.

32. Brent Waters, Dirk Balfanz, Glenn Durfee, and Diana Smetters. Building an encrypted and searchable audit log. In *Proceedings of NDSS 2004*, 2004.

33. Danfeng Yao, Nelly Fazio, Yevgeniy Dodis, and Anna Lysyanskaya. ID-based encryption for complex hierarchies with applications to forward security and broadcast encryption. In *ACM Conference on Computer and Communications Security—CCS 2004*, pages 354–63, 2004.

Fast Algorithms for the Free Riders Problem in Broadcast Encryption

Zulfikar Ramzan[1,*] and David P. Woodruff[2,3,*]

[1] Symantec, Inc.
zulfikar_ramzan@symantec.com
[2] MIT
dpwood@mit.edu
[3] Tsinghua University

Abstract. We provide algorithms to solve the *free riders* problem in broadcast encryption. In this problem, the broadcast server is allowed to choose some small subset F of the revoked set R of users to allow to decrypt the broadcast, despite having been revoked. This may allow the server to significantly reduce network traffic while only allowing a small set of non-privileged users to decrypt the broadcast.

Although there are worst-case instances of broadcast encryption schemes where the free riders problem is difficult to solve (or even approximate), we show that for many specific broadcast encryption schemes, there are efficient algorithms. In particular, for the complete subtree method [25] and some other schemes in the subset-cover framework, we show how to find the optimal assignment of free riders in $O(|R||F|)$ time, which is independent of the total number of users. We also define an approximate version of this problem, and study specific distributions of R for which this relaxation yields even faster algorithms.

Along the way we develop the first approximation algorithms for the following problem: given two integer sequences $a_1 \geq a_2 \geq \cdots \geq a_n$ and $b_1 \geq b_2 \geq \cdots \geq b_n$, output for all i, an integer j' for which $a_{j'} + b_{i-j'} \leq (1 + \epsilon) \min_j (a_j + b_{i-j})$. We show that if the differences $a_i - a_{i+1}, b_i - b_{i+1}$ are bounded, then there is an $O(n^{4/3}/\epsilon^{2/3})$-time algorithm for this problem, improving upon the $O(n^2)$ time of the naive algorithm.

Keywords: free riders, broadcast encryption, applications.

1 Introduction

Broadcast Encryption: Broadcast encryption schemes allow a single center to transmit encrypted data over a broadcast channel to a large number of users U such that only a select subset $\mathcal{P} \subseteq U$ of *privileged users* can decrypt it. Such schemes aim to provide one-to-many secure communication services to a large user base without incurring scalability costs. One important application of broadcast encryption is to provide the users in a group with a common cryptographic key that they can then use to communicate amongst each other.

---

* Part of this research was done while both authors were at DoCoMo Labs.

C. Dwork (Ed.): CRYPTO 2006, LNCS 4117, pp. 308–325, 2006.

Other traditional applications of broadcast encryption include secure multi-cast of privileged content such as premium video, audio, and text content as well as protection on external storage devices such as the hard disk of a mobile phone, USB storage devices, Flash cards, CD and DVD ROMs, etc.

Broadcast encryption schemes typically involve a series of pre-broadcast transmissions at the end of which the users in P can compute a broadcast session key bk. The remainder of the broadcast is then encrypted using bk. There are a number of variations on this general problem:

- Privileged Sets: Privileged sets may be determined arbitrarily, have fixed size, or contain some type of hierarchical structure. They can be static across all broadcasts or dynamic. In other words, a user may be revoked permanently or only have his privileges temporarily revoked for a particular broadcast.
- Key Management: User keys may be fixed at the onset, be updated each time period, or be a function of previous transmissions.
- Coalition Resistance: The scheme may be secure against any revoked user coalition, or there may be some bound on the size of the tolerable coalition.

The most widely applicable scenario (and the one we consider in this paper) is where the revocation set is dynamic, the users are only given a collection of set-up keys at the onset (these keys are used for the lifetime of the system), and the system can tolerate an arbitrary number of colluders.

The following performance parameters are of interest: the number of pre-broadcast transmissions t made by the center, the amount of keying material k the receivers must persistently store, and the amount of time τ_d the receiver must spend to derive the broadcast session key bk from the pre-broadcast transmissions. Let $R = U \setminus P$ be the set of revoked users, let $r = |R|$, and let $n = |U|$.

Related Work: Berkovits introduced the concept of broadcast encryption [6]. In his scheme $t = O(n)$, although $k = 1$; the scheme is also insecure if used repeatedly. Later, Fiat and Naor [14] formalized the basic problem. The topic has since been studied extensively (e.g., see [4, 7, 10, 15, 20, 21, 22, 23, 25, 17]). We limit our discussion to the work most relevant to us. Naor et al. [25] proposed the *Complete Subtree* (CS) scheme, which is stateless and requires a pre-broadcast transmission of $t = O(r \log \frac{n}{r})$ ciphertexts and storage of $k = O(\log n)$ keys per user. The scheme is simple and provides information-theoretic security. In [25], the *Subset Difference* (SD) scheme is also proposed. Here, t is reduced to $O(r)$, and is therefore independent of the total number of users n. On the other hand, the scheme only provides computational security and a user must store $k = O(\log^2 n)$ keys. Furthermore, the receiver must perform $\tau_d = O(\log n)$ computation to compute the broadcast key bk. Halevy and Shamir [17] improved upon this construction with their *Layered Subset Difference* (LSD) scheme. For any $\epsilon > 0$, they achieve $k = O(\log^{1+\epsilon} n)$ without substantially increasing t.

The CS, SD, LSD, are all examples of schemes in the subset-cover framework. Here various subsets of the users are formed and a cryptographic key is assigned to each subset. Each user is provided with all the keys for subsets he is a member of. To encrypt to a privileged set, the center finds a collection of subsets whose union is exactly the privileged set, and encrypts content using just those keys.

Free Riders in Broadcast Encryption: Most of the prior work on broadcast encryption assumes a stringent security requirement. In particular, any member of the privileged set $\mathcal{P}$ can decrypt a broadcast, but *no one* outside of this set can. For many applications this might be excessive. In particular, the system might be able to tolerate some number of non-privileged users being able to decrypt a broadcast. Such a user is termed a *free rider*.

Here the center chooses some (preferably small) subset F of R who can decrypt the broadcast, despite having been revoked. Let $f = |F|$. In general, this concept is useful in commercial applications of cryptography (e.g., by network operators) since the negative fiscal impact of allowing some small number of free riders may be dwarfed by the savings incurred by reducing other operational systems costs, such as the cost of transmitting traffic on the network, etc. We will demonstrate the benefits of choosing free riders in Section 3.2.

Abdalla, Shavit, and Wool (ASW) [1] introduced the notion of free riders. They demonstrate that allowing free riders can reduce the amount of data communicated in many situations. Next, they consider the problem of determining the free rider assignment that minimizes traffic for a given number of free riders. Via simple reductions from SetCover, they observe that this problem is NP complete in the worst case, and is unlikely to have a polynomial-time approximation scheme (PTAS). Further, they suggest some heuristic algorithms that apply to broadcast encryption schemes in the subset-cover framework. Finally, they experimentally analyze their heuristics on the CS scheme.

Our Contributions: In this paper we first show that for the CS scheme, one can provably obtain the *optimal solution* (with respect to reducing the overall traffic) to the free riders problem in worst-case time $O(rf + r \log \log n)$. Our techniques are also applicable to other schemes in the subset-cover framework.

In contrast, ASW [1] only give heuristics with no performance guarantees.[1] Thus, our work *provably* demonstrates the positive aspects of free riders in broadcast encryption. Moreover, our running time is almost *independent* of n, and as $r, f << n$ in practice, is likely to be extremely efficient. Note that neither us nor ASW consider the role of free riders on improving other aspects of performance, like the amount of storage required on end-user devices. In fact, our storage requirements are the same as that of the underlying broadcast encryption scheme, but the overall traffic is reduced.

Second, we relax the free riders problem to allow an algorithm to output a free rider assignment with cost at most $(1 + \epsilon)$ times that of the optimal assignment. We show how this relaxation is likely to be useful for *specific distributions* of revoked users R. In particular, we show that when R is uniformly chosen from sets of size r, it results in a running time of $O(rf^{1/3} \text{polylog } n)$ for constant ϵ.

---

[1] Even though our algorithms have provable performance guarantees, our work does not contradict the ASW NP-hardness or inapproximability results. In particular, their results only apply to worst-case instances. In our case, we give optimal solutions for *specific* set-covering schemes. Interestingly, our algorithms apply to the very same broadcast encryption scheme (the complete subtree method) that ASW used in experimentally analyzing their heuristics.

Third, consider the following MinSum problem: given two arrays of integers $a_1 \geq a_2 \geq \cdots \geq a_{m_1} \geq 0$ and $b_1 \geq b_2 \geq \cdots b_{m_2} \geq 0$ such that for all i, $a_i - a_{i+1}$ and $b_i - b_{i+1}$ are bounded by L, output an array c such that for all $2 \leq i \leq m_1 + m_2$, $a_{c[i]} + b_{i-c[i]} \leq (1 + \epsilon) \min_{j=1}^{i-1}(a_j + b_{i-j})$. Note that $c[i]$ is an integer between 1 and $i - 1$, inclusive. There is a trivial algorithm for this which runs in time $O(m_1 m_2)$. We show a direct connection between this problem and the running time of our algorithm for specific distributions on R. Using tools from computational geometry, the running time of MinSum was recently improved [8, 13] to $O(m_1 + m_2 + m_1 m_2 / \log(m_1 + m_2))$, which we can apply to our problem (for $\epsilon = 0$) to obtain faster solutions to the free riders problem.

More importantly, as far as we are aware, we provide the first approximation (e.g., $\epsilon > 0$) algorithms for MinSum which asymptotically beat the trivial $O(m_1 m_2)$ running time. Namely, we achieve $O(m_1^{1/3} m_2 L^{2/3} / \epsilon^{2/3})$ time, which for constant ϵ and small L (as is the case for the CS, SD, and LSD schemes), is about $m_1^{1/3} m_2$. This algorithm may be of independent interest since the MinSum problem occurs naturally in computational geometry, see, e.g., [13], where the problem is compared with computing all-pairs-shortest-paths.

Our algorithms extend to other broadcast encryption schemes that use set covering mechanisms with a hierarchical tree-like structure, such as the natural extension of the CS scheme to k-ary trees. We suspect they also extend to the SD [25] and LSD [17] schemes, and the natural extensions of these schemes to k-ary trees. Further, our techniques apply to multi-certificate revocation/validation and group key management, which we discuss below.

Other Applications of Free Riders: The idea of designing systems that permit some number of non-privileged users to receive the same services as privileged users to lower system costs is widely applicable. We mention other settings where the techniques in this paper apply (that do not seem to have been considered previously). One such application is multi-certificate revocation/validation [2]. Here a certificate authority (CA) in a public-key infrastructure might wish to revoke a number of user certificates, but could be willing to permit a subset of these users to get away with using revoked credentials. Multi-certificate revocation/validation schemes allow the CA to issue a single proof that validates the non-revoked certificates for a specified time period. Permitting free riders decreases the proof size, which reduces communication complexity.

Another application is dynamic group key management [3, 27, 19, 24, 30, 29], which is closely related to broadcast encryption. Here a user group communicates among each other using a common cryptographic key. As users join or leave the group at different times the key is updated by sending out a new (encrypted) group session key. If one were willing to allow a small number of former group members to be free riders and decrypt this key-update message, then the overall communication costs could be lowered. Popular schemes for multi-certificate revocation and group key management use the same set-covering mechanisms encountered in broadcast encryption, and thus our techniques apply.

Organization: Section 2 discusses preliminaries and gives an overview of our algorithm, while section 3 gives more detail. Section 4 discusses our algorithmic improvements to the MinSum problem. Due to space constraints, we defer many proofs to the full version of this paper [28].

2   Preliminaries and Overview of Our Algorithm

Preliminaries: Let U denote the users, with $n = |U|$. Let R denote the revoked set at a given time, with $r = |R|$. Let F denote the set of free riders with $f = |F|$.

We now define the complete subtree (CS) scheme [25], which falls into the subset-cover framework mentioned in Section 1. W.l.o.g., assume n is a power of 2. Users are associated with the n leaves of a complete binary tree. The server creates a key for each node v of the binary tree, and distributes it to all users in the subtree rooted at v. One can find $O(r \log n/r)$ keys (nodes) such that every user in $U \setminus R$ has one such key, but each user in R lacks all such keys [25].

Benefits of Free Riders: We give an example to demonstrate the benefits of choosing the optimal set of free riders for the CS scheme. Suppose $r = \sqrt{n}$ and $f = cr = c\sqrt{n}$ for some constant $0 < c < 1$. For simplicity suppose that $r, r - f$ are powers of 2. Consider the level L of the complete binary tree, as used in the CS scheme, with exactly $\sqrt{n}$ nodes. Put $r - f$ revoked users in a *complete* subtree of one node in level L, and for each other node L, put at most 1 revoked user (so exactly f nodes in L have at least one revoked user in their subtree). Then the optimal F contains the f isolated users. The revoked users not in F constitute all the leaves of a complete subtree, and we only need $O(\log n)$ traffic to transmit to the set $U \setminus (R \setminus F)$. This is accomplished by including a key at each sibling along the path from the root of the complete subtree to the root of the complete binary tree.

If, however, we choose the free riders randomly, then with overwhelming probability $\Omega(f)$ isolated revoked users (i.e., those not included in the complete subtree) remain. For each such user u, we must include a key at each sibling along the path from u to L (not including the node in level L). Since this path length is $\log n - \log \sqrt{n} = \Omega(\log n)$, and since the subtrees rooted in level L are disjoint, the total network traffic is $\Omega(f(\log n - \log \sqrt{n})) = \Omega(\sqrt{n} \log n)$. Therefore, by choosing F sensibly (as opposed to randomly) we save a $\sqrt{n}$ factor in network traffic.

The reader is referred to ASW [1] for a more extensive analysis of the benefits of free riders on the network traffic. Here we concentrate on the efficiency of algorithms for finding F.

Notation: For a vertex v in a tree, let $T(v)$ be the subtree rooted at v, and let $L(v)$ and $R(v)$ denote the subtrees rooted at the left and right child of v, respectively. We use a standard identification of nodes of a complete binary tree. Namely, we define the root to be 1. Then, for a node identified with integer i, its left child is $2i$ and its right child is $2i + 1$. We define the height of a node v to be $\lfloor \log_2 v \rfloor$, so the root has height 0. For nodes u, v, we define $a(u, v)$ to be the common ancestor of u and v of largest height. We say that node u is *to the left* of node v if $u \in L(a(u,v))$ while $v \in R(a(u,v))$. Let $\log()$ denote the base-2 logarithm. We work in the RAM model where arithmetic operations on $O(\log n)$ size words take $O(1)$ time.

Cost Function: Let $F \subseteq R$ be subject to $|F| \leq f$. Let V be a subset of vertices of the complete binary tree for which

$$(R \setminus F) \cap \cup_{v \in V} T(v) = \emptyset \text{ and } (U \setminus R) \subseteq \cup_{v \in V} T(v).$$

Then the cost of F is the size of a minimal such V, and the optimal asignment F is that with the smallest cost. This cost function captures the network traffic in any scheme within the subset-cover framework, since in these schemes the number of ciphertexts the server transmits equals $|V|$.

We first show how to quickly compute an optimal assignment of free riders for the CS scheme. We focus on this scheme for its simplicity and to contrast our results with the original free riders paper [1]. For these schemes we prove the following about our main algorithm Freerider-Approx.

Theorem 1. *Algorithm* Freerider-Approx(R, f) *outputs an optimal assignment of f free riders in $O(rf + r \log \log n)$ time.*

Remark 1. Although we use the terminology Freerider-Approx, the algorithm we describe is also capable of producing an optimal answer (i.e., no approximation).

To appreciate the techniques of Freerider-Approx, we first sketch a simple dynamic-programming algorithm achieving $O(nf)$ time. This already shows that many natural broadcast encryption schemes admit theoretically-efficient (poly(n) time) algorithms for the free riders problem, despite the fact that the problem is NP-hard in general.

Recall that in the CS scheme, users are associated with leaves of a complete binary tree, and the revoked set R determines some set of r leaves. For each node v in the tree, we can use dynamic programming to compute the optimal traffic $opt(i, T(v))$ attained by assigning at most i free riders to its subtree $T(v)$, for each $0 \leq i \leq f$. Note that if $i \geq |T(v) \cap R|$ then either every leaf of $T(v)$ is in R, in which case $opt(i, T(v)) = 0$, else $opt(i, T(v)) = 1$ since we can just let everyone in $T(v)$ receive the broadcast using a single message. Now, suppose for each $0 \leq i \leq f$ we have $opt(i, L(v))$ and $opt(i, R(v))$, together with indicator bits stating whether $L(v) \cap R$ and $R(v) \cap R$ are empty. Then the optimal cost of assigning at most i free riders to $T(v)$ is 1 if both indicator bits are set. This is because, in this case, and only in this case, we can use a single key at v. In this case we set the indicator bit at v to be 1. Otherwise the cost is $\min_j (opt(j, L(v)) + opt(i - j, R(v)))$. We can find this collection of minima in $O(f^2)$ time, for $0 \leq i \leq f$. Finally, by tracking which indices realize the minima, we can backtrack to find the optimal free-rider assignment itself. This running time is $O(nf^2)$, which we can reduce to $O(nf)$ by observing that finding the minima for each v only requires $O(|L(v) \cap R| \cdot |R(v) \cap R|)$ time. Unfortunately, this running time depends linearly on n, which may be huge in practice.

On the other hand, r is likely to be small, which we can exploit by observing that it suffices to perform the dynamic programming step for only a very small set of internal nodes called the *meeting points*. These are defined as follows. Recall that a Steiner tree of a subset U of vertices of an unweighted tree is a minimal-sized tree containing each vertex of U. Suppose we compute the Steiner tree T of the subset R of the complete binary tree, where the ith revoked user is identified with the ith leaf. Then the *meeting points* of T are the leaves of T together with the internal nodes v of the binary tree for which both $L(v)$ and $R(v)$ are non-empty (i.e., they contain a revoked user). We refer to the latter nodes as *internal meeting points*. A simple induction shows there are $2r - 1$ meeting points, $r - 1$ of which are internal. We give a very efficient $O(r \log \log n)$-time algorithm to output the list of meeting points.

Now, if a node v is not a meeting point, then there is no reason to perform the dynamic programming step on v, as its optimal assignments can be immediately inferred from those of its only child. The next question is how to perform the dynamic programming step, which computes $c_i = \min_j(a_j + b_{i-j})$ for all i, where a_k, b_k, c_k are the smallest costs of assigning at most k free riders to the trees $L(v), R(v)$, and $T(v)$, respectively. This is an instance of the MinSum problem. Suppose v is the meeting point joining meeting points x and y, and let $r_x = |L(v) \cap R|$ and $r_y = |R(v) \cap R|$. Then there is an $O(\min(f, r_x) \min(f, r_y))$ time algorithm for the MinSum problem, which works by computing sums $a_j + b_{i-j}$ for all i, j, and then taking the appropriate minima.

In fact, MinSum can be solved in time

$$O\left(\min(r_x, f) + \min(r_y, f) + \frac{\min(r_x, f) \min(r_y, f)}{\log(\min(f, r_x) + \min(f, r_y))}\right)$$

using ideas from computational geometry [8, 13]. It turns out that there are two properties of our problem which allow us to significantly reduce this running time. The first is that for the schemes we consider, $a_i - a_{i+1}$ and $b_i - b_{i+1}$ are very small (i.e., $\log n$ for for the CS scheme and $O(1)$ for the SD scheme). The second is that for many applications, it suffices to output an *approximate* solution to the free riders problem, that is, an assignment of at most f free riders with resulting cost at most $(1 + \epsilon)$ times that of the optimal. To this end, we relax MinSum to just require that we output an array d with $d_i \le (1 + \epsilon)c_i$ for all i.

We show in section 3.1 that if $a_i - a_{i+1}, b_i - b_{i+1}$ are bounded by L, we can output d in $O((r_x)^{1/3} r_y L^{2/3}/\epsilon^{2/3})$ time. For the small values of L that we consider, this shows that a constant-factor approximation can be computed in about $(r_x)^{1/3} r_y$ time, significantly improving the time of the naive algorithm.

Unfortunately, it turns out that for a worst-case choice of R, our improvements for MinSum do not asymptotically improve the running time of Freerider-Approx, though they do so for *specific distributions* of R that may occur in practice. In particular, for $f = \Omega(\log n)$ we show that when R is uniformly chosen from sets of size r, then with probability at least $1 - 1/n$ Freerider-Approx outputs a constant-factor approximation in $O(rf^{1/3}\text{polylog } n)$ time, whereas if Freerider-Approx were to use the naive MinSum algorithm, its running time would be

$\Omega(rf)$. The more general version of our theorem, for arbitrary f and arbitrary approximation ratios, can be found in Section 3.

For a worst-case analysis, we analyze Freerider-Approx in terms of the naive algorithm for MinSum and show by a careful choice of parameters that the algorithm outputs a $(1 + \epsilon)$-approximation. Showing that it only takes $O(rf)$ time requires an analysis over Steiner trees, which although elementary, is non-trivial.

3   Details of the Algorithm

3.1   Finding the Meeting Points

We assume $0 < f < r < n$, as otherwise the problem is trivial. Thus, we can root T at an internal meeting point. We root T at the meeting point of smallest height in the complete binary tree. This node is unique, since if there were two such nodes their common ancestor would be a meeting point of smaller height.

We assume that we are given R as a list of integers in the range $\{n, n + 1, \ldots, 2n - 1\}$. We first sort R. The best known integer sorting algorithms run in expected time $O(r\sqrt{\log \log r})$ [18] and worst-case $O(r \log \log r)$ [16], though more practical ones running in time $O(r \log r)$ or $O(r \log \log n)$ can be based on textbook algorithms [9] or van Emde Boas trees [11, 12].

One way to visualize our algorithm for computing the meeting points is to think of a line containing r points. The ith point is u_i, the ith revoked user in the given sorted list. The edge (u_i, u_{i+1}) is labeled as follows: we find $a = a(u_i, u_{i+1})$, and we label (u_i, u_{i+1}) with both a and $\mathsf{height}(a)$, referring to the latter as the *edge weight*. Note that a and its height can be found using $O(\log \log n)$ arithmetic operations via a binary search on the binary representations of u_i and u_{i+1} (recall that we are assuming the unit-cost RAM model). Then a is an internal meeting point with $u_i \in L(a)$ and $u_{i+1} \in R(a)$.

We maintain a sorted list E of edge weights and an output list MP of meeting points. We initialize MP to R. We then find the largest-weight edge $e = \{u, v\}$ (i.e., $a(u, v)$ is of largest height), breaking ties arbitrarily, add $a(u, v)$ to the end of MP, and delete e from E. This corresponds to collapsing the edge $\{u, v\}$ on the line and replacing it with the vertex $a(u, v)$. We then update the weight and ancestor label of the other edges in E containing u or v (e.g., the neighbors of the collapsed edge on the line). We show that the "line structure" of the remaining edges E is preserved across iterations, so any vertex u on the line can belong to at most 2 edges, and thus this update step is efficient. After $r - 1$ deletions, the entire line is collapsed into a single vertex, the root of the Steiner tree, and we are done. The running time is $O(r \log \log n)$ plus the time to sort E.

A list of meeting points is *ordered* if for each meeting point v, its unique closest meeting points (under the shortest path distance) p_l and p_r, in $L(v)$ and $R(v)$ respectively, precede v in the list. In the full version of this paper [28], we prove:

Theorem 2. Meeting-Points *outputs an ordered list of meeting points in time* $O(r \log \log n)$.

3.2   The Free Rider Approximation Algorithm

Our free rider approximation algorithm Freerider-Approx makes use of the subroutine MeetingPoints and also the subroutine Min-Sum. We defer the description of Min-Sum to Section 4, but we summarize its relevant properties as follows.

Theorem 3. *Let* $\epsilon \geq 0$, *and let* $a_1 \geq a_2 \geq \cdots \geq a_{m_1} \geq 0$ *and* $b_1 \geq b_2 \geq \cdots \geq b_{m_2} \geq 0$ *be integer sequences such that for all* i, $a_i - a_{i+1}$ *and* $b_i - b_{i+1}$ *are bounded by* L. *Then* Min-Sum$((a_i), (b_i), \epsilon)$ *outputs an array* c *such that for all* $2 \leq i \leq m_1 + m_2$, *we have* $a_{c[i]} + b_{i-c[i]} \leq (1 + \epsilon) \min_j (a_j + b_{i-j})$.

1. *If* $\epsilon > 0$, *the time complexity is* $O\left(\frac{m_1^{1/3} m_2 L^{2/3}}{\epsilon^{2/3}}\right)$.

2. *If* $\epsilon = 0$, *the time complexity is* $O\left(\frac{m_1 m_2}{\log(m_1+m_2)} + m_1 + m_2\right)$.

The basic idea behind our algorithm Freerider-Approx for computing an assignment of free riders is to use dynamic programming on the ordered list of meeting points provided by MeetingPoints, invoking Min-Sum for each dynamic programming step. Note that MeetingPoints returns a list of *ordered* meeting points (see Section 3.1), so we can do dynamic programming by walking through the list in order. This effectively replaces the factor of n in the time complexity of the naive dynamic programming algorithm with a factor of r.

Now consider the approximation error. Suppose we want a free-rider assignment whose resulting traffic cost is at most $(1 + \epsilon)$ times that of the optimal assignment. Suppose for each dynamic programming step we invoke Min-Sum with parameter $\epsilon' = \epsilon/(2 \min(r, \log n))$. When we combine two such approximations, we obtain a $(1 + \epsilon')^2$ approximation. Since any path in the Steiner tree contains at most $\min(r, \log n)$ meeting points, the total error is at most $(1 + \epsilon)$.

For each internal meeting point v, we assume the two meeting points x, y for which $v = a(x, y)$ can be determined in constant time. This is easily achieved by keeping track of pointers in the implementation of MeetingPoints. Let MPleft(v) be the operation returning x, and MPright(v) the operation returning y.

In our algorithm each meeting point u has data fields r_u and A_u. Here r_u denotes the number of revoked users in the tree $T(u)$. A_u is an array, indexed from $0 \leq i \leq \min(f, r_u)$, which records the cost found by our algorithm of assigning at most i free riders to $T(u) \cap R$. Finally, for internal meeting points u, there is a field B_u which is a companion array to A_u, indexed from $0 \leq i \leq \min(f, r_u)$, such that $A_u[i] = A_{\mathsf{MPleft}(u)}[B_u[i]] + A_{\mathsf{MPright}(u)}[i - B_u[i]]$. Thus, B_u records the assignments of free riders realizing the costs found by A_u.

By a simple inductive argument (given in the full version), one can show that $|MP| = 2r - 1$. Our main algorithm is described below.

Freerider-Approx (R, f, ϵ):

1. $MP \leftarrow \mathsf{MeetingPoints}(R)$, $\epsilon' = \epsilon/(2\min(r, \log n))$.
2. For each revoked user u, set $r_u = 1$ and $A_u \leftarrow (0, 0)$.
3. For $i = r + 1$ to $2r - 1$,
   (a) $v \leftarrow MP[i]$, $x \leftarrow \mathsf{MPleft}(v)$, $y \leftarrow \mathsf{MPright}(v)$.
   (b) Add $\mathsf{height}(x) - \mathsf{height}(v) - 1$ to each entry of A_x.
       If $r_x \leq f$ and $A_x[r_x] \neq 0$, set $A_x[r_x] = 1$.
   (c) Add $\mathsf{height}(y) - \mathsf{height}(v) - 1$ to each entry of A_y.
       If $r_y \leq f$ and $A_y[r_y] \neq 0$, set $A_y[r_y] = 1$.
   (d) $B_v \leftarrow \mathsf{MinSum}(A_x, A_y, \epsilon')$.
   (e) $r_v = r_x + r_y$.
   (f) For each $0 \leq i \leq \min(f, r_v)$, set $A_v[i] = A_x[B_v[i]] + A_y[i - B_v[i]]$.
       If $r_v \leq f$ and $A_v[r_v] \neq 0$, set $A_v[r_v] = 1$.
4. Output $A_{MP[2r-1]}[f] + \mathsf{height}(MP[2r - 1])$ as the cost.
5. Use the B_v's to do a Depth-First-Search on MP to find/output the assignment.

Due to space constraints, we defer the formal proof of the following theorem to the full version of the paper [28], though we give some intuition and bound its time complexity here.

Theorem 4. Freerider-Approx *outputs an assignment of free riders with cost at most* $(1 + \epsilon)$ *times that of the optimum.*

The algorithm can be explained as follows. For each revoked user u, if we assign no free riders to $T(u) = u$, then $T(u)$'s cost is 0 since $T(u)$ contains no privileged users. Thus the cost of assigning at most one free rider to $T(u)$ is also 0, so in step (2) we set $A_u \leftarrow (0, 0)$, consistent with the definition of A_u.

In step (3b) or (3c), it may be the case that x or y is not a child of v. Suppose this is true of x. Now consider the optimal assignment of at most i free riders to $L(v)$ for some i. Then, unless $r_x \leq f$, there must be a revoked user in $T(x)$ that cannot be made a free rider. Let P be the shortest-path from x to v, not including v. It is then easy to see that the optimal assignment of at most i free riders to $L(v)$ is the optimal assignment of at most i free riders to $T(x)$ together with one extra key for each sibling of a node on P. Thus, we should add the length of P to each entry of A_x, and $|P| = \mathsf{height}(x) - \mathsf{height}(v) - 1$. This logic also explains step (4). Indeed, in this case we consider the path from $\mathsf{MP}[2r - 1]$ to the root of the complete binary tree, which has length $\mathsf{height}(MP[2r - 1])$.

There are two exceptions though. The first is if $r_x \leq f$, in which case we should reset $A_x[r_x]$ to 1 since we can turn the r_x revoked users into free riders and use a single key at the root of $L(v)$ to cover all of v. There is, however, one more case for which this may not be optimal. This occurs if $A_x[r_x] = 0$ (after adding $\mathsf{height}(x) - \mathsf{height}(v) - 1$ to each entry of A_x) in which case every leaf of $T(x)$ is a revoked user, and thus there is 0 cost if we do not assign any free riders. This can only occur if x is in fact a child of v. In this case, we should leave $A_x[r_x] = 0$. This logic also explains the if statement in step (3f).

We now elaborate on step (5). If, for instance, v is the root of the Steiner tree with $x = \mathsf{MPleft}(v)$ and $y = \mathsf{MPright}(v)$ with $B_v[f] = i$, then one assigns at most i free riders to $T(x)$ and at most $f - i$ to $T(y)$. If $A_x[i] = 0$, it means that the leaves of $T(x)$ are all revoked users, and we should not assign any of them to be free riders. Otherwise, if $i \geq r_x$, we make every revoked user in $T(x)$ a free rider. Otherwise, we recurse on $\mathsf{MPleft}(x)$ and $\mathsf{MPright}(x)$. The analysis for y is similar, and it is easy to see the overall time complexity of this step is $O(r)$.

Let us now look at the time complexity of Freerider-Approx. We first note that as we have stated the algorithm, we cannot hope to do better than $O(rf)$.

Lemma 1. *For any ϵ, the worst-case time complexity of* Freerider-Approx *in the unit-cost RAM model is $\Omega(rf)$.*

Proof. Suppose the revoked set has the form

$$u_1 = 2n - 1, u_2 = 2n - 2, u_3 = 2n - 4, u_4 = 2n - 8, \ldots, u_i = 2n - 2^{i-1}, \ldots$$

In this case the Steiner tree consists of vertices $u_1, \ldots, u_r, v_1, \ldots, v_{r-1}$, where the v_i are the internal meeting points satisfying, $L(v_i) \cap R = u_{i+1}$ and $R(v_i) \cap R = \{u_1, u_2, \ldots, u_i\}$. Note that the time complexity of Min-Sum is at least $|A_y|$ for any ϵ. Then in this example, for $i \geq f$, there are at least f revoked users in $R(v_i)$, and thus the time complexity of MinSum at meeting point v_i is at least f. It follows that the time complexity of Freerider-Approx is at least $\Omega((r - f)f) = \Omega(rf)$.

It turns out that the above lemma is indeed a worst-case.

Theorem 5. *The time complexity of* Freerider-Approx *is $O(rf + r \log \log n)$.*

Proof. By Theorem 2, steps (1-2) can be done in $O(r \log \log n)$ time. The time complexity of $\mathsf{MinSum}(A_x, A_y, \epsilon')$ is at least $|A_x| + |A_y|$, so the time complexity of step (3) is dominated by step (3d). Note also that step (4) can be done in $O(1)$ time and step (5) in $O(r)$ time, so the total time of the algorithm is $O(r \log \log n)$ plus the total time spent in step (3d). If $C(|A_x|, |A_y|)$ denotes the time complexity of $\mathsf{MinSum}(A_x, A_y, \epsilon')$, then the total time spent in step (3d) is

$$\sum_{i=r+1}^{2r-1} C(|A_{\mathsf{MPleft}\ (MP[i])}|, |A_{\mathsf{MPright}\ (MP[i])}|). \tag{1}$$

To bound this sum, observe that for any v, $|A_v| = 1 + \min(f, r_v)$. To solve MinSum, we will consider the naive $O(m_1 m_2)$-time algorithm, as it turns out in the *worst-case* our faster algorithms of Section 4 do not yield an asymptotic improvement. Then, up to a constant factor, sum (1) is

$$\sum_v \min(f, r_{\mathsf{MPleft}(v)}) \min(f, r_{\mathsf{MPright}(v)}), \tag{2}$$

where the sum is over all internal meeting points v.

It will be convenient to define the *collapsed Steiner tree* on the revoked set R. This tree is formed by taking the rooted Steiner tree on R, and then repeatedly collapsing edges for which either endpoint is not a meeting point (if one endpoint is a meeting point, the collapsed edge is labeled by that endpoint). We are left with a binary tree on $2r - 1$ nodes (the meeting points) containing r leaves (the revoked users). Denote this tree by T.

We first consider the contribution S to (2) from v with $r_{\mathsf{MPleft}(v)}, r_{\mathsf{MPright}(v)} < f$, i.e.,

$$S = \sum_{v \mid r_{\mathsf{MPleft}(v)}, r_{\mathsf{MPright}(v)} < f} \min(f, r_{\mathsf{MPleft}(v)}) \min(f, r_{\mathsf{MPright}(v)})$$

$$= \sum_{v \mid r_{\mathsf{MPleft}(v)}, r_{\mathsf{MPright}(v)} < f} r_{\mathsf{MPleft}(v)} r_{\mathsf{MPright}(v)}.$$

Create a formal variable X_u for each $u \in R$. Then by writing $|L(v) \cap R|$ as $\sum_{u \in L(v) \cap R} X_u$ and $|R(v) \cap R|$ as $\sum_{u \in R(v) \cap R} X_u$, we may express S as a degree-2 multilinear polynomial $S(\{X_u\}_{u \in R})$, where the actual value S corresponds to setting $X_u = 1$ for all u. We claim that in $S(\{X_u\}_{u \in R})$, for each u there can be at most $f - 1$ different v for which $X_u \cdot X_v$ appears, and further, the coefficient of each of these monomials is 1.

To see this, let $a \in T$ be the ancestor of u of maximal height for which $r_{\mathsf{MPleft}(a)}$, $r_{\mathsf{MPright}(a)} < f$. Consider the path $u = u_0, u_1, \ldots, u_k = a$ from u to a in T. Let $v_0, v_1, \ldots, v_{k-1}$ be the siblings of $u_0, u_1, \ldots, u_{k-1}$. Then the subtrees $T(v_i)$ and $T(v_j)$ are disjoint for all $i \neq j$. Thus, $X_u \cdot X_v$ can appear at most once in S for a given $v \in T(a)$, and the number of such v is bounded by $r_a < 2f$.

As there are r different u, setting each X_u to 1 shows that $S = O(rf)$.

Next, consider the subgraph T' of T on nodes v for which either $r_{\mathsf{MPleft}(v)} \geq f$ or $r_{\mathsf{MPright}(v)} \geq f$ (or both). Observe that T' is a tree.

Consider any vertex v for which exactly one of $r_{\mathsf{MPleft}(v)}, r_{\mathsf{MPright}(v)}$ is at least f. In this case we say that v is *lop-sided*. Suppose $c(v)$ is v's child for which $r_{c(v)} < f$. Then the contribution from v to (2) is $f \cdot |T(c(v)) \cap R|$. We claim that $T(c(v)) \cap T(c(v')) = \emptyset$ for all lop-sided $v \neq v'$. Indeed, as T' is a tree, there is a path P from v to v' in T'. Moreover, as both $T(c(v))$ and $T(c(v'))$ are trees, if $T(c(v)) \cap T(c(v')) \neq \emptyset$, then there is a path P' from $c(v')$ to $c(v)$ in $T(c(v)) \cup T(c(v'))$. As T' and $T(c(v)) \cup T(c(v'))$ are disjoint, the path $Pc(v')P'v$ is a cycle in T, contradicting that T is a tree. Therefore,

$$\sum_{lop-sided\ v} \min(f, r_{\mathsf{MPleft}(v)}) \min(f, r_{\mathsf{MPright}(v)})$$

$$= f \sum_{lop-sided\ v} |T(c(v)) \cap R| \leq f \cdot r.$$

Thus, to bound (2), it remains to account for those vertices v for which both $r_{\mathsf{MPleft}(v)} \geq f$ and $r_{\mathsf{MPright}(v)} \geq f$. Observe that each such v contributes $O(f^2)$ to (2). So, if we can show that there are at most $O(r/f)$ such v, it will follow that (2) is bounded by $O(rf)$.

To this end, let T'' be the subgraph of T consisting of all vertices v for which $r_{\mathsf{MPleft}(v)} \geq f$ and $r_{\mathsf{MPright}(v)} \geq f$. Next, create a binary tree T''' by first adding

T'' to T''', then adding MPleft(v) and MPright(v) to T''' for each $v \in T''$. Now, for any two leaves $v_1 \neq v_2$ in T''', we have $T(v_1) \cap T(v_2) = \emptyset$, as otherwise T would contain a cycle. Moreover, by definition of the vertices in T'', we have $|T(v_1)| \geq |T(v_1) \cap R| \geq f$ and $|T(v_2)| \geq |T(v_2) \cap R| \geq f$. Thus, there can be at most r/f leaves in T'''. As T''' is a binary tree, it follows that it can contain at most $2r/f = O(r/f)$ nodes, and thus T'' also contains at most this many nodes. This completes the proof.

Thus, we have proven Theorem 1. We now consider the case when the revoked set R is chosen uniformly at random from sets of size r. This might often be the case in practice, if say, the revoked users are uncorrelated with each other. This example exploits our algorithmic improvements in Section 4. In the full version of the paper [28], we prove the following theorem.

Theorem 6. *If the revoked set R is uniformly chosen from sets of size r, then with probability at least $1 - 1/n$, if* Freerider-Approx *uses our* MinSum $(1 + \epsilon)$-*approximation algorithm, then it runs in time*

$$O\left(\frac{1}{\epsilon^{2/3}}\left(r \log^{8/3} n + r f^{1/3} \log^{5/3} n\right)\right),$$

which is $O(r f^{1/3}$polylog $n)$ for $\epsilon > \frac{1}{\text{polylogn}}$. On the other hand, if $r \geq \frac{4}{3}\log n$, then with probability at least $1 - 1/n$, if Freerider-Approx *uses the naive algorithm for* MinSum, *then it takes time*

$$\Omega\left(r \min\left(f, \frac{f^2}{\log n}\right)\right),$$

which is $\Omega(rf)$ for $f = \Omega(\log n)$.

The intuition behind the proof is as follows. For a randomly chosen R of size r, one can use a Chernoff-type bound for negatively-correlated random variables [26] to show that with probability at least $1 - 1/n$, for all nodes v in the complete binary tree, $|T(v) \cap R| \leq 2t(v)\frac{r}{n} + 2\log n$, where $t(v)$ is the number of leaves in $T(v)$. Note that, $\mathbf{E}[|T(v) \cap R|] = t(v)\frac{r}{n}$, so this variable is tightly concentrated.

To get the lower bound of $\Omega(rf)$ for the naive MinSum algorithm, we first consider nodes with height $\lfloor\frac{r}{2f}\rfloor$, assuming first that $f \leq r/4$. Using the above bound on $|T(v) \cap R|$, one can show that $\Omega(\frac{r}{f})$ of them must have $|T(v) \cap R| \geq f$, as otherwise there is some v with height $\lfloor\frac{r}{2f}\rfloor$ for which $|T(v) \cap R|$ is too large. It follows that there are $\Omega(\frac{r}{f})$ nodes with height less than $\lfloor\frac{r}{2f}\rfloor$ each contributing $\Omega(f^2)$ to sum (2), giving total cost $\Omega(rf)$. If $f > \frac{r}{4}$, one can show the root of the Steiner tree itself contributes $\Omega(rf)$ to sum (2).

Finally, to get the upper bound of $O(rf^{1/3}$polylog $n)$, one modifies sum (2) to,

$$\sum_v \min(f, r_{\mathsf{MPleft}(v)})^{1/3} \min(f, r_{\mathsf{MPright}(v)}) \left(\frac{L \log n}{\epsilon}\right)^{2/3},$$

where $L = \max_{i,v} opt(i, T(v)) - opt(i + 1, T(v))$. One can use the bound on $|T(v) \cap R|$ above for each v to bound this sum. Using the definition of the CS scheme, it is easy to show that $opt(i, T(v)) - opt(i+1, T(v)) \leq \log n$ for any i, v.

Finally, we note that it is straightforward to extend Freerider-Approx to broadcast encryption schemes based on complete k-ary trees for general k. The idea is to binarize the tree by replacing a degree-k node with a binary tree on k nodes, and then run Freerider-Approx on the virtual tree.

We suspect that one can extend Freerider-Approx to the SD scheme of [25] with the same time complexity, though we have not worked out the details. The only change is the dynamic programming step, steps (3-4), for which one must be careful if a key involving the joining node is part of an optimal assignment of at most i free riders at a given subtree. One can run MinSum as before, but needs to compare the output to a few extra cases involving the key at the joining node (which MinSum does not account for).

4   Min-sum Problem

In this section we give our MinSum algorithms and analyses. Recall that we are given $\epsilon \geq 0$, and integer sequences $a_1 \geq a_2 \geq \cdots \geq a_{m_1} \geq 0$ and $b_1 \geq b_2 \geq \cdots \geq b_{m_2} \geq 0$ with $a_i - a_{i+1}, b_i - b_{i+1} \leq L$ for all i. The goal is to output an array c such that for all $2 \leq i \leq m_1 + m_2$, we have $a_{c[i]} + b_{i-c[i]} \leq (1+\epsilon) \min_j (a_j + b_{i-j})$. W.l.o.g., we will assume that $m_1 \leq m_2$. Thus, $m_1 + m_2 = O(m_2)$. The naive algorithm for this problem which, for all i, computes all pairs of sums $a_j + b_{i-j}$, runs in $O(m_1(m_1 + m_2)) = O(m_1 m_2)$ time. We now show how to do much better.

4.1   Algorithm Overview

The essence of our algorithm is the following balancing technique. The first idea is that, since the lists a_j and b_j are sorted, we can quickly find (via binary search) the largest "level" i^* for which $\min_j(a_j + b_{i^*-j}) \geq s$ for some parameter s to be optimized. Our algorithm then has two parts: its solution to levels $i \leq i^*$ and its solution to levels $i > i^*$.

To solve levels $i \leq i^*$, we have two ideas. For these levels, we crucially exploit the fact that the differences $a_j - a_{j+1}$ and $b_j - b_{j+1}$ are bounded by L and that $\epsilon > 0$. The first idea is that to solve some such level i, instead of computing all sums $a_j + b_{i-j}$ we can get by with only computing a small fraction of sums. Indeed, since the differences are bounded, we have $a_{j-1} + b_{i-(j-1)}$ is approximately equal to $a_j + b_{i-j}$ for all j. Since we only need an approximate answer, we can just take the minimum found over the small fraction of sums that we compute. This fraction depends on ϵ, L, and s. Note, the dependence on s arises since we want a *multiplicative approximation*, and thus, the smaller s is, the larger the fraction of sums we will need to compute.

The next idea is that we do not even need to solve all the levels $i \leq i^*$. Indeed, since the differences are bounded and we are content with an approximation, if we solve some level i' then we can use this solution for levels "close" to i', i.e., those levels in a range $[i' - d, i' + d]$ for some parameter d that depends on ϵ, L, and s. In other words, if $a_j + b_{i'-j}$ is our solution for level i', then for a level $i \in [i' - d, i' + d]$ we can bound its cost by $a_j + b_{i'-j}$.

To solve levels $i > i^*$ we have one main idea, which allows us to compute the *exact* solutions for these levels. Suppose $(j^*, i^* - j^*)$ realizes the optimum in level i^*. Then $a_{j^*} + b_{i^*-j^*} \geq s$. It follows that a_{j^*} and $b_{i^*-j^*}$ must both be less than $s + L$. Indeed, since the a_j and b_j are non-negative and the differences are bounded by L, then if say, $a_{j^*} \geq s + L$, then $a_{j^*+1} \geq s$ and thus the minimum in level $i^* + 1$ is at least s, contradicting the maximality of i^*. Thus, in the sequences $a_{j^*} \geq a_{j^*+1} \geq \cdots \geq a_{m_1}$ and $b_{i^*-j^*} \geq b_{i^*-j^*+1} \geq \cdots \geq b_{m_2}$ there are a lot of identical elements, i.e., we cannot have $a_\ell > a_{\ell+1}$ too often since on the one hand $a_{j^*} \leq s + L$, and on the other hand $a_{m_1} \geq 0$ (similarly for the b_j).

Then the key observation is that for any level $i > i^*$, the minimum value $a_j + b_{i-j}$ satisfies the following property:

$$\text{Either } j = j^*, \text{ or } a_{j-1} > a_j, \text{ or } b_{i-j-1} > b_{i-j}.$$

Indeed, suppose not, i.e., $a_{j-1} = a_j$ and $b_{i-j-1} = b_{i-j}$ for some $j \neq j^*$. Then, since $i > i^*$, either $j > j^*$ or $i - j > i - j^*$. W.l.o.g., suppose that $j > j^*$. Then, since the b_j values are non-increasing, we have that $a_{j-1} + b_{i-j+1}$ is also a minimum for level j. Repeating this "pushing" argument, we can push a_j until j either equals j^* or satisfies $a_{j-1} > a_j$. This is a contradiction.

Now as observed before, there cannot be too many different $j > j^*$ for which $a_{j-1} > a_j$. In fact, there can be at most $s + L$ of them. So we can just compute all valid pairs of sums in this case, and take the appropriate minima.

Finally, we choose s so as to balance the time complexities of these two parts.

4.2   Algorithm Details

First consider $\epsilon > 0$. Our main algorithm MinSum will call upon a subroutine MinSumLevel$((a), (b), i)$ which outputs $\min_j(a_j + b_{i-j})$ and an index j realizing this minimum. MinSumLevel can be implemented in $O(m_1)$ time by computing all the sums in the ith level and taking a minimum.

We need the notion of a *hop*. We say that (a_i, a_{i+1}) is a hop if $a_i > a_{i+1}$. In this case we say that a_{i+1} is a *hop-stop*. We similarly define hops and hop-stops for the sequence (b). For any integer z, if $a_i \leq z$, then since $a_{m_1} \geq 0$, there can be at most z hops in the sequence $a_i \geq a_{i+1} \geq a_{i+2} \geq \cdots \geq a_{m_1}$.

Let s be an integer parameter to be determined. MinSum first performs a binary search on level i to find the largest i for which $\min_j(a_j + b_{i-j}) \geq s$ (which works since the sequence $M_i = \min_j(a_j + b_{i-j})$ is non-increasing). This can be done in $O(m_1 \log(m_1 + m_2)) = O(m_1 \log m_2)$ time by using MinSumLevel at each level of a binary search on i. Let i^* be the largest such level, and let j^* be the index MinSumLevel$((a), (b), i^*)$ returns realizing this minimum. By the maximality of i^*, it follows that $a_{j^*}, b_{i^*-j^*} \leq s + L$, so there can be at most $s + L$ hops to the right of a_{j^*} (including a_{j^*}), and at most $s + L$ hops to the right of $b_{i^*-j^*}$ (including $b_{i^*-j^*}$). Suppose $p_1 < p_2 < \cdots < p_\alpha$ are the indices of the hop-stops to the right of a_{j^*}, and let $q_1 < q_2 < \cdots < q_\beta$ be the indices of the hop-stops to the right of $b_{i^*-j^*}$. Note that $\alpha, \beta \leq s + L$.

For levels $i > i^*$, MinSum computes all sums $a_x + b_y$ with $x \in \{j^*\} \cup \{p_1, \ldots, p_\alpha\}$ and $y \in [m_2]$, together with all sums $a_x + b_y$ with $x \in [m_1]$ and $y \in \{i^* - j^*\} \cup \{q_1, \ldots, q_\beta\}$. The number of sums computed is $O((s + L)(m_1 + m_2)) = O((s + L)m_2)$. Each sum is labeled by the level it falls in, and for each level, the minimum value is computed. Note that this can be done in $O((s+L)m_2)$ time simply by walking through the pairs and keeping track of the minimum for each level. Thus, in $O((s + L)m_2)$ time we have solved all levels greater than i^*.

Now let ϵ' be $\Theta(\epsilon s/(Lm_1))$, and fix any level $i \leq i^*$. Then to solve level i exactly, there are at most m_1 pairs of indices we must take a minimum over:

$$(z, i - z), (z + 1, i - z - 1), (z + 2, i - z - 2), \ldots,$$

where $z = 1$ if $i \leq m_2 + 1$, and otherwise $z = i - m_2$.

Instead of computing all these sums, we only compute those within the list:

$$(z, \ i - z), (z + \lfloor \epsilon'm_1 \rfloor, \ i - z - \lfloor \epsilon'm_1 \rfloor), (z + \lfloor 2\epsilon'm_1 \rfloor, \ i - z - \lfloor 2\epsilon'm_1 \rfloor),$$

$$(z + \lfloor 3\epsilon'm_1 \rfloor, \ i - z - \lfloor 3\epsilon'm_1 \rfloor), \ \ldots$$

We then take a minimum only over the sums with indices in this list. Note that for all i, $\lfloor (i + 1)\epsilon'm_1 \rfloor - \lfloor i\epsilon'm_1 \rfloor \leq \epsilon'm_1 + 1$. Since $a_j - a_{j+1}, b_j - b_{j+1} \leq L$, it follows that one of the sums from this list will be at most $\epsilon'm_1L \leq \epsilon s/2$ (for an appropriate choice of ϵ') more than the value of the true minimum for level i. Moreover, the number of sums computed is only $O(1/\epsilon')$, so the total time for level i is $O(Lm_1/(\epsilon s))$.

Finally, we observe that we don't have to solve every level $i \leq i^*$, but rather we can use the output for level i as the output for levels "close" to i. Let ϵ'' be $\Theta(\epsilon s/(Lm_2))$. We want to solve levels $1, 2, \ldots, i^*$. Suppose we obtain an additive $\epsilon s/2$ approximation for each level in the following list:

$$1, \ \lfloor \epsilon''i^* \rfloor, \ \lfloor 2\epsilon''i^* \rfloor, \ \lfloor 3\epsilon''i^* \rfloor, \ \ldots, \ i^*.$$

To output the minimum for some level i not appearing in the list, we find the nearest level $i' \geq i$ in the list. Then, suppose $a_j + b_{i'-j}$ is our solution for level i'. Then in constant time, we can find integers $d_1 \leq j$ and $d_2 \leq i' - j$ such that $d_1 + d_2 = i$. We can then output the pair (d_1, d_2) as our solution for level i (or more precisely, we just output d_1 since d_2 can be inferred). Since the sequences (a) and (b) are non-increasing, $a_{d_1} + b_{d_2} \leq a_j + b_{i'-j}$.

Note that i' is at most $O(\epsilon''i^*)$ levels away from i, so its minimum can be at most an additive

$$O(\epsilon''Li^*) = O(\epsilon''L(m_1 + m_2)) \leq \epsilon s/2$$

more than that of i (for an appropriate choice of ϵ''). Thus, as the error in level i' is at most $\epsilon s/2$, and the minimum in level i' is at most $\epsilon s/2$ larger than that of level i, our output has additive error at most ϵs. Since the minimum in all levels $i \leq i^*$ is at least s, our output is a $(1 + \epsilon)$-approximation.

Note that we need only solve $O(1/\epsilon'') = O(Lm_2/(\epsilon s))$ levels, each taking $O(Lm_1/(\epsilon s))$ time, so the time over all levels $i \leq i^*$ is $O(L^2 m_1 m_2/(\epsilon^2 s^2))$. Thus, the total time is $O(L^2 m_1 m_2/(\epsilon^2 s^2) + (s + L)m_2 + m_1 \log m_2)$. Setting $s = \Theta(m_1^{1/3} L^{2/3}/\epsilon^{2/3})$ gives total time complexity

$$O\left(\frac{m_1^{1/3} m_2 L^{2/3}}{\epsilon^{2/3}} + Lm_2\right).$$

The first term dominates this expression whenever $L \leq m_1$. On the other hand, if $L > m_1$, then $m_1^{1/3} m_2 L^{2/3} \epsilon^{-2/3} = \Omega(m_1 m_2)$, and we can switch to the trivial $O(m_1 m_2)$-time algorihm which works just by computing $a_j + b_{i-j}$ for all i, j, and taking the appropriate minima. Thus, the total time is $O\left(\frac{m_1^{1/3} m_2 L^{2/3}}{\epsilon^{2/3}}\right)$.

For the binary-tree scheme, $L = O(\log n)$, and for the NNL scheme $L = O(1)$. Thus, for constant ϵ, the running time is close to $m_1^{1/3} m_2$, greatly improving the $O(m_1 m_2)$ time of the naive algorithm. We conclude that,

Theorem 7. *For any* $\epsilon > 0$, MinSum *can be solved in* $O\left(\frac{m_1^{1/3} m_2 L^{2/3}}{\epsilon^{2/3}}\right)$ *deterministic time.*

There is an alternative algorithm when $\epsilon = 0$ described in [13] (which is essentially a direct application of a technique given in [8]) that uses ideas from computational geometry. We defer the description to the full version of this paper [28], but state the main theorem:

Theorem 8. *[13] If* $\epsilon = 0$, MinSum *can be solved in deterministic time*

$$O\left(\frac{m_1 m_2}{\log m_2} + m_1 + m_2\right).$$

Acknowledgment. The second author would like to thank Andrew Yao and Tsinghua University for their support while conducting part of this research.

References

[1] M. Abdalla, Y. Shavitt, and A. Wool. Key Management for Restricted Multicast Using Broadcast Encryption. In *ACM Trans. on Networking*, vol. 8, no. 4. 2000.

[2] W. Aiello, S. Lodha, and R. Ostrovsky. Fast digital identity revocation. In *Proc. of Crypto 1998*.

[3] J. Anzai, N. Matsuzaki and T. Matsumoto. A Quick Group Key Distribution Scheme with Entity Revocation. In *Proc. of Asiacrypt 1999*, LNCS 1716.

[4] T. Asano. A Revocation Scheme with Minimal Storage at Receivers. In *Proc. Of Asiacrypt 2002*, LNCS 2501, pages 433–450. Springer-Verlag, 2002.

[5] K. Azuma. Weighted Sums of Certain Dependent Random Variables. *Tôhoku Math. Journ.* **19** (1967) pp. 357-367.

[6] S. Berkovits. How to Broadcast a Secret. In *Proc. of Eurocrypt 1991*, LNCS 547, pages 535–541. Springer-Verlag, 1991.

[7] R. Canetti, T. Malkin and K. Nissim. Efficient Communication-Storage Tradeoffs for Multicast Encryption. In *Proc. of Eurocrypt 1999*, Springer-Verlag.

[8] T. M. Chan. *All-pairs shortest paths with real weights in $O(n^3/\log n)$ time*. In *Proc 9th WADS*, LNCS 3608, pp. 318-324, 2005.

[9] T. H. Cormen, C. E. Leiserson, and R. L. Rivest. Introduction to Algorithms, 1990, The MIT Press/McGraw-Hill.

[10] *Content Protection for Pre-recorded Media Specification* and *Content Protection for Recordable Media Specification*, available from http://www.4centity.com/tech/cprm.

[11] P. van Emde Boas. *Preserving order in a forest in less than logarithmic time and linear space*. Inform. Process. Lett. **6**(3) 1977, 80-82.

[12] P. van Emde Boas, R. Kaas, and E. Zijlstra. *Design and implementation of an efficient priority queue*. Math Systems Theory **10**(2) 1976/1977. Also, FOCS 1975.

[13] J. Erickson. Blog. Post at
http://3dpancakes.typepad.com/ernie/2005/08/chans_technique.html

[14] A. Fiat and M. Naor. Broadcast Encryption. In *Proc. of Crypto 1993*, LNCS 773, pages 480–491. Springer-Verlag, 1994.

[15] C. Gentry and Z. Ramzan. RSA Accumulator Based Broadcast Encryption. In *Proc. Information Security Conference*, 2004.

[16] Y. Han. *Deterministic sorting in $O(n \log \log n)$ time and linear space*. Journal of Algorithms, 2004, pp. 96-105.

[17] D. Halevy and A. Shamir. The LSD Broadcast Encryption Scheme. In *Proc. of Crypto 2002*, LNCS 2442, pages 47–60. Springer-Verlag, 2002.

[18] Y. Han and M. Thorup. *Sorting in $O(n\sqrt{\log \log n})$ expected time and linear space*. FOCS, 2002, pp. 135-144.

[19] Y. Kim, A. Perrig and G. Tsudik. Simple and Fault-Tolerant Key Agreement for Dynamic Collaborative Groups. In *Proc. Of ACM CCS*, 2000.

[20] R. Kumar, S. Rajagopalan and A. Sahai. Coding Constructions for Blacklisting Problems. In *Proc. of Crypto 1999*, LNCS 1666, pages 609–623. Springer-Verlag.

[21] M. Luby and J. Staddon. Combinatorial Bounds for Broadcast Encryption. In *Proc. of Eurocrypt 1998*, LNCS 1403, pages 512–526. Springer-Verlag, 1998.

[22] N. Matsuzaki, J. Anzai and T. Matsumoto. Light Weight Broadcast Exclusion Using Secret Sharing. In *Proc. of ACISP 2000*, LNCS 1841, Springer-Verlag.

[23] S. Mitra. Iolus: A Framework for Scalable Secure Multicasting. In *Proc. of ACM SIGCOMM*, September 1997.

[24] D.A. McGrew and A.T.Sherman. Key Establishment in Large Dynamic Groups Using One-Way Function Trees. Manuscript available at http://www.csee.umbc.edu/~Sherman/Papers/itse.ps.

[25] D. Naor, M. Naor and J. Lotspiech. Revocation and Tracing Schemes for Stateless Receivers. In *Proc. Of Crypto 2001*. Full version: ECCC Report No. 43, 2002.

[26] A. Panconesi and A. Srinivasan. *Randomized Distributed Edge Coloring via an Extension of the Chernoff-Hoeffding Bounds*, SICOMP, 1997.

[27] R. Poovendran and J.S. Baras. An Information Theoretic Analysis of Rooted-Tree Based Secure Multicast Key Distribution Schemes. In *Proc. of Crypto 1999*.

[28] Z. Ramzan and D. P. Woodruff. Fast Algorithms for the Free Riders Problem in Broadcast Encryption. IACR ePrint Archive, http://eprint.iacr.org, 2006.

[29] D.M. Wallner, E.J. Harder, and R.C. Agee. Key Management for Multicast: Issues and Architectures. Internet Draft, September 1998.

[30] C.K. Wong, M. Gouda and S.S. Lam. Secure Group Communications Using Key Graphs. In *Proc. of SIGCOMM 1998*, 1998.

The Number Field Sieve
in the Medium Prime Case

Antoine Joux[1,3], Reynald Lercier[1,2], Nigel Smart[4], and Frederik Vercauteren[5,*]

[1] DGA
[2] CELAR
Route de Laillé, 35170 Bruz, France
Reynald.Lercier@m4x.org
[3] Université de Versailles St-Quentin-en-Yvelines
PRISM
45, avenue des Etats-Unis, 78035 Versailles Cedex, France
Antoine.Joux@m4x.org
[4] Dept. Computer Science, University of Bristol
MVB, Woodland Road, Bristol, BS8 1UB, United Kingdom
nigel@cs.bris.ac.uk
[5] Department of Electrical Engineering, University of Leuven
Kasteelpark Arenberg 10, B-3001 Leuven-Heverlee, Belgium
frederik.vercauteren@esat.kuleuven.be

Abstract. In this paper, we study several variations of the number field sieve to compute discrete logarithms in finite fields of the form $\mathbb{F}_{p^n}$, with p a medium to large prime. We show that when n is not too large, this yields a $L_{p^n}(1/3)$ algorithm with efficiency similar to that of the regular number field sieve over prime fields. This approach complements the recent results of Joux and Lercier on the function field sieve. Combining both results, we deduce that computing discrete logarithms have heuristic complexity $L_{p^n}(1/3)$ in all finite fields. To illustrate the efficiency of our algorithm, we computed discrete logarithms in a 120-digit finite field $\mathbb{F}_{p^3}$.

1 Introduction

Today's public key cryptosystems usually rely on either integer factorisation or discrete logarithms in finite fields or on elliptic curves. In this paper, we consider discrete logarithm computations in finite fields of the form $\mathbb{F}_{p^n}$, with $n > 1$ and p a medium to large prime. For a long time, these fields have been mostly ignored in cryptosystems, and the security of their discrete logarithm problems are much less known than the security in prime fields $\mathbb{F}_p$ or fields of fixed characteristic such as $\mathbb{F}_{2^n}$ or $\mathbb{F}_{3^n}$. Indeed, in $\mathbb{F}_p$ it is well-known that the best available algorithm is the number field sieve, first introduced in [7] and further studied in [18,20,21,12,19]. Similarly, in fixed characteristic and n tending to infinity, the best available algorithm is the function field sieve, first introduced in [3] and further studied in [4,11,8]. Both algorithms have complexity $L_{p^n}(1/3)$.

---

* Postdoctoral Fellow of the Research Foundation - Flanders (FWO - Vlaanderen).

C. Dwork (Ed.): CRYPTO 2006, LNCS 4117, pp. 326–344, 2006.

However, until very recently for the intermediate case $\mathbb{F}_{p^.}$, no general $L_{p^.}(1/3)$ algorithm was known and the best available approach had complexity $L_{p^.}(1/2)$ (see [2,1]). With the advent of pairing-based and torus-based cryptography, the security of discrete logarithms in $\mathbb{F}_{p^.}$ is a growing concern. Recently, two very different methods have been proposed to deal with some of these fields. The first approach is based on rational torus representation and was proposed by Granger and Vercauteren [9]. It was effectively used by Lercier and Vercauteren in [15]. The second approach, by Joux and Lercier [13], proposes a family of algorithms which are based on the function field sieve and yields complexity $L_{p^.}(1/3)$ where applicable.

In this paper, we introduce several variations of the number field sieve algorithm that depend on the relative size of p and n. The main difference with existing algorithms lies in the construction of the number fields and the choice of sieving space: for p large compared to $L_{p^.}(2/3)$, we use a new polynomial selection algorithm and for p small compared to $L_{p^.}(2/3)$, we sieve over elements of degree higher than one. Furthermore, we show that these variations can fill the gap which was left open by [13]. As a consequence, we conclude that discrete logarithm computation has heuristic complexity $L_{p^.}(1/3)$ for all finite fields, which greatly improves the $L_{p^.}(1/2)$ complexity of [2,1].

The paper is organised as follows: in Section 2 we describe our setup and the basic number field sieve variation that we are considering. In Section 3 we show how the basic algorithm can be modified in order to cover all the finite fields left out by [13]. In Section 4 we describe the mathematical background required to make this description rigorous and in Section 5 we give a precise heuristic analysis of the algorithm. Finally in Section 6 we report on experiments obtained with an actual implementation of this algorithm.

2   Our Basic Variation on the Number Field Sieve

The number and function field sieve algorithms are both index calculus algorithms that search for multiplicative identities using smooth objects over well-chosen smoothness bases. In the number field sieve, the smoothness bases contain ideals of small norm. Since we are using ideals, some important mathematical technicalities arise when transforming these multiplicative identities into linear equations involving logarithms. For the sake of simplicity, we defer these technicalities to Section 4. When dealing with index calculus algorithms, the complexities are usually expressed using the following notation:

$$L_q(\alpha, c) = \exp((c + o(1))(\log q)^{\alpha}(\log \log q)^{1-\alpha}),$$

where log denotes natural logarithm. When the constant c is not given explicitly, the notation $L_q(\alpha)$ is often used. In particular, for the prime field $\mathbb{F}_p$ and for extension fields $\mathbb{F}_{p^.}$ with fixed characteristic p, the number field sieve and the function field sieve respectively yield $L_p(1/3, (64/9)^{1/3})$ and $L_{p^.}(1/3, (32/9)^{1/3})$ algorithms. Moreover, in [13] it was shown that variations of the function field sieve yield $L_{p^.}(1/3)$ algorithms for $\mathbb{F}_{p^.}$ as long as $p \leq L_{p^.}(1/3)$. In this paper, we consider the complementary case: $p \geq L_{p^.}(1/3)$.

2.1   Setup, Sieving and Linear Algebra

Our basic variation addresses finite fields $\mathbb{F}_{p^n}$ with $p = L_{p^n}(2/3, c)$ and c near $2 \cdot (1/3)^{1/3}$. Recall that the regular number field sieve algorithm over $\mathbb{F}_p$ starts by choosing two polynomials $f_1, f_2 \in \mathbb{Z}[X]$ with a common root in $\mathbb{F}_p$. In our basic variation, we generalise this to $\mathbb{F}_{p^n}$ as follows: first, f_1 is chosen as a degree n polynomial, with very small coefficients and irreducible over $\mathbb{F}_p$. Then, we set f_2 equal to the polynomial $f_1 + p$. Since $f_2 \equiv f_1 \bmod p$, both polynomials clearly have a common root in $\mathbb{F}_{p^n}$ (in fact all of them are equal).

Let $K_1 \simeq \mathbb{Q}[X]/(f_1(X)) \cong \mathbb{Q}[\theta_1]$ and $K_2 \cong \mathbb{Q}[X]/(f_2(X)) \cong \mathbb{Q}[\theta_2]$ be the two number fields defined respectively by f_1 and f_2, i.e. θ_1 and θ_2 are roots of f_1 and f_2 in $\mathbb{C}$. Choose a smoothness bound B and a sieve limit S and consider all pairs (a, b) of coprime integers, with $|a| \le S$ and $|b| \le S$, such that $a - b\theta_1$ and $a - b\theta_2$ both have B-smooth norms. After some post-processing described in Section 4, which involves adding unit contributions or Schirokauer maps, each such pair yields a linear equation between "logarithms of ideals" in the smoothness bases. Since the number of small norm ideals involved on each side is smaller than $n\pi(B)$, with $\pi(x)$ the number of primes smaller than x, it suffices to collect $2n\pi(B)$ equations during the sieving phase. Once the sieving phase is complete, we solve the resulting sparse system of linear equations modulo the cardinality of $\mathbb{F}_{p^n}^*$, or more precisely, modulo a large factor of this cardinality and recover logarithms of ideals of small norm. There are two options for this large factor: either we take it prime, but then we need to factor $p^n - 1$ (which would not increase the total complexity) or we simply take it composite without small factors, since all operations we perform modulo a large prime factor remain valid for such a composite factor. Finally, small prime factors of $p^n - 1$ can be dealt with separately using a combination of the Pohlig-Hellman and Pollard rho algorithm.

2.2   Individual Discrete Logarithms

Once the two steps described above, sieving and linear algebra, have been performed, we obtain the logarithms of the ideals of the smoothness bases. In order to complete the computation, an additional step is required, namely computing the logarithm of random elements in the finite field. We propose a classical approach based on "special-q" descent, which is similar to the approach in [12] for the case of logarithms over a prime field.

Represent $\mathbb{F}_{p^n}$ as $\mathbb{F}_p[t]/(f_1(t))$ and assume we want to compute the discrete logarithm of some element $y \in \mathbb{F}_{p^n}$. First, we search for an element of the form $z = y^i t^j$ for some $i, j \in \mathbb{N}$ with the following additional properties:

1. after lifting z to the number field K_1 (also denoted by z), its norm factors into primes smaller than some bound $B_1 \in L_{p^n}(2/3, 1/3^{1/3})$,
2. the norm of the lift of z should be squarefree, implying that only degree one prime ideals will appear in the factorisation of (z).

Assuming that z does not belong to a strict subfield[1] of $\mathbb{F}_{p^{.}}$, nothing prevents such a squarefree factorisation to occur. Therefore, we make the usual heuristic hypothesis and assume that the norm has a squarefree factorisation into small primes with probability similar to that of a random number of the same size. According to [10], this probability can be expressed as a product of two terms. The first term is the usual probability of smoothness, without the squarefree condition. The second term, comes from the squarefree condition and quickly tends to $6/\pi^2$. Since the constant $6/\pi^2$ vanishes into the $o(1)$ of the L notation, we simply ignore this technicality in the remainder of the paper.

Whilst the first condition on the element z is standard, the second condition is required to guarantee that only degree one prime ideals will appear in the factorisation of (z). This condition is necessary since during the sieving phase, we only computed logarithms of degree one prime ideals. Since squared factors in the norm could correspond to higher degree ideals, we would not know the corresponding logarithms.

After finding an adequate candidate, we factor the principal ideal generated by z into degree one prime ideals of small norm. Note that there will be several prime ideals not contained in the factor base, since their norm is allowed to be bigger than the smoothness bound B. To compute the logarithm of such an ideal $\mathfrak{q}$, we start a special-$\mathfrak{q}$ descent as follows: we sieve on pairs (a, b), where a and b are chosen to ensure that $\mathfrak{q}$ divides $a - b\theta_1$ in the number field K_1. After finding a pair (a, b) such that the norm of $a - b\theta_1$ (divided by the norm of $\mathfrak{q}$) and the norm of $a - b\theta_2$ factor into primes smaller than $B_2 < B_1$, we iterate the descent lowering the bound at each step until it becomes lower than B. At this point, all the discrete logarithms are known and we can backtrack to recover the logarithm of each of the initial $\mathfrak{q}$ and consequently the logarithm of z. Note that during the descent, we will need to consider special-$\mathfrak{q}$ ideals in both number fields.

2.3 Practical Improvements

Galois extensions. If possible, we should choose f_1 such that K_1 is Galois over $\mathbb{Q}$. Let $\mathrm{Gal}(K_1/\mathbb{Q})$ be its Galois group, then since p is inert in K_1, we obtain an isomorphism $\mathrm{Gal}(K_1/\mathbb{Q}) \simeq \mathrm{Gal}(\mathbb{F}_q/\mathbb{F}_p)$ (see [17, Chapter I, 9.6]) which implies that K_1 has to be a cyclic number field of degree n. The first major improvement for Galois K_1, is that the factor basis associated with K_1 can be n times smaller. We refer to Section 4.3 for a detailed description. The second improvement is related to finding an adequate z in the individual logarithm phase. Assuming that n is prime, the prime ideals lying above a rational prime l are of two types only: either (l) is inert or splits into degree one prime ideals. In the former case, we simply have to compute logarithms in $\mathbb{F}_p$, which is a much smaller finite field, so we can neglect its cost. In the latter case, we proceed as before. For Galois K_1 of prime extension degree n, we therefore only need to find a z with sufficiently smooth norm and can ignore the squarefree condition.

---

[1] Of course, when z belongs to a strict subfield, we are in an easier case of the algorithm, since it then suffices to compute the logarithm in $\mathbb{F}_p^*$ of z w.r.t. the norm of a generator of $\mathbb{F}_{p^{.}}$.

Choice of polynomials. Instead of simply setting $f_2 = f_1 + p$, we can in fact take any polynomial f_2 such that $f_1 | f_2$ over the finite field $\mathbb{F}_p$. In particular, if f_1 defines a cyclic number field, f_2 should be chosen to maximise the automorphism group of K_2. For instance, for all primes $p \equiv 2, 5 \bmod 9$, $f_1 = x^6 + x^3 + 1$ is irreducible over $\mathbb{F}_p$ and by choosing $f_2 = x^6 + (p+1)x^3 + 1$, K_2 will have a non-trivial automorphism group of order 2. Constructing a K_2 with large automorphism group for general p remains an open problem.

Since f_1 normally has tiny coefficients, the coefficients of the polynomial f_2 will be much larger than those of f_1. To balance the size of the norms we compute during the algorithm, there are basically two approaches: unbalance the size of a and b of the elements we sieve over or change the polynomial selection such that the coefficients of both polynomials are of the same size. One possibility is to choose f_1 such that (at least) one of its coefficients, c, is of order $\sqrt{p}$. Write $c \equiv c_1/c_2 \bmod p$ with c_1 and c_2 also of order $\sqrt{p}$ and define $f_2 \equiv c_2 f_1 \bmod p$. The polynomial f_2 is no longer monic, but its coefficients are $O(\sqrt{p})$ instead of $O(p)$.

Individual Logarithms. Instead of directly decomposing z as a product of ideals, it is useful in practice to start by writing z as a fraction of the form:

$$\frac{\sum a_i t^i}{\sum b_i t^i},$$

where the coefficients a_i and b_i are of the order of $\sqrt{p}$.

More precisely, this is done by reducing the following lattice L, where the generating vectors are in columns and where the algebraic numbers in the first line stand for the subcolumns of their coordinates:

$$
L = \begin{pmatrix}
\mathbf{z} & \mathbf{tz} & \mathbf{t^2 z} & \cdots & \mathbf{t^{n-1} z} & \mathbf{p} & \mathbf{pt} & \mathbf{pt^2} & \cdots & \mathbf{pt^{n-1}} \\
1 & 0 & 0 & \cdots & 0 & 0 & 0 & 0 & \cdots & 0 \\
0 & 1 & 0 & \cdots & 0 & 0 & 0 & 0 & \cdots & 0 \\
0 & 0 & 1 & \cdots & 0 & 0 & 0 & 0 & \cdots & 0 \\
\vdots & \vdots & \vdots & \ddots & \vdots & \vdots & \vdots & \vdots & \ddots & \vdots \\
0 & 0 & 0 & \cdots & 1 & 0 & 0 & 0 & \cdots & 0
\end{pmatrix}.
$$

Clearly, any vector in this lattice represents a fraction equal to z, where the numerator is encoded by the upper half of the vector and the denominator by the lower half. Since the determinant of the lattice is p^n and the dimension $2n$, we expect that lattice reduction can find a short vector of norm $\sqrt{p}$. Note that this is only a practical expectation in small dimension that does not hold when n becomes too large.

3   Extension to Other Fields

The analysis (cf. Section 5) of the previous algorithm yields a $L_{p'}(1/3)$ complexity when p has the right size compared to p^n. In this section, we discuss the possible adaptations to larger and smaller values of p.

3.1 Smaller Values of p

When p is smaller than in the basic case, we encounter a major problem, when sieving over the (a, b) pairs. Indeed, we would like to keep the norms of $a - b\theta_1$ and $a - b\theta_2$ below $L_{p'}(2/3)$, however, since p is small, n is larger than before and there are less than $L_{p'}(1/3)$ relations. To summarise, the possible sieving space is too small and we cannot collect enough relations. Thus, we need to enlarge the sieving space. A simple technique is to sieve over elements of higher degree. Instead of pairs (a, b), we consider $(t + 1)$-tuples $(a_0, \cdots, a_t)$ and compute the norms of $\sum_{i=0}^{t} a_i \theta_1^i$ and $\sum_{i=0}^{t} a_i \theta_2^i$. Each of these norms can be obtained through a resultant computation of two polynomials $A(x) = \sum_{i=0}^{t} a_i x^i$ and $f_1(x)$ (resp. $f_2(x)$). It is well-known that the resultant can be obtained as the determinant of an $(n + t) \times (n + t)$ matrix formed of t columns containing the coefficients of f_1 and n columns containing the coefficients of A. Thus, we can clearly bound the norm by $(n + t)^{n+t} B_a{}^n B_f{}^t$, where B_a is an upper bound on the absolute values of the a_i and B_f a similar bound on the coefficients of f_1 (resp. f_2).

3.2 Larger Values of p

When p is larger than in the basic case, the basic polynomial construction is no longer usable, because due to the addition of p, a coefficient of f_2 is larger than $L_{p'}(2/3)$. In order to lower the size of the coefficients, we need to change the polynomial construction and to use polynomials with higher total degree. This is not completely surprising, since the number field sieve in the prime field case uses polynomials with total degree increasing with p. The most difficult part is to find a polynomial construction which does not somehow fail in the individual logarithm phase. We propose the following simple construction. Start from a polynomial f_0 of degree n with small coefficients (and optionally cyclic Galois group when possible). Then, we choose a constant W and let $f_1(x) = f_0(x + W)$, the largest coefficient of f_1 is around W^n. Then, using lattice reduction, we search for a polynomial f_2 of degree D and coefficients smaller than W^n, such that $f_1 | f_2 \bmod p$. This can be done by reducing the lattice L, where the generating vectors are in columns

$$L = \big(\mathbf{f_1(x)} \ \mathbf{xf_1(x)} \ \mathbf{x^2 f_1(x)} \cdots \mathbf{x^{D-n} f_1(x)} \ \mathbf{p} \ \mathbf{px} \ \mathbf{px^2} \cdots \mathbf{px^D} \big)$$

It is clear that every linear combination of the columns results in a polynomial which is divisible by f_1 modulo p. Furthermore, the dimension of the lattice L is $D + 1$ and its determinant is p^{D+1}. We know that for lattices of dimension s and determinant $\mathcal{D}$, LLL [14] outputs a vector shorter than $2^{s/4} \mathcal{D}^{1/s}$. We want this vector to encode a polynomial different from f_1. This can be ensured if the vector is smaller than the encoding of f_1. Thus, we need:

$$2^{(D+1)/4} p^{n/(D+1)} \le W^n.$$

Since the term $2^{(D+1)/4}$ can be hidden in the $o(1)$ of the L representation, the optimal size of the coefficients in f_1 and f_2 is $R = W^n \approx p^{n/(D+1)}$.

4   Mathematical Aspects

In this section, we describe the mathematical background required to make our algorithm rigorous. As before, let $\mathbb{F}_q$ be a finite field with $q = p^n$ elements, with p prime. Given $h = g^x \in \mathbb{F}_q^*$ and a large prime divisor l of $q - 1$, we show how to recover $x \bmod l$.

We will work in several number fields of the form $K = \mathbb{Q}[\theta]$ for some algebraic integer θ with minimal polynomial $f(X) \in \mathbb{Z}[X]$. In the basic case, all these number fields will be of degree n over $\mathbb{Q}$, such that p remains prime in the ring of integers $\mathcal{O}_K$ of K, i.e. $\mathcal{O}_K/(p) \cong \mathbb{F}_q$. In general, we also need number fields of degree $m > n$, such that there exists a prime ideal $\mathfrak{Q}$ of degree n lying above p, i.e. $\mathcal{O}_K/\mathfrak{Q} \cong \mathbb{F}_q$. To construct such a number field, it simply suffices for $f(X) \bmod p$ to have an irreducible factor of degree n. Denote by $\varphi_{\mathfrak{Q}}$ the surjective ring homomorphism obtained by dividing out the ideal $\mathfrak{Q}$, then for any element $x \in \mathcal{O}_K$ we define $\bar{x} = \varphi_{\mathfrak{Q}}(x)$.

4.1   Factoring Ideals in $\mathcal{O}_K$

Recall that in the algorithm we look for coprime pairs of integers (a, b) such that the principal ideal $(a - b\theta)$ factors into prime ideals of small norm, i.e. of norm smaller than some predefined bound B. The possible prime ideals occurring in the factorisation of $(a - b\theta)$ are given by the following lemma [6, Lemma 10.5.1].

Lemma 1. *Let $K = \mathbb{Q}[\theta]$ and (a, b) coprime integers, then any prime ideal $\mathfrak{p}$ which divides $a - b\theta$ either divides the index $f_\theta = [\mathcal{O}_K : \mathbb{Z}[\theta]]$ or is of degree one.*

Therefore, let $\mathcal{F}$ consist of degree one prime ideals of norm smaller than B and the finitely many prime ideals that divide the index f_θ, then we try to decompose $(a - b\theta)$ over $\mathcal{F}$. Each degree one prime ideal is generated by $(p_i, \theta - c_{p.})$ with p_i a rational prime smaller than B and $c_{p.}$ a root of $f(X) \bmod p_i$.

Furthermore, finding the decomposition of the ideal $(a - b\theta)$ into prime ideals is equally easy. First compute its norm $N_{K/\mathbb{Q}}(a - b\theta) = b^{\deg(f)} f(a/b)$ and test whether $N_{K/\mathbb{Q}}(a - b\theta)$ is B-smooth. If it is, write $N_{K/\mathbb{Q}}(a - b\theta) = \prod_i p_i^{e_i}$; now we need to make the following distinction:

- For rational primes $p_i \nmid f_\theta$, we obtain that $\mathfrak{p}_i = (p_i, \theta - c_{p.})$ with $c_{p.} \equiv a/b \bmod p_i$ will occur in the ideal factorisation of $(a - b\theta)$ and that the $\mathfrak{p}_i$-adic valuation is precisely e_i.
- For rational primes $p_i | f_\theta$, use Algorithm 4.8.17 in Cohen [6].

At the end, we therefore have several pairs of coprime integers (a, b) with corresponding ideal decompositions

$$(a - b\theta) = \prod_i \mathfrak{p}_i^{e_i} . \tag{1}$$

For p smaller than in the basic case, we will need to sieve over $(t + 1)$-tuples of coprime integers $(a_0, a_1, \ldots, a_t)$ such that the principal ideal $(\sum_{i=0}^{t} a_i \theta^i)$ factors into prime ideals of norm smaller than B. The following lemma is an easy generalisation of Lemma 1.

Lemma 2. *Let $K = \mathbb{Q}[\theta]$ and $(a_0, \ldots, a_t)$ a $(t+1)$-tuple of coprime integers, then any prime ideal $\mathfrak{p}$ that divides $(\sum_{i=0}^{t} a_i \theta^i)$ either divides the index $f_\theta = [\mathcal{O}_K : \mathbb{Z}[\theta]]$ or is of degree $\leq t$.*

To find the decomposition of the ideal $(\sum_{i=0}^{t} a_i \theta^i)$, we compute its norm as $N_{K/\mathbb{Q}}(\sum_{i=0}^{t} a_i \theta^i) = \mathrm{Res}(\sum_{i=0}^{t} a_i X^i, f(X))$, with Res the resultant. For each prime factor p_i not dividing the index f_θ, we proceed as follows: each irreducible factor of $\gcd(\sum_{i=0}^{t} a_i X^i, f(X))$ over $\mathbb{F}_{p_i}$, corresponds to an ideal $\mathfrak{p}_i$ lying above p occurring in the ideal decomposition. Furthermore, if the gcd itself is irreducible, the $\mathfrak{p}_i$-adic valuation is simply $m/\deg \mathfrak{p}_i$; if not, we use Algorithm 4.8.17 in Cohen [6]. For prime factors dividing the index, we proceed as described above.

4.2 From Ideals to Elements

We now show how to transform the relations involving ideals into multiplicative relations involving elements only. After mapping these relations to $\mathbb{F}_q^*$ using $\varphi_\mathfrak{Q}$ and taking logarithms, we obtain linear equations between logarithms of elements in $\mathbb{F}_q^*$ modulo the prime factor l.

We make a distinction between two cases: the simple case where K has class number one and computable unit group, the general case.

K with Class Number One and Computable Unit Group. Since the class number equals one, every ideal I in $\mathcal{O}_K$ is principal, i.e. there exists $\gamma_I \in \mathcal{O}_K$ with $I = (\gamma_I)$. The ideal decomposition $(a - b\theta) = \prod_i \mathfrak{p}_i^{e_i}$ then leads to

$$a - b\theta = u \prod_i \gamma_i^{e_i}$$

with $(\gamma_i) = \mathfrak{p}_i$ and u a unit in $\mathcal{O}_K$. Let (r_1, r_2) be the signature of K, then the unit group $\mathcal{O}_K^* \cong \mu(K) \times \mathbb{Z}^{r_1+r_2-1}$, with $\mu(K) = \langle u_0 \rangle$ a finite cyclic group of order v. Assuming we can compute a system of fundamental units $u_1, \ldots, u_r$ with $r = r_1 + r_2 - 1$, we can write $u = u_0^{n_0} u_1^{n_1} \cdots u_r^{n_r}$. In this setting we thus obtain r logarithmic maps λ_i for $i = 1, \ldots, r$ defined by

$$\lambda_i : \mathcal{O}_K^* \to \mathbb{Z} : u \mapsto n_i,$$

and a logarithmic map $\lambda_0 : \mathcal{O}_K^* \mapsto \mathbb{Z}/v\mathbb{Z} : u \mapsto n_0$. Finally, we obtain the decomposition

$$a - b\theta = \prod_{i=0}^{r} u_i^{\lambda_i(u)} \prod_i \gamma_i^{e_i}.$$

Applying $\varphi_\mathfrak{Q}$ and taking logarithms of both sides then leads to

$$\log_g(a - b\bar{\theta}) \equiv \sum_{i=0}^{r} \lambda_i(u) \log_g \bar{u}_i + \sum_i e_i \log_g \bar{\gamma}_i \bmod q - 1. \tag{2}$$

General K. Here we assume that the large prime factor l of $q - 1$ does not divide the class number $h(K)$, which constitutes a very minor restriction. To obtain a relation between elements, we raise both sides of the ideal decomposition $(a - b\theta) = \prod_i \mathfrak{p}_i^{e_i}$ to the power $h = h(K)$ and obtain

$$(a - b\theta)^h = u \prod_i \delta_i^{e_i} \tag{3}$$

with u a unit and $\delta_i \in \mathcal{O}_K$ with $(\delta_i) = \mathfrak{p}_i^h$. Note that the left hand side denotes the h-th power of the element (not the ideal) $a - b\theta$. Furthermore, we will never compute the elements δ_i, but only need that these exist and correspond to the ideals $\mathfrak{p}_i$.

It is tempting to apply φ_Ω to both sides and take logarithms as in the easy case, leading to

$$h \log_g(a - b\bar{\theta}) \equiv \log_g \bar{u} + \sum_i e_i \log_g \bar{\delta}_i \bmod q - 1 \,.$$

The problem with this approach is that for each equation, the unit u will be different, so $\log_g \bar{u}$ is a new unknown for every equation, and thus we will never obtain a linear system of full rank. Note that in the easy case, we could express u as a product of fundamental units, which effectively circumvented this problem.

To solve this problem we follow Schirokauer [18]: since it is sufficient to compute the discrete logarithm modulo a large prime $l | q - 1$, we need not work with the whole unit group $\mathcal{O}_K^*$, but only modulo l-th powers of units, i.e. $\mathcal{O}_K^*/(\mathcal{O}_K^*)^l$. Clearly, for $u \in (\mathcal{O}_K^*)^l$, we have $\log_g \bar{u} \equiv 0 \bmod l$. Instead of defining logarithmic maps on the whole unit group, Schirokauer defines such maps on $\mathcal{O}_K^*/(\mathcal{O}_K^*)^l$.

For simplicity we will assume that l does not ramify in K. Consider the set $\Gamma_K = \{\gamma \in \mathcal{O}_K \mid N_{K/\mathbb{Q}}(\gamma) \neq 0 \bmod l\}$, and note that the elements $a - b\theta$ are in Γ, since they are smooth and thus $l \nmid N_{K/\mathbb{Q}}(a - b\theta)$. Also note that Γ_K is multiplicative and contains the unit group $\mathcal{O}_K^*$.

Let $\varepsilon = l^D - 1$, with D the least common multiples of the irreducible factors of $f(X) \bmod l$, then for all $\gamma \in \Gamma_K$ we have

$$\gamma^\varepsilon \equiv 1 \bmod l \,.$$

Define the map λ from Γ_K to $l\mathcal{O}_K/l^2\mathcal{O}_K$ by $\lambda(\gamma) = (\gamma^\varepsilon - 1) + l^2\mathcal{O}_K$. Fix a basis $\{lb_i + l^2\mathcal{O}_K : i = 1, \ldots, n\}$ for $l\mathcal{O}_K/l^2\mathcal{O}_K$ with $\{lb_i + l^2\mathcal{O}_K : i = 1, \ldots, r\}$ a basis for $\lambda(\mathcal{O}_K^*)$ and define maps $\lambda_i : \Gamma_K \to \mathbb{Z}/l\mathbb{Z}$ by

$$(\gamma^\varepsilon - 1) \equiv l \sum_{i=1}^n \lambda_i(\gamma)b_i \bmod l^2 \,.$$

Note that $\lambda(\gamma \cdot \gamma') = \lambda(\gamma) + \lambda(\gamma')$ and similarly $\lambda_i(\gamma \cdot \gamma') = \lambda_i(\gamma) + \lambda_i(\gamma')$, so these maps are logarithmic on Γ_K. As a consequence, the maps λ_i for $i = 1, \ldots, n$ are in fact homomorphisms from $\mathcal{O}_K^*$ to $\mathbb{Z}/l\mathbb{Z}$. Consider the homomorphism

$$\bar{\lambda} : \mathcal{O}_K^*/(\mathcal{O}_K^*)^l \to (\mathbb{Z}/l\mathbb{Z})^r : u \mapsto (\lambda_1(u), \ldots, \lambda_r(u)) \,.$$

We have the following trivial lemma which is a special case of Schirokauer's theorem.

Lemma 3. *Assuming that $\overline{\lambda}$ is injective, $u \in \mathcal{O}_K^*$ is an l-th power if and only if $\overline{\lambda}(u) = 0$.*

If we furthermore assume that $\mathcal{O}_K^*$ does not contain the primitive l-th roots of unity, then $\mathcal{O}_K^*/(\mathcal{O}_K^*)^l$ has l^r elements and thus $\overline{\lambda}$ defines an isomorphism. Therefore, there exist units $u_1, \ldots, u_r \in \mathcal{O}_K^*$ such that $\lambda_i(u_i) = 1$ and $\lambda_i(u_j) = 0$ for $i \neq j$. Note that these units are not unique, since they are only determined modulo $(\mathcal{O}_K^*)^l$. Any unit $u \in \mathcal{O}_K$ can thus be written as

$$u = \prod_{i=1}^{r} u_i^{\lambda_i(u)} \cdot \xi^l \tag{4}$$

for some unit ξ.

In Equation (3) we now modify the elements δ_i by multiplying with a well defined unit

$$\delta_i' = \delta_i \cdot \prod_{j=1}^{r} u_j^{-\lambda_i(\delta_i)} ,$$

such that $\lambda_j(\delta_i') = 0$ for $j = 1, \ldots, r$. Note that δ_i' still generates $\mathfrak{p}_i^h$.

Rewriting Equation (3) then leads to $(a - b\theta)^h = u' \prod_i (\delta_i')^{e_i}$, with u' a unit. Since we have constructed δ_i' such that $\lambda_j(\delta_i') = 0$ for $j = 1, \ldots, r$, we have $h\lambda_j(a - b\theta) = \lambda_j(u')$. Rewriting the unit u' as in (4) finally leads to the equation

$$(a - b\theta)^h = \xi_{a,b}^l \cdot \prod_{i=1}^{r} u_i^{h\lambda_i(a-b\theta)} \cdot \prod_i (\delta_i')^{e_i} ,$$

for some $\xi_{a,b} \in \mathcal{O}_K^*$. Applying $\varphi_{\mathfrak{Q}}$ and taking logarithms of both sides modulo l then gives

$$\log_g(a - b\overline{\theta}) \equiv \sum_{i=0}^{r} \lambda_i(a - b\theta) \log_g \overline{u}_i + \sum_i e_i h^{-1} \log_g \overline{\delta}_i' \bmod l. \tag{5}$$

Note that $h^{-1} \log_g \overline{\delta}_i'$ and $\log_g \overline{u}_i$ correspond precisely to the virtual logarithms of ideals and Schirokauer maps introduced in [12] and [19], but are now given as logarithms of the reduction of specific elements in $\mathfrak{p}_i$ and $\mathcal{O}_K^*$. Furthermore, note that these values are well defined modulo l.

4.3   Exploiting Automorphisms

In this section we show that if the number field K has a non-trivial automorphism group $\text{Aut}(K)$, the size of the factor basis $\mathcal{F}$ can be reduced by a factor of $\#\text{Aut}(K)$.

Let K be a number field of degree n with non-trivial automorphism group $\text{Aut}(K)$ and assume p is inert in K. Clearly, the prime ideal (p) is invariant under each $\phi \in \mathrm{Aut}(K)$. Therefore, each automorphism ϕ defines an automorphism $\overline{\phi}$ of $\mathbb{F}_q$ by reduction modulo p. Furthermore, since p is unramified, this map will be injective, so $\overline{\mathrm{Aut}}(K) \subset \mathrm{Gal}(\mathbb{F}_q/\mathbb{F}_p)$, implying that $\mathrm{Aut}(K)$ has to be cyclic and $\#\mathrm{Aut}(K)|n$.

Let $A = \#\mathrm{Aut}(K)$ and $d = n/A$ and denote by $\psi \in \mathrm{Aut}(K)$ the unique automorphism such that $\psi(x) \equiv x^{p^d} \bmod p$ for all $x \in \mathcal{O}_K$. Then $\mathrm{Aut}(K) = \langle \psi \rangle$ and define $\psi_k = \psi^k$ for $k = 0, \dots, A-1$.

Write $\mathcal{F} = \mathcal{F}_e \cup \mathcal{F}_u$ where $\mathcal{F}_u$ contains all unramified degree one prime ideals and $\mathcal{F}_e$ the others. Note that $\mathcal{F}_e$ is tiny, since these prime ideals have to divide the discriminant of f. Since $\psi_k(\mathfrak{p})$ is a prime ideal of the same degree and ramification index as $\mathfrak{p}$, we can partition $\mathcal{F}_u$ into A disjunct sets $\mathcal{F}_{u,k}$ for $k = 0, \dots, A-1$ such that $\mathfrak{p} \in \mathcal{F}_{u,0} \Leftrightarrow \psi_k(\mathfrak{p}) \in \mathcal{F}_{u,k}$. The decomposition (1) can then be rewritten as

$$(a - b\theta) = \prod_{\mathfrak{p}_i \in \mathcal{F}_e} \mathfrak{p}_i^{e_i} \prod_{k=0}^{A-1} \prod_{\mathfrak{p}_i \in \mathcal{F}_{u,0}} \psi_k(\mathfrak{p}_i)^{e_{i\cdots}},$$

where most of the e_i and $e_{i,k}$ will be zero.

Let δ_i be a generator of the principal ideal $\mathfrak{p}_i^h$, then clearly $\psi_k(\delta_i)$ is a generator of $(\psi_k(\mathfrak{p}_i))^h$ and $\log_g \overline{\psi_k(\delta_i)} \equiv p^{dk} \log_g \overline{\delta_i} \bmod q - 1$. Rewrite Equation (3) as

$$(a - b\theta)^h = u' \prod_{i \in [\mathcal{F}_e]} \delta_i^{e_i} \prod_{k=0}^{A-1} \prod_{i \in [\mathcal{F}_{u,0}]} \psi_k(\delta_i)^{e_{i\cdots}},$$

for some unit u' and with $[\cdot]$ denoting the index set of a set. Note that we no longer have the equality $\lambda_j(u') = h\lambda_j(a - b\theta)$, since $\lambda_j(\psi_k(\delta_i))$ will not be zero. However, we can explicitly compute

$$\lambda_j(u') \equiv h\lambda_j(a - b\theta) - \sum_{i \in [\mathcal{F}_e]} e_i\lambda_j(\delta_i) - \sum_{k=0}^{A-1} \sum_{i \in [\mathcal{F}_{u,0}]} e_{i,k}\lambda_j(\psi_k(\delta_i)) \bmod l.$$

The equivalent of Equation (5) then becomes

$$
\begin{aligned}
\log_g(a - b\overline{\theta}) \equiv{} & \sum_{j=0}^{r} h^{-1}\lambda_j(u') \log_g \overline{u}_j + \sum_{i \in [\mathcal{F}_e]} h^{-1} e_i \log_g \overline{\delta}_i \\
& + \sum_{i \in [\mathcal{F}_{u,0}]} \left(h^{-1} \sum_{k=0}^{A-1} p^{dk} e_{i,k} \right) \log_g \overline{\delta}_i \bmod l.
\end{aligned}
\tag{6}
$$

Note that the number of unknowns in the right hand side of the above equation now only amounts to $\#\mathcal{F}_{u,0} + \#\mathcal{F}_e + r$, which is roughly $\#\mathrm{Aut}(K)$ times smaller than in the previous section. However, this approach is only feasible when we are able to compute the class number h. Furthermore, for each ideal $\mathfrak{p}_i \in \mathcal{F}_{u,0}$ we now have to explicitly compute a δ_i that generates $\mathfrak{p}_i^h$ and all the Schirokauer maps $\lambda_j(\psi_k(\delta_i))$ for $j = 1, \dots, r$, $k = 0, \dots, A-1$ and all i. Clearly none of these issues arise when K has class number one and computable unit group.

5 Asymptotic Heuristic Complexity

In this section we summarise the heuristic complexity of the different variations described in Sections 2 and 3. The details of this complexity analysis are given in Appendix A. Note that the basic algorithm is a special case of the algorithm given in Section 3.1, so we do not treat it separately.

The precise complexity depends on the ratio of p compared to p^n and we have the following cases:

- p can be written as $L_q(l_p, c)$ with $1/3 < l_p < 2/3$, then the algorithm given in 3.1 has complexity
$$L_q(1/3, (128/9)^{1/3}).$$

- p can be written as $L_q(2/3, c)$ for a constant c, then the algorithm given in 3.1 has complexity
$$L_q(1/3, 2c') \quad \text{with} \quad c' = \frac{4}{3}\left(\frac{3t}{4(t+1)}\right)^{1/3},$$
where we sieve over $(t+1)$-tuples where the degree t is taken as the natural number closest to the real root of $3c^3t(t+1)^2 - 32 = 0$.

- p can be written as $L_q(2/3, c)$ for a constant c, then the algorithm given in 3.2 has complexity
$$L_q(1/3, 2c') \quad \text{with} \quad 9c'^3 - \frac{6}{c}c'^2 + \frac{1}{c^2}c' - 8 = 0.$$

- p can be written as $L_q(l_p, c)$ with $l_p > 2/3$, then the algorithm given in 3.2 has complexity
$$L_q(1/3, (64/9)^{1/3}).$$

In Figure 1 we have plotted the constant c' which determines the complexity $L_q(1/3, 2c')$ as a function of the constant c. The plot also shows for which c to switch from algorithm 3.1 to algorithm 3.2.

6 Numerical Experiments

We modified Joux and Lercier's discrete logarithm C-implementation in order to handle a 120-digit realistic experiment. More precisely, we consider a finite field of the form $\mathbb{F}_{p^3}$ with
$$p = \lfloor 10^{39}\pi \rfloor + 2622 = 3141592653589793238462643383279502886819,$$
the smallest 40-digit integer than $10^{39}\pi$ such that $(p-1)/2$ is prime. Its multiplicative group has $p^3 - 1$ elements, which factors as

$2 \times 13 \times 19 \times 199 \times 4177 \times 212145751 \times 547465820443 \times 337091202666800863 \times l \times (p-1)/2 \,,$

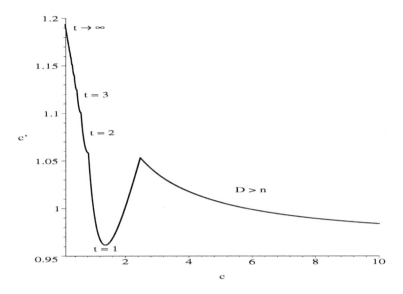

Fig. 1. Asymptotic complexity $L_q(1/3, 2c')$ (note factor 2) as a function of c with $p \in L_q(1/3, c)$. Also indicated are the degree t of the elements in the sieving space and the degree $D \geq n$ of the number field K_2.

where $l = 1227853277188311599267416539617839$. The hardest part is of course to compute discrete logarithms modulo the 110-bit factor l since smaller factors can be easily handled with the Pollard-rho algorithm and discrete logarithms modulo $p - 1$ are the same as discrete logarithms of norms in $\mathbb{F}_p$. To solve this problem, let $\mathbb{Q}[\theta_1]$ and $\mathbb{Q}[\theta_2]$ be the two number fields defined respectively by

$$f_1(X) = X^3 + X^2 - 2X - 1 \quad \text{and} \quad f_2(X) = f_1(X) + p,$$

where we have $\mathbb{F}_{p^3} \simeq \mathbb{F}_p[t]/(f_1(t))$. With these settings, $\mathbb{Q}[\theta_1]$ is a cubic cyclic number field, the Galois group of which is given by $\text{Aut}(\mathbb{Q}[\theta_1]) = \{\theta_1 \mapsto \theta_1, \theta_1 \mapsto \theta_1^2 - 2, \theta_1 \mapsto -\theta_1^2 - \theta_1 + 1\}$, and a system of fundamental units by $u_1 = \theta_1 + 1$ and $u_2 = \theta_1^2 + \theta_1 - 1$. $\mathbb{Q}[\theta_2]$ has signature $(1, 1)$, thus we will have to apply one single Schirokauer logarithmic map λ_1 to our relations to deal with obstructions in the $\mathbb{Q}[\theta_2]$ side.

We construct smoothness bases with $1\,000\,000$ prime ideals as follows,

- in the $\mathbb{Q}[\theta_1]$ side, we include $899\,999$ prime ideals, but only $300\,000$ are meaningful due to the Galois action,
- in the $\mathbb{Q}[\theta_2]$ side, we include $700\,000$ prime ideals.

The lattice sieving considers only algebraic integers $a + b\theta_2$ which are already divisible by a prime ideal in $\mathbb{Q}[\theta_2]$, such that norms to be smoothed in $\mathbb{Q}[\theta_2]$ are 150 bit integers (once removed the known prime ideal) and norms in $\mathbb{Q}[\theta_1]$ are 110 bit integers (an important tip to improve the running time with such polynomials is to unbalance the size of a versus the size of b). The sieving took 12 days on a 1.15 GHz 16-processors HP AlphaServer GS1280.

We then had to compute the kernel of a $1\,163\,482 \times 793\,188$ matrix, the coefficients of which are mostly equal modulo ℓ to ± 1, $\pm p$ or $\pm p^2$. We thus first modified the Joux-Lercier structured gaussian elimination step to reduce such a system to $450\,246$ equations in $445\,097$ unknowns with $44\,544\,016$ non null entries. Time needed for this on one processor was only a few minutes. Then, the critical phase was the final computation via Lanczos's algorithm. Our parallelised version of this algorithm (modified for such matrices too) took one additional week. At the end, since $\mathbb{Q}[\theta_1]$ is principal, we may check that we have discrete logarithms of generators of ideals in the smoothness bases, for instance,

$$(t^2 + t + 1)^{(p^3-1)/l} = G^{29406688645015596112746712243217 1},$$
$$(t - 3)^{(p^3-1)/l} = G^{36422456363509538073334012349071 9},$$
$$(3\,t - 1)^{(p^3-1)/l} = G^{46887658774739638067572350292825 7},$$

where $G = g^{(p^3-1)/1159268202574177739715462155841484\,l}$ and $g = -2t + 1$.

In the final step, we took as a challenge $\gamma = \sum_{i=0}^{2} (\lfloor \pi \times p^{i+1} \rfloor \bmod p) t^i$. We first easily computed its discrete logarithm in basis g modulo $(p^3 - 1)/l$,

38895389158901518975845922936941184677534991099612214604576972713861472869108247732 8.

To obtain a complete result, we expressed

$$\cdot = \frac{-90987980355959529347 \cdot^2 \; - \; 114443008248522156910 \cdot \; + \; 154493664373341271998}{94912764441570771406 \cdot^2 \; - \; 120055569809711861965 \cdot \; - \; 81959619964446352567} \cdot$$

where numerator and denominator are both smooth once considered as algebraic integers in $\mathbb{Q}[\theta_1]$. Using a three level tree with 80 special-$\mathfrak{q}$ ideals, we recovered the discrete logarithm modulo l, namely $110781190155780903592153105706975$. Each special-$\mathfrak{q}$ sieving took 10 minutes or a total of 14 hours.

7   Conclusion

In this paper, we have presented a new variation of the number field sieve algorithm to compute discrete logarithms in $\mathbb{F}_{p^n}^*$. For $p > L_{p^n}(1/3)$, our variation yields a complexity of $L_{p^n}(1/3)$. For smaller values of p, the function field sieve algorithm described in [13] also gives $L_{p^n}(1/3)$ complexity. As a consequence, we have $L_{p^n}(1/3)$ heuristic algorithms to compute discrete logarithms in all finite fields $\mathbb{F}_{p^n}$. This should be compared to the previous $L_{p^n}(1/2)$ algorithms given in [1,2]. Another major advantage is that this algorithm has a fast postprocessing phase for individual logarithms, once the main algorithm terminates. This is extremely useful for applications which require the computation of many logarithms in the same field. To give an example, this was required by the identity based cryptosystem of Maurer and Yacobi [16].

Acknowledgement

The authors would like to thank Henri Cohen for providing extensive references on the existence and construction of cyclic number fields.

References

1. L. Adleman and J. DeMarrais. A subexponential algorithm for discrete logarithms over all finite fields. In D. Stinson, editor, *Proceedings of CRYPTO'93*, volume 773 of *Lecture Notes in Comput. Sci.*, pages 147–158. Springer, 1993.
2. L. Adleman and J. DeMarrais. A subexponential algorithm for discrete logarithms over all finite fields. *Math. Comp.*, 61(203):1–15, 2003.
3. L.M. Adleman. The function field sieve. In *Algorithmic Number Theory, Proceedings of the ANTS-I conference*, volume 877 of *Lecture Notes in Comput. Sci.*, pages 108–121, 1994.
4. L.M. Adleman and M.A. Huang. Function field sieve method for discrete logarithms over finite fields. In *Information and Computation*, volume 151, pages 5–16. Academic Press, 1999.
5. E. R. Canfield, Paul Erdős, and Carl Pomerance. On a problem of Oppenheim concerning "factorisatio numerorum". *J. Number Theory*, 17(1):1–28, 1983.
6. H. Cohen. *A course in computational algebraic number theory*, volume 138 of *Graduate Texts in Mathematics*. Springer-Verlag, Berlin, 1993.
7. D.M. Gordon. Discrete logarithms in GF(p) using the number field sieve. *SIAM J. Discrete Math.*, 6(1):124–138, 1993.
8. R. Granger, A. Holt, D. Page, N. Smart, and F. Vercauteren. Function field sieve in characteristic three. In D. Buell, editor, *Algorithmic Number Theory, Proceedings of the ANTS-VI conference*, volume 3076 of *Lecture Notes in Comput. Sci.*, pages 223–234. Springer, 2004.
9. R. Granger and F. Vercauteren. On the discrete logarithm problem on algebraic tori. In V. Shoup, editor, *Proceedings of CRYPTO'2005*, volume 3621 of *Lecture Notes in Comput. Sci.*, pages 66–85. Springer, 2005.
10. A. Ivić and G. Tenenbaum. Local densities over integers free of large prime factors. *Quart. J. Math. Oxford Ser. (2)*, 37(148):401–417, 1986.
11. A. Joux and R. Lercier. The function field sieve is quite special. In C. Fieker and D. Kohel, editors, *Algorithmic Number Theory, Proceedings of the ANTS-V conference*, volume 2369 of *Lecture Notes in Comput. Sci.*, pages 431–445. Springer, 2002.
12. A. Joux and R. Lercier. Improvements to the general number field sieve for discrete logarithms in prime fields. A comparison with the gaussian integer method. *Math. Comp.*, 72:953–967, 2003.
13. A. Joux and R. Lercier. The function field sieve in the medium prime case. In S. Vaudenay, editor, *Proceedings of EUROCRYPT'2006*, volume 4004 of *Lecture Notes in Comput. Sci.*, pages 254–270. Springer, 2006.
14. A. K. Lenstra, H. W. Lenstra, Jr., and L. Lovász. Factoring polynomials with rational coefficients. *Math. Ann.*, 261(4):515–534, 1982.
15. R. Lercier and F. Vercauteren. Discrete logarithms in $\mathbb{F}_{p^{18}}$ - 101 digits. NMBRTHRY mailing list, June 2005.
16. U.M. Maurer and Y. Yacobi. A non-interactive public-key distribution system. *Des. Codes Cryptogr.*, 9(3):305–316, 1996.
17. J. Neukirch. *Algebraic number theory*, volume 322 of *Grundlehren der Mathematischen Wissenschaften*. Springer-Verlag, Berlin, 1999.
18. O. Schirokauer. Discrete logarithms and local units. *Philos. Trans. Roy. Soc. London Ser. A*, 345(1676):409–423, 1993.
19. O. Schirokauer. Virtual logarithms. *J. Algorithms*, 57(2):140–147, 2005.

20. O. Schirokauer, D. Weber, and T. Denny. Discrete logarithms: the effectiveness of the index calculus method. In *Algorithmic Number Theory, Proceedings of the ANTS-II conference*, volume 1122 of *LNCS*, pages 337–361. Springer, Berlin, 1996.

21. D. Weber. Computing discrete logarithms with the general number field sieve. In *Algorithmic Number Theory, Proceedings of the ANTS-II conference*, volume 1122 of *LNCS*, pages 391–403. Springer, Berlin, 1996.

A   Asymptotic Heuristic Complexity

A.1   Basic Algorithm

In order to analyse the heuristic asymptotic complexity of the basic algorithm described in Section 2 we first assume that we can write the following relations between $q = p^n$, p and n, for some constant c to be determined later on:

$$n = \frac{1}{c} \cdot \left(\frac{\log q}{\log \log q} \right)^{1/3}, \qquad p = \exp\left(c \cdot \sqrt[3]{\log^2 q \cdot \log \log q} \right).$$

We further assume that the sieve limit S and the smoothness bound B are chosen equal and expressed as:

$$S = B = \exp\left(c' \cdot \sqrt[3]{\log q \cdot \log^2 \log q} \right),$$

for some constant c'. Since f_1 has small coefficients, say bounded by b_0, the norm of $a - b\theta_1$ is smaller than $nb_0 B^n = B^{n+o(1)}$. Similarly, the norm of $a - b\theta_2$ is smaller than $p \cdot B^{n+o(1)}$. Thus, the product of these two norms is smaller than $p \cdot B^{2n+o(1)} = L_q(2/3, c + 2c'/c)$. We make the usual heuristic hypothesis and assume that the probability for each value of this product to split into primes lower than B follows the well-known theorem of Canfield, Erdös and Pomerance [5]: a random number below $L_q(r, \gamma)$ is $L_q(s, \delta)$-smooth with probability $L_q(r - s, -\gamma(r - s)/\delta)$. As explained above, in the squarefree factorisation case, this probability is lower, by a factor $6/\pi^2$ which we can neglect in the L-notation. Plugging in our values, we find that the product of the norms is smooth with probability $L_q(1/3, -(1/3) \cdot (c/c' + 2/c))$. In order to minimise the total runtime, we want to roughly balance the complexities of the sieving phase and of the linear algebra phase. This is achieved when $c' = (1/3) \cdot (c/c' + 2/c)$ and thus:

$$c' = \frac{1}{3} \left(\frac{1}{c} + \sqrt{3c + c^{-2}} \right).$$

The heuristic complexity is $L_q(1/3, 2c')$ and it varies depending on c. It is minimal for $c = 2 \cdot (1/3)^{1/3}$, where $c' = 2 \cdot (1/3)^{2/3}$. At this minimal point, the complexity is $L_q(1/3, (64/9)^{1/3})$, thus identical to the complexity of the number field sieve over prime fields.

A.2   Algorithm for Smaller p

When considering our algorithmic variants for smaller values of p, two cases are possible. Either p is written $L_q(2/3, c)$ as before or it is of the form $L_q(l_p, c)$ with $1/3 < l_p < 2/3$. In the first case, the overall complexity is $L_q(1/3)$ with a constant which depends on c, in the second case, the complexity is $L_q(1/3, (128/9)^{1/3})$. The parameters of the algorithms are the smoothness bound B, the degree t of the elements we are sieving over and the bound on the coefficients of these elements S. Clearly, the norm on the f_1 side is bounded by $S^{n+o(1)}$ and the norm on the f_2 side is bounded by $S^{n+o(1)}p^{t+o(1)}$. Thus the product of the norms is bounded by $S^{2n+o(1)}p^{t+o(1)}$. Choose for t the nearest integer to:

$$\frac{c_t}{c} \cdot \left(\frac{\log q}{\log \log q} \right)^{2/3 - l_\bullet},$$

for a constant c_t to be determined later on. Choose the sieve bound as:

$$S = \exp \left(c_S c \cdot \log^{l_\bullet -1/3} q \cdot \log^{4/3-l_\bullet} \log q \right).$$

Then, the total sieving space contains S^t elements, where

$$S^t = \exp \left((c_S c_t + o(1)) \cdot \log^{1/3} q \cdot \log^{2/3} \log q \right)$$

when $l_p < 2/3$. For the $l_p = 2/3$ case, the round-off error in t is no longer negligible, which explains why the complexity varies with c. We continue the analysis for $l_p < 2/3$, the $l_p = 2/3$ case is discussed in the next subsection and summarised in Figure 1. Let the smoothness bound be $B = L_q(1/3, c')$ as before. Rewrite the bound on the product of norms as $L_q(2/3, 2c_S + c_t)$, then the smoothness probability is

$$Pr = L_q(1/3, -(1/3) \cdot (2c_S/c' + c_t/c')).$$

In order to choose c_S and c_t, we need to equate the runtime of the sieving, i.e. the size of the sieving space S^t, with the time of the linear algebra, i.e. B^2, and also with B/Pr. This implies $S^{t/2} = B = 1/Pr$. Translating this into equations involving the various constants, we find:

$$c' = \frac{1}{3} \left(\frac{2c_S}{c'} + \frac{c_t}{c'} \right) = \frac{c_S c_t}{2}.$$

As a consequence, we deduce:

$$c' = \sqrt{\frac{2c_S + c_t}{3}} = \frac{c_S c_t}{2}.$$

We now write $c_t = x$, $c_S = \mu x$ and minimise c'. With this notation:

$$\frac{(2\mu + 1)x}{3} = \frac{\mu^2 x^4}{4} \quad \text{or equivalently } x^3 = \frac{4(2\mu + 1)}{3\mu^2}.$$

Since $c' = \mu x^2/2$, we write:

$$c'^3 = \frac{2(2\mu + 1)^2}{9\mu}.$$

Clearly, minimising c' is equivalent to minimising c'^3 and implies:

$$\frac{8(2\mu + 1)}{9\mu} = \frac{2(2\mu + 1)^2}{9\mu^2}, \text{ or } 4\mu = 1 + 2\mu, \text{ and finally } \mu = \frac{1}{2}.$$

As a consequence $x = (32/3)^{1/3}$, $c' = x^2/4 = (16/9)^{1/3}$. Thus, the complexity of the algorithm is $L_q(1/3, 2c') = L_q(1/3, (128/9)^{1/3})$.

The case of $l_p = 2/3$. In this case, it is easier to consider a family of algorithms, indexed by t and to compute the complexity of each algorithm. We then choose $B = L_q(1/3, c')$ and $S = L_q(1/3, 2c'/(t + 1))$, the total size of the sieving space and the runtime of the linear algebra are $L_q(1/3, 2c')$, the product of the norms is $L_q(2/3, 4c'/(t + 1)c + tc)$ and the probability of smoothness is $L_q(1/3, (-1/3) \cdot (4/(t + 1)c + tc/c'))$. As usual, we equalise c' with the constant in the probability expression and find:

$$c' = \frac{1}{3}\left(\frac{4}{(t + 1)c} + \frac{tc}{c'}\right).$$

This implies:

$$3c'^2 - \frac{4c'}{(t + 1)c} - tc = 0.$$

Thus:

$$c' = \frac{1}{3}\left(\frac{2}{(t + 1)c} + \sqrt{\frac{4}{(t + 1)^2c^2} + 3tc}\right).$$

Note that for $t = 1$, we recover the formula of the basic case. This comes to a minimum when:

$$\frac{2}{(t+1)c^2} = \frac{-8(t+1)^{-2}c^{-3} + 3t}{2\sqrt{4(t+1)^{-2}c^{-2} + 3tc}}$$

$$\Leftrightarrow$$

$$64(t + 1)^{-2}c^{-2} + 48tc = \left(-8(t + 1)^{-1}c^{-1} + 3t(t + 1)c^2\right)^2$$

$$\Leftrightarrow$$

$$48tc = -48tc + 9t^2(t + 1)^2c^4.$$

Thus, we take:

$$c = 2\left(\frac{4}{3t(t + 1)^2}\right)^{1/3}.$$

As a consequence:

$$c' = \frac{4}{3}\left(\frac{3t}{4(t + 1)}\right)^{1/3}.$$

We recover the basic complexity $L_q(1/3, (64/9)^{1/3})$ when $t = 1$ and find the expected limit

$$L_q(1/3, (128/9)^{1/3})$$

when t tends to infinity. We show how the complexity constant varies with c in Figure 1.

A.3   Algorithm for Larger p

For larger p, as for smaller p, two cases arise. Either p is written $L_q(2/3, c)$ or it is of the form $L_q(l_p, c)$ with $l_p > 2/3$. In the first case, the overall complexity is $L_q(1/3)$ with a constant which depends on c, in the second case, the complexity is $L_q(1/3, (64/9)^{1/3})$. The parameters in this case are the smoothness bound B and the degree D. The degree of f_1 is n, the bound on the coefficients when sieving is $S = B$ and the size of the coefficients of both f_1 and f_2, according to the construction of polynomials described above is $R = p^{n/(D+1)} = q^{1/(D+1)}$. Choose:

$$D = c_D \left(\frac{\log q}{\log \log q} \right)^{1/3}$$

and write $B = L_q(1/3, c')$. Bound the norms on the f_1 side by $RS^{n+o(1)}$ and the norms on the f_2 side by $RS^{D+1+o(1)}$. Thus the product of the norms is bounded by $B^{n+D+1+o(1)}q^{2/(D+1)}$. When $l_p > 2/3$, n is negligible compared to D and the product of the norms can be rewritten as $L_q(2/3, c'c_D + 2/c_D)$. This is minimised when $c_D = (2/c')^{1/2}$, and becomes $L_q(2/3, 2\sqrt{2c'})$. The probability of smoothness is $L_q(1/3, -(1/3) \cdot 2\sqrt{2/c'})$. As usual, we equalise the (opposite of) this constant with c'. This yields $c' = (8/9)^{1/3}$ and finally the expected $L_q(1/3, (64/9)^{1/3})$ complexity.

When $l_p = 2/3$, matters are more complicated. The product of the norms is now rewritten as $L_q(2/3, c'(c_D + 1/c) + 2/c_D)$, again this is minimised at $c_D = (2/c')^{1/2}$. However, it now becomes, $L_q(2/3, c'/c+2\sqrt{2c'})$. The probability of smoothness is $L_q(1/3, -(1/3) \cdot (1/c + 2\sqrt{2/c'}))$. Equating this constant to c' yields:

$$(3c' - 1/c)^2 = 8/c' \quad \text{or} \quad 9c'^3 - \frac{6}{c}c'^2 + \frac{1}{c^2}c' - 8 = 0.$$

When c tends to infinity, we recover the $(64/9)^{1/3}$ constant in the complexity.

Inverting HFE Is Quasipolynomial*

Louis Granboulan[1], Antoine Joux[2,3], and Jacques Stern[1]

[1] École normale supérieure
Département d'Informatique 45, rue d'Ulm
75230 Paris Cedex 05, France
Louis.Granboulan@ens.fr, Jacques.Stern@ens.fr
[2] DGA
[3] Université de Versailles St-Quentin-en-Yvelines
PRISM
45, avenue des Etats-Unis
78035 Versailles Cedex, France
Antoine.Joux@m4x.org

Abstract. In the last ten years, multivariate cryptography has emerged as a possible alternative to public key cryptosystems based on hard computational problems from number theory. Notably, the HFE scheme [17] appears to combine efficiency and resistance to attacks, as expected from any public key scheme. However, its security is not yet completely understood. On one hand, since the security is related to the hardness of solving quadratic systems of multivariate binary equations, an NP complete problem, there were hopes that the system could be immune to subexponential attacks. On the other hand, several lines of attacks have been explored, based on so-called relinearization techniques [12,5], or on the use of Gröbner basis algorithms [7]. The latter approach was used to break the first HFE Challenge 1 in 96 hours on a 833 MHz Alpha workstation with 4 Gbytes of memory. At a more abstract level, Faugère and Joux discovered an algebraic invariant that explains why the computation finishes earlier than expected. In the present paper, we pursue this line and study the asymptotic behavior of these Gröbner basis based attacks. More precisely, we consider the complexity of the decryption attack which uses Gröbner bases to recover the plaintext and the complexity of a related distinguisher. We show that the decryption attack has a quasipolynomial complexity, where quasipolynomial denotes an subexponential expression much smaller than the classical subexponential expressions encountered in factoring or discrete logarithm computations. The same analysis shows that the related distinguisher has provable quasipolynomial complexity.

1 Introduction

In the last ten years, multivariate cryptography has emerged as a possible alternative to public key cryptosystems based on hard computational problems from

---

* This work is supported in part by the French government through X-Crypt, in part by the European Commission through ECRYPT.

number theory. The public key of multivariate schemes is a system of multivariate quadratic (MQ) equations over a finite field, and the underlying question of finding a solution to such systems, a well known NP-complete problem, seems to form a basis for the security of the schemes, the same way RSA-like cryptosystems have their security based on the hardness of factoring, and ElGamal-like cryptosystems rely on the discrete logarithm problem.

Although all MQ schemes are more or less built on the same pattern, mixing the equations and the unknowns coming from a trapdoor *internal MQ function* using invertible affine transforms, there exist many such schemes, each relying on its own specific trapdoor. We refer the reader to the survey [18] for details. We simply recall that the original proposal of Matsumoto and Imai [15] could be viewed as a multivariate variation on RSA and used as its internal MQ function a bijective monomial in some extension field. The resulting scheme was broken by Patarin in [16]. In order to repair the scheme, he later proposed the HFE cryptosystem [17], using a low degree polynomial as internal MQ function. In the same paper, he also proposed a wide range of variations on the HFE cryptosystem. We would like to remark that our result focuses on the basic scheme and that its extension to the variations is an interesting open problem.

As already mentioned, finding a solution of a generic system of MQ equations is NP-complete. For this reason, there were hopes that MQ schemes, HFE in particular, might very well be immune to subexponential attacks, square root attacks or even quantum computers. Indeed, these attacks are a common drawback of number theoretic cryptosystem and overcoming them is a worthy goal. To illustrate the possibilities, MQ schemes could achieve post-quantum computer security or yield extremely short signatures thanks to the lack of a square root attack. Such signatures could be, for a comparable security level, twice as short as a pairing based short signature of [3].

On the other hand, despite these hopes, several lines of attacks have been explored, such as the so-called relinearization techniques [12,5], or the use of Gröbner basis algorithms [7]. The former was of a rather theoretical flavor, reducing the cryptanalysis of HFE to the resolution of an overdefined system of quadratic equations in the extension field with many excess equations, and describing a technique called relinearization to solve such overdefined systems in general. However the complexity of this attack remains unclear: despite the fact that relinearization was claimed to succeed in polynomial time, a close look at the claim shows that it only makes sense in a setting where the degree (called d in [17]) of the internal HFE polynomial is fixed. The latter approach followed a more experimental path since it used implementation of very generic Gröbner basis algorithms to solve the above quadratic system, thus breaking the first HFE Challenge 1 in 96 hours on a 833 MHz Alpha workstation with 4 Gbytes of memory. At a more abstract level, Faugère and Joux discovered an algebraic invariant that explains why the computation finishes much earlier than expected for a quadratic system of this size. Surprisingly, the authors did not try to derive a complexity bound for the problem of inverting HFE. This may be due to the lack of complexity estimates in the original paper [17] itself. This paper does not fully adhere to the current trend in cryptography, that defines a key generation algorithm with input a security parameter. In [17], the two main parameters of the schemes, the dimension n of the extension field and the degree d of the hidden polynomial are somehow unrelated. It is clear however that, in order to allow polynomial time decryption, any instantiation of HFE requires both n and d to be polynomial in the security parameter. Moreover, inverting HFE using exhaustive search has complexity 2^n. As a consequence, it is natural to use n itself as the security parameter. Thus d must be polynomial in n and we assume throughout the sequel that $d = O(n^\alpha)$, for some constant α.

This sheds some light on the hope that HFE might be immune to subexponential attacks. This hope stems from a remark in [17], which notes that the complexity of the so-called *affine multiple attack* is $O(n^{O(d)})$, thus exponential in the security parameter. Incidentally, this shows that the existence of polynomial time attacks for fixed d has been known right from the beginning. Similarly, the affine multiple attack is subexponential whenever d is small enough, say $d = O(\log n)$. In fact, in order to hope for full exponential security, we clearly need to choose $\alpha \geq 1$.

Another approach against MQ schemes uses the rank of the differential of the public key and has been proven successful to break the PMI scheme [11]. This technique also allows to build a quasipolynomial distinguisher for HFE [6], with complexity $O(\exp(c(\log n)^2))$, which happens to be the same as for our attack. As far as we know, this approach does not lead to a decryption attack against the HFE cryptosystem.

1.1   Our Results

The main result of this paper is the following: there exists a heuristic quasipolynomial decryption attack against HFE. In fact, we do not actually propose a new attack but revisit the method described by Joux and Faugère in [7], which performs a Gröbner basis computation [4], with the efficient algorithms of Lazard and Faugère [14,8,9]. The efficiency of this approach was already shown by experiments from [7], and had been partly supported by mathematical arguments. Here, we give a more thorough theoretical analysis that allows to conclude that the attack is asymptotically efficient for any instantiation of HFE.

More accurately, our estimate yields complexity $O(n^{O(\log d)})$, that is $\exp(O(\log n)^2)$, for any instantiation of HFE where n is chosen as the security parameter, and where $d = O(n^\alpha)$, for some constant α. This greatly improves on the exponential estimate from [17]. This heuristic complexity estimate should also be compared with the complexity of factorization and integer discrete logarithm. In that case, we let n denote the size of the problem, i.e. $n = \log(N)$ to factor N or $n = \log(p)$ to compute discrete logarithm modulo p. The complexity is subexponential and its expression is $\exp(O(n^{1/3}(\log n)^{2/3}))$, which is clearly higher than what we obtain for HFE. In order to make this distinction clear, we say that the complexity of our attack is quasipolynomial rather than subexponential. Another widely used hard problem is the elliptic curve discrete logarithm problem, where the best attack has exponential complexity $O(\exp(n/2))$.

1.2 Organization of the Paper

The paper is organized as follows: we first recall the definition of the HFE cryptosystem. Next, we survey known facts about Gröbner bases and their computation, focusing on the so-called degree of regularity of algebraic systems which is an extremely important parameter during the execution of Gröbner basis algorithms. Then comes the main contribution of the paper, where we bound the degree of regularity of the algebraic system arising from an attempt to directly invert HFE. This is done by showing that another system with a much smaller number of unknowns is in fact hidden into this algebraic system. Finally, we use this bound to show that the distinguishing and decryption attacks, obtained by applying a Gröbner basis computation to HFE systems, respectively have provable quasipolynomial and heuristic quasipolynomial complexities.

2 The HFE Cryptosystem

Although the HFE cryptosystem was originally defined using any finite field as a base field, we restrict ourselves to the simpler case, where the base field is the two elements field.

2.1 Notations

Fields. We denote by $\mathbb{F}_2$ the finite field with two elements and by $\mathbb{F}_{2^n}$ the extension field with 2^n elements, which is isomorphic (as a vector space) to $(\mathbb{F}_2)^n$. A normal basis of $\mathbb{F}_{2^n}$ is defined by an element θ such that θ, θ^2, θ^4, ..., $\theta^{2^{n-1}}$ generate $\mathbb{F}_{2^n}$. It is well known that such a basis always exists. Note that the original description of HFE used a polynomial basis, however, since change of bases are linear and since arbitrary linear transforms are already used during the HFE construction, using a normal basis involves no loss of generality. Moreover, this approach greatly simplifies the exposition of our attack.

Monomials. Let $f(x) = x^d$ be a monomial over $\mathbb{F}_{2^n}$. The binary decomposition of the exponent reads $d = \sum_{0 \leq i < n} d_i 2^i$. Using the linearity of the Frobenius operator: $x \longrightarrow x^2$, it is easily seen that the Hamming weight of the d_i sequence is exactly the degree of the representation of f over $(\mathbb{F}_2)^n$, as a system of multivariate polynomial functions.

Polynomials. For any polynomial $P \in \mathbb{F}_{2^n}[X]$ we denote by $d^\circ P$ the degree of P, that is the maximal degree of its monomials. We let $w^\circ P$ be the maximal Hamming weight of the exponents of P's monomials, as defined in the previous paragraph, and call $w^\circ P$ the Hamming weight of P. It is well known that systems of MQ equations over $(\mathbb{F}_2)^n$ are in bijection with polynomials such that $w^\circ P = 2$. Also, affine functions over $(\mathbb{F}_2)^n$ are in bijection with polynomials such that $w^\circ P = 1$.

2.2   The HFE Cryptosystem

The cryptosystem HFE is defined from a polynomial $f \in \mathbb{F}_{2^n}[X]$, with $w°f = 2$ and $d°f < d$, where n and d are (usually implicitly) defined from a security parameter. In the sequel, t will denote the smallest number such that $2^t > d$. The public key of HFE is obtained by composing f with two affine invertible functions, S, Y, thus yielding the polynomial $P = T \circ f \circ S$.

Encrypting with HFE is straightforward, it suffices to evaluate the public polynomial P on the input to be encrypted. Decryption is harder and uses the fact that it is easy to compute the inverses of S, T, and also to solve a polynomial equation of low degree $d < 2^t$ in time polynomial in d and n. Therefore, provided that $d = O(n^\alpha)$ and thus $t = O(\log n)$, a polynomial time decryption algorithm is available from the trapdoor.

3   Gröbner Bases Computations

Gröbner basis algorithms compute an algebraic basis of an ideal in a multivariate polynomial ring, given an ordering of the monomials. The output is such that any element f of the ideal can efficiently be written as an algebraic combination of the resulting basis $f_1, \cdots, f_m$ by repeating a sequence of simple reductions. Each reduction decreases the degree w.r.t. the ordering by suitably withdrawing from the current polynomial a multiple of some f_i by a monomial, until the zero polynomial is found. The original algorithm for computing Gröbner bases is due to Buchberger [4] and is based on maintaining a sequence of polynomials, and repeatedly using reduction, an operation that reduces the degree, and the so-called S-polynomial operation, an operation that increases the degree by computing an element of the ideal from a so-called critical pair of elements. In the early eighties, Lazard [13] realized that Gröbner basis computations could be achieved by applying Gaussian elimination to a specific matrix, called the Macaulay matrix, which is obtained by indexing the columns by all monomials with n variables of degree at most an integer r, and filling all rows with the coefficients of all multiples by a monomial of a family of polynomials generating the ideal, provided they remain of degree at most r. The main problem with this approach, is that the complete Macaulay matrix contains many "obvious" dependencies, which arise from generic properties and could be predicted in advance. Later Faugère [9] gave a simple criterion, that permit the construction of a reduced version of the Macaulay matrix that does not contain these obvious dependencies. This yielded a extremely efficient algorithm, called F_5, for the computation of Gröbner bases. In the sequel, we denote the variant of the Macaulay matrix containing all the multiples of degree r by the Macaulay-Faugère matrix of degree r.

The F_5 algorithm works by constructing Macaulay-Faugère matrices of increasing degree and by performing linear algebra on those. Its main goal is to find a linear combination of rows encoding a polynomial of degree smaller than r. By definition, the degree of regularity of a sequence $f_1, \cdots, f_m$ of polynomials is the minimal degree where such a linear combination exists. Each such linear combination encodes a new polynomial which needs to be added to the original sequence to get the Gröbner bases, except when the polynomial is zero in which case it might be necessary to remove some polynomial in the current ideal basis, thus simplifying it.

The degree of regularity D is a very important parameter of Gröbner basis computation using F_5, since it leads to a decomposition in two steps. During the first step, up to the degree of regularity, the computation behaves nicely and its complexity can be easily predicted. What happens during the second step, when non-trivial combinations have appeared is much harder to predict in general. However, for random systems, the behavior is quite simple, an extremely large number of new polynomials appear in the Macaulay-Faugère matrices of degrees D or $D+1$, and after that the computation quickly terminates. Moreover, most real-life systems of equations have a similarly tame behavior and rarely need to construct Macaulay-Faugère matrices beyond the degree of regularity plus a small constant. On the other hand, it is possible to cook up wild systems with a very bad behavior.

For any system of polynomial equations in n unknowns with degree of regularity D, the first step of F_5 involves the construction of Macaulay-Faugère matrices up to degree D, thus of dimension at most n^D. Performing the linear algebra on these matrices costs at most n^{3D} operations. If we let $\mathcal{D}$ denote the largest degree of Macaulay-Faugère matrices occurring during the rest of the algorithm, the total cost is $n^{3\mathcal{D}}$. For well behaved systems of equations, $\mathcal{D}$ is not much larger than D and the overall complexity is $n^{O(D)}$.

3.1   Known Bounds on Degrees of Regularity

Previous work shows that for a quadratic system of equations in n variables, the degree of regularity cannot become too large. Moreover, for random systems of equations, the known bound on the degree of regularity is reached. The general analysis that we need was done by Bardet, Faugère and Salvy [2] and is neatly described in Bardet's thesis [1, chap. 4].

The result given there is that for a system of τn quadratic equations in n unknowns, the degree of regularity is at most:

$$D_\tau(n) = \left(\tau - \frac{1}{2} - \sqrt{\tau(\tau-1)}\right)n + \frac{-a_1}{2(\tau(\tau-1))^{1/6}}n^{1/3} \qquad (1)$$

$$-\left(2 - \frac{2\tau-1}{4\sqrt{\tau(\tau-1)}}\right) + O(n^{-1/3}), \qquad (2)$$

where $a_1 \approx -2.33811$ is the first real zero of the Airy function.

Moreover, for random systems the probability of having a smaller degree of regularity is negligible. Furthermore, experiments show that such random systems are well behaved in the sense that the F_5 computation do not involve Macaulay-Faugère matrices of much higher degree. To formalize this observation, we propose the following conjecture:

Conjecture 1. For all $\tau > 1$, there exists a constant K such that for a large enough random system S of $\lceil \tau n \rceil$ quadratic equations in n unknowns, a Gröbner basis for the ideal generated by S can be computed in time n^{Kn}, with overwhelming probability.

4   Systems of Equations Arising from HFE Instances

In this section, our goal is to study the complexity of a direct Gröbner basis approach to the resolution of HFE systems. This direct approach consists in writing down that each public polynomial, belonging to the encryption key, when evaluated on the (unknown) plaintext yields the corresponding ciphertext bit. This approach was first described in [7].

4.1   Outline of the Strategy

Our strategy is to bound the degree of regularity D of the system of polynomials stemming from directly attempting to invert HFE from the description of its public key through a Gröbner basis algorithm. From such a bound, which is smaller than the bound for a random system in the same number of unknowns, we then derive our main results. First, when the first simplification in a Macaulay-Faugère matrix is encountered, we deduce the effective degree of regularity of the polynomial system under consideration. If this degree is small enough, our distinguisher knows that the system has no chance to be random and asserts that it is an HFE instance. No heuristic is required for this attack, whose runtime is bounded by n^{3D}. Second, for the decryption attack, we assume that HFE based systems of equations behave nicely and that $\mathcal{D}$ is not much larger than D. Under this heuristic assumption, which is supported by the experiments described in [10,7,1], where the computations never constructed Macaulay-Faugère matrices beyond the degree of regularity plus 1, we claim that the complexity of the decryption attack remains $n^{O(D)}$.

The key idea that allows us to bound D is to apply a sequence of transformations involving the unknown secret trapdoor, in order to show that D does not exceed the degree of regularity of a much smaller system. More precisely, this other system involves $l+t+l-2 = (2\lambda+1)\cdot t-2$ equations in $(\lambda+1)\cdot t$ unknowns, where 2^t is, as defined in section 2.2 a bound for the degree of the internal HFE polynomial and λ is an appropriate constant. Since this system has a much smaller number of unknowns, we obtain a much better bound on D than for generic systems in n unknowns.

4.2   Reducing the Number of Unknowns

Let θ, θ^2, θ^4, ..., $\theta^{2^{n-1}}$ be our normal basis for the finite field $\mathbb{F}_{2^n}$. Let $f(X)$ be the secret polynomial of an HFE instance over $\mathbb{F}_{2^n}$, of degree $d < 2^t$. The corresponding public key polynomials in n unknowns x_0, ..., x_{n-1} over $\mathbb{F}_2$ are P_1, P_2, ..., P_n. They are obtained by writing X over the normal basis as:

$$X = \sum_{0 \leq i < n} \theta^{2^i} x_i,$$

by taking the coordinates of $f(X)$, viewed as polynomials in $x_0, \ldots, x_{n-1}$, in the normal basis and, finally, by applying two invertible linear transforms, as explained in section 2.2.

The resulting polynomials $P_1, P_2, \ldots, P_n$ are quadratic. Thus, cryptanalyzing a message encrypted with the HFE cryptosystem requires solving a quadratic system of equations over $\mathbb{F}_2$ defined by fixing the target values of the f_i polynomials. In order to decrypt an HFE instance using a generic Gröbner basis approach, this is the system of equations we need to consider. Thus, following our general strategy, we want to bound its degree of regularity.

First of all, since the degree of regularity only depends on the high degree homogeneous parts of each equation in the system, the target value has no influence on this parameter. Moreover, assuming that the quadratic parts of the P_i, together with the quadratic parts of the field equations $x_i^2 - x_i = 0$, are linearly independent, the degree of regularity remains the same if we remove the secret linear transformations before and after f. At this point, we are left with computing (or more precisely bounding) the degree of regularity of the system of secret internal equations directly given by the coordinates of f over the normal basis, together with the field equations $x_i^2 - x_i = 0$. Let $f_0, f_1, \ldots,$ f_{n-1} denote these secret polynomials, which are related to f by means of the equation:

$$f\left(\sum_{0 \leq i < n} \theta^{2^i} x_i \right) = \sum_{0 \leq i < n} \theta^{2^i} f_i.$$

Another System with Higher Degree of Regularity Than the Internal System. In order to bound the degree of regularity of the ideal generated by $(f_0, \cdots, f_{n-1}, x_0^2 - x_0, \cdots, x_{n-1}^2 - x_{n-1})$, we first remark that this degree is left unchanged when solving this system over $\mathbb{F}_{2^n}$ instead of $\mathbb{F}_2$. Moreover, thanks to the field equations, the solutions are the same in both cases. Over $\mathbb{F}_{2^n}$, we can transform the system by writing:

$$F_j(x_0, \cdots, x_{n-1}) = \sum_{0 \leq i < n} \theta^{2^{i+j}} f_i.$$

In fact, F_0 is just a representation of f in terms of $x_0, \ldots, x_{n-1}$, F_1 is a representation of f^2, F_2 a representation of f^4 and so on $\ldots$

Clearly, replacing the f_i by the F_j is an invertible linear transform, which for the same reasons as before, does not affect the degree of regularity. The next step is to make a linear change of variables, replacing the x_i by y_j, where:

$$y_j = \sum_{0 \leq i < n} \theta^{2^{i+j}} x_i.$$

This change corresponds to setting $y_0 = X$, $y_1 = X^2$, $\ldots$, $y_{n-1} = X^{2^{n-1}}$. It cleanly expresses each F_i as a quadratic equation in terms of X and its Frobenius images. However, the field equations are not yet in a nice form. Luckily, a final linear transform turns them into:

$$y_1 = y_0^2,$$
$$y_2 = y_1^2,$$
$$\vdots$$
$$y_{n-1} = y_{n-2}^2,$$
$$y_0 = y_{n-1}^2.$$

Finally, in order to bound the degree of regularity, it is enough to remark that due to the degree bound on f, F_0 is a function of $y_0, \ldots, y_{t-1}$ and that $y_t, \ldots, y_{n-1}$ are not used. Likewise, F_1 is a function of $y_1, \ldots, y_t$ and F_j a function of $y_j, \ldots, y_{t+j-1}$ (when j is small enough). Thanks to this observation, we can focus on a subset of the equations and variables. Assume that we restrict ourselves to $F_0, F_1, \ldots, F_{l-1}$ then we need only use the variables y_0 to y_{t+l-1}. Moreover, among these variables, we keep $t + l - 2$ field equations of the form $y_{j+1} = y_j^2$. Of course, restricting ourselves to such a subset can only increase the degree of regularity, since any non trivial relation among the equations of the smaller system clearly holds in the larger one. Setting $l = \lambda t$, for an adequately chosen constant λ, we now obtain a system of $l + t + l - 2 = (2\lambda + 1) \cdot t - 2$ equations in $(\lambda + 1) \cdot t$ unknowns.

Note. In the case where $d = 2^t$, we can slightly improve the above description to reduce the complexity of the attack. This is due to the fact that the variable y_t in F_0 only appears as a linear term. Thus, the additional variable y_{t+l} only appears linearly. Since the degree of regularity only depends on the high order (quadratic) terms, we can safely ignore this extra variable. This confirms the practical observation made in [7], in connection with the degree of the polynomial complexity with fixed degree d. They observed that this degree slowly increased with d and that the increase steps occurred immediately after increasing d beyond a power of 2.

4.3   Bound on the Degree of Regularity of the Internal System

To upper bound the degree of regularity arising with HFE systems, it is now sufficient to apply the generic bound for random systems on the internal system with a reduced number of variables, expressed as a function of $t = \lceil \log_2 d \rceil$ as in section 4.2.

We apply the bound from section 3.1 with $\tau = (2\lambda + 1)/(\lambda + 1)$, fixing $\lambda = 1$. This is not the optimal choice and it would be better to let λ grow with n in order to make τ close to its limit 2, thus finding a tighter bound. However, the simple choice $\lambda = 1$ is sufficient to fulfill our purpose. With this choice, we find that ignoring low order terms the degree of regularity of $3t$ quadratic equations in $2t$ unknowns is:

$$D = 2(1 - \sqrt{3/4})t + O(t^{1/3}).$$

This can be summarized by the following theorem:

Theorem 1. *For basic HFE instances, defined by a secret polynomial of degree* d *over* $\mathbb{F}_{2^n}$, *the maximum possible degree of regularity* $D(n, d)$ *is asymptotically upper bounded by*

$$(2 + \epsilon)(1 - \sqrt{3/4}) \min(n, \log_2 d),$$

for all ϵ.

4.4   Complexity of the Attacks

The above study has shown that the degree of regularity of the HFE system is an integer D upper bounded by $2(1 - \sqrt{3/4})t + O(t^{1/3})$. As noted in section 3, the complexity of computing a Gröbner basis with F_5 is bounded by the cost of linear algebra on a matrix whose columns are indexed by the monomials of degree $\mathcal{D}$ in n unknowns, where $\mathcal{D}$ is the degree of the largest Macaulay-Faugère matrix that is used. Moreover, during a first phase the algorithm only builds Macaulay-Faugère matrices of degree up to D. Moreover, the end of the first phase is easily detected. As a consequence, we easily measure the effective value of D. Since we do not know the HFE secret key, we need to perform our Gröbner basis computation in the public world, using a system with n unknowns. Thus the respective runtimes of the first phase and of the full algorithm are bounded by $O(n^{3D})$ and $O(n^{3\mathcal{D}})$, assuming that all linear algebra is done using ordinary Gaussian elimination.

An alternative way of viewing this result is to make the heuristic assumption that the reduced system described in section 4.2 behaves as a random system in t unknowns. Thus using conjecture 1, a Gröbner basis for this hidden system can be found in time t^{Kt}. Further assuming that this can be achieved with the F_5 algorithm, the Gröbner computation in the public world requires a running time n^{Kt}.

Distinguishing Attack. At the end of the first phase, we end up with the effective value of D. We know that for HFE instances, this value is quite small and that for random systems, it is much larger (with overwhelming probability). This simple fact yields a simple distinguisher which at the end of the first phase can tell whether the original system is an HFE instance or not. Of course, when working on a random system, the distinguisher should use an early abort strategy and stop as soon as the current Macaulay-Faugère matrix has degree larger than the expected D. Thanks to the early abort strategy, replacing D by its bound, we find the complexity of the distinguisher is $2^{O(\log(n)^2)}$ even when the input is a random system. This distinguisher was first mentioned by Faugère when describing his HFE challenge experiment. It offers an alternative to the distinguisher described in [6].

Note that for a random system of quadratic equations, we need a different variant of the theorem of Bardet than described in section 3.1. This variant, which holds for quadratic systems of n equations in n unknowns over $\mathbb{F}_2$, is also given in [1] gives a formula for the degree of regularity which is similar to equation 2 for $\tau = 2$, with slightly different constants. It implies that the degree of regularity of a random system is $O(n)$ and thus much higher than the $O(\log n)$ value we obtained for D in the case of an HFE instance.

Decryption Attack. In the case of the decryption attack, we let the Gröbner basis algorithm terminate and from the result, we find the corresponding plaintext as in [7]. Here the runtime is $n^{3\mathcal{D}}$. Using our heuristic assumption about the good behavior of the system of equations, $n^{3\mathcal{D}}$ is $n^{O(D)}$. As a consequence, the overall heuristic runtime is $2^{O(\log(n)^2)}$ as announced previously.

5 Conclusion

In this paper, we analyzed the behavior of the Gröbner basis based attack on HFE systems as proposed in [7]. We showed that this attack takes quasipolynomial time for any practical instantiation of the basic HFE cryptosystem. The runtime analysis of the distinguisher part of the attack gives a provable complexity, while the decryption part of the attack only leads to a heuristic complexity. Comparing the result with the best existing subexponential algorithms for factoring and computing discrete logarithms, we find that for comparable key and/or ciphertext sizes, breaking the basic HFE scheme is asymptotically much easier than breaking RSA or discrete logarithm based systems.

References

1. M. Bardet. *Étude des systèmes algébriques surdéterminés. Applications aux codes correcteurs et à la cryptographie.* PhD thesis, Université Paris 6, Dec. 2004. http://www-calfor.lip6.fr/~bardet/.
2. M. Bardet, J.-C. Faugère, and B. Salvy. On the complexity of Gröbner basis computation of semi-regular overdetermined algebraic equations. In *Proc. ICPSS International Conference on Polynomial System Solving*, 2004.
3. D. Boneh, B. Lynn, and H. Shacham. Short signatures from the Weil pairing. In C. Boyd, editor, *Proceedings of ASIACRYPT'2001*, volume 2248 of *Lecture Notes in Comput. Sci.*, pages 514–532. Springer, 2001.
4. B. Buchberger. Gröbner bases : an algorithmic method in polynomial ideal theory. In N.-K. Bose, editor, *Multidimensional systems theory*, number 16 in Mathematics and its Applications, chapter 6, pages 184–232. D. Reidel Pub. Co., 1985.
   Based on his PhD thesis: *Ein Algorithmus zum Auffinden der Basiselemente des Restklassenringes nach einem nulldimensionalen Polynomideal*, U. Innsbruck, Austria, 1965.
5. N. Courtois. The security of Hidden Field Equations (HFE). In *CT-RSA'01*, volume 2020 of *Lecture Notes in Comput. Sci.*, pages 266–281. Springer-Verlag, 2001.
6. V. Dubois, L. Granboulan, and J. Stern. An efficient provable distinguisher for HFE. In *Proceedings of ICALP*, Lecture Notes in Comput. Sci., 2006. To appear.
7. J.-C. Faugère and A. Joux. Algebraic cryptanalysis of Hidden Field Equation (HFE) cryptosystems using Gröbner Bases. In *Crypto'03*, volume 2729 of *Lecture Notes in Comput. Sci.*, pages 44–60. Springer-Verlag, 2003.

8. J.-C. Faugère. A new efficient algorithm for computing Gröbner bases (F4). *J. Pure Appl. Algebra*, 139(1-3):61–88, 1999. Effective methods in algebraic geometry (Saint-Malo, 1998).

9. J.-C. Faugère. A new efficient algorithm for computing Gröbner bases without reduction to zero (F5). In T. Mora, editor, *ISSAC 2002*, pages 75–83, 2002.

10. J.-C. Faugère. Algebraic cryptanalysis of HFE using Gröbner bases. Technical Report 4738, INRIA, Feb. 2003. `ftp://ftp.inria.fr/INRIA/tech-reports/dienst/RR-4738.pdf`.

11. P.-A. Fouque, L. Granboulan, and J. Stern. Differential cryptanalysis for Multivariate Schemes. In *Eurocrypt'05*, volume 3386 of *Lecture Notes in Comput. Sci.*, pages 341–353, 2005.

12. A. Kipnis and A. Shamir. Cryptanalysis of the HFE Public Key Cryptosystem. In *Crypto'99*, volume 1666 of *Lecture Notes in Comput. Sci.*, pages 19–30. Springer-Verlag, 1999.

13. D. Lazard. Gröbner bases, gaussian elimination and resolution of systems of algebraic equations. In *Computer algebra (London, 1983)*, volume 162 of *Lecture Notes in Comput. Sci.*, pages 146–156, 1983.

14. D. Lazard. Solving systems of algebraic equations. *ACM SIGSAM Bulletin*, 35(3):11–37, Sept. 2001.

15. T. Matsumoto and H. Imai. Public Quadratic Polynomial-tuples for efficient signature-verification and message encryption. In *Eurocrypt'88*, volume 330 of *Lecture Notes in Comput. Sci.*, pages 419–453. Springer-Verlag, 1988.

16. J. Patarin. Cryptanalysis of the Matsumoto and Imai Public Key Scheme of Eurocrypt'88. In *Crypto'95*, volume 963 of *Lecture Notes in Comput. Sci.*, pages 248–261. Springer-Verlag, 1995.

17. J. Patarin. Hidden Field Equations (HFE) and Isomorphisms of Polynomials (IP): two families of asymetric algorithms. In *Eurocrypt'96*, volume 1070 of *Lecture Notes in Comput. Sci.*, pages 33–46. Springer-Verlag, 1996.

18. C. Wolf and B. Preneel. Taxonomy of Public Key Schemes based on the problem of Multivariate Quadratic equations. Cryptology ePrint Archive, Report 2005/077, 2005. `http://eprint.iacr.org/`.

Cryptanalysis of 2R⁻ Schemes

Jean-Charles Faugère[1] and Ludovic Perret[2]

[1] LIP6, 8 rue du Capitaine Scott, Paris 75015, France
Jean-Charles.Faugere@inria.fr
[2] UCL, Crypto Group, Microelectronic Laboratory
Place du Levant, 3
Louvain-la-Neuve, B 1348, Belgium
ludovic.perret@uclouvain.be

Abstract. In this paper, we study the security of 2R⁻ schemes [17,18], which are the "minus variant" of two-round schemes. This variant consists in removing some of the n polynomials of the public key, and permits to thwart an attack described at Crypto'99 [25] against two-round schemes. Usually, the "minus variant" leads to a real strengthening of the considered schemes. We show here that this is actually not true for 2R⁻ schemes. We indeed propose an efficient algorithm for decomposing 2R⁻ schemes. For instance, we can remove up to $\lfloor \frac{n}{2} \rfloor$ equations and still be able to recover a decomposition in $O(n^{12})$. We provide experimental results illustrating the efficiency of our approach. In practice, we have been able to decompose 2R⁻ schemes in less than a handful of hours for most of the challenges proposed by the designers [18]. We believe that this result makes the principle of two-round schemes, including 2R⁻ schemes, useless.

Keywords: Cryptanalysis, Functional Decomposition Problem (FDP), Gröbner bases, F₅ algorithm.

1 Introduction

Last years a new kind of cryptanalysis has made its entrance in cryptography: the so-called *algebraic cryptanalysis*. A fundamental issue of this cryptanalysis consists in finding zeroes of algebraic systems. Gröbner bases, which are a fundamental tool of commutative algebra, constitute the most elegant and efficient way for solving this problem. They provide an algorithmic solution for solving several problems related to algebraic systems (some of them can be found in [1]). We present here a new application of Gröbner bases. More precisely, we propose a new algorithm for solving the *Functional Decomposition Problem* (FDP). The problem is as follows:

Functional Decomposition Problem (FDP)
Input : multivariate polynomials $h_1, \ldots, h_u$.
Find : – if any – multivariate polynomials $f_1, \ldots, f_u$, and $g_1, \ldots, g_n$, such that:

$$(h_1, \ldots, h_u) = \big(f_1(g_1, \ldots, g_n), \ldots, f_u(g_1, \ldots, g_n)\big).$$

This problem is related to security of 2R⁻ schemes [17,18].

C. Dwork (Ed.): CRYPTO 2006, LNCS 4117, pp. 357–372, 2006.
© International Association for Cryptologic Research 2006

1.1   Related Works

As stated by E. Biham [6], "the design of this scheme (2R) is unique as it uses techniques from symmetric ciphers in designing public key cryptosystems, while still claiming security based on relation to the difficulty of decomposing compositions of multivariate ... functions". Anyway, the security of 2R schemes has been already carefully investigated [6,25,26]. E. Biham proposed in [6] a successful cryptanalysis of 2R schemes with S-Boxes. This attack exploits the birthday paradox, but can be avoided by increasing the security parameters of 2R schemes [18]. At Crypto'99 [25], D.F. Ye, Z.D. Dai, and K.Y. Lam have presented a quite efficient method for solving the Functional Decomposition Problem. The security of 2R schemes is indeed related to this problem. To thwart this last attack, L. Goubin and J. Patarin have proposed [18] to use a general technique for repairing multivariate schemes, namely keeping secret some polynomials of the public key. The resulting schemes are called 2R$^-$ schemes. Note that V. Carlier, H. Chabanne, and E. Dottax [9] have described a method for protecting the confidentiality of block ciphers design exploiting the principle of 2R$^-$ schemes. Usually, the "minus modification" leads to a real strengthening of the considered schemes. For instance, C^* is broken [22] while C^{*--} is the basis of Sflash [10], the signature scheme recommended for low-cost smart cards by the European consortium Nessie[1]. Here, we show that 2R$^-$ is not more secure than 2R.

1.2   Organization of the Paper and Main Results

The paper is organized as follows. We begin in Section 2 by introducing our notations and defining essential tools used in this paper, namely ideals, Gröbner bases, and several operations on ideals (sum, intersection, quotient, ...). Section 3 gives a brief review of one-round, 2R and 2R$^-$ schemes. We also present the *Functional Decomposition problem* (FDP) in a more formal manner, which is at the basis of the security of 2R and 2R$^-$ schemes. An algorithm for solving this problem efficiently would allow to decompose the public key of 2R and 2R$^-$ schemes into two independent quadratic systems, making thereby the principle of these cryptosystems useless. In Section 4, we present a general algorithm for solving FDP. Our method is inspired on the algorithm of D.F. Ye, Z.D. Dai, and K.Y. Lam [25]. Note that their algorithm only works for particular instances of FDP, namely when $u = n$, or $u = n - 1$. Briefly, our algorithm works as follows. Let $(h_1, \ldots, h_u) = (f_1(g_1, \ldots, g_n), \ldots, f_u(g_1, \ldots, g_n))$ be an instance of FDP. We first construct the ideal $\partial \mathcal{I}_h = \left\langle \frac{\partial h_i}{\partial x_j} : 1 \leq i \leq u, 1 \leq j \leq n \right\rangle$ generated by the partial derivatives of the h_is. We then show that for all $i, 1 \leq i \leq n, x_n^{d+1} g_i \in \partial \mathcal{I}_h$, for some $d \geq 0$. In most cases, this allows to recover a basis of the vector space $\mathcal{L}(g) = \text{Vect}(g_1, \ldots, g_n)$ generated by $g_1, \ldots, g_n$. This is the most difficult part of our algorithm. The f_is being indeed recovered from the knowledge of $\mathcal{L}(g)$ by solving a linear system. The complexity of this algorithm depends on the ratio n/u. For example, our algorithm runs in $O(n^{12})$, if $n/u < 1/2$. More generally,

---

[1] https://www.cosic.esat.kuleuven.be/nessie/deliverables/decision-final.pdf

we provide a global analysis of the theoretical complexity of our method. As a
side effect, we give several insights into the theoretical behavior of the algorithm
of D.F. Ye, Z.D. Dai, and K.Y. Lam. We conclude this section by providing
experimental results illustrating the efficiency of our approach. We have been
able to solve in few hours instances of FDP used in 2R⁻ schemes for most of the
challenges proposed in [18].

2   Preliminaries

Throughout this paper, we denote by $\mathbb{K}[x_1,\ldots,x_n]$ the polynomial ring in the
n indeterminate $x_1,\ldots,x_n$ over a finite field $\mathbb{K}$ with $q = p^r$ elements (p a prime,
and $r \geq 1$). The set of polynomials $p_1,\ldots,p_s$ of $\mathbb{K}[x_1,\ldots,x_n]$ can be regarded
as a mapping $\mathbb{K}^n \to \mathbb{K}^s$:

$$(v_1,\ldots,v_n) \mapsto \big(p_1(v_1,\ldots,v_n),\ldots,p_s(v_1,\ldots,v_n)\big).$$

We will call these polynomials *components*. We will also denote by $\mathcal{I} = \langle p_1,\ldots,p_s\rangle$
$= \{\sum_{k=1}^{s} p_k u_k \colon u_1,\ldots,u_s \in \mathbb{K}[x_1,\ldots,x_n]\}$ the *ideal generated* by $p_1,\ldots,p_s$. We
define now essential notions used in this paper. For a more thorough introduction
to these tools, we refer to classical books on commutative algebra, such as [1,11].
Most of the results presented in this part are well known in commutative algebra,
and thus given without proofs. For these proofs, we also refer to [1,11]. The reader
already familiar with Gröbner bases and quotient ideals can skip this part.

2.1   Gröbner Bases

Informally, a Gröbner basis of an ideal is a generatring set of this ideal with
"good" algorithmic properties. These bases are defined with respect to *monomial orders*. Here, we will use the lexicographic (LEX) and degree reverse lexicographical (DRL) orders, which are definedas follows:

Definition 1. *Let $\alpha = (\alpha_1,\ldots,\alpha_n)$ and $\beta = (\beta_1,\ldots,\beta_n) \in \mathbb{N}^n$. Then:*
*$- x_1^{\alpha_1}\cdots x_n^{\alpha_n} \prec_{LEX} x_1^{\beta_1}\cdots x_n^{\beta_n}$, if the left-most nonzero entry of the vector $\alpha - \beta$
is positive.*
*$- x_1^{\alpha_1}\cdots x_n^{\alpha_n} \prec_{DRL} x_1^{\beta_1}\cdots x_n^{\beta_n}$, if $\sum_{i=1}^{n} \alpha_i > \sum_{i=1}^{n} \beta_i$, or $\sum_{i=1}^{n} \alpha_i = \sum_{i=1}^{n} \beta_i$
and the right-most nonzero entry of $\alpha - \beta$ is negative.*

To define Gröbner bases, we have to introduce the following definitions.

Definition 2. *For any n-uple $\alpha = (\alpha_1,\ldots,\alpha_n) \in \mathbb{N}^n$, we denote by x^α the
monomial $x_1^{\alpha_1}\cdots x_n^{\alpha_n}$. We define the total degree of this monomial by the sum
$\sum_{i=1}^{n} \alpha_i$. The **leading monomial** of a polynomial $f \in \mathbb{K}[x_1,\ldots,x_n]$ is the
largest monomial – w.r.t some monomial ordering $\prec$ – among the monomials of
f. This leading monomial will be denoted by $\mathrm{LM}(f,\prec)$. The **leading coefficient**
of f, denoted by $\mathrm{LC}(f,\prec)$, is the coefficient of $\mathrm{LM}(f,\prec)$ in f. The **degree** of f
– denoted $deg(f)$ – is the total degree of $\mathrm{LM}(f,\prec)$. Finally, the **maximal total
degree** of f is the maximal total degree of the monomials occurring in f.*

We are now ready to define one of the main objects of this paper. Indeed:

Definition 3. *A set of polynomials G is a* **Gröbner basis** *– w.r.t. a monomial ordering* $\prec$ *– of an ideal* $\mathcal{I}$ *in* $\mathbb{K}[x_1, \ldots, x_n]$, *if for all* $f \in \mathcal{I}$ *there exists* $g \in G$ *such that* $\mathrm{LM}(g, \prec)$ *divides* $\mathrm{LM}(f, \prec)$. *This Gröbner basis is called* **reduced** *if, for all* $g \in G$, $\mathrm{LC}(g, \prec) = 1$, *and any monomial of* $g \in G$ *is not divisible by any element of* $\mathrm{LM}(G \setminus \{g\}, \prec)$. *Let* G *be Gröbner basis – w.r.t. a monomial ordering* $\prec$ *– of an ideal* $\mathcal{I}$ *in* $\mathbb{K}[x_1, \ldots, x_n]$, *and* d *be a positive integer. We call* d-**Gröbner basis** *(or* **truncated Gröbner basis***) of an homogeneous ideal* $\mathcal{I}$ *the set:*

$$\{g \in G : deg(g) = d\}.$$

A Gröbner basis of a given ideal is not unique in general. The reduced Gröbner basis allows to achieve uniqueness. A reduced Gröbner basis can be obtained from a Gröbner basis in polynomial-time. Gröbner bases are a fundamental tool to study algebraic systems in theory and practice. They provide an algorithmic solution for solving several problems related to polynomial systems (some of them can be found in [1]). The historical method for computing Gröbner bases is Buchberger's algorithm [8,7]. Recently, more efficient algorithms have been proposed. To date, F_5 [13] is the most efficient for computing Gröbner bases (a brief description of this algorithm is given in Appendix A). Here we will concentrate on Gröbner bases w.r.t. lexicographical and degree reverse lexicographical orders.

LEX and DRL Gröbner Bases

Lexicographical Gröbner bases (LEX Gröbner bases) offer a way for eliminating variables.

Theorem 1 (Elimination Theorem). *Let* $\mathcal{I}$ *be an ideal in* $\mathbb{K}[x_1, \ldots, x_n]$, *and* $k \in \{1, \ldots, n\}$. *If* G *is a LEX Gröbner basis of* $\mathcal{I}$, *then* $G \cap \mathbb{K}[x_{k+1}, \ldots, x_n]$ *is a Gröbner basis of* $\mathcal{I} \cap \mathbb{K}[x_{k+1}, \ldots, x_n]$.

The shape of degree reverse lexicographical Gröbner bases (DRL Gröbner bases) is much more complicated. However, DRL Gröbner bases have several interesting properties. For instance, the polynomials of lowest degree of an ideal $\mathcal{I}$ appear in a DRL Gröbner bases of this ideal. More precisely:

Theorem 2. *Let* $\mathcal{I} \subset \mathbb{K}[x_1, \ldots, x_n]$, $d = min\{deg(f) : f \in \mathcal{I}\}$, *and* G *be a DRL Gröbner basis of* $\mathcal{I}$. *Then:*

$$\mathrm{Vect}\big(g \in G : deg(g) = d\big) = \mathrm{Vect}\big(g \in \mathcal{I} : deg(g) = d\big).$$

Proof. A proof of this theorem can be found in [3]. □

We should mention that the variable x_n has a special role for the DRL order.

Lemma 1. *Let* $f \in \mathbb{K}[x_1, \ldots, x_n]$, *and* m *be a positive integer. Then:*

$$x_n^m | f \iff x_n^m | \mathrm{LM}(f, \prec_{DRL}).$$

Sum, Intersection, and Quotient of Ideals
We now go over the definitions of several operations on ideals.

Definition 4. *Let $\mathcal{I}$ and $\mathcal{J}$ be ideals in $\mathbb{K}[x_1, \ldots, x_n]$. Then:*
- *the **sum** of $\mathcal{I}$ and $\mathcal{J}$, noted $\mathcal{I} + \mathcal{J}$, is the $\mathcal{I} + \mathcal{J} = \{f + g : f \in \mathcal{I} \text{ and } g \in \mathcal{J}\}$.*
- *the **intersection** of $\mathcal{I}$ and $\mathcal{J}$, is defined as $\mathcal{I} \cap \mathcal{J} = \{f \in \mathbb{K}[x_1, \ldots, x_n] : f \in \mathcal{I} \text{ and } f \in \mathcal{J}\}$.*
- *$\mathcal{I} + \mathcal{J}$ and $\mathcal{I} \cap \mathcal{J}$ are ideals.*

Given two ideals and their generators, we would like to compute a set of generators for the intersection. This is actually much more delicate than the analogous problem for sums, which is straightforward. Indeed, $\mathcal{I} = \langle p_1, \ldots, p_s \rangle + \mathcal{J} = \langle g_1, \ldots, g_r \rangle = \langle p_1, \ldots, p_s, g_1, \ldots, g_r \rangle$. The following result permits to solve the problem for intersections.

Theorem 3. *Let $\mathcal{I}, \mathcal{J}$ be ideals in $\mathbb{K}[x_1, \ldots, x_n]$, and t be a new variable. Then:*

$$\mathcal{I} \cap \mathcal{J} = \big(t \cdot \mathcal{I} + (1 - t) \cdot \mathcal{J} \big) \cap \mathbb{K}[x_1, \ldots, x_n],$$

where $t \cdot \mathcal{I} = \{t \cdot h : h \in \mathcal{I}\}$, and $(1 - t) \cdot \mathcal{J} = \{(1 - t) \cdot h : h \in \mathcal{J}\}$ are in $\mathbb{K}[t, x_1, \ldots, x_n]$.

This result, together with the Elimination Theorem (i.e. Theorem 1), provide a method for computing intersections of ideals. Given ideals $\mathcal{I} = \langle p_1, \ldots, p_s \rangle$ and $\mathcal{J} = \langle g_1, \ldots, g_r \rangle$ in $\mathbb{K}[x_1, \ldots, x_n]$, we consider the ideal $\langle t \cdot p_1, \ldots, t \cdot p_s, (1 - t) \cdot g_1, \ldots, (1 - t) \cdot g_r \rangle \subset \mathbb{K}[t, x_1, \ldots, x_n]$. Those elements of a LEX Gröbner basis (with $t \succ_{LEX} x_1 \succ_{LEX} \cdots \succ_{LEX} x_n$) that do not contain the variable t will exactly form a Gröbner basis for $\mathcal{I} \cap \mathcal{J}$.

Definition 5. *Let $\mathcal{I}$ and $\mathcal{J}$ be ideals in $\mathbb{K}[x_1, \ldots, x_n]$. The **ideal quotient** of $\mathcal{I}$ by $\mathcal{J}$, denoted $\mathcal{I} : \mathcal{J}$, is the set*

$$\mathcal{I} : \mathcal{J} = \{f \in \mathbb{K}[x_1, \ldots, x_n] : fg \in \mathcal{I}, \text{ for all } g \in \mathcal{J}\}.$$

The following proposition relates the quotient operation to the sum and intersection operations.

Proposition 1. *Let $\mathcal{I}$, and $\{\mathcal{I}_k\}_{1 \leq k \leq r}$ be ideals in $\mathbb{K}[x_1, \ldots, x_n]$. Then:*

$$\text{i)} \ \left(\bigcap_{k=1}^{r} \mathcal{I}_k \right) : \mathcal{I} = \bigcap_{k=1}^{r} (\mathcal{I}_k : \mathcal{I})$$
$$\text{ii)} \ \mathcal{I} : \left(\sum_{k=1}^{r} \mathcal{I}_k \right) = \bigcap_{k=1}^{r} (\mathcal{I} : \mathcal{I}_k)$$

If f is a polynomial and $\mathcal{I}$ an ideal, we shall write $\mathcal{I} : f$ instead of $\mathcal{I} : \langle f \rangle$. A special case of ii) is:

$$\mathcal{I} : \langle f_1, \ldots, f_r \rangle = \bigcap_{k=1}^{r} (\mathcal{I} : f_k).$$

We now address the question of computing generators of the ideal quotient $\mathcal{I} : \mathcal{J}$. The following observation is crucial:

Theorem 4. *Let $\mathcal{I}$ be an ideal in $\mathbb{K}[x_1, \ldots, x_n]$, and $f \in \mathbb{K}[x_1, \ldots, x_n]$. If $\langle g_1, \ldots, g_p \rangle = \mathcal{I} \cap \langle f \rangle$, then*

$$\langle g_1/f, \ldots, g_p/f \rangle = \mathcal{I} : f.$$

In order to construct a basis of an ideal quotient, we proceed as follows. Given ideals $\mathcal{I} = \langle p_1, \ldots, p_s \rangle$ and $\mathcal{J} = \langle g_1, \ldots, g_r \rangle$ in $\mathbb{K}[x_1, \ldots, x_n]$, we compute a basis for the intersections $\mathcal{I} \cap \langle g_1 \rangle, \ldots, \mathcal{I} \cap \langle g_n \rangle$ by using the above described method. For each i, we divide by f each element of a basis of $\mathcal{I} \cap \langle g_i \rangle$. This leads to a basis for $\mathcal{I} : g_i$. We then obtain a basis for $\mathcal{I} : \mathcal{J}$ by computing the intersections $\bigcap_{k=1}^{r}(\mathcal{I} : g_i)$.

3   2R⁻ Schemes

In [20], T. Matsumoto and H. Imai proposed one of the first examples of PKCs using compositions of multivariate polynomials. The public key of one of them, called C* ([21]), represented by "$t \circ \psi \circ s$", where t, s are two secret linear mappings over $GF(2)^n$, and ψ is the multivariate representation of $GF(2^n) \rightarrow GF(2^n), x \mapsto x^{1+2^{\theta}}$. This scheme has been broken by J. Patarin at Crypto'95 [22].

One-round schemes [17,18] are generalizations of C*. The public key of these schemes is indeed of the form "$t \circ \psi \circ s$", where t, s are two affine mappings over $\mathbb{K}^n$, and a $\psi : \mathbb{K}^n \rightarrow \mathbb{K}^n$ is a bijective mapping given by n multivariate polynomials of degree two. J. Patarin and L. Goubin [17,18] propose several constructions for ψ:

1. S-box functions: $(a_1, \ldots, a_n) \mapsto$

$$\left(S_1(a_1 \ldots, a_{n_1}), S_2(a_{n_1+1} \ldots, a_{n_1+n_2}), \ldots, S_b(a_{n_1+n_2+\cdots+n_{\bullet -1}+1}, \ldots, a_n) \right),$$

where $n = \sum_i n_i$, and each $S_i : K^{n_\bullet} \rightarrow K^{n_\bullet}$ is quadratic.
2. Triangular functions:

$$(a_1, \ldots, a_n) \mapsto \left(a_1, a_2 + q_1(a_1), a_3 + q_2(a_1, a_2, a_3), \ldots, a_n + q_{n-1}(a_1, \ldots, a_{n-1}) \right),$$

where each q_i is quadratic.
3. Combinations of S-box and triangular functions.

They showed that all these constructions are insecure [17,18]. To circumvent attacks, they introduce *two-round schemes* whose public key is the composition of two one-round schemes. The secret key of two-round schemes consists of:

Three affine bijections $r, s, t : K^n \rightarrow K^n$.
Two applications $\psi, \phi : K^n \rightarrow K^n$, given by n quadratic polynomials.

The public key is composed of n polynomials $p_1, \ldots, p_n$ of total degree 4 describing:

$$p = t \circ \psi \circ s \circ \phi \circ r, K^n \rightarrow K^n.$$

When all the polynomials are given, this scheme is called *2R scheme*. If only some of them are given, it is called *2R⁻ scheme*. The public-key part of the computation is merely an application of the mapping p (for encrypting a message, or checking the validity of a signature). For the secret-key computations, we need to invert the mappings ψ and ϕ. The authors then propose to choose the mappings among the constructions $1, 2, 3$ described above and also:

4. C* functions: monomials over an extension of degree n over K,
5. D* functions [16].

In [17,18], it has been proved that when ψ is chosen in the classes 2. and 4., then the resulting 2R scheme is weak. It is not clear that a similar result holds for 2R⁻ schemes.

Anyway, does composing two weak one-round schemes leads to a secure scheme ? The answer is closely related to the difficulty of the following problem:

Functional Decomposition Problem (FDP)
Input : $h = (h_1, \ldots, h_u) \in \mathbb{K}[x_1, \ldots, x_n]^u$.
Find : – if any – $f = (f_1, \ldots, f_u) \neq h \in \mathbb{K}[x_1, \ldots, x_n]^u$, and $g = (g_1, \ldots, g_n) \in \mathbb{K}[x_1, \ldots, x_n]^n$, such that:

$$\big(h_1(x), \ldots, h_u(x)\big) = \big(f_1(g_1(x), \ldots, g_n(x)), \ldots, f_u(g_1(x), \ldots, g_n(x))\big),$$

noted $h(x) = (f \circ g)(x)$ hereafter, where $x = (x_1, \ldots, x_n)$.

During the last years several results have been obtained on the univariate polynomial decomposition area [15,23,24]. However, multivariate decomposition problem has not been studied so much. Particular instances (multi-univariate,...) of FDP have been investigated in [23,19]. In [12], M. Dickerson provided several insights into the theoretical complexity of FDP. However, this kind of results solely guarantees the difficulty of the worst-case. In the cryptographic context, D.F. Ye, Z.D. Dai, and K.Y. Lam presented in [25,26] a quite efficient method for solving instances of FDP used in 2R. Note that their method only works when $u = n$, or $u = n - 1$ [26]. To the best of our knowledge, there exists no previously known algorithm for solving FDP when $u < n - 1$. An efficient method for solving FDP in this case would permit to decompose 2R⁻ schemes into two independent schemes given by quadratic polynomials. To break these schemes, we then would only have to solve two quadratic systems. As mentioned by J. Patarin and L. Goubin [18], this would make the principle of two-round schemes, including 2R⁻, useless.

4   A General Algorithm for Solving FDP

In this part, we present a new algorithm for solving FDP. Our approach is inspired on the works of D.F. Ye, Z.D. Dai, and K.Y. Lam [25,26]. According to these authors, we can restrict our attention to homogeneous instances of FDP [25]. The *homogenization* of a polynomial $p \in \mathbb{K}[x_1, \ldots, x_n]$, denoted p^*, is defined by $p^*(x_0, x_1, \ldots, x_n) = x_0^{deg(p)} p(x_1/x_0, \ldots, x_n/x_0)$, where x_0 is a new variable. For any mapping $f : \mathbb{K}^n \to \mathbb{K}^u$, given by the polynomials $f_1, \ldots, f_u$, we define its *homogenization* by $f^* = (x_0^{deg(f)}, f_1^*, \ldots, f_n^*)$. The *dehomogenization* of f^* is then $f = (f_1^*(1, x_1, \ldots, x_n), \ldots, f_n^*(1, x_1, \ldots, x_n))$. We have:

Lemma 2 ([25]). *Let $f : \mathbb{K}^n \to \mathbb{K}^u$ and $g : \mathbb{K}^n \to \mathbb{K}^n$ be two mappings, then:*

$$(f \circ g)^* = f^* \circ g^*.$$

Note 1. In [25], it is stated that this lemma is correct only if $deg(f)deg(g) > |K|$. We no longer need this condition over $\mathbb{K}[x_1, \ldots, x_n]$.

Thus, if we can decompose $h^* = f^* \circ g^*$, then a decomposition of $h = f \circ g$ is simply obtained by dehomogenization of f^* and g^*[25]. Now, we assume that $f : \mathbb{K}^n \to \mathbb{K}^u$ and $g : \mathbb{K}^n \to \mathbb{K}^n$ are two homogeneous functions of degree two. Finally, let $h = f \circ g$, and $\{h_i\}_{1 \leq i \leq u}, \{f_i\}_{1 \leq i \leq u}, \{g_i\}_{1 \leq i \leq n}$ be the components of h, f, g respectively.

4.1   Description of the Algorithm

The aim of our algorithm is to find the vector space $\mathcal{L}(g) = \text{Vect}(g_1, \ldots, g_n)$ generated by $g_1, \ldots, g_n$. More precisely, this vector space will be recovered from a DRL Gröbner basis of a suitable ideal. Note that the knowledge of $\mathcal{L}(g)$ is sufficient for decomposing h. Indeed, any bijective linear combination A of the g_is leads to a decomposition of h since:

$$h = (f \circ A^{-1}) \circ (A \circ g).$$

Let us first assume that we know the vector space $\mathcal{L}(g)$. For all $i, 1 \leq i \leq u$:

$$f_i = \sum_{1 \leq k, \ell \leq n} f_{k,\ell}^{(i)} x_k x_\ell \in \mathbb{K}[x_1, \ldots, x_n],$$
$$g_i = \sum_{1 \leq k, \ell \leq n} g_{k,\ell}^{(i)} x_k x_\ell \in \mathbb{K}[x_1, \ldots, x_n].$$

Therefore, for all $i, 1 \leq i \leq u$:

$$h_i = f_i(g_1, \ldots, g_n) = \sum_{1 \leq k, \ell \leq n} f_{k,\ell}^{(i)} g_k g_\ell. \tag{1}$$

By comparing the coefficients in the right-most and left-most parts of these equalities, we obtain a linear system of $O(u C_{n+2}^2)$ equations in the $u C_{n+2}^2$ unknown coefficients of the f_is. It seems difficult to rigorously evaluate the rank of this linear system, a question that has been avoided in the previous works on FDP [25,26]. However, it is very likely that this linear system is of full rank when the f_is are dense polynomials. For the instances of FDP used in $2R^-$ schemes, we experimentally only obtain linear systems of full rank. The difficult part is actually to determine the vector space $\mathcal{L}(g)$. For this, we observe that:

$$\frac{\partial h_i}{\partial x_j} = \sum_{1 \leq k, \ell \leq n} f_{k,\ell} \left(\frac{\partial g_k}{\partial x_j} g_\ell + g_k \frac{\partial g_\ell}{\partial x_j} \right). \tag{2}$$

The polynomials $g_1, \ldots, g_n$ being of degree two, their partial derivatives are of degree one. Hence:

$$\partial \mathcal{I}_h = \left\langle \frac{\partial h_i}{\partial x_j} : 1 \leq i \leq u, 1 \leq j \leq n \right\rangle \subseteq \langle x_k g_\ell \rangle_{1 \leq k, \ell \leq n} = \mathcal{V}.$$

This ideal $\partial \mathcal{I}_h$ usually provides enough information for recovering the polynomials $g_1, \ldots, g_n$.

Theorem 5. *Let $M(d)$ be the set of monomials of degree $d \geq 0$ in $x_1, \ldots, x_n$, and*

$$V_d = \mathrm{Vect}\, (mg_k : m \in M(d+1), \ and \ 1 \leq k \leq n),$$
$$\tilde{V}_d = \mathrm{Vect}\, \left(\left\{ m\frac{\partial h_i}{\partial x_j} : m \in M(d), 1 \leq i \leq u, \ and \ 1 \leq j \leq n \right\} \right).$$

Then, for all $i, 1 \leq i \leq n$:

$$x_n^{d+1} g_i \in \partial \mathcal{I}_h, \text{if } \dim(\tilde{V}_d) \geq n|M(d+1)|,$$

where $\dim(\tilde{V}_d)$ is the dimension of $\tilde{V}_d$ as a vector space over V_d.

Proof. We first study the case $d = 0$. Let $\tilde{V} = \tilde{V}_0$ be the linear space generated by the partial derivatives of the h_is, i.e.:

$$\tilde{V}_0 = \tilde{V} = \mathrm{Vect}\, \left(\left\{ \frac{\partial h_i}{\partial x_j} \right\}_{1 \leq j \leq n}^{1 \leq i \leq u} \right) \subset \partial \mathcal{I}_h.$$

According to (2), each element of $\tilde{V}$ can be written as a sum of $\{x_k g_\ell\}_{1 \leq k, \ell \leq n}$. Now let $A_{\tilde{V}} \in \mathcal{M}_{n^2 \times n^2}(\mathbb{K})$ be a matrix associated to the linear transformation $\mathrm{Vect}(\{x_k g_\ell\}_{1 \leq k, \ell \leq n}) \mapsto \tilde{V}$. For some basis:

$$
A_{\tilde{V}} = \begin{array}{c}
\\
\frac{\partial h_1}{\partial x_1} \\
\vdots \\
\frac{\partial h_1}{\partial x_\cdot} \\
\vdots \\
\frac{\partial h_\cdot}{\partial x_\cdot} \\
\vdots \\
\frac{\partial h_\cdot}{\partial x_1} \\
\vdots \\
\frac{\partial h_\cdot}{\partial x_\cdot}
\end{array}
\begin{array}{ccccccccc}
x_1 g_1 & \cdots & x_n g_1 & \cdots & x_k g_\ell & \cdots & x_1 g_n & \cdots & x_n g_n \\
\left(\right. & & & & \cdots & & & & \left. \right) \\
& & & & \cdots & & & & \\
& & & & \cdots & & & & \\
& & & & \cdots & & & & \\
& & & & \cdots & & & & \\
& & & & \cdots & & & & \\
& & & & \cdots & & & & \\
& & & & \cdots & & & & \\
\end{array}
$$

One can see at once that the $x_n g_i$s lie in $\tilde{V}$ if the number of linearly independent rows of this matrix is at least equal to its number of columns. That is, $x_n g_i \in \partial \mathcal{I}_h$, for all $i, 1 \leq i \leq n$, if:

$$\dim(\tilde{V}) \geq n|M(1)| = n^2.$$

Observe that $\dim(\tilde{V})$ is upper-bounded by un. Thus, $\dim(\tilde{V}) \geq n|M(1)| = n^2$ only holds if $u = n$. This explains why the method proposed in [25,26] is limited to 2R schemes. To circumvent this problem, we have to consider a vector space of higher dimension. This is the motivation for considering:

$$\tilde{V}_d = \mathrm{Vect}\left(\left\{m\frac{\partial h_i}{\partial x_j} : m \in M(d), 1 \leq i \leq u, \text{ and } 1 \leq j \leq n\right\}\right).$$

From (2), we deduce that each polynomial of $\tilde{V}_d$ can be written as a sum of elements of:

$$V_d = \mathrm{Vect}\left(mg_k : m \in M(d+1), \text{ and } 1 \leq k \leq n\right).$$

Let then $A_{\tilde{V}.}$ be a matrix associated to $V_d \mapsto \tilde{V}_d$. For some basis:

$$A_{\tilde{V}.} = m\frac{\partial h.}{\partial x.}\begin{array}{c} \cdots \quad \cdots \quad mg_k \quad \cdots \quad \cdots \\ \begin{pmatrix} & & \cdots & & \\ & & \cdots & & \\ & & \cdots & & \\ & & & & \\ & & \cdots & & \end{pmatrix} \end{array}$$

Thus, $x_n^{d+1}g_i \in \tilde{V} \subset \partial \mathcal{I}_h$, for all $i, 1 \leq i \leq n$, if $\dim(\tilde{V}_d)$ is at least equal to the number of columns of $A_{\tilde{V}.}$. That is, if $\dim(\tilde{V}_d) \geq n|M(d+1)|$.   □

Remark 1. *At the end of this part, we will provide an explicit value of d in function of the ratio n/u.*

According to Theorem 5, the polynomials g_is are contained, up to some power of x_n, in $\partial \mathcal{I}_h$. Therefore, the quotient of this ideal by a suitable power of x_n contains the polynomials $g_1, \ldots, g_n$.

Corollary 1. *Using the same notations as in Theorem 5. If $\dim(\tilde{V}_d) \geq n|M(d+1)|$, then:*

$$\mathcal{L}(g) \subset \langle g_1, \ldots, g_n \rangle \subseteq \partial \mathcal{I}_h : (x_n^{d+1}).$$

Proof. The proof of this corollary is obviously deduced from Theorem 5, and very definition of the quotient.   □

Thus each element of $\mathcal{L}(g)$ is included in $\partial \mathcal{I}_h : (x_n^{d+1})$. Let then G be a (reduced) DRL Gröbner basis of this ideal. It is then natural to consider the set $B_g = \mathrm{Vect}\left(g \in G : deg(g) = 2\right)$, since according to Theorem 2:

$$\mathcal{L}(g) = B_g, \text{if } \#B_g = n, \text{ and } min\left(deg(g) : g \in G\right) = 2.$$

If these conditions are not fulfilled, then one can not recover efficiently $\mathcal{L}(g)$ from B_g. Observe that the condition $\#B_g = n$ implies that there exists a unique decomposition (up to bijective linear combinations). To get away with these problems, we can apply several heuristics such as computing $\partial\mathcal{I}_h : (x_1^{d+1}), \ldots, \partial\mathcal{I}_h : (x_{n-1}^{d+1})$. In practice, it has been always sufficient to compute $\partial\mathcal{I}_h : (x_n^d)$, for a suitable d (i.e. $\dim(\tilde{V}_d) \geq n|M(d+1)|$).

4.2   The Algorithm Algo$_{\mathrm{FDP}}$

We describe now our algorithm for general instances of FDP, i.e. we no longer suppose here that h is given by homogeneous polynomials.

Algo$_{\mathrm{FDP}}$
Input: $h = f \circ g : \mathbb{K}^n \to \mathbb{K}^u$, given by u polynomials $h_1, \ldots, h_u \in \mathbb{K}[x_1, \ldots, x_n]$ of degree 4
Output : $f_1', \ldots, f_u', g_1', \ldots, g_n'$, such that $(h_1, \ldots, h_u) = (f_1'(g_1', \ldots, g_n'), \ldots, f_u'(g_1', \ldots, g_n'))$
$h_0^*(x_0, x_1, \ldots, x_n) \leftarrow x_0^4$
$h_i^*(x_0, x_1, \ldots, x_n) \leftarrow x_0^4 h_i(x_1/x_0, \ldots, x_n/x_0)$, for all $i, 1 \leq i \leq u$
$\partial\mathcal{I}_h^* \leftarrow \left\langle \frac{\partial h_i^*}{\partial x.} : 0 \leq i \leq u, 0 \leq j \leq n \right\rangle$
Let d be the smallest integer such that $\dim(\tilde{V}_d^*) \geq n|M(d+1)|$, with:
$\tilde{V}_d^* = \mathrm{Vect}\left(\left\{m\frac{\partial h_i^*}{\partial x.} : m \in M(d), 0 \leq i \leq u, \text{ and } 0 \leq j \leq n\right\}\right)$.
Compute a reduced 2-DRL Gröbner basis G of $\partial\mathcal{I}_h^* : (x_n^{d+1})$
$B_{g^*} \leftarrow \{g^* \in G, \deg(g^*) = 2\}$
If $\#B_{g^*} \neq n+1$ or $\min(\deg(g) : g \in G) \neq 2$ **then** Return Fail
Recover a basis B_{f^*} of $\mathrm{Vect}(f^*)$ by solving the system of linear equations given by (1)
Return $\{g^*(1, x_1, \ldots, x_n) \in B_{g^*}\}$ and $\{f^*(1, x_1, \ldots, x_n) \in B_{f^*}\}$

Remark 2. *In practice, our algorithm never returned* Fail *for instances of FDP used in* 2R$^-$.

Theorem 6. *Let* $g_0^*(x_0, x_1, \ldots, x_n) = x_0^2$, *and* $g_i^*(x_0, x_1, \ldots, x_n) = x_0^2 g_i(x_1/x_0, \ldots, x_n/x_0)$, *for all* $i, 1 \leq i \leq n$. *Moreover, let* $M(d)$ *be the set of monomials of degree* $d \geq 0$ *in* $x_0, x_1, \ldots, x_n$, *and*

$$V_d^* = \mathrm{Vect}\,(mg_k^* : m \in M(d+1), \text{ and } 0 \leq k \leq n),$$
$$\tilde{V}_d^* = \mathrm{Vect}\left(\left\{m\frac{\partial h_i^*}{\partial x_j} : m \in M(d), 0 \leq i \leq u, \text{ and } 0 \leq j \leq n\right\}\right).$$

Algo$_{\mathrm{FDP}}$ returns a solution of FDP (and not Fail*) in:*

$$O(n^{3(d+3)}),$$

where d is the smallest integer such that $\dim(\tilde{V}_d^*) \geq (n+1)|M(d+1)|$.

Proof. Let us suppose that our algorithm returns a solution (and not Fail). According to Corollary 1, we know that for all $i, 0 \leq i \leq n, g_i^* \in \partial\mathcal{I}_h^* : (x_n^{d+1})$. The complexity of Algo$_{\mathrm{FDP}}$ is then dominated by the cost of computing a reduced DRL Gröbner basis G of $\partial\mathcal{I}_h^* : (x_n^{d+1})$. This step can be done as explained in Section 2. However, an alternative method can be used in this particular situation. This is due to the particular role of x_n in a DRL order. From Lemma 1, we know that if x_n^{d+1} divides the leading monomial of a polynomial, then it also divides the entire polynomial. Thus, we can restrict our attention to polynomials of a DRL Gröbner Bases G' of $\partial \mathcal{I}_h^*$ whose leading monomials contain x_n^{d+1}. One can see directly that:

$$\left(g \in G : deg(g) = 2 \right) = \left(\frac{g'}{x_n^{d+1}} : g' \in G', \text{and } x_n^{d+1} | \mathrm{LM}(g', \prec_{DRL}) \right).$$

More precisely, it is sufficient to compute a reduced $(d+3)$-DRL Gröbner basis of $\partial \mathcal{I}_h^*$. According to Appendix A, this can be done with the F_5 algorithm in $O(n^{3(d+3)})$. From a practical point of view, the two methods proposed for computing G are similar. But the last one is more suitable for evaluating the complexity. □

Remark 3. *It should be noticed that our algorithm can easily be adapted for polynomials f of degree greater that 2.*

Comparison with Previous Approach

In short, our method can be viewed as a generalization of the approach of D.F. Ye, Z.D. Dai, and K.Y. Lam [25,26]. When $u = n$, it is sufficient to consider the ideal $\partial \mathcal{I}_h^* : (x_n^1)$ for recovering $\mathcal{L}(g)$. This is a simplified description of the method described in [25,26]. When $u < n$, $\partial \mathcal{I}_h^* : (x_n^1)$ no longer provides enough information for recovering $\mathcal{L}(g)$. To overcome this difficulty, we proposed here to consider ideals of the form $\partial \mathcal{I}_h^* : (x_n^{d+1})$. We then proved that $\mathcal{L}(g)$ is contained in this ideal as soon as d is sufficiently large.

It is important to know the exact value of the parameter d. This value can be lower-bounded in fonction of the ratio n/u. For this, we observe that $(n+1)|M(d+1)| = (n+1)C_{n+1+d}^{d+1}$ and $\dim(\tilde{V}_d^*)$ is very likely to be equal $(u+1)(n+1)C_{n+d}^d$. We then obtain that d should verify:

$$d \geq \frac{n}{u} - 1.$$

For instance, if the number of equations removed (i.e. $n-u$) is smaller than $\lfloor \frac{n}{2} \rfloor$, this yields a complexity of $O(n^{12})$, and $O(n^9)$ if $u = n$. We will show now that this approximation is perfectly coherent with our experimental results.

4.3   Experimental Results

Generation of the Instances

We have only considered instances $h = f \circ g$ of FDP admitting a solution. We constructed these instances in the following way:

$- f = t \circ \psi \circ s$ and $g = \phi \circ r$, with $r, s, t \circ \psi \circ s : K^n \to K^n$ are random affine bijections, and $\psi, \phi : K^n \to K^n$ are S-box functions contructed as explained in Section 3. We then remove $r \geq 0$ polynomials of h.

Programming language – Workstation

The experimental results have been obtained with a Xeon bi-processor 3.2 Ghz, with 6 Gb of Ram. The instances of FDP have been generated using the Maple software. We used our own implementation (in language C) of F$_5$ for computing truncated Gröbner bases.

Table Notations

The following notations are used in the table below:
- n, the number of variables,
- b, the number of blocks (as defined in Section 3),
- n_i, the number of variables in each block (see Section 3),
- q, the size of the field,
- r, the number of polynomials removed,
- $d_{theo} = \lceil \frac{n}{u} - 1 \rceil$, the predicted (see 4.2) value of d for which Algo$_{FDP}$ returns a solution
- d_{real}, the real value of d for which Algo$_{FDP}$ returns a solution
- T, the total time taken by our algorithm,
- $\sqrt{q^n}$, the current security bound [18,6] for 2R$^-$ schemes.

Practical Results

Let us now present results obtained with our algorithm.

n	b	n_i	r	q	d_{theo}	d_{real}	T	$\sqrt{q^n}$
8	4	2	0	65521	0	0	0.0 s.	
8	4	2	4	65521	1	1	0.0 s.	$\approx 2^{64}$
8	4	2	5	65521	2	2	0.3 s.	$\approx 2^{64}$
8	4	2	6	65521	3	3	1.9 s.	$\approx 2^{64}$
10	5	2	5	65521	1	1	0.2 s.	$\approx 2^{80}$
10	5	2	6	65521	2	2	3.2 s.	$\approx 2^{80}$
10	5	2	7	65521	3	3	21.4 s.	$\approx 2^{80}$
10	5	2	8	65521	4	4	180.8 s.	$\approx 2^{80}$
12	3	4	0	65521	1	1	0.1 s.	
12	3	4	5	65521	1	1	0.9 s.	$\approx 2^{96}$
12	3	4	6	65521	1	1	0.9 s.	$\approx 2^{96}$
12	3	4	7	65521	2	2	20.5 s.	$\approx 2^{96}$
12	3	4	8	65521	2	2	25.2 s.	$\approx 2^{96}$
12	3	4	9	65521	3	3	414 s.	$\approx 2^{96}$
20	5	4	0	65521	0	0	1.6 s.	
20	5	4	5	65521	1	1	55.2 s.	$\approx 2^{160}$
20	5	4	10	65521	1	1	78.9 s.	$\approx 2^{160}$
20	10	2	10	65521	1	1	78.8 s.	$\approx 2^{160}$
20	2	10	10	65521	1	1	78.7 s.	$\approx 2^{160}$
24	6	4	0	65521	0	0	4.9 s.	
24	6	4	12	65521	1	1	376.1 s.	$\approx 2^{192}$

30	15	2	15	65521	1	1	2910.5 s.	$\approx 2^{160}$
32	8	4	0	65521	0	0	31.3 s.	
32	8	4	10	65521	1	1	3287.9 s.	$\approx 2^{256}$
32	8	4	16	65521	1	1	4667.9 s.	$\approx 2^{256}$
36	18	2	15	65521	1	1	13427.4 s.	$\approx 2^{256}$

Interpretation of the Results

Let us mention that $n = 16$ and $n = 32$ were two challenges proposed by the designers of $2R^-$ schemes. First it should be observed that the parameters b and n_i of the S-box functions seem irrelevant for the complexity of our algorithm. We also tested our algorithm for instances of FDP constructed with various forms of ψ, ϕ (C^*+S-Box functions, Triangular+S-Box functions,...) and several values of q. These results are very similar to the ones obtained for S-Box functions, and thus not quoted here. The major observation is that our algorithm behaves exactly as predicted. That is, $d_{theo} = \lceil \frac{n}{u} - 1 \rceil$ is exactly equal to the d_{real} observed in practice.

Acknowledgements

We thank Lilian Bohy, Jintai Ding and anonymous referees for numerous comments which improved the presentation of the results.

References

1. A.W. Adams and P. Loustaunau. *An Introduction to Gröbner Bases.* Graduate Studies in Mathematics, Vol. 3, AMS, 1994.
2. G. Ars, J.-C. Faugère, H. Imai, M. Kawazoe, and M. Sugita. *Comparison Between XL and Gröbner Basis Algorithms.* Advances in Cryptology – ASIACRYPT 2004, Lecture Notes in Computer Science, vol. 3329, pp. 338-353, 2004.
3. G. Ars, and J.-C. Faugère. *Algebraic Immunities of functions over finite fields.* Proceedings of BFCA'05, Rouen, 2005. Also available at http://eprint.iacr.org/2004/188.ps.
4. M. Bardet, J-C. Faugère, B. Salvy and B-Y. Yang. *Asymptotic Behaviour of the Degree of Regularity of Semi-Regular Polynomial Systems.* In MEGA 2005, Eighth International Symposium on Effective Methods in Algebraic Geometry, 15 pages, 2005.
5. M. Bardet, J-C. Faugère, and B. Salvy. *On the Complexity of Gröbner Basis Computation of Semi-Regular Overdetermined Algebraic Equations.* In Proc. of International Conference on Polynomial System Solving (ICPSS), pp. 71–75, 2004.
6. E. Biham. *Cryptanalysis of Patarin's 2-Round Public Key System with S Boxes (2R).* Advances in Cryptology – CRYPTO 2000, Lecture Notes in Computer Science, vol. 1807, Springer–Verlag, pp. 408-416, 2000.
7. B. Buchberger. *Gröbner Bases : an Algorithmic Method in Polynomial Ideal Theory.* Recent trends in multidimensional systems theory. Reider ed. Bose, 1985.
8. B. Buchberger, G.-E. Collins, and R. Loos. *Computer Algebra Symbolic and Algebraic Computation.* Springer-Verlag, second edition, 1982.

9. V. Carlier, H. Chabanne, and E. Dottax *Grey Box Implementation of Block Ciphers Preserving the Confidentiality of their Design.* Proceedings of BFCA'05, Rouen, 2005. Also available at http://eprint.iacr.org/2004/188.ps.

10. N. Courtois, L. Goubin, and J. Patarin. *SFLASH, a Fast Asymmetric Signature Scheme for low-cost Smartcards – Primitive Specification and Supporting Documentation.* Available at http://www.minrank.org/sflash-b-v2.pdf.

11. D. A. Cox, J.B. Little, and D. O'Shea. *Ideals, Varieties, and Algorithms: an Introduction to Computational Algebraic Geometry and Commutative Algebra.* Undergraduate Texts in Mathematics. Springer-Verlag. New York, 1992.

12. M. Dickerson. *The functional Decomposition of Polynomials.* Ph.D Thesis, TR 89-1023, Departement of Computer Science, Cornell University, Ithaca, NY, July 1989.

13. J.-C. Faugère. *A New Efficient Algorithm for Computing Gröbner Casis without Reduction to Zero:* F$_5$. Proceedings of ISSAC, pp. 75–83. ACM press, July 2002.

14. J. C. Faugère, P. Gianni, D. Lazard, and T. Mora. *Efficient Computation of Zero-Dimensional Gröbner Bases by Change of Ordering.* Journal of Symbolic Computation, 16(4), pp. 329–344, 1993.

15. D. Kozen, and S. Landau. *Polynomial decomposition algorithms.* J. Symb. Comput. (7), pp 445–456, 1989.

16. L. Goubin, and J. Patarin. *Trapdoor One-way Permutations and Multivariate Polynomials.* Information and Communication Security, First International Conference (ICICS'97), Lecture Notes in Computer Science vol. 1334, Springer–Verlag, pp. 356–368, 1997.

17. L. Goubin, and J. Patarin. *Asymmetric Cryptography with S-Boxes.* Information and Communication Security, First International Conference (ICICS'97), Lecture Notes in Computer Science vol. 1334, Springer–Verlag, pp. 369–380, 1997.

18. L. Goubin, and J. Patarin. *Asymmetric Cryptography with S-Boxes – Extended Version.* Available at http://citeseer.ist.psu.edu/patarin97asymmetric.html.

19. J. Gutierrez, R. Rubio, J. von zur Gathen. *Multivariate Polynomial Decomposition.* Algebra in Engineering, Communication and Computing, 14 (1), pp. 11–31.

20. T. Matsumoto, and H. Imai. *Algebraic Methods for Constructing Asymmetric Cryptosystems.* Algebraic and Error-Correcting Codes. Prod. Third Intern. Conf., Grenoble, France, Springer-Verlag, pp. 108–119, 1985.

21. T. Matsumoto, and H. Imai. *Public Quadratic Polynomial-tuples for efficient signature-verification and message-encryption.* Advances in Cryptology – EUROCRYPT 1988, Lecture Notes in Computer Science, vol. 330, Springer–Verlag, pp. 419–453, 1988.

22. J. Patarin. *Cryptanalysis of the Matsumoto and Imai Public Key Scheme of Eurocrypt'88.* Advances in Cryptology – CRYPTO 1995, Lecture Notes in Computer Science, Springer-Verlag, vol. 963, pp. 248-261, 1995.

23. J. von zur Gathen. *Functional decomposition of polynomials: the tame case.* J. Symb. Comput. (9), pp. 281–299, 1990.

24. J. von zur Gathen. *Functional decomposition of polynomials: the wild case.* J. Symb. Comput. (10), pp. 437–452, 1990.

25. D.F. Ye, K.Y. Lam, Z.D. Dai. *Cryptanalysis of "2R" Schemes,* Advances in Cryptology – CRYPTO 1999, Lecture Notes in Computer Science, vol. 1666, Springer–Verlag, pp. 315–325, 1999.

26. D.F. Ye, Z.D. Dai and K.Y. Lam. *Decomposing Attacks on Asymmetric Cryptography Based on Mapping Compositions,* Journal of Cryptology (14), pp. 137–150, 2001.

Appendix A

The F5 Algorithm

The historical method for computing Gröbner bases is Buchberger's algorithm [8,7]. Recently, more efficient algorithms have been proposed. To date, F5 [13] is the most efficient for computing Gröbner bases. In a nutshell, this algorithm constructs incrementally the following matrices in degree d:

$$
A_d = \begin{array}{c} \\ t_1 f_1 \\ t_2 f_2 \\ t_3 f_3 \\ \ldots \end{array} \begin{array}{c} m_1 \succ m_2 \succ m_3 \ldots \\ \begin{bmatrix} \cdots & \cdots & \cdots & \cdots \\ \cdots & \cdots & \cdots & \cdots \\ \cdots & \cdots & \cdots & \cdots \\ \cdots & \cdots & \cdots & \cdots \end{bmatrix} \end{array}
$$

where the indices of the columns are monomials sorted for the admissible ordering $\prec$ and the rows are product of some polynomials f_i by some monomials t_j such that $\deg(t_j f_i) \le d$. For a regular system [13] the matrices A_d are of full rank. In a second step, row echelon forms of theses matrices are computed, i.e.

$$
A'_d = \begin{array}{c} \\ t_1 f_1 \\ t_2 f_2 \\ t_3 f_3 \\ \ldots \end{array} \begin{array}{c} m_1 \ m_2 \ m_3 \ldots \\ \begin{bmatrix} 1 & 0 & 0 & \ldots \\ 0 & 1 & 0 & \ldots \\ 0 & 0 & 1 & \ldots \\ 0 & 0 & 0 & \ldots \end{bmatrix} \end{array}
$$

Note that for each d, A'_d contains a d-Gröbner basis of the ideal considered. Important parameters to evaluate the complexity of F5 is the maximal degree d occurring in the computation and the size of the matrix A_d. The overall cost is thus dominated by $(\#A_d)^3$. Very roughly, $(\#A_d)$ can be approximated by $O(n^d)$. A more precise complexity analysis can be found in [4,5].

Receipt-Free Universally-Verifiable Voting with Everlasting Privacy[*]

Tal Moran and Moni Naor[**]

Department of Computer Science and Applied Mathematics,
Weizmann Institute of Science, Rehovot, Israel

Abstract. We present the first universally verifiable voting scheme that can be based on a general assumption (existence of a non-interactive commitment scheme). Our scheme is also the first receipt-free scheme to give "everlasting privacy" for votes: even a computationally unbounded party does not gain any information about individual votes (other than what can be inferred from the final tally).

Our voting protocols are designed to be used in a "traditional" setting, in which voters cast their ballots in a private polling booth (which we model as an untappable channel between the voter and the tallying authority). Following in the footsteps of Chaum and Neff [7,16], our protocol ensures that the integrity of an election cannot be compromised *even if the computers running it are all corrupt* (although ballot secrecy may be violated in this case).

We give a generic voting protocol which we prove to be secure in the Universal Composability model, given that the underlying commitment is universally composable. We also propose a concrete implementation, based on the hardness of discrete log, that is slightly more efficient (and can be used in practice).

Keywords: receipt-free voting human protocol universally composable formal proof.

1   Introduction

One of the earliest secret election protocols is the source of the word "ostracism": when the citizens of ancient Athens believed a politician had too much power, they held a vote to determine if he should be exiled for a period of ten years. The vote was conducted by having each citizen write the name of the person he most hated on a pot shard (these shards were called *ostraca*) and place it in a vase. After the votes were cast, the shards were taken out of the vases and counted. Many modern voting systems follow a very similar protocol, replacing clay with paper and vases with ballot boxes.

---

[*] This work was partially supported by the Minerva Foundation.
[**] Incumbent of the Judith Kleeman Professorial Chair.

C. Dwork (Ed.): CRYPTO 2006, LNCS 4117, pp. 373–392, 2006.

1.1   Challenges in Designing Voting Protocols

One of the main problems with traditional systems is that the accuracy of the election is entirely dependent on the people who count the votes. In modern systems, this usually consists of fairly small committees: if an entire committee colludes, they can manufacture their own results. Even worse, depending on the exact setup it may be feasible to stuff ballot boxes, destroy votes or perform other manipulations.

The problems with assuring election integrity were a large factor in the introduction of mechanical voting machines, and more recently, optical scan and "Direct Recording Electronic" (DRE) machines. These perform a function identical to a ballot box and paper ballots, using a different medium: the basic protocol remains the same. While alleviating some of the problems (such as ballot stuffing), in some cases they actually aggravate the main one: instead of relying on a large number of election committees (each of which has a limited potential for harm), their security relies on a much smaller number of programmers (who may be able to undetectably change the results of the entire election).

There has also been a large amount of more theoretical research, aimed at using cryptographic tools to define and solve the problems inherent in conducting secure elections. The most important advantage of cryptographic voting protocols over their physical counterparts is the potential for *universal verifiability*: the possibility that every voter (and even interested third parties) can verify that the ballot-counting is performed correctly. The challenge, of course, is satisfying this property while still maintaining the secrecy of individual ballots.

A problem that was first introduced with mechanical voting machines, and exacerbated in DRE and many cryptographic systems, is that part of the protocol must be performed by a machine (or computer), whose inner workings are opaque to most voters. This can have a serious impact on the trust a voter places in the results of the election (e.g., "how do I know that when I pushed the button next to candidate A the machine didn't cast a vote for B?"). One of the targets recently identified in the cryptographic literature is to design systems that can be trusted by human voters *even if the election computers are running malicious code*.

Another attack on both traditional and cryptographic voting systems is vote-buying and coercion of voters. To prevent this, we would like a voter to be unable to convince a third party of her vote *even if she wants to do so*. This property, called *receipt-freeness*, is strictly stronger than ballot secrecy, and seems even harder to achieve simultaneously with universal-verifiability. As is the case for election integrity, it is much more difficult to design a receipt-free protocol if the voter is required to perform secret calculations on a computer: the voter may be forced to use an untrusted computer to perform the calculations (or even one provided by the coercer), in which case the coercer can learn the secret.

There are also problems specific to the cryptographic case. One of these is that cryptographic protocols are often based on computational assumptions (e.g., the infeasibility of solving a particular problem). Unfortunately, some computational assumptions may not hold forever (e.g., Adi Shamir estimated that existing public-key systems will remain secure for less than thirty years [21]).

A voting protocol is said to provide *information-theoretic privacy* if a computationally unbounded adversary does not gain any information about individual votes (apart from the final tally). If the privacy of the votes depends on computational assumptions, we say the protocol provides *computational privacy*. Note that to coerce a voter, it is enough that the voter *believe* there's a good chance of her privacy being violated, whether or not it is actually the case (so even if Shamir's estimate is unduly pessimistic, the fact that such an estimate was made by an expert may be enough to allow voter coercion). Therefore, protocols that provide computational privacy may not be proof against coercion: the voter may fear that her vote will become public some time in the future.

While integrity that depends on computational assumptions only requires the assumptions to hold during the election, privacy that depends on computational assumptions requires them to hold forever. To borrow a term from Aumann et al. [1], we can say that information-theoretic privacy is *everlasting* privacy.

A related problem is that we would like to base our voting schemes on assumptions that are as strong as possible. Existing voting schemes generally require public-key encryption (or very specific computational assumptions, such as the hardness of computing discrete log in certain groups).

1.2 Our Results

In this paper, we present the first universally verifiable voting scheme that can be based on a general assumption (existence of a non-interactive commitment scheme).

Our protocol also satisfies the following properties:

- It has everlasting privacy (provided the commitment scheme is statistically hiding). To the best of our knowledge, only one published protocol has this property [9], and this protocol is not receipt-free.
- The protocol does not require human voters to perform any complex operations (beyond choosing a random string and comparing two strings)
- The integrity of the election is guaranteed even if the DRE is corrupted.
- It is receipt-free. We use a technique from Neff's voting scheme [16] to achieve receipt-freeness without requiring complex computation on the voter's part.

We give a rigorous proof that our protocol is secure in the Universally Composable model (given a universally-composable commitment scheme). This is a very strong notion of security. We also give a slightly more efficient protocol based on Pedersen Commitments (this protocol is secure, but not in the UC model, since Pedersen Commitments are not UC secure).

An additional contribution of this paper is a formal definition of receipt-freeness in the general multi-party computation setting (we also prove that our protocol satisfies this definition). Our definition is a generalization of Canetti and Gennaro's definition for an incoercible computation [4]. To the best of our knowledge, this is the first definition to capture receipt-freeness in the general case (most previous papers that deal with receipt-freeness do not provide a formal definition at all).

1.3   Previous Work on Voting Protocols

The first published electronic voting scheme was proposed by Chaum [5], based on *mixes*. Loosely speaking, a *mix* is a device that hides the correspondence between its inputs and outputs by accepting (encrypted) inputs in large batches and mixing them before output. This can be used to hide the correspondence between voters and their votes, allowing each voter to make sure her vote was counted (ensuring the integrity of the election) while preserving the secrecy of the vote. A strong advantage of this scheme over previous voting systems (e.g., putting paper slips in ballot boxes) is that the integrity of the vote no longer rests in the hands of a few trustees: every voter can verify that their vote was counted (i.e. it has *individual* verification). The *privacy* of the votes does depend on a small number of trustees (the mixing centers), though. Other advantages are convenience and speed: a voter can vote from any location with network access, and the votes are tabulated by computers immediately after they were all cast.

Many additional protocols were suggested since Chaum's. Almost all can be classified into one of three categories:

Mix-Type. These are protocols based on mixes, such as Chaum's original protocol.

Blind Signatures. These are protocols based on "blind signatures", introduced by Chaum in [6]. A blind signature allows a signer to digitally sign a document without knowing what was signed. In a voting scheme based on blind signatures, the general idea is that the voter has her ballot blindly signed by the voting authority, and later publishes the ballot using an anonymous channel. Although Chaum suggested the use of blind signatures for voting in his original paper, the first published protocol that makes use of blind signatures was by Fujioka et al. [11]. A major problem of blind signature schemes is that they require *anonymous channels* (so that the voter can publish her signed vote linking the vote to the voter).

Homomorphic. A function E is homomorphic if for any x and y in its domain it satisfies $E(x)E(y) = E(x+y)$. The general idea of a homomorphic voting scheme is for each voter to encrypt her vote using a public-key homomorphic encryption function, where the public key is published before the election. Each voter must prove that her encrypted vote is an encryption of a *valid* vote (the voting schemes differ on the exact way in which this is done). The encrypted votes are summed using the homomorphic property of the encryption function (without decrypting them). Finally, a set of trustees cooperate to decrypt the final tally (the secret key for the encryption scheme is divided between the trustees). The advantages of using homomorphic schemes are efficiency and verifiability: many operations can be carried out on the encrypted votes, in public, so they are both verifiable and can be performed during the voting process (without interaction between the voting authorities). The first protocol of this type was devised by Cohen (Benaloh) and Fischer [8]. Additional examples of this type of scheme are [2,9,10,12].

Receipt-Free Voting. Only a small fraction of the proposed voting schemes satisfy the property of receipt-freeness. Benaloh and Tuinstra [2] were the first to define this concept, and to give a protocol that achieves it (it turned out that their full protocol was not, in fact, receipt free, although their single-authority version was [12]). Their protocol was based on homomorphic encryption rather than mixes. To satisfy receipt-freeness, Benaloh and Tuinstra also required a physical "voting booth": completely untappable channels between the voting authority and the voter. Sako and Kilian showed that a one-way untappable channel is enough [20], and gave a receipt-free mix-type voting scheme based on this assumption (our protocol makes this assumption as well). Other protocols were also devised, however the minimal assumption required by protocols that do not use a trusted third party device (e.g., a smart card) is the one-way untappable channel.

Human Considerations. Almost all the existing protocols require complex computation on the part of the voter (infeasible for an unaided human). Thus, they require the voter to trust that the computer actually casting the ballot on her behalf is accurately reflecting her intentions. Chaum [7], and later Neff [16], proposed universally-verifiable receipt-free voting schemes that overcome this problem. Recently, Reynolds proposed another protocol similar to Neff's [18].

All three schemes are based in the "traditional" setting, in which voters cast their ballots in the privacy of a voting booth. Instead of a ballot box the booth contains a DRE. The voter communicates her choice to the DRE (e.g., using a touch-screen or keyboard). The DRE encrypts her vote and posts the encrypted ballot on a public bulletin board. It then proves to the voter, in the privacy of the voting booth, that the encrypted ballot is a truly an encryption of her intended vote. After all the votes have been cast, the votes are shuffled and decrypted using mix-type schemes.

Chaum's protocol uses a two-part ballot. Together, both parts define the vote in a manner readable to a human. Either part separately, however, contains only an encrypted ballot. The voter chooses one part at random (after verifying that the ballot matches her intended choice), and this part becomes her receipt. The other part is destroyed. The ballots are constructed so that in an invalid ballot, at least one of the two parts must be invalid (and so with probability at least $\frac{1}{2}$ this will be caught at the tally stage). Chaum's original protocol used Visual Cryptography [15] to enable the human voter to read the complete (two-part) ballot, and so required special printers and transparencies. Bryans and Ryan showed how to simplify this part of the protocol to use a standard printer [3,19].

Neff's protocol makes ingenious use of zero-knowledge arguments. The idea is that a zero-knowledge argument system has a simulator that can output "fake" proofs indistinguishable from the real ones. The DRE performs an interactive zero-knowledge protocol with the voter to prove that the encrypted ballot corresponds to the correct candidate. The DRE uses the simulator to output a zero-knowledge proof for *every other* candidate. The proofs are of the standard "cut-and-choose" variety. In a "real" proof, the DRE commits, then the voter gives a random challenge and the DRE responds. In the "fake" proofs, the voter first gives the random challenge and then the DRE commits and responds. The voter only has to make sure that she gave the challenge for the real candidate *after* the DRE was committed, and that the challenge printed on the receipt matches what she gave. Everything else can be publicly checked outside the voting booth. Since no one can tell from the receipt in which order the commitments and challenges were made, the zero-knowledge property ensures that they cannot be convinced which of the proofs is the real one.

2   The Model

The physical setup of our system is very similar to many existing (non-electronic) voting schemes. Voters cast their ballots at polling stations. The votes are tabulated for each station separately, and the final tally is computed by summing the results for all stations.

2.1   Basic Assumptions

Human Capability. An important property of our protocol is that its security is maintained even if the computers running the elections are corrupt (and only some of the human voters remain honest). Thus, we must define the operations we expect a human to perform. We make three requirements from human voters:

1. They can send messages to the DRE (e.g., using a keyboard). We require voters to send a few short phrases. This should be simple for most humans (but may be a slight problem for disabled voters).
2. They can verify that two strings are identical (one of which they chose themselves)
3. They can choose a random string. This is the least obvious of the assumptions we make. Indeed, choosing truly uniformly random bits is probably not something most humans can do. However, using physical aids (coins, dice, etc.) and techniques for randomness extraction it may be possible. In our security proofs, in order to clarify the presentation, we will ignore these subtleties and assume the voters can choose uniformly random strings.

Physical Commitment. In order to achieve receipt-freeness, our protocol requires a commitment with an extremely strong hiding property: The verifier's view at the end of the commit stage is a deterministic function of her view before the commit stage (i.e., not only can the view not contain any information about the committed string, it cannot even contain randomness added by the committer). Such a commitment is not possible in the "plain" cryptographic model (even with computational assumptions), but can be easily implemented by physical means (for example, by covering part of the printer's output with an opaque shield, so that the voter can see that something has been printed but not what). Note that the security of the protocol does not depend on physical commitment, only the receipt-freeness.

2.2   Participating Parties

In our description, we consider only a single polling booth (there is no interaction between booths in our system, apart from the final, public summation of results). Formally, we consider a few classes of participants in our voting protocol:

Voters. There are an arbitrary number of voters participating in the protocol (we will denote the number of voters by n). Each voter has a secret input (the candidate she wants to vote for).

DRE. The protocol has only a single DRE party. The DRE models the ballot box: it receives the votes of all the voters and outputs the final tally at the end.

Verifier. A Verifier is a party that helps verify that the voting protocols are being followed correctly. Although there can be many verifiers (and voters can be verifiers as well) the verifiers are deterministic and use only public information, so we model them as a single party.

Adversary. The adversary attempts to subvert the voting protocol. We detail the adversarial model in Sections 2.4 and 2.5.

2.3   Protocol Structure and Communication Model

As is customary in universally-verifiable voting protocols, we assume the availability of a public *Bulletin Board*: a broadcast channel with memory. All parties can read from the board and all messages from the DRE are sent to the bulletin board.

Our voting protocols consist of three phases:

1. *Casting a Ballot.* In this phase, each voter communicates directly with the DRE over a private, untappable, channel (inside the voting booth). All communication from the DRE to the voter is through the bulletin board.

   In practice, the voter will not be able to access the bulletin board while in the voting booth. Thus, we assume there is a separate channel between the DRE and the voter, also with memory (e.g., a printer). The DRE outputs its messages both to the printer (the printed messages form the voter's receipt), and to the bulletin board. This implementation adds an additional stage in which the voter verifies that the contents of her receipt match the contents on the bulletin board.

   We need the physical commitment only at one point in the Ballot-Casting phase. One possible implementation is by a shield covering part of the printer's output (so that the voter can see something has been printed under the shield, but not what).

2. *Tallying the results.* This phase begins after all voters have cast their ballots, and in this phase the results of the vote are revealed. The tallying consists of an interactive protocol between the DRE and a random beacon, whose only messages are uniformly random strings. In practice, we can use the Fiat-Shamir heuristic, replacing the beacon's messages with a secure hash of the entire transcript to that point (in the analysis the hash function is modeled

as a random oracle). The entire transcript of the tally phase is sent to the bulletin board.

3. *Universal Verification.* This phase consists of verifying the consistency of the messages on the bulletin board. This can be performed by any interested party, and is a deterministic function of the information on the bulletin board.

2.4　Universal Composability

We consider a number of different adversarial models. In the basic model, the adversary can adaptively corrupt the DRE and the voters (since there are arbitrarily many verifiers, and all are running the same deterministic function of public information, it is reasonably to assume not all of them can be corrupted). We formalize the capabilities of the adversary by defining them in the Universally Composable model, using the ideal voting functionality, $\mathcal{F}_V$. In the ideal world, The DRE does nothing unless it is corrupted. When the DRE is corrupt, ballots no longer remain secret (note that this must be the case in any protocol where the voter divulges her vote to the DRE). The integrity of the vote is always maintained. The verifiers have no input, but get the final output from the functionality (or $\perp$ if the functionality was halted by the adversary).

The adversary is allowed to intercept every outgoing message from the functionality, and has the choice of either delivering the message or sending the **Halt** command to the functionality (in which case the message will not be received by other parties).

The ideal functionality realized by our voting scheme accepts only one "honest" command and one "cheating" command (beyond the special **Halt** and **Corrupt** commands that can be sent by the adversary)

Vote c. On receiving this command from voter v, the functionality verifies that this is the only **Vote** command sent by v. It then:
1. Increments counter c.
2. If the DRE is corrupt, the functionality outputs **Voted** v, c to the adversary.
3. Broadcasts the message **Voted** v.
4. If all n voters have sent a **Vote** command, the functionality outputs the final value of the counters to the verifiers.

ChangeVote c, c'. This command can only be sent by a corrupt voter v and only if the DRE is also corrupt. On receiving this command, the functionality verifies that a **Vote** c command was previously sent by v. It then decrements counter c and increments counter c'. This command can be sent after the last voter has voted and before the final tally is output.

The security, privacy and robustness of our protocol is proven by showing that any attack against the protocol in the real world can be performed against the ideal functionality in the ideal world (where the possible attacks are explicitly defined). Due to space contraints, the formal description of the protocol and proof of security will appear only in the full version.

2.5   Receipt-Freeness

The property of receipt-freeness is not adequately captured by the UC model. In this model, in addition to corrupting parties, the adversary can *coerce* parties. A coercion attack models the real-life scenario in which voters are bribed or threatened to act according the adversaries wishes. The adversary can interrogate coerced parties and give them commands, but does not completely control them (a formal definition of receipt freeness can be found in Section 4). When considering the receipt-freeness of our protocol, we do not allow the adversary to coerce or corrupt the DRE. The latter is because corrupting the DRE reveals the voter's inputs, and so the protocol is trivially coercible. The former is because the DRE is a machine, so it does not make sense to bribe or threaten it.

It may make sense to coerce or corrupt the DRE's *programmers*, however. The difference between this situation and a corrupt DRE is that a corrupt DRE can communicate freely with the adversary, while a "maliciously programmed" DRE can communicate with the adversary only through the public communication channel (in one direction) and the voter's input (in the other direction). We briefly discuss this problem in Section 5.

2.6   Timing Attacks

Like any cryptographic proof, the security of our protocol is guaranteed only as far as the real-world matches the model on which our proof is based. One point we think is important to mention is the "timing" side-channel. Our model does not specify timing information for messages appearing on the bulletin board — only the order of the messages. However, in a real life implementation it may be possible to time the messages sent by the DRE. If the DRE actually does send messages simultaneously to the bulletin board and the voter, this timing information can be used to determine the voter's input (since the time it takes the voter to respond will be different). To prevent this attack, we specify that the DRE sends its output to the bulletin board only after the voter leaves the booth. One possible implementation (that also guarantees that the DRE can't leak information using timing, is that the DRE is not connected to a network at all. Instead, it prints the output to be sent to the bulletin board. The printout is given by the voter to the election officials on exiting the booth, who can scan it and upload the information to the bulletin board.

3   Informal Protocol Description

3.1   Overview

At the highest level, our voting scheme is extremely simple: the voter enters the voting booth and selects a candidate. The DRE uses a statistically-hiding commitment scheme to publicly commit to the candidate (e.g., by posting the commitment on a public bulletin board). It then proves privately to the voter that the commitment is really to the voter's candidate. After all voters have cast their ballots, the DRE publishes the final tally. It then proves, using a zero knowledge proof of knowledge, that the tally corresponds to the commitments published for each voter.

Since we know how to construct a ZK proof of knowledge for any NP language, and in particular we can construct such a proof system for any string commitment scheme, it would appear that we could use any such system for the private proof (the one that convices the voter that her ballot is being cast as she intended). The zero-knowledge property would ensure that the voter cannot use the proof to convince any third party of her vote.

The problem is that the voter is human, and the general zero-knowledge proof systems require complex computations that are infeasible to perform without the help of computers. Since the scheme must remain secure even if the DRE is malicious, the voter cannot trust the DRE to make these calculations. Allowing voters to use their own computers is not much better. Most voters do not know how to verify that their computer is actually running the correct code. Even worse, a coercive adversary could require a voter to use a computer supplied by the adversary, in which case it could easily learn the identity of the voter's candidate.

Our solution is that used in Neff's voting scheme: Neff observed that a standard cut-and-choose zero knowledge proof of some statement S has the following structure: the prover commits to two proofs P_0, P_1, the verifier makes a choice b, the prover reveals proof P_b and finally the verifier makes sure the revealed proof is correct. The protocol is constructed so that if both P_0 and P_1 are valid proofs, the statement S holds but, given b, anyone can construct a pair P_0', P_1' so that P_b' is correct (even if S does not hold). The insight is that a human can easily make the choice b without the aid of a computer. To keep the proof private, the DRE constructs a *fake* proof for all the other candidates by running the ZK simulator. The only difference between the real and fake proofs in this case is that in the real proof the DRE first commits, and then the voter chooses b, while in the fake proof the voter reveals b before the DRE commits. This *temporal* information cannot be inferred from the receipt. However, since the receipt is public, anyone can check (using a computer) that both P_b and P_b' are valid proofs. Even if the voter does not trust her own computer, it is enough that someone with a good implementation of the verification algorithm perform the check.

In the following subsections we present a concrete implementation of our generic protocol. This implementation is based on Pedersen commitments. In Section 3.2 we describe an example of a hypothetical voter's interaction with the system. Section 3.3 goes behind the scenes, and describes what is actually occurring during this session (as well as giving a brief description of the protocol itself).

3.2   A Voter's Perspective

Figure 3.1 shows what Dharma, our hypothetical voter, would see while casting her ballot in an election between candidates Alice, Betty and Charlie. Dharma identifies herself to the election official at the entrance to the polling station and enters the booth.

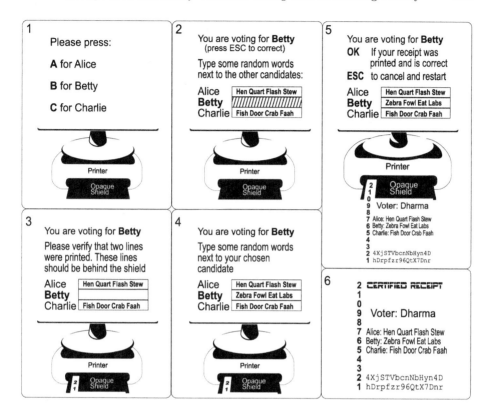

Fig. 3.1. Ballot for Betty

Inside the booth is the DRE: a computer with a screen, keyboard and an ATM-style printer

1. The screen presents the choice of candidates: "Press A for Alice, B for Betty and C for Charlie". Dharma thinks Betty is the best woman for the job, and presses B.
2. The DRE now tells Dharma to enter some random words next to each of the other candidates (Alice and Charlie). For an even simpler experience, the DRE can prefill the random words, and just give Dharma the option of changing them if she wants. At any time until the final stage, Dharma can change her mind by pressing **ESC**. In that case, the DRE spits out whatever has been printed so far (this can be discarded), and returns to stage 1.
3. The DRE prints two rows. The actual printed information is hidden behind a shield, but Dharma can verify that the two rows were actually printed.
4. Dharma enters a random challenge for her chosen candidate.
5. The DRE prints out the rest of the receipt. Dharma verifies that the challenges printed on the receipt are identical to the challenges she chose. If everything is in order, she presses **OK** to finalize her choice. If something is

wrong, or if she changed her mind and wishes to vote for a different candidate, she presses **ESC** and the DRE returns to stage 1.

6. The DRE prints a "Receipt Certified" message on the final line of the receipt. Dharma takes her receipt and leaves the voting booth. At home, she verifies that the public bulletin board has an exact copy of her receipt, including the two lines of "gibberish" (the bulletin board can be viewed from the internet). Alternatively, she can give her receipt to an organization she trusts (e.g., "Betty's Popular People's Front"), who will perform this verification for her.

After all the voters have cast their ballots, the protocol moves to the Final Tally phase. Voters are not required to participate in this phase — it consists of a single broadcast from the DRE to the public bulletin board (here we assume we are using the Fiat-Shamir heuristic to make the final tally non-interactive). Note that we do not actually require the DRE to be connected to a network. The DRE can store its output (e.g., on a removable cartridge). After the DRE has written the final tally message, the contents of the cartridge can be uploaded to the internet. Anyone tampering with the cartridge would be detected.

Anyone interested in verifying the election results participates in the Universal Verification phase. This can include voters, candidates and third parties. Voters do not have to participate, as long as they made sure that a copy of their receipt appears on the bulletin board, and they trust that at least one of the verifying parties is honest.

Receipt-Freeness. To get an intuitive understanding for why this protocol is receipt-free, suppose Eve tries to bribe Dharma to vote for Alice instead. There are only two things Dharma does differently for Alice and Betty: she presses B in the first step, and she fills in Alice's random words before the DRE prints the first two lines of the receipt, while filling in Betty's afterwards. Eve has no indication of what Dharma pressed (since the receipt looks the same either way). The receipt also gives no indication in what order the candidate's words were filled (since the candidates always appear in alphabetical order). Because the first two lines of the receipt are hidden behind the shield when Dharma enters the challenge for her chosen candidate, she doesn't gain any additional information as a result of filling out the challenges for Alice and Charlie; so whatever Eve asks her to do, she can always pretend she filled out the challenge for Alice after the challenge for Betty.

3.3 Behind the Scenes: An Efficient Protocol Based on the Discrete Log Assumption

Pedersen Commitments. The concrete protocol we describe in this section is based on Pedersen commitments [17]; statistically hiding commitments whose security is based on the hardness of discrete-log. We briefly describe the Pedersen commitment, assuming discrete log is hard in some cyclic group G of order q, and $h, g \in G$ are generators such that the committer does not know $\log_g h$. To commit to $a \in \mathbb{Z}_q$, the committer chooses a random element $r \in \mathbb{Z}_q$, and sends $h^a g^r$.

Note that if the committer can find $a' \neq a$ and r' such that $h^{a'} g^{r'} = h^a g^r$, then $h^{a'-a} = g^{r-r'}$, and so the commiter can compute $\log_g h = \frac{r-r'}{a'-a}$ (in contradiction to the assumption that discrete log is hard in G). Therefore the commitment is computationally binding. If r is chosen uniformly at random from $\mathbb{Z}_q$, then g^r is a uniformly random element of G (since g is a generator), and so for any a the commitment $h^a g^r$ is also a uniformly random element of G. So the commitment is perfectly hiding.

We'll now show what happened behind the scenes, assuming the parameters G, h and g of the Pedersen commitment scheme are decided in advance and known to all parties. For simplicity, we also assume a collision-resistant hash function $H : \{0,1\}^* \mapsto \mathbb{Z}_q$. This allows us to commit to an arbitrary string $a \in \{0,1\}^*$ by committing to $H(a)$ instead. Denote $P(a,r) = h^{H(a)} g^r$. To open $P(a,r)$, the committer simply sends a, r.

The security of the scheme depends on a security parameter, k. The probability that the DRE can change a vote without being detected is $2^{-k+1} + O(nk\epsilon)$, where n is the number of voters and ϵ is the probability of successfully breaking the commitment scheme.

Casting the Ballot. We'll go over the stages as described above.

1. (Dharma chose Betty). The DRE computes a commitment: $v = P(\text{Betty}, r)$ (where r is chosen randomly), and prepares the first step in a proof that this commitment is really a commitment to Betty. This step consists of computing, for $1 \leq i \leq k$, a "masked" copy of v: $b_i = vg^{r \cdots} = P(\text{Betty}, r + r_{B,i})$, where $r_{B,i}$ is chosen randomly.
2. (Dharma enters fake challenges). The DRE translates each challenge to a k bit string using a predetermined algorithm (e.g., by hashing). Let A_i be the ith bit of the challenge for Alice. For each bit i such that $A_i - 0$ the DRE computes a commitment to Alice: $a_i = P(\text{Alice}, r_{A,i})$, while for each bit such that $A_i = 1$ the DRE computes a masked copy of the real commitment to Betty: $a_i = vg^{r \cdots}$. Note that $a_i = P(\text{Betty}, r + r_{A,i})$. The set of commitments $a_1, \ldots, a_k$ will form a fake proof that v is a commitment to Alice (we'll see why we construct them in this way in the description of universal verification phase step 3.3). The DRE also computes $c_1, \ldots, c_k$ in the same way for Charlie.
   The DRE now computes a commitment to everything it has calculated so far: $x = P([v, a_1, \ldots, a_k, b_1, \ldots, b_k, c_1, \ldots, c_k], r_x)$. It prints x on the receipt (this is what is printed in the first two lines).
3. (Dharma enters the real challenge). The DRE translates this challenge into a k bit string as in the previous step. Denote B_i the ith bit of the real challenge.
4. (The DRE prints out the rest of the receipt). The DRE now computes the answers to the challenges: For every challenge bit i such that is $X_i = 0$ (where $X \in \{A, B, C\}$), the answer to the challenge is $s_{X,i} = r_{X,i}$. For $X_i = 1$, the answer is $s_{X,i} = r + r_{X,i}$. The DRE stores the answers. It then prints the candidates and their corresponding challenges (in alphabetical order), and the voter's name (Dharma).

5. (Dharma accepts the receipt). The DRE prints a "Receipt Certified" message on the final line of the receipt. (the purpose of this message is to prevent voters from changing their minds at the last moment, taking the partial receipt and then claiming the DRE cheated because their receipt does not appear on the bulletin board). It then sends a copy of the receipt to the public bulletin board, along with the answers to the challenges and the information needed to open the commitment x: $(s_{A,1}, \ldots, s_{A,k}, s_{B,1}, \ldots, s_{B,k}, s_{C,1}, \ldots, s_{C,k})$ and $([v, a_1, \ldots, a_k, b_1, \ldots, b_k, c_1, \ldots, c_k], r_x)$.

Final Tally. The DRE begins the tally phase by announcing the final tally: how many voters voted for Alice, Betty and Charlie. Denote the total number of voters by n, and $v_i = P(X_i, r_i)$ the commitment to voter i's choice (X_i) that was sent in the Ballot Phase. The DRE now performs the following proof k times:

1. The DRE chooses a random permutation π of the voters, and n "masking numbers" $m_1, \ldots, m_n$. It then sends the permuted, masked commitments of the voters:
$$v_{\pi(1)} g^{m_{\pi(1)}}, \ldots, v_{\pi(n)} g^{m_{\pi(n)}}$$
2. The random beacon sends a challenge bit b
3. If $b = 0$, the DRE sends π and $m_1, \ldots, m_n$ (unmasking the commitments to prove it was really using the same commitments it output in the Ballot-Casting phase). If $b = 1$, the DRE opens the masked commitments (without revealing π, the correspondence to the original commitments). It sends:
$$(X_{\pi(1)}, r_{\pi(1)} + m_{\pi(1)}), \ldots, (X_{\pi(n)}, r_{\pi(n)} + m_{\pi(n)})$$

Universal Verification (and security proof intuition). The purpose of the universal verification stage is to make sure that the DRE sent well-formed messages and correctly opened all the commitments. For the messages from the Ballot-Casting phase, the verifiers check that:

1. $x = P([v, a_1, \ldots, a_k, b_1, \ldots, b_k, c_1, \ldots, c_k], r_x)$ This ensures that the DRE committed to v and $b_1, \ldots, b_k$ (in Dharma's case) before Dharma sent the challenges $B_1, \ldots, B_k$ (because x was printed on the receipt before Dharma sent the challenges).
2. For every commitment x_i (where $x \in \{a, b, c\}$), its corresponding challenge X_i, and the response $s_{X,i}$, the verifiers check that x_i is a good commitment to X when $X_i = 0$ (i.e., $x_i = P(X, s_X.)$) and that x_i is a masked version of v if $X_i = 1$ (i.e., $v s_X. = x_i$). Note that if x_i is both a masked version of v *and* a good commitment to X, then v must be a good commitment to X (otherwise the DRE could open v to two different values, contradicting the binding property of the commitment). This means that if v is a commitment to some value other than the voter's choice, the DRE will be caught with probability at least $1 - 2^{-k}$: every commitment x_i can be either a good masked version of v or a good commitment to X, but not both. So for each of the k challenges (which are not known in advance to the DRE), with probability $\frac{1}{2}$. The DRE will not be able to give a valid response.

For the final tally phase, the verifiers also check that all commitments were opened correctly (and according to the challenge bit). As in the Ballot-Casting phase, if the DRE can correctly answer a challenge in both directions (i.e., the commitments are a permutation of good masked versions of commitments to the voter's choices, and also when opened they match the tally), then the tally must be correct. So the DRE has probability at least $\frac{1}{2}$ of getting caught for each challenge if it gave the incorrect tally. If the DRE wants to change the election results, it must either change the final tally, change at least one of the voter's choices or break the commitment scheme. Since the protocol uses $O(nk)$ commitments (note that cheating on the commitments in the fake proofs doesn't matter), the total probability that it can cheat is bounded by $2 \cdot 2^{-k} + O(nk\epsilon)$.

Protocol Complexity. We can consider both the time and communication complexity of the protocol. In terms of time complexity, the DRE must perform $O(knm)$ commitments in the Ballot Casting phase (where m is the number of candidates), and $O(kn)$ commitments in the Final Tally phase (the constants hidden in the O notation are not large in either case). Verifiers have approximately the same time complexity (they verify that all the commitments were opened).

The total communication complexity is also of the same order. In this case, the important thing is to minimize the DRE's communication to the voter (since this must fit on a printed receipt). Here the situation is much better: the receipt only needs to contain a single commitment and the challenges sent by the voter (each challenge has k bits). Note that we do not use any special properties of the commitment on the receipt (in practice, this can be the output of a secure hash function rather than a Pedersen commitment).

3.4   Using Generic Commitment

The protocol we described above makes use of a special property of Pedersen commitment: the fact that we can make a "masked" copy of a commitment. The essence of our zero knowledge proof is that on the one hand, we can prove that a commitment is a masked copy of another without opening the commitment. On the other hand, just by seeing two commitments there is no way to tell that they are copies, so opening one does not give us information about the other.

Our generic protocol uses the same idea, except that we implement "masked copies" using an arbitrary commitment scheme. The trick is to use a double commitment. Denote $C(a, r)$ a commitment to a with randomness r.

The idea is to create all the "copies" in advance (we can do this since we know how many copies we're going to need): a "copyable" commitment to a, assuming we're going to need k copies, consists of $v = C(C(a, r_1), s_1), \ldots, C(C(a, r_k), s_k)$. The ith copy of the commitment is $C(a, r_i)$. The hiding property of C ensures that there is no way to connect $C(a, r_i)$ to v. On the other hand, the binding property of C ensures that we cannot find any s' such that $C(C(a, r_i), s')$ is in v unless this was the original commitment.

Unlike the case with Pedersen commitments, when using the double commitment trick an adversary *can* "copy" a commitment to a different value (the adversary can always use different values in the inner commitments). The insight here is that the adversary still has to commit in advance to the locations of the errors. After the DRE commits, we randomly permute the indices (so $v = C(C(a, r_{\sigma(1)}), s_{\sigma(1)}), \ldots, C(C(a, r_{\sigma(k)}), s_{\sigma(k)})$, for some random permutation σ). If the DRE did commit to the same value at every index, this permutation will not matter. If the DRE committed to different values, we show that it will be caught with high probability. Intuitively, we can consider each commitment in the tally phase as a row of a matrix whose columns correspond to the voters. By permuting the indices of the copies, we are effectively permuting the columns of the matrix (since in the ith tally step we force the DRE to use the ith copy. The DRE can pass the test only if all the rows in this matrix have the same tally. But when the columns are not constant, this occurs with small probability[1]. The formal protocol description and proof of security, as well as a proof that our protocol is receipt-free, will appear in the full version.

4   Incoercibility and Receipt-Freeness

Standard "secure computation" models usually deal with two types of parties: honest parties with a secret input that follow the protocol, and corrupt parties that are completely controlled by the adversary. In voting protocols, we often need to consider a third type of player: a party that has a secret input, but is threatened (or bribed) by the adversary to behave in a different manner. Such a "coerced" player differs from a corrupt party in that she doesn't do what the adversary wishes if she can help it; if she can convince the adversary that she's following its instructions while actually following the protocol using her secret input, she will.

Benaloh and Tuinstra [2] were the first to introduce this concept. Most papers concerning receipt-free voting (including Benaloh and Tuinstra), do not give a rigorous definition of what it means for a protocol to be receipt-free, only the intuitive one: "the voter should not be able to convince anyone else of her vote".

Canetti and Gennaro considered this problem for general multiparty computation (of which voting is a special case), and gave a formal definition of *incoercibility* [4]. Their definition is weaker than receipt-freeness, however: the adversary is only allowed to coerce a player after the protocol is complete (i.e., it cannot require a coerced player to follow an arbitrary strategy, or even specify what randomness to use).

Juels, Catalano and Jakobsson also give a formal definition of *coercion-resistance* [13]. Their definition has a similar flavor, but is specifically tailored to voting in a public-key setting. It is stronger than receipt-freeness in that a coercion-resistant protocol must also prevent an *abstention-attack* (preventing a coerced voter from voting). However, this strong definition requires anonymous channels from voters to tallying authorities (otherwise an abstention-attack is always possible).

---

[1] For technical reasons, the technique we use in practice is slightly different, however the idea is the same.

Our formalization of receipt-freeness is a generalization of Canetti and Gennaro's definition (and so can be used for any secure function evaluation), and is strictly stronger (i.e., any protocol that is receipt-free under our definition is incoercible as well). Loosely, both definitions require that any "coercion" the adversary can do in the real world it can also do in an ideal world (where the parties only interaction is sending their input to an ideal functionality that computes the function). The difference is the adversarial model we consider in the real world. Canetti and Gennaro allow the adversary to give coerced players an alternative input. In our definition, the adversary can give coerced players an arbitrary strategy (i.e. commands they should follow instead of the real protocol interactions).

4.1   Formal Definition

Formally, a receipt-free protocol must specify, in addition to the strategy for honest parties, a *coercion strategy*. When a party is coerced by the adversary, it begins following the coercion strategy instead of the honest strategy. The coercion strategy tells the party how to fake its responses to the adversary.

Ideal Setting. Consider the "ideal setting" for computing a function f. In this setting we have n players, $P_1, \ldots, P_n$, with inputs $x_1, \ldots, x_n$ A trusted party collects the inputs from all the players, computes $f(x_1, \ldots, x_n)$ and broadcasts the result. In this setting, the ideal adversary $\mathcal{I}$ is limited to the following options:

1. corrupt a subset of the parties. In this case the adversary learns the parties' real inputs and can replace them with inputs of its own choosing.
2. coerce a subset of the parties. The only thing an adversary can do with a coerced party is to force it to use the input $\perp$ (signifying a forced abstention).

$\mathcal{I}$ can perform these actions iteratively (i.e., adaptively corrupt or coerce parties based on information gained from previous actions), and when it is done the ideal functionality computes the function. $\mathcal{I}$'s view in the ideal case consists its own random coins, the inputs of the corrupted parties and the output of the ideal functionality $f(x_1, \ldots, x_n)$ (where for corrupted parties x_i is the input chosen by the adversary).

Real-World Setting. In the "real-world setting", we also have the n parties, $P_1, \ldots, P_n$, with inputs $x_1, \ldots, x_n$. Each party has a protocol specification, consisting of a pair of interactive turing machines (H_i, C_i). There is also a "real" adversary $\mathcal{A}$.

The protocol proceeds in steps: In each step $\mathcal{A}$ can choose:

1. to corrupt a subset of the parties. In this case the adversary learns the entire past view of the party and completely controls its actions from that point.
2. to coerce a subset of the parties. The adversary supplies each coerced party with a "fake" input x_i'. From this point on it can interactively query and send commands to the coerced party. The coerced party switches to the coercion strategy C_i, which specifies how to respond to $\mathcal{A}$'s queries and commands.

390 T. Moran and M. Naor

$\mathcal{A}$ performs these actions iteratively, adaptively coercing, corrupting and inter-acting with the parties. $\mathcal{A}$'s view in the real-world consists of its randomness, the inputs and all communication of corrupted parties, its communications with the coerced parties and all public communication.

Definition 1. *A protocol is receipt-free if there exists an ideal adversary $\mathcal{I}$, such that for any real adversary $\mathcal{A}$ (with which $\mathcal{I}$ can communicate as a black-box) and any input vector $x_1, \ldots, x_n$, $\mathcal{I}$'s output in the ideal world is indistinguishable from $\mathcal{A}$'s view of the protocol in the real world with the same input vector (where the distributions are over the random coins of $\mathcal{I}$, $\mathcal{A}$ and the parties).*

It is important to note that even though a protocol is receipt-free by our defini-tion, it may still be possible to coerce players (a trivial example is if the function f consists of the player's inputs). What the definition does promise is that if it is possible to coerce a party in the real world, it is also possible to coerce that party in the ideal world (i.e. just by looking at the output of f).

5 Discussion

Splitting the Vote. Our scheme suffers a large drawback compared to many of the published universally-verifiable schemes: we do not know how to distribute the vote between multiple authorities. In our protocol, the DRE acts as the only tallying authority. Thus, a corrupt DRE can reveal the voters' ballots. This is also the case with Chaum and Neff's schemes, however (as well as traditional DRE schemes in use today). It seems a fairly difficult problem to solve while taking into account the limited abilities of voters (e.g., most secret sharing schemes cannot be performed by unaided humans).

Robustness of the Voting Scheme. The *robustness* of a voting scheme is its ability to recover from attacks without requiring the election to be canceled. Because the DRE is the sole tallying authority, a corrupt DRE can clearly disrupt the elections (e.g., by failing to output the tally). The UC proof shows that voters cannot disrupt the elections just by interacting with the DRE (the only thing a voter can do in the ideal model is either vote or abstain). However, our model does not prevent *false* accusations against the DRE. For example, a corrupt voter that is able to fake a receipt can argue that the DRE failed to publish it on the bulletin board. Using special paper for the receipts may help, but preventing this and similar attacks remains an open problem.

Traditional Paper Trail as Backup. Many critics of non-cryptographic DRE sys-tems are pushing for a "voter verified paper-trail": requiring the DRE to print a plaintext ballot that the voter can inspect, which is then placed in a real bal-lot box. In a non-cryptographic system, the paper trail can help verify that a DRE is behaving honestly, and act as a recovery mechanism when it is not. In our system, a paper trail can be used for the recovery property alone: if the DRE is misbehaving, our protocol ensures it will be detected with high proba-bility (without requiring random audits of the paper trail). In that case, we can perform a recount using the paper ballots.

Randomness and Covert channels. One problem that also plagues Neff's scheme, and possibly Chaum's [14], is that a corrupt DRE can convey information to an adversary using subliminal channels. In this case, an adversary only needs access to the bulletin board in order to learn all the voters choices. The source of the problem is that the DRE uses a lot of randomness (e.g., the masking factors for the commitments and the random permutations in the final tally phase).

In our scheme based on Pedersen commitments, we have a partial solution to this problem. It requires the output of the DRE to be sent through a series of "filters" before reaching the printer or bulletin board. The idea is that for each Pedersen commitment of the form $x = h^a g^r$ received by a filter, it will choose a random masking factor s, and output xg^s. If the DRE opens x by sending (a, r), the filter will send instead $(a, r + s)$. In a similar way the filter can mask the permutations used in the final-tally phase by choosing its own random permutations and composing them. Note that the filter does not need to know the value of the commitments or the original permutations in order to perform its operation. If the DRE is honest, the filter receives no information and so cannot covertly send any. If the filter is honest, any randomness sent by the DRE is masked, so no information embedded in that randomness can leak. By using a series of filters, each from a different trustee, we can ensure that the DRE does not utilize covert channels (as long as at least one of the filters is honest). A general solution to this problem is still an interesting open problem.

An even stronger adversary may be able to both coerce voters and maliciously program the DRE. Our protocol is vulnerable in this case. For example, a coercer can require a voter to use a special challenge, which is recognized by the DRE (a coercer can verify that the voter used the required challenge, since it appears on the public bulletin board). Once it knows a voter is coerced, the DRE can change the vote as it wishes (since the coerced voter will not be able to complain). Possibly mitigating the severity of this attack is the fact that, in order to significantly influence the outcome of an election, the adversary must coerce many voters. This makes it much harder to keep the corruption secret.

Separate Tallying. Our scheme requires the tally to be performed separately for each DRE. This reveals additional information about voter's choices (in many real elections this information is also available, however). An open problem is to allow a complete tally without sacrificing any of the other properties of our scheme (such as receipt-freeness and everlasting privacy).

References

1. Y. Aumann, Y. Z. Ding, and M. O. Rabin. Everlasting security in the bounded storage model. *IEEE Transactions on Information Theory*, 48(6):1668–1680, 2002.
2. J. Benaloh and D. Tuinstra. Receipt-free secret-ballot elections. In *STOC '94*, pages 544–553, 1994.
3. J. W. Bryans and P. Y. A. Ryan. A simplified version of the Chaum voting scheme. Technical Report CS-TR 843, University of Newcastle, 2004.

4. R. Canetti and R. Gennaro. Incoercible multiparty computation. In *FOCS '96*, pages 504–513, 1996.
5. D. Chaum. Untraceable electronic mail, return addresses, and digital pseudonyms. *Communications of the ACM*, 24(2):84–88, 1981.
6. D. Chaum. Blind signature systems. In *CRYPTO '83*, page 153, 1983.
7. D. Chaum. E-voting: Secret-ballot receipts: True voter-verifiable elections. *IEEE Security & Privacy*, 2(1):38–47, Jan./Feb. 2004.
8. J. D. Cohen(Benaloh) and M. J. Fischer. A robust and verifiable cryptographically secure election scheme. In *FOCS '85*, pages 372–382, 1985.
9. R. Cramer, M. Franklin, B. Schoenmakers, and M. Yung. Multi-authority secret-ballot elections with linear work. In *EUROCRYPT '96*, pages 72–83, 1996.
10. R. Cramer, R. Gennaro, and B. Schoenmakers. A secure and optimally efficient multi-authority election scheme. In *Eurocrypt '97*, pages 103–118, 1997.
11. A. Fujioka, T. Okamoto, and K. Ohta. A practical secret voting scheme for large scale elections. In *AUSCRYPT '92*, volume 718 of *LNCS*, pages 244–251, 1993.
12. M. Hirt and K. Sako. Efficient receipt-free voting based on homomorphic encryption. In *Eurocrypt 2000*, volume 1807 of *LNCS*, pages 539+, 2000.
13. A. Juels, D. Catalano, and M. Jakobsson. Coercion-resistant electronic elections. In *WPES '05*, pages 61–70, 2005.
14. C. Karlof, N. Sastry, and D. Wagner. Cryptographic voting protocols: A systems perspective. In *USENIX Security '05*, pages 33–50, 2005.
15. M. Naor and A. Shamir. Visual cryptography. In *Eurocrypt '94*, volume 950 of *LNCS*, pages 1–12, 1995.
16. C. A. Neff. Practical high certainty intent verification for encrypted votes, October 2004. http://www.votehere.net/vhti/documentation/vsv-2.0.3638.pdf.
17. T. P. Pedersen. Non-interactive and information-theoretic secure verifiable secret sharing. In *CRYPTO '91*, volume 576 of *LNCS*, pages 129–140, 1991.
18. D. J. Reynolds. A method for electronic voting with coercion-free receipt. FEE '05. Presentation: http://www.win.tue.nl/~berry/fee2005/presentations/reynolds.ppt.
19. P. Y. A. Ryan. A variant of the Chaum voter-verifiable scheme. In *WITS '05*, pages 81–88, 2005.
20. K. Sako and J. Kilian. Receipt-free mix-type voting schemes. In *EUROCRYPT '95*, volume 921 of *LNCS*, pages 393–403, 1995.
21. A. Shamir. Cryptographers panel, RSA conference, 2006. Webcast: http://media.omediaweb.com/rsa2006/1_5/1_5_High.asx.

Cryptographic Protocols for Electronic Voting

David Wagner

UC Berkeley

Abstract. Electronic voting has seen a surge of growth in the US over the past five years; yet many questions have been raised about the trustworthiness of today's e-voting systems. One solution that has been proposed involves use of sophisticated cryptographic protocols to prove to voters that their vote has been recorded and counted correctly. These protocols are of interest both for their use of many fundamental tools in cryptography (e.g., zero-knowledge proofs, mixnets, threshold cryptosystems) as well as because of the social importance of the problem. I will survey some promising recent work in this area, discuss several open problems for future research, and suggest some ways that cryptographers might be able to contribute to improving the integrity of public elections.

C. Dwork (Ed.): CRYPTO 2006, LNCS 4117, p. 393, 2006.

Asymptotically Optimal Two-Round Perfectly Secure Message Transmission

Saurabh Agarwal[1], Ronald Cramer[2], and Robbert de Haan[3]

[1] Basic Research in Computer Science,
funded by Danish National Research Foundation
saurabh@daimi.au.dk
http://www.brics.dk
[2] CWI, Amsterdam & Mathematical Institute, Leiden University, The Netherlands
http://www.cwi.nl/~cramer, http://www.math.leidenuniv.nl/~cramer
[3] CWI, Amsterdam, The Netherlands
http://www.cwi.nl/~haan

Abstract. The problem of perfectly secure message transmission concerns two synchronized non-faulty processors sender (S) and receiver (R) that are connected by a synchronous network of $n \geq 2t + 1$ noiseless 2-way communication channels. Their goal is to communicate privately and reliably, despite the presence of an adversary that may actively corrupt at most t of those channels. These properties should hold information theoretically and without error.

We propose an asymptotically optimal solution for this problem. The proposed protocol consists of two communication rounds, and a total of $O(\ell n)$ bits are exchanged in order to transmit a message of ℓ bits. Earlier, at CRYPTO 2004, an equally optimal solution has been claimed. However, we give a counter-example showing that their result is not perfectly reliable. The flaw seems to be fundamental and non-trivial to repair. Our approach is overall entirely different, yet it also makes essential use of their neat communication efficient technique for reliably transmitting conflict graphs.

What distinguishes our approach from previous ones is a technique that allows to identify *all actively corrupted channels*, initially trading it off against privacy. A perfectly secure and reliable secret key is then distilled by privacy amplification.

Keywords: Reliable and private transmission, information theoretic security, zero-error protocols, communication efficiency.

1 Introduction

The problem of perfectly secure message transmission (PSMT) was first introduced in [2]. In its more general description, it concerns two synchronized non-faulty processors sender (S) and receiver (R) that are connected by a synchronous network of n noiseless 2-way communication channels. The goal is for S to communicate a secret message M, drawn from a finite field K, to R. This

C. Dwork (Ed.): CRYPTO 2006, LNCS 4117, pp. 394–408, 2006.

should be done in such a way that for any set of at most t channels that is controlled and coordinated by an active adversary ($\mathcal{A}$), the adversary is neither able to disrupt the transmission of M to $\mathcal{R}$, nor is he able to obtain any new information about M. Moreover, these properties should hold information theoretically and without error. Of course $\mathcal{S}$ and $\mathcal{R}$ have no a priori knowledge of which particular channels are under the control of $\mathcal{A}$.

In general, such perfect communication is not possible for every selection of t and n. The good values for t and n depend on whether communication is 1-way (only from $\mathcal{S}$ to $\mathcal{R}$) or 2-way ($\mathcal{S}$ and $\mathcal{R}$ converse). It has been established that $n \geq 3t + 1$ is necessary and sufficient for 1-way communication and $n \geq 2t + 1$ is necessary and sufficient for 2-way communication [2].

The efficiency of any protocol solving the PSMT problem is typically measured in three parameters; the number of channels t that can be controlled by the adversary, the number of rounds[1] r of the protocol and the number of bits sent to reliably and privately communicate one bit of actual message from $\mathcal{S}$ to $\mathcal{R}$. The last parameter is also known as the communication complexity of the protocol.

Clearly a protocol which can tolerate the strongest adversary, uses the minimum number of rounds and which has a minimal communication complexity is preferred. For $r = 1$ and $n = 3t + 1$, a protocol with optimal communication complexity is known [2,5]. On the other hand, no protocol with optimal communication complexity for $r > 1$ and $n = 2t + 1$ is known. In this paper we give an asymptotically optimal protocol for $r = 2$ and $n = 2t + 1$. The authors of an earlier paper [5] claim to have found an optimal protocol for $r = 2$ and $n = 2t + 1$, but as we note in Section 3.3 the protocol of [5] is not perfectly reliable and therefore not a perfectly secure message transmission protocol.[2]

1.1 Organization of the Paper

In Section 2 we isolate some of the known basic techniques that are used throughout the paper. In Section 3, we give an overview of prior work in this area and in particular give a counter-example that breaks the perfect reliability of the protocol proposed in [5]. In Section 4, we introduce the new techniques which lead to our asymptotically optimal protocol using the communication efficient technique for reliably transmitting conflict graphs from [5]. The latter is then described in Section 5, where also the communication complexity is worked out.

2 Preliminaries

2.1 Shamir Sharing

Let K be a finite field with $|K| > n$. A selection $\{s_1, s_2, \ldots, s_n\}$ of shares according to Shamir's (t, n)-threshold secret sharing scheme [4] over the field K has the following properties,

---

[1] A round or a phase is a single sided communication from $\mathcal{S}$ to $\mathcal{R}$ or vice versa.

[2] This paper also presents a protocol which ends in a single round with high probability. However as the authors of [5] have noted in their presentation at CRYPTO 2004, this protocol is incorrect. Failure can be enforced with probability close to 1.

(1) any $t + 1$ shares fix all other shares in the selection and
(2) given any subset of $t + 1 - e$ shares, no information can be obtained about any disjoint subset of e shares.

Definition 1. *We say that a set of shares is* consistent *when there is a selection in the Shamir secret sharing scheme that leads to this set of shares. A set of shares which is not consistent is called* inconsistent.

Definition 2. *We say that a channel is* corrupted *if information sent on the channel is changed before it reaches its destination in any round of the protocol. Otherwise the channel is called* uncorrupted. *In other words, the information sent on an uncorrupted channel is unchanged in any round of the protocol.*

Definition 3. *We say that a value is* broadcast *if it is simultaneously sent over all communication channels. Since the value can then be correctly determined using majority voting on the other end, such values are always perfectly reliably transmitted. Since corruptions that occur during broadcasts are easy to detect, in the sequel we assume without loss of generality that broadcasts occur without any corruptions on the channels.*

2.2   Protocol Π_i

There is a two-round subprotocol Π_i that has implicitly been used both in the protocol of [3] and also in the incorrect protocol of [5]. During this protocol, $\mathcal{R}$ attempts to privately transmit a value $s \in K$ to $\mathcal{S}$ over the ith channel and obtains feedback about the result afterwards. It has the following functionality:

– If channel i is corrupted, it is disqualified by $\mathcal{R}$ after the second round.
– If channel i is not corrupted, $\mathcal{R}$ is certain that $\mathcal{S}$ correctly received s.
– If channel i is not under the control of $\mathcal{A}$, $\mathcal{A}$ obtains no information about the value s.

We now briefly describe the details of the protocol. Assume that $|K| > n$. First, $\mathcal{R}$ selects an arbitrary set $\{s_1, s_2, \ldots, s_n\}$ of shares in Shamir's (t, n)-threshold secret sharing scheme where the ith share is s, corresponding to a randomly chosen secret. $\mathcal{R}$ then sends all the shares over channel i and the share s_j over every other channel j. We denote the shares received on channel i by $\{s'_1, s'_2, \ldots, s'_n\}$ and the shares received on the other channels j by t'_j. This completes the first round. If the received set $\{s'_1, s'_2, \ldots, s'_n\}$ of shares is not consistent, $\mathcal{S}$ disqualifies channel i and broadcasts a notification to $\mathcal{R}$. Otherwise, for every pair of values such that $s'_j \neq t'_j$, $\mathcal{S}$ broadcasts[3] j and s'_j. Finally, $\mathcal{R}$ verifies for all received values whether $s'_j = s_j$ and disqualifies channel i if this is not the case or if $\mathcal{S}$ disqualified channel i. The properties now follow from a straightforward application of (1) and (2) for the value s.

---

[3] In [5] these shares are actually not transmitted using broadcast, but the functionality is the same.

The Symmetry of Conflicts in Π_i. We now describe an interesting property of the first round of the protocol Π_i that is used to break the perfect reliability of the protocol from [5] in Section 3.3. Let $\{s_1, s_2, \ldots, s_n\}$ be the set of shares in Shamir's (t, n)-threshold secret sharing scheme that has been transmitted by $\mathcal{R}$ over the ith channel in the first round of protocol Π_i. We denote the set of shares received on channel i by $\{u_1, u_2, \ldots, u_n\}$ and the shares received over the other channels j by v_j. We say that channel j *conflicts with* channel i if $u_j \neq v_j$.

Assume that channel i is under the control of the adversary and that the set $\{u_1, u_2, \ldots, u_n\}$ is a consistent set of shares that differs from the original set $\{s_1, s_2, \ldots, s_n\}$. If t of the original values s_j that $\mathcal{R}$ sent over the uncontrolled channels were kept intact and the shares sent over the remaining $t - 1$ controlled channels were changed to u_j, there will be only one pair u_j, v_j for which $u_j \neq v_j\ (= s_j)$, i.e., there will be only one uncontrolled channel j that conflicts with channel i. Note that this is also the minimal number of conflicts possible when the set of shares sent over channel i is altered, since the shares sent over the $t + 1$ uncontrolled channels completely fix the original set of shares and therefore cannot all be consistent with the altered set of shares.

Now consider the situation where channel i is not under the control of the adversary and where $\{u_1, u_2, \ldots, u_n\}$ is the original set of shares. Furthermore, assume that only the share on the controlled channel j is modified, resulting in an altered share s_j. Then $u_j \neq s_j$ and the received shares are exactly the same as in the previous situation. This implies that $\mathcal{S}$ cannot distinguish between the two situations.

2.3   Information Reconciliation

In this section we describe an information reconciliation technique that is based on an idea by Sayeed and Abu-Amara [3]. We assume that $\mathcal{S}$ has a vector consisting of $n = 2t + 1$ uniformly random values and that at least $t + 1$ of these values are known by $\mathcal{R}$. Furthermore, suppose that the adversary $\mathcal{A}$ knows at most t of these values and nothing else about the other values. The goal is for $\mathcal{S}$ to transmit enough information to allow $\mathcal{R}$ to recover the random vector, without allowing $\mathcal{A}$ to do the same.

Concretely, let K be a finite field with $|K| > n + t$ and assume that $\mathcal{S}$ has a uniformly random vector $\underline{v} = (v_1, v_2, \ldots, v_n) \in K^n$. We now consider the vector $\underline{v}$ as a set of values for the first n shares in Shamir's $(n - 1, n + t + 1)$-threshold secret sharing scheme. By property (1), these shares fix the remaining $t+1$ shares in the scheme. Let $\mathcal{S}$ broadcast t of these remaining shares to $\mathcal{R}$.

$\mathcal{R}$ now knows at least n different shares in the scheme, which completely fix all the shares in the scheme and in particular the first n shares. Since $\mathcal{A}$ can learn the value of at most $n - 1$ different shares, it follows from (2) that $\mathcal{A}$ cannot completely reconstruct the first n shares in the scheme. Therefore, the requirements are met.

2.4   Privacy Amplification

We now describe a well-known technique for perfect privacy amplification, that is very well-suited for use in PSMT protocols. Suppose S and R share b uniformly random elements in K and that it is promised that $a < b$ of these elements are completely unknown to the adversary. Then there is a simple technique that allows S and R to non-interactively generate a random elements about which A has no information.

Assume that $|K| > a + b$. Then we can view the b shared random elements as the first b shares in a Shamir's $(b-1, a+b)$-threshold secret sharing scheme. Again these shares fix all the other shares in the selection and by property (2), A has no information about the values of the a 'new' shares. These shares can therefore be taken as the outcome of the privacy amplification.

3   Earlier Protocols for PSMT

3.1   Overview

The main known protocols for perfectly secure message transmission all roughly have the same structure.

1. First S and R interact in such a manner that they obtain sufficiently correlated information, about which A has sufficient uncertainty.
2. Then S and R perform information reconciliation, i.e., they agree upon certain information that is not completely known to A.
3. Subsequently S and R non-interactively perform privacy amplification on this information and obtain a random string which is completely unknown to A.
4. This string now serves as a one-time pad, S encrypts the *actual* message with it and communicates the result to R.

For simplicity we leave out the encryption part of the procedure in the sequel and only focus on establishing the one-time pad. The first known two round protocol for PSMT is due to Dolev et al. [2]. The communication complexity of this protocol is $O(2^n)$, and therefore it is not efficient in terms of communication complexity. The article by Sayeed and Abu-Amara [3] presents the first efficient two round PSMT protocol, which achieves a communication complexity of $O(n^3)$. In [5], the authors claim to present a protocol with a communication complexity of $O(n)$, which can be shown to be optimal. However, the protocol in [5] is not perfectly reliable as is shown in Section 3.3. Furthermore, it seems to be nontrivial to repair the protocol, indicating that some new techniques may be required. As a consequence, the problem of finding two-round PSMT protocols with better communication complexity is still open. In the sequel we demonstrate some new techniques, and show that at least *asymptotically* a linear communication complexity can be obtained.

3.2   Protocol by Sayeed and Abu-Amara

The protocol due to Sayeed and Abu-Amara [3] is easily explained in terms of the techniques described in Section 2. Initially, Π_i is executed in parallel for every channel $i \in \{1, 2, \ldots, n\}$. This results in n random values $\{v_1, v_2, \ldots, v_n\}$ that are received by $\mathcal{S}$, of which at least $t + 1$ are equal to values that were originally transmitted by $\mathcal{R}$. Furthermore, $\mathcal{R}$ finds out in the second round which values were correctly received. Also, $\mathcal{A}$ knows at most t of the values received by $\mathcal{S}$, which correspond to the channels that are under his control. $\mathcal{S}$ and $\mathcal{R}$ can now apply the information reconciliation technique from Section 2.3 and the privacy amplification technique from Section 2.4 to obtain a completely secret element v, which can then be used as a one-time pad.

3.3   Protocol of CRYPTO 2004

We demonstrate in this section how the protocol Π_i is used to obtain a two round protocol in [5]. To make it clear that the protocol is not perfectly reliable, we show a strategy for $\mathcal{A}$ so that $\mathcal{R}$ cannot decide without error probability between two different random pads at the end of the protocol. Since the adversary can guess the information sent over the channels that are not under his control and the protocol should have zero error probability, we can assume without loss of generality that $\mathcal{A}$ also has full knowledge of the information that is transmitted over the uncontrolled channels.

To further simplify matters we only discuss the core functionality of the protocol, and only for the situation where $> 2t/3$ of the channels are corrupted in the first round. We can do this, since the protocol in [5] applies different techniques depending on whether $\leq 2t/3$ of the channels are corrupted in the first round or more channels are corrupted. Furthermore, we discuss the protocol using the notation and techniques of this paper.

The protocol from [5] starts by executing the protocol Π_i in parallel for every channel i. The protocol then continues in the second round as follows:

1. An arbitrary set $\{q_1, q_2, \ldots, q_n\}$ of shares in Shamir's $(\lfloor 4t/3 \rfloor, n)$-threshold secret sharing scheme is selected according to a randomly chosen secret.
2. Privacy amplification is applied to these shares, leading to a random vector $y = (y_1, \ldots, y_{\lfloor t/3 \rfloor})$ that is later used as a one-time pad to mask and send a message $M \in K^{\lfloor t/3 \rfloor}$.
3. For every channel i, the first round of Π_i is executed again with an arbitrary set $U_i = \{(u_1)_i, (u_2)_i, \ldots, (u_n)_i\}$ of shares for which $(u_i)_i = q_i$, where this time $\mathcal{S}$ is the party transmitting.
4. For every conflict $(s'_j)_i \neq (t'_j)_i$ that occurred during the first execution of Π_i, the value $(u_j)_i$ is broadcast by $\mathcal{S}$. [4]

We now set $t = 4$ and $n = 9$ and demonstrate the claimed strategy for the adversary so that $\mathcal{R}$ cannot decide without error probability at the end of the

---

[4] In [5] these shares are actually not transmitted using broadcast, but the functionality is the same.

protocol between two different one-time pads y_1 and y_2. It then follows that the protocol is not perfectly reliable.

Assume that the channels 1, 7, 8 and 9 are under control of the adversary. Let Q_1 and Q_2 be two sets of shares in Shamir's $(5, 9)$-threshold secret sharing scheme that lead to different pads y_1 and y_2 after privacy amplification, but for which the shares are the same in the 3th, 4th, 5th and 6th position. It is straightforward to show that such sets always exist regardless of the choice of Q_1 and in fact such sets can be found for any choice of y_1 and y_2. We demonstrate a strategy for the adversary such that $\mathcal{R}$ cannot determine whether Q_1 or Q_2 was the original set of shares, due to the fact that $\mathcal{R}$ cannot determine at the end of the protocol whether channel 1 has been corrupted during the protocol or channel 2. Let Q_1 be the set of shares that is selected by $\mathcal{S}$ in step 1. The strategy, which consists of a number of instances of the technique demonstrated in Section 2.2, then works as follows:

- In the first round, the adversary precisely creates conflicts in two directions between the channels 6 and 7, 6 and 8 and 6 and 9 using the technique described in Section 2.2.
- In the second round, the adversary precisely creates conflicts in two directions between the channels 1 and 2, 5 and 7, 5 and 8 and 5 and 9. This is done in such a way that the shares sent on the channels 1, 3, 4, 5 and 6 are consistent with the set Q_2. Since the shares on the uncontrolled channels are not altered, the shares sent on the channels 2, 3, 4, 5 and 6 remain consistent with the set Q_1. This can be done as follows.

  - For $j \in \{1, 2, 7, 8, 9\}$, choose a share $(u'_1)_j$ in such a way that the set of shares that is defined by the shares $(u'_1)_j$, $(u_3)_j$, $(u_4)_j$, $(u_5)_j$ and $(u_6)_j$ contains as the jth share the jth share of Q_2. This is going to ensure that the shares sent on the channels 1, 3, 4, 5 and 6 are consistent with the set Q_2.
  - Let U'_1 be the set of shares that will replace the set U_1, defined by the shares $(u'_1)_1$, $(u_3)_1$, $(u_4)_1$, $(u_5)_1$ and $(u_6)_1$. The corresponding shares sent over the controlled channels are replaced by the new shares in this set. This causes channel 2 to conflict with channel 1.
  - Replace the share $(u_1)_2$ sent over channel 1 by $(u'_1)_2$. This causes channel 1 to conflict with channel 2.
  - For $j \in \{7, 8, 9\}$, let U'_j be the set of shares that will replace the set U_j, defined by the shares $(u'_1)_j$, $(u_2)_j$, $(u_3)_j$, $(u_4)_j$ and $(u_6)_j$. The corresponding shares sent over the controlled channels are replaced by the new shares in these sets. This causes channel 5 to conflict with the channels 7, 8 and 9.
  - Replace the shares $(u_7)_5$, $(u_8)_5$, $(u_9)_5$ sent over the channels 7, 8 and 9 by arbitrary different values. This causes the channels 7, 8 and 9 to conflict with channel 5.

In the first round more than $2t/3$ of the channels are corrupted, since the sets of shares are changed on the channels 7, 8 and 9, so the example matches the assumed setting. These three channels will be disqualified at the end of the second round. However, as described in Section 2.2, $\mathcal{R}$ cannot determine after the second application of the first round of Π_i whether channel 1 has been corrupted or channel 2. By design, the received shares on the channels could have come from the set Q_2 if the channels 2, 7, 8 and 9 had been corrupted and the shares on the actual uncorrupted channels are consistent with Q_1. Therefore, $\mathcal{R}$ cannot distinguish between these cases. As a final remark, note that the broadcast shares $(u_j)_i$ from step 4 are superfluous in this scenario, as these shares were already correctly received in the second round.

4   Alternative Two Round PSMT

In this section we describe the new two round protocol. Using the technique from Section 5.1 and given a large enough message M, it introduces a communication cost of $O(\ell n)$ bits. To this end, we introduce the protocol $\hat{\Pi}_i$, that is based on some completely new techniques. As already noted in Section 3.1, we leave out the encryption part of the procedure and only focus on establishing the one-time pad.

4.1   Protocol $\hat{\Pi}_i$

The main contribution of this paper is the replacement of the protocol Π_i from Section 2.2 by a stronger two-round protocol $\hat{\Pi}_i$ with the following functionality:

- $\mathcal{S}$ and $\mathcal{R}$ both obtain a uniformly random vector $\mathcal{Z}_i = (z_1, z_2, \ldots, z_d) \in K^d$. However, they do not necessarily control which vector this is.
- If channel i is not under the control of $\mathcal{A}$, $\mathcal{A}$ obtains no information about the vector $\mathcal{Z}_i$.

Here $d \in \mathbb{Z}_{>0}$ is some constant value that can be selected before the start of the protocol.

We compare the protocol $\hat{\Pi}_i$ with the protocol Π_i. As shown in the protocol due to Sayeed and Abu-Amara (Section 3.2), after Π_i has been invoked once for every channel, up to t of the values that were actually received by $\mathcal{S}$ may be unknown to $\mathcal{R}$. Therefore, almost all privacy had to be sacrificed during information reconciliation. However, when the protocol $\hat{\Pi}_i$ finishes, information reconciliation has already occurred.

Furthermore, we will show that by choosing a message M of sufficiently large size, the relative amount of privacy that has to be given up during the information reconciliation in the new protocol can be made arbitrarily small, whereas in the protocol due to Sayeed and Abu-Amara this amount is always proportional to the message size. Additionally incorporating the technique from Section 5.1 allows us to obtain the desired communication complexity.

4.2   Sketch of the Techniques Used

From the discussion in Section 2.2, we see that controlled channels can change their share in order to match an altered set of shares sent over another channel in the protocol Π_i, while these channels cannot be detected due to insufficient feedback. We demonstrate a completely different technique that allows *all* corrupted channels to be detected, even if only a single share has been altered on such a channel. In Section 4.3 we then demonstrate that if $\mathcal{S}$ sends some appropriate additional information, which is similar to the conflict information that is broadcast during the protocol Π_i, $\mathcal{R}$ can reconstruct the altered set $\{s_1', s_2', \ldots, s_n'\}$ of shares that $\mathcal{S}$ received.

The key to detecting all corrupted channels is the fact that there are always $t+1$ channels that are not controlled by the adversary. The shares corresponding to these channels completely determine the original set of shares and therefore any combination of shares consisting of one altered share together with the shares corresponding to the $t + 1$ uncontrolled channels has to be inconsistent. Such an inconsistent combination allows $\mathcal{R}$ to detect a corrupted channel, since $\mathcal{S}$ can send the altered share to $\mathcal{R}$ and $\mathcal{R}$ can verify whether this share has been altered. However, since it is not known which channels are under the control of $\mathcal{A}$, we need to perform this procedure for all subsets consisting of $t + 1$ channels to make sure that the proper subset of $t + 1$ channels has been attempted.

4.3   Details of Protocol $\hat{\Pi}_i$

Assume that $|K| > n + t$ and let $\mathcal{N} = \{1, 2, \ldots, n\}$. The first round is just as in the protocol Π_i, except that instead of one random set of shares, m sets of shares are initially selected and transmitted.

Round 1. In the first round, $\mathcal{R}$ selects m arbitrary sets of shares according to Shamir's (t, n)-threshold secret sharing scheme for m randomly chosen secret values. We use notation (s_j) to denote the vector of all j^{th} shares and denote the set consisting of all shares by $\{(s_1), (s_2), \ldots, (s_n)\}$. Furthermore, the notation $(s_j)_l$ is used to denote the jth share in the lth set of shares. $\mathcal{R}$ sends the set $\{(s_1), (s_2), \ldots, (s_n)\}$ over the channel i and the vector (s_j) over every other channel j. This completes the first round.

Round 2. Assume that $\mathcal{S}$ receives the set $\{(s_1'), (s_2'), \ldots, (s_n')\}$ on channel i and vectors (t_j') on the other channels j, where we define $(t_i') := (s_i')$. Without loss of generality, we may again assume that all the m sets of received shares are consistent, since otherwise $\mathcal{S}$ can disqualify channel i and notify $\mathcal{R}$.

We now perform the verification as described in Section 4.2, where for every channel j and every combination of $t + 1$ channels $j_1, j_2, \ldots, j_{t+1}$ not including channel j we try to find an index l for which the combination of the shares $(t_{j_1}')_l, (t_{j_2}')_l, \ldots, (t_{j_{t+1}}')_l$ and $(t_j')_l$ is not consistent. For every such selection of channels, if such an index l exists $\mathcal{S}$ broadcasts one such share $(t_j')_l$ and its index l to $\mathcal{R}$, who can then verify whether this share is correct. In Lemma 1, we show that this approach allows $\mathcal{R}$ to identify all corrupted channels j.

It is clear that in the above procedure many shares are broadcast that were not initially known to $\mathcal{A}$. In order to remove all information that $\mathcal{A}$ may have gained due to this, all sets of shares corresponding to the broadcast shares are discarded. Of the remaining sets, the ith shares are then kept. Due to the similarities with the protocol Π_i, it now follows that these shares are known to $\mathcal{A}$ only if channel i is under his control, whereas $\mathcal{A}$ gains no information about these shares otherwise.

Concretely, for $j = 1, 2, \ldots, n$, let the set $\mathcal{Q}_j = \{V_{j1}, V_{j2}, \ldots, V_{jw}\}$ consist of all combinations of $t+1$ channels that do not include channel j, i.e., $\mathcal{Q}_j = \{V \subset \mathcal{N}\backslash\{j\} : |V| = t+1\}$. Then for all members of the set $\mathcal{Q}_j$ $(1 \leq j \leq n)$ the corresponding shares determine the full set of shares and every set $\mathcal{Q}_j$ has the same number of elements (namely $w = \binom{n-1}{t+1}$ elements).

The following protocol now specifies the verification step that is performed after the first round.

Protocol 1 (Classify channels)

1. Let $j \in \mathcal{N}$ and $k \in \{1, \ldots, w\}$. Then either the received shares in the $(t+2)$-sized set of shares $W_{jkl} := \{(t'_j)_l\} \cup \{(t'_z)_l : z \in V_{jk}\}$ are consistent for every $l \in \{1, \ldots, m\}$, or there is a smallest integer l_{jk} such that the shares in $W_{jkl_{jk}}$ are not consistent.

   Taking $l_{jk} = 0$ when the shares in W_{jkl} are consistent for every l, we let $L_j = (l_{j1}, \ldots, l_{jw})$ be the vector containing all such smallest indices, $I_j = \{l_{j1}, \ldots, l_{jw}\}\backslash\{0\}$ be the corresponding set of indices and define

   $$E_j := ((t'_j)_{l_{jm}})_{m \in \{1, \ldots, w\}:l_{jm} \neq 0}.$$

2. For $j = 1, \ldots, n$, $\mathcal{S}$ broadcasts L_j and E_j. Furthermore, $\mathcal{S}$ defines

   $$\mathcal{Z}_i := ((s'_i)_l)_{l \in \{1, \ldots, m\}\backslash(\bigcup_{j=1}^n I_j)}.$$

During Protocol 2 almost the same 'conflict information' is transmitted as in the second round of Π_i, with as its main difference that now whenever $(s'_j)_l \neq (t'_j)_l$ both conflicting values are returned instead of only the value $(s'_j)_l$. However, this information is used in a completely different way. Whereas in previous protocols this information was required to discover channels that have been corrupted, that functionality is now completely superfluous due to the previous protocol. Instead, the information transmitted during Protocol 2 is exactly sufficient to allow for complete information reconciliation by $\mathcal{R}$, in the sense that it helps $\mathcal{R}$ to completely determine what $\mathcal{S}$ received in the first round.

Protocol 2 (Gather reconciliation information)

1. Define $C_i := \{(l, j, (s'_j)_l, (t'_j)_l) : (s'_j)_l \neq (t'_j)_l, j \in \mathcal{N}\backslash\{i\}, l \in \{1, \ldots, m\}\}$.
2. $\mathcal{S}$ broadcasts C_i.

After $\mathcal{S}$ has finished transmitting, $\mathcal{R}$ can now execute the following protocol to reconstruct $\mathcal{Z}_i$. Note that nothing needs to be transmitted anymore at this point.

Protocol 3 (Reconcile)
According to Lemma 1, a channel j has been corrupted iff there exists an entry $(t'_j)_l$ in E_j such that $(t'_j)_l \neq (s_j)_l$. This allows $\mathcal{R}$ to completely split up the set of channels in a set U_c of channels that have been corrupted and a set U_u of uncorrupted channels.

1. *First assume that $i \in U_u$. Then*

$$\mathcal{Z}_i = ((s'_i)_l)_{l \in \{1,\dots,m\} \setminus (\bigcup_{j=1}^n I_j)} = ((s_i)_l)_{l \in \{1,\dots,m\} \setminus (\bigcup_{j=1}^n I_j)},$$

   which is a vector known to $\mathcal{R}$.
2. *Now assume that $i \in U_c$. Fix $l \in \{1,\dots,m\}$, let $H \subset U_u$ be a set of $t+1$ uncorrupted channels and take $j \in H$. Then either $(t'_j)_l = (s'_j)_l$ or $(l, j, (s'_j)_l, (t'_j)_l) \in \mathcal{C}_i$. In the first case, $(t'_j)_l$ (which is then equal to $(s_j)_l$, which is known to $\mathcal{R}$) gives a share in the lth set of shares $\{(s'_1)_l, \dots, (s'_n)_l\}$, and in the second case $(s'_j)_l$ gives a share in the lth set of shares.*

   Since the lth set of shares is consistent, the $t+1$ shares that can be obtained in this way fix all the shares in this set and in particular the share $(s'_i)_l$. Therefore, $\mathcal{R}$ can obtain $\mathcal{Z}_i$ in this case as well.

It follows that $\mathcal{R}$ and $\mathcal{S}$ both obtain the same vector $\mathcal{Z}_i$ at the end of the protocol. If $\mathcal{Z}_i$ is not empty, the values in the vector are either completely known to the adversary or he has no information about these values, depending only on whether the adversary controls channel i or not. This completes the description of protocol $\hat{\Pi}_i$.

Two Round PSMT. Assume that we execute the protocol $\hat{\Pi}_i$ in parallel for all n channels. Without loss of generality we may assume that all vectors $\mathcal{Z}_i$ have the same length, since otherwise $\mathcal{S}$ and $\mathcal{R}$ can just remove entries according to some predetermined method. Also, it should be clear that for any i that the protocol is executed for, the set $\bigcup_{j=1}^n I_j$ can contain at most $nw = n\binom{n-1}{t+1}$ indices. Therefore, by choosing m large enough, the length of the vectors $\mathcal{Z}_i$ can in fact be fixed to any nonzero value, so we can assume that the vectors $\mathcal{Z}_i$ have nonzero length.

At most t of the vectors $\mathcal{Z}_i$ are known to the adversary at the end of the n parallel executions of $\hat{\Pi}_i$, whereas he has no information about the remaining vectors. Therefore, applying a parallel version of the privacy amplification technique from Section 2.4 on these n vectors gives $t+1$ completely secret vectors. The values in these vectors can then be used in the second round by $\mathcal{S}$ to one-time-pad encrypt message elements from K.

Proofs. We now provide the results that support our claims. The following lemma shows that $\mathcal{R}$ can determine for a channel j whether it has been corrupted in the first round of $\hat{\Pi}_i$ by comparing the received values in the set E_j with the original values that were transmitted on channel j in the first round.

Lemma 1. *Fix any* $j \in \mathcal{N}$. *Then* $(t'_j)_l \neq (s_j)_l$ *for some* $l \in \{1, \ldots, m\}$ *iff there is a value* $l' \in \{1, \ldots, m\}$ *such that* $(t'_j)_{l'} \neq (s_j)_{l'}$ *and* $(t'_j)_{l'}$ *is an entry of* E_j.

Proof. ($\Leftarrow$) Trivial.

($\Rightarrow$) If $(t'_j)_l \neq (s_j)_l$ for some $l \in \{1, \ldots, m\}$, then there is a set V_{jk} of $t+1$ uncontrolled channels that does not contain channel j. In particular, the shares in the set W_{jkl} are inconsistent, since the shares corresponding to these uncontrolled channels lead to share $(s_j)_l$ for j, which is different from $(t'_j)_l$.

Now let l' be the smallest value for which the shares in the set $W_{jkl'}$ are inconsistent. Since the channels in V_{jk} are not under control of the adversary, the shares in the set $W_{jkl'} \setminus \{(t'_j)_{l'}\}$ lead to share $(s_j)_{l'}$ for channel j, where $(t'_j)_{l'} \neq (s_j)_{l'}$ since otherwise the shares would be consistent. Since $(t'_j)_{l'}$ is an entry of E_j by definition, the lemma follows.

At first sight, it may seem that $\mathcal{A}$ can deduce information from the minimum collision indices l_{jk} that are broadcast during Protocol 2. However, the lemma below shows that this is not the case.

Lemma 2. *The values* l_{jk} *in Protocol 1 completely depend on the actions of the adversary in the first round. In particular, these values are known to the adversary even before they are broadcast in the second round.*

Proof. By Lagrange's theorem, a unique linear relation $\sum_{i=1}^{t+1} \lambda_i s_i = s_{t+2}$ necessarily holds for a consistent set $\{s_1, \ldots, s_{t+2}\}$ of shares in Shamir's (t, n)-threshold secret sharing scheme, where the λ_i's are publicly known constants that only depend on the a priori fixed evaluation points on the used secret sharing polynomial. Lets assume that the first $e <= t$ shares are replaced by values s'_i. It is straightforward to verify that the new set $\{s'_1, \ldots, s'_e, s_{e+1}, \ldots, s_{t+2}\}$ is consistent iff $\sum_{i=1}^{e} \lambda_i s'_i + \sum_{i=e+1}^{t+1} \lambda_i s_i = s_{t+2}$. This is the case iff $\sum_{i=1}^{e} \lambda_i (s_i - s'_i) = 0$. However, the values $s_i - s'_i$ are chosen by and known to the adversary. Therefore, the adversary already knows beforehand whether any particular received subset of $t+2$ shares is consistent or not and can in particular predict all the minimum indices l_{jk}.

Proposition 1. *If* $(l, j, (s'_j)_l, (t'_j)_l) \in \mathcal{C}_i$, *then at least one of the channels* i *and* j *has been corrupted. Furthermore, both* $(s'_j)_l$ *and* $(t'_j)_l$ *were already known to the adversary at the end of the first round.*

Assume that $\mathcal{A}$ does not control channel i. Since the first round of $\hat{\Pi}_i$ is a parallel version of the first round of Π_i, it is clear that $\mathcal{A}$ obtains no information about the ith shares in the sets of shares in the first round. The following lemma states that the adversary does not learn anything new about the values of the entries of $\mathcal{Z}_i$ (provided that $\mathcal{Z}_i$ has any entries at all) in the second round of $\hat{\Pi}_i$. This shows that the proposed protocol is perfectly private.

Lemma 3. *If channel* i *is not under control of the adversary and* $\mathcal{Z}_i$ *contains a nonzero number of entries, then the adversary obtains no new information about the values of the entries of* $\mathcal{Z}_i$ *in the second round.*

Proof. According to Lemma 2, the indices that are transmitted during Protocol 1 are selected by (and therefore known to) the adversary before the execution of the second round. Furthermore, the shares that are transmitted during Protocol 1 are completely uncorrelated with the vector $\mathcal{Z}_i$, since the corresponding sharings are discarded before the vector $\mathcal{Z}_i$ is constructed. Finally, Proposition 1 shows that only information that is already known to the adversary is transmitted during Protocol 2. Therefore, the adversary does not learn anything new about the values of the entries of $\mathcal{Z}_i$ in the second round.

5   Towards Linear Overhead

5.1   Improved Reliable Transmission

The most expensive transmission in the described protocols so far, both for the previous protocols as well as for the new one, is the transmission of the collisions in the second round, i.e., the transmission of the values s'_j and/or t'_j for which $s'_j \neq t'_j$. In this section we demonstrate a technique used in [5], which can reduce the communication cost by switching from broadcast to a combination of broadcast and error correcting. We describe this technique for a single value of $l \in \{1, \ldots, m\}$ in the new protocol, where we group all shares received on the channels that relate to a set of shares that has index l. It is straightforward to improve the technique we describe by taking all shares into account at once, but this has no impact on the obtained communication complexity.

We start by regrouping the vectors in the sets $\mathcal{C}_1, \ldots, \mathcal{C}_n$ as follows: For $l = 1, \ldots, m$, define $C_l := \{(i, j, (s'_j)_l, (t'_j)_l) : (l, j, (s'_j)_l, (t'_j)_l) \in C_i, i \in \{1, 2, \ldots, n\}\}$. Then the set C_l contains all conflicts arising from the distribution of the n l^{th} sets of shares. We then move to a more efficient method to reliably transmit the sets C_l. Define the undirected graph $\mathcal{G}_l = (\mathcal{P}, \mathcal{E}_l)$ by

$$(i, j) \in \mathcal{E}_l \Leftrightarrow (i, j, (s'_j)_l, (t'_j)_l) \in C_l \vee (j, i, (s'_i)_l, (t'_i)_l) \in C_l$$

and let M_l be the size of a maximum matching on $\mathcal{G}_l$. From Proposition 1 it follows that every edge in this graph involves at least one channel that has been corrupted. Therefore there are at least M_l channels that have been corrupted in total.

These channels can all be detected by $\mathcal{R}$ using the information from Protocol 1. This implies that $\mathcal{R}$ will be able to discard the shares received on at least M_l channels during the reliable transmission. Therefore, we can use an error correcting code with codewords of length n that can handle M_l erasures and $t - M_l$ errors for the transmission. Using codewords of length n, $n - M_l - 2(t - M_l) = M_l + 1$ shares can now be transmitted.

Since every edge in the graph involves at least one channel that is in the maximum matching, there can be at most $2M_l n$ edges in the graph. In particular, this implies that every set C_l can contain at most $4M_l n$ vectors. Using an error correcting code, every set C_l can thus be transmitted by sending $O(n^2)$ shares over the channels.

5.2   Complexity Analysis

Choose a field K such that its elements can be represented using bit strings of length $O(\log(n))$ and assume that m is such that $m > nw \log_n(m)$. The length $\max\{0, m - nw\}$ of the vectors $\mathcal{Z}_i$ can be chosen to be of size $\geq cm$ for any constant c in the interval $(0, 1)$, by enlarging m as necessary. As the privacy amplification section shows, this implies that we can obtain a secret key of size $(t+1)cm = O(nm)$, i.e., of $O(mn \log(n))$ bits. In order to have a PSMT-protocol with a linear communication complexity, the total number of shares transmitted in each round should be $O(mn^2)$ or, stated equivalently, the total number of bits transmitted in each round should be $O(mn^2 \log(n))$. Let us now analyze the communication complexity of the parts of the new protocol.

The First Round. For every i, $\mathcal{R}$ sends mn elements over channel i and m elements over every other channel j. This sums up to a total of $O(mn)$ shares that are sent over each channel and therefore to $O(mn^2)$ shares in total that are transmitted in the first round.

Protocol: Classify Channels. For every $i \in \{1, 2 \ldots, n\}$, at most $O(w)$ indices and field elements are broadcast at the end of Protocol 1 for every j, so that in total $O(n^2 w(\log(m) + \log(n)))$ bits have to be broadcast. This gives $O(n^3 w(\log(m) + \log(n)))$ bits that are transmitted during Protocol 1. Our assumption implies that $m > n$, so that this can be rewritten to $O(n^3 w \log(m))$. Furthermore, by assumption $m > nw \log_n(m)$, so that $m \log(n) > nw \log(m)$, i.e., $n^3 w \log(m) = O(mn^2 \log(n))$.

Protocol: Gather Reconciliation Information. Assume that we regroup the sets C_i during the Protocol 2 as described in Section 5.1 and obtained the sets C_l. Then, using some appropriate padding between the vectors C_l, all information can be transmitted by communicating only $O(mn^2)$ shares.

This completes the analysis.

6   Conclusion

In this paper we have given an asymptotically optimal two round PSMT protocol for $n = 2t + 1$. It is not difficult to extend the protocol for $n > 2t + 1$ as well. In particular, there exists a protocol that asymptotically achieves a constant communication overhead when $t = cn$ for any $0 < c < 1/2$.

The main difference when compared to earlier two round PSMT protocols is the ability to completely isolate corrupted channels. However, this comes at the expense of a high computational cost to both sender and receiver. We do not know whether a similar protocol can exist where sender and receiver are restricted to polynomial time computations (in terms of the number of channels) only.

Acknowledgements

The authors would like to thank Serge Fehr, Dennis Hofheinz, Eike Kiltz, Jesper Buus Nielsen and Carles Padró for useful discussions and comments.

References

1. Y. Desmedt and Y. Wang. Perfectly Secure Message Transmission Revisited. In *EUROCRYPT'02*, volume 2332 of *LNCS*, pages 502–517. Springer-Verlag, 2002.
2. D. Dolev, C. Dwork, O. Waarts, and M. Yung. Perfectly Secure Message Transmission. *JACM*, 40(1):17–47, January 1993.
3. H. Sayeed and H. Abu-Amara. Efficient Perfectly Secure Message Transmission in Synchronous Networks. *Information and Computation*, 126(1):53–61, 1996.
4. A. Shamir. How to Share a Secret. *Communications of the ACM*, 22:612–613, 1979.
5. S. Srinathan, A. Narayanan, and C. Pandu Rangan. Optimal Perfectly Secure Message Transmission. In *CRYPTO'04*, pages 545–561, Santa Barbara, California, 2004.

Random Selection with an Adversarial Majority[*]

Ronen Gradwohl[1,**], Salil Vadhan[2,***], and David Zuckerman[3,†]

[1] Department of Computer Science and Applied Math, Weizmann Institute of Science
ronen.gradwohl@weizmann.ac.il
[2] Division of Engineering & Applied Sciences, Harvard University
salil@eecs.harvard.edu
[3] Department of Computer Science, University of Texas at Austin
diz@cs.utexas.edu

Abstract. We consider the problem of random selection, where p players follow a protocol to jointly select a random element of a universe of size n. However, some of the players may be adversarial and collude to force the output to lie in a small subset of the universe. We describe essentially the first protocols that solve this problem in the presence of a dishonest majority in the full-information model (where the adversary is computationally unbounded and all communication is via non-simultaneous broadcast). Our protocols are nearly optimal in several parameters, including the round complexity (as a function of n), the randomness complexity, the communication complexity, and the tradeoffs between the fraction of honest players, the probability that the output lies in a small subset of the universe, and the density of this subset.

1 Introduction

Suppose p players wish to jointly make a random choice from a universe of size n. They follow some protocol, and if all parties play honestly, the output is indeed a uniformly random one. However, some of the players may form a coalition and deviate arbitrarily from the protocol, in an attempt to force some output. The problem of random selection is that of designing a protocol in which the influence of coalitions of dishonest players is somehow limited.

Random selection is a very useful building block for distributed algorithms and cryptographic protocols, because it allows one to first design protocols assuming a public source of randomness, which is often an easier task, and then replace public randomness with the output of a random selection protocol. Of

---

[*] The full version of this paper is available on *ECCC* [22].

[**] Research supported by US-Israel Binational Science Foundation Grant 2002246.

[***] Supported by US-Israel BSF grant 2002246, NSF grants CNS-0430336 and CCR-0133096, and ONR Grant N00014-04-1-0478.

[†] Most of this work was done while visiting Harvard University, and was supported in part by a Radcliffe Institute for Advanced Study Fellowship, a John Simon Guggenheim Memorial Foundation Fellowship, a David and Lucile Packard Fellowship for Science and Engineering, and NSF Grant CCR-0310960.

C. Dwork (Ed.): CRYPTO 2006, LNCS 4117, pp. 409–426, 2006.

course, for this to work, there must be a good match between the guarantees of the random selection protocol and the requirements of the application at hand. Nevertheless, this general paradigm has been applied successfully numerous times in the past in various settings, e.g., [42,20,17,32,12,13,30,21,25,3,23]. This motivates a systematic study of random selection in its own right, like the one we undertake in this paper.

The Setting. The problem of random selection has been widely studied in a variety of settings, which differ in the following respects:

Adversary's Computational Power. In some work on random selection, such as Blum's 'coin-tossing over the telephone' [4], the adversary is assumed to be computationally bounded (e.g., probabilistic polynomial time). Generally, in this setting one utilizes one-way functions and other cryptographic primitives to limit the adversary's ability to cheat, and thus the resulting protocols rely on complexity assumptions. In this paper, we study the *information-theoretic* setting, where the adversary is computationally unbounded (so complexity assumptions are useless).

Communication Model and the Adversary's Information. There is a choice between having point-to-point communication channels, a broadcast channel, or both. In the case of point-to-point communication, one can either assume private channels, as in [6,10], or allow the adversary full access to all communication, as in the *full-information model* of Ben-Or and Linial [7]. We allow a broadcast channel and work in the full-information model (so there is no benefit from point-to-point channels). We do not assume simultaneous communication, and thus consider a 'rushing' adversary, which can send its messages in a given round after receiving those of the honest players.

Number of Players. There has been work specifically studying two-party protocols where one of the players is adversarial; examples in the full-information model include [17,39]. Other works study p-player protocols for large p, such as the large body of work on collective coin-flipping (random selection where the universe is of size $n = 2$) and leader election [7,38,1,11,31,8,43,37,15]. In this paper, we focus on the latter setting of p-*player protocols*, but some of our results are significant even for $p = 2$.

To summarize, here we study general multiparty protocols for random selection in the full-information model (with a broadcast channel). This is the first work in this setting to focus on the case that a majority of the players may be dishonest.[1] It may be surprising that protocols exist for this case, as the other two other well-studied problems in this setting, leader election and collective coin-flipping, are provably impossible to solve with an adversarial majority [38].

The Goal: Construct p-player protocols for selecting an element of $[n]$ such that even if a β fraction of players are cheating, the probability that the output lands in any small subset of $[n]$, of density μ, is at most ε.

---

[1] We note that dishonest majorities have been studied extensively in the settings of computationally bounded parties and private channels, both for Byzantine agreement and secure computation, e.g., [20,26,19].

Particular applications of random selection protocols often have special additional requirements, such as "simulatability." However, all of the existing work on random selection with information-theoretic security, such as [7,38,17,28,1,12,13,21,37,15,14,39], seem to include at least some variant of our requirement above. Thus it is of interest to understand this requirement on its own, in particular the tradeoffs between the parameters p, n, β, μ, and ε, as well as the efficiency of protocols meeting the requirement.

As these five parameters vary, we have a very general class of problems, which includes many previously studied problems as special cases (See Section 2.2.). Some natural settings of parameters are n being exponentially large in the security parameter (e.g., choosing a random k-bit string), p being constant or polynomial, β being a constant in $(0,1)$ (we are particularly interested in $\beta \geq 1/2$), and μ, ε either being constants in $(0,1)$ or tending to zero.

Regarding protocol efficiency, we focus primarily on information-theoretic measures, such as the communication and round complexities, but we also provide some computationally efficient versions of our protocols.

Our Results. In this paper, we give several protocols for random selection that tolerate an arbitrarily large fraction of cheating players $\beta < 1$. The protocols are nearly optimal in many of the parameters, for example:

- One of our protocols achieves an error probability of $\varepsilon = \tilde{O}(\mu^{1-\beta})$, when the number of players is constant and the density μ of bad outcomes is arbitrary. This comes close to the lower bound of $\varepsilon \geq \mu^{1-\beta}$ proven by Goldreich, Goldwasser, and Linial [17]. For a nonconstant number of players, we can come polynomially close to the lower bound, achieving $\varepsilon = \mu^{\Omega(1-\beta)}$, provided that the fraction β of cheating players is bounded away from 1.
- One of our protocols can handle any density μ of bad outcomes that is smaller than the fraction $\alpha = 1 - \beta$ of honest players while achieving an error probability ε that is bounded away from 1. More generally, we can handle any constants α, μ such that $\lfloor 1/\alpha \rfloor \leq \lceil 1/\mu \rceil - 1$, which is a tight tradeoff by a lower bound of Feige [15].
- In our protocols, the total number of coins tossed by the honest parties is $\log n + o(\log n)$ (when the other parameters are constant), which almost equals the lower bound of $\log n - O(1)$. As the only bits communicated in our protocols are the random coin tosses, the communication complexity is also nearly optimal.
- As a function of n, the round complexity of our protocols is $\log^* n + O(1)$ (when the other parameters are constant). This is within a factor of essentially 2 of the $(1/2 - o(1)) \log^* n$ lower bound proven by Sanghvi and Vadhan [39], which applies whenever $\beta \geq 1/2$, and $\mu > 0$ and $\varepsilon < 1$ are constants.

Techniques. Our protocols build upon recent work on round-efficient leader election [37,15] and round-efficient two-party random selection [39]. Specifically, the leader election protocols of Russell and Zuckerman [37] and Feige [15] work by iterating a one-round protocol that reduces the task of electing a leader from

p players to that of electing from polylog(p) players. Similarly, the two-party random selection protocol of Sanghvi and Vadhan [39] utilizes a one-round protocol that reduces selecting from a universe of size n to selecting from one of size polylog(n). We combine these approaches, iteratively reducing both the number of players and the universe size in parallel. To do this, we construct new one-round universe reduction protocols that work for many parties (instead of just two, as in [39]). We obtain these by establishing a connection between randomness extractors [29] (or, equivalently, randomness-efficient samplers) and universe reduction protocols. Optimizing parameters of the underlying extractors then translates to optimizing parameters of the universe reduction protocols, resulting in the near-optimal bounds we achieve in our final protocols. Our main results, as outlined above, refer to protocols that use optimal extractors, as proven to exist via the probabilistic method, and thus are not explicit or computationally efficient. In the full version of this work [22], we also give computationally efficient versions of our protocols, using some of the best known explicit constructions of extractors. Any additional deficiencies in these protocols are due to limitations in the state-of-the-art in constructing extractors, which we view as orthogonal to the issues we study here. Indeed, if the loss turns out to be too much for some application, then that would provide motivation for further research on explicit constructions of extractors.

Organization. Section 2 includes definitions, a more detailed description of previous work and how it relates to this paper, and our results. Section 3 contains the one-round selection protocols that are the final ingredient in our protocols, and in Section 4 we give protocols that reduce the number of players and the size of the universe. In Section 5 we informally describe how the different pieces fit together to form our final protocols, and defer details and formal proofs from this section to the full version [22]. Finally, in Section 6, we state known and new lower bounds on the various parameters of random selection.

2 Definitions and Results

2.1 Random Selection Protocols

We now define random selection protocols, the model, and the complexity measures in which we are interested.

A (p, n)-*selection protocol* is a p-player protocol for selecting an element of $[n]$. In each round of the protocol, the players broadcast messages that they may base on the messages sent by all players in previous rounds, as well as their own internal coin tosses. The players may not legally base their outputs in round i on the outputs of other players in round i. However, since we can not guarantee simultaneity within a round, we allow the dishonest players to base their outputs on the outputs of other players from the same round (but not from later rounds). This is known as rushing. At the end, a predetermined function of all sent messages is computed, outputting an element of $[n]$.

Given this definition, we have the following notion of security.

Definition 1. *A (p,n)-selection protocol is called $(\beta, \mu, \varepsilon)$-resilient if when at most a β fraction of players are cheating and S is any subset of $[n]$ of density at most μ, the probability that the output lands in S is at most ε. We refer to S as a* bad set.

We will be interested in the asymptotic behavior of protocols, so when we discuss (p,n)-selection protocols, we are implicitly referring to a family of protocols, one for each value of p and n (or some infinite set of pairs (p,n)). We are then interested in optimizing a variety of complexity measures:

The *computation time* of a (p,n)-selection protocol is the maximum total time spent by all (honest) players (to compute their messages and the final function) in an execution of the protocol. We call a protocol *explicit* if its computation time is poly$(\log n, p)$. The *round complexity* is the total number of rounds of the protocol. The *randomness complexity* of a protocol is the (maximum possible) total number of random bits used by the honest players. [2] (Typically this maximum is achieved when all players are honest.) The *communication complexity* of a protocol is the total number of bits communicated by the honest players.[3]

All our protocols are public-coin, in the sense that the honest players flip their random coins and broadcast the results. Thus, the communication complexity is equal to the randomness complexity. By convention, we assume that if a player sends a message that deviates from the protocol in some syntactically obvious way (e.g. the player outputs more bits than requested), then its message is replaced with some canonical string of the correct form (e.g. the all-zeroes string).

2.2   Previous Work

We now discuss the relationship of the above definitions, specifically of $(\beta, \mu, \varepsilon)$-resilient (p,n)-selection protocols, to existing notions and results in the literature.

Two-Party Random Selection. This is the special case where $p = 2$ and $\beta = 1/2$, and attention in previous work has focused on the tradeoff between μ and ε as well as the round complexity. Specifically,

- Goldreich, Goldwasser, and Linial [17] constructed, for every $n = 2^i$, an explicit $(2,n)$-selection protocol that is $(1/2, \mu, O(\sqrt{\mu}))$-resilient for every $\mu > 0$. The protocol takes $2 \log n$ rounds. They also prove that the bound of $\varepsilon = O(\sqrt{\mu})$ is tight (as a special case of a more general result mentioned later).

---

[2] Actually, it will be convenient to allow the players to pick elements uniformly at random from $\{1, \ldots, m\}$ where m is determined during the protocol and may not be a power of 2, and in such a case we view this as costing $\log_2 m$ random bits.

[3] As with randomness complexity, it will be convenient to allow players to send elements of $\{1, \ldots, m\}$, in which case we charge $\log_2 m$ bits of communication.

- Sanghvi and Vadhan [39] constructed, for every constant $\delta > 0$ and every n, an explicit $(2, n)$-selection protocol that is $(1/2, \mu, O(\sqrt{\mu + \delta}))$-resilient for every $\mu > 0$. Their protocol takes $\log^* n + O(1)$ rounds. They also prove that $(\log^* n - \log^* \log^* n - O(1))/2$ rounds are necessary for any $(2, n)$-selection protocol that is $(1/2, \mu, \varepsilon)$-resilient for constants $\mu > 0$ and $\varepsilon < 1$.

Collective Coin-Flipping [7]. This is the special case when $n = 2$ and $\mu = 1/2$. Attention in the literature has focused on constructing efficient protocols that are $(\beta, 1/2, \varepsilon)$-resilient where β and ε are constants (independent of p), β is as large as possible, and $\varepsilon < 1$. Such a protocol exists for every constant $\beta < 1/2$ [8] and can be made explicit [43]. Conversely, it is impossible to achieve $\beta = 1/2$ and $\varepsilon < 1$ [38]. Efficient constructions of such protocols have been based on leader election (described below).

Leader Election

Definition 2. *A p-player leader election protocol is a (p, p)-selection protocol. It is (β, ε)-resilient if when at most a β fraction of players are cheating, the probability that the output is the index of a cheating player is at most ε.*

- Every $(\beta, \beta, \varepsilon)$-resilient (p, p)-selection protocol is a (β, ε)-resilient p-player leader election protocol. The converse does not hold because the the former considers *each* subset $S \subset [p]$ of density at most β as a potential bad set of outcomes, but the latter only considers the subset consisting of the cheating players.
- Nevertheless, a p-player leader election protocol can be used to construct a (p, n)-selection protocol for any n by having the elected leader choose a uniform, random element of $[n]$ as the output. If the election protocol is (β, ε)-resilient, then the resulting selection protocol will be $(\beta, \mu, \varepsilon + (1-\varepsilon)\cdot\mu)$-resilient for every $\mu \geq 0$.
- By the impossibility result for collective coin-flipping mentioned above [38] and the previous bullet, it is impossible to have an election protocol that is (β, ε)-resilient for $\beta = 1/2$ and $\varepsilon < 1$.
- A long line of work [1,11,31,43,37,15] on optimizing the resilience and round complexity for leader election has culminated in the following result of Russell and Zuckerman [37].[4] For every constant $\beta < 1/2$, there exists an $\varepsilon < 1$ such that for all p, there is an explicit (β, ε)-resilient p-player leader election protocol of round complexity $\log^* p + O(1)$. Consequently, for all constants $\beta < 1/2$ and $\mu > 0$, there is a constant $\varepsilon < 1$ such that for all p and n, there is an explicit $(\beta, \mu, \varepsilon)$-resilient (p, n)-selection protocol.

Multi-Party Random Selection. This is the general problem that encompasses the previous special cases.

- Goldreich, Goldwasser, and Linial [17] constructed, for every $n = 2^i$ and every p, an explicit (p, n)-selection protocol that is $(\beta, \mu, \mu^{1-O(\beta)})$-resilient

---

[4] A very recent paper [2] aims to optimize ε as a function of β, obtaining efficient leader election protocols with $\varepsilon = O(\beta)$.

for all sufficiently small β and every $\mu > 0$. The protocol runs in polylog(n) rounds. They also showed that any $(\beta, \mu, \varepsilon)$-resilient protocol must satisfy $\varepsilon \geq \mu^{1-\beta}$.

- Russell and Zuckerman [37] constructed, for every n and p such that $n \geq p^c$ for a constant c, an explicit one-round (p, n)-selection protocol that is $(\beta, \mu, \mu \cdot n/n^{\Omega(1-\beta)})$-resilient for every $\mu > 0$ and $1 > \beta > 0$.

Notice that all but the last of the above results require that the fraction β of bad players satisfies $\beta \leq 1/2$.[5] For collective coin-flipping and leader election, this is supported by impossibility results showing that $\beta \geq 1/2$ is impossible. For 2-party random selection, it does not make sense to discuss $\beta > 1/2$. The only result which applies to $\beta \geq 1/2$ is the last one (of [37]). However, the resilience $\mu \cdot n/n^{\Omega(1-\beta)}$ is quite weak and only interesting when the density μ of the bad set is close to $1/n$.[6] Our work is the first to show strong results for the case $\beta > 1/2$.

2.3   Our Results

In this section, we present our main results. All of our protocols utilize certain kinds of randomness-efficient samplers (equivalently, randomness extractors). Here we present the versions of our results obtained by using optimal samplers, proven to exist via the probabilistic method. We also have explicit (i.e., computationally efficient) versions of our protocols, obtained by using best known explicit constructions of samplers. One such protocol is given by Theorem 7, and the rest are deferred to the full version of this work [22].

The first main result of this paper is the following:

Theorem 3. *For all constants $k \in \mathbb{N}$, $k > 0$ and $\delta > 0$, there exists a constant $\varepsilon < 1$ and a (p, n)-selection protocol with the following properties:*

(i)   The protocol has $\max(\log^* p, \log^* n) + O(1)$ rounds.
(ii) The protocol is $(1 - \alpha, \mu, \varepsilon)$-resilient for $\alpha = 1/(k+1) + \delta$ and $\mu = 1/k - \delta$.
(iii) The randomness complexity of the protocol is $(\log n)/\alpha + o(\log n) + O(p \log p)$.

The tradeoff between α and μ in the above theorem is optimal up to the slackness parameter δ. This is shown in Corollary 27, as a consequence of a lower bound of Feige [15]. Furthermore, the round and randomness complexity are nearly optimal as functions of n, as shown by Corollary 25 and Theorem 28.

Setting $p = 2$ and $\alpha = 1/2$, we obtain the following two-party protocol:

Corollary 4. *For every constant $\delta > 0$, there exists a constant $\varepsilon < 1$ and a $(2, n)$-selection protocol with the following properties:*

---

[5] The hidden constant in the protocol of [17] is larger than 2.

[6] The significance of the [37] protocol is that it is one round and only requires n polynomial in p; in fact, there is a trivial protocol with somewhat better parameters when n is exponential in p (Lemma 10).

(i) *The protocol has* $\log^* n + O(1)$ *rounds.*
(ii) *The protocol is* $(1/2, 1/2 - \delta, \varepsilon)$-*resilient.*
(iii) *The randomness complexity of the protocol is* $2\log n + o(\log n)$.

This protocol improves upon the two-party protocol of [39][7] in two ways: first, the randomness complexity is a nearly optimal $2\log n + o(\log n)$, and not polylog(n). Second, their protocol is $(1/2, \nu, \varepsilon')$-resilient for some small constant ν, and not for the nearly optimal $\frac{1}{2} - \delta$. In other words, their resilience is not optimal in the density of the bad set. On the other hand, the error probability ε' of their protocol is smaller than that of ours. However, a special case of our second theorem below gives the parameters of [39] with the added benefit of optimal randomness complexity.

Our next two results optimize the error probability ε as a function of the density μ of the bad set and fraction β of cheating players. The first achieves a near-optimal tradeoff when the number of players is small (e.g., constant).

Theorem 5. *For all* $\mu, \alpha > 0$ *and* $p, n \in \mathbb{N}$, *there exists a* (p, n)-*selection protocol with the following properties:*

(i) *The protocol has* $\log^* n - \log^*(1/\mu) + O(1)$ *rounds.*
(ii) *The protocol is* $(1 - \alpha, \mu, \varepsilon)$-*resilient for*

$$\varepsilon = \mu^\alpha \cdot O\left(\frac{1}{\alpha} \cdot \log \frac{1}{\mu} + (1 - \alpha)p\right)^{1-\alpha} \cdot 2^{(1-\alpha)p}.$$

(iii) *The randomness complexity is* $[\log n + o(\log n) + O(p + \log(1/\mu))]/\alpha + \log(1/(1 - \alpha)) + O(p \log p)$.

Note that when the number p of players and the fraction α of honest players are constants, the bound becomes $\varepsilon = \tilde{O}(\mu^\alpha)$, which nearly matches the lower bound of $\varepsilon \geq \mu^\alpha$ proven in [17] (see Theorem 26). However, the bound on ε grows exponentially with p. This is removed in the following theorem, albeit at the price of achieving a slightly worse error probability of $\mu^{\Omega(\alpha)}$ (for constrained values of α).

Theorem 6. *There is a universal constant* c *such that for all* $p, n \in \mathbb{N}$, $\mu, \alpha > 0$ *satisfying* $\alpha \geq \sqrt{c \log \log(1/\mu)/\log(1/\mu)}$, *there exists a* (p, n)-*selection protocol with the following properties:*

(i) *The protocol has* $\max\{\log^* p, \log^* n\} - \log^*(1/\mu) + O(1)$ *rounds.*
(ii) *The protocol is* $(1 - \alpha, \mu, \mu^{\Omega(\alpha)})$-*resilient.*
(iii) *The randomness complexity is* $[\log n + o(\log n) + O(p)]/\alpha + O(p \log p) + \text{poly}(1/\alpha, \log(1/\mu))$.

---

[7] Note that in [39], the claimed round complexity is $2\log^* n + O(1)$, but this difference from our claim is only a difference of convention: in their model, only one player may communicate in each round, whereas we use the convention of multi-party protocols, in which all players may communicate simultaneously in one round.

One disadvantage of the above two theorems (as compared to, say, the honest-majority protocols of [17]) is that the protocols require an a priori upper-bound μ on the density of the bad set. However, we also benefit from this, in that the round complexity improves as μ tends to zero. In particular, if $\mu \leq 1/\log^{(k)} n$ for some constant k, where $\log^{(k)}$ denotes k iterated logarithms, then the round complexity is *constant*.

An explicit version of the protocol of Theorem 3 is the following theorem:

Theorem 7. *For all constants $k \in \mathbb{N}$, $k > 0, \gamma > 0$ and $\delta > 0$, there exists a constant $\varepsilon < 1$ and an explicit (p, n)-selection protocol with the following properties:*

(i) *The protocol has $\max(\log^* p, \log^* n) + O(1)$ rounds.*
(ii) *The protocol is $(1 - \alpha, \mu, \varepsilon)$-resilient for $\alpha = 1/(k+1) + \delta$ and $\mu = 1/k - \delta$.*
(iii) *The randomness complexity of the protocol is $(\log n)^{1+\gamma} + O(p \log p)$.*

Apart from its explicitness, note that the randomness complexity of the above theorem is now $(\log n)^{1+\gamma}$ for an arbitrarily small constant γ, rather than $(1 + o(1))(\log n)/\alpha$. Intuitively, this occurs because the explicit sampler we use (based on an extractor of [34]) only has randomness complexity polynomially close to optimal. It is possible to remedy this and obtain a randomness complexity of $(1 + o(1)) \log n$ by using other samplers (e.g. based on the extractors of [33]) for the first few rounds of universe reduction, but this creates some messy constraints on the other parameters, so we omit a formal statement.

We also have explicit versions of Theorem 5 and Theorem 6.

Theorem 8. *For every constant $\gamma > 0$, and every $p, n \in \mathbb{N}$, $\mu, \alpha > 0$, there exists a (p, n)-selection protocol with the following properties:*

(i) *The protocol has $\log^* n - \log^*(1/\mu) + O(1)$ rounds.*
(ii) *The protocol is $(1 - \alpha, \mu, \varepsilon)$-resilient for*

$$\varepsilon = \mu^\alpha \cdot O\left(\frac{1}{\alpha} \cdot \log \frac{1}{\mu} + (1 - \alpha)p\right)^{1-\alpha} \cdot 2^{(1-\alpha)p}.$$

(iii) *The randomness complexity is $[(\log n)^{1+\gamma} + O(p + \log(1/\mu))]/\alpha + O(\log(1/(1-\alpha))) + O(p \log p)$.*
(iv) *The protocol is explicit given appropriate samplers of size*

$$s = \text{poly}(2^p, 1/\mu, \log^{(3)} n)^{1/(\alpha)}.$$

which can be obtained probabilistically in time $O(s)$ and deterministically in time $2^{O(s)}$.

Theorem 9. *There is a universal constant c such that for every constant $\gamma > 0$ and every $p, n \in \mathbb{N}$, $\mu, \alpha > 0$ satisfying $\alpha \geq \sqrt{c \log \log(1/\mu)/\log(1/\mu)}$, there exists a (p, n)-selection protocol with the following properties:*

(i) The protocol has $\max\{\log^* p, \log^* n\} - \log^*(1/\mu) + O(1)$ rounds.
(ii) The protocol is $(1 - \alpha, \mu, \mu^{\Omega(\alpha)})$-resilient.
(iii) The randomness complexity is $[(\log n)^{1+\gamma} + O(p)]/\alpha + O(p \log p) + \text{poly}(1/\alpha, \log(1/\mu))$.
(iv) The protocol is explicit given appropriate samplers of size

$$s = \text{poly}(1/\mu, 1/\alpha, \log^{(3)} n)^{1/\alpha}.$$

which can be obtained probabilistically in time $O(s)$ and deterministically in time $2^{O(s)}$.

Note that the protocols are explicit whenever $s = O(\log \log n)$ (in particular, when μ and α are constants). Due to space constraints, details of these explicit constructions are deferred to the full version [22].

3   One-Round Protocols

We start with some simple one-round protocols that will play a role in our later constructions.

Lemma 10. *For every $p, \ell \in \mathbb{N}$ and $n = \ell^p$, there is an explicit (p, n)-selection protocol that is $(\beta, \mu, n^\beta \cdot \mu)$-resilient for every $\beta, \mu > 0$.*

Proof sketch. Each player outputs a random element of $[\ell]$, and we take the concatenation of the player's outputs.   □

The above protocol has two main disadvantages. First, the size of the universe $n = \ell^p$ must be at least exponential in the number of players. (We note that Russell and Zuckerman [37] showed how to reduce this requirement to be only polynomial, at the price of a somewhat worse resilience. We will avoid this difficulty in a different manner, by first reducing the number of players.) Second, in terms of resilience, a bad set of density μ gets multiplied by a factor that grows polynomially with the universe size (namely, n^β). However, when the number of players is small (e.g. a fixed constant) and the universe is small (e.g. $n = O(1/\mu)$), it can achieve a nearly optimal bound on ε as a function of β and μ (cf. Theorem 26).

Lemma 11 ([15], Cor. 5). *For every $p, n \in \mathbb{N}$ and $\alpha, \mu \in [0, 1]$ such that $\lfloor 1/\alpha \rfloor \leq \lceil 1/\mu \rceil - 1$, there exists an $\varepsilon < 1$ and a (p, n)-selection protocol that is $(1 - \alpha, \mu, \varepsilon)$-resilient. Specifically, one can take $\varepsilon = 1 - \exp(-\Omega(\alpha \cdot (1 - \mu) \cdot np))$.*

Proof sketch. Every player chooses a random subset of $[n]$ of density at least $1 - \mu$, and the output is the first element of $[n]$ that is contained in every set S that was picked by at least an α fraction of players. Such an element exists because there exist at most $\lfloor 1/\alpha \rfloor \leq \lceil 1/\mu \rceil - 1$ such sets S, but any $\lceil 1/\mu \rceil - 1$ sets of density at least $1 - \mu$ must have a common intersection.   □

The advantage of the above protocol is that it achieves an optimal tradeoff between α and μ (cf. Theorem 27). The main disadvantage is that ε can depend on p and n (this time with exponentially bad dependence), and that it is not sufficiently explicit — even the communication is of length $\Theta(n)$ (rather than polylog(n)).

4   Universe and Player Reduction

The simple 1-round protocols of the previous section behave well when the number of players p and universe size n are small. Thus, as in previous work, our main efforts will be aimed at giving protocols to reduce p and n while approximately preserving the fraction β of bad players and the density μ of the bad set. Roughly speaking, in one round we will reduce p and n to polylog(p) and polylog(n), respectively. For this, we consider protocols that select a subset of the universe (or a subset of the players).

4.1   Definitions

Definition 12. A $[(p,n) \mapsto n']$-universe reduction protocol is a p-player protocol whose output is a sequence $(s_1, \ldots, s_{n'})$ of elements of $[n]$. Such a protocol is $[(\beta,\mu) \overset{\gamma}{\mapsto} \mu']$-resilient if when at most a β fraction of players are cheating and S is any subset of $[n]$ of density at most μ, the probability that at most a μ' fraction of the output sequence is in S is at least γ. It is explicit if the players' strategies are computable in time poly($\log n, p$), and given the protocol transcript and $i \in [n']$, the i'th element of the output sequence is computable in time poly($\log n, p$).

Notice that a (p,n)-selection protocol is equivalent to a $[(p,n) \mapsto n']$-universe reduction protocol with $n' = 1$, and the former is $(\beta, \mu, \varepsilon)$-resilient if and only if the latter is $[(\beta,\mu) \overset{1-\varepsilon}{\mapsto} 0]$-resilient.

Definition 13. A $[p \mapsto p']$-player reduction protocol is a $[(p,p) \mapsto p']$-universe reduction protocol. It is $[\beta \overset{\gamma}{\mapsto} \beta']$-resilient if when at most a β fraction of players are cheating, the probability that at most a β' fraction of the output sequence are indices of cheating players is at least γ.

Definition 14. A $[(p,n) \mapsto (p',n')]$-universe+player reduction protocol is a p-player protocol whose output is a sequence $(s_1, \ldots, s_{n'})$ of elements of $[n]$ and a sequence $(t_1, \ldots, t_{p'})$ of elements of $[p]$. Such a protocol is $[(\beta,\mu) \overset{\gamma}{\mapsto} (\beta',\mu')]$-resilient if when at most a β fraction of players are cheating and S is any subset of $[n]$ of density at most μ, the probability that at most a β' fraction of the first output sequence are cheating players and at most a μ' fraction of the second output sequence is in S is at least γ. It is explicit if the players' strategies are computable in time poly($\log n, p$), and given the protocol transcript and $i \in [n']$ (resp., $j \in [p']$), the i'th (resp., j'th) element of the first (resp., second) output sequence is computable in time poly($\log n, p$).

4.2   One-Round Reduction Protocols

In the following one-round protocols, think of $\theta = 1/\text{polylog}(n)$ and $\varepsilon = 1/\text{poly}(n)$. We use both a one-round player-reduction protocol and a one-round universe-reduction protocol.

Theorem 15 ([37,15]). *For every $p \in \mathbb{N}$, $\varepsilon > 0$, and $\theta > 0$, there is an explicit, one-round $[p \mapsto p']$-player reduction protocol with*

$$p' = O\left(\frac{1-\beta}{\theta^2} \cdot \log \frac{p}{\varepsilon}\right),$$

that is $[\beta \overset{1-\varepsilon}{\mapsto} \beta + \theta]$-resilient for all $\beta > 0$. Moreover, the randomness complexity is $p \cdot \log(p/p')$.

The starting point for our universe reduction protocol is the simple protocol of Lemma 10. That protocol has the property that a β fraction of cheating players cannot make any outcome in $[n]$ appear with probability more than $1/n^{1-\beta}$. (This can be seen by taking $\mu = 1/n$.) Thus the output can be viewed as a source with "min-entropy rate" at least $1 - \beta$.[8] To get a higher quality output, it is natural to try applying a *randomness extractor*, a function that extracts almost-uniform bits from sources with sufficient min-entropy. However, randomness extractors require an additional random *seed* to do such extraction. Thus we will enumerate over all seeds of the extractor, and the resulting sequence will be the output of our universe reduction protocol. Fortunately, there exist extractors where the number of seeds is only polylogarithmic in n, the domain of the source.

Actually, it is more convenient for us to work with an object that is essentially equivalent to extractors, namely (averaging) samplers (cf., [9,43,16]). Samplers are functions that output sample points of a given universe, with the property that the fraction of samples from any particular subset of the universe is roughly equal to the density of that subset. In the following definition, U_r denotes an element of $[r]$ chosen uniformly at random.

Definition 16. *A function $\text{Samp} : [r] \to [n]^t$ is a (θ, ε) sampler if for every set $S \subseteq [n]$,*

$$\Pr_{(i_1,\ldots,i_t) \leftarrow \text{Samp}(U_r)} \left[\frac{\#\{j : i_j \in S\}}{t} > \frac{|S|}{n} + \theta\right] \leq \varepsilon.$$

We say that Samp is explicit if for every $x \in [r]$ and every $j \in [t]$, the j'th component of $\text{Samp}(x)$ can be computed in time $\text{poly}(\log r, \log n)$.[9]

---

[8] The *min-entropy* of a random variable X is defined as $H_\infty(X) = \max_x \Pr[X = x]$. If X takes values in a universe U, then its *min-entropy rate* is defined to be $H_\infty(X)/\log|U|$.

[9] Often the definition of samplers also requires that the fraction of samples that lie in S is also not much larger than the density of S. However, this follows from our definition (paying a factor of 2 in ε) by considering $\overline{S}$. Moreover, we will only need the one-sided version, and below, in Definition 20 we will consider a variant which is not symmetric with respect to approximation from above and below.

Zuckerman [43] showed that samplers (as defined above) are essentially equivalent to randomness extractors.

Given $p, \ell \in \mathbb{N}$ and a sampler Samp : $[r] \to [n]^{n'}$ with $r = \ell^p$, we obtain a $[(p, n) \mapsto n']$-universe reduction protocol Π_{Samp} as follows: the players use the protocol of Lemma 10 to select an element $x \in [\ell^p]$, and then output the sequence Samp(x).

Lemma 17. *If* Samp *is a* (θ, ε) *averaging sampler, then for every* $\mu, \beta > 0$, Π_{Samp} *is* $[(\beta, \mu) \overset{\gamma}{\mapsto} \mu + \theta]$-*resilient for* $\gamma = 1 - r^\beta \cdot \varepsilon$. *Moreover, the randomness complexity is* $\log r$.

Proof. Call $x \in [r]$ "bad" if $\#\{j : i_j \in S\}/t > |S|/n + \theta$ when $(i_1, \ldots, i_t) \leftarrow$ Samp(x), and note that the number of bad x's is at most $\varepsilon \cdot r$ by the properties of the sampler. The players use the protocol of Lemma 10 to select an element x from a universe of size r, where the fraction of bad elements is ε. This is a (p, r)-selection protocol that is $(\beta, \varepsilon, r^\beta \cdot \varepsilon)$-resilient, and so the probability of selecting a good x is at least $\gamma = 1 - r^\beta \cdot \varepsilon$. If a good x is selected, then the fraction of bad elements is increased by at most θ. □

Notice that for this to be useful, we need the error probability ε of the sampler to be smaller than $r^{-\beta}$, and in fact we will be interested in β that are arbitrarily close to 1. Fortunately, we have samplers that achieve this. (This is equivalent to the fact that we have extractors that work for min-entropy rate arbitrarily close to 0.)

Lemma 18 (nonconstructive samplers [36,43]). *There is a universal constant c such that for every $n \in \mathbb{N}, \theta > 0, \varepsilon > 0$ and $r \geq c \cdot n/(\varepsilon\theta^2)$, there exists a (θ, ε) sampler* Samp : $[r] \to [n]^t$ *with* $t = O(\log(1/\varepsilon)/\theta^2)$.

It is important to note that the lower bound on r depends linearly on $1/\varepsilon$; this means that we can make the error $\varepsilon \leq r^{-\beta}$ for any $\beta < 1$. Combining the above two lemmas, we have:

Theorem 19 (nonconstructive 1-round universe reduction). *For every $p, n \in \mathbb{N}, \beta, \varepsilon, \theta > 0$, there exists a 1-round $[(p, n) \mapsto n']$-universe reduction protocol that is $[(\beta, \mu) \overset{1-\varepsilon}{\mapsto} \mu + \theta]$-resilient for every $\mu > 0$, with*

$$n' = O\left(\frac{\log(1/\varepsilon) + (\beta/(1 - \beta)) \cdot \log n + \beta \cdot p}{\theta^2}\right).$$

Moreover, the randomness complexity is $p + (\log n + \log(1/\varepsilon) + 2\log(1/\theta))/(1 - \beta) + O(1)$.

Proof. First note that without loss of generality, $\theta \geq 1/n$, otherwise we can use the trivial protocol that outputs the entire universe. So now choose $r \in [(cn/(\varepsilon\theta^2))^{1/(1-\beta)}, 2^p \cdot (cn/(\varepsilon\theta^2))^{1/(1-\beta)}]$ such that r is the p'th power of some natural number, and apply Lemma 17 with $\varepsilon' = \varepsilon/r^\beta$. □

Thus, for $p = \text{polylog}(n)$, $\theta = 1/\text{polylog}(n)$, $\varepsilon = 1/\text{poly}(n)$, and $\beta = 1 - 1/\text{polylog}(n)$, we can reduce the universe size from n to $\text{polylog}(n)$. If the number of players is constant, then we can iterate this $\log^* n$ times to reduce the universe size to a constant. However, if the number of players p is large, then the above will not reduce the universe size below βp. Therefore, we will combine this with the player reduction of Theorem 15.

Notice that if we want to preserve the density μ of the bad set up to a constant factor, then we can set $\theta = 1/\mu$ and the above protocol will reduce the universe size to n' depending polynomially on $1/\mu$. However, to obtain some of our results (namely, Theorems 6, 5, and their explicit versions), it will be beneficial to reduce to a universe size that depends almost-linearly on $1/\mu$. To achieve this, we use a variant of our sampler-based protocol that is tailored to a particular value of μ.

Definition 20. *A function* $\text{Samp} : [r] \to [n]^t$ *is a* $(\mu, \theta, \varepsilon)$ *density-tailored sampler if for every set* $S \subseteq [n]$ *with* $|S| \leq \mu \cdot n$,

$$\Pr_{(i_1,\ldots,i_t) \leftarrow \text{Samp}(U_r)} \left[\frac{\#\{j : i_j \in S\}}{t} > \mu + \theta \right] \leq 1 - \varepsilon.$$

We say that Samp *is* explicit *if for every* $x \in [r]$ *and every* $j \in [t]$, *the* j'*th component of* $\text{Samp}(x)$ *can be computed in time* $\text{poly}(\log r, \log n)$.

Density-tailored samplers are essentially equivalent to 'slice extractors', defined in [36]. As in Lemma 17, these density-tailored samplers also induce selection protocols.

Lemma 21. *If* Samp *is a* $(\mu, \theta, \varepsilon)$ *density-tailored sampler, then for every* $\beta > 0$, Π_{Samp} *is* $[(\beta, \mu) \xmapsto{\gamma} \mu + \theta]$*-resilient for* $\gamma = 1 - r^\beta \cdot \varepsilon$. *Moreover the randomness complexity is* $\log r$.

The reason we are interested in these density-tailored samplers is that they exist with slightly better parameters for certain values of μ.

Lemma 22 (nonconstructive density-tailored samplers [40]). *There is a universal constant* c *such that for every* $n \in \mathbb{N}, \mu > 0, \theta > 0, \varepsilon > 0, t \geq c \cdot \log(1/\varepsilon) \cdot \max\{1/\mu, \mu/\theta^2\}$, *and* $r \geq c \cdot n \cdot (\mu \log(1/\mu))/(\varepsilon \log(1/\varepsilon))$, *there exists a* $(\mu, \theta, \varepsilon)$ *density-tailored sampler* $\text{Samp} : [r] \to [n]^t$.

Note that the number of samples t in these samplers depends linearly on $1/\mu$ (if $\theta = \Omega(\mu)$) and not polynomially as in Lemma 18. Combining the above lemma with Lemma 17 we get a nonconstructive 1-round universe reduction protocol with different parameters from those of Theorem 19:

Theorem 23 (nonconstructive, density-tailored 1-round universe reduction). *There is a universal constant* c *such that for every* $p, n \in \mathbb{N}, \beta, \mu, \varepsilon, \theta > 0$ *and every*

$$n' \geq c \cdot \max\left\{ \frac{\mu}{\theta^2}, \frac{1}{\theta} \right\} \cdot \left(\log \frac{1}{\varepsilon} + \frac{\beta}{1-\beta} \cdot \left(\log n + \log \frac{1}{\beta} \right) + \beta \cdot p \right),$$

there exists a 1-round $[(p, n) \mapsto n']$-universe reduction protocol that is $[(\beta, \mu) \stackrel{1-\varepsilon}{\mapsto} \mu + \theta]$-resilient. Moreover, the randomness complexity is $p + [\log n + \log(1/\varepsilon) - \log \log(1/\varepsilon) - \log(1/\mu) + \log \log(1/\mu) + \log(1/\beta)]/(1 - \beta) + O(1)$.

Proof. We can choose $r \in [r', 2^p \cdot r']$ such that r is the p'th power of some natural number and

$$r' = \left(\frac{cn \cdot \mu \log \frac{1}{\mu}}{\beta \cdot \varepsilon \log \frac{1}{\varepsilon}} \right)^{\frac{1}{1-\beta}}.$$

We then apply Lemmas 21 and 22 with $\varepsilon' = \varepsilon/r^\beta$.   □

5   Putting It Together

In this section we give a sketch of how our main results are obtained by combining the various building blocks described above. Due to space constraints, details and formal proofs are deferred to the full version of this work [22].

First we construct a sub-protocol that reduces the size of the universe and the number of players to $n', p' = \text{poly}(\log(1/\varepsilon), 1/\theta)$ (Theorem 24). This is accomplished by iterating the 1-round player reduction protocol of Theorem 15 to reduce the number of players, and iterating the 1-round universe reduction protocol of Theorem 19 to reduce the size of the universe. To save on the round complexity, the player reduction and universe reduction can be done in parallel. This yields the following theorem, which is the main component of our protocols:

Theorem 24 (many-round universe+player reduction). *For every $n, p \in \mathbb{N}$ and every $\beta, \theta, \varepsilon > 0$, there exists a $[(p, n) \mapsto (p', n')]$-universe+player reduction protocol that is $[(\beta, \mu) \stackrel{1-\varepsilon}{\mapsto} (\beta + \theta, \mu + \theta)]$-resilient for every $\mu > 0$, with*

$$n' = \text{poly}(\log(1/\varepsilon), 1/\theta)$$
$$p' = \text{poly}(\log(1/\varepsilon), 1/\theta).$$

Moreover, the number of rounds is $t = \max\{\log^* n, \log^* p\} - \log^* n' + O(1)$ and the randomness complexity is $[\log n + o(\log n) + O(p + t \cdot \log(1/\varepsilon) + t \cdot \log(1/\theta))]/(1 - \beta) + O(p \log p)$.

For Theorem 3, the size of the universe and the number of players are both reduced to constant ($\text{poly}(1/\delta)$, where δ is the constant in the statement of the theorem). We can then run the protocol of Lemma 11, and since all parameters are constant, the error probability of the final output is also bounded by a constant less than 1.

For Theorem 5, we wish to use the protocol of Lemma 10 as the final protocol, and thus obtain a near optimal probability of error. For this to work well, however, we require that n' be exponential in p, and that $n' = \tilde{O}(1/\mu)$. This requires a more delicate reduction than the one above.

First, we reduce the size of the universe to poly$(1/\mu)$ in the same way as above (Theorem 24, ignoring the player reduction). Next we further reduce the size of the universe to $n' = \tilde{O}(1/\mu)$, using the density-tailored 1-round universe reduction protocol of Theorem 23. This use of a density-tailored reduction is critical, as it allows the reduction of n' to $1/\mu$. Once this is accomplished, we can use the protocol of Lemma 10.

The protocol of Theorem 6 is the same as that of Theorem 5, except that it is combined with an appropriate player-reduction protocol (which can be run in parallel).

6   Lower Bounds

The are several known and new lower bounds for the different parameters of random selection, and here we state the relevant theorems. A lower bound on the round complexity is a corollary of a theorem of [39]:

Corollary 25. *For any* (p, n)-*selection protocol that is* $(\beta, \mu, \varepsilon)$-*resilient with* $\beta \geq 1/2$ *and for constants* $\mu > 0$ *and* $\varepsilon < 1$, *the round complexity is at least* $(\log^* n - \log^* \log^* n - O(1))/2$.

The lower bound on the error probability ε is given by a theorem of [17]:

Theorem 26 ([17]). *For any* (p, n)-*selection protocol that is* $(1-\alpha, \mu, \varepsilon)$-*resilient,* $\varepsilon \geq \mu^\alpha$.

The optimal tradeoff between μ and α is given by the following corollary of [15, Thm. 4]:

Corollary 27. *For any* (p, n)-*selection protocol that is* $(1-\alpha, \mu, \varepsilon)$-*resilient with* $\varepsilon < 1$, $\lfloor 1/\alpha \rfloor \leq \lceil 1/\mu \rceil - 1$.

Finally, in the full version of this work [22], we prove a lower bound on the randomness and communication complexities.

Theorem 28. *For any* (p, n)-*selection protocol that is* $(1 - \alpha', \mu, \varepsilon)$-*resilient for* $\varepsilon < 1$, *the randomness and communication complexities are at least*

$$\max\left\{(1 - \alpha')p, \frac{1 - \varepsilon}{\alpha} \log \frac{\mu n}{\varepsilon}\right\},$$

where $\alpha = \lfloor \alpha' p \rfloor / p$.

Acknowledgements

We thank Omer Reingold for many useful discussions, and the anonymous referees for helpful comments.

References

1. N. Alon and M. Naor. Coin-flipping games immune against linear-sized coalitions. *SIAM J. Computing*, 22(2):403–417, April 1993.
2. S. Antonakopoulos. Fast leader-election protocols with bounded cheaters' edge. In *Proc. 38th STOC*, 187–196, 2006.
3. B. Barak. Constant-round coin-tossing with a man in the middle or realizing the shared random string model . In *43rd FOCS*, 2002.
4. M. Blum. Coin flipping by telephone. In *IEEE Spring COMPCOM*, 1982.
5. D. Beaver and S. Goldwasser. Multiparty computation with faulty majority. In *CRYPTO 89*, Springer LNCS 435, 589–590, 1989.
6. M. Ben-Or, S. Goldwasser, and A. Wigderson. Completeness theorems for non-cryptographic fault-tolerant distributed computation. In *20th STOC*, 1–10, 1988.
7. M. Ben-Or and N. Linial. Collective coin fliping. In *Advances in Computing Research*, volume 5: Randomness and Computation, JAI Press, 1989, 91–115.
8. R. Boppana and B. Narayanan. Perfect-information leader election with optimal resilience. *SIAM J. Computing*. 29(4): 1304-1320 (2000).
9. M. Bellare and J. Rompel. Randomness-efficient oblivious sampling. In *35th FOCS*, 1994.
10. D. Chaum, C. Crépeau, and I. Damgård. Multiparty unconditionally secure protocols. In *20th STOC*, 11–19, 1988.
11. J. Cooper and N. Linial. Fast perfect-information leader-election protocols with linear immunity. *Combinatorica*, 15:319-332, 1995.
12. I. Damgård. Interactive hashing can simplify zero-knowledge protocol design without computational assumptions (extended abstract). In *CRYPTO*, 1993.
13. I. Damgård, Oded Goldreich, and Ave Wigderson. Hashing functions can simplify zero-knowledge protocol design (too). TR RS-94-39. BRICS, 1994.
14. Y. Z. Ding, D. Harnik, A. Rosen, and R. Shaltiel. Constant-round oblivious transfer in the bounded storage model. In *1st TCC*, 2004.
15. U. Feige. Noncryptographic selection protocols. In *40th FOCS*, 142–152, 1999.
16. O. Goldreich. A sample of samplers - a computational perspective on sampling (survey). Report 97-020, *Electronic Colloquium on Computational Complexity*, 1997.
17. O. Goldreich, S. Goldwasser, N. Linial. Fault-tolerant computation in the full information model. *SIAM J. Computing* 27(2), 1998.
18. S. Goldwasser and L. Levin. Fair computation of general functions in presence of immoral majority. In *CRYPTO 90*, Springer LNCS 537, 77–93, 1990.
19. S. Goldwasser and Y. Lindell. Secure computation without agreement. In *DISC 2002*: 17-32
20. O. Goldreich, S. Micali, and A. Wigderson. How to play ANY mental game. In *19th STOC*, 218–229, 1987.
21. O. Goldreich, A. Sahai, and S. Vadhan. Honest-verifier statistical zero-knowledge equals general statistical zero-knowledge. In *30th STOC*, 1998.
22. R. Gradwohl, S. Vadhan, and D. Zuckerman. Random Selection with an Adversarial Majority. Report TR06-26, *Electronic Colloquium on Computational Complexity*, Feb. 2006.
23. J. Katz and R. Ostrovsky. Round-optimal secure two-party computation. In *CRYPTO*, 2004.
24. J. Katz, R. Ostrovsky, and A. Smith: Round efficiency of multi-party computation with a dishonest majority. In *EUROCRYPT 2003*: 578-595.

25. Y. Lindell. Parallel coin-tossing and constant-round secure two-party computation. In *CRYPTO*, 2001.

26. M. Pease, R. Shostak, and L. Lamport. Reaching agreement in the presence of faults. *J. ACM*, 27(2):228–234, 1980.

27. C. Lu, O. Reingold, S. Vadhan, and A. Wigderson. Extractors: Optimal up to constant factors. In *35th STOC*, 2003.

28. M. Naor, R. Ostrovsky, R. Venkatesan, and M. Yung. Perfect zero-knowledge arguments for NP can be based on general complexity assumptions. *J. Cryptology* 11, 1998.

29. N. Nisan and D. Zuckerman. Randomness is linear in space. *J. Computer and System Sci.*, 52(1): 43–52, 1996.

30. T. Okamoto. On relationships between statistical zero-knowledge proofs. *J. Computer and System Sci.*, 60(1): 47–108, 2000.

31. R. Ostrovsky, S. Rajagopalan, and U. Vazirani. Simple and efficient leader election in the full information model. In *Proc. 26th STOC*, 234–242, 1994.

32. R. Ostrovsky, R. Venkatesan. and M. Yung. Interactive hashing simplifies zero-knowledge protocol design. In *EUROCRYPT*, 1993.

33. R. Raz, O. Reingold, and S. Vadhan. Extracting all the randomness and reducing the error in Trevisan's extractors. *J. Computer and System Sc.*, 65(1):97–128, 2002.

34. R. Raz, O. Reingold, and S. Vadhan. Error Reduction for Extractors. In *40th FOCS*, 1999.

35. Alexander Russell, Michael Saks and David Zuckerman. Lower bounds for leader election and collective coin- flipping in the perfect information model. *SIAM J. Computing*, 31:1645–1662, 2002.

36. J. Radhakrishnan and A. Ta-Shma. Bounds for dispersers, extractors, and depth-two superconcentrators. *SIAM J. Discrete Math.*, 13(1):2–24, 2000.

37. A. Russell and D. Zuckerman. Perfect-information leader election in $\log^* n + O(1)$ rounds. *J. Computer and System Sci.*, 63:612–626, 2001.

38. M. Saks. A robust noncryptographic protocol for collective coin flipping. *SIAM J. Discrete Math.*, 2(2):240-244, 1989.

39. S. Sanghvi and S. Vadhan. The round complexity of two-party random selection. In *37th STOC*, 2005.

40. S. Vadhan. Constructing locally computable extractors and cryptosystems in the bounded-storage model. *J. Cryptology* 17(1), 43–77, 2004.

41. A. Wigderson and D. Zuckerman, Expanders that beat the eigenvalue bound: explicit construction and applications. *Combinatorica*, 19 (1999) 125–138. 17

42. A. Yao. How to generate and exchange secrets. In *Proc. 27th FOCS*, 1986.

43. D. Zuckerman. Randomness-optimal oblivious sampling. *Random Structures and Algorithms*, 11(4): 345-367 (1997).

Oblivious Transfer and Linear Functions

Ivan B. Damgård[1], Serge Fehr[2,*], Louis Salvail[1], and Christian Schaffner[1,**]

[1] BRICS, FICS, Aarhus University, Denmark
{ivan, salvail, chris}@brics.dk
[2] CWI Amsterdam, The Netherlands
fehr@cwi.nl

Abstract. We study unconditionally secure 1-out-of-2 Oblivious Transfer (*1-2 OT*). We first point out that a standard security requirement for *1-2 OT* of bits, namely that the receiver only learns one of the bits sent, holds if and only if the receiver has no information on the XOR of the two bits. We then generalize this to *1-2 OT* of strings and show that the security can be characterized in terms of binary linear functions. More precisely, we show that the receiver learns only one of the two strings sent if and only if he has no information on the result of applying any binary linear function (which non-trivially depends on both inputs) to the two strings.

We then argue that this result not only gives new insight into the nature of *1-2 OT*, but it in particular provides a very powerful tool for analyzing *1-2 OT* protocols. We demonstrate this by showing that with our characterization at hand, the reducibility of *1-2 OT* (of strings) to a wide range of weaker primitives follows by a very simple argument. This is in sharp contrast to previous literature, where reductions of *1-2 OT* to weaker flavors have rather complicated and sometimes even incorrect proofs.

1 Introduction

1-2 Oblivious-Transfer, *1-2 OT* for short, is a two-party primitive which allows a sender to send two bits (or, more generally, strings) B_0 and B_1 to a receiver, who is allowed to learn one of the two according his choice C. Informally, it is required that the receiver only learns B_C but not B_{1-C} (*obliviousness*), while at the same time the sender does not learn C (*privacy*). *1-2 OT* was introduced in [28] (under the name of "multiplexing") in the context of quantum cryptography, and, inspired by [25] where a different flavor was introduced, later re-discovered in [19].

1-2 OT turned out to be very powerful in that it was shown to be sufficient for secure general two-party computation [22]. On the other hand, it is quite easy to see that unconditionally secure *1-2 OT* is not possible without any assumption. Even with the help of quantum communication and computation, unconditionally secure *1-2 OT* remains impossible [23,24]. As a consequence, much effort

* Supported by the Dutch Organization for Scientific Research (NWO).
** Supported by the EC-Integrated Project SECOQC, No: FP6-2002-IST-1-506813.

C. Dwork (Ed.): CRYPTO 2006, LNCS 4117, pp. 427–444, 2006.

has been put into constructing unconditionally secure protocols for *1-2 OT* using physical assumptions like (various models for) noisy channels [8,16,12,9], or a memory bounded adversary [6,17,18]. Similarly, much effort has been put into reducing *1-2 OT* to (seemingly) weaker flavors of *OT*, like *Rabin OT*, *1-2 XOT*, etc. [7,3,5,29,4,10].

In this work, we focus on a slightly modified notion of *1-2 OT*, which we call *Randomized 1-2 OT*, *Rand 1-2 OT* for short, where the bits (or strings) B_0 and B_1 are not *in*put by the sender, but generated uniformly at random during the *Rand 1-2 OT* and then *out*put to the sender. It is still required that the receiver only learns the bit (or string) of his choice, B_C, whereas the sender does not learn any information on C. It is obvious that a *Rand 1-2 OT* can easily be turned into an ordinary *1-2 OT* simply by using the generated B_0 and B_1 to mask the actual input bits (or strings). Furthermore, all known constructions of unconditionally secure *1-2 OT* protocols make (implicitly) the detour via a *Rand 1-2 OT*.

In a first step, we observe that the obliviousness condition of a *Rand 1-2 OT* of *bits* is equivalent to requiring the XOR $B_0 \oplus B_1$ to be (close to) uniformly distributed from the receiver's point of view. The proof is very simple, and it is kind of surprising that (to the best of our knowledge) this has not been realized before. We then ask and answer the question whether there is a natural generalization of this result to *Rand 1-2 OT* of *strings*. Note that requiring the bitwise XOR of the two strings to be uniformly distributed is obviously not sufficient. We show that the obliviousness condition for *Rand 1-2 OT* of strings can be characterized in terms of *non-degenerate linear functions* (bivariate binary linear functions which non-trivially depend on both arguments, as defined in Definition 2): obliviousness holds if and only if the result of applying any non-degenerate linear function to the two strings is (close to) uniformly distributed from the receiver's point of view.

We then show the usefulness of this new understanding of *1-2 OT*. We demonstrate this on the problem of reducing *1-2 OT* to weaker primitives. Concretely, we show that the reducibility of an (ordinary) *1-2 OT* to weaker flavors via a non-interactive reduction follows by a trivial argument from our characterization of the obliviousness condition. This is in sharp contrast to the current literature: The proofs given in [3,29,4] for reducing *1-2 OT* to *1-2 XOT*, *1-2 GOT* and *1-2 UOT* (we refer to Section 5 for a description of these flavors of *OT*) are rather complicated and tailored to a particular class of privacy-amplifying hash functions; whether the reductions also work for a less restricted class is left as an open problem [4, page 222]. And, the proof given in [5] for reducing *1-2 OT* to one execution of a general *UOT* is not only complicated, but also incorrect, as we will point out. Thus, our characterization of the obliviousness condition allows to simplify existing reducibility proofs (and, along the way, to solve the open problem posed in [4], as well as to improve the reduction parameters in most cases), but it also allows for new (respectively until now only incorrectly proven) reductions. Furthermore, our techniques are useful for the construction and analysis of *1-2 OT* protocols in other settings, for instance in the bounded quantum-storage model [13].

2   Notation

Let P and Q be two probability distributions over the same domain $\mathcal{X}$. The *variational distance* $\delta(P, Q)$ is defined as $\delta(P, Q) := \frac{1}{2} \sum_{x \in \mathcal{X}} |P(x) - Q(x)|$. Note that this definition makes sense also for *non-normalized* distributions, and indeed we define and use $\delta(P, Q)$ for arbitrary positive-valued functions P and Q with common domain. In case $\mathcal{X}$ is of the form $\mathcal{X} = \mathcal{U} \times \mathcal{V}$, we can expand $\delta(P, Q)$ to $\delta(P, Q) = \sum_u \delta(P(u, \cdot), Q(u, \cdot)) = \sum_v \delta(P(\cdot, v), Q(\cdot, v))$. We write $P \approx_\varepsilon Q$ to denote that P and Q are ε-close, i.e., that $\delta(P, Q) \le \varepsilon$.

For a random variable X it is common to denote its distribution by P_X. We adopt this notation. Alternatively, we also write $[X]$ for the distribution P_X of X. For two random variables X and Y, whereas $[XY]$ naturally denotes the joint distribution P_{XY}, we write $[X][Y]$ to denote the independent distribution $P_{XY} : (x, y) \mapsto P_X(x)P_Y(y)$. Using this notation, X and Y are (close to) *independent* if and only if $[XY] = [X][Y]$ (respectively $[XY] \approx_\varepsilon [X][Y]$). We feel that this notation is sometimes easier to read as it refrains from putting the crucial information into the subscript.

By UNIF we denote a uniformly distributed binary random variable (independent of anything else), such that $P_{\text{UNIF}}(b) = \frac{1}{2}$ for both $b \in \{0, 1\}$, and UNIF^ℓ stands for ℓ independent copies of UNIF.

3   Defining *1-2 OT*

3.1   (Randomized) *1-2 OT* of Bits

Formally capturing the intuitive understanding of the security of *1-2 OT* is a non-trivial task. We adopt the security definition of [11], where it is argued that this definition is the "right" way to define unconditionally secure *1-2 OT*. In their model, a secure *1-2 OT* protocol is as good as an ideal *1-2 OT* functionality.

In this paper, we will mainly focus on a slight modification of *1-2 OT*, which we call *Randomized 1-2 OT* (although *Sender*-randomized *1-2 OT* would be a more appropriate, but also rather lengthy name). A Randomized *1-2 OT*, or *Rand 1-2 OT* for short, essentially coincides with an (ordinary) *1-2 OT*, except that the two bits B_0 and B_1 are not *input* by the sender but generated uniformly at random during the protocol and *output* to the sender. This is formalized in Definition 1 below.

There are two main justifications for focusing on *Rand 1-2 OT*. First, an ordinary *1-2 OT* can easily be constructed from a *Rand 1-2 OT*: the sender can use the randomly generated B_0 and B_1 to one-time-pad encrypt his input bits for the *1-2 OT*, and send the masked bits to the receiver (as first realized in [1]). For a formal proof of this we refer to the full version of [11].[1] And second, all information-theoretically secure constructions of *1-2 OT* protocols we are aware

---

[1] The definition of *Rand 1-2 OT* in [11] differs syntactically slightly from our Definition 1, in particular in that it involves an auxiliary input Z, but it is not too hard to see that the two definitions are equivalent. We discuss this in more detail in [15].

of in fact do implicitly build a *Rand 1-2 OT* and use the above reduction to achieve *1-2 OT*.

We formalize *Rand 1-2 OT* in such a way that it minimizes and simplifies as much as possible the security restraints, while at the same time remaining sufficient for *1-2 OT*.

Definition 1 (Rand 1-2 OT). *An ε-secure Rand 1-2 OT is a protocol between sender* S *and receiver* R, *with* R *having input $C \in \{0,1\}$ (while* S *has no input), such that for any distribution of C, the following properties hold:*

ε-*Correctness: For honest* S *and* R, S *has output $B_0, B_1 \in \{0,1\}$ and* R *has output B_C, except with probability ε.*

ε-*Privacy: For honest* R *and any (dishonest)* $\tilde{\mathsf{S}}$ *with output[2] V, $[CV] \approx_\varepsilon [C][V]$.*

ε-*Obliviousness: For honest* S *and any (dishonest)* $\tilde{\mathsf{R}}$ *with output W, there exists a binary random variable D such that $[B_{1-D} W B_D D] \approx_\varepsilon [\text{UNIF}][W B_D D]$.*

The privacy condition simply says that $\tilde{\mathsf{S}}$ learns no information on C, and obliviousness requires that there exists a choice bit D (supposed to be C) such that when given the choice bit D and the corresponding bit B_D, then the other bit B_{1-D} is independent and random from $\tilde{\mathsf{R}}$'s point of view.

3.2   (Randomized) *1-2 OT* of Strings

In a *1-2 String OT* the sender inputs two *strings* (of the same length), and the receiver is allowed to learn one of the two and only one of the two. Formally, for any positive integer ℓ, *1-2 OT$^\ell$* and *Rand 1-2 OT$^\ell$* can be defined along the same lines as *1-2 OT* and *Rand 1-2 OT* of *bits*; for instance for *Rand 1-2 OT$^\ell$* the binary random variables B_0 and B_1 (as well as UNIF) in Definition 1 are simply replaced by random variables S_0 and S_1 (and UNIF$^\ell$) with range $\{0,1\}^\ell$.

4   Characterizing Obliviousness

4.1   The Case of Bit *OT*

It is well known (and it follows from the obliviousness condition) that in a (*Rand*) *1-2 OT* the receiver R should in particular learn (essentially) no information on the XOR $B_0 \oplus B_1$ of the two bits. The following proposition shows that this is not only necessary for the obliviousness condition but also *sufficient*.

Theorem 1. *The ε-obliviousness condition for a Rand 1-2 OT is satisfied for a particular (possibly dishonest) receiver* $\tilde{\mathsf{R}}$ *with output W if and only if*

$$[(B_0 \oplus B_1)W] \approx_\varepsilon [\text{UNIF}][W].$$

---

[2] Note that $\tilde{\mathsf{S}}$'s output V may consist of $\tilde{\mathsf{S}}$'s complete view on the protocol.

Before going into the proof (which is surprisingly simple), consider the following example. Assume a candidate protocol for *Rand 1-2 OT* and a dishonest receiver $\tilde{R}$ which is able to output $W = 0$ if $B_0 = 0 = B_1$, $W = 1$ if $B_0 = 1 = B_1$ and $W = 0$ or 1 with probability $1/2$ each in case $B_0 \neq B_1$. Then, it is easy to see that conditioned on, say, $W = 0$, (B_0, B_1) is $(0, 0)$ with probability $\frac{1}{2}$, and $(0, 1)$ and $(1, 0)$ each with probability $\frac{1}{4}$, such that the condition on the XOR from Theorem 1 is satisfied. On the other hand, neither B_0 nor B_1 is uniformly distributed (conditioned on $W = 0$), and it appears as if the receiver has some joint information on B_0 and B_1 which is forbidden by a *(Rand) 1-2 OT*. But that is not so. Indeed, the same view can be obtained when attacking an *ideal Rand 1-2 OT*: submit a random bit C to obtain B_C and output $W = B_C$. And in the light of Definition 1, if $W = 0$ we can split the event $(B_0, B_1) = (0, 0)$ into two disjoint subsets (subevents) $\mathcal{E}_0$ and $\mathcal{E}_1$ such that each has probability $\frac{1}{4}$, and we define D by setting $D = 0$ if $\mathcal{E}_0$ or $(B_0, B_1) = (0, 1)$, and $D = 1$ if $\mathcal{E}_1$ or $(B_0, B_1) = (1, 0)$. Then, obviously, conditioned on $D = d$, the bit B_{1-d} is uniformly distributed, even when given B_d. The corresponding holds if $W = 1$.

Proof. The "only if" implication is well known and straightforward. For the "if" implication, we first argue the perfect case where $[(B_0 \oplus B_1) W] = [\text{UNIF}][W]$. For any value w with $P_W(w) > 0$, the non-normalized distribution $P_{B_0 B_1 W}(\cdot, \cdot, w)$ can be expressed as depicted in the left table of Figure 1, where we write a for $P_{B_0 B_1 W}(0, 0, w)$, b for $P_{B_0 B_1 W}(0, 1, w)$, c for $P_{B_0 B_1 W}(1, 0, w)$ and d for $P_{B_0 B_1 W}(1, 1, w)$. Note that $a+b+c+d = P_W(w)$ and, by assumption, $a+d = b+c$. Due to symmetry, we may assume that $a \leq b$. We can then define D by extending $P_{B_0 B_1 W}(\cdot, \cdot, w)$ to $P_{B_0 B_1 DW}(\cdot, \cdot, \cdot, w)$ as depicted in the right two tables in Figure 1: $P_{B_0 B_1 DW}(0, 0, 0, w) = P_{B_0 B_1 DW}(0, 1, 0, w) = a$, $P_{B_0 B_1 DW}(1, 0, 0, w) = P_{B_0 B_1 DW}(1, 1, 0, w) = c$ etc. Important to realize is that $P_{B_0 B_1 DW}(\cdot, \cdot, \cdot, w)$ is indeed a valid extension since by assumption $c + (b - a) = d$.

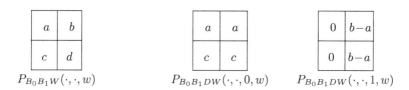

$$P_{B_0 B_1 W}(\cdot, \cdot, w) \qquad P_{B_0 B_1 DW}(\cdot, \cdot, 0, w) \qquad P_{B_0 B_1 DW}(\cdot, \cdot, 1, w)$$

Fig. 1. Distributions $P_{B_0 B_1 W}(\cdot, \cdot, w)$ and $P_{B_0 B_1 DW}(\cdot, \cdot, \cdot, w)$

It is now obvious that $P_{B_0 B_1 DW}(\cdot, \cdot, 0, w) = \frac{1}{2} P_{B_0 DW}(\cdot, 0, w)$ as well as $P_{B_0 B_1 DW}(\cdot, \cdot, 1, w) = \frac{1}{2} P_{B_1 DW}(\cdot, 1, w)$. This finishes the perfect case.

Concerning the general case, the idea is the same as above, except that one has to take some care regarding the error parameter $\varepsilon \geq 0$. As this does not give any new insight, and we anyway state and fully prove a more general result in Theorem 2, we skip this part of the proof.[3]   □

---

[3] Although the special case $\ell = 1$ in Theorem 2 is quantitatively slightly weaker than Theorem 1.

4.2   The Case of String OT

The obvious question after the previous section is whether there is a natural generalization of Theorem 1 to $1\text{-}2\,OT^\ell$ for $\ell \geq 2$. Note that the straightforward generalization of the XOR-condition in Theorem 1, requiring that any receiver has no information on the bit-wise XOR of the two strings, is clearly too weak, and does not imply the obliviousness condition for $Rand\ 1\text{-}2\,OT^\ell$: for instance the receiver could know the first half of the first string and the second half of the second string.

The Characterization.   Let ℓ be an arbitrary positive integer.

Definition 2. *A function* $\beta : \{0,1\}^\ell \times \{0,1\}^\ell \to \{0,1\}$ *is called a* non-degenerate linear function *(NDLF) if it is of the form* $\beta : (s_0, s_1) \mapsto \langle a_0, s_0 \rangle \oplus \langle a_1, s_1 \rangle$ *for two non-zero* $a_0, a_1 \in \{0,1\}^\ell$, *i.e., if it is linear and non-trivially depends on both input strings.*

In case $\ell = 1$, the XOR is a NDLF, and it is the *only* NDLF. Based on this notion, the obliviousness condition of $Rand\ 1\text{-}2\,OT^\ell$ can be characterized as follows.

Theorem 2. *The ε-obliviousness condition for a $Rand\ 1\text{-}2\,OT^\ell$ is satisfied for a particular (possibly dishonest) receiver $\tilde{\mathsf{R}}$ with output W if*

$$[\beta(S_0, S_1)\,W] \approx_{\varepsilon/2^{2\ell+1}} [\mathrm{UNIF}]\,[W]$$

for every NDLF β, and, on the other hand, the ε-obliviousness condition may be satisfied only if $[\beta(S_0, S_1)\,W] \approx_\varepsilon [\mathrm{UNIF}]\,[W]$ for every NDLF β.

Note that the number of NDLFs is exponential in ℓ, namely $(2^\ell - 1)^2$. Nevertheless, we show in Section 5 that this characterization turns out to be very useful. There, we will also argue that an exponential overhead (in ℓ) in the sufficient condition is unavoidable. The proof of Theorem 2 also shows that the set of NDLFs forms a minimal set of functions among all sets that imply obliviousness. In this sense, our characterization is tight.

We would like to point out that Theorem 4 in [4] also provides a tool to analyze the obliviousness condition of $1\text{-}2\,OT$ protocols in terms of linear functions; however, the condition that needs to be satisfied is much stronger than for our Theorem 2: it additionally requires that one of the two strings is *a priori* uniformly distributed (from the receiver's point of view).[4] This difference is crucial, because showing that one of the two strings is uniform (conditioned on the receiver's view) is usually technically involved and sometimes not even possible, as the example given after Theorem 1 shows. This is also demonstrated by the fact that the analysis in [4] of the considered $1\text{-}2\,OT$ protocol is tailored to one particular class of privacy-amplifying hash functions, and it is stated as

---

[4] Concretely, it is additionally required that every non-trivial parity of that string is uniform, but by the XOR-Lemma this is equivalent to the whole string being uniform.

an open problem how to prove their construction secure when a different class of hash functions is used. The condition for Theorem 2, on the other hand, is naturally satisfied for typical constructions of *1-2 OT* protocols, as we shall see in Section 5. As a result, Theorem 2 allows for much simpler and more elegant security proofs for *1-2 OT* protocols, and, as a by-product, allows to solve the open problem from [4]. We explain this in detail in Section 5, and the interested reader may well jump ahead and save the proof of Theorem 2 for later.

The proof for the "only if" part of Theorem 2 is given in the full version of this paper [15]; in fact, a slightly stronger statement is shown, namely that the ε-obliviousness condition implies $[\beta(S_0, S_1)W] \approx_\varepsilon [\mathrm{UNIF}][W]$ for any *2-balanced* function (as defined in [15]). The "if" part, which is the interesting direction, is proven below.

Proof of Theorem 2 ("if" part). First, we consider the perfect case: if $[\beta(S_0, S_1)W]$ *equals* $[\mathrm{UNIF}][W]$ for every NDLF β, then the obliviousness condition for *Rand 1-2 OT$^\ell$* holds (perfectly).

THE PERFECT CASE: As the case $\ell = 1$ is already settled, we assume that $\ell \geq 2$.

Fix an arbitrary output w of the receiver, and consider the non-normalized probability distribution $P_{S_0 S_1 W}(\cdot, \cdot, w)$. We use the variable p_{s_0, s_1} to refer to $P_{S_0 S_1 W}(s_0, s_1, w)$, and we write **o** for the all-zero string $(0, \ldots, 0) \in \{0, 1\}^\ell$. We assume that $p_{\mathbf{o}, \mathbf{o}} \leq p_{\mathbf{o}, s_1}$ for any $s_1 \in \{0, 1\}^\ell$; we show later that we may do so. We extend this distribution to $P_{S_0 S_1 DW}(\cdot, \cdot, \cdot, w)$ by setting

$$P_{S_0 S_1 DW}(s_0, s_1, 0, w) = p_{s_0, \mathbf{o}} \quad \text{and} \quad P_{S_0 S_1 DW}(s_0, s_1, 1, w) = p_{\mathbf{o}, s_1} - p_{\mathbf{o}, \mathbf{o}} \quad (1)$$

for any strings $s_0, s_1 \in \{0, 1\}^\ell$, and we collect the equations resulting from the condition that $P_{S_0 S_1 W}(\cdot, \cdot, w) = P_{S_0 S_1 DW}(\cdot, \cdot, 0, w) + P_{S_0 S_1 DW}(\cdot, \cdot, 1, w)$ needs to be satisfied: for any two $s_0, s_1 \in \{0, 1\}^\ell \setminus \{\mathbf{o}\}$

$$p_{s_0, \mathbf{o}} + p_{\mathbf{o}, s_1} = p_{\mathbf{o}, \mathbf{o}} + p_{s_0, s_1}. \quad (2)$$

If all these equations do hold (for any w) then as in the case of $\ell = 1$, the random variable D is well defined and $[S_{1-D} S_D W D] = [\mathrm{UNIF}^\ell][S_D W D]$ holds, since $P_{S_0 S_1 DW}(s_0, s_1, 0, w)$ does not depend on s_1 and $P_{S_0 S_1 DW}(s_0, s_1, 1, w)$ not on s_0.

Before moving on, we first justify the assumption that $p_{\mathbf{o}, \mathbf{o}} \leq p_{\mathbf{o}, s_1}$ for any $s_1 \in \{0, 1\}^\ell$. In general, we choose $t \in \{0, 1\}^\ell$ such that $p_{\mathbf{o}, t} \leq p_{\mathbf{o}, s_1}$ for any $s_1 \in \{0, 1\}^\ell$, and we set $P_{S_0 S_1 DW}(s_0, s_1, 0, w) = p_{s_0, t}$ and $P_{S_0 S_1 DW}(s_0, s_1, 1, w) = p_{\mathbf{o}, s_1} - p_{\mathbf{o}, t}$, resulting in the equations $p_{s_0, t} + p_{\mathbf{o}, s_1} = p_{\mathbf{o}, t} + p_{s_0, s_1}$ for $s_0 \in \{0, 1\}^\ell \setminus \{\mathbf{o}\}$ and $s_1 \in \{0, 1\}^\ell \setminus \{t\}$. However, these equations follow from the equations given by (2): subtract equation (2) with s_1 replaced by t from equation (2). Therefore, it suffices to focus on the equations given by (2).

We proceed by showing that the equations provided by the assumed uniformity of $\beta(S_0, S_1)$ for any β imply the equations given by (2). Consider an arbitrary pair $a_0, a_1 \in \{0, 1\}^\ell \setminus \{\mathbf{o}\}$ and let β be the associated NDLF, i.e., such that $\beta(s_0, s_1) = \langle a_0, s_0 \rangle \oplus \langle a_1, s_1 \rangle$. By assumption, $\beta(S_0, S_1)$ is uniformly distributed, independent of W. Thus, for any fixed w, and writing p_{σ_0,σ_1} for $P_{S_0 S_1 W}(s_0, s_1, w)$, this can be expressed as

$$\sum_{\substack{\sigma_0,\sigma_1: \\ \langle a_0,\sigma_0\rangle=\langle a_1,\sigma_1\rangle}} p_{\sigma_0,\sigma_1} = \sum_{\substack{\sigma_0,\sigma_1: \\ \langle a_0,\sigma_0\rangle\neq\langle a_1,\sigma_1\rangle}} p_{\sigma_0,\sigma_1}, \tag{3}$$

where both summations are over all $\sigma_0, \sigma_1 \in \{0,1\}^\ell$ subject to the indicated respective properties. Recall, that this equality holds for any pair $a_0, a_1 \in \{0,1\}^\ell \setminus \{\mathbf{o}\}$. Thus, for fixed $s_0, s_1 \in \{0,1\}^\ell \setminus \{\mathbf{o}\}$, if we add up over all such pairs a_0, a_1 subject to $\langle a_0, s_0\rangle = \langle a_1, s_1\rangle = 1$, we get the equation

$$\sum_{\substack{a_0,a_1: \\ \langle a_0,s_0\rangle=\langle a_1,s_1\rangle=1}} \sum_{\substack{\sigma_0,\sigma_1: \\ \langle a_0,\sigma_0\rangle=\langle a_1,\sigma_1\rangle}} p_{\sigma_0,\sigma_1} = \sum_{\substack{a_0,a_1: \\ \langle a_0,s_0\rangle=\langle a_1,s_1\rangle=1}} \sum_{\substack{\sigma_0,\sigma_1: \\ \langle a_0,\sigma_0\rangle\neq\langle a_1,\sigma_1\rangle}} p_{\sigma_0,\sigma_1},$$

which, after re-arranging the terms of the summations, leads to

$$\sum_{\sigma_0,\sigma_1} \sum_{\substack{a_0,a_1: \\ \langle a_0,s_0\rangle=\langle a_1,s_1\rangle=1 \\ \langle a_0,\sigma_0\rangle=\langle a_1,\sigma_1\rangle}} p_{\sigma_0,\sigma_1} = \sum_{\sigma_0,\sigma_1} \sum_{\substack{a_0,a_1: \\ \langle a_0,s_0\rangle=\langle a_1,s_1\rangle=1 \\ \langle a_0,\sigma_0\rangle\neq\langle a_1,\sigma_1\rangle}} p_{\sigma_0,\sigma_1}. \tag{4}$$

We are now going to argue that, up to a constant multiplicative factor, equation (4) coincides with equation (2).

First, it is straightforward to verify that the variables $p_{\mathbf{o},\mathbf{o}}$ and p_{s_0,s_1} occur only on the left hand side, both with multiplicity $2^{2(\ell-1)}$ (the number of pairs a_0, a_1 such that $\langle a_0, s_0\rangle = \langle a_1, s_1\rangle = 1$), whereas $p_{s_0,\mathbf{o}}$ and $p_{\mathbf{o},s_1}$ only occur on the right hand side, with the same multiplicity $2^{2(\ell-1)}$.

Now, we argue that any other p_{σ_0,σ_1} equally often appears on the right and on the left hand side, and thus vanishes from the equation. Note that the set of pairs a_0, a_1, over which the summation runs on the left respectively the right hand side, can be understood as the set of solutions to a binary (non-homogeneous) linear equations system:

$$\begin{pmatrix} s_0 & 0 \\ 0 & s_1 \\ \sigma_0 & \sigma_1 \end{pmatrix} \begin{pmatrix} a_0 \\ a_1 \end{pmatrix} = \begin{pmatrix} 1 \\ 1 \\ 0 \end{pmatrix} \text{ respectively } \begin{pmatrix} 1 \\ 1 \\ 1 \end{pmatrix}.$$

Also note that the two linear equation systems consist of three equations and involve at least 4 variables (as $a_0, a_1 \in \{0,1\}^\ell$ and $\ell \geq 2$). Therefore, using basic linear algebra, one is tempted to conclude that they both have solutions, and, because they have the same homogeneous part, they have the same number of solutions (equal to the number of homogeneous solutions). However, this is only guaranteed if the matrix defining the homogeneous part has full rank. But here this is precisely the case if and only if $(\sigma_0, \sigma_1) \notin \{(\mathbf{o}, \mathbf{o}), (s_0, \mathbf{o}), (\mathbf{o}, s_1), (s_0, s_1)\}$, where those four exceptions have already been treated above.

It follows that the equations (3), which are guaranteed by assumption, imply the equations (2). This concludes the proof for the perfect case.

THE GENERAL CASE: Now, we consider the general case where there exists some $\varepsilon > 0$ such that $\delta\big([\beta(S_0, S_1)W], [\text{UNIF}][W]\big) \leq 2^{-2\ell-1}\varepsilon$ for any NDLF β. We use the observations from the perfect case, but additionally we keep track of the "error term".

For any w with $P_W(w) > 0$ and any NDLF β, set

$$\varepsilon_{w,\beta} = \delta\big(P_{\beta(S_0,S_1)W}(\cdot, w), P_{\text{UNIF}}P_W(w)\big).$$

Note that $\sum_w \varepsilon_{w,\beta} = \delta\big([\beta(S_0, S_1)W], [\text{UNIF}][W]\big) \leq 2^{-2\ell-1}\varepsilon$, independent of β. Fix now an arbitrary w with $P_W(w) > 0$. Then, (3) only holds up to an error of $2\varepsilon_{w,\beta}$, where β is the NDLF associated to a_0, a_1. As a consequence, equation (4) only holds up to an error of $2\sum_\beta \varepsilon_{w,\beta}$ and thus (2) holds up to an error of $\delta_{s_0,s_1} = \frac{2}{2^{2\ell-2}}\sum_\beta \varepsilon_{w,\beta}$, where the sum is over the $2^{2\ell-2}$ functions associated to the pairs a_0, a_1 with $\langle a_0, s_0 \rangle = \langle a_1, s_1 \rangle = 1$. Note that δ_{s_0,s_1} depends on w, but the set of β's, over which the summation runs, does not. Adding up over all possible w's gives

$$\sum_w \delta_{s_0,s_1} = \frac{2}{2^{2\ell-2}}\sum_w \sum_\beta \varepsilon_{w,\beta} = \frac{2}{2^{2\ell-2}}\sum_\beta \sum_w \varepsilon_{w,\beta} \leq 2^{-2\ell}\varepsilon.$$

Since (2) only holds approximately, $P_{S_0S_1DW}$ as in (1) is not necessarily a valid extension, but close. This can obviously be overcome by instead setting

$$P_{S_0S_1DW}(s_0, s_1, 0, w) = p_{s_0,\mathbf{o}} \pm \delta'_{s_0,s_1} \text{ and } P_{S_0S_1DW}(s_0, s_1, 1, w) = p_{\mathbf{o},s_1} - p_{\mathbf{o},\mathbf{o}} \pm \delta''_{s_0,s_1}$$

with suitably chosen $\delta'_{s_0,s_1}, \delta''_{s_0,s_1} \geq 0$ with $\delta'_{s_0,s_1} + \delta''_{s_0,s_1} = \delta_{s_0,s_1}$, and with suitably chosen signs "+" or "−".[5] Using that every $P_{S_0S_1DW}(s_0, s_1, 0, w)$ differs from $p_{s_0,\mathbf{o}}$ by at most δ'_{s_0,s_1}, it follows from a straightforward computation that $\delta\big(P_{S_{1-D}S_DDW}(\cdot, \cdot, 0, w), P_{\text{UNIF}}P_{S_DDW}(\cdot, 0, w)\big) \leq \sum_{s_0,s_1} \delta'_{s_0,s_1}$. The corresponding holds for $P_{S_0S_1DW}(\cdot, \cdot, 1, w)$. It follows that

$$\delta\big(P_{S_{1-D}S_DWD}, P_{\text{UNIF}}P_{S_DWD}\big) \leq \sum_w \sum_{s_0,s_1}(\delta'_{s_0,s_1} + \delta''_{s_0,s_1}) = \sum_{s_0,s_1} \sum_w \delta_{s_0,s_1} \leq \varepsilon$$

which concludes the proof. $\square$

5   Applications

In this section we will show the usefulness of Theorem 2 for the construction of *1-2 OT$^\ell$*, based on weaker primitives (like a noisy channel, a quantum uncertainty relation or other flavors of *OT*). In particular, we will show that the reducibility of *1-2 OT* to any weaker flavor of *OT* follows as a simple argument using Theorem 2.

---

[5] Most of the time, it probably suffices to correct one of the two, say, choose $\delta'_{s_0,s_1} = \delta_{s_0,s_1}$ and $\delta''_{s_0,s_1} = 0$; however, if for instance $p_{s_0,\mathbf{o}}$ and $p_{\mathbf{o},s_1} - p_{\mathbf{o},\mathbf{o}}$ are both positive but $P_{S_0S_1W}(s_0, s_1, w) = 0$, then one has to correct both.

5.1 Reducing *1-2 OT$^\ell$* to Independent Repetitions of Weak *1-2 OT*'s

Background. A great deal of effort has been put into constructing protocols for *1-2 OT$^\ell$* based on physical assumptions like (various models for) noisy channels [8,16,12,9] or a memory bounded adversary [6,17,18], as well as into reducing *1-2 OT$^\ell$* to (seemingly) weaker flavors of *OT*, like *Rabin OT*, *1-2 XOT*, *1-2 GOT* and *1-2 UOT* [7,3,5,29,4,10]. Note that the latter three flavors of *OT* are weaker than *1-2 OT* in that the (dishonest) receiver has more freedom in choosing the sort of information he wants to get about the sender's input bits B_0 and B_1: B_0, B_1 or $B_0 \oplus B_1$ in case of *1-2 XOT*, $g(B_0, B_1)$ for an arbitrary one-bit-output function g in case of *1-2 GOT*, and an arbitrary (probabilistic) Y with mutual information $I(B_0B_1; Y) \leq 1$ in case of *1-2 UOT*.[6]

All these reductions of *1-2 OT* to weaker versions follow a specific construction design (which is also at the core of the *1-2 OT* protocols based on noisy channels or a memory-bounded adversary). By repeated (independent) executions of the underlying primitive, S transfers a randomly chosen bit string $X = (X_0, X_1) \in \{0,1\}^n \times \{0,1\}^n$ to R such that: (1) depending on his choice bit C, the honest R knows either X_0 or X_1, (2) any S̃ has no information on which part of X R learned, and (3) any R̃ has some uncertainty in X. Then, this is completed to a *Rand 1-2 OT* by means of privacy amplification [2]: S samples two functions f_0 and f_1 from a universal-two class $\mathcal{F}$ of hash functions, sends them to R, and outputs $S_0 = f_0(X_0)$ and $S_1 = f_1(X_1)$, and R outputs $S_C = f_C(X_C)$. Finally, the *Rand 1-2 OT* is transformed into an ordinary *1-2 OT* in the obvious way.

Correctness and privacy of this construction are clear, they follow immediately from (1) and (2). How easy or hard it is to prove obliviousness depends heavily on the underlying primitive. In case of *Rabin OT* it is rather straightforward. In case of *1-2 XOT* and the other weaker versions, this is non-trivial. The problem is that since R might know $X_0 \oplus X_1$, it is not possible to argue that there exists $d \in \{0, 1\}$ such that R's uncertainty on X_{1-d} is large when given X_d. This, though, would be necessary in order to finish the proof by simply applying the privacy amplification theorem [2]. This difficulty is overcome in [3,4] by tailoring the proof to a particular universal-two class of hash functions (namely the class of all *linear* hash functions). Whether the reduction also works for a less restricted class of hash functions is left in [3,4] as an open problem, which we solve here as a side result. Using a smaller class of hash functions would allow for instance to reduce the communication complexity of the protocol.

In [10], the difficulty is overcome by giving up on the simplicity of the reduction. The cost of two-way communication allowing for interactive hashing is traded for better reduction parameters. We would like to emphasize that

---

[6] As a matter of fact, reducibility has been proven for any bound on $I(B_0B_1; Y)$ strictly smaller than 2. Note that there is some confusion in the literature in what a *universal OT*, *UOT*, should be: In [3,29,4], a *UOT* takes as input two *bits* and the receiver is doomed to have at least one bit (or any other non-trivial amount) of *Shannon* entropy on them; we denote this by *1-2 UOT*. Whereas in [5], a *UOT* takes as input two *strings* and the receiver is doomed to have some *Renyi* entropy on them. We address this latter notion in more detail in Section 5.2.

these parameters are incomparable to ours, because a different reduction is used, whereas our approach provides a *better analysis* of the non-interactive reductions.

The New Approach. We argue that, independent of the underlying primitive, obliviousness follows as a simple consequence of Theorem 2, in combination with a straightforward observation regarding the composition of NDLFs with strongly universal-two hash functions (Proposition 1 below). Recall that a class $\mathcal{F}$ of hash functions from, say, $\{0,1\}^n$ to $\{0,1\}^\ell$ is *strongly universal-two* [27] if for any distinct $x, x' \in \{0,1\}^n$ the two random variables $F(x)$ and $F(x')$ are independent and uniformly distributed (over $\{0,1\}^\ell$), where the random variable F represents the random choice of a function in $\mathcal{F}$.

Proposition 1. *Let $\mathcal{F}_0$ and $\mathcal{F}_1$ be two classes of strongly universal-two hash functions from $\{0,1\}^{n_0}$ respectively $\{0,1\}^{n_1}$ to $\{0,1\}^\ell$, and let β be a fixed NDLF $\beta : \{0,1\}^\ell \times \{0,1\}^\ell \to \{0,1\}$. Consider the class $\mathcal{F}$ of all composed functions $f : \{0,1\}^{n_0} \times \{0,1\}^{n_1} \to \{0,1\}$ with $f(x_0, x_1) = \beta(f_0(x_0), f_1(x_1))$ where $f_0 \in \mathcal{F}_0$ and $f_1 \in \mathcal{F}_1$. Then, $\mathcal{F}$ is strongly universal-two.*

The proof is straightforward; for completeness, it is given in the full version [15].[7]

Now, briefly, obliviousness for a construction as sketched above can be argued as follows. The only restriction is that $\mathcal{F}$ needs to be *strongly* universal-two. From the independent repetitions of the underlying weak *OT* (*Rabin OT, 1-2 XOT, 1-2 GOT* or *1-2 UOT*) it follows that $\tilde{R}$ has "high" collision entropy in X. Hence, for any NDLF β, we can apply the privacy amplification theorem [2] (respectively the version given in Appendix A) to the (strongly) universal-two hash function $\beta(f_0(\cdot), f_1(\cdot))$ and argue that $\beta(f_0(X_0), f_1(X_1))$ is close to uniform for randomly chosen f_0 and f_1. Obliviousness then follows immediately from Theorem 2.

We save the quantitative analysis (Theorem 3) for next section, where we consider a reduction of *1-2 OT* to the weakest kind of *OT*: to *one* execution of a *UOT*. Based on this, we compare in Appendix B the quality of the analysis of the above reductions based on Theorem 2 with the results in [4]. It turns out that our analysis is tighter for *1-2 GOT* and *1-2 UOT*, whereas the analysis in [4] is tighter for *1-2 XOT*; but in all cases, our analysis is much simpler and, we believe, more elegant.

5.2 Reducing *1-2 OT$^\ell$* to One Execution of *UOT*

We assume the reader to be somewhat familiar with the notion of *Renyi entropy* H_α of order α. Definition and some elementary properties needed in this section are given in Appendix A. We also refer to Appendix A for the slightly non-standard notion of *average conditional* Renyi entropy $H_\alpha(X|Y)$ we are using.

---

[7] The claim does not hold in general for ordinary (as opposed to strongly) universal-two classes: if $n_0 = n_1 = \ell$ and $\mathcal{F}_0$ and $\mathcal{F}_1$ both only contain the identity function $id : \{0,1\}^\ell \to \{0,1\}^\ell$ (and thus are universal-two), then $\mathcal{F}$ consisting of the function $f(x_0, x_1) = \beta(id(x_0), id(x_1)) = \beta(x_0, x_1)$ is not universal-two.

Universal Oblivious Transfer.   Probably the weakest flavor of OT is the *Universal OT* (UOT) as it was introduced in [5], in that it gives the receiver the most freedom in getting information on the string X. Formally, for a finite set $\mathcal{X}$ and parameters $\alpha \geq 0$ (allowing $\alpha = \infty$) and $r > 0$, an (α, r)-$UOT(\mathcal{X})$ works as follows. The sender inputs $x \in \mathcal{X}$, and the receiver may choose an arbitrary conditional probability distribution $P_{Y|X}$ with the only restriction that for a uniformly distributed X it must satisfy $H_\alpha(X|Y) \geq r$.[8] The receiver then gets as output y, sampled according to the distribution $P_{Y|X}(\cdot|x)$, whereas the sender gets no information on the receiver's choice for $P_{Y|X}$. Note that a *1-2 UOT* is a special case of this kind of UOT since "*1-2 UOT* $= (1,1)$-$UOT(\{0,1\}^2)$".

The crucial property of such an UOT is that the input is not restricted to two bits, but may be two bit-*strings*; this potentially allows to reduce *1-2 OT* to *one* execution of a UOT, rather than to many independent executions of the same primitive as for the 1-2 flavors of OT mentioned above. Indeed, following the design principle discussed in Section 5.1, it is straightforward to come up with a candidate protocol for *1-2 OT$^\ell$* which uses *one* execution of a (α, r)-$UOT(\mathcal{X})$ with $\mathcal{X} = \{0,1\}^n \times \{0,1\}^n$. The protocol is given in Figure 2, where $\mathcal{F}$ is a (strongly) universal-two class of hash functions from $\{0,1\}^n$ to $\{0,1\}^\ell$.

$OT2UOT(c)$:
 1. S and R run (α, r)-$UOT(\mathcal{X})$: S inputs a random $x = (x_0, x_1) \in \mathcal{X} = \{0,1\}^n \times \{0,1\}^n$, R inputs $P_{Y|X}$ with $P_{Y|X}(x_c'|(x_0', x_1')) = 1$ for any (x_0', x_1'), and as a result R obtains $y = x_c$.
 2. S samples independent random $f_0, f_1 \in \mathcal{F}$, sends f_0 and f_1 to R, and outputs $s_0 = f_0(x_0)$ and $s_1 = f_1(x_1)$.
 3. R computes and outputs $s_c = f_c(y)$.

Fig. 2. Protocol $OT2UOT$ for *Rand 1-2 OT$^\ell$*

In [5] it is claimed that, for appropriate parameters, protocol $OT2UOT$ is a secure *Rand 1-2 OT$^\ell$* (respectively, the resulting protocol for *1-2 OT* is secure). However, we argue below that the proof given is not correct (and it is not obvious how to fix it). In Theorem 3 we then show that its security follows easily from Theorem 2.

A Flaw in the Security Proof.   In [5] the security of protocol $OT2UOT$ is argued as follows. Using (rather complicated) *spoiling-knowledge techniques*, it is shown that, conditioned on the receiver's output (which we suppress to simplify the notation) at least one out of $H_\infty(X_0)$ and $H_\infty(X_1|X_0 = x_0)$ is "large" (for any x_0), and, similarly, at least one out of $H_\infty(X_1)$ and $H_\infty(X_0|X_1 = x_1)$. Since collision entropy is lower bounded by min-entropy, it then follows from the privacy amplification theorem that at least one out of $H(F_0(X_0)|F_0)$ and $H(F_1(X_1)|F_1, X_0 = x_0)$ is close to ℓ, and similarly, one out of $H(F_1(X_1)|F_1)$ and $H(F_0(X_0)|F_0, X_1 = x_1)$. It is then claimed that this proves $OT2UOT$ secure.

---

[8] This notion of UOT is even slightly weaker than what is considered in [5], where $H_\alpha(X|Y = y) \geq r$ for all y is required.

We argue that this very last implication is not correct. Indeed, what is proven about the entropy of $F_0(X_0)$ and $F_1(X_1)$ does not exclude the possibility that both entropies $H(F_0(X_0)|F_0)$ and $H(F_1(X_1)|F_1)$ are maximal, but that $H(F_0(X_0) \oplus F_1(X_1)|F_0, F_1) = 0$. This would allow the receiver to learn the (bitwise) XOR $S_0 \oplus S_1$, which is clearly forbidden by the obliviousness condition.

Also note that the proof does not use the fact that the two functions F_0 and F_1 are chosen *independently*. However, if they are chosen to be the same, then the protocol is clearly insecure: if the receiver asks for $Y = X_0 \oplus X_1$, and if $\mathcal{F}$ is a class of *linear* universal-two hash functions, then $\tilde{\mathsf{R}}$ obviously learns $S_0 \oplus S_1$.

Reducing 1-2 OT$^\ell$ to UOT. The following theorem guarantees the security of *OT2UOT* (for an appropriate choice of the parameters). The only restriction we have to make is that $\mathcal{F}$ needs to be a *strongly* universal-two class of hash function.

Theorem 3. *Let $\mathcal{F}$ be a strongly universal-two class of hash functions from $\{0,1\}^n$ to $\{0,1\}^\ell$. Then OT2UOT reduces a $2^{-\kappa}$-secure Rand 1-2 OT$^\ell$ to a (perfect) $(2,r)$-UOT($\{0,1\}^{2n}$) with $n \geq r \geq 4\ell + 3\kappa + 4$.*

Using the bounds from Lemma 2 (in Appendix A) on the different orders of Renyi entropy, the reducibility of *1-2 OT$^\ell$* to (α, r)-UOT($\mathcal{X}$) follows immediately for *any* $\alpha > 1$.

Informally, obliviousness for protocol *OT2UOT* is argued as for the reduction of *1-2 OT* to *Rabin OT*, *1-2 XOT* etc., discussed in Section 5.1, simply by using Proposition 1 in combination with the privacy amplification theorem, and applying Theorem 2. The formal proof given in Appendix C additionally keeps track of the "error term". From this proof it also becomes clear that the exponential (in ℓ) overhead in Theorem 2 is unavoidable. Indeed, a sub-exponential overhead would allow ℓ in Theorem 3 to be super-linear (in n), which of course is nonsense.

6   Generalizations and Further Applications

The general technique described in this section also comes in handy in a quantum setting. The fact that we do not need to know *how* the entropy is distributed over X is fundamental to prove secure a protocol for *1-2 OT$^\ell$* in the bounded quantum-storage model as introduced in [14]. In upcoming work [13], we present a protocol for *Rand 1-2 OT* for which we can use a quantum uncertainty relation to show a lower bound on the min-entropy of the $2n$-bit string X transmitted by the sender using a quantum encoding. We prove a quantum version of Theorem 2 which enables us to use the result about privacy amplification against quantum adversaries [26] to conclude that our protocol is oblivious against adversaries with bounded quantum memory. This application motivates further the use of (strongly) universal-two hashing, because up to date, no other means of privacy amplification have been shown secure against quantum adversaries.

In [15], we show that it is also possible to generalize Theorem 2 to *1-n OT*: it then states that the condition for *Rand 1-n OT* is satisfied if for any NDLF β and for any $0 \leq i < j \leq n - 1$ it holds that $\beta(S_i, S_i)$ is (essentially) uniform, conditioned on the receiver's output W and on all S_k with $k \neq i, j$. This comes in handy for the construction and analysis of *1-n OT* schemes, as demonstrated in [13], where also *1-n OT* schemes in the bounded quantum-storage model are considered.

7   Conclusion

We have established a characterization of the obliviousness condition for (a slightly modified version of) *1-2 OT$^\ell$* (Theorem 2). Using this characterization in combination with a composition result about strongly universal-two hash functions (Proposition 1), it follows by a very simple argument that when starting with a $2n$-bit string X with enough (collision) entropy, arbitrarily splitting up X into two n-bit strings X_0, X_1 followed by strongly universal-two hashing yields obliviousness as required by a *1-2 OT*. This allows for easy analyses whenever this design principle is used or can be applied, like reductions of *1-2 OT$^\ell$* to weaker flavors, or *1-2 OT$^\ell$* in the bounded (quantum) storage model, but possibly also in other contexts like in a computational setting when unconditional obliviousness is required.

Acknowledgments

We would like to thank Renato Renner for bringing up the idea of characterizing obliviousness in terms of the XOR, and Jürg Wullschleger for observing that our earlier results, which were expressed in terms of balanced functions, can also be expressed in terms of NDLFs. We are also grateful to Claude Crépeau and George Savvides for enlightening discussions regarding the formal definition of *1-2 OT*.

References

1. D. Beaver.   Precomputing oblivious transfer.   In *Advances in Cryptology—CRYPTO '95*, volume 963 of *Lecture Notes in Computer Science*. Springer, 1995.
2. C. H. Bennett, G. Brassard, C. Crépeau, and U. M. Maurer. Generalized privacy amplification. *IEEE Transactions on Information Theory*, 41(6), 1995.
3. G. Brassard and C. Crépeau. Oblivious transfers and privacy amplification. In *Advances in Cryptology—CRYPTO '97*, volume 1294 of *Lecture Notes in Computer Science*. Springer, 1997.
4. G. Brassard, C. Crépeau, and S. Wolf. Oblivious transfer and privacy amplification. *Journal of Cryptology*, 16(4), 2003.
5. C. Cachin. On the foundations of oblivious transfer. In *Advances in Cryptology—EUROCRYPT '98*, volume 1403 of *Lecture Notes in Computer Science*. Springer, 1998.

6. C. Cachin, C. Crépeau, and J. Marcil. Oblivious transfer with a memory-bounded receiver. In *39th Annual IEEE Symposium on Foundations of Computer Science (FOCS)*, pages 493–502, 1998.
7. C. Crépeau. Equivalence between two flavours of oblivious transfers. In *Advances in Cryptology—CRYPTO '87*, volume 293 of *Lecture Notes in Computer Science*. Springer, 1987.
8. C. Crépeau and J. Kilian. Achieving oblivious transfer using weakened security assumptions (extended abstract). In *29th Annual IEEE Symposium on Foundations of Computer Science (FOCS)*, 1988.
9. C. Crépeau, K. Morozov, and S. Wolf. Efficient unconditional oblivious transfer from almost any noisy channel. In *International Conference on Security in Communication Networks (SCN)*, volume 4 of *Lecture Notes in Computer Science*, 2004.
10. C. Crépeau and G. Savvides. Optimal reductions between oblivious transfers using interactive hashing. In *Advances in Cryptology—EUROCRYPT '06*, Lecture Notes in Computer Science. Springer, 2006.
11. C. Crépeau, G. Savvides, C. Schaffner, and J. Wullschleger. Information-theoretic conditions for two-party secure function evaluation. In *Advances in Cryptology—EUROCRYPT '06*, Lecture Notes in Computer Science. Springer, 2006. Full version: http://eprint.iacr.org.
12. I. B. Damgård, S. Fehr, K. Morozov, and L. Salvail. Unfair noisy channels and oblivious transfer. In *Theory of Cryptography Conference (TCC)*, volume 2951 of *Lecture Notes in Computer Science*. Springer, 2004.
13. I. B. Damgård, S. Fehr, R. Renner, L. Salvail, and C. Schaffner. A tight high-order entropic uncertainty relation with applications in the bounded quantum-storage model. In preparation, 2006.
14. I. B. Damgård, S. Fehr, L. Salvail, and C. Schaffner. Cryptography in the bounded quantum-storage model. In *46th Annual IEEE Symposium on Foundations of Computer Science (FOCS)*, 2005.
15. I. B. Damgård, S. Fehr, L. Salvail, and C. Schaffner. Oblivious transfer and linear functions (full version). Available at http://eprint.iacr.org/2005/349, 2006.
16. I. B. Damgård, J. Kilian, and L. Salvail. On the (im)possibility of basing oblivious transfer and bit commitment on weakened security assumptions. In *Advances in Cryptology—EUROCRYPT '99*, volume 1592 of *Lecture Notes in Computer Science*. Springer, 1999.
17. Y. Z. Ding. Oblivious transfer in the bounded storage model. In *Advances in Cryptology—CRYPTO '01*, volume 2139 of *Lecture Notes in Computer Science*. Springer, 2001.
18. Y. Z. Ding, D. Harnik, A. Rosen, and R. Shaltiel. Constant-round oblivious transfer in the bounded storage model. In *Theory of Cryptography Conference (TCC)*, volume 2951 of *Lecture Notes in Computer Science*, pages 446–472. Springer, 2004.
19. S. Even, O. Goldreich, and A. Lempel. A randomized protocol for signing contracts. In *Advances in Cryptology: Proceedings of CRYPTO 82*. Plenum Press, 1982.
20. J. Håstad, R. Impagliazzo, L. A. Levin, and M. Luby. A pseudorandom generator from any one-way function. *SIAM Journal on Computing*, 28(4), 1999.
21. R. Impagliazzo, L. A. Levin, and M. Luby. Pseudo-random generation from one-way functions. In *21st Annual ACM Symposium on Theory of Computing (STOC)*, 1989.
22. J. Kilian. Founding cryptography on oblivious transfer. In *20th Annual ACM Symposium on Theory of Computing (STOC)*, 1988.
23. H.-K. Lo and H. F. Chau. Is quantum bit commitment really possible? *Physical Review Letters*, 78(17):3410–3413, 1997.

24. D. Mayers. Unconditionally secure quantum bit commitment is impossible. *Physical Review Letters*, 78(17):3414–3417, 1997.
25. M. O. Rabin. How to exchange secrets by oblivious transfer. Technical Report TR-81, Harvard Aiken Computation Laboratory, 1981.
26. R. Renner and R. Koenig. Universally composable privacy amplification against quantum adversaries. In *Theory of Cryptography Conference (TCC)*, volume 3378 of *Lecture Notes in Computer Science*, pages 407–425. Springer, 2005. Also available at http://arxiv.org/abs/quant-ph/0403133.
27. M. N. Wegman and J. L. Carter. New classes and applications of hash functions. In *20th Annual IEEE Symposium on Foundations of Computer Science (FOCS)*, 1979.
28. S. Wiesner. Conjugate coding. *ACM Special Interest Group on Automata and Computability Theory (SIGACT News)*, 15, 1983. Original manuscript written circa 1970.
29. S. Wolf. Reducing oblivious string transfer to universal oblivious transfer. In *IEEE International Symposium on Information Theory (ISIT)*, 2000.

A   (Conditional) Renyi Entropy

Let $\alpha \geq 0$, $\alpha \neq 1$. The *Renyi entropy of order* α of a random variable X with distribution P_X is defined as

$$H_\alpha(X) = \frac{1}{1-\alpha} \log \left(\sum_x P_X(x)^\alpha \right) = -\log \left(\left(\sum_x P_X(x)^\alpha \right)^{\frac{1}{\alpha-1}} \right).$$

The limit for $\alpha \to 1$ is the *Shannon entropy* $H(X) = -\log \left(\sum_x P_X(x) \log P_X(x) \right)$ and the limit for $\alpha \to \infty$ the *min-entropy* $H_\infty(X) = -\log \left(\max_x P_X(x) \right)$. Another important special case is the case $\alpha = 2$, also known as *collision entropy* $H_2(X) = -\log \left(\sum_x P_X(x)^2 \right)$.

The *conditional* Renyi entropy $H_\alpha(X|Y=y)$ for two random variables X and Y is naturally defined as $H_\alpha(X|Y=y) = \frac{1}{1-\alpha} \log \left(\sum_x P_{X|Y}(x|y)^\alpha \right)$. Furthermore, in the literature $H_\alpha(X|Y)$ is often defined as $\sum_y P_Y(y) H_\alpha(X|Y=y)$, like for Shannon entropy. However, for our purpose, a slightly different definition will be useful. For $1 < \alpha < \infty$, we define the *average conditional* Renyi entropy $H_\alpha(X|Y)$ as

$$H_\alpha(X|Y) = -\log \left(\sum_y P_Y(y) \left(\sum_x P_{X|Y}(x|y)^\alpha \right)^{\frac{1}{\alpha-1}} \right),$$

and as $H_\infty(X|Y) = -\log \left(\sum_y P_Y(y) \max_x P_{X|Y}(x|y) \right)$ for $\alpha = \infty$. This notion is useful in particular because it has the property that if the *average* conditional Renyi entropy is large, then the conditional Renyi entropy is large with high probability:

Lemma 1. *Let* $\alpha > 1$ *(allowing* $\alpha = \infty$*) and* $t \geq 0$*. Then with probability at least* $1 - 2^{-t}$ *(over the choice of* y*)* $H_\alpha(X|Y=y) \geq H_\alpha(X|Y) - t$.

The proof is straightforward and thus omitted. The following lemma follows from well known properties of the Renyi entropy which are easily seen to translate to the average conditional Renyi entropy.

Lemma 2. *For any* $1 < \alpha < \infty$: $H_2(X|Y) \geq H_\infty(X|Y) \geq \frac{\alpha-1}{\alpha} H_\alpha(X|Y)$.

Finally, our notion of average conditional Renyi entropy is such that the privacy amplification theorem of [2] still provides a lower bound on the average conditional collision entropy as we define it (as can easily be seen from the proof given in [2]). However, for us it is convenient to express the smoothness in terms of variational distance rather than entropy, as in [21,20]:

Theorem 4 ([20]). *Let X be a random variable over $\mathcal{X}$, and let F be the random variable corresponding to the random choice of a member of a universal-two class $\mathcal{F}$ of hash functions from $\mathcal{X}$ to $\{0,1\}^\ell$. Then*

$$\delta\big([F(X)F], [\text{UNIF}^\ell][F]\big) \leq 2^{-\frac{1}{2}(H_2(X)-\ell)-1} .$$

B   Quantitative Comparison

We compare the simple reduction of *1-2 OT$^\ell$* to n executions of *1-2 XOT*, *1-2 GOT* and *1-2 UOT*, respectively, using our analysis based on Theorem 2 as discussed in Section 5.1 (together with the quantitative statement given in Theorem 3), with the results achieved in [4].[9] The quality of (the analysis of) a reduction is given by the *reduction parameters* c_{len}, c_{sec} and c_{const} such that the *1-2 OT$^\ell$* is guaranteed to be $2^{-\kappa}$-secure as long as $n \geq c_{\text{len}} \cdot \ell + c_{\text{sec}} \cdot \kappa + c_{\text{const}}$. The smaller these constants are, the better is the (analysis of the) reduction. The comparison of these parameters is given in Figure 3 (we focus on c_{len} and c_{sec} since c_{const} is not really relevant, unless very large).

	1-2 XOT		*1-2 GOT*		*1-2 UOT*	
	c_{len}	c_{sec}	c_{len}	c_{sec}	c_{len}	c_{sec}
BCW [4]	2	2	4.8	4.8	14.6	14.6
this work	4	3	4	3	13.2	10.0

Fig. 3. Comparison of the reduction parameters

The parameters in the first line can easily be extracted from Theorems 5, 7 and 9 of [4] (where in Theorem 9 $p_e \approx 0.19$). The parameters in the second line corresponding to the reductions to *1-2 XOT* and *1-2 GOT* follow immediately from Theorem 3, using the fact that in *one* execution of a *1-2 XOT* or a *1-2 GOT* the receivers average conditional collision entropy (as defined in Appendix A) on the sender's two input bits is at least 1 (in case of *1-2 XOT* this is trivial, and in case of *1-2 GOT* this can easily be computed). The parameters for *1-2 UOT* follow from Theorem 3 and the following observation. If for one execution of the *1-2 UOT* the receiver's average (Shannon) entropy is at least 1, then it follows from Fano's Inequality that his average guessing probability is at most $1 - p_e$

---

[9] As mentioned earlier, these results are incomparable to the parameters achieved in [10], where *interactive* reductions are used.

(with p_e as above), and thus his average conditional min-entropy, which lower bounds the collision entropy, is at least $-\log(1-p_e) \approx 0.3$. c_{len} and c_{sec} are then computed as $c_{\text{len}} \approx 4/0.3$ and $c_{\text{sec}} \approx 3/0.3$.

C   Proof of Theorem 3

Define the event $\mathcal{E} = \{y : H_2(X|Y=y) \geq H_2(X|Y) - \kappa - 1\}$. By Lemma 1 $P[\mathcal{E}] \geq 1 - 2^{-\kappa-1}$. We will show below that conditioned on $\mathcal{E}$, the obliviousness condition of Definition 1 holds with "error term" $2^{-\kappa-1}$. It then follows that

$$\delta([B_{1-D}\,B_D\,W\,D], [\text{UNIF}]\,[B_D\,W\,D])$$
$$\leq \delta\big(P_{B_{1-D}B_DWD\mathcal{E}}, P_{\text{UNIF}}P_{B_DWD\mathcal{E}}\big) + \delta\big(P_{B_{1-D}B_DWD\bar{\mathcal{E}}}, P_{\text{UNIF}}P_{B_DWD\bar{\mathcal{E}}}\big)$$
$$= \delta\big(P_{B_{1-D}B_DWD|\mathcal{E}}, P_{\text{UNIF}}P_{B_DWD|\mathcal{E}}\big)P[\mathcal{E}] + \delta\big(P_{B_{1-D}B_DWD|\bar{\mathcal{E}}}, P_{\text{UNIF}}P_{B_DWD|\bar{\mathcal{E}}}\big)P[\bar{\mathcal{E}}]$$
$$\leq 2^{-\kappa-1} + 2^{-\kappa-1} = 2^{-\kappa}.$$

It remains to prove the claimed obliviousness when conditioning on $\mathcal{E}$. To simplify notation, instead of conditioning on $\mathcal{E}$ we consider a distribution $P_{Y|X}$ with $H_2(X|Y=y) \geq H_2(X|Y) - \kappa - 1$ for all y. Note that $H_2(X|Y) - \kappa - 1 \geq 4\ell + 2\kappa + 3$. Fix an arbitrary y. Consider an arbitrary NDLF $\beta : \{0,1\}^\ell \times \{0,1\}^\ell \to \{0,1\}$. Let F_0 and F_1 be the random variables that represent the random choices of f_0 and f_1, and set $B = \beta(F_0(X_0), F_1(X_1))$. In combination with Proposition 1, privacy amplification (Theorem 4) guarantees that

$$\delta\big(P_{BF_0F_1|Y=y}, P_{\text{UNIF}}P_{F_0F_1|Y=y}\big) \leq 2^{-\frac{1}{2}(H_2(X|Y=y)+1)} \leq 2^{-\frac{1}{2}(4\ell+2\kappa+4)} = 2^{-2\ell-\kappa-2}.$$

It now follows that

$$\delta\big([\beta(S_0,S_1)\,W], [\text{UNIF}]\,[W]\big) = \delta\big(P_{BF_0F_1Y}, P_{\text{UNIF}}P_{F_0F_1Y}\big)$$
$$= \sum_y \delta\big(P_{BF_0F_1|Y=y}, P_{\text{UNIF}}P_{F_0F_1|Y=y}\big)P_Y(y) \leq 2^{-2\ell-\kappa-2}.$$

Obliviousness as claimed now follows from Theorem 2. $\square$

On Expected Constant-Round Protocols
for Byzantine Agreement

Jonathan Katz* and Chiu-Yuen Koo

Dept. of Computer Science, University of Maryland, College Park, USA
{jkatz, cykoo}@cs.umd.edu

Abstract. In a seminal paper, Feldman and Micali (STOC '88) show an n-party Byzantine agreement protocol tolerating $t < n/3$ malicious parties that runs in expected constant rounds. Here, we show an expected constant-round protocol for *authenticated* Byzantine agreement assuming *honest majority* (i.e., $t < n/2$), and relying only on the existence of a secure signature scheme and a public-key infrastructure (PKI). Combined with existing results, this gives the first expected constant-round protocol for secure computation with honest majority in a point-to-point network assuming only one-way functions and a PKI. Our key technical tool — a new primitive we introduce called *moderated VSS* — also yields a simpler proof of the Feldman-Micali result.

We also show a simple technique for sequential composition of protocols without simultaneous termination (something that is inherent for Byzantine agreement protocols using $o(n)$ rounds) for the case of $t < n/2$.

1 Introduction

When designing cryptographic protocols, it is often convenient to abstract away various details of the underlying communication network. As one noteworthy example,[1] it is often convenient to assume the existence of a *broadcast channel* that allows any party to send the same message to all other parties (and all parties to be assured they have received identical messages) in a single round. With limited exceptions (e.g., in a small-scale wireless network), it is understood that the protocol will be run in a network where only point-to-point communication is available and the parties will have to "emulate" the broadcast channel by running a broadcast sub-protocol. Unfortunately, this "emulation" typically increases the round complexity of the protocol substantially.

Much work has therefore focused on reducing the round complexity of broadcast or the related task of *Byzantine agreement* (BA) [30,26]; we survey this work

---

* This research was supported by NSF Trusted Computing grant #0310751, NSF CAREER award #0447075, and US-Israel Binational Science Foundation grant #2004240.
[1] Other "abstractions" include the assumptions of private and/or authenticated channels. For non-adaptive, computationally-bounded adversaries these can be realized without affecting the round complexity using public-key encryption and digital signatures, respectively. See also Section 2.

C. Dwork (Ed.): CRYPTO 2006, LNCS 4117, pp. 445–462, 2006.

in Section 1.2. As discussed there, a seminal result of Feldman and Micali [17] is a protocol for Byzantine agreement in a network of n parties tolerating $t < n/3$ malicious parties that runs in an expected *constant* number of rounds. This resilience is the best possible — regardless of round complexity — unless additional assumptions are made. The most common assumption is the existence of a public-key infrastructure (PKI) such that each party P_i has a public key pk_i for a digital signature scheme that is known to all other parties (a more formal definition is given in Section 2); broadcast or BA protocols in this model are termed *authenticated*. Authenticated broadcast protocols are known for $t < n$ [30,26,13], but all existing protocols that assume only a PKI and secure signatures require $\Theta(t)$ rounds.[2] Recent work of Fitzi and Garay [20] gives an authenticated BA protocol beating this bound, but using specific number-theoretic assumptions; see Section 1.2 for further discussion.

1.1   Our Contributions

As our main result, we extend the work of Feldman and Micali and show an authenticated BA protocol tolerating $t < n/2$ malicious parties and running in expected constant rounds. Our protocol assumes only the existence of signature schemes and a PKI, and is secure against a rushing adversary who adaptively corrupts up to t parties. For those unfamiliar with the specifics of the Feldman-Micali protocol, we stress that their approach does *not* readily extend to the case of $t < n/2$. In particular, they rely on a primitive termed *graded VSS* and construct this primitive using in an essential way the fact that $t < n/3$. We take a different approach: we introduce a new primitive called *moderated VSS* (mVSS) and use this to give an entirely self-contained proof of our result.

We suggest that mVSS is a useful alternative to graded VSS *in general*, even when $t < n/3$. For one, mVSS seems easier to construct: we show a generic construction of mVSS *in the point-to-point model* from any VSS protocol that relies on a broadcast channel, while a generic construction of this sort for graded VSS seems unlikely. Perhaps more importantly, mVSS provides what we believe to be a conceptually-simpler approach to the problem at hand: in addition to our authenticated BA protocol for $t < n/2$, our techniques give a BA protocol (in the plain model) for $t < n/3$ which is more round-efficient than the Feldman-Micali protocol and which also admits a self-contained proof (cf. the full version of this work [24]) that we believe is significantly simpler than that of [17].

Since mVSS is impossible for $t \geq n/2$, our techniques do not extend to the case of dishonest majority. It remains an interesting open question as to whether it is possible to achieve broadcast for $n/2 \leq t < n$ in expected constant rounds.

As mentioned earlier, cryptographic protocols are often designed under the assumption that a broadcast channel is available; when run in a point-to-point network, these protocols must "emulate" the broadcast channel by running a

---

[2] Feldman and Micali claim an expected $O(1)$-round solution for $t < n/2$ in the conference version of their paper, but this claim no longer appears in either the journal version of their work or Feldman's thesis [15].

broadcast protocol as a sub-routine. If the original (outer) protocol uses multiple invocations of the broadcast channel, and these invocations are each emulated using a *probabilistic* broadcast protocol, subtle issues related to the parallel and sequential *composition*[3] of the various broadcast sub-protocols arise; see the detailed discussion in Section 4. Parallel composition can be dealt with using existing techniques [4,20]. There are also techniques available for handling sequential composition of protocols without simultaneous termination [4,28]; however, this work applies only to the case $t < n/3$ [4] or else is rather complex [28]. As an additional contribution, we show an extension of previous work [4] that enables sequential composition for $t < n/2$ (assuming digital signatures and a PKI) in a simpler and more round-efficient manner than [28]. See Section 5.

The above results, in combination with prior work [2,12], yield the first expected constant-round protocol for secure computation in a point-to-point network that tolerates an adversary corrupting a minority of the parties and is based only on the existence of one-way functions and a PKI. (The constant-round protocol of Goldwasser and Lindell [23], which does not assume a PKI, achieves a weaker notion of security that does not guarantee output delivery.)

1.2 Prior Work on Broadcast/Byzantine Agreement

In a synchronous network with pairwise authenticated channels and no additional set-up assumptions, BA among n parties is achievable iff the number of corrupted parties t satisfies $t < n/3$ [30,26]; furthermore, in this case the corrupted parties may be computationally unbounded. In this setting, a lower bound of $t+1$ rounds for any deterministic BA protocol is known [18]. A protocol with this round complexity — but with exponential message complexity — was shown by Pease, et al. [30,26]. Following a long sequence of works, Garay and Moses [21] show a fully-polynomial BA protocol with optimal resilience and round complexity.

To circumvent the above-mentioned lower bound, researchers beginning with Rabin [32] and Ben-Or [3] explored the use of randomization to obtain better round complexity. This line of research [6,10,16,14] culminated in the work of Feldman and Micali [17], who show a randomized BA protocol with optimal resilience $t < n/3$ that runs in an expected *constant* number of rounds. Their protocol requires channels to be both private and authenticated (but see footnote 1 and the next section).

To achieve resilience $t \geq n/3$, additional assumptions are needed even if randomization is used. The most widely-used assumptions are the existence of digital signatures and a public-key infrastructure (PKI); recall that protocols in this setting are termed *authenticated*. Implicit in this setting is that the adversary is computationally bounded (so it cannot forge signatures), though if information-theoretic "pseudo-signatures" [31] are used security can be obtained even against an unbounded adversary. Pease, et al. [30,26] show an authenticated broadcast

---

[3] These issues are unrelated to those considered in [27] where the different executions are oblivious of each other: here, there is an outer protocol scheduling all the broadcast sub-protocols. For the same reason, we do not consider *concurrent* composition since we are interested only in "stand-alone" security of the outer protocol.

protocol for $t < n$, and a fully-polynomial protocol achieving this resilience was given by Dolev and Strong [13]. These works rely only on the existence of digital signature schemes and a PKI, and do not require private channels.

The $(t + 1)$-round lower bound for deterministic protocols holds in the authenticated setting as well [13], and the protocols of [30,26,13] meet this bound. Some randomized protocols beating this bound for the case of $n/3 \leq t < n/2$ are known [34,6,36], but these are only partial results:

- Toueg [34] gives an expected $O(1)$-round protocol, but assumes a trusted dealer. After the dealing phase the parties can only run the BA protocol a *bounded* number of times.

- A protocol by Bracha [6] implicitly requires a trusted dealer to ensure that parties agree on a "Bracha assignment" in advance (see [16]). Furthermore, the protocol only achieves expected round complexity $\Theta(\log n)$ and tolerates (slightly sub-optimal) $t \leq n/(2 + \epsilon)$ for any $\epsilon > 0$.

- Waidner [36], building on [6,16], shows that the dealer in Bracha's protocol can be replaced by an $\Omega(t)$-round pre-processing phase during which a broadcast channel is assumed. The expected round complexity (after the pre-processing) is also improved from $\Theta(\log n)$ to $\Theta(1)$.

The latter two results assume private channels.

Fitzi and Garay [20], building on [34,7,29], give the first full solution to this problem: that is, they show the first authenticated BA protocol with optimal resilience $t < n/2$ and expected constant round complexity that does not require any trusted dealer or pre-processing (other than a PKI). Even assuming private channels, however, their protocol requires specific number-theoretic assumptions (essentially, some appropriately-homomorphic public-key encryption scheme) and not signatures alone. We remark that, because of its reliance on additional assumptions, the Fitzi-Garay protocol cannot be adapted to the information-theoretic setting using pseudo-signatures.

2   Model and Technical Preliminaries

By a *public-key infrastructure* (PKI) in a network of n parties, we mean that prior to any protocol execution all parties hold the same vector $(pk_1, \ldots, pk_n)$ of public keys for a digital signature scheme, and each honest party P_i holds the honestly-generated secret key sk_i associated with pk_i. Malicious parties may generate their keys arbitrarily, even dependent on keys of honest parties.

Our results are in the *point-to-point* model (unless otherwise stated), by which we mean the standard model in which parties communicate in synchronous rounds using pairwise *private* and *authenticated* channels. Authenticated channels are necessary for our work, but can be realized using signature schemes if one is willing to assume a PKI. (Note that signatures are stronger than authenticated channels since they allow third-party verifiability.) For static adversaries, private channels can be realized using one additional round by having each party P_i send to each party P_j a public key $PK_{i,j}$ for a semantically-secure public-key encryption scheme (using a different key for each sender avoids issues of malleability). For adaptive adversaries, more complicated solutions are available [1,9] but we do not discuss these further. For simplicity, we assume *unconditional* private/authenticated channels with the understanding that these guarantees hold only computationally if the above techniques are used.

When we say a protocol (for Byzantine agreement, VSS, etc.) tolerates t malicious parties, we always mean that it is secure against a *rushing* adversary who may *adaptively* corrupt up to t parties during execution of the protocol and coordinate the actions of these parties as they deviate from the protocol in an arbitrary manner. Parties not corrupted by the adversary are called *honest*. Our definitions always implicitly cover both the "unconditional" and "authenticated" cases, in the following way: For $t < n/3$ we allow a computationally-unbounded adversary (this is the unconditional case). For $t < n/2$ we assume a PKI and also assume that the adversary cannot forge a new valid signature on behalf of any honest party (this is the authenticated case). Using a standard hybrid argument and assuming the existence of one-way functions, this implies that authenticated protocols are secure against computationally-bounded adversaries. Using pseudo-signatures, authenticated protocols can be made secure even against a computationally-unbounded adversary (though in a statistical, rather than perfect, sense).

When we describe signature computation in authenticated protocols we often omit for simplicity additional information that must be signed along with the message. Thus, when we say that party P_i signs message m and sends it to P_j, we implicitly mean that P_i signs the concatenation of m with additional information such as: (1) the identity of the recipient P_j, (2) the current round number, (3) an identifier for the message (in case multiple messages are sent to P_j in the same round); and (4) an identifier for the (sub-)protocol (in case multiple sub-protocols are being run; cf. [27]).

Byzantine Agreement. We focus on Byzantine agreement, which readily implies broadcast. The standard definitions of BA and broadcast follow.

Definition 1. *(Byzantine agreement): A protocol for parties $P_1, \ldots, P_n$, where each party P_i holds initial input v_i, is a* Byzantine agreement protocol tolerating t malicious parties *if the following conditions hold for any adversary controlling at most t parties:*

- *All honest parties output the same value.*
- *If all honest parties begin with the same input value v, then all honest parties output v.*

If the $\{v_i\}$ are restricted to binary values, the protocol achieves binary Byzantine agreement. ◇

Definition 2. *(Broadcast): A protocol for parties $\mathcal{P} = \{P_1, \ldots, P_n\}$, where a distinguished dealer $P^* \in \mathcal{P}$ holds an initial input M, is a* broadcast protocol tolerating t malicious parties *if the following conditions hold for any adversary controlling at most t parties:*

- *All honest parties output the same value.*
- *If the dealer is honest, then all honest parties output M.*

3   Byzantine Agreement in Expected Constant Rounds

In this section, we construct expected constant-round protocols for Byzantine agreement in both the unconditional ($t < n/3$) and authenticated ($t < n/2$) settings. Our main result is the protocol for the case $t < n/2$ (which is the first such construction assuming only a PKI and digital signatures); however, we believe our result for the case $t < n/3$ is also interesting as an illustration that our techniques yield a conceptually-simpler and more efficient protocol in that setting as compared to [17]. We develop both protocols in parallel so as to highlight the high-level similarities in each.

We begin by reviewing the notions of *gradecast* and *VSS*:

Gradecast. *Gradecast*, a relaxed version of broadcast, was introduced by Feldman and Micali [17]; below we provide a definition which is slightly weaker than the one in [17, Def. 11], but suffices for our purposes.

Definition 3. *(Gradecast): A protocol for parties* $\mathcal{P} = \{P_1, \ldots, P_n\}$, *where a distinguished dealer* $P^* \in \mathcal{P}$ *holds an initial input* M, *is a* gradecast *protocol* tolerating t malicious parties *if the following conditions hold for any adversary controlling at most t parties:*

- *Each honest party* P_i *eventually terminates the protocol and outputs both a message* m_i *and a grade* $g_i \in \{0, 1, 2\}$.
- *If the dealer is honest, then the output of every honest party* P_i *satisfies* $m_i = M$ *and* $g_i = 2$.
- *If there exists an honest party* P_i *who outputs a message* m_i *and grade* $g_i = 2$, *then the output of every honest party* P_j *satisfies* $m_j = m_i$ *and* $g_j \geq 1$.   ◇

The first result that follows is due to [17]. A proof of the second result can be found in the full version of this paper [24].

Lemma 1. *There exists a constant-round gradecast protocol tolerating* $t < n/3$ *malicious parties.*

Lemma 2. *There exists a constant-round authenticated gradecast protocol tolerating* $t < n/2$ *malicious parties.*

Verifiable Secret Sharing (VSS). *VSS* [11] extends the concept of secret sharing [5,33] to the case of Byzantine faults. Below we provide what is essentially the standard definition except that for technical reasons (namely, security under parallel composition) we explicitly incorporate a notion of "extraction."

Definition 4. *(Verifiable secret sharing): A two-phase protocol for parties* $\mathcal{P} = \{P_1, \ldots, P_n\}$, *where a distinguished* dealer $P^* \in \mathcal{P}$ *holds initial input* s, *is a* VSS protocol tolerating t malicious parties *if the following conditions hold for any adversary controlling at most t parties:*

Validity *Each honest party P_i outputs a value s_i at the end of the second phase (the* reconstruction phase*). Furthermore, if the dealer is honest then $s_i = s$.*

Secrecy *If the dealer is honest during the first phase (the* sharing phase*), then at the end of this phase the joint view of the malicious players is independent of the dealer's input s.*

Extraction *At the end of the sharing phase the joint view of the honest parties defines a value s' (which can be computed in polynomial time from this view) such that all honest parties will output s' at the end of the reconstruction phase.* ◇

The first result that follows is well known (see, e.g., [22,19]). The second result follows readily from existing results; see the full version for a proof [24].

Lemma 3. *There exists a constant-round VSS protocol tolerating $t < n/3$ malicious parties that relies on a broadcast channel only during the sharing phase.*

Lemma 4. *There exists a constant-round authenticated VSS protocol tolerating $t < n/2$ malicious parties that relies on a broadcast channel only during the sharing phase.*

3.1 Moderated VSS

We introduce a variant of VSS called *moderated VSS*, in which there is a distinguished party (who may be identical to the dealer) called the *moderator*. Roughly speaking, the moderator "simulates" a broadcast channel for the other parties. At the end of the sharing phase, parties output a boolean flag indicating whether or not they trust the moderator. If the moderator is honest, all honest parties should set this flag to 1. Furthermore, if any honest party sets this flag to 1 then the protocol should achieve all the properties of VSS (cf. Def. 4). A formal definition follows.

Definition 5. *(Moderated VSS): A two-phase protocol for parties $\mathcal{P} = \{P_1, \ldots, P_n\}$, where there is a distinguished dealer $P^* \in \mathcal{P}$ who holds an initial input s and a moderator $P^{**} \in \mathcal{P}$ (who may possibly be the dealer), is a* moderated VSS *protocol tolerating t malicious parties if the following conditions hold for any adversary controlling at most t parties:*

- *Each honest party P_i outputs a bit f_i at the end of the first phase (the* sharing phase*), and a value s_i at the end of the second phase (the* reconstruction phase*).*
- *If the moderator is honest during the sharing phase, then each honest party P_i outputs $f_i = 1$ at the end of this phase.*
- *If there exists an honest party P_i who outputs $f_i = 1$ at the end of the sharing phase, then the protocol achieves VSS; specifically: (1) if the dealer is honest then all honest parties output s at the end of the reconstruction phase, and the joint view of all the malicious parties at the end of the sharing phase is independent of s, and (2) the joint view of the honest parties at the*

end of the sharing phase defines an efficiently-computable value s' such that all honest parties output s' at the end of the reconstruction phase.   ◇

We stress that if all honest parties P_i output $f_i = 0$ at the end of the sharing phase, then no guarantees are provided: e.g., honest parties may output different values at the end of the reconstruction phase, or the malicious players may learn the dealer's secret in the sharing phase.

The main result of this section is the following, which holds for any $t < n$:

Theorem 1. *Assume there exists a constant-round VSS protocol Π, using a broadcast channel in the sharing phase only, which tolerates t malicious parties. Then there exists a constant-round moderated VSS protocol Π', using a gradecast channel, which tolerates t malicious parties.*

Proof. We show how to "compile" Π so as to obtain the desired Π'. Essentially, Π' is constructed by replacing each broadcast in Π with two invocations of gradecast: one by the party who is supposed to broadcast the message, and one by the moderator P^{**}. In more detail, Π' is defined as follows: At the beginning of the protocol, all parties set their flag f to 1. The parties then run an execution of Π. When a party P is directed by Π to send message m to P', it simply sends this message. When a party P is directed by Π to broadcast a message m, the parties run the following "simulated broadcast" subroutine:

1. P gradecasts the message m.
2. The moderator P^{**} gradecasts the message it output in the previous step.
3. Let (m_i, g_i) and (m'_i, g'_i) be the outputs of party P_i in steps 1 and 2, respectively. Within the underlying execution of Π, party P_i will use m'_i as the message "broadcast" by P.
4. Furthermore, P_i sets $f_i := 0$ if either (or both) of the following conditions hold: (1) $g'_i \neq 2$, or (2) $m'_i \neq m_i$ and $g_i = 2$.

Party P_i outputs f_i at the end of the sharing phase, and outputs whatever it is directed to output by Π at the end of the reconstruction phase.

We now prove that Π' is a moderated VSS protocol tolerating t malicious parties. First, note that if the moderator is honest during the sharing phase then no honest party P_i ever sets $f_i := 0$. To see this, note that if P^{**} is honest then $g'_i = 2$ each time the simulated broadcast subroutine is executed. Furthermore, if P_i outputs some m_i and $g_i = 2$ in step 1 of that subroutine then, by definition of gradecast, P^{**} also outputs m_i in step 1. Hence $m'_i = m_i$ and f_i remains 1.

To show the second required property of moderated VSS, consider any execution of the simulated broadcast subroutine. We show that if there exists an honest party P_i who holds $f_i = 1$ upon completion of that subroutine, then the functionality of broadcast was achieved (in that execution of the subroutine). It follows that if P_i holds $f_i = 1$ at the end of the sharing phase, then Π' provided a faithful execution of all broadcasts in Π and so the functionality of VSS is achieved.

If P_i holds $f_i = 1$, then $g'_i = 2$. (For the remainder of this paragraph, all variables are local to a particular execution of the broadcast subroutine.) Since

$g'_i = 2$, the properties of gradecast imply that any honest party P_j holds $m'_j = m'_i$ and so all honest parties agree on the message that was "broadcast." Furthermore, if the "dealer" P (in the simulated broadcast subroutine) is honest then $g_i = 2$ and $m_i = m$. So the fact that $f_i = 1$ means that $m'_i = m_i = m$, and so all honest parties use the message m "broadcast" by P in their underlying execution of Π. ∎

By applying the above theorem to the VSS protocol of Lemma 3 (resp., Lemma 4) and then instantiating the gradecast channel using the protocol of Lemma 1 (resp., Lemma 2), we obtain:

Corollary 1. *There exists a constant-round protocol for moderated VSS tolerating $t < n/3$ malicious parties.*

Corollary 2. *There exists a constant-round protocol for authenticated moderated VSS tolerating $t < n/2$ malicious parties.*

3.2   From Moderated VSS to Oblivious Leader Election

In this section, we construct an oblivious leader election (OLE) protocol based on any moderated VSS protocol. The following definition of oblivious leader election is adapted from [20]:

Definition 6. *(Oblivious leader election): A two-phase protocol for parties P_1, ..., P_n is an* oblivious leader election protocol with fairness δ tolerating t malicious parties *if each honest party P_i outputs a value $v_i \in [n]$, and the following condition holds with probability at least δ (over random coins of the honest parties) for any adversary controlling at most t parties:*

> *There exists a $j \in [n]$ such that (1) each honest party P_i outputs $v_i = j$, and (2) P_j was honest at the end of the first phase.*

If the above event happens, we say an honest leader was elected. ◇

Our construction of OLE uses a similar high-level approach as in the construction of an oblivious common coin from graded VSS [17]. However, we introduce different machinery and start from *moderated* VSS. Intuitively, we generate a random coin $c_i \in [n^4]$ for each party P_i. This is done by having each party P_j select a random value $c_{j,i} \in [n^4]$ and then share this value using moderated VSS with P_i acting as moderator. The $c_{j,i}$ are then reconstructed and c_i is computed as $c_i = \sum_j c_{j,i} \bmod n^4$. An honest party then outputs i minimizing c_i. Since moderated VSS (instead of VSS) is used, each party P_k may have a different view regarding the value of the $\{c_i\}$. However:

- If P_i is honest then (by the properties of moderated VSS) all honest parties reconstruct the same values $c_{j,i}$ (for any j) and hence compute an identical value for c_i.
- If P_i is dishonest, but there exists an honest party P_j such that P_j outputs $f_j = 1$ in all invocations of moderated VSS where P_i acts as the moderator, then (by the properties of moderated VSS) all honest parties compute an identical value for c_i.

Relying on the above observations, we devise a way such that all honest parties output the same i (such that P_i is furthermore honest) with constant probability.

Theorem 2. *Assume there exists a constant-round moderated VSS protocol tolerating t malicious parties. Then there exists a constant-round OLE protocol with fairness $\delta = \frac{n-t}{n} - \frac{1}{n^2}$ tolerating t malicious parties. Specifically, if $n \geq 3$ and $t < n/2$ then $\delta \geq 1/2$.*

Proof. Each party P_i begins with $\mathsf{trust}_{i,j} = 1$ for $j \in \{1, \ldots, n\}$.

Phase 1: Each party P_i chooses random $c_{i,j} \in [n^4]$ for $1 \leq j \leq n$. The following is executed n^2 times in parallel for each ordered pair (i, j):

All parties execute the sharing phase of a moderated VSS protocol in which P_i acts as the dealer with initial input $c_{i,j}$, and P_j acts as the moderator. If a party P_k outputs $f_k = 0$ in this execution, then P_k sets $\mathsf{trust}_{k,j} := 0$.

Upon completion of the above, let $\mathsf{trust}_k \overset{\mathrm{def}}{=} \{j : \mathsf{trust}_{k,j} = 1\}$.

Phase 2: The reconstruction phase of the moderated VSS protocol is run n^2 times in parallel to reconstruct the secrets previously shared. Let $c_{i,j}^k$ denote P_k's view of the value of $c_{i,j}$. (If a reconstructed value lies outside $[n^4]$, then $c_{i,j}^k$ is assigned some default value in the correct range.) Each party P_k sets $c_j^k := \sum_{i=1}^n c_{i,j}^k \bmod n^4$, and outputs $j \in \mathsf{trust}_k$ that minimizes c_j^k.

We prove that the protocol satisfies Definition 6. Following execution of the above, define:

$$\mathsf{trusted} = \left\{ k : \begin{array}{c} \text{there exists a } P_i \text{ that was honest at the end of phase 1} \\ \text{for which } k \in \mathsf{trust}_i \end{array} \right\}.$$

If P_i was honest in phase 1, then $i \in \mathsf{trusted}$. Furthermore, by the properties of moderated VSS, if $k \in \mathsf{trusted}$ then for any honest P_i, P_j and any $1 \leq \ell \leq n$, we have $c_{\ell,k}^i = c_{\ell,k}^j$ and hence $c_k^i = c_k^j$; thus, we may freely omit the superscript in this case. We claim that for $k \in \mathsf{trusted}$, the coin c_k is uniformly distributed in $[n^4]$. Let $c_k' = \sum_{\ell : P_\ell \text{ malicious in phase 1}} c_{\ell,k} \bmod n^4$ (this is the contribution to c_k of the parties that are malicious in phase 1), and let P_i be honest. Since $k \in \mathsf{trusted}$, the properties of VSS hold for all secrets $\{c_{\ell,k}\}_{\ell=1}^n$ and thus c_k' is independent of $c_{i,k}$. (If we view moderated VSS as being provided unconditionally, independence holds trivially. When this is instantiated with a protocol for moderated VSS, independence follows from the information-theoretic security of moderated VSS.[4]) It follows that c_k is uniformly distributed in $[n^4]$.

By union bound, with probability at least $1 - \frac{1}{n^2}$, all coins $\{c_k : k \in \mathsf{trusted}\}$ are distinct. Conditioned on this event, with probability at least $\frac{n-t}{n}$ the party

---

[4] Formally, one could define an appropriate ideal functionality within the UC framework [8] and show that any protocol for moderated VSS implements this functionality (relying on the extraction property in Definition 4); security under parallel composition then follows. One could also appeal to a recent result showing security of statistically-secure protocols under parallel self composition when inputs are not chosen adaptively (as is the case here) [25].

with the minimum c_j among the set trusted corresponds to a party which was honest in phase 1. This concludes the proof.   ∎

Combining Theorem 2 with Corollaries 1 and 2, we obtain:

Corollary 3. *There exists a constant-round protocol for OLE with fairness 2/3 tolerating $t < n/3$ malicious parties. (Note that when $n < 4$ the result is trivially true.)*

Corollary 4. *There exists a constant-round protocol for authenticated OLE with fairness 1/2 tolerating $t < n/2$ malicious parties. (Note that when $n < 3$ the result is trivially true.)*

3.3   From OLE to Byzantine Agreement

For the unconditional case (i.e., $t < n/3$), Feldman and Micali [17] show how to construct an expected constant-round binary Byzantine agreement protocol based on any constant-round *oblivious common coin* protocol. We construct a more round-efficient protocol based on oblivious leader election. This also serves as a warmup for the authenticated case.

Theorem 3. *If there exists a constant-round OLE protocol with fairness $\delta = \Omega(1)$ tolerating $t < n/3$ malicious parties, then there exists an expected constant-round binary Byzantine agreement protocol tolerating t malicious parties.*

Proof. We describe a protocol for binary Byzantine agreement, assuming the existence of an OLE protocol tolerating $t < n/3$ malicious parties. Each party P_i uses local variables $b_i \in \{0, 1\}$ (which is initially P_i's input), $\texttt{lock}_i$ (initially set to ∞), and $\texttt{accept}_i$ (initially set to $\texttt{false}$).

Step 1: Each P_i sends b_i to all parties. Let $b_{j,i}$ be the bit P_i receives from P_i. (When this step is run at the outset of the protocol, a default value is used if P_i does not receive anything from P_j. In subsequent iterations, if P_i does not receive anything from P_j then P_i leaves $b_{j,i}$ unchanged.)

Step 2: Each party P_i sets $\mathcal{S}_i^b := \{j : b_{j,i} = b\}$ for $b \in \{0, 1\}$. If $|\mathcal{S}_i^0| \geq t + 1$, then P_i sets $b_i := 0$. If $|\mathcal{S}_i^0| \geq n - t$, then P_i sets $\texttt{lock}_i := 0$.
Each P_i sends b_i to all parties. If P_i receives a bit from P_j, then P_i sets $b_{j,i}$ to that value; otherwise, $b_{j,i}$ remains unchanged.

Step 3: Each party P_i defines $\mathcal{S}_i^b$ as in step 2. If $|\mathcal{S}_i^1| \geq t + 1$, then P_i sets $b_i := 1$. If $|\mathcal{S}_i^1| \geq n - t$, then P_i sets $\texttt{lock}_i := 0$.
Each P_i sends b_i to all parties. If P_i receives a bit from P_j, then P_i sets $b_{j,i}$ to that value; otherwise, $b_{j,i}$ remains unchanged.
If $\texttt{lock}_i = \infty$, then P_i sets $\texttt{accept}_i := \texttt{true}$.

Step 4: Each party P_i defines $\mathcal{S}_i^b$ as in step 2. If $|\mathcal{S}_i^0| \geq t + 1$, then P_i sets $b_i := 0$. If $|\mathcal{S}_i^0| \geq n - t$, then P_i sets $\texttt{accept}_i := \texttt{false}$.
Each P_i sends b_i to all parties. If P_i receives a bit from P_j, then P_i sets $b_{j,i}$ to that value; otherwise, $b_{j,i}$ remains unchanged.

Step 5: Each party P_i defines $\mathcal{S}_i^b$ as in step 2. If $|\mathcal{S}_i^1| \geq t + 1$, then P_i sets $b_i := 1$. If $|\mathcal{S}_i^1| \geq n - t$, then P_i sets $\mathtt{accept}_i := \mathtt{false}$.

Each P_i sends b_i to all parties. If P_i receives a bit from P_j, then P_i sets $b_{j,i}$ to that value; otherwise, $b_{j,i}$ remains unchanged.

Step 6: All parties execute the OLE protocol; let ℓ_i be the output of P_i. Each P_i does the following: if $\mathtt{accept}_i = \mathtt{true}$, then P_i sets $b_i := b_{\ell_i,i}$. If $\mathtt{lock}_i = 0$, then P_i outputs b_i and terminates; otherwise, P_i goes to step 1.

We refer to an execution of step 1 through 6 as an *iteration*. First we claim that if an honest P_i sets $\mathtt{lock}_i := 0$ in step 2 or 3 of some iteration, then all honest parties P_j hold $b_j = b_i$ by the end of step 3 of that same iteration. Consider the case when P_i sets $\mathtt{lock}_i := 0$ in step 2. (The case when P_i sets $\mathtt{lock}_i := 0$ in step 3 is exactly analogous.) This implies that $|S_i^0| \geq n - t$ and hence $|S_j^0| \geq n - 2t \geq t + 1$ and $b_j = 0$. Since this holds for all honest players P_j, it follows that in step 3 we have $|S_j^1| \leq t$ and so b_j remains 1.

Next, we show that if — immediately prior to any given iteration — no honest parties have terminated and there exists a bit b such that $b_i = b$ for all honest P_i, then by the end of step 3 of that iteration all honest parties P_i hold $b_i = b$ and $\mathtt{lock}_i = 0$. This follows easily (by what we have argued in the previous paragraph) once we show that there exists an honest party who sets $\mathtt{lock}_i := 0$ while holding $b_i = b$. Consider the case $b = 0$ (the case $b = 1$ is exactly analogous). In this case $|\mathcal{S}_i^0| \geq n - t$ in step 2 for any honest P_i. Thus, any honest P_i sets $\mathtt{lock}_i := 0$ and holds $b_i = 0$ by the end of this step.

Arguing exactly as in the previous two paragraphs, one can similarly show: (i) if — immediately prior to any given iteration — there exists a bit b such that $b_i = b$ for all honest P_i, then by the end of step 5 of that iteration all honest parties P_i hold $b_i = b$, $\mathtt{lock}_i = 0$, and $\mathtt{accept}_i = \mathtt{false}$ (and hence all honest parties output b and terminate the protocol in that iteration). (ii) If an honest party P_i sets $\mathtt{lock}_i := 0$ in some iteration, then all honest parties P_j hold $b_j = b_i$ and $\mathtt{accept}_j = \mathtt{false}$ by the end of step 5 of that iteration. (iii) If an honest party P_i sets $\mathtt{accept}_i := \mathtt{false}$ in some iteration, then all honest parties P_j hold $b_j = b_i$ by the end of step 5 of that same iteration.

Next, we show that if an honest party P_i outputs $b_i = b$ (and terminates) in some iteration, then all honest parties output b and terminate by the end of the next iteration. Note that if P_j fails to receive b_i from P_i, then $P_{i,j}$ is unchanged; thus, if P_i terminates with output $b_i = b$, it can be viewed as if P_i keeps on sending $b_i = b$ in the next iteration. (In particular, note that P_i must have sent $b_i = b$ in step 5.) Hence it suffices to show that by the end of the (current) iteration, $b_j = b$ for all honest parties P_j. But this is implied by (ii), above.

Finally, we show that if an honest leader[5] P_ℓ is elected in step 6 of some iteration, then all honest parties P_i terminate by the end of the next iteration. By (i), it is sufficient to show that $b_i = b_{\ell,i} = b_\ell$ at the end of step 6 of the current iteration. Consider two sub-cases: if all honest P_j hold $\mathtt{accept}_j = \mathtt{true}$ then this is immediate. Otherwise, say honest P_i holds $\mathtt{accept}_i = \mathtt{false}$. By (iii), $b_\ell = b_i$

---

[5] This implies that P_ℓ was uncorrupted in step 5 of the iteration in question.

at the end of step 5, and hence all honest parties P_j have $b_j = b_i$ by the end of step 6.

If the OLE protocol elects an honest leader with constant probability, it follows that the above protocol is an expected constant-round binary Byzantine agreement protocol.   ∎

When $t < n/3$, any binary Byzantine agreement protocol can be transformed into a (multi-valued) Byzantine agreement protocol using two additional rounds [35]. In Section 4, we show how parallel composition can also be used to achieve the same result without using any additional rounds. Using either approach in combination with the above and Corollary 3, we have:

Corollary 5. *There exists an expected constant-round protocol for Byzantine agreement (and hence also broadcast) tolerating $t < n/3$ malicious parties.*

For the authenticated case (i.e., $t < n/2$), Fitzi and Garay [20] show a construction of expected constant-round binary Byzantine agreement based on OLE; however, they do not explicitly describe how to achieve termination. We construct a multi-valued Byzantine agreement protocol based on OLE and explicitly show how to achieve termination. In the proof below, we assume a gradecast channel for simplicity of exposition; in the full version, we improve the round complexity of our protocol by working directly in the point-to-point model.

Theorem 4. *Assume there exist a constant-round authenticated gradecast protocol and a constant-round authenticated OLE protocol with fairness $\delta = \Omega(1)$, both tolerating $t < n/2$ malicious parties. Then there exists an expected constant-round authenticated Byzantine agreement protocol tolerating t malicious parties.*

Proof. Let V be the domain of possible input values, let the input value for party P_i be $v_i \in V$, and let $\phi \in V$ be some default value. Each P_i begins with an internal variable lock_i set to ∞.

Step 1: Each P_i gradecasts v_i. Let $(v_{j,i}, g_{j,i})$ be the output of P_i in the gradecast by P_j.

Step 2: For any v such that $v = v_{j,i}$ for some j, party P_i defines the sets $\mathcal{S}_i^v := \{j : v_{j,i} = v \wedge g_{j,i} = 2\}$ and $\tilde{\mathcal{S}}_i^v := \{j : v_{j,i} = v \wedge g_{j,i} \geq 1\}$. If $\text{lock}_i = \infty$, then:

    1. If there exists a v such that $|\tilde{\mathcal{S}}_i^v| > n/2$, then $v_i := v$; otherwise, $v_i := \phi$.

    2. If $|\mathcal{S}_i^{v_i}| > n/2$, then $\text{lock}_i := 1$.

Step 3: Each P_i gradecasts v_i. Let $(v_{j,i}, g_{j,i})$ be the output of P_i in the gradecast by P_j.

Step 4: For any v such that $v = v_{j,i}$ for some j, party P_i defines $\mathcal{S}_i^v$ and $\tilde{\mathcal{S}}_i^v$ as in step 2. If $\text{lock}_i = \infty$, then:

    – If there exists a v such that $|\tilde{\mathcal{S}}_i^v| > n/2$, then $v_i := v$; otherwise, $v_i := \phi$.

    P_i sends v_i to all parties. Let $v_{j,i}$ be the value P_i receives from P_j.

Step 5: All parties execute the OLE protocol; let ℓ_i be the output of P_i.

Step 6: Each P_i does the following: if $\text{lock}_i = \infty$ and $|\mathcal{S}_i^{v_i}| \leq n/2$, then P_i sets $v_i := v_{\ell_i, i}$.

Step 7: If $\texttt{lock}_i = 0$, then P_i outputs v_i and terminates the protocol. If $\texttt{lock}_i = 1$, then P_i sets $\texttt{lock}_i := 0$ and goes to step 1. If $\texttt{lock}_i = \infty$, then P_i goes to step 1.

We refer an execution of steps 1 through 7 as an *iteration*. An easy observation is that once an honest party P_i sets $\texttt{lock}_i := 1$, then v_i remains unchanged for the remainder of the protocol and P_i outputs v_i (and terminates) at the end of the next iteration. We first claim that if — immediately prior to any given iteration — there exists a value v such that $v_i = v$ for all honest P_i, then after step 2 of that iteration $v_i = v$ and $\texttt{lock}_i \neq \infty$ for all honest P_i. To see this, note that in this case all honest parties gradecast v in step 1. Furthermore, by the properties of gradecast, all honest parties will be in $\mathcal{S}_i^v$ and $\tilde{\mathcal{S}}_i^v$ for any honest P_i. Assuming honest majority, it follows that $v_i = v$ and $\texttt{lock}_i \neq \infty$ for all honest P_i after step 2, and the claim follows.

Consider the first iteration in which an honest party P_i sets $\texttt{lock}_i := 1$ (in step 2), and let P_j be a different honest party. We show that $v_j = v_i$ by the end of that same iteration (regardless of the outcome of the OLE protocol). To see this, note that by definition of gradecast $\mathcal{S}_j^{v_i} \subseteq \tilde{\mathcal{S}}_i^{v_i}$ and so $v_j = v_i$ after step 2 (whether or not P_j also sets $\texttt{lock}_j := 1$ at this point). Since this holds for all honest parties, every honest party gradecasts v_i in step 3 and thus $|\mathcal{S}_j^{v_i}| > n/2$. It follows that $v_j = v_i$ at the end of that iteration.

Combining the above arguments, it follows that if an honest P_i sets $\texttt{lock}_i := 1$ in some iteration, then all honest P_j will have $\texttt{lock}_j \neq \infty$ by the end of next iteration, and all honest parties will output an identical value and terminate by the end of the iteration after that.

What remains to show is that if an honest leader[6] P_ℓ is elected in some iteration, then all honest parties P_i will hold the same value v_i by the end of that iteration. By what we have argued already above, we only need to consider the case where $\texttt{lock}_i = \infty$ for all honest parties P_i in step 4. Now, if in step 6 all honest P_i have $|\mathcal{S}_i^{v_i}| \leq n/2$, it follows easily that each honest P_i sets $v_i := v_{\ell,i} = v_\ell$ (using the fact that P_ℓ was honest in step 4) and so all honest parties hold the same value v_i at the end of step 6. So, say there exists an honest party P_i such that $|\mathcal{S}_i^{v_i}| > n/2$ in step 6. Consider another honest party P_j:

- If $|\mathcal{S}_j^{v_j}| > n/2$, then $\mathcal{S}_i^{v_i} \cap \mathcal{S}_j^{v_j} \neq \emptyset$ and (by properties of gradecast) $v_i = v_j$.
- If $|\mathcal{S}_j^{v_j}| \leq n/2$, then P_j sets $v_j := v_{\ell,j}$. But, using properties of gradecast, $\mathcal{S}_i^{v_i} \subseteq \tilde{\mathcal{S}}_\ell^{v_i}$ and so P_ℓ set $v_\ell := v_i$ in step 4 (since P_ℓ was honest at that point). Hence $v_{\ell,j} = v_\ell = v_i$.

This concludes the proof. ∎

Combining Lemma 2, Corollary 4, and Theorem 4 we obtain our main result:

Corollary 6. *There exists an expected constant-round protocol for authenticated Byzantine agreement tolerating $t < n/2$ malicious parties.*

We refer the reader to the full version of this work [24] for an exact calculation of the expected round complexities of our BA protocols.

---

[6] This implies that P_ℓ was uncorrupted in step 4 of the iteration in question.

4   Secure MPC in Expected Constant Rounds

Beaver, et al. [2] and Damgård and Ishai [12] show computationally-secure constant-round protocols Π for secure multi-party computation tolerating dishonest minority, assuming the existence of one-way functions and a broadcast channel (the solution of [12] is secure against an adaptive adversary). To obtain an expected constant-round protocol Π' in the point-to-point model (assuming one-way functions and a PKI), a natural approach is to replace each invocation of the broadcast channel in Π with an invocation of an expected constant-round (authenticated) broadcast protocol bc; i.e., to set $\Pi' = \Pi^{\mathsf{bc}}$. There are two subtle problems with this approach that we must deal with:

Parallel Composition. In protocol Π, all n parties may access the broadcast channel in the same round; this results in n parallel executions of bc in protocol Π^{bc}. Although the expected round complexity of *each* execution of bc is constant, the expected number of rounds for *all* n executions of bc to terminate may no longer be constant.

A general technique for handling this issue is proposed by [4]; their solution is somewhat complicated. In our case, however, we may rely on an idea of Fitzi and Garay [20] that applies to OLE-based protocols such as ours. The main idea is that when multiple broadcast sub-routines are run in parallel, only a *single* leader election (per iteration) is required for all of these sub-routines. Using this approach, the expected round complexity for n parallel executions will be identical to the expected round complexity of a single execution. We defer to the full version a detailed exposition of their method as applied to our protocol.

Sequential Composition. A second issue is that protocol bc does not provide simultaneous termination. (As noted in [28], this is inherent for any expected constant-round broadcast protocol.) This may cause problems both in the underlying protocol Π as well as in sequential executions of bc within Π^{bc}.

Existing methods for dealing with this issue either apply only in the case $t < n/3$ [4] or else are rather complex [28]. In the full version [24], we show a simple way to deal with sequential composition for the case of $t < n/2$ (assuming digital signatures and a PKI); a high-level overview is given in Section 5.

Taking the above into account, we obtain the following (security without abort is the standard notion of security in the case of honest majority; see [23]):

Theorem 5. *Assuming the existence of one-way functions, for every probabilistic polynomial-time functionality f there exists an expected constant-round protocol for computing f in a point-to-point network with a PKI. The protocol is secure **without** abort, and tolerates $t < n/2$ adaptive corruptions.*

5   Sequential Composition

We have already noted that certain difficulties arise due to sequential composition of protocols without simultaneous termination (see also the discussion in [28]). As an example of what can go wrong, assume protocol Π (that relies on a broadcast channel) requires a party P_i to broadcast values in rounds 1 and 2. Let $\mathsf{bc}_1, \mathsf{bc}_2$ denote the corresponding invocations of broadcast within the composed protocol Π^{bc} (which runs in a point-to-point network). Then, because honest parties in bc_1 do not terminate in the same round, honest parties may begin execution of bc_2 in different rounds. But security of bc_2 is no longer guaranteed in this case!

At a high level, we can fix this by making sure that bc_2 remains secure as long as all honest parties begin execution of bc_2 within a certain number of rounds. Specifically, if honest parties are guaranteed to terminate bc_1 within g rounds of each other, then bc_2 must remain secure as long as all honest parties start within g rounds. We now show how to achieve this for an arbitrary number of sequential executions of bc, without blowing up the round complexity too much.

Let us first formally define the *staggering gap* of a protocol:

Definition 7. *A protocol Π has* staggering gap *g if any honest parties P_i, P_j are guaranteed to terminate Π within g rounds of each other.*

Let $\mathsf{rc}(\Pi)$ denote the round complexity of a protocol Π. A proof of the following appears in the full version [24]:

Lemma 5. *Let* bc *be a protocol for (authenticated) broadcast with staggering gap g. Then for any constant $c \geq 0$ there exists a protocol* $\mathsf{Expand}'(\mathsf{bc})$ *which achieves the same security as* bc *as long as all honest parties begin execution of* $\mathsf{Expand}'(\mathsf{bc})$ *within c rounds of each other. Furthermore,*

$$\mathsf{rc}\left(\mathsf{Expand}'(\mathsf{bc})\right) = (2c + 1) \cdot \mathsf{rc}(\mathsf{bc}) + 1,$$

and the staggering gap of $\mathsf{Expand}'(\mathsf{bc})$ *is 1.*

This result holds unconditionally for the case of $t < n/3$ malicious parties, and under the assumption of a PKI and secure digital signatures for $t < n/2$.

To see that the above solves our problem, suppose we want to sequentially compose protocols $\mathsf{bc}_1, \ldots, \mathsf{bc}_\ell$, each having staggering gap g. We simply run ℓ sequential executions of $\mathsf{bc}'_i = \mathsf{Expand}'(\mathsf{bc}_i)$ (setting $c = 1$) instead. Each bc'_i has staggering gap 1, meaning that honest parties terminate within 1 round of each other. But each bc'_{i+1} is secure as long as honest parties begin execution within 1 round of each other, so things are ok.

Applying this technique to compile a protocol Π (which uses a broadcast channel in each of its ℓ rounds) to a protocol Π' (running in a point-to-point network), we obtain $\mathsf{rc}(\Pi') = \ell \cdot (3 \cdot \mathsf{rc}(\mathsf{bc}) + 1)$. This can be improved slightly [24].

References

1. D. Beaver and S. Haber. Cryptographic protocols provably secure against dynamic adversaries. In *Advances in Cryptology — Eurocrypt '92*.
2. D. Beaver, S. Micali, and P. Rogaway. The round complexity of secure protocols. In *22nd Annual ACM Symposium on Theory of Computing (STOC)*, 1990.

3. M. Ben-Or. Another advantage of free choice: Completely asynchronous agreement protocols. In *2nd Annual ACM Symposium on Principles of Distributed Computing (PODC)*, 1983.

4. M. Ben-Or and R. El-Yaniv. Resilient-optimal interactive consistency in constant time. *Distributed Computing*, 16(4):249–262, 2003.

5. G. Blakley. Safeguarding cryptographic keys. In *National Computer Conference*, volume 48, pages 313–317. AFIPS Press, 1979.

6. G. Bracha. An $O(\log n)$ expected rounds randomized Byzantine generals protocol. *J. ACM*, 34(4):910–920, 1987.

7. C. Cachin, K. Kursawe, and V. Shoup. Random oracles in Constantinople: Practical asynchronous Byzantine agreement using cryptography (extended abstract). In *19th Annual ACM Symposium on Principles of Distributed Computing (PODC)*, 2000.

8. R. Canetti. Universally composable security: A new paradigm for cryptographic protocols. In *42nd Annual IEEE Symposium on Foundations of Computer Science (FOCS)*, 2001.

9. R. Canetti, U. Feige, O. Goldreich, and M. Naor. Adaptively secure multiparty computation. In *28th Annual ACM Symposium on Theory of Computing (STOC)*, 1996.

10. B. Chor and B. Coan. A simple and efficient randomized Byzantine agreement algorithm. *IEEE Trans. Software Engineering*, 11(6):531–539, 1985.

11. B. Chor, S. Goldwasser, S. Micali, and B. Awerbuch. Verifiable secret sharing and achieving simultaneity in the presence of faults. In *26th Annual IEEE Symposium on the Foundations of Computer Science (FOCS)*, 1985.

12. I. Damgård and Y. Ishai. Constant-round multiparty computation using a blackbox pseudorandom generator. In *Advances in Cryptology — Crypto 2005*.

13. D. Dolev and H. Strong. Authenticated algorithms for Byzantine agreement. *SIAM J. Computing*, 12(4):656–666, 1983.

14. C. Dwork, D. Shmoys, and L. Stockmeyer. Flipping persuasively in constant time. *SIAM J. Computing*, 19(3):472–499, 1990.

15. P. Feldman. *Optimal Algorithms for Byzantine Agreement*. PhD thesis, Massachusetts Institute of Technology, 1988.

16. P. Feldman and S. Micali. Byzantine agreement in constant expected time (and trusting no one). In *26th Annual IEEE Symposium on the Foundations of Computer Science (FOCS)*, 1985.

17. P. Feldman and S. Micali. An optimal probabilistic protocol for synchronous Byzantine agreement. *SIAM J. Computing*, 26(4):873–933, 1997.

18. M. J. Fischer and N. A. Lynch. A lower bound for the time to assure interactive consistency. *Information Processing Letters*, 14(4):183–186, 1982.

19. M. Fitzi, J. Garay, S. Gollakota, C. P. Rangan, and K. Srinathan. Round-optimal and efficient verifiable secret sharing. In *3rd Theory of Cryptography Conference (TCC)*, 2006.

20. M. Fitzi and J. A. Garay. Efficient player-optimal protocols for strong and differential consensus. In *22nd Annual ACM Symposium on Principles of Distributed Computing (PODC)*, 2003.

21. J. A. Garay and Y. Moses. Fully polynomial Byzantine agreement for $n > 3t$ processors in $t + 1$ rounds. *SIAM J. Comput.*, 27(1):247–290, 1998.

22. R. Gennaro, Y. Ishai, E. Kushilevitz, and T. Rabin. The round complexity of verifiable secret sharing and secure multicast. In *33rd Annual ACM Symposium on Theory of Computing (STOC)*, 2001.

23. S. Goldwasser and Y. Lindell. Secure computation without agreement. *J. Cryptology*, 18(3):247–287, 2005.

24. J. Katz and C.-Y. Koo. On expected constant-round protocols for Byzantine agreement. Available at `http://eprint.iacr.org/2006/065`.

25. E. Kushilevitz, Y. Lindell, and T. Rabin. Information-theoretically secure protocols and security under composition. STOC 2006, to appear.

26. L. Lamport, R. Shostak, and M. Pease. The Byzantine generals problem. *ACM Trans. Program. Lang. Syst.*, 4(3):382–401, 1982.

27. Y. Lindell, A. Lysyanskaya, and T. Rabin. On the composition of authenticated Byzantine agreement. In *34th Annual ACM Symposium on Theory of Computing (STOC)*, 2002.

28. Y. Lindell, A. Lysyanskaya, and T. Rabin. Sequential composition of protocols without simultaneous termination. In *21st Annual ACM Symposium on Principles of Distributed Computing (PODC)*, 2002.

29. J. B. Nielsen. A threshold pseudorandom function construction and its applications. In *Advances in Cryptology — Crypto 2002*.

30. M. Pease, R. Shostak, and L. Lamport. Reaching agreement in the presence of faults. *J. ACM*, 27(2):228–234, 1980.

31. B. Pfitzmann and M. Waidner. Information-theoretic pseudosignatures and Byzantine agreement for $t \geq n/3$. Technical Report RZ 2882 (#90830), IBM Research, 1996.

32. M. Rabin. Randomized Byzantine generals. In *24th Annual IEEE Symposium on Foundations of Computer Science (FOCS)*, 1983.

33. A. Shamir. How to share a secret. *Comm. ACM*, 22(11):612–613, 1979.

34. S. Toueg. Randomized Byzantine agreements. In *3rd Annual ACM Symposium on Principles of Distributed Computing (PODC)*, 1984.

35. R. Turpin and A. B. Coan. Extending binary Byzantine agreement to multivalued Byzantine agreement. *Information Processing Letters*, 18(2):73–6, 1984.

36. M. Waidner. *Byzantinische Verteilung ohne Kryptographische Annahmen trotz Beliebig Vieler Fehler (in German)*. PhD thesis, University of Karlsruhe, 1991.

Robust Multiparty Computation with Linear Communication Complexity

Martin Hirt[1] and Jesper Buus Nielsen[2]

[1] ETH Zurich, Switzerland
hirt@inf.ethz.ch
[2] University of Aarhus, Denmark
buus@daimi.au.dk

Abstract. We present a robust multiparty computation protocol. The protocol is for the cryptographic model with open channels and a poly-time adversary, and allows n parties to actively securely evaluate any poly-sized circuit with resilience $t < n/2$. The total communication complexity in bits over the point-to-point channels is $\mathcal{O}(Sn\kappa + n\,\mathcal{BC})$, where S is the size of the circuit being securely evaluated, κ is the security parameter and $\mathcal{BC}$ is the communication complexity of one broadcast of a κ-bit value. This means the average number of bits sent and received by a single party is $\mathcal{O}(S\kappa + \mathcal{BC})$, which is almost independent of the number of participating parties. This is the first robust multiparty computation protocol with this property.

1 Introduction

Efficient Multiparty Computation. A multiparty computation (MPC) protocol allows n parties to compute an agreed-upon function of their inputs in such a way that every party learns the correct function output, but no party learns any additional information about the other parties' inputs. A protocol is said to be t-robust if this holds even if up to t of the parties, called the corrupted parties, in a coordinated manner try to falsify the protocol output by sending wrong messages, and try to learn extra information by pooling the values that they learned in the protocol.

Since the publication of the first MPC protocols [Yao82, GMW87, CDG87, BGW88, CCD88, RB89, Bea91b] a lot of research went into improving the communication complexity [GV87, BB89, BMR90, BFKR90, Bea91a, FH96, GRR98, CDD+99, CDM00, CDD00, HMP00]. In [HM01] it was shown that any circuit with S gates can be computed unconditionally secure and t-robust for $t < n/3$ with communication complexity $\mathcal{O}(Sn^2\kappa + n^2\,\mathcal{BC})$, where κ is the size of the elements of the field secret-sharing is done over and $\mathcal{BC}$ is the communication complexity of broadcasting a κ-bit value. Recently, similar communication complexities were achieved for $t < n/2$, once with cryptographic security ($\mathcal{O}(Sn^2\kappa + n\,\mathcal{BC})$, where κ is the security parameter, [HN05]), and once with information-theoretic security ($\mathcal{O}(Sn^2\kappa + n^3\,\mathcal{BC})$, [BH05]). In these protocols there is an *overall* $\mathcal{O}(n)$ to $\mathcal{O}(n^3)$ broadcasts, and besides this the only communication is from each party sending an average of $\mathcal{O}(\kappa)$ bits to each other party per gate to be evaluated.

C. Dwork (Ed.): CRYPTO 2006, LNCS 4117, pp. 463–482, 2006.

Our Contributions. In this paper we show that, maybe somewhat surprisingly, a circuit with S gates can be computed cryptographically secure and t-robust for $t < n/2$ with communication complexity $\mathcal{O}(Sn\kappa + n\,\mathcal{BC})$. This means that the average number of bits sent and received by a single party is $\mathcal{O}(S\kappa + \mathcal{BC})$, which is almost independent of the number of participating parties. In particular, each party does not send messages to each of the other parties to evaluate a gate.

In contrast to many other efficient MPC protocols, the stated complexity holds also for circuits with many inputs and outputs, i.e., we give realizations with linear complexity for input, addition, multiplication, random, local output, and global output gates.

Related Work. In concurrent and independent work [DI06], it was showed that a circuit with S gates can be computed t-robust with communication complexity $\mathcal{O}(Sn\,\mathrm{poly}(\kappa))$. Their protocol is constant-round and can be proven adaptively secure, in contrast to our protocol which requires linear rounds and is proven only static secure. However, their protocol only achieves sub-optimal resilience (essentially $t < n/5$), whereas our protocol achieves optimal resilience ($t < n/2$).

2 Protocol Overview

Our MPC protocol follows the approach with homomorphic encryption of [CDN01]: We assume that a homomorphic encryption scheme is available, where the encryption key is known to all parties, and the decryption key is shared among the parties such that we have verifiable threshold decryption. The function to be computed is represented as an arithmetic circuit with input, addition (resp. linear), multiplication, random, and output gates. The evaluation proceeds gate-by-gate, i.e., for each gate an encryption of its value is computed.

Furthermore, we use segmentation and player elimination [HMP00]: The circuit is divided into n segments of (almost) equal size, and the parties evaluate one segment after each other. The evaluation of a segment may fail, but in this case, two parties, (at least) one of them being corrupted, are identified and eliminated, and the segment is repeated. Note that at most t times a segment can fail and must be repeated, hence the adversary can at most double the overall costs. Also note that, as we have an honest majority in the original party set, we also have honest majority in the party set resulting from a sequence of eliminations.

When a segment is to be evaluated, an arbitrary (non-eliminated) party is selected to be the king for this segment [HNP05]. Most sub-protocols are so-called king-fail-detect protocols, i.e., when cheating occurs, then the sub-protocol may fail, but then this must be detected by at least one honest party, and the king must be able to identify a cheater.

The actual evaluation of a segment follows the circuit-randomization approach of [Bea91b]: First, the parties generate one multiplication triple (A, B, C) for each multiplication and each random gate in the segment, where A and B are encryptions of random values and C is an encryption of their product. Then the parties evaluate the segment gate by gate, where for evaluating a multiplication or a random gate, one multiplication triple is consumed. The generation of the random values is based on super-invertible matrices: This technique allows us to transform n encryptions of random values, t of them known to the adversary, into $n - t$ encryptions of random values unknown to the adversary. Furthermore, we use homomorphic proof systems, i.e., proofs that allow to combine several proofs into a single, compact proof. Finally we introduce the notion of strong soundness for a Σ-protocol, which allows to turn Σ-protocols into zero-knowledge proofs by challenging with an unpredictable, but deterministic value.

3   Preliminaries

The Security Model. Our protocols can be proven secure in the UC model [Can01] against a static, actively corrupted minority. The page limit is however far from allowing us full simulation proofs, and we therefore do not need to introduce the UC model in detail. However we briefly sketch the used communication model.

We assume that at the beginning a party set $\mathcal{P} = \{P_1, \ldots, P_n\}$ is given, and all parties $P_i \in \mathcal{P}$ agree on this set. Furthermore, the parties $P_i \in \mathcal{P}$ have access to synchronous, authenticated point-to-point channels. The corrupted parties are modeled by a probabilistic poly-time (PPT) Turing machine A, the adversary. Before the execution of the protocol A gets to pick a subset $C \subset \mathcal{P}$ of size $|C| < n/2$, the corrupted parties. Then the protocol is executed. The adversary decides the scheduling of the honest parties, with the only restriction that in each round each honest party is scheduled exactly once. The adversary sees all messages sent, and can chooses the messages to be sent by corrupted parties based on this.

Broadcast. We assume a protocol BROADCAST for broadcasting values. Some party $P_i \in \mathcal{P}$ has input (P_i, m) and the remaining parties have input (P_i). The output of all parties will be (P_i, m). The parties will output a common value (P_i, m') even when P_i is corrupted. We will only broadcast $\mathcal{O}(\kappa)$-bit values, where κ is the security parameter, and we use $\mathcal{BC}$ to denote the communication complexity of broadcasting $\mathcal{O}(\kappa)$-bit values. In the analysis of the communication complexity we assume that $\mathcal{BC} \geq \mathcal{O}(n^2\kappa)$, as this is the complexity of the most efficient broadcast protocol for the cryptographic model [Nie03].

Threshold Signature Schemes. We also need a threshold signature scheme. Here all parties know the verification key vk and the signing key sk is shared among the n parties, with P_i holding a secret signing-key share sk_i. A party P_i can compute a signature share $\sigma_i = \text{sig}_{sk_i}(m)$ and prove in zero-knowledge that σ_i is a valid signature share from P_i on m. There is a threshold $t < n$ such that given a signature share σ_i on m from $t + 1$ parties one can compute $\text{sig}_{sk}(m) = \sigma = \text{combine}(\{\sigma_i\}_i)$, and given t of the keys sk_i a poly-time adversary cannot compute a signature on any message m on which it did not receive $\text{sig}_{sk_i}(m)$ from at least one honest P_i.

Super-Invertible Matrices. We use super-invertible matrices over $\mathbb{Z}_N$. Given a matrix $M = \{m_{i,j} \in \mathbb{Z}_N\}_{i=1,\ldots,r}^{j=1,\ldots,c}$ with r rows and c columns and a subset $C \subset$

$\{1,\ldots,c\}$ we let $M_C = \{m_{i,j}\}_{i=1,\ldots,r}^{j\in C}$ denote the matrix consisting of columns $j \in C$ from M. If $c = r$ we call M invertible if its columns are linear independent over $\mathbb{Z}_N$. If $c \geq r$ we call M super-invertible if M_C is invertible for all $C \subset \{1,\ldots,c\}$ with $|C| = r$. If $\{1,\ldots,c\}$ all are mutually prime with N, as will be the case later, a simple construction exists. For $i = 1,\ldots,r$, let f_i be a polynomial of degree at most $r - 1$ with $f_i(i) = 1$ and $f_i(j) = 0$ for $j \in \{1,\ldots,r\} \setminus \{i\}$, and let $M = \{m_{i,j} = f_i(j)\}_{i=1,\ldots,r}^{j=1,\ldots,c}$. Then $M_{\{1,\ldots,r\}}$ is the identity matrix and thus invertible. Furthermore, any matrix M_C with $C \subset \{1,\ldots,c\}$ and $|C| = r$ can be mapped onto $M_{\{1,\ldots,r\}}$ using an invertible matrix given by Lagrange interpolation over points from $\{1,\ldots,c\}$.

We use the following simple fact about super-invertible matrices. Pick any $C \subset \{1,\ldots,c\}$ with $|C| = r$. First sample x_j for $j \notin C$ using any joint distribution. Then for each $j \in C$ sample $x_j \in_R \mathbb{Z}_N$ uniformly at random, and independent of $\{x_j\}_{j\notin C}$. Let $y = (y_1,\ldots,y_r) = M(x_1,\ldots,x_c)$. Then y is uniformly random in $(\mathbb{Z}_N)^r$.

4   Paillier's Encryption Scheme

We will use Paillier's public-key encryption scheme, along with a number of known and new tools for this encryption scheme.

The public key is $N = pq$ where p and q are $\mathcal{O}(\kappa)$-bit primes. In addition $p = 2p' + 1$ and $q = 2q' + 1$ where p' and q' are primes such that p, q, p', q' are distinct.

A plaintext $m \in \mathbb{Z}_N$ is encrypted as $M = \mathcal{E}(m; r) = G^m r^N \bmod N^2$, where $G = N + 1$ and $r \in_R \mathbb{Z}_N^*$ is uniformly random. The element $N + 1$ has order N in $\mathbb{Z}_{N^2}^*$ and it can be showed that $\mathcal{E}$ is a homomorphism from $\mathbb{Z}_N \times \mathbb{Z}_N^*$ to $\mathbb{Z}_{N^2}^*$, where $\mathcal{E}(m_0; r_0)\mathcal{E}(m_1; r_1) \bmod N^2 = \mathcal{E}(m_0 + m_1 \bmod N; r_0 r_1 \bmod N)$.

By the choice of p and q the integers $N = pq$ and $\phi(N) = 2p'2q'$ are mutually prime, so one can compute an integer d such that $d \bmod N = 1$ and $d \bmod \phi(N) = 0$. It then follows that $\mathcal{E}(m; r)^d \bmod N^2 = \mathcal{E}(md \bmod N; r^d \bmod N) = \mathcal{E}(m \bmod N; 1)$, as $d \bmod \phi(N) = 0$ and the order of $r \in \mathbb{Z}_N^*$ divides $\phi(N)$. So, $\mathcal{E}(m; r)^d = (N + 1)^m \bmod N^2$. It can then be showed that m can be computed efficiently from $(N + 1)^m \bmod N^2$, *without knowing* d. Therefore d acts as the private key and decryption is equivalent to computing $\mathcal{E}(m; r)^d \bmod N^2$. In the following we use $m = \mathcal{D}(M)$ to denote the decryption function.

4.1 Σ-Protocols with Strong Soundness.

We will use a large number of zero-knowledge proofs, all based on Σ-protocols. A Σ-protocol is a three-move proof system. The verifier knows an instance x and the prover knows a witness w for x. The prover sends a first message a sampled using x and w. The verifier returns a challenge $e \in E$ for some challenge set E, and then the prover returns a reply c depending on e. Based on the conversation (x, a, e, c) the verifier either accepts or rejects. A conversation is called valid if it is accepted by the verifier. To be called a Σ-protocol the system must be complete and have the following two properties. Special soundness: Given two valid conversations (x, a, e, c) and (x, a, e', c') with the the same first message a and different challenges $e \neq e'$ one can *efficiently* compute a witness w for x. Special honest-verifier zero-knowledge: Given any (x, e), where x is a correct instance, meaning that there exists a witness for x, one can *efficiently* compute (a, c) such that the conversation (x, a, e, c) has a distribution statistically close to that generated by the protocol when run with instance x and challenge e, and the prover knowing a witness for x.

For any pair (x, a), being an instance and a first message, we say that (e, c) is a valid pair if (x, a, e, c) is a valid conversation, and we call e a valid challenge for (x, a) if there exists a valid pair (e', c) for (x, a) with $e' = e$. It follows from the special soundness that a Σ-protocol has the following property: if x is an incorrect instance, meaning an instance which has no witness, then for any (x, a) there exists at most one valid challenge e. We introduce the notion of strong soundness for a Σ-protocol, by which we will mean that if x is an incorrect instance, then for all a for which there exists a valid challenge e for (x, a), this (unique) challenge can be computed *efficiently* from (x, a). We call a Σ-protocol with strong soundness a strong Σ-protocol.

As we will see later, a Σ-protocol with strong soundness can be turn into a zero-knowledge proof in a particularly simple manner, by letting $e = \mathrm{sig}_{sk}(nonce)$ be a signature on some fresh value *nonce*.

Proof of Plaintext Knowledge. The first strong Σ-protocol is for proving plaintext knowledge. Normally a Paillier encryption is computed as $A = G^m r^N \bmod N^2$. For generality, we need to consider the case where some other arbitrary element $H \in \mathbb{Z}_{N^2}^*$ was used in place of G. I.e., the encryption has been computed as $A = H^m r^N \bmod N^2$. In this case m might be an arbitrary integer, but we assume that a public bound ℓ is known such that $m \leq 2^\ell$.

Proof of plaintext knowledge

1. The verifier knows an instance $x = (H, A)$ and the prover knows a witness $w = (m, r)$ such that $A = H^m r^N \bmod N^2$.
2. The prover samples $n \in_R \mathbb{Z}_{2^{\ell+\kappa}}$ and $s \in_R \mathbb{Z}_N^*$, computes $B = H^n s^N \bmod N^2$ and sends the first message $a = B$.
3. The verifier sends a challenge $e \in \mathbb{Z}_{\mathsf{Bound}}$, where Bound is a lower bound on the primes p', q'.
4. The prover computes $z = em + n$ and $t = r^e s \bmod N$ and sends the reply $c = (z, t)$.
5. The verifier accepts the conversation if $H^z t^N \equiv_{N^2} A^e B$.

If $\ell \in \mathcal{O}(\kappa)$, as will always be the case, the bit-length of a conversation is seen to be $\mathcal{O}(\kappa)$.

This proof is known to be a Σ-protocol (see e.g. [CDN01]). We argue that it also has strong soundness. For this we have to consider an incorrect instance $x = (H, A)$, where $A = H^m r^N \bmod N^2$ has no solution for (m, r), and a first message $a = B$ for which there exists $(e, (z, t))$ such that $H^z t^N \equiv_{N^2} A^e B$. Then we must show how to compute e efficiently from (x, a). The algorithm for doing this uses the factorization (p, q) of N.

Since $H, A \in \mathbb{Z}_{N^2}^*$, they are ciphertexts and can be written as $A = \mathcal{E}(m_A; r_A)$ and $H = \mathcal{E}(m_H; r_H)$. The equivalence $A \equiv_{N^2} H^m r^N$ having no solution for (m, r) is then easily seen to be equivalent to $\mathcal{E}(m_A; r_A) \equiv_{N^2} \mathcal{E}(m_H; r_H)^m \mathcal{E}(0; r)$ having no solution for (m, r), which in turn is equivalent to $\mathcal{E}(0; 1) \equiv_{N^2} \mathcal{E}(m_H m - m_A \bmod N; r_H^m r r_A^{-1} \bmod N)$ having no solution for (m, r). Since setting $r = r_H^{-m} r_A \bmod N$ guarantees that $1 = r_H^m r r_A^{-1} \bmod N$, it follows that $A \equiv_{N^2} H^m r^N$ having no solution for (m, r) is equivalent to $m_H m \equiv_N m_A$ having no solution for m. This implies that there is a prime factor of N, p say, such that $m_H m \equiv_p m_A$ has no solution for m. From this it easily follows that $m_A \not\equiv_p 0$ and $m_H \equiv_p 0$. We assumed that $H^z t^N \equiv_{N^2} A^e B$. If we write $B = \mathcal{E}(m_B; r_B)$, then using arguments as those above it follows that $z m_H \equiv_p e m_A + m_B$. From $m_H \equiv_p 0$ it then follows that $e m_A \equiv_p -m_B$, and from $m_A \not\equiv_p 0$ it follows that $e \equiv_p -m_B m_A^{-1}$. Since $e < \mathsf{Bound} \le p$ we therefore have that $e = -m_B m_A^{-1} \bmod p$. Since $m_B = \mathcal{D}(B)$ and $m_A = \mathcal{D}(A)$ can be computed efficiently from (x, a) given the factorization of N, it follows that $e = -m_B m_A^{-1} \bmod p$ can be computed efficiently from (x, a) given the factorization of N. This establishes the strong soundness.

Proof of Identical Encryptions. The next strong Σ-protocol is for proving identical encryptions.

Proof of identical encryption

1. The verifier knows an instance $x = (H_0, H_1, A_0, A_1)$ and the prover knows a witness $w = (m, r_0, r_1)$ with $A_0 = H_0^m r_0^N \bmod N^2$ and $A_1 = H_1^m r_1^N \bmod N^2$.
2. The prover samples $n \in_R \mathbb{Z}_{2^\cdot +}$ and $s_0, s_1 \in_R \mathbb{Z}_N^*$, computes $B_0 = H_0^n s_0^N \bmod N^2$ and $B_1 = H_1^n s_1^N \bmod N^2$ and sends the first message $a = (B_0, B_1)$.
3. The verifier sends a challenge $e \in \mathbb{Z}_{\mathsf{Bound}}$.
4. The prover computes $z = em + n$ and $t_0 = r_0^e s_0 \bmod N$ and $t_1 = r_1^e s_1 \bmod N$, and sends the reply $c = (z, t_0, t_1)$.
5. The verifier accepts the proof if $H_0^z t_0^N \equiv_{N^2} A_0^e B_0$ and $H_1^z t_1^N \equiv_{N^2} A_1^e B_1$.

The bit-length of a conversation is $\mathcal{O}(\kappa)$ bits. Again the proof is known to be a Σ-protocol. The strong soundness can be argued using techniques very similar to those used for the proof of plaintext knowledge. In particular, if for a value $(x, a) = ((H_0, H_1, A_0, A_1), (B_0, B_1))$ there does not exist (m, r_0, r_1) such that $A_0 = H_0^m r_0^N \bmod N^2$ and $A_1 = H_1^m r_1^N \bmod N^2$, but still there exists a valid challenge e, then e can be computed efficiently given (x, a) and (p, q).

4.2   Homomorphic Conversations

Another notion that we use for efficiency purposes is that of homomorphic conversations. We look at the proof of identical encryption for an example.

Let $x = (H_0, H_1, A_0, A_1)$ be an instance for the proof system. If the instance is correct there exists (m, r_0, r_1) such that $A_0 = H_0^m r_0^N \bmod N^2$ and $A_1 = H_1^m r_1^N \bmod N^2$. Assume then that we have ℓ correct instances $x^{(l)} = (H_0, H_1, A_0^{(l)}, A_1^{(l)})$ with witnesses $(m^{(l)}, r_0^{(l)}, r_1^{(l)})$. If we let $A_0 = \prod_{i=1}^\ell A_0^{(l)} \bmod N^2$ and $A_1 = \prod_{i=1}^\ell A_1^{(l)} \bmod N^2$, then (H_0, H_1, A_0, A_1) is again a correct instance, as it has the witness (m, r_0, r_1) with $m = \sum_{i=1}^{\ell} m^{(l)}$, $r_0 = \prod_{i=1}^{\ell} r_0^{(l)} \bmod N$ and $r_1 = \prod_{i=1}^{\ell} r_1^{(l)} \bmod N$. We call (H_0, H_1, A_0, A_1) the combined instance.

A similar property holds for valid conversations. A value $(x, a, e, c) = \big((H_0, H_1, A_0, A_1), (B_0, B_1), e, (z, t_0, t_1)\big)$ is a valid conversation iff $H_0^z t_0^N \equiv_{N^2} A_0^e B_0$ and $H_1^z t_1^N \equiv_{N^2} A_1^e B_1$.

Assume now that we have ℓ valid conversations $(x^{(l)}, a^{(l)}, e, c^{(l)}) = \big((H_0, H_1, A_0^{(l)}, A_1^{(l)}), (B_0^{(l)}, B_1^{(l)}), e, (z^{(l)}, t_0^{(l)}, t_1^{(l)})\big)$, with the same (H_0, H_1) and the same challenge e. Define A_0 and A_1 as above, and let $B_0 = \prod_{i=1}^{\ell} B_0^{(l)} \bmod N^2$, $B_1 = \prod_{i=1}^{\ell} B_1^{(l)} \bmod N^2$, $z = \sum_{l=1}^{\ell} z^{(l)}$, $t_0 = \prod_{i=1}^{\ell} t_0^{(l)} \bmod N$ and $t_1 = \prod_{l=1}^{\ell} t_1^{(l)} \bmod N$. Then it can be seen that $(x, a, e, c) = \big((H_0, H_1, A_0, A_1), (B_0, B_1), e, (z, t_0, t_1)\big)$ is again a valid conversation, now for the combined instance (H_0, H_1, A_0, A_1) computed above. We call (x, a, e, c) the combined conversation.

When combining instances and conversations in the following we will simply write $a = \text{combine}(\{a_i\}_{l=1}^{\ell})$ and $c = \text{combine}(\{c_i\}_{l=1}^{\ell})$ for the above way to compute the combined first message a and the combined reply c. We later use combined conversations for efficiency purposes. Note that the only value in $(x, a, e, c) = \big((H_0, H_1, A_0, A_1), (B_0, B_1), e, (z, t_0, t_1)\big)$ which is larger than the corresponding value in a normal conversation is z, where $z = \sum_{l=1}^{\ell} z^{(l)}$. We will however only combine polynomially many proofs, so z will be an $\mathcal{O}(\kappa)$-bit value, even in combined proofs.

4.3   Distributed Decryption

We then sketch how the decryption function can be distributed. Recall that when $M = \mathcal{E}(m; r) = G^m r^N \bmod N^2$, then $M^d \bmod N^2 = (N+1)^m \bmod N^2$, from which m can be computed efficiently without knowing d. Distributing the decryption function is therefore equivalent to distributing the function $M \mapsto M^d \bmod N^2$, letting each P_i hold a share d_i of d.

The key distribution is done by a trusted server which generates N and d and hands N to all parties. It furthermore samples an integer secret sharing $(d_1, \ldots, d_n)$ such that $d = \sum_{i=1}^{n} d_i$ and hands d_i secretly to party P_i, for $i = 1, \ldots, n$. To distributedly decrypt a ciphertext $M = \mathcal{E}(m; r)$, each party P_i contributes with the decryption share $M_i = M^{d_i} \bmod N^2$. Then the parties compute $M_0 = \prod_{i=1}^{n} M_i \bmod N^2 = M^d \bmod N^2$. Since $M^d = (N+1)^m \bmod N^2$ all parties can now efficiently compute m from M_0. In the following we write this as $m = \text{combine}(M_1, \ldots, M_n)$. It can be showed [CDN01] that this distributed decryption protocol does not leak information about d, as the decryption shares M_i can be simulated given just M and m, without using the values d_i.

To add robustness to the protocol the actual encryption key distributed by the trusted server is of the form $ek = (N, com_1, \ldots, com_n)$, where com_i is a commitment to d_i. Furthermore, the decryption key share given to P_i is of the form $dk_i = (d_i, o_i)$, where o_i is an opening of com_i to d_i. After sending the decryption share $M_i = M^{d_i} \bmod N^2$ the party P_i will then prove that com_i

contains a commitment to a value d_i for which it holds that $M_i = M^{d_i} \bmod N^2$. This guarantees that if all parties give acceptable proofs for their decryption share M_i, then $m = \text{combine}(M_1, \ldots, M_n)$ is the plaintext of M, except with negligible probability.

The proof that com_i is a commitment to a value d_i such that $M_i = M^{d_i} \bmod N^2$ we will call the proof of correct decryption share. The proof is done using a Σ-protocol. We denote the instance by $x = (ek, M, M_i)$ and we call a valid conversation $\big((ek, M, M_i), a, e, c\big)$ a valid conversation for M_i being the decryption share of M from P_i. The details of this Σ-protocol have been removed from this extended abstract due to lack of space.

The Σ-protocol for proving correct decryption share has total bit-length $\mathcal{O}(\kappa)$. Furthermore, it has homomorphic conversations, in the following sense: Given n decryption shares $M_1, \ldots, M_n$ along with a valid conversation $\big((ek, M, M_i), a_i, e, c_i\big)$ for each M_i being the decryption share of M from P_i, it is possible to compute $m = \text{combine}(M_1, \ldots, M_n)$, $a = \text{combine}(a_1, \ldots, a_n)$ and $c = \text{combine}(c_1, \ldots, c_n)$ such that $\big((ek, M, m), a, e, c\big)$ is a valid conversation for a Σ-protocol for proving that m is the plaintext of M when ek is the public key. In addition, the Σ-protocol for proving that m is the plaintext of M has strong soundness in the following sense. Given (ek, M, m) where m is not the plaintext of M (when the N in ek is the encryption key) and given any first message a, if there exists a valid challenge e for $\big((ek, M, m), a\big)$, then this e can be computed efficiently from $\big((ek, M, m), a\big)$ given the factorization of N.

5   Setup, Key-Share Backup and Party Elimination

We use $\mathcal{P}_{\text{ori}} = \{P_1, \ldots, P_n\}$ to denote the original party set which agreed on running the protocol. Before our MPC protocol is run we assume that the following key-distribution has been made. An encryption-key ek for Paillier's encryption scheme has been generated and made public, and P_i has been given a decryption key share dk_i, as described in Section 4. This means that the decryption key is shared with threshold $n - 1$. A verification-key/signing-key pair (vk, sk) for a threshold signature scheme has been generated and sk is shared among $\mathcal{P}_{\text{ori}}$ with threshold $n - 1$ and P_i holding sk_i. A verification-key/signing-key pair $(vkzk, skzk)$ for a threshold signature scheme has been generated and $skzk$ is shared among $\mathcal{P}_{\text{ori}}$ with threshold $t = \lfloor (n-1)/2 \rfloor$ and P_i holding $skzk_i$. Finally an auxiliary string has been setup as described in [Dam00] to allow transforming all Σ-protocols into concurrent zero-knowledge proofs.[1]

Because the decryption key dk and the signing key sk are shared with threshold $n-1$ among the parties in $\mathcal{P}_{\text{ori}}$ one corrupted party can halt the protocol. To deal with this issue we use the technique of key-share backup by Rabin [Rab98]. As part of the protocol setup, the key shares sk_i and dk_i of each $P_i \in \mathcal{P}_{\text{ori}}$ are verifiably secret shared among the parties in $\mathcal{P}_{\text{ori}}$, with threshold $t = \lfloor (n-1)/2 \rfloor$.

---

[1] The proofs of identical encryption and correct decryption share, with strong soundness, will not be made zero-knowledge using [Dam00], but by challenging with a signature, as described in Section 6.

The parties then hold an active party set, originally $\mathcal{P}_{\mathtt{act}} = \mathcal{P}_{\mathtt{ori}}$. Then, whenever some protocol fails, the parties will run a detection protocol, DETECT-ELIMINATE, where all parties in $\mathcal{P}_{\mathtt{ori}}$ end up agreeing on a subset $D \subset \mathcal{P}_{\mathtt{act}}$ that can be thought of as the active parties causing the failure. Then each $P_i \in \mathcal{P}_{\mathtt{ori}}$ will for each $P_j \in D$ send its shares of dk_j and sk_j to all parties, and all parties reconstruct dk_j and sk_j. Then the parties set $\mathcal{P}_{\mathtt{act}} := \mathcal{P}_{\mathtt{act}} \setminus D$. This ensures that all parties at all times hold dk_j and sk_j for all eliminated parties $P_j \in \mathcal{P}_{\mathtt{ori}} \setminus \mathcal{P}_{\mathtt{act}}$.

As opposed to [Rab98] our detection protocol will not detect only corrupted parties. Our detection protocol will however guarantee that there are at least as many corrupted parties as honest parties in the detected set D. Therefore $\mathcal{P}_{\mathtt{act}}$ will continue to have honest majority. This is an idea from the party elimination framework from [HMP00, HM01].

6   Some Sub-protocols

6.1   Transferable Zero-Knowledge Proofs

We need a protocol TRANS-ZK for a party P_i to give a short transferable zero-knowledge proof. We assume that there exists a two-party protocol for a prover P_i to prove some instance x using a witness w, and we assume that the protocol is zero-knowledge when run concurrently. Such protocols exists for all the problems that we need to prove in zero-knowledge later, as we based all proofs on Σ-protocols which can be made concurrent zero-knowledge using [Dam00]. The protocol TRANS-ZK uses the threshold signature scheme $(vkzk, skzk)$ with threshold $t = \lfloor (n-1)/2 \rfloor$.

TRANS-ZK

1. $P_i \in \mathcal{P}_{\mathtt{ori}}$ has input (sid, P_i, x, w), where $sid \in \{0,1\}^\kappa$ is a unique session identifier, x is an instance for some proof system and w is a witness for x. Each $P_j \in \mathcal{P}_{\mathtt{ori}} \setminus \{P_i\}$ has input (sid, P_i).
2. P_i: Send (sid, x) to each $P_j \in \mathcal{P}_{\mathtt{ori}}$ and use the witness w to run the two-party zero-knowledge proof for x with each P_j.
3. Each $P_j \in \mathcal{P}_{\mathtt{ori}}$: If P_i sends a value x and gives an accepting zero-knowledge proof for x, then send $\sigma_j = \mathrm{sig}_{skzk_j}(sid, i, x)$ to P_i.[a]
4. P_i: Collect $t+1$ valid signature shares σ_j on (sid, i, x) and compute a signature $\sigma = \mathrm{combine}(\{\sigma_j\}_j)$ on (sid, i, x). Then output $proof(sid, i, x) = (sid, \sigma)$.

---

[a] Here, and for the rest of the paper, whenever a party sends a signature share under sk_j or $skzk_j$ it will give a zero-knowledge proof of correctness to the recipient, and we will not mentioning this explicitly from now on.

The value $proof(sid, i, x)$ is then the transferable proof. If $\mathrm{ver}_{vkzk}((sid, i, x), \sigma) = 1$, then any $P_j \in \mathcal{P}_{\mathtt{ori}}$ will accept (sid, σ) as evidence that x was proved by P_i in session sid. The proof is clearly zero-knowledge when the underlying zero-knowledge proof is concurrent zero-knowledge. As for soundness, a party P_i can only construct a valid transferable proof $proof(sid, i, x)$ by obtaining $t+1$ signature shares on (sid, i, x). Since we have assumed that at most a minority of the parties in $\mathcal{P}_{\text{ori}}$ are corrupted, it follows that a corrupted prover P_i must give an accepting proof for x to at least one honest party to construct a valid transferable proof $proof(sid, i, x)$. This guarantees the soundness. Since $t = \lfloor (n-1)/2 \rfloor$ and we have assumed that a majority of the parties in $\mathcal{P}_{\text{ori}}$ are honest, an honest prover will always receive $t + 1$ signature shares on (sid, i, x). This guarantees the completeness of the proof system.

For all the proof systems that we use, the communication complexity of one proof is $\mathcal{O}(\kappa)$. Therefore the total communication complexity of one run of TRANS-ZK is $\mathcal{O}(n\kappa)$, and proofs (sid, σ) have length $\mathcal{O}(\kappa)$.

6.2   King-Fail-Detect Protocols

The rest of our sub-protocols will be so-called King-Fail-Detect (KFD) protocols, where all parties have inputs, only active parties have outputs and a designated active party $P_{\text{king}} \in \mathcal{P}_{\text{act}}$ acts as King. Each party $P_i \in \mathcal{P}_{\text{ori}}$ has some input $(sid, P_{\text{king}}, x_i)$, where sid is a unique session identifier.[2] The output of each active party $P_i \in \mathcal{P}_{\text{act}}$ will be some value (sid, s_i, y_i), where $s_i \in \{\texttt{ok!}, \texttt{failed!}\}$ is a termination status. If $s_i = \texttt{ok!}$ for all honest $P_i \in \mathcal{P}_{\text{act}}$, then we say that the session succeeded. In that case the values y_i constitute the outputs of the protocol. Otherwise, we say that the session failed. If an honest $P_i \in \mathcal{P}_{\text{act}}$ has output $s_i = \texttt{failed!}$, then $y_i = P_j$ for some $P_j \in \mathcal{P}_{\text{act}}$, meaning that P_i accuses P_j of being corrupted. In words: A KFD protocol is said to fail if some honest, active party considers it failed, and if a KFD protocol fails, then at least one honest active party will accuse some active party of causing the failure.

We put some restrictions on what accusations are allowed by the honest parties. Specifically we require that a KFD protocol has the following two properties. King Awareness: When P_{king} is honest and at least one honest party $P_i \in \mathcal{P}_{\text{act}}$ outputs $s_i = \texttt{failed!}$ and $y_i = P_{\text{king}}$, then P_{king} outputs $s_{\text{king}} = \texttt{failed!}$. Sound Detection: If P_{king} has outputs $s_{\text{king}} = \texttt{failed!}$ and $y_{\text{king}} = P_j \in \mathcal{P}_{\text{act}}$, then P_j is corrupted, and if some honest $P_i \in \mathcal{P}_{\text{act}} \setminus \{P_{\text{king}}\}$ has outputs $s_i = \texttt{failed!}$ and $y_i = P_j \in \mathcal{P}_{\text{act}} \setminus \{P_{\text{king}}\}$, then P_j is corrupted. In words: No honest party accuses another honest party, except maybe the King. And, if some honest party accuses an honest King, then the King will in turn accuse some corrupted party.

After a KFD protocol the below protocol DETECT-ELIMINATE will sometimes be run to detect failing parties and eliminate them. The protocol uses the threshold signature scheme (vk, sk) with threshold $n - 1$.

In Step 2 in DETECT-ELIMINATE P_{king} is instructed to compute a signature σ on $(sid, \texttt{ok!})$ when it received correct signature shares $\sigma_i = \text{sig}_{sk_i}(sid, \texttt{ok!})$ from all $P_i \in \mathcal{P}_{\text{act}}$. This is possible as P_{king} knows sk_i for all $P_i \notin \mathcal{P}_{\text{act}}$, as these were reconstructed in Step 5 in earlier runs of DETECT-ELIMINATE.[3]

---

[2] Typically the eliminated parties $P_i \in \mathcal{P}_{\text{ori}} \setminus \mathcal{P}_{\text{act}}$ have no input x_i, but in some case they do.

[3] Only DETECT-ELIMINATE will change the value of $\mathcal{P}_{\text{act}}$.

DETECT-ELIMINATE

1. Each $P_i \in \mathcal{P}_{\text{act}}$: If $s_i = \text{ok!}$, then send $\sigma_i = \text{sig}_{sk_i}(sid, \text{ok!})$ to P_{king}.
2. P_{king}: If $s_{\text{king}} = \text{failed!}$, then BROADCAST $(\text{corrupt!}, y_{\text{king}})$ to $\mathcal{P}_{\text{ori}}$. Otherwise, if some $P_i \in \mathcal{P}_{\text{act}}$ did not send $\sigma_i = \text{sig}_{sk_i}(sid, \text{ok!})$, then broadcast $(\text{complained!}, P_i)$ to $\mathcal{P}_{\text{ori}}$, for *one* of these parties. Otherwise, compute a signature σ on $(sid, \text{ok!})$ and broadcast σ to $\mathcal{P}_{\text{ori}}$.
3. Each $P_i \in \mathcal{P}_{\text{act}}$: If P_{king} broadcast $(\text{complained!}, P_i)$, then BROADCAST (s_i, y_i) to $\mathcal{P}_{\text{ori}}$.
4. Each $P_k \in \mathcal{P}_{\text{ori}}$: If P_{king} broadcast $(\text{corrupt!}, P_j)$ with $P_j \in \mathcal{P}_{\text{act}}$, then let $D = \{P_{\text{king}}, P_j\}$. If P_{king} broadcast $(\text{complained!}, P_i)$ and P_i did not broadcast $s_i = \text{failed!}$ and $y_i = P_j \in \mathcal{P}_{\text{act}}$, then let $D = \{P_{\text{king}}, P_i\}$. If P_{king} broadcast $(\text{complained!}, P_i)$ and P_i broadcast $s_i = \text{failed!}$ and $y_i \in \mathcal{P}_{\text{act}}$, then let $D = \{P_i, y_i\}$. If P_{king} broadcast $\sigma = \text{sig}_{sk}(sid, \text{ok!})$, then let $D = \emptyset$. Otherwise, let $D = \{P_{\text{king}}\}$.
5. Each $P_k \in \mathcal{P}_{\text{ori}}$: Let $\mathcal{P}_{\text{act}} := \mathcal{P}_{\text{act}} \setminus D$, and for each $P_i \in D$, reconstruct the key shares dk_i, sk_i.

Note that if any honest $P_i \in \mathcal{P}_{\text{act}}$ has $s_i = \text{failed!}$, then $D \neq \emptyset$, and note that King Awareness and Sound Detection guarantee that there are always at least as many corrupted parties as honest parties in D. So, if the session fails, at least one corrupted party is eliminated, and $\mathcal{P}_{\text{act}}$ will keep having honest majority. One run of DETECT-ELIMINATE can be seen to have communication complexity $\mathcal{O}(n\kappa + \mathcal{BC}) = \mathcal{O}(\mathcal{BC})$.

6.3 KFD Signing

We use the below KFD protocol for signing under sk. It is straight-forward to verify that this is a KFD protocol, with King Awareness and Sound Detection. When the protocol succeeds it is clear that all parties received a signature σ on (sid, m_{king}). Since sk is shared with threshold $n-1$ and sid is used only once, it is clear that at most one value (sid, m) is signed for each sid. The communication complexity is $\mathcal{O}(n\kappa)$.

KFD-SIGN

1. Each $P_i \in \mathcal{P}_{\text{act}}$ has input $(sid, P_{\text{king}}, m_i)$ for $P_{\text{king}} \in \mathcal{P}_{\text{act}}$, and if P_{king} is honest, then $m_i = m_{\text{king}}$ for all honest $P_i \in \mathcal{P}_{\text{act}}$.
2. Each $P_i \in \mathcal{P}_{\text{act}}$: Send $\sigma_i = \text{sig}_{sk_i}(sid, m_i)$ to P_{king}.
3. P_{king}: If some $P_i \in \mathcal{P}_{\text{act}}$ did not send a valid σ_i, then output $(sid, \text{failed!}, P_i)$. Otherwise, compute $\sigma_i = \text{sig}_{sk_i}(sid, m_{\text{king}})$ for $P_i \in \mathcal{P}_{\text{ori}} \setminus \mathcal{P}_{\text{act}}$, compute $\sigma = \text{combine}(\sigma_1, \ldots, \sigma_n)$, and send σ to all parties in $\mathcal{P}_{\text{act}}$.
4. Each $P_i \in \mathcal{P}_{\text{act}}$: If receiving σ such that $\text{ver}_{vk}((sid, m_i), \sigma) = 1$, then output $(sid, \text{ok!}, (m_i, \sigma))$. Otherwise, output $(sid, \text{failed!}, P_{\text{king}})$.

6.4 KFD Broadcast

We use the KFD protocol KFD-BROADCAST for broadcasting values. A party $P_j \in \mathcal{P}_{\text{ori}}$ has a message m which should be sent to all active parties. If m is an instance of some proof system, then the party P_j also has a witness w for m which allows P_j to give a zero-knowledge proof for m. In that case the broadcast only succeeds if P_j can give a proof of m. If the proof fails, the output is some dummy value D.

KFD-BROADCAST

1. Some $P_j \in \mathcal{P}_{\text{ori}}$ has input $(sid, P_{\text{king}}, P_j, m, w, D)$ for $P_{\text{king}} \in \mathcal{P}_{\text{act}}$, and each $P_i \in \mathcal{P}_{\text{act}} \setminus \{P_j\}$ has input $(sid, P_{\text{king}}, P_j, D)$.
2. P_j: Run TRANS-ZK(sid, P_j, m, w) to get $proof(sid, j, m)$. The other parties have input (sid, P_j).
3. P_j: Send $\big(sid, m, proof(sid, j, m)\big)$ to each $P_i \in \mathcal{P}_{\text{act}}$.
4. Each $P_i \in \mathcal{P}_{\text{act}}$: If receiving $\big(sid, m', proof(sid, j, m')\big)$ from P_j, then send it to P_{king}.
5. P_{king}: If receiving any $\big(sid, m', proof(sid, j, m')\big)$ from any $P_i \in \mathcal{P}_{\text{act}}$, then pick one of them, $\big(sid, m, proof(sid, j, m)\big)$, and send it to each $P_i \in \mathcal{P}_{\text{act}}$.
6. Each $P_i \in \mathcal{P}_{\text{act}}$: If $\big(sid, m', proof(sid, j, m')\big)$ was sent to P_{king} in Step 4, but P_{king} did not return any $\big(sid, m, proof(sid, j, m)\big)$ in Step 5, then output $(sid, \text{failed!}, P_{\text{king}})$. Otherwise, run KFD-SIGN$(sid, P_{\text{king}}, m_i)$ (with P_{king} as King)[a], where m_i is determined as follows. If P_{king} sent $\big(sid, m, proof(sid, j, m)\big)$, then $m_i = m$, and otherwise $m_i = D$.
7. Each $P_i \in \mathcal{P}_{\text{act}}$: If KFD-SIGN outputs ok!, then output $(sid, \text{ok!}, m_i)$. Otherwise, output $(sid, \text{failed!}, P_{\text{king}})$.

---

[a] For the rest of the paper, when a KFD protocol runs another KFD protocol, it calls it with its own King, and we do not mention this explicitly.

It is easy to verify that KFD-BROADCAST is a KFD protocol. Assume then that the protocol succeeds. In that case KFD-SIGN terminated without failure, so all honest $P_i \in \mathcal{P}_{\text{act}}$ had the same input m_i. Therefore all honest $P_i \in \mathcal{P}_{\text{act}}$ output the same m_i. It is easy to see that if in addition P_j is honest, then $m_i = m$, were m is the input of P_j. Finally, observe that if P_j cannot prove m, then the output of all honest $P_i \in \mathcal{P}_{\text{act}}$ will be the dummy value D. Let $\ell = |m|$. If $\ell \geq \kappa$, the communication complexity is seen to be $\mathcal{O}(n\ell)$.

6.5   KFD Promote

The following KFD sub-protocol allows to promote a vector $x_0, \ldots, x_{\ell-1}$ of κ-bit values, known only by active parties, to *all* parties. This can be done with communication complexity $\mathcal{O}(n\ell\kappa + n^2\kappa) = \mathcal{O}(n\ell\kappa + \mathcal{BC})$ following an idea of [FH06].

6.6   KFD Decryption

The KFD sub-protocol KFD-DECRYPT allows to decrypt an agreed ciphertext M toward all active parties. The decryption is performed through a King P_{king}. The protocol assumes that the parties in $\mathcal{P}_{\text{act}}$ agree on M, and does not guarantee that all active parties receive the decryption m (this will be detected in DETECT-ELIMINATE), but it guarantees that no party outputs a wrong decryption m. The protocol uses the homomorphic proofs for correct decryption share.

KFD-PROMOTE

1. Each $P_i \in \mathcal{P}_{\mathtt{act}}$ has input $(sid, P_{\mathtt{king}}, (x_0, \ldots, x_{\ell-1}))$ and each $P_j \in \mathcal{P}_{\mathtt{ori}} \setminus \mathcal{P}_{\mathtt{act}}$ has input $(sid, P_{\mathtt{king}}, \ell)$.

2. Each $P_j \in \mathcal{P}_{\mathtt{ori}}$: Let $f(z) = \sum_{l=0}^{\ell-1} x_l z^l$ be the polynomial constituted of the coefficients x_l, and let $\vec{m}_i$ be the vector with the ith block of $2\lceil \ell/|\mathcal{P}_{\mathtt{act}}| \rceil$ values on $f(z)$, i.e., $\vec{m}_i = (f(i \cdot 2\lceil \ell/|\mathcal{P}_{\mathtt{act}}| \rceil), \ldots, f((i+1) \cdot 2\lceil \ell/|\mathcal{P}_{\mathtt{act}}| \rceil - 1))$.

3. Each $P_j \in \mathcal{P}_{\mathtt{act}}$: Invoke KFD-SIGN, to compute signatures σ_i on $\vec{m}_i$ for every $P_i \in \mathcal{P}_{\mathtt{act}}$.

4. Each $P_i \in \mathcal{P}_{\mathtt{act}}$: Send $(\vec{m}_i, \sigma_i)$ to every $P_j \in \mathcal{P}_{\mathtt{ori}} \setminus \mathcal{P}_{\mathtt{act}}$.

5. Each $P_j \in \mathcal{P}_{\mathtt{ori}} \setminus \mathcal{P}_{\mathtt{act}}$: Accept those vectors $\vec{m}_i$ with good signature σ_i, and uses them to interpolate $f(z)$ and compute $(x_0, \ldots, x_{\ell-1})$.

KFD-DECRYPT

1. Each $P_i \in \mathcal{P}_{\mathtt{act}}$ has input $(sid, P_{\mathtt{king}}, ek, M, dk_i)$.

2. Each $P_i \in \mathcal{P}_{\mathtt{act}}$: Compute decryption share $M_i = \mathcal{D}_{dk_i}(M)$ and send M_i to $P_{\mathtt{king}}$, along with the first message a_i in a proof that M_i is the correct decryption share from P_i.

3. $P_{\mathtt{king}}$: For $P_i \in \mathcal{P}_{\mathtt{ori}} \setminus \mathcal{P}_{\mathtt{act}}$, use dk_i to compute $M_i = \mathcal{D}_{dk_i}(M)$ and the first message a_i in a proof that M_i is the correct decryption share from P_i. Compute $m = \mathrm{combine}(M_1, \ldots, M_n)$ and $a = \mathrm{combine}(a_1, \ldots, a_n)$ and KFD-BROADCAST (m, a) to each $P_i \in \mathcal{P}_{\mathtt{act}}$.

4. Each $P_i \in \mathcal{P}_{\mathtt{act}}$: Run KFD-SIGN on (sid) to receive $e = \mathrm{sig}_{sk}(sid)$.

5. Each $P_i \in \mathcal{P}_{\mathtt{act}}$: Using e as challenge, compute the reply c_i for the proof begun in Step 2. Send c_i to $P_{\mathtt{king}}$.

6. $P_{\mathtt{king}}$: If for some $P_i \in \mathcal{P}_{\mathtt{act}}$ the value $((ek, M, M_i), a_i, e, c_i)$ is not a valid conversation for M_i being the decryption share of M from P_i, then output $(sid, \mathtt{failed!}, P_i)$. Otherwise, for each $P_i \in \mathcal{P}_{\mathtt{ori}} \setminus \mathcal{P}_{\mathtt{act}}$, use dk_i to compute the reply c_i of the proof begun in Step 3, using e as challenge. Then compute $c = \mathrm{combine}(c_1, \ldots, c_n)$ and send c to all parties.

7. Each $P_i \in \mathcal{P}_{\mathtt{act}}$: If $((ek, M, m), a, e, c)$ is not a valid conversation for m being the plaintext of M, then output $(sid, \mathtt{failed!}, P_{\mathtt{king}})$. Otherwise, output $(sid, \mathtt{ok!}, m)$.

The protocol is a KFD protocol, with King Detection and Sound Detection: If any KFD sub-protocol fails, the parties adopt the accusation from the first one to fail. If all KFD sub-protocols succeed, but some party $P_i \in \mathcal{P}_{\mathtt{act}}$ does not give a valid conversation $((ek, M, M_i), a_i, e, c_i)$, then $P_{\mathtt{king}}$ outputs $(sid, \mathtt{failed!}, P_i)$. Otherwise, $P_{\mathtt{king}}$ will, by the homomorphic-conversations property, always compute a valid conversation $((ek, M, m), a, e, c)$ for m being the plaintext of M. Therefore no honest party will output $(sid, \mathtt{failed!}, P_{\mathtt{king}})$.

We then consider the correctness of the protocol, by which we mean that no honest party outputs (sid, m) with $m \neq \mathcal{D}(M)$. Assume for the sake of contradiction that some PPT adversary can control the corrupted parties in such a way that some honest party outputs (sid, m) with $m \neq \mathcal{D}(M)$. Consider the proof given by $P_{\mathtt{king}}$ from the viewpoint of an honest $P_i \in \mathcal{P}_{\mathtt{act}}$. In Step 3 $P_{\mathtt{king}}$ sends (m, a). Then $e = \mathrm{sig}_{sk}(sid)$ is generated, and in Step 6 $P_{\mathtt{king}}$ then sends c such that $((ek, M, m), a, e, c)$ is a valid conversation for m

being the plaintext of M. Since m is *not* the plaintext of M, it follows from the strong soundness that using the factorization (p, q) of N one can compute from (m, a) the one challenge $e' = e(p, q, M, m, a)$ for which there exists c' such that $\big((ek, M, m), a, e', c'\big)$ is a valid conversation. Since $\big((ek, M, m), a, e = \mathrm{sig}_{sk}(sid), c\big)$ *is* a valid conversation (otherwise, P_i would not output (sid, m)), it follows that $e(p, q, M, m, a) = \mathrm{sig}_{sk}(sid)$. Since the use of KFD-BROADCAST ensures that *all* $P_i \in \mathcal{P}_{\mathtt{act}}$ receive the same (m, a), we can consider the point were (m, a) is received by the *first* honest $P_i \in \mathcal{P}_{\mathtt{act}}$. At this point one can use (p, q) to compute $\mathrm{sig}_{sk}(sid) = e(p, q, M, m, a)$. However, at this point in the protocol no honest party $P_i \in \mathcal{P}_{\mathtt{act}}$ has yet input (sid) to KFD-SIGN, so by the security of KFD-SIGN, it should be infeasible to compute $\mathrm{sig}_{sk}(sid)$. Since the security of KFD-SIGN does not depend on (p, q) being hidden, we have a contradiction. This argument can easily be turned into a formal reduction to the unforgeability of sig_{sk}.

We then consider the privacy of the protocol. Revealing the decryption shares leaks no additional information as they can be simulated from M and the plaintext m. We therefore just have to ensure that the proofs of correct decryption share do not leak any additional information. This follows from the fact that the proofs can be simulated given only the decryption shares M_i and the plaintext m, and especially without using d_i. Instead of d_i the simulator will use the *signing key* sk. Recall namely that the proofs of correct decryption shares are performed using a Σ-protocol, and observe that knowing sk allows the simulator to compute the challenge $e = \mathrm{sig}_{sk}(sid)$ *before* the proof is run. Using the instance (ek, M, M_i) and the challenge e the simulator can apply the special honest-verifier zero-knowledge property to compute a first message a_i and a reply c_i for each honest $P_i \in \mathcal{P}_{\mathtt{act}}$, such that $\big((ek, M, M_i), a_i, e, c_i\big)$ has a distribution statistically close to that in the protocol. Then the simulator sends a_i to $P_{\mathtt{king}}$. When $e = \mathrm{sig}_{sk}(sid)$ is output by KFD-SIGN, the simulator then simply sends the preprocessed c_i as reply.

6.7   Random Encrypted Elements

The next KFD sub-protocol, KFD-RANDOM, allows the parties to generate a common ciphertext B, where the plaintext $\mathcal{D}(B)$ is computationally indistinguishable to the corrupted parties from a uniformly random value from $\mathbb{Z}_N$. For efficiency many such values are generated in parallel.

Let H consist of the indices i of the honest $P_i \in \mathcal{P}_{\mathtt{act}}$, and let C consist of the indices j of the corrupted $P_j \in \mathcal{P}_{\mathtt{act}}$. Note that by the assumption that $\mathcal{P}_{\mathtt{act}}$ has honest majority we have that $|H| \geq t'+1$. Consider the ciphertexts $(R_1, \ldots, R_{n'})$. For $i \in H$, the value $r_i = \mathcal{D}(R_i)$ is computationally indistinguishable to the corrupted parties from a uniformly random value from $\mathbb{Z}_N$, by semantic security. For $j \in C$, the proof of plaintext knowledge (along with the choice of dummy value) guarantees that the value $r_j = \mathcal{D}(R_j)$ is known by P_j, and thus the corrupted parties. Since the corrupted parties have no knowledge on r_i for $i \in H$ and they know r_j for $j \in C$, it follows that to the corrupted parties, the vector $(r_1, \ldots, r_{n'})$ looks as a vector with $|C|$ values chosen by themselves and then

---

KFD-RANDOM

1. Each $P_i \in \mathcal{P}_{\texttt{act}}$ has input $(sid, P_{\texttt{king}})$, where $P_{\texttt{king}} \in \mathcal{P}_{\texttt{act}}$. Let $n' = |\mathcal{P}_{\texttt{act}}|$ and $t' = \lfloor (n'-1)/2 \rfloor$. For notational convenience assume $\mathcal{P}_{\texttt{act}} = \{P_1, \ldots, P_{n'}\}$.
2. Each $P_i \in \mathcal{P}_{\texttt{act}}$: Compute $R_i = \mathcal{E}(r_i; \rho_i)$ for uniformly random $r_i \in \mathbb{Z}_N$, $\rho_i \in \mathbb{Z}_N^*$. Then KFD-BROADCAST R_i along with a proof of plaintext knowledge for R_i. The dummy value is $D = \mathcal{E}(0; 1)$.
3. Each $P_k \in \mathcal{P}_{\texttt{act}}$: Now n' values R_i were received. Using some fixed agreed upon scheme, pick a super-invertible matrix $S = \{s_{j,i}\}_{j=1,\ldots,t'+1}^{i=1,\ldots,n'}$ with $t' + 1$ rows and n' columns. Then for $j = 1, \ldots, t' + 1$, let $B_j = \prod_{i=1}^{n'} R_i^{s\cdots} \bmod N^2$. Then output $(B_1, \ldots, B_{t'+1})$.

---

filled in with $|H|$ independent, uniformly random values r_i. Since S is a super-invertible matrix with n' columns and $t' + 1 \leq |H|$ rows, it follows that to the corrupted parties the vector $(b_1, \ldots, b_{t'+1}) = S(r_1, \ldots, r_{n'})$ looks as a uniformly random vector from $(\mathbb{Z}_N)^{t'+1}$. By the homomorphic properties of the encryption function it follows that $\mathcal{D}(B_j) = b_j$ for $j = 1, \ldots, t' + 1$. It therefore follows that KFD-RANDOM outputs encrypted values which to the corrupted parties look independent and uniformly random. This is exactly what we need from the protocol.

The communication complexity is given by the n' runs of KFD-BROADCAST on $\mathcal{O}(\kappa)$-bit values, giving a total communication complexity of $\mathcal{O}(n'n\kappa)$. The protocol generates $t' + 1 = \Theta(n')$ outputs. To generate ℓ outputs the protocol will be run $\lceil \ell/(t'+1) \rceil \leq \ell/(t'+1) + 1$ times in parallel. In that case the communication complexity is $\mathcal{O}(n\ell\kappa + n^2\kappa) = \mathcal{O}(n\ell\kappa + \mathcal{BC})$.

Recall that each output B was generated as $B = \prod_{i=1}^{n'} R_i^{s} \bmod N^2$ for a row $(s_1, \ldots, s_{n'})$ in S, and P_i knows r_i and ρ_i such that $R_i = \mathcal{E}(r_i; \rho_i)$. Therefore all parties can compute $B_i = R_i^{s} \bmod N^2$ for $i = 1, \ldots, n'$ such that $B = \prod_{i=1}^{n'} B_i \bmod N^2$, and P_i can compute $b_i = s_i r_i \bmod N$ and $\beta_i = \rho_i^{s} \bmod N$ such that $B_i = \mathcal{E}(b_i; \beta_i)$. We use this in the next sub-protocol.

6.8 Random Encrypted Multiplication Triples

Our last KFD sub-protocol, KFD-TRIPLES, allows the parties to generate a triple (A, B, C), where $\mathcal{D}(C) = \mathcal{D}(A)\mathcal{D}(B) \bmod N$, and where $\mathcal{D}(A)$ and $\mathcal{D}(B)$ are computationally indistinguishable to the corrupted parties from independent, uniformly random values from $\mathbb{Z}_N$. The protocol uses the homomorphic proofs of identical encryption.

The protocol is a KFD protocol, with King Detection and Sound Detection: If any KFD sub-protocol fails, the parties adopt the accusation from the first one to fail. If all KFD sub-protocol succeed, but some party $P_i \in \mathcal{P}_{\texttt{act}}$ does not give a valid conversation $((G, A, B_i, C_i), a_i, e, c_i)$, then $P_{\texttt{king}}$ outputs $(sid, \texttt{failed!}, P_i)$. Otherwise, $P_{\texttt{king}}$ can, by the homomorphic-conversations property, always compute a valid $((G, A, B, C), a, e, c)$. Therefore no honest party will output $(sid, \texttt{failed!}, P_{\texttt{king}})$.

As for the correctness, notice that the use of KFD-BROADCAST guarantees that when KFD-TRIPLES succeeds, then the honest $P_i \in \mathcal{P}_{\texttt{act}}$ output a common

---

KFD-TRIPLES

1. First KFD-RANDOM is run to generate random encryptions. Two of these are taken and renamed to A and B. All parties know $\{B_i\}_{P_i \in \mathcal{P}_{\text{act}}}$ such that $B = \prod_{P_i \in \mathcal{P}_{\text{act}}} B_i \bmod N^2$, and P_i knows (b_i, β_i) such that $B_i = \mathcal{E}(b_i; \beta_i) = G^{b_i} \beta_i^N \bmod N^2$.

2. Each $P_i \in \mathcal{P}_{\text{act}}$: Compute $C_i = A^{b_i} \gamma_i^N \bmod N^2$ for uniformly random $\gamma_i \in_R \mathbb{Z}_N^*$, and send C_i to P_{king}, along with the first message a_i in a proof that there exists (b_i, β_i, γ_i) such that $B_i = G^{b_i} \beta_i^N \bmod N^2$ and $C_i = A^{b_i} \gamma_i^N \bmod N^2$.

3. P_{king}: Compute $C = \prod_{P_i \in \mathcal{P}_{\text{act}}} C_i \bmod N^2$ and $a = \text{combine}(\{a_i\}_{P_i \in \mathcal{P}_{\text{act}}})$ and KFD-BROADCAST (C, a).

4. Each $P_i \in \mathcal{P}_{\text{act}}$: On (C, a), run KFD-SIGN on (sid) to receive $e = \text{sig}_{sk}(sid)$.

5. Each P_i: Compute the reply c_i in the proof begun in Step 2, using e as challenge. Send c_i to P_{king}.

6. P_{king}: If for any $P_i \in \mathcal{P}_{\text{act}}$ the value $((G, A, B_i, C_i), a_i, e, c_i)$ is not a valid conversation, then output $(sid, \texttt{failed!}, P_i)$. Otherwise, compute $c = \text{combine}(\{c_i\}_{P_i \in \mathcal{P}_{\text{act}}})$ and send c to all parties.

7. Each $P_i \in \mathcal{P}_{\text{act}}$: If $((G, A, B, C), a, e, c)$ is not a valid conversation for the claim that there exists (b, β, γ) such that that $B = G^b \beta^N \bmod N^2$ and $C = A^b \gamma^N \bmod N^2$, then output $(sid, \texttt{failed!}, P_{\text{king}})$. Otherwise, output $(sid, \texttt{ok!}, (A, B, C))$.

---

value (A, B, C). And, as for KFD-DECRYPT, it follows from the strong soundness and the unforgeability of sig_{sk} that when the protocol succeeds, then except with negligible probability there exists (b, β, γ) such that that $B = G^b \beta^N \bmod N^2$ and $C = A^b \gamma^N \bmod N^2$. Therefore $\mathcal{D}(B) = \mathcal{D}(G^b \beta^N) = b$ and $\mathcal{D}(C) = \mathcal{D}(A^b \gamma^N) = b\mathcal{D}(A) \bmod N = \mathcal{D}(B)\mathcal{D}(A) \bmod N$, as required.

As for the privacy, the encryption $C_i = A^{b_i} \gamma_i^N \bmod N^2$ hides b_i, and the proofs can be simulated without b_i using the simulation technique from KFD-DECRYPT.

To generate ℓ triples, the protocol is run ℓ times in parallel. In that case the communication complexity in bits is $\mathcal{O}(n\ell\kappa + \mathcal{BC})$. Namely, the communication complexity for generating 2ℓ random values using KFD-RANDOM is $\mathcal{O}(n\ell\kappa + \mathcal{BC})$. The extra communication complexity per triple is then given by the run of KFD-BROADCAST and KFD-SIGN, each $\mathcal{O}(n\kappa)$, and sending a constant number of $\mathcal{O}(\kappa)$-bit values between P_{king} and each of the other parties.

7   The MPC Protocol

In the MPC protocol EVAL the parties are given an acyclic circuit $\text{Circ} = \{G_{gid}\}$, where each G_{gid} is a gate with gate identifier $gid \in \{0, 1\}^\kappa$. Each gate is of one of the following types:

Input gate (gid, in, i): Let $P_i \in \mathcal{P}_{\text{ori}}$ give a secret input x_{gid}.

Multiplication gate $(gid, \text{mul}, gid_1, gid_2)$: $x_{gid} \leftarrow x_{gid_1} x_{gid_2} \bmod N$.

Linear gate $(gid, \text{lin}, a_0, a_1, gid_1, \ldots, a_\ell, gid_\ell)$: $x_{gid} \leftarrow a_0 + \sum_{l=1}^\ell a_l x_{gid_l}$.

Random gate (gid, ran): A secret $x_{gid} \in \mathbb{Z}_N$ is sampled uniformly at random.

Global output gate $(gid, \mathtt{gout}, gid_1)$: Let every $P_i \in \mathcal{P}_{\mathtt{ori}}$ learn x_{gid_1}.

Local output gate $(gid, \mathtt{lout}, gid_1, i)$: Let $P_i \in \mathcal{P}_{\mathtt{ori}}$, and only P_i, learn x_{gid_1}.

We assume that we are given a segmentation $(\mathrm{Seg}_1, \ldots, \mathrm{Seg}_m)$ of the circuit, such that the value of every gate in Seg_i only depends on values of gates in Seg_j with $j \le i$. Furthermore, we require that if a segment contains an output gate, then it does not contain an input or a random gate. This requirement is made to ensure that if a segment is evaluated twice in a row, then all outputs will be the same. This is needed for privacy reasons. If in addition it should not be possible for the parties to pick inputs based on earlier outputs, all segments containing input gates should appear before all segments containing output gates.

During the protocol a party $P_i \in \mathcal{P}_{\mathtt{act}}$ stores a representation (gid, X_{gid}) for some gates gid. These values will be consistent in the sense that all parties storing a value (gid, X_{gid}) for some gid agree on X_{gid}. When gid is an output gate, then $X_{gid} \in \mathbb{Z}_N$ is a plaintext, and otherwise $X_{gid} \in \mathbb{Z}_{N^2}^*$ is a ciphertext. For an input gate $(gid, \mathtt{in}, i)$ to be called correct we require that P_i gave a proof of plaintext knowledge for X_{gid}. When P_i is honest, we also require that $\mathcal{D}(X_{gid}) = x_{gid}$. When $(gid, \mathtt{in}, i)$ is correct we define $\mathcal{V}(gid) = \mathcal{D}(X_{gid})$. We extend $\mathcal{V}(\cdot)$ to the rest of the circuit by $\mathcal{V}(gid) = \mathcal{V}(gid_1)\mathcal{V}(gid_2) \bmod N$ for multiplication gates and $\mathcal{V}(gid) = a_0 + \sum_{l=1}^{\ell} a_l \mathcal{V}(gid_l) \bmod N$ for linear gates. We then call a multiplication or linear gate correct if $\mathcal{D}(X_{gid}) = \mathcal{V}(gid)$, and we say that a global output gate is correct if $X_{gid} = \mathcal{V}(gid_1)$. The goal is to end up with all gates in Circ being correct and all parties $P_i \in \mathcal{P}_{\mathtt{ori}}$ holding (gid, X_{gid}) for all output gates.

It is straightforward to see that the inputs to DETECT-ELIMINATE in Step 2e have King Awareness and Sound Detection: If any of the KFD sub-protocols fail, the parties adopt the accusation from the first one to fail, and if no KFD sub-protocol fails and yet some active honest party does not have a representation of all output gates at the end of the protocol, then clearly $P_{\mathtt{king}}$ itself is corrupted.

The use of KFD-BROADCAST ensures that all parties receiving values receive the same values. Therefore the stored representations are at all times consistent.

That the circuit is at all times correct follows from the homomorphic properties of $\mathcal{E}$. For a linear gate it holds that $\mathcal{D}(X_{gid}) = a_0 + \sum_{l=1}^{\ell} a_l \mathcal{D}(X_{gid_l}) \bmod N$. So, if $\mathcal{D}(X_{gid_l}) = \mathcal{V}(gid_l)$ for $l = 1, \ldots, \ell$, then $\mathcal{D}(X_{gid}) = \mathcal{V}(gid)$. For a multiplication gate it holds that $\mathcal{D}(X_{gid}) = \alpha\beta - \alpha\mathcal{D}(B_{gid}) - \beta\mathcal{D}(A_{gid}) + \mathcal{D}(C_{gid}) \bmod N = (\alpha - \mathcal{D}(A_{gid}))(\beta - \mathcal{D}(B_{gid})) \bmod N$, as $\mathcal{D}(C_{gid}) = \mathcal{D}(A_{gid})\mathcal{D}(B_{gid}) \bmod N$. Since $\alpha = \mathcal{D}(X_{gid_1}A_{gid}) = \mathcal{D}(X_{gid_1}) + \mathcal{D}(A_{gid}) \bmod N$ and $\beta = \mathcal{D}(X_{gid_2}) + \mathcal{D}(B_{gid}) \bmod N$, it follows that $\mathcal{D}(X_{gid}) = \mathcal{D}(X_{gid_1})\mathcal{D}(X_{gid_2}) \bmod N$, as desired. The correctness of global output gates follows from the correctness of KFD-DECRYPT, the correctness of local output gates follows from the correctness of global output gates and from the fact that any (also eliminated) parties can input a blinding b_{gid}.

Since the circuit is at all times correct and a segment only succeeds if all active parties end up with a representation of all gates in the segment, it follows that when a segment succeeds, then $X_{gid} = \mathcal{V}(gid)$ for all output gates gid at all active parties. The correctness of KFD-PROMOTE then guarantees that all parties

---

EVAL

1. Each $P_i \in \mathcal{P}_{\texttt{ori}}$ has input Circ $= (\text{Seg}_1, \ldots, \text{Seg}_m)$ and $x_{gid} \in \mathbb{Z}_N$ for every $(gid, \texttt{in}, i) \in$ Circ. Let $\mathcal{P}_{\texttt{act}} = \mathcal{P}_{\texttt{ori}}$ and let cur $= 1$.

2. Terminate if cur $> m$, otherwise, let Seg $= \text{Seg}_{\text{cur}}$ and let $P_{\texttt{king}}$ denote the party in $\mathcal{P}_{\texttt{act}}$ with the smallest index.

   (a) *(Prepare multiplication triples)* Run KFD-TRIPLES with $P_{\texttt{king}}$ to generate triples $(A_{gid}, B_{gid}, C_{gid})$ for each $(gid, \texttt{mul}, \cdot, \cdot), (gid, \texttt{ran}) \in$ Seg.

   (b) *(Evaluate segment)* Every gate $G_{gid} \in$ Seg is evaluated (in parallel) as soon as its source-gates have been evaluated.

   Input: For $(gid, \texttt{in}, i)$, $P_i \in \mathcal{P}_{\texttt{ori}}$ generates $X_{gid} \leftarrow \mathcal{E}(x_{gid})$ and KFD-BROADCAST the encryption X_{gid} along with a proof of plaintext knowledge, with $P_{\texttt{king}}$ as King and $D = \mathcal{E}(0; 1)$ as dummy value. Every P_j stores the value X_{gid} broadcasted by P_i as (gid, X_{gid}).

   Random: For $(gid, \texttt{ran})$, $P_i \in \mathcal{P}_{\texttt{act}}$ stores (gid, A_{gid}).

   Linear function: For $(gid, \texttt{lin}, a_0, a_1, gid_1, \ldots, a_\ell, gid_\ell)$, let $X_{gid} = \mathcal{E}(a_0; 1)\left(\prod_{l=1}^{\ell} X_{gid_l}^{a_l}\right) \bmod N^2$, and store (gid, X_{gid}).

   Multiplication: For $(gid, \texttt{mul}, gid_1, gid_2)$, run KFD-DECRYPT on $X_{gid_1} A_{gid} \bmod N^2$ and $X_{gid_2} B_{gid} \bmod N^2$. Let α and β be the respective outputs, let $X_{gid} = \mathcal{E}(\alpha\beta; 1) B_{gid}^{-\alpha} A_{gid}^{-\beta} C_{gid} \bmod N^2$, and store (gid, X_{gid}).

   Global output: For $(gid, \texttt{gout}, gid_1)$, let $X_{gid} = X_{gid_1}$, run KFD-DECRYPT on X_{gid}, and let x_{gid} be the output. Store (gid, x_{gid}).

   Local output: For $(gid, \texttt{lout}, gid_1, i)$, $P_i \in \mathcal{P}_{\texttt{ori}}$ selects a random blinding $b_{gid} \in_R \mathbb{Z}_N$, and the parties first evaluate $(gid, \texttt{in}, i)$ for input b_{gid}, resulting in an encryption B_{gid}, then evaluate $(gid, \texttt{gout}, \cdot)$ for the encryption $X_{gid} = X_{gid_1} B_{gid} \bmod N^2$, and store the blinded result (gid, x_{gid}).

   (c) *(Result promotion)* Let ℓ be the number of global and local outputs in Seg. Invoke KFD-PROMOTE with input $(x_{gid_1}, \ldots, x_{gid_\ell})$ to promote these values from $\mathcal{P}_{\texttt{act}}$ to $\mathcal{P}_{\texttt{ori}}$. Parties $P_i \in \mathcal{P}_{\texttt{ori}} \setminus \mathcal{P}_{\texttt{act}}$ store (gid, x_{gid}) for every output x_{gid}.

   (d) *(Result unblinding)* Every party $P_i \in \mathcal{P}_{\texttt{ori}}$ with local output $(gid, \texttt{lout}, \cdot, i) \in$ Seg stores $(gid, x_{gid} - b_{gid} \bmod N)$.

   (e) *(Detect and eliminate)* Each $P_i \in \mathcal{P}_{\texttt{act}}$: Run DETECT-ELIMINATE with input (s_i, y_i) derived as follows. If some KFD sub-protocol terminated with output $s_i = \texttt{failed!}$, then use the accusation y_i from the first one to fail. Otherwise, if all KFD sub-protocols output $s_i = \texttt{ok!}$, but a value $(gid, \cdot)$ was not stored for all gates $G_{gid} \in$ Seg, then use $s_i = \texttt{failed!}$ and $y_i = P_{\texttt{king}}$. Otherwise, use $s_i = \texttt{ok!}$.

   (f) *(Conclude segment)* If no parties were eliminated in DETECT-ELIMINATE, then let cur $:=$ cur $+ 1$. Go to Step 2.

---

end up storing $(gid, \mathcal{V}(gid))$ for all output gates $(gid, \texttt{gout}, gid_1) \in$ Seg, and the correctness of local outputs is guaranteed by the correctness of the (blinded) global outputs.

The privacy of the protocol follows as for previous MPC protocols based on threshold homomorphic encryption (see e.g. [CDN01]). The important point being that e.g. $\alpha = \mathcal{D}(X_{gid_1}) + \mathcal{D}(A_{gid}) \bmod N$ leaks no information to the corrupted parties, as $\mathcal{D}(A_{gid})$ looks uniformly random to them and thus blinds

$\mathcal{D}(X_{gid_1})$. The privacy of local output is guaranteed by the blinding chosen by the output party.

As for the communication complexity it can be seen that evaluating Seg_{cur} generates communication $\mathcal{O}(|\text{Seg}_{\text{cur}}|n\kappa + \mathcal{BC})$ bits. We can divide the circuit Circ of any function into $m = \mathcal{O}(n)$ segments of size at most $\lceil|\text{Circ}|/n\rceil$, and evaluating Circ requires evaluation of at most $m + t = \mathcal{O}(n)$ segments. Hence, the total communication complexity is $\mathcal{O}(|\text{Circ}|n\kappa + n\mathcal{BC})$ bits.

8   Conclusions

Any function can be securely evaluated by n parties with communication $\mathcal{O}(|\text{Circ}|n\kappa + n\mathcal{BC})$ bits, where Circ is a circuit computing the function, κ is the security parameter, and $\mathcal{BC}$ is the communication complexity of broadcast. The communication complexity is linear in the number of parties, for all types of gates, including multiplication, input and output gates.

References

[BB89]   J.Bar-Ilan and D.Beaver. Non-cryptographic fault-tolerant computing in constant number of rounds of interaction. *PODC'89*, 1989.

[Bea91a]   D.Beaver. Efficient multiparty protocols using circuit randomization. In *Crypto'91*, LNCS 576, 1991.

[Bea91b]   D.Beaver. Secure multi-party protocols and zero-knowledge proof systems tolerating a faulty minority. *Journal of Cryptology*, 4(2):75–122, 1991.

[BFKR90]   D.Beaver, J.Feigenbaum, J.Kilian, and P.Rogaway. Security with low communication overhead (extended abstract). *Crypto'90*, LNCS 537, 1990.

[BGW88]   M.Ben-Or, S.Goldwasser, and A.Wigderson. Completeness theorems for non-cryptographic fault-tolerant distributed computation (extended abstract). *20th STOC*, 1988.

[BH05]   Z.Beerliova-Trubiniova and M.Hirt. Efficient multi-party computation with dispute control. *TCC'06*, LNCS 3876, 2006.

[BMR90]   D.Beaver, S.Micali, and P.Rogaway. The round complexity of secure protocols (extended abstract). *22nd STOC*, 1990.

[Can01]   R.Canetti. Universally composable security: A new paradigm for cryptographic protocols. *42nd FOCS*, 2001.

[CCD88]   D.Chaum, C.Crépeau, and I.Damgård. Multiparty unconditionally secure protocols (extended abstract). *20th STOC*, 1988.

[CDD+99]   R.Cramer, I.Damgård, S.Dziembowski, M.Hirt, and T.Rabin. Efficient multiparty computations secure against an adaptive adversary. *EuroCrypt'99*, LNCS 1592, 1999.

[CDD00]   R.Cramer, I.Damgård, and S.Dziembowski. On the complexity of verifiable secret sharing and multiparty computation. *32nd STOC*, 2000.

[CDG87]   D.Chaum, I.Damgård, and J. van de Graaf. Multiparty computations ensuring privacy of each party's input and correctness of the result. *Crypto'87*, LNCS 293, 1987.

[CDM00]   R.Cramer, I.Damgård, and U.Maurer. General secure multi-party computation from any linear secret-sharing scheme. *EuroCrypt'00*, LNCS 1807, 2000.

[CDN01]   R.Cramer, I.Damgaard, and J.B.Nielsen. Multiparty computation from threshold homomorphic encryption. *EuroCrypt'01*, LNCS 2045, 2001.

[Dam00]   I.Damgård. Efficient concurrent zero-knowledge in the auxiliary string model. *EuroCrypt'00*, LNCS 1808, 2000.

[DI06]   I.Damgård and Y.Ishai. Scalable secure multiparty computation. *Crypto'06*, 2006.

[FH06]   M.Fitzi and M.Hirt. Optimally efficient multi-valued byzantine agreement. *25th PODC*, 2006.

[FH96]   M.Franklin and S.Haber. Joint encryption and message-efficient secure computation. *Journal of Cryptology*, 9(4):217–232, 1996.

[GMW87]   O.Goldreich, S.Micali, and A.Wigderson. How to play any mental game or a completeness theorem for protocols with honest majority. *19th STOC*, 1987.

[GRR98]   R.Gennaro, M.Rabin, and T.Rabin. Simplified VSS and fast-track multi-party computations with applications to threshold cryptography. *PODC'98*, 1998.

[GV87]   O.Goldreich and R.Vainish. How to solve any protocol problem - an efficiency improvement. *Crypto'87*, LNCS 293, 1987.

[HM01]   M.Hirt and U.Maurer. Robustness for free in unconditional multi-party computation. *Crypto'01*, LNCS 2139, 2001.

[HMP00]   M.Hirt, U.Maurer, and B.Przydatek. Efficient secure multi-party computation. *Asiacrypt'00*, LNCS 1976, 2000.

[HN05]   M.Hirt and J.B.Nielsen. Upper bounds on the communication complexity of optimally resilient cryptographic multiparty computation. *Asiacrypt'05*, LNCS 3788, 2005

[HNP05]   M.Hirt, J.B.Nielsen, and B.Przydatek. Cryptographic asynchronous multi-party computation with optimal resilience. *EuroCrypt'05*, LNCS 3494, 2005.

[Nie03]   J.B.Nielsen. On Protocol Security in the Cryptographic Model. PhD Thesis. Department of Computer Science, University of Aarhus, 2003.

[Pai99]   P.Paillier. Public-key cryptosystems based on composite degree residue classes. *EuroCrypt'99*, LNCS 1592, 1999.

[Rab98]   T.Rabin. A simplified approach to threshold and proactive RSA. *Crypto'98*, LNCS 1462, 1998.

[RB89]   T.Rabin and M.Ben-Or. Verifiable secret sharing and multiparty protocols with honest majority. *21th STOC*, 1989.

[Yao82]   A. C.-C. Yao. Protocols for secure computations (extended abstract). *23rd FOCS*, 1982.

On Combining Privacy with Guaranteed Output Delivery in Secure Multiparty Computation[*]

Yuval Ishai[1], Eyal Kushilevitz[1], Yehuda Lindell[2,**], and Erez Petrank[1]

[1] Technion
{yuvali, eyalk, erez}@cs.technion.ac.il
[2] Bar-Ilan University
lindell@cs.biu.ac.il

Abstract. In the setting of multiparty computation, a set of parties wish to jointly compute a function of their inputs, while preserving security in the case that some subset of them are corrupted. The typical security properties considered are privacy, correctness, independence of inputs, guaranteed output delivery and fairness. Until now, all works in this area either considered the case that the corrupted subset of parties constitutes a strict minority, or the case that a half or more of the parties are corrupted. Secure protocols for the case of an honest majority achieve full security and thus output delivery and fairness are guaranteed. However, the security of these protocols is *completely compromised* if there is no honest majority. In contrast, protocols for the case of no honest majority do not guarantee output delivery, but do provide privacy, correctness and independence of inputs for *any number* of corrupted parties. Unfortunately, an adversary controlling only *a single party* can disrupt the computation of these protocols and prevent output delivery.

In this paper, we study the possibility of obtaining general protocols for multiparty computation that *simultaneously* guarantee security (allowing abort) in the case that an arbitrary number of parties are corrupted *and* full security (including guaranteed output delivery) in the case that only a minority of the parties are corrupted. That is, we wish to obtain the best of both worlds in a single protocol, depending on the corruption case. We obtain both positive and negative results on this question, depending on the type of the functionality to be computed (standard or reactive) and the type of dishonest majority (semi-honest or malicious).

1 Introduction

1.1 Background

Secure multiparty computation (MPC) [43, 28, 6, 13] allows a set of mutually distrusting parties to compute a function in a distributed way while guaranteeing (to the extent possible) the privacy of their local inputs and the correctness of

---

[*] Research supported by grant 36/03 from the Israel Science Foundation.
[**] Much of this work was carried out while the author was visiting the Technion.

C. Dwork (Ed.): CRYPTO 2006, LNCS 4117, pp. 483–500, 2006.

the outputs. To be more exact, security is typically formulated by comparing a real execution of a protocol to an ideal execution where the parties just send their inputs to a trusted party and receive back their outputs. A real protocol is said to be secure if an adversary can do no more harm in a real execution than in an ideal execution (which is secure by definition). The main security properties that have been identified, and are implied by this formulation, are: privacy (parties learn nothing more than their output), correctness (the outputs are correctly computed), independence of inputs (parties must choose their inputs independently of the other parties), fairness (if one party receives its output, all parties receive their output), and guaranteed output delivery (the honest parties are guaranteed to successfully complete the computation).

The vast body of research in this area can be divided into two, almost disjoint, lines of work: one dealing with the case of an *honest majority* and another dealing with the case of *no honest majority*. There are inherent differences between the two settings both in terms of the type of security guarantees one can hope to obtain and in terms of the techniques used for implementing protocols. We elaborate on this below.

Secure Computation with Honest Majority. If a majority of the participants are honest, then it is possible to obtain the strongest type of security one could hope for (including all of the properties described above). In fact, it is even possible to obtain this level of security unconditionally, without relying on unproven complexity assumptions (as long as we do assume secure point-to-point channels and either the existence of a broadcast channel [41] or of a trusted preprocessing phase [40][1]). Protocols for the honest-majority case include unconditionally secure protocols from [6, 13, 41, 3, 17, 23, 15, 31] and computationally secure protocols from [28, 5, 18, 19, 20, 25]. In this setting, n parties can compute an arbitrary polynomial-time computable function of their inputs, while providing unconditional security against $s < n/2$ malicious parties [41, 15]. Moreover, settling for computational security, the same can be done in a constant number of rounds (assuming that one-way functions exist) [5, 20]. It is known that the bound $s < n/2$ is tight even for simple functionalities such as coin-flipping [14] and even if one settles for computational security. We refer to protocols with security against $s < n/2$ malicious parties as being of type $\mathcal{P}_{n/2}$.

Secure Computation with No Honest Majority. If a half or more of the participants *may* be corrupted, then it is impossible to construct protocols that guarantee (full) fairness, let alone output delivery. Rather, all protocols must allow the adversary to *abort* the protocol (possibly, even after first learning its output, but before the honest parties receive their output) [14].[2] We will refer to this type of security as "security with abort" (and will sometimes call security for the

---

[1] The assumptions of broadcast or trusted preprocessing are not needed in the case where more than two thirds of the parties are honest [6, 13].

[2] In fact, the protocols can guarantee that the latter fairness problem occurs only if some specified distinguished party, say the first, is corrupted (cf. [27, 30]).

case of an honest majority full security in order to clarify which level of security is discussed). Unlike the honest majority case, protocols in the case of no honest majority can only guarantee *computational* security and need to rely on cryptographic assumptions. Accordingly, the techniques used for obtaining such protocols involve cryptographic tools such as zero-knowledge proofs, commitment, and oblivious transfer. Protocols in this setting include the two-party protocols of [43, 28, 36, 39, 33] and the multi-party protocols of [28, 4, 29, 34, 30, 12, 38, 2]. In this setting, n parties can compute any polynomial-time computable function of their inputs, while providing security with abort against $t < n$ malicious parties [43, 28] (assuming enhanced trapdoor permutations exist).[3] This can also be done in a constant number of rounds (assuming that trapdoor permutations and collision-resistant hash functions exist) [36, 34, 38]. We refer to protocols that are secure with abort against $t < n$ malicious parties as being of type $\mathcal{P}_n$. We stress that, although output delivery and fairness are not guaranteed, there is no compromise on privacy, correctness and independence of inputs.

A major *qualitative advantage* of type $\mathcal{P}_n$ protocols is that they provide security (albeit with abort) against an *arbitrary* number of corrupted parties. In particular, each party is assured that the *privacy* of its input is maintained even if *all the rest of the world* conspires against it (the same holds for correctness and independence of inputs as well). Thus, the privacy of each party is entirely in its own hands. On the other hand, a major *qualitative disadvantage* of type $\mathcal{P}_n$ protocols is that they fail to be resilient even against a single malicious party. In particular, a single malicious party can abort the protocol and prevent honest parties from receiving any outputs. This disadvantage becomes more severe as the number of parties grows, to the point of rendering the guarantee of security with abort essentially useless in a system with many parties. (Consider, for instance, a voting system in which each voter can disrupt the elections.)

The above state of affairs raises the following natural question:

> *Is it possible to construct a single protocol for general MPC that will provide security with abort (or even just privacy) in the case that an arbitrary number of parties are corrupted and, simultaneously, will provide full security (including guaranteed output delivery) in the case that only a minority (or even just one) of the parties are corrupted?*

In other words, is it possible to combine the main qualitative advantages of the two types of protocols? To the best of our knowledge, the above fundamental question has not been studied before. This is a major gap in our current understanding of the feasibility of secure computation.

1.2 Our Results

We initiate the study of the above question, obtaining both negative and positive results. We distinguish between standard and reactive secure computation tasks.

---

[3] Here too, one either needs to rely on the availability of a broadcast channel, or alternatively rely on some setup assumption such as PKI. Another alternative is to relax the requirement of a *unanimous* abort, allowing a situation where only some of the honest parties abort [30, 22].

A standard (or non-reactive) functionality receives a single input from every party and provides a single output to every party. This corresponds to the usual notion of secure function evaluation. In contrast, a reactive functionality can receive inputs and provide outputs in multiple rounds, maintaining a state information between subsequent invocations. (For example, a commitment functionality is reactive because it has distinct "commit" and "reveal" phases.) We also distinguish between two main variants of the above question, depending on the type of adversary that is present. We will consider semi-honest and malicious adversaries. A semi-honest adversary follows the protocol specification but may try to infer additional information about the honest parties' inputs from what it sees. In contrast, a malicious adversary may instruct the corrupted parties under its control to arbitrarily deviate from the protocol specification. In order to strengthen our negative results, we will relax the requirements and allow the protocol to achieve only privacy in the case of no honest majority (without requiring the full definition of security-with-abort).

A Negative Result for Reactive Functionalities. We observe that the answer to the above question is negative for the case of reactive functionalities. This is proved by considering a type of *commitment* or *secret-sharing* functionality and showing that it is impossible to achieve the requirements for this specific functionality. The intuition behind this result is that if a secret is to be shared between n parties with maximal privacy (i.e., so that no strict subset of the parties can recover the secret), then even a single party can prevent a subsequent reconstruction of the secret. This is due to the fact that any one party can just withhold its share and refuse to participate in the secret reconstruction phase. Since the remaining $n - 1$ parties should not be able to recover the secret in the case that they are corrupted, they cannot recover it when they are honest here as well. Thus, it is impossible to construct a protocol for secret sharing that achieves security (or security with abort) against $n-1$ *semi-honest* (and of course malicious) parties, and full security (including guaranteed output delivery) against just a single malicious party. Our proof is motivated by the above-described intuition, but the details are somewhat different. In particular, our proof uses a non-standard formalization of secret-sharing as a *reactive* functionality. We note that some standard formulations of verifiable secret-sharing from the literature (e.g., the one from [27]) are non-reactive, and our proof does not hold for these formulations.

A Positive Result for Standard Functionalities. In light of the above, we restrict our attention from now on to the evaluation of standard (non-reactive) functionalities. In this case, we show the following positive result: any n-party functionality f can be computed by a protocol that *simultaneously* achieves the following two security properties: (1) full security in the presence of any minority of *malicious* parties; and (2) full security in the presence of any number of *semi-honest* parties (we obtain full security here even in the case of no honest majority because the adversary is semi-honest and so never aborts). Moreover, such a protocol can be implemented in a constant number of rounds. (The above results can be based on standard cryptographic assumptions.) Thus, in contrast to the case of reactive functionalities, our main question can be answered affirmatively for the case where privacy is only required to hold against semi-honest parties.

The above positive result is shown via the following simple protocol: first, the parties use a type $\mathcal{P}_n$ protocol to evaluate a randomized functionality $\hat{f}$ that outputs an *authenticated secret-sharing* of the output of f for which the shares of $< n/2$ colluding parties yields no information about the secret. If this stage terminates successfully (without aborting), then the parties exchange their outputs of $\hat{f}$, from which they can recover the output of f. In case the adversary aborted the computation of $\hat{f}$, the parties use a protocol of type $\mathcal{P}_{n/2}$ to evaluate f. This combination of the two types of protocols combines their relative advantages (in the sense discussed above), but falls short of providing privacy against $t < n$ *malicious* parties. We also describe an interesting variant of this protocol that can provide a relaxed form of full privacy even in the malicious case. Specifically, in addition to security against $s < n/2$ malicious parties, it can be guaranteed that any coalition of $t < n$ malicious parties can learn no more information than is implied by $t + 1$ invocations of the functionality. For some natural functionalities (like a voting functionality), this relaxed notion of privacy is equivalent to the standard notion of privacy (see the end of Section 3 for discussion). Thus, in some interesting cases, our main question can be answered affirmatively even for malicious adversaries, as long as only privacy (and not security) is sufficient.

A Negative Result for Standard Functionalities. We finally turn to the most challenging remaining case of standard functionalities and where the adversary can be *malicious* both in the case of an honest majority and in the case of no honest majority. Arguably, as far as the corruption model is concerned, this is the most interesting case. Unlike the previous cases, here we do not completely settle the question. Instead, we prove that there do not exist *constant round* protocols for general secure multiparty computation that are secure for the case of an honest majority and private in the case of no honest majority (we actually show that the number of rounds must grow with the security parameter). We note that constant-round protocols *do exist* for type $\mathcal{P}_{n/2}$ protocols and for type $\mathcal{P}_n$ protocols alone, as well as for the setting of our above positive result. Thus, our negative result here provides a nontrivial separation in terms of round complexity. We leave open the possibility of protocols that require a non-constant number of rounds in this setting.

Summary. We stress that all of our results apply to the *computational* model of security. They are also quite insensitive to whether or not a broadcast primitive is given. In particular, all negative results apply even when a broadcast channel is assumed, and a variant of the positive results (replacing $s < n/2$ by $s < n/3$) can be obtained also without broadcast or any setup assumptions. Finally, we stress again that we strengthen our negative results by showing that even privacy alone (rather than security-with-abort) cannot be achieved. Our main results are summarized in Table 1.

Table 1. Summary of main results: the existence and non-existence of protocols that simultaneously guarantee security against $s < n/2$ malicious parties and privacy against $t < n$ malicious or semi-honest parties. Negative results apply whenever $s + t \geq n$, and in particular when $s = 1$ and $t = n - 1$. Our positive result actually provides security for the semi-honest case.

	Standard functionalities	Reactive functionalities
Privacy against $t < n$ semi-honest parties and security against $s < n/2$ malicious parties	Yes (constant-round) Theorem 1	No Theorem 4
Privacy against $t < n$ malicious parties and security against $s < n/2$ malicious parties	No constant-round Theorem 2	No Theorem 4

1.3 Related Work

There is a significant body of work that deals with extending the basic feasibility results for MPC in various directions. We mention some of this work below and explain the differences from ours.

Multi-threshold MPC. Several previous works consider the possibility of refining known feasibility results for MPC by allowing different security thresholds for different types of corruption [23, 22, 24]. The focus of these works is very different from ours. In particular, they all fall into the setting of an honest majority, while we study the situation where honest majority cannot be guaranteed. Moreover, in contrast to [23, 24], in this work we only deal with *computational* security and do not insist on any kind of unconditional security.

Fair MPC. Recall that a secure MPC protocol is said to be *fair* if at the end of the protocol either all parties learn the output or no party learns the output. (See [30] a for a comprehensive discussion of fairness and its variants.) It is known that even simple functionalities such as coin-flipping or XOR cannot be fairly computed in the case of no honest majority [14]. In contrast, in this work, in the setting of no honest majority we settle for privacy or security-with-abort which do not imply fairness. Thus, our negative results are not implied by impossibility results that relate to fairness. In light of the fact that "full fairness" cannot be achieved without an honest majority, there has been a considerable amount of work suggesting possible alternatives. The *gradual release* technique [21, 7, 26] can guarantee the following partial form of fairness: if the adversary quits the protocol early, then the adversary and the honest parties must invest comparable amounts of time in order to recover the outputs. This approach departs from traditional cryptographic protocols in that it does not set an a-priori bound on the resources invested by *honest* parties. In contrast, in this work, we consider the traditional setting in which the running time of honest parties is bounded by some fixed polynomial in the security parameter. Other approaches for getting around the impossibility of fairness involve relaxed notions of fairness [37, 4, 29], or employing semi-trusted third parties or physical assumptions [1, 8, 35]. None of these approaches applies to the standard model we consider here.

Organization. In Section 2, we provide some necessary definitions. In Section 3, we present our positive result for the case of standard (non-reactive) functionalities. In Section 4, we present a negative result for standard functionalities and, in Section 5, we present stronger negative results for the case of reactive functionalities. Due to lack of space, some of the material (including proofs) were omitted from this extended abstract and can be found in the full version.

2 Preliminaries

The Model. Our default network model consists of n parties, $P_1, \ldots, P_n$, who interact in synchronous rounds via authenticated secure point-to-point channels and a broadcast medium. We note, however, that the question we consider and our results are quite insensitive to the network model and in particular to the availability of broadcast, as discussed in the end of Section 1.2. Similarly, our positive results hold even when the adversary is allowed *rushing* (i.e., sending its messages at a given round only after receiving all messages sent by honest parties in the same round), whereas our negative results hold even if all messages of a given round are delivered simultaneously. This should be contrasted with the known impossibility result for coin-flipping [14], which crucially depends on the adversary's rushing capability. Indeed, flipping coins becomes a trivial task if the network can enforce simultaneous message exchange.

Throughout the paper, we consider *computational* security against computationally bounded adversaries. The security parameter will be denoted by k. We assume for simplicity that the adversary is *static*, namely that the set of corrupted parties is chosen at the onset of the protocol in a non-adaptive manner. This strengthens our negative results, and is not essential for our positive results.

Finally, we consider both *malicious* adversaries, who have total control over the behavior of corrupted parties and may thus arbitrarily deviate from the protocol specification, and *semi-honest* adversaries, who can record all information viewed by corrupted parties but do not otherwise modify their behavior. We will sometimes also refer to *fail-stop* adversaries which can only deviate from a semi-honest behavior by aborting, namely stopping to send messages. Unless stated otherwise, adversaries are assumed to be malicious.

Defining Security. We assume some familiarity of the reader with the basic simulation-based approach for defining secure computation, as described in detail in [9, 27]. Generally speaking, this definitional approach compares the *real-life* interaction of an adversary with the protocol and its interaction with an *ideal* function evaluation process in which a trusted party is employed. Security is then defined by requiring that whatever can be achieved in the real model could have also been achieved (or *simulated*) in the ideal model. In other words, for every adversary A attacking the real execution of the protocol there exists an adversary A', sometimes referred to as a *simulator*, which can "achieve the same effect" by attacking the ideal function evaluation process. The two main types of security considered in this work differ in how the ideal model is defined.

IDEAL MODEL – WITHOUT ABORT. This variant of the ideal model corresponds to the strong notion of security (with guaranteed output delivery) that is achieved by type $\mathcal{P}_{n/2}$ protocols. In this case, the interaction of the simulator A' with the ideal evaluation of f is very simple:

- Honest parties send their input x_i to the trusted party. Corrupted parties may send the trusted party arbitrary inputs as instructed by A'. Denote by x_i' the value sent by party P_i. In the case of a semi-honest adversary, we require that $x_i' = x_i$.
- The trusted party computes $f(x_1', \ldots, x_n') = (y_1, \ldots, y_n)$ and sends the value y_i to party P_i. (If f is randomized, this computation involves random coins that are generated by the trusted party.) Any missing or "invalid" value x_i' is substituted by a valid default value (say 0) before evaluating f.

IDEAL MODEL – WITH ABORT. In this case, a malicious (or even a fail-stop) adversary can abort the computation in the ideal model *after learning its outputs.*[4] We use $\perp$ to represent the output of honest parties resulting from the adversary aborting the protocol and let $I \subset [n]$ denote the set of corrupted parties. The following definition of the ideal model follows that of [34]:

- The parties send their inputs to the trusted party. As before, we let x_i' denote the (possibly modified) value sent by party P_i.
- The trusted party computes $f(x_1', \ldots, x_n') = (y_1, \ldots, y_n)$ and sends to each corrupted party P_i, $i \in I$, its output y_i.
- Based on the values $(y_i)_{i \in I}$, received in the previous step, the adversary chooses whether to continue the ideal function evaluation process or to abort. In the former case, the trusted party sends to each uncorrupted party P_i, $i \notin I$, its output value y_i. In the latter case, each party P_i such that $i \notin I$ receives the special value $\perp$ indicating that the adversary chose to abort.

When referring to constant-round protocols, we allow simulators to run in *expected* polynomial time (cf. [34]); our negative results hold even if the simulator's running time is unbounded.

Security Versus Privacy. In the literature on secure computation, privacy is often synonymous with security against a semi-honest adversary. Here we also refer to privacy against *malicious* adversaries, in which case we consider the privacy property in isolation. Privacy requires that the adversary's view when attacking the real model can be simulated in the ideal model. This should be contrasted with the (stronger) standard security definitions, which consider the joint distribution of the adversary's view and the outputs of uncorrupted

---

[4] It is possible to realize a somewhat stronger notion of security in which the adversary can abort the protocol *after learning the outputs* only if it corrupts some predefined party, say P_1. In this case, the adversary should be given the opportunity to abort the protocol *before learning the output* even when the corrupted parties do not include P_1. Since the distinction between these two variants does not effect our results, we prefer the simpler notion.

parties in both the real and ideal models. We note that considering privacy on its own is often not sufficient. In particular, as discussed in [9], privacy alone for randomized functionalities does not capture our intuition of what privacy *should* mean. Moreover, privacy against a *malicious* adversary (in contrast to privacy against a semi-honest adversary) does not compose. Nevertheless, it is a meaningful notion of security that captures the adversary's inability to learn more than it should about the *inputs* of the honest parties. Furthermore, in this work we mainly use privacy for our negative results, thereby strengthening them.

Authenticated Secret Sharing. Our positive results will make use of *authenticated secret sharing* [41, 16] (also known as "honest-dealer VSS" or "weak VSS"). An s-secure authenticated secret sharing scheme allows an honest dealer to distribute a secret σ among n players such that an adversary corrupting up to s players learns nothing about σ from its shares. Moreover, even by modifying its shares, the adversary cannot prevent honest players from later reconstructing the correct value σ, except with negligible probability. An authenticated secret sharing scheme is defined by a pair $\mathcal{S} = (\mathcal{D}, \mathcal{R})$ of efficient algorithms, where $\mathcal{D}(1^k, 1^n, \sigma)$ is a randomized secret distribution algorithm outputting an n-tuple of shares, and $\mathcal{R}(\sigma_1, \ldots, \sigma_n)$ is a reconstruction function recovering the secret from the (possibly modified) shares. We will also need to rely on the existence of an efficient *share completion* algorithm which, given a secret σ and a set of $s' \leq s$ shares, can sample the remaining $n - s'$ shares conditioned on the given information. A formal definition of authenticated secret-sharing can be found in the full version. We rely on the following fact.

Fact 1. *(implicit in [41]) There exists an s-secure n-party authenticated secret sharing scheme for any $s < n/2$.*

3   Positive Result for Standard Functionalities

In this section, we present a protocol demonstrating the possibility of obtaining the best of both worlds for standard (non-reactive) functionalities, as long as the adversary is semi-honest in the case of no honest majority. That is, we obtain a protocol that provides *full security* against a minority of malicious parties, and *full security* against any number of semi-honest parties.

Theorem 1. *Assume the existence of enhanced trapdoor permutations. Then, for any probabilistic polynomial-time computable (non-reactive) functionality f, there exists a single protocol π such that:*

1. *π securely computes f in the presence of an s-bounded malicious adversary, for any $s < n/2$;*
2. *π securely computes f in the presence of a t-bounded semi-honest adversary, for any $t < n$.*

Furthermore, if collision resistant hash functions exist, the protocol π can be constructed with just a constant number of rounds.

We prove the theorem by presenting a protocol that achieves the desired security features. Assume, without loss of generality, that the (possibly randomized) functionality f has a single output that is given to all parties (see Remark 1 below). We also assume without loss of generality that the output length of f, denoted by ℓ, is determined by k. We use the fact that known protocols that securely compute f with abort in the presence of an $(n-1)$-bounded malicious adversary are fully secure in the presence of an $(n-1)$-bounded semi-honest adversary. Our protocol proceeds as follows.

Protocol 1. Multiparty computation secure against a minority of malicious parties and against any number of semi-honest parties:

- **Inputs:** *Each party P_i, $1 \le i \le n$, has an input x_i. All parties hold a security parameter k.*
- **Output:** *Each party P_i would like to obtain $y = f(x_1, \ldots, x_n)$.*
- **Parameters:** *Security thresholds $s = \lfloor (n-1)/2 \rfloor < n/2$ and $t = n - 1$.*
- **The protocol:**
  Phase I:
    1. *The parties begin by computing an authenticated secret-sharing of $f(x_1, \ldots, x_n)$. Specifically, let $\mathcal{S} = (\mathcal{D}, \mathcal{R})$ be an s-secure authenticated secret sharing scheme (see Section 2), and let $\hat{f}$ be the randomized functionality defined by*

$$\hat{f}(1^k, x_1, \ldots, x_n) = \mathcal{D}(1^k, 1^n, f(x_1, \ldots, x_n)).$$

    Let π_n be a protocol that securely computes $\hat{f}$ with abort in the presence of t-bounded malicious adversaries (and is fully secure in the presence of t-bounded semi-honest adversaries). The parties invoke π_n on their inputs x_i, obtaining local outputs y_i.
    2. *If Phase I terminated successfully (with honest parties outputting values $y_i \ne \bot$), each party P_i sends y_i to all other parties and outputs $y = \mathcal{R}(y_1, \ldots, y_n)$. Otherwise (if Phase I terminated with "abort"), proceed to Phase II below.*
  Phase II:
    Let $\pi_{\mathrm{n}/2}$ be protocol that securely computes f in the presence of s-bounded malicious adversaries. The parties invoke $\pi_{\mathrm{n}/2}$ on their original inputs x_i, and output their outputs in this invocation.

Notice that even a single malicious party can abort the computation of π_n in the first phase. However, in such a case all the honest parties output $y_i = \bot$ and proceed to the next phase where the protocol is fully secure in the presence of s-bounded malicious adversaries. The key observation used in constructing the protocol is that *either* the adversary is semi-honest, in which case the protocol will terminate in phase I, *or* the adversary is malicious, but then we know that it is s-bounded. Thus, it will learn nothing in the first phase, and the protocol in the second phase will yield the desired result.

Remark 1 (Single output vs. multiple outputs). In Protocol 1, we assume that the functionality f has a single output y which is given to all parties. To handle general functionalities, we can use the standard reduction where the output includes a concatenation of n values $y_i \oplus r_i$, where y_i is the local output of P_i, and r_i is a mask randomly selected by P_i and included in its input. It is not difficult to show that this reduction is secure for any number of corruptions.

In the full version of this paper, we prove the following:

Lemma 1. *Protocol 1 is secure against semi-honest t-bounded adversaries for any $t < n$.*

Lemma 2. *Protocol 1 is secure against malicious s-bounded adversaries, for any $s < n/2$.*

To conclude the proof of Theorem 1, we note that protocols $\pi_{n/2}, \pi_n$ as needed indeed exist. That is, we can take $\pi_{n/2}$ to be the protocol of [28] and π_n to be the protocol of [41] (obtaining a non-constant number of rounds), or we can take $\pi_{n/2}$ to be the protocol of [38] and π_n to be the protocol of [5] (obtaining a constant number of rounds, but additionally requiring the existence of collision-resistant hash functions). $\qquad\qquad\Box$

In the full version we discuss extensions of Theorem 1 to the cases of adaptive security and UC security.

Remark 2 (On strengthening semi-honest security). In the case of no honest majority $(n/2 \leq t < n)$, Theorem 1 only refers to security against a semi-honest adversary. However, the security guarantee provided by Protocol 1 in this case is somewhat stronger. In particular, for any *non-aborting adversary* (i.e., an adversary that does not cause the honest parties to abort), the computation will be completed in Phase I, and privacy is thus preserved in this case. Since an adversary that aborts may be identified as misbehaving, real adversarial parties may have a lot to lose by causing an abort in Phase I. Note, however, that Protocol 1 *does not* generally satisfy security-with-abort, or even privacy, against $t < n$ *malicious* parties. Indeed, a malicious majority can learn the entire input in Phase II. In Section 4 we show that a similar type of insecurity is inherent to constant-round protocols.

An Interesting Protocol Variant. A useful property of most known protocols of type $\mathcal{P}_n$ (e.g., the protocols in [27] or [34]) is that if the protocol aborts, then a corrupted party is identified (where the same corrupted party is identified by all honest parties). This feature gives rise to the following variant of Protocol 1, that obtains a meaningful notion of security even against a *majority of malicious* parties (in addition to full security against a minority of malicious parties):

> Repeatedly run Phase I, disqualifying a bad party at each iteration until no aborts occur. (If a party is disqualified, its input is replaced by a default value.) Then, all remaining parties exchange their authenticated shares and reconstruct the output of f as in Step 2 of Protocol 1.

We first note that this protocol may run at most $t + 1$ iterations, as one corrupted party is eliminated in each repetition. For this variant of Protocol 1, an interesting security guarantee can be made. First, similarly to Protocol 1, the protocol is secure against $s < n/2$ malicious parties. Moreover, even if a malicious adversary corrupts $t < n$ parties, this adversary can be simulated in a relaxed ideal model where the simulator can invoke the functionality up to $t + 1$ times. In each such invocation, the simulator gets to learn the outputs first, and then decides whether to abort (and continue to the next invocation) or to deliver the outputs to the uncorrupted parties (in which case the protocol terminates). If it gets to invocation $t + 1$, then it must deliver the output.

For many natural functionalities the above type of security is quite reasonable, and in particular may imply privacy. For instance, in the case of the basic voting functionality (sum of $0/1$ values), the above protocol is fully private against a malicious adversary, because t invocations of the functionality are equivalent to a single invocation in terms of the information learned by the simulator. This holds because in a single invocation the adversary can learn exactly how many honest parties voted 1 and how many voted 0. Future invocations, where the adversary changes the votes of the corrupted parties, will yield no additional information. Thus, using this approach, we actually get a voting protocol with security against $s < n/2$ and privacy against $t < n$ *malicious* parties. We stress that even for this specific functionality the above protocol falls short of providing full security, or even security with abort, against a malicious majority. For instance, corrupted parties can force the vote tally to be even, or force a tie between the candidates. Note, however, that obtaining such an unfair advantage is possible only when a *majority* of the voters are maliciously corrupted. In such a case, protecting the *privacy* of the honest minority is typically the most important concern. This concern is completely resolved by our protocol, without compromising full security in the (more likely) case of a malicious minority.

4   Negative Result for Standard Functionalities

In this section we prove our main negative result for standard (non-reactive) functionalities. Specifically, we demonstrate the existence of a 3-party functionality for which there does not exist any *constant-round* protocol that is both secure against one malicious party and private against two malicious parties. (A generalization to n parties is presented later.) The proof of this result is quite subtle; we first describe the general idea at a high level.

Denote the 3 parties by A, B, C and their respective inputs by a, b, c. We consider a functionality f which, on randomly chosen inputs, allows a coalition of B and another party (A or C) to learn *partial* information about the input of the remaining party, but not everything. Let q be a round number such that the following holds: After q rounds, A and B can learn from the honest execution of the protocol the *maximal* amount of information about c allowed by the the ideal evaluation of f. On the other hand, after $q - 1$ rounds B and C can only obtain a substantially smaller amount of information about a. (If no such such q exists, we can exchange the roles of A and C and the rest of the argument proceeds symmetrically.) An adversary corrupting A and B can now break the privacy of the protocol as follows. It acts honestly for the first $q - 1$ rounds. Starting from round q, it makes A stop sending any messages but keeps B acting honestly. At this stage, the adversary has already learned the maximal amount of information about c allowed by the ideal model. However, the protocol must continue as if only A is corrupted, in which case the outputs of B and C should correspond to substituting the input of A by some default value. (Here we use the fact that the protocol should be fully secure in the case that only A is corrupted.) The crucial point is that B and C do not have enough information about a to guarantee that it is always used as the default value. Thus, with some noticeable probability, B learns an output of f on (a', b, c) for some $a' \neq a$. This new information can be used by the adversary to learn more about c than is allowed in the ideal model.

We now proceed with the formal statement of the result and its proof.

Theorem 2. *There exists a finite 3-party functionality f for which there is no constant-round protocol π such that:*

- *π securely computes f in the presence of a malicious adversary corrupting one party;*
- *π privately computes f in the presence of a malicious adversary corrupting two parties.*

This holds even if the malicious adversary is restricted to be of a fail-stop type.

The proof of Theorem 2 relies on a technical definition of information yielding defined below. We will consider honest executions of the protocol on random inputs, and would like to measure the "amount of information" some partial view yields on an input. For this, it will be convenient to use the following measurement. Intuitively, it measures the maximal probability of guessing the input correctly *without being wrong* (except with probability that tends to 0 as k tends to infinity), but possibly outputting "don't know".

Definition 1 *Let $(X(k), V(k))$ be an ensemble of (possibly) correlated random variables and let $0 \leq p \leq 1$ be a constant. We say that V yields X with probability (at least) p if there exists a polynomial-size prediction algorithm P such that $\Pr[P(V(k)) = X(k)] \geq p - \epsilon(k)$, and $\Pr[P(V(k)) \notin \{X(k), \text{"don't know"}\}] \leq \epsilon(k)$ for some $\epsilon(k)$ that tends to 0 as k tends to infinity.*

In the following proof we will use Definition 1 with X being an input to a protocol and V a corresponding view obtained by running the protocol with security parameter k. (In our specific example, X will not depend on k.)

Proof Sketch: We prove Theorem 2 by exhibiting a specific functionality f as required. This functionality involves three parties, A, B, C, where A's input is denoted a, C's input is c and B has no input. The functionality is defined by $f(a, \perp, c) = (\perp, out, \perp)$, where $out = c$ if $a = c$, and $out = *$ otherwise. The inputs a, c are taken from the set $\{0, 1, 2\}$. We will argue that there is no constant-round protocol that computes f with full security against a single malicious party and privacy against two malicious parties.

Suppose, by way of contradiction, that π is a protocol for f that obtains both privacy against two parties and security against one. Now, consider the honest execution of π on inputs a and c that are selected uniformly and independently at random in the domain $\{0, 1, 2\}$. For an honest run of the protocol, we denote by q_{AB} the first round after which the view of the coalition of A and B on the protocol yields c (in the sense of Definition 1) with probability $1/3$. We know that $q_{AB} > 0$, since initially it is impossible for A and B to guess c without error. We also know that q_{AB} is no larger than the (constant) number of rounds in the protocol, since eventually even the view of B alone must yield c with probability $1/3$. Thus, q_{AB} is well defined. Define q_{BC} symmetrically, i.e., the first round in which the view of B and C yields a with probability $1/3$. Assume, without loss of generality, that $q_{AB} \leq q_{BC}$. In other words, A and B learn c at no later than B and C learn a. The proof for the case $q_{AB} \geq q_{BC}$ proceeds symmetrically.

Let $q = q_{AB}$, i.e., after q rounds the joint views of A and B yields c with probability $p_1 = 1/3$, where the probability is taken over the choice of a, c and the random coin tosses of all parties. Let P be a corresponding prediction algorithm (of c from the view of A and B), and denote by G_1 the event that P does not output "don't know", given the view of A and B after q rounds. By the definition of q and P, and relying only on the fact that $\Pr[P(V(k)) = X(k)] \geq p - \epsilon(k)$, it follows that

$$\Pr[G_1] \geq 1/3 - \epsilon(k) \tag{1}$$

After $q - 1$ rounds, the view of B and C does not yield the value of a with probability $1/3$. This means that, for any poly-size predictor P' that is wrong with probability at most $\epsilon'(k)$ (for some ϵ' that tends to 0 as k tends to infinity), there is some constant $p_2 < 1/3$ and infinitely many k's such that the probability that P''s output is not "don't know", based on the view of B and C in the first $q - 1$ rounds, is at most p_2. (To be consistent with Definition 1, note that p_2 equals some $p' - \epsilon'(k)$ where $p' < 1/3$.)

Note that the assumption that p_2 is bounded away from $1/3$ is where we use the assumption that the protocol has a constant number of rounds. A more fundamental reason for the restriction on the number of rounds is the following. We will need the information obtained in round q to be significantly better than the information obtained in round $q - 1$, and furthermore will need the difference between the two amounts of information to be bigger than the difference between round q and the maximal amount of information. The difficult case of non-constant-round protocols, for which the above property cannot be met, is when after m rounds the view yields the input with probability $1/3 - 2^{-m}$. This suggests that our general proof approach cannot go beyond a logarithmic number of rounds.

From here on, we consider the following experiment. Pick the two inputs a, c uniformly and independently at random. The adversary attacks the protocol, when invoked on uniformly random inputs, as follows. It corrupts A and B, and after $q - 1$ rounds makes A abort. (That is, the last message sent by A is that sent in Round $q - 1$.) Define the event G_1 as above, based on the view of A and B in the first q rounds (note that the joint view of A and B depends only on C's messages to A and B in the first q rounds and these are all obtained even though A halts after $q - 1$ rounds). We now define another event G_2 on the probability space induced by the same execution of the protocol. By the security guarantee, and since B is still following the protocol, the protocol must continue as if only A is corrupted. In particular, B should eventually output c with probability (roughly) $1/3$. Let G_2 denote the event that B outputs some value in $\{0, 1, 2\}$ (in which case, as follows from the definition of f, it also learns the value of the input c). Note, however, that there is no guarantee that when G_2 occurs, the output of B is necessarily equal to a (since A is corrupted).

In the following, $\epsilon(k)$ should be interpreted as *some* function that tends to 0 as k tends to infinity (not necessarily negligible in k) as in Definition 1. Using the above notations, the following facts can now be deduced. First,

$$\Pr[G_1 \text{ and } a \neq c] < \epsilon(k). \tag{2}$$

This is because when $a \neq c$ then even the full view does not allow A and B to predict c with "small" error (where from here on, small error means with error $\epsilon(k)$ for some function $\epsilon(\cdot)$ that tends to 0 as its input tends to infinity). Thus, in the ideal model, it is impossible to predict with small error. This means that if Eq. (2) did not hold, then the assumed predictor could be used to distinguish the output of the adversary in a real protocol execution from its output in an ideal one.

In addition to the above, we have that in the ideal model B must output c with probability $1/3$. Thus, by the security guarantee

$$\Pr[G_2] \geq 1/3 - \epsilon(k). \tag{3}$$

(Note that Eq. (3) actually holds for a *negligible* function $\epsilon(k)$. Nevertheless, it suffice to use a function $\epsilon(k)$ that tends to 0 as k tends to infinity.) Finally, we claim that the probability that B generates output in $\{0, 1, 2\}$, but the predictor outputs "don't know" is at most $\epsilon(k)$; that is:

$$\Pr[G_2 \text{ and } \operatorname{not}(G_1)] \leq \epsilon(k). \tag{4}$$

This last inequality is shown by way of contradiction. If Equation 4 does not hold, then there is some constant $p > 0$ such that $\Pr[G_2 \text{ and } \operatorname{not}(G_1)] > p$ for infinitely many k. If that happens, we can derive a contradiction to the guaranteed privacy of the protocol π. To show that privacy does not hold, we show that there is some constant $p' > 1/3$ such that the adversary (that corrupts A and B) can guess c with probability greater than p' for infinitely many k's (while outputting a wrong guess with vanishingly small probability). This cannot happen in the ideal model. The adversary uses the following strategy: Corrupt A and B; Run the protocol normally for $q - 1$ rounds; Invoke the predictor P on the partial view of A and B, and if it outputs some value c then output the same value and halt; Otherwise, pretend as if A is dead and continue to run the protocol until it terminates (with B acting honestly); If B outputs some value c then output the same value and halt; Otherwise, output "don't know". Note that a guess c output by the algorithm is wrong with only small probability, and the event of outputting some value c is the union of G_1 and G_2 which, using Equation 1 and the assumption above, is at least $1/3 - \epsilon(k) + p$ for infinitely many k's. Since p is a positive constant and $\epsilon(k)$ goes to zero, we obtain a contradiction.

So, we assume Equation 4 does hold and obtain a contradiction to the property of round q; i.e., to the fact that the view of B and C does not yield the value a with probability $1/3$ after $q - 1$ rounds. We construct a new predictor that yields the value of a from the view of B and C at round $q - 1$ in the following manner. It takes the view of B and C and continues to run the protocol until the end assuming that A aborted. Since A is not involved from Round q and on, this can be done based on the partial view of B and C alone (in the first $q - 1$ rounds). Note that this run is distributed exactly like the experiment previously described, where A aborts and B and C continue to run the protocol on their own. The experiment and distribution remain identical even though we previously thought of A and B as trying to obtain the value c and now we think of B and C as trying to obtain the value a. The predictor outputs whatever B outputs. By Equation 3, the probability that B outputs a value in $\{0, 1, 2\}$ is roughly $1/3$. By Equation 4, whenever this happens it is almost always the case that G_1 occurs. Namely, if we had run the predictor P on the view of A and B, it would have predicted the value c correctly. But if G_1 occurs then, by Equation 2, we get that $a = c$. To sum up, the output of B generated from the view of B and C in round $q - 1$ yields a with probability (almost) $1/3$ and is wrong with only small probability, contradicting the definition of the round q.   □

Note that while we allow the simulator an arbitrary malicious behavior in the ideal world, the real-world adversary described above is only a fail-stop one. This strengthens our result.

Using standard player-partitioning arguments, the result for the above 3-party functionality f can be generalized to rule out general constant-round protocols for any s, t such that $s + t \geq n$.

Theorem 3. *Let n, s, t be such that $s + t \geq n$. Then there is an n-party functionality for which there is no constant-round protocol that is simultaneously:*

- *secure against an s-bounded malicious adversary;*
- *private against a t-bounded malicious adversary.*

5   Negative Result for Reactive Functionalities

For the case of reactive functionalities, we obtain the following strong negative result.

Theorem 4. *There exists a finite, reactive 3-party functionality f for which there does not exist a protocol π such that:*

- *π securely computes f in the presence of a malicious adversary corrupting one party;*
- *π privately computes f in the presence of a semi-honest adversary corrupting two parties.*

This holds even if the malicious adversary is restricted to be of a fail-stop type.

In the full version of the paper we prove Theorem 4 using a two-phase functionality, which roughly corresponds to the task of committing to a secret and later decommitting it. Similarly to Theorem 3, this negative result can also be generalized to any s, t such that $s + t \geq n$.

References

[1] N. Asokan, V. Shoup, and M. Waidner. Optimistic Fair Exchange of Digital Signatures (Extended Abstract). In *Proc. EUROCRYPT 1998*, pages 591-606.

[2] B. Barak, and A. Sahai. How To Play Almost Any Mental Game Over The Net - Concurrent Composition via Super-Polynomial Simulation. In *Proc. of 46th FOCS*, pp. 543-552, 2005.

[3] J. Bar-Ilan and D. Beaver. Non-cryptographic fault-tolerant computing in a constant number of rounds. In *Proc. 8th ACM PODC*, pages 201–209, 1989.

[4] D. Beaver, S. Goldwasser. Multiparty Computation with Faulty Majority. In *Proc. of FOCS 1989*, pp. 468-473.

[5] D. Beaver, S. Micali and P. Rogaway. The Round Complexity of Secure Protocols. In *22nd STOC*, pages 503–513, 1990.

[6] M. Ben-Or, S. Goldwasser and A. Wigderson. Completeness Theorems for Non-Cryptographic Fault-Tolerant Distributed Computation. In *20th STOC*, pages 1–10, 1988.

[7] D. Boneh and M. Naor. Timed Commitments. In *CRYPTO 2000*, pp. 236-254.

[8] C. Cachin and J. Camenisch. Optimistic Fair Secure Computation. In *Proc. CRYPTO 2000*, pages 93-111.

[9] R. Canetti. Security and Composition of Multiparty Cryptographic Protocols. *Journal of Cryptology*, 13(1):143–202, 2000.

[10] R. Canetti. Universally Composable Security: A New Paradigm for Cryptographic Protocols. FOCS 2001: 136-145.

[11] R. Canetti, U. Feige, O. Goldreich, and M. Naor. Adaptively Secure Multi-Party Computation. In *28th STOC*, pages 639-648, 1996.

[12] R. Canetti, Y. Lindell, R. Ostrovsky and A. Sahai. Universally Composable Two-Party and Multi-Party Computation. In *34th STOC*, pages 494–503, 2002.

[13] D. Chaum, C. Crépeau and I. Damgård. Multi-party Unconditionally Secure Protocols. In *20th STOC*, pages 11–19, 1988.

[14] R. Cleve. Limits on the Security of Coin Flips when Half the Processors Are Faulty. In *Proc. of STOC 1986*, pp. 364-369.

[15] R. Cramer, I. Damgård, S. Dziembowski, M. Hirt, and T. Rabin. Efficient Multiparty Computations Secure Against an Adaptive Adversary. In *Proc. EUROCRYPT 1999*, pages 311-326.

[16] R. Cramer, I. Damgård, and S. Fehr. On the Cost of Reconstructing a Secret, or VSS with Optimal Reconstruction Phase. In *Proc. CRYPTO 2001*, pages 503-523.

[17] R. Cramer, I. Damgård, and U. Maurer. General secure multi-party computation from any linear secret-sharing scheme. In *EUROCRYPT '00*, pp. 316-334.

[18] R. Cramer, I. Damgård, and J. Nielsen. Multiparty computation from threshold homomorphic encryption. In *EUROCRYPT '01*, pp. 280-299.

[19] I. Damgård and J. Nielsen. Multiparty Computation from Threshold Homomorphic Encryption. CRYPTO 2003: 247-264.

[20] I. Damgård and Y. Ishai. Constant-Round Multiparty Computation Using a Black-Box Pseudorandom Generator. In *CRYPTO 2005*, pp. 378–394.

[21] S. Even, O. Goldreich and A. Lempel. A Randomized Protocol for Signing Contracts. In *Communications of the ACM,* 28(6):637–647, 1985.

[22] M. Fitzi, M. Hirt, T. Holenstein, and J. Wullschleger. Two-Threshold Broadcast and Detectable Multi-party Computation. EUROCRYPT 2003: 51-67.

[23] M. Fitzi, M. Hirt, and U. M. Maurer. Trading Correctness for Privacy in Unconditional Multi-Party Computation (Extended Abstract). CRYPTO 1998: 121-136

[24] M. Fitzi, T. Holenstein, and J. Wullschleger. Multi-party Computation with Hybrid Security. EUROCRYPT 2004: 419-438

[25] J. A. Garay, P. D. MacKenzie, and K. Yang. Efficient and Universally Composable Committed Oblivious Transfer and Applications. TCC 2004.

[26] J. A. Garay, P. D. MacKenzie, M. Prabhakaran, and K. Yang. Resource Fairness and Composability of Cryptographic Protocols. *Proc. 3rd TCC*, 2006. Also appears in Cryptology ePrint Archive, Report 2005/370.

[27] O. Goldreich. *Foundations of Cryptography: Volume 2 – Basic Applications.* Cambridge University Press, 2004.

[28] O. Goldreich, S. Micali and A. Wigderson. How to Play any Mental Game – A Completeness Theorem for Protocols with Honest Majority. In 19*th STOC,* pages 218–229, 1987. For details see [27].

[29] S. Goldwasser and L. Levin. Fair Computation of General Functions in Presence of Immoral Majority. In *CRYPTO '90*, pp. 77–93.

[30] S. Goldwasser, and Y. Lindell. Secure Multi-Party Computation without Agreement. J. Cryptology 18(3): 247-287 (2005). Preliminary version in DISC 2002.

[31] M. Hirt and U. M. Maurer. Robustness for Free in Unconditional Multi-party Computation. CRYPTO 2001: 101-118.

[32] Y. Ishai and E. Kushilevitz. Randomizing polynomials: A new representation with applications to round-efficient secure computation. In *Proc. 41st FOCS*, pp. 294–304, 2000.

[33] J. Katz and R. Ostrovsky. Round-Optimal Secure Two-Party Computation. In *CRYPTO 2004*, pages 335-354.

[34] J. Katz, R. Ostrovsky, and A. Smith. Round Efficiency of Multi-party Computation with a Dishonest Majority. In *EUROCRYPT 2003*, pages 578-595.

[35] M. Lepinski, S. Micali, C. Peikert, and A. Shelat. Completely fair SFE and coalition-safe cheap talk. In *Proc. PODC 2004*, pages 1-10.

[36] Y. Lindell. Parallel Coin-Tossing and Constant-Round Secure Two-Party Computation. *J. Cryptology* 16(3): 143-184 (2003). Preliminary version in Crypto 2001.

[37] M. Luby, S. Micali, and C. Rackoff. How to Simultaneously Exchange a Secret Bit by Flipping a Symmetrically-Biased Coin. In *24th FOCS*, pp. 11-21, 1983.

[38] R. Pass. Bounded-Concurrent Secure Multi-Party Computation With a Dishonest Majority. In *Proc. STOC 2004*, pages 232-241.

[39] R. Pass and A. Rosen. Bounded-Concurrent Secure Two-Party Computation in a Constant Number of Rounds. In *Proc. FOCS 2003.*, pp. 404-413, 2005.

[40] B. Pfitzmann and M. Waidner. Information-Theoretic Pseudosignatures and Byzantine Agreement for $t \geq n/3$. IBM Research Report RZ 2882 (#90830), IBM Research Division, Zurich, 1996.

[41] T. Rabin and M. Ben-Or. Verifiable Secret Sharing and Multiparty Protocols with Honest Majority. In *Proc. 21st STOC*, pages 73–85. ACM, 1989.

[42] A. Shamir. How to share a secret. *Commun. ACM*, 22(6):612–613, June 1979.

[43] A. Yao. How to Generate and Exchange Secrets. In 27*th FOCS*, pages 162–167, 1986.

Scalable Secure Multiparty Computation

Ivan Damgård[1],[*] and Yuval Ishai[2],[**]

[1]Aarhus University
ivan@daimi.au.dk
[2]Technion
yuvali@cs.technion.ac.il

Abstract. We present the first general protocol for secure multiparty computation which is *scalable*, in the sense that the amortized work per player does not grow, and in some natural settings even vanishes, with the number of players. Our protocol is secure against an *active* adversary which may *adaptively* corrupt up to some *constant fraction* of the players. The protocol can be implemented in a constant number rounds assuming the existence of a "computationally simple" pseudorandom generator, or in a small non-constant number of rounds assuming an arbitrary pseudorandom generator.

1 Introduction

We consider the complexity of general secure multiparty computation (MPC) [37, 23, 9, 12] involving a large number of players. This problem is motivated by real-life applications that require a large-scale cooperation between many mutually distrustful entities. Alternatively, one may wish to distribute a critical role of a single "dealer" (distributing keys or other forms of correlated randomness) among many untrusted players.

As another motivating example, consider the case where a *small* number of players wish to securely perform some complex distributed computation on their inputs. In this case, the players could directly interact with each other using known methods from the MPC literature. However, all these methods have some inherent weaknesses. The first type of methods (for MPC with honest majority) will completely fail if a majority of the players become corrupted. Even if players trust each others' "honorable intentions", they may still be subject to hacking and other external attacks. Since attacking successfully a small number of players is sufficient to corrupt a majority, one may need to invest heavily in physical and other security measures to make this solution work. The second type of methods (for MPC with no honest majority) cannot provide important security features such as fairness or guaranteed output delivery [13]. Moreover, protocols

---

[*] Supported by BRICS, Basic research in Computer Science, Center of the Danish National Research Foundation and FICS, Foundations in Cryptography and Security, funded by the Danish Natural Sciences Research Council.
[**] Supported by grant 2004361 from the U.S.-Israel Binational Science Foundation.

of the latter type make use of expensive zero-knowledge proofs and public-key primitives.

One could hope to avoid the weaknesses discussed above by distributing the computation among a *large* number of players with diverse platforms and security measures. The event that a substantial fraction, say 25%, of this large number of players become corrupted may be considered much less likely than the adversary breaking into a majority of the original small set of players.

For all of the application scenarios described above, one could in principle employ any of a large variety of general MPC protocols from the literature. However, a major drawback of all these protocols is that their efficiency *does not scale well* with the number of players. There has been a long line of work attempting to improve this state of affairs by reducing the communication complexity of general MPC protocols [20, 21, 25, 24, 29, 15, 17, 26, 8]. Still, even in the most efficient protocols to date [24, 26, 8], the amount of communication involving *each player* grows linearly with the number of other players.

1.1   Our Results

In this work, we show that this is not an inherent state of affairs by presenting the first *scalable* MPC protocol. In this protocol, the amortized work per player does not grow, and in some cases even vanishes, with the number of players, while still tolerating a constant fraction of maliciously corrupted players.

It is convenient to formulate our results in the clients-servers model of [14, 16]. This model refines the standard MPC model by separating between *clients*, who provide inputs and receive outputs, and *servers* who perform the actual computation. (Note that the same party can play both roles, as is the case in the standard model of secure computation.) The main advantage of this refinement is that it allows to decouple the number of "consumers" (clients) from the expected "level of security" (which depends on the number of servers and the security threshold). In all cases, we require security against a computationally-bounded adversary that may (actively, adaptively) corrupt at most some *constant fraction* of the servers and *any* strict subset of the clients.

Our main result comes in two flavors:

- In the standard MPC setting, where every player can hold an input and receive an output, the amount of work involving each player is $\tilde{O}(|C|)$, where $|C|$ is the size of the circuit(s) being evaluated. Here the $\tilde{O}(\cdot)$ notation ignores factors that depend polynomially on the security parameter k and polylogarithmically on the number of players,[1] as well as additive terms that depend on the number of players and the size of the inputs, but not on $|C|$ (see more below). The security of this protocol can be based either on the existence of a

---

[1] Since the security parameter needs to be (at least) polylogarithmic in the number of players, the $\text{polylog}(n)$ term can be viewed as being absorbed in the $\text{poly}(k)$ term. Also, the $\text{polylog}(n)$ term only applies to the computation and not to the communication.

"computationally simple" pseudorandom generator (PRG)[2] (with a constant number of rounds) or on an arbitrary PRG (with poly(k) rounds).

- In the case where the number of clients is constant, as in the above motivating example, the *total* amount of work of all clients and servers is $\tilde{O}(|C|)$. Security in this case is based on the existence of a computationally simple PRG.

The $\tilde{O}(|C|)$ communication complexity of the latter protocol is essentially the best one could hope for given the current state of the art in the area of secure computation. Indeed, it is not known how to break this bound even in the case of security against a *single, semi-honest* player, with the only exceptions being protocols that apply to special classes of circuits (e.g., constant-depth circuits [4]) or protocols that involve an exponential amount of computation [6, 34]. Thus, even in a much more liberal setting than the one addressed here, significantly improving the $\tilde{O}(|C|)$ bound would be considered a major breakthrough.

The main disadvantage of our protocol compared to most previous works along this line is its non-optimal security threshold.[3] However, in our scenario we view the improved efficiency as being *qualitatively* more significant than the difference between different "constant" levels of resilience. The latter is very important in the case of computations involving *few* players. For instance, improving the resilience threshold from $t < n/3$ to $t < n/2$ enables three players to perform tasks that otherwise could not be performed at all. This difference is less significant in our scenario, where the number of players is large: in practice, it seems unlikely that increasing the number of players would automatically allow an adversary to corrupt more players. Therefore, the clients can compensate for the degraded level of security by employing a larger number of servers. Using our protocol, this can be done at a low extra cost.

A major advantage of our protocol over some previous related works (e.g., [15, 26]) is its provable *adaptive* security. Insisting on adaptive security is very significant when the number of players is large, since otherwise it would be possible to fool the (non-adaptive) adversary by assigning the computation to a small, randomly chosen, subset of the players. We note, however, that in contrast to the above simple approach, it is not clear if the protocols of [15, 26] are actually insecure in any "harmful" sense in the presence of an adaptive adversary.

On the Amortized Measure of Complexity. We stress again that our asymptotic complexity measurements ignore *additive* terms that depend (polynomially) on the number of players and the security parameter, but do not depend on the size of the circuitry to be evaluated. Thus, the scalability of our protocol only holds in an *amortized* sense, i.e., when amortizing over (one or several)

---

[2] Specifically, we assume the existence of a PRG in NC^1 or similar complexity classes. This is a mild assumption, which in particular is implied by virtually all standard *concrete* intractability assumptions used in cryptography (see [1] for discussion).

[3] We did not attempt to optimize the exact resilience of our protocol. In particular, in some cases we give away resilience in order to simplify subprotocols and their analysis. We note, however, that giving up on *some* resilience is inherent to our technique.

"large" circuits whose total size is much larger than the number of servers. The latter scenario is very realistic even for one-time computations (as in the contexts of distributed data-mining on large databases, or distributed generation of cryptographic keys), let alone when the computation is used to generate resources for a life-time of future interactions. Similar conventions were used in previous works along this line [25, 15, 24, 17, 26]. Similarly to previous works, we also ignore the cost of broadcasting the inputs held by the input clients to the n servers. This cost is inevitable in the general case (consider the functionality which outputs to all players the concatenation of all inputs) and can be *entirely avoided* in the case of a small number of clients. Moreover, in many applications the inputs to the computation are either short or nonexistent (as in the case of emulating a trusted dealer of correlated randomness). Thus, the amortized complexity of our protocol is typically quite insensitive to whether a broadcast channel is given,[4] and in many natural scenarios reflects their actual cost.

1.2   Techniques

Our protocol employs a variety of previous tools and techniques, as well as new ones. One source for improvement over previous works in the area is the use of the "share-packing" technique of Franklin and Yung [21]. To employ this technique in our context we need to combine it with an efficient procedure for generating many packed random secrets that are guaranteed to satisfy some given (possibly complex) replication pattern. The latter procedure extends previous techniques of Hirt et al. [25, 24]. Another source for improvement is the use of the "easy PRG" technique of Applebaum et al. [2], which in turn relies on Yao's garbled circuit technique [37] together with information-theoretic randomization techniques [28]. These previous tools are described in Section 3.

At a very high level, our protocol proceeds roughly as follows. Each client concatenates to its inputs a vector of replicated random bits, i.e., certain pairs of these bits must be equal. The replication structure (that is, which pairs have to be equal) is derived from the circuit C being evaluated. This random input vector is of roughly the same size as C. The client then divides this string into blocks of size $\Theta(n)$, applies to each block an efficient (constant-rate) Reed-Solomon encoding, and sends to each server a single entry of the encoding of each block. Now the servers engage in an efficient interactive testing procedure, verifying that the encoded blocks distributed by each client are not far from being valid encodings of vectors that are consistent with the structure of C. This procedure may result in eliminating some of the blocks. Next, each server locally applies a *low-degree* computation to its local information and sends the result back to the clients. (This step does not depend on the structure of C.) Finally, the output of the protocol is obtained by each client by first applying error correction of Reed-Solomon codes and then applying a decoding procedure that essentially mimics the computation of C. The intuition on how the previous tools are used in the protocol will be explained towards the end of Section 3.

---

[4] The constant rounds feature of our protocol can also be maintained (settling for *expected* constant rounds) without the availability of broadcast [19, 31, 30].

Related Work. In a concurrent and independent work, Hirt and Nielsen [27] present a different MPC protocol whose (amortized) communication complexity per player is linear in the circuit size and independent of the number of other players. This work significantly differs from ours both in the features of the protocol and in the underlying techniques. The main advantage of the protocol from [27] is that it achieves an optimal security threshold ($t < n/2$). However, unlike our protocol, it cannot be proved to be adaptively secure. Moreover, it does not match the complexity of our protocol in the case of a constant number of clients (as in a distributed implementation of two-party computation). Recall that in the latter case, the *total* complexity of our protocol can be made linear in the circuit size. The two works differ also in the underlying cryptographic assumptions and primitives. The protocol of [27] employs a threshold homomorphic public-key encryption, whereas our protocol involves primitives from the "private-key" world. Thus, the current work and [27] are quite complementary.

Organization. Following some preliminaries (Section 2), in Section 3 we describe previous tools and techniques on which our protocol relies. In Section 4 we describe a scalable MPC protocol under some setup assumptions, which are eliminated in Section 5. For lack of space, some of the material was omitted from this extended abstract and can be found in the full version.

2   Preliminaries

We consider a system consisting of several players, who interact in synchronous rounds via authenticated secure point-to-point channels and a broadcast medium. (The availability of broadcast is not a necessary assumption, and the number of broadcasts employed by our protocols will be independent of the size of the circuit being evaluated.)

Players can be designated three different roles: *input clients* who hold inputs, *output clients* who receive outputs, and *servers* who may be involved in the actual computation. This is just a generalization of the standard model, where each player can play all three roles. We denote the number of servers by n. We will sometimes refer to a server or a client as a "player" and denote server i by P_i. The functionalities we wish to compute only receive inputs from input clients and only provide outputs to output clients. For simplicity we will only explicitly consider deterministic functionalities providing all output clients with the same output, though an extension of our results to the general case is straightforward. (In contrast, we will employ subprotocols that compute randomized functionalities and provide servers and input clients with outputs as well.)

We assume by default an active, adaptive, rushing adversary corrupting at most t servers, where t is bounded by some constant fraction of n that will be either explicitly specified or understood from the context. The adversary may also corrupt an arbitrary strict subset of the clients. We refer the reader to, e.g., [10] for the standard definition of security in this model. Our protocol heavily builds on previous work in the area of *information-theoretic* multiparty computation. We assume some familiarity of the reader with basic techniques in this area, such as Shamir's secret-sharing [36] and the "BGW protocol" [9].

We will use secret-sharing over a finite field K of characteristic 2, where $|K| = O(n)$. We let $\omega_1, \ldots, \omega_n$ denote n distinct elements of K, where ω_i will be an interpolation point assigned to server P_i.

The Linear Preprocessing Model. It will be convenient to first describe and analyze our protocols in the *linear preprocessing model*, where we allow some restricted trusted setup as described below. We later show how the protocols can be converted to the *plain model*, where no setup assumptions are made.

In the linear preprocessing model, we assume a dealer who initially gives to each player a set of values in K or in its subfield GF(2). The values distributed by the dealer are restricted to be "linearly correlated". Specifically, the dealer picks a random codeword in a linear code defined over K or over GF(2), and then hands to each player a subset of the coordinates in the codeword. It is public which subsets are used, but the values themselves are private. This procedure can be repeated multiple times, possibly using different linear codes.

Of course, we do not expect that a dealer as above would exist in practice. This is only a convenient abstraction, that can be formalized as an ideal functionality. Much of the effort in the paper is devoted to finding efficient methods for amortizing the cost of multiple invocations of this functionality, for the types of distributed linear codes employed by our high level protocol.

A linear correlation pattern as above can be generally specified via a *distributed linear code* $L = (V, \pi_1, \ldots, \pi_n)$, where V is a linear code (vector space) over K or over GF(2), and π_i is the set of coordinates assigned to P_i. For $v \in V$ we use the notation $v(i)$ to denote v_{π_i}, the restriction of v to coordinates belonging to player i, which naturally extends to $v(H)$ for a set $H \subseteq [n]$ of players.

3   Basic Tools

In this section we present two tools from previous work on which our protocol relies: the share packing technique of Franklin and Yung [21] (Section 3.1) and the "easy PRG" technique of Applebaum et al. [2] (Section 3.2).

3.1   Share Packing

Most MPC protocols from the literature distribute the inputs and the random inputs via the use of Shamir's secret-sharing scheme [36]. This scheme is *ideal*, and in particular its share size is the best possible using any information-theoretic threshold scheme. However, viewed as an encoding of the shared secret, the rate achieved by Shamir's scheme is still very low: the size of the encoding is n times larger than the size of the secret. This suggests that better efficiency could be obtained via the use of better encodings.

A first and major step in this direction was taken by Franklin and Yung [21].[5] They observe that for the purpose of MPC, one does not necessarily have to rely

---

[5] The approach of Franklin and Yung was recently generalized, in the case of a passive adversary, to non-threshold adversary structures [38].

on a strict threshold scheme in which every t players learn nothing about the secret and every $t + 1$ players can fully reconstruct it. Instead, the latter requirement can be relaxed: it is only required that the encoding be "sufficiently robust". (Such relaxed secret-sharing schemes are known as *ramp schemes*.) Specifically, the following natural *multi-secret* generalization of Shamir's scheme is used. Let t be a security threshold and $\ell \geq 1$ be a parameter specifying a number of secrets to be jointly shared between the players. Assume that the field K has more than $\ell + n$ elements, and associate with each secret index $1 \leq i \leq \ell$ a distinct field element $\mu_i \in K$, different from all elements ω_j assigned to players. A block of ℓ secrets $(s_1, \ldots, s_\ell) \in K^\ell$ is distributed as follows.

Procedure ShareBlock($K, \ell, t, (s_1, \ldots, s_\ell)$):

- Pick a random polynomial p of degree (at most) $T = t + \ell - 1$ over K, subject to the constraints $p(\mu_i) = s_i$, $1 \leq i \leq \ell$.
- Send to each player P_j the share $p(\omega_j)$.

It is easy to verify that every t players learn nothing about the secret block $\mathbf{s} = (s_1, \ldots, s_\ell)$ from the shares they receive in ShareBlock. Moreover, this type of secret-sharing supports the homomorphic properties required by MPC protocols. Indeed, suppose that two blocks of secrets $\mathbf{s}, \mathbf{s}'$ have been distributed using polynomials p, p', respectively. Then, by locally adding their shares the players hold points on the degree-T polynomial $p + p'$ whose value at point μ_i is $s_i + s_i'$. Thus, two shared blocks can be locally added. Similarly, by locally multiplying their shares the players hold points on the degree-$2T$ polynomial $p \cdot p'$ whose value at point μ_i is $s_i \cdot s_i'$. Given these properties, it was suggested in [21] to generalize the standard MPC protocols by applying them to packed shares instead of standard shares. This allows to simultaneously perform distributed addition and multiplication on blocks of ℓ packed secrets, thereby evaluating a function f on a batch of ℓ (independent) inputs. The new protocol requires $O(\ell)$ more players than the original one, to compensate for the higher degree of the polynomials used for sharing the secrets.

From here on, we will take ℓ to be some constant fraction of n, and apply the share packing technique to protocols that tolerate a constant fraction of corrupted players. In this setting of the parameters, the resulting protocol uses $\Omega(n)$ times less communication than ℓ repetitions of the original protocol, and still tolerates a (smaller) constant fraction of corrupted players.

The problem encountered by [21] when trying to go beyond multiple evaluations of the same function is that "splitting" a vector of packed secrets is an expensive operation. Thus, the above savings could not be directly translated to the case of evaluating a complex function on a single input. We explain below how one can get around this problem using the garbled circuit technique, once we have explained a second source of savings.

3.2   The "Easy PRG" Technique

All variants of our protocol make use of a pseudorandom generator (PRG). The main variant of the protocol further requires an "easy PRG", namely a PRG computable in NC^1 or similar complexity classes. (Here by PRG we refer to any nontrivial PRG that extends its seed by just a single bit.) Such a PRG is implied by most standard cryptographic intractability assumptions such as those related to factoring, discrete logarithms, and lattice problems (see [1]). Efficient constructions of PRGs with a low algebraic degree that expand their seed by a constant factor can make our protocol quite efficient in practice. Candidates for such PRGs can be found in [32, 3].

Our protocol will use the PRG in order to reduce the functionality f we wish to evaluate to a *low-degree* randomized functionality $\hat{f}$. (Here and in the following, the degree of a randomized functionality is its degree over $GF(2)$, counting both inputs and random inputs towards the degree.) More precisely, we would like to obtain a low-degree randomized functionality $\hat{f}(x, r)$ with the following properties: (1) given the output of $\hat{f}(x, r)$ it is possible to efficiently *decode* $f(x)$; and (2) given the output $f(x)$ alone it is possible to efficiently *simulate* the output distribution of $\hat{f}(x, r)$ induced by a uniform choice of r, up to computational indistinguishability. The above properties of $\hat{f}$ allow to obtain a non-interactive *secure reduction* from f to $\hat{f}$: to evaluate f the players first invoke $\hat{f}$, and then apply the efficient decoder (from property (1) above) to recover the output of f.

The question of securely reducing general functionalities to low-degree functionalities was initially studied in [28] in an information-theoretic setting and was recently studied in a computational setting in [2]. The idea of [2] is to combine the "computational" circuit garbling technique of Yao [37] with "information-theoretic" randomization techniques of [28, 1]. The former reduces the evaluation of a circuit C to an evaluation of a degree-3 randomized functionality $\tilde{C}$ having an oracle access to a PRG; the latter then uses the easiness of the PRG for reducing $\tilde{C}$ to a degree-3 randomized functionality $\hat{C}$ which does not make use of a PRG oracle. We now describe the high-level details of these reductions that are useful for our purposes.

Computational Reduction. Yao's technique (e.g., using its variants described in [35] or [2]) reduces the functionality $C(x)$ defined by a circuit C to a randomized functionality $\tilde{C}(x, \tilde{r})$ having the following structure. The random input $\tilde{r}$ includes $O(|C|)$ wire keys $s_{w,b} \in \{0, 1\}^k$ (where w ranges over the wires and $b \in \{0, 1\}$), as well as wire masks $\lambda_w \in \{0, 1\}$. The functions computing the output bits of $\tilde{C}$ can be divided into three classes:

I. For each input bit of C, we have a function taking the input bit x_i and corresponding wire keys $s_{i,0}, s_{i,1}$ and outputting the key s_{i,x_i}. Overall, we have $|x|$ functions of type I, each with $O(k)$ input bits and $O(k)$ output bits, where the outputs can be viewed as degree-2 polynomials in the inputs.

II. For each NAND gate of C, with input wires α, β and output wire γ, we have a function mapping the 6 wire keys $s_{\alpha,b}, s_{\beta,b}, s_{\gamma,b}$ and the 3 wire masks $\lambda_\alpha, \lambda_\beta, \lambda_\gamma$ to $O(k)$ outputs. Each of these outputs is a degree-3 polynomial acting on the $O(k)$ input bits mentioned above as well as the outputs of the PRG applied to the wire keys. (The PRG is applied here to expand the wire keys and use the result to encrypt other keys and wire masks.[6]) Overall, we have $|C|$ functions of type II, each with $O(k)$ input bits and $O(k)$ output bits, where the outputs can be viewed as degree-3 polynomials in the inputs given an oracle to a PRG.

III. For each output wire w of C, we have a function taking a wire mask λ_w and outputting the same value.

The crucial feature of the functionality $\tilde{C}$ is that it consists of $|C|$ parallel copies of *few* types of degree-3 functionalities, each having a *short* input and output. The inputs to these functionalities have to be replicated in a way that reflects the structure of the circuit C. Note, however, that $\tilde{C}$ makes use of a PRG oracle, which we would like to eliminate. This is done via the following information-theoretic reduction.

Information-Theoretic Reduction. Using information-theoretic randomization techniques of [28, 1], it is possible to reduce the randomized functionality $\tilde{C}(x, \tilde{r})$ defined above to a similar randomized functionality $\hat{C}(x, \hat{r})$ whose outputs are degree-3 polynomials in $x, \hat{r}$. This reduction relies on the underlying PRG being computationally easy. The functionality $\hat{C}$ has a similar high level structure to that of $\tilde{C}$ described above, except that the random input $\hat{r}$ of $\hat{C}$ is longer and includes $\text{poly}(k)$ additional bits for every gate. Moreover, functions of type II now have $\text{poly}(k)$ many inputs and outputs instead of k. Again, the inputs to all of the $O(|C|)$ functions are a suitable replication of the functionality's inputs $x, \hat{r}$, where the replication pattern depends on the structure of C.

We summarize the above with the following lemma:

Lemma 1. *[2] Suppose an "easy PRG" (e.g., a PRG in NC^1) exists. Then, the secure computation of an arbitrary circuit $C(x)$ can be reduced to the secure computation of a degree-3 randomized functionality $\hat{C}(x, \hat{r})$, where $\hat{C}$ outputs a concatenation of $|C|$ functions of types I, II, and III as above. This reduction is secure against an active, adaptive adversary corrupting any set of players. It is a non-interactive reduction, in the sense that computing C involves only a single invocation to $\hat{C}$ and local computation, without further communication.*

We are now ready to explain at a high level how the tools described above can be combined. Intuitively, Yao's garbled circuit technique transforms each gate in the circuit C to a table of encrypted values, such that with appropriately encrypted inputs, one can evaluate C on these inputs while learning only the intended result. The lemma above essentially says that the encrypted circuit can be created by evaluating (securely and in parallel) $|C|$ degree-3 functions, namely the functions of type II above. Moreover, these functions are all of the same form because we may assume that all gates are NAND-gates, and the original garbling procedure treats every gate in the same way.

Now recall that evaluating many instances of the same function is exactly what Franklin and Yung's share packing technique allows us to do efficiently. There is

---

[6] Here we assume that the PRG has a sufficiently high stretch. This assumption can be eliminated with an $O(k)$ multiplicative overhead to the total complexity; see [2].

one technical difficulty that needs to be overcome, however: the random inputs for the instances of the function have to be correlated (in fact, replicated) in a way that reflects the structure of C. Thus, in the plain malicious model (without preprocessing) we will need to devise a protocol for efficiently enforcing such complex correlations. This problem will be addressed in Section 5.

4   Scalable MPC in the Linear Preprocessing Model

In this section we show how the previous tools can be combined to obtain scalable MPC protocols in the linear preprocessing model. In fact, in this model we can get the *overall* complexity to be $\tilde{O}(|C|)$, assuming a constant number of output clients[7] and regardless of the number of input clients. For simplicity, from here on we will mostly address the *communication* complexity. A treatment of the computational complexity (along with relevant optimizations) appears in the full version. We also assume for simplicity that the functionality outputs the same value $C(x)$ to all output clients. Our main result in this setting is captured by the following theorem.

Theorem 1 (Scalable MPC in the linear preprocessing model). *Suppose an easy PRG exists. Then, in the linear preprocessing model, there is a general two-round MPC protocol tolerating $t = \Omega(n)$ malicious servers, in which the communication complexity for computing a circuit C involves $O(|C| \cdot \text{poly}(k))$ bits sent to each output client, each input client must broadcast its (masked) inputs to the n servers, and there is no communication between the servers.*

Proof sketch: Let $\ell = \Theta(n)$ be the share packing parameter, as described in Section 3.1. The high level idea is to evaluate $\hat{C}(x, r)$ on a packed representation of its inputs and random inputs produced using ShareBlock. That is, we would like to evaluate ℓ copies of each type of output (I, II, or III) *in parallel* using the approach of [21]. To set up such an evaluation, we partition the (at most $|C|$) functions of each type into blocks of size ℓ. (For simplicity, assume wlog that the number of functions of each type is an exact multiple of ℓ, and so is the number of input bits held by each input client.) To evaluate ℓ instances of each function in parallel we replace each of the poly(k) input bits on which it depends by a block of ℓ bits, one from each instance. (That is, block i contains the i-th input bit from each instance.) We assume that the input bits x_i are all packed into the first $|x|/\ell$ blocks, referred to as input blocks, and that the o wire masks λ_w of the output wires are packed into the last o/ℓ blocks, referred to as output blocks. (We stress that in order to secret-share the bits belonging to a single block B, we form a block B' of ℓ elements in K where the i-th element equals the i-th bit in B, and secret-share B' using the procedure ShareBlock from Section 3.1; this should not be confused with packing several bits into one element of K.)

---

[7] This assumption can also be dispensed with by letting a single output client receive the (encrypted and authenticated) outputs of all clients, and then send to every other output client its share of the output. However, the resulting protocol only satisfies a relaxed notion of security, allowing the adversary to abort the computation.

In addition to the input and output blocks, we have $O(|C| \cdot \mathrm{poly}(k)/\ell)$ blocks containing bits of the random input $\hat{r}$, suitably replicated for allowing the parallel evaluation of the above 3 types of functions. A bit more precisely, this replicated assignment of random bits into blocks is formed as follows. First, the output bits of $\hat{C}$ are partitioned into groups of size ℓ, where the outputs in each group are computed by applying the same function g on some set of $k' = \mathrm{poly}(k)$ bits of $\hat{r}$. (The function g is the same for each of the ℓ outputs in a group, but the sets of k' input bits may be different; note that since $\hat{C}$ employs only $\mathrm{poly}(k)$ distinct functions g, we can fit all outputs of $\hat{C}$ into $O(|C| \cdot \mathrm{poly}(k)/\ell)$ groups.) The relevant bits of $\hat{r}$ are then placed into k' blocks so that the i-th output from the group is obtained by applying g to the i-th input bit from each block. Overall, each bit of $\hat{r}$ is placed into as many slots as the number of (type I, II, or III) functions in which it is involved. Note that the replication pattern of the bits of $\hat{r}$ depends on the circuit topology via the functions of type II.

Let $m = O(|C| \cdot \mathrm{poly}(k)/\ell)$ denote the total number of blocks. To represent the replication requirements induced by the circuit structure it is convenient to view the $m\ell$ elements of the blocks as being colored according to their partition classes. That is, we have a coloring function $\phi_C : [m] \times [\ell] \to [\ell m]$ such that $\phi_C(i,j) = \phi_C(i',j')$ iff entry j of block i should be identical to entry j' of block i' (otherwise they are assumed to be independent). The entries of the input blocks will all be colored by unique colors (since each input is used only by a single function of type I), but all other entries (including those of the output wires) will be replicated.

We are now ready to describe the scalable protocol in the linear preprocessing model. We assume that the field K is of characteristic 2, and is of minimal size subject to $|K| > n$ (the requirement $|K| > n$ is imposed by ShareBlock). Let $\ell = \Theta(n)$ and $T = t + \ell - 1$. To enable non-interactive computation of degree-3 polynomials on blocks of ℓ packed secrets, it suffices to use $n > 2T$ players in the semi-honest case and $n > 5T$ players in the malicious case (as required for evaluating degree-3 polynomials with minimal interaction, cf. [16], Section 2.2). The protocol proceeds as follows.

Protocol 1. Scalable MPC protocol in linear preprocessing model:

1. *Setup. Using linear preprocessing, the players obtain packed shares of replicated random binary secrets, where the replication pattern is specified by ϕ_C. Moreover, each input client obtains the (random) secrets belonging to the blocks of inputs it owns. Finally, the players obtain packed shares of $O(m)$ independent 0-blocks, namely blocks whose ℓ secrets $(s_0, \dots, s_{\ell-1})$ are all set to 0. These 0-blocks are secret-shared using polynomials of degree $3T$ and will be used in Round 3 to support secure multiplication.*

   Note that each of the above distribution patterns can be expressed as a random codeword in a distributed linear code over $\mathrm{GF}(2)$, and thus can be implemented within the linear preprocessing framework. (The distributed codes used above are only linear over $\mathrm{GF}(2)$ and not over the extension field K, since the random secrets are restricted to take $0/1$ values.)

2. Round 1. *Each input client broadcasts to the servers a correction vector of length ℓ for each input block it owns, namely the exclusive-or of this block with a corresponding block of random secrets obtained during setup.*

3. Round 2. *First, each server locally corrects its share of each input block according to the corrections broadcasted in Round 1. For instance, if the correction for the first block is $c \in \{0,1\}^\ell$, server P_j finds the (unique) polynomial p of degree T such that $p(\mu_i) = c_i$ for $1 \le i \le \ell$ and $p(\omega_{j'}) = 0$ for $1 \le j' \le t$, and adds $p(\omega_j)$ to its share of the first block as received during setup. Following this stage, all inputs to $\hat{C}$ (including actual inputs x and random inputs $\hat{r}$) are shared between the servers.*

    Now the servers locally compute random shares of the outputs of $\hat{C}$ and send them to the output clients. This local computation is done by simply evaluating the corresponding degree-3 polynomials on the local (packed) shares and adding the (packed) shares of the 0-blocks. (Note that the careful structuring of the blocks, defined by the replication pattern ϕ_C, ensures that the computation applied to each set of $k' = \text{poly}(k)$ blocks is the same for each position i, $1 \le i \le \ell$, within the blocks.)

4. Output. *Each output client reconstructs the outputs of $\hat{C}$ from the shares it received, using error-correction of Reed-Solomon codes to recover from possible errors caused by malicious servers. It then applies the efficient decoding procedure of $\hat{C}$ to recover the output of C.*

The security of Protocol 1 follows by combining (1) the security of the reduction from C to $\hat{C}$ (Lemma 1) and (2) the security of the above procedure for computing $\hat{C}$ (namely, the one defined by Protocol 1 without the final decoding step). The latter procedure is analogous to a similar protocol in [16], Section 2.2, except for the use of packed shares instead of standard shares. Thus, a formal proof of (2) can be easily obtained by generalizing the corresponding proof in [16]. Finally, the security of Protocol 1 is implied by (1) and (2) using a suitable composition theorem (e.g., from [10, 11]).

The communication complexity includes broadcasting the (masked) inputs by the clients, and n field elements communicated to each output client (one from each server) for each ℓ-tuple of output bits of $\hat{C}$. Overall, each output client receives $O(|\hat{C}| \cdot n/\ell) = O(|\hat{C}|) = O(|C| \cdot \text{poly}(k))$ field elements, as required. $\square$

We note that a variant of Protocol 1 (and also of our main protocol) can be based on an arbitrary PRG, using techniques of [5, 24] to directly evaluate the PRG-based functionality $\tilde{C}$ defined in Section 3.2 instead of the degree-3 functionality $\hat{C}$. This variant requires $\text{poly}(k)$ rounds of interaction, and cannot go below the $\tilde{O}(n|C|)$ communication bound.

5   Scalable MPC in the Plain Model

In this section we implement Protocol 1 in the plain model. This amounts to securely emulating Step 1, namely the generation of replicated random secrets, without the trusted setup of linear preprocessing. In the semi-honest model, this can be done in a straightforward way by letting each player share random (packed) secrets replicated according to ϕ_C, and adding up the contributions of $t + 1$ different players. Thus, for the semi-honest model we immediately get a scalable protocol with a total of $O(n|C| \cdot \text{poly}(k))$ communication. In fact, it suffices to let only the *clients* participate in the generation of the random packed shares. (Indeed, even one uncorrupted client would suffice to keep the shared secrets hidden from the adversary.) So when the number of clients is constant, the communication is only $O(|C| \cdot \text{poly}(k))$.

We turn to the question of securely generating the required linear correlations in the malicious model. We will begin by describing an efficient procedure for generating many *independent* instances of the *same* linear correlation. Later, in Section 5.1, we will extend this basic procedure to deal with the more challenging case of correlations induced by an arbitrary replication pattern ϕ_C, as required by the setup phase of Protocol 1.

The basic procedure we present below is similar to a subprotocol of Hirt and Maurer [24], and in particular uses their ideas of player elimination (also used, in different ways, in [25, 22, 26, 8]) and the idea of verifying a large batch of linearly-shared secrets by using randomized linear combinations. We will improve its efficiency by using distributed coin-flipping and a pseudorandom generator to succinctly generate the coefficients of a long (pseudo-)random linear combination. In the full version we describe an optimization of the asymptotic *computational* complexity of the protocol. (The corresponding subprotocol from [24] does not rely on distributed coin flipping nor a PRG; instead, it requires each player to separately communicate a long linear combination vector. Moreover, this subprotocol was implicitly analyzed in the non-adaptive case; an extension of its analysis to the adaptive case would involve an additional overhead.)

Generating Independent Instances of a Linear Correlation. Consider the randomized functionality $\mathsf{RandDLC}(L)$, which distributes to the players shares of a random codeword from a distributed linear code $L = (V, \pi_1, \ldots, \pi_n)$. We want to securely compute m independent instances of $\mathsf{RandDLC}(L)$ at a low cost.

To get the desired efficiency with a minimal amount of interaction, we will not insist on securely computing the desired functionality in the strict sense, but rather settle for the following relaxation. At the end of the protocol, we can tolerate a small set B of (say, at most $2t$) "eliminated" players, such that only the outputs of honest players *outside* B should be consistent with the functionality. Moreover, this set B is publicly known. (Jumping ahead, we will require that the same global set B apply jointly to all of the subprotocols we use.) This notion of security can be formally captured by relaxing the standard simulation requirement to consider, in addition to the view of corrupted players, the joint outputs of all honest players outside the set B (instead of *all* honest players in the standard simulation-based definitions). For instance, consider the case of applying $\mathsf{RandDLC}$ to secret-share a block of random secrets. In this case, such a definition would imply that the outputs of all honest players *outside* B are consistent with a valid sharing of the same block, and moreover that the adversary learns nothing about the secrets in the block.

This relaxed notion of security is implicit in previous protocols (e.g., [24, 22]). It does not compromise the security of our high level protocols, since: (1) all relevant functionalities are *robust*, in the sense that a "correct" output can be efficiently decoded from a partial output; and (2) such a decoding procedure will be employed by the output clients in the high level protocol. Thus, from the point of view of the higher level application, the missing shares of B do not make any difference.

We turn to describe our amortized protocol for $\mathsf{RandDLC}(L)$. In fact, we start with an even simpler *deterministic* functionality we denote by $\mathsf{DLC}^P(L)$.

Definition 1 (Functionality $\mathsf{DLC}^P(L)$). *Let $L = (V, \pi_1, \ldots, \pi_n)$ be a distributed linear code, where V is spanned by the columns of a matrix M. Let P be a player (an input client or a server) referred to as a* dealer. *The functionality $\mathsf{DLC}^P(L)$ is defined as follows:*

- **Inputs:** *P holds an input y, defining a vector $v = My \in V$. Other players have no input.*
- **Outputs:** *Each server P_j outputs $v(j)$.*

The secure computation of $\mathsf{DLC}^P(L)$ is trivial in the semi-honest model, since P can simply send to each server j its share $v(j)$. It is nontrivial in the malicious model, since we need to ensure that the outputs of honest servers are consistent with L, and that they are furthermore consistent with v when P is uncorrupted. Our motivation for securely computing $\mathsf{DLC}^P(L)$ is that this will later be used as a building block for computing $\mathsf{RandDLC}(L)$. Specifically, $\mathsf{RandDLC}(L)$ can be reduced to multiple instances of $\mathsf{DLC}^P(L)$ by adding together random vectors contributed by different dealers P. Thus, from here on we can focus on the goal of amortizing the cost of securely computing m instances of $\mathsf{DLC}^P(L)$, on input vectors $v_1, \ldots, v_m$. This is done as follows.

Protocol 2. Efficient amortized implementation of $\mathsf{DLC}^P(L)$:

- **Inputs:** *P holds m inputs $y_1, \ldots, y_m$, defining m vectors $v_i = My_i \in V$. Other players have no inputs.*
- **Outputs:** *Each server P_j outputs $v_1(j), \ldots, v_m(j)$.*

1. Share distribution. *P sends to each server P_j, $1 \le j \le n$, its share $v_i(j)$, $1 \le i \le m$, of each vector v_i. In addition, it similarly distributes σ random vectors $r^1, \ldots, r^\sigma \in_R L$ (which will be used as "blinding" vectors to protect the privacy of $v_1, \ldots, v_m$). To resist an adaptive adversary, we will need to let $\sigma = n + k$, where k is the security parameter.*

2. Coin-flipping. *The servers use a constant-round multiparty coin-flipping protocol (e.g., a 3-round protocol based on the 2-round VSS protocol from [22]) to pick a σ-tuple of random k-bit seeds $(s^1, \ldots, s^\sigma)$. Each seed s^h is locally expanded by each server, via a PRG, to a vector R^h in $\{0,1\}^m$, specifying a (pseudo-)random linear combination of the v_i's.*

3. Proving. *P computes and broadcasts the σ linear combinations specified by the vectors R^h, masking each with a different blinding vector r^h. That is, it*

computes and broadcasts the σ codewords $u^h = r^h + \sum_{i=1}^{m} R_i^h v_i$. (The vectors u^h serve as a succinct distributed witness for the validity of the vectors v_i.) Each P_j verifies that all u^h are valid codewords; otherwise P is disqualified and all honest servers output 0.

4. Complaining. *Each P_j verifies that every u^h is consistent with the linear combination R^h and the $m + \sigma$ shares it received. If some inconsistency is found, P_j broadcasts a complaint.*

5. Output. *Let B be the set of servers who broadcasted a complaint. If $|B| > t$, the dealer P is disqualified and all honest servers output 0. Otherwise, each P_j outputs its shares of $v_1, \ldots, v_m$ and discards its shares of $r^1, \ldots, r^\sigma$ (which were used for blinding).*

Lemma 2. *Protocol 2 satisfies the following:*

1. *If P is honest then it is never disqualified, the outputs of the honest servers are consistent with the v_i's, and the set B includes only corrupted servers.*
2. *Suppose P is corrupted. If P is not disqualified, then (except with negligible probability) all outputs of honest servers outside B are consistent with L. In other words, the event that P is not disqualified and the outputs of honest servers outside B are inconsistent occurs with negligible probability in k.*
3. *If P is honest, the adversary learns nothing about the vectors v_i except the shares of corrupted servers. (Moreover, its view can be efficiently simulated given these shares.)*

Proof sketch: Fact (1) follows by inspection of the protocol, and (3) follows from the use of the blinding vectors r^h. It remains to show that (2) holds. For this, we bound the probability that there is *some* set H of honest servers whose shares are inconsistent yet pass all tests. Suppose first that the linear combinations R^h were entirely random, and fix a possible set $H \subseteq [n]$ of honest servers. If some share vector received by H, wlog $v_1(H)$, is inconsistent with L, then a random linear combination R^h of all vectors is inconsistent with L with probability at least $1/2$. (Indeed, conditioned on the linear combination v' of the remaining vectors, there are two cases: (a) if $v'(H)$ is consistent with L then $[1 \cdot v_1 + v'](H)$ is inconsistent with L; (b) if $v'(H)$ is inconsistent with L then $[0 \cdot v_1 + v'](H)$ is inconsistent with L.) It follows that H will pass all σ consistency tests with at most $2^{-\sigma}$ probability. Since $\sigma = n + k$, it follows by a union bound that there is no set H of honest servers who receive inconsistent shares and do not complain, except with negligible probability. Specifically, this failure probability is bounded by $2^n \cdot 2^\sigma = 2^{-k}$. Finally, by the security of the PRG the above analysis remains the essentially the same when each R is pseudorandom (e.g., replace $1/2$ in the above analysis by 0.4).   □

Note that when m is large, the communication complexity of the protocol is dominated by the length of the inputs and is thus essentially optimal. Also note that one could instantiate the PRG with an (unconditionally secure) ϵ-biased generator [33].

The Multiple-Dealer Case. To implement RandDLC(L) we need to use multiple (parallel) invocations of the above DLC$^{P_i}(L)$ protocol with different dealers

P_i. A major concern in this case is that each invocation may result in a different set B_i of eliminated players. Indeed, the adversary can arrange that for each corrupted P_i, the set B_i will include a disjoint set of t honest players. Our goal is to obtain a publicly known set B of at most $2t$ players, whose elimination would result in all honest players holding consistent shares from all remaining invocations of DLC. Moreover, to achieve this goal we can afford to disqualify up to t honest dealers, overriding their inputs with 0. This further relaxation will not affect the security of our applications, in which DLC will be invoked with random and independent codewords rather than ones that depend on the inputs. Specifically, when using n invocations of DLC for implementing RandDLC, each player will distribute a random codeword and output the sum of the received shares. In this case, eliminating a few honest players will have no effect on security. (In fact, it would suffice that $2t + 1$ players act as dealers.)

The above is easy to achieve using the following player elimination approach. Let B_i denote the union of all sets B induced by DLC invocations in which P_i acts as the dealer. It follows from Lemma 2 that either P_i is corrupted or all players in B_i are corrupted. Thus, we can eliminate both P_i and some $P_j \in B_i$ and include them in B. As a result, the size of B grows by 2 and the number of remaining corrupted players is reduced by at least 1. After no more than t such elimination steps all public inconsistencies are resolved and B contains at most $2t$ players. At the end of this procedure, all dealers belonging to the (publicly known) B are eliminated. We are guaranteed that the outputs of all non-eliminated honest players are consistent in all DLC invocations in which one of them serves as a dealer.

In the setting where the number of clients is constant, we can avoid the above procedure by letting only the clients act as dealers. (This suffices because there is no need to protect against an adversary that corrupts all clients.) In this case we can afford to let the final B be the union of all B_i associated with clients that were not disqualified, while still tolerating a constant fraction of corrupted servers $(t = \Omega(n))$. Since this approach does not involve an elimination of honest clients, it can also be used to verify shares of the *inputs*, thus avoiding the need to broadcast all inputs.

5.1   Generating Packed Replicated Secrets

The above RandDLC protocol suffices for generating the packed shares of 0-blocks required by the setup phase of Protocol 1. Indeed, to generate these blocks, let L be the space of degree-T polynomials over K which evaluate to 0 on all points μ_i, $1 \leq i \leq \ell$, where each P_j is assigned the value of the polynomial at point ω_j. The challenge is to efficiently generate m blocks of random binary secrets *subject to the replication pattern defined by* ϕ_C. Our main observation is that testing if a shared m-tuple of blocks is consistent with the "complex" replication pattern ϕ_C can be reduced, via local computation (and no communication), to testing $O(\ell^2)$ instances of a "repetitive" replication pattern. The latter can be efficiently handled using the testing procedure in Protocol 2. The extra cost of $O(\ell^2)$ elementary tests will be amortized over a large number of blocks $m \gg \ell^2$, and thus will not effect the amortized communication. We stress that the above holds for an arbitrary replication pattern ϕ_C, and thus does not depend on the circuit topology.

The idea for this reduction is the following. Consider the case where a single dealer P distributes m blocks of ℓ secrets. We want to verify that these are indeed *binary* secrets that are *consistent with* ϕ_C. The replication condition ϕ_C can be expressed as the conjunction of at most $m\ell$ atomic conditions of the form "$\phi_C(i,j) = \phi_C(i',j')$". We refer to such a condition as being of type (j, j'), indicating that the j-th secret in one block is compared with the j'-th secret in another, not necessarily distinct, block. There are at most ℓ^2 different types of atomic conditions, where each block is involved in at most $O(\ell)$ types (assuming C has constant fan-out).

The key idea is that each condition of type (j, j') can be tested by letting the servers locally concatenate their two shares from the two relevant blocks (i, i'), and then (jointly) test that the combined shares belong to some linear space $L_{j,j'}$. More specifically, let v, v' represent share vectors corresponding to blocks i, i' respectively, and let (v, v') be the distributed vector obtained by letting each server locally concatenate its shares of v, v' in that order. Then, the set of vectors (v, v') satisfying the above atomic replication constraint can be expressed as a test of membership in a distributed linear code $L_{j,j'}$ over $GF(2)$. Indeed, if both (u, u') and (v, v') satisfy the atomic replication constraint, then so does $(u + v, u' + v')$. Finally, we can batch together all atomic conditions of the same type and test them together using the testing procedure defined by Steps 2-5 of Protocol 2. This gives a procedure for verifying the validity of the m blocks distributed by a dealer P_i via ℓ^2 (parallel) invocations of the previous testing procedure. This procedure will either disqualify the dealer, or output a set B_i of at most t servers such that all shares of honest servers outside B_i are consistent packed shares of replicated binary secrets satisfying ϕ_C.

More concretely, for each player P, acting as a dealer, we run in parallel the following procedure.

Protocol 3. Generating packed random secrets consistent with replication pattern ϕ_C:

1. *P generates m blocks of random binary secrets subject to the replication constraints ϕ_C, picks a random degree-T polynomial encoding each block, and sends to each P_j the value of each of the m polynomials at point ω_j. In addition, a random "blinding" vector $r^h \in_R L_{j,j'}$ should be distributed for each type (j, j').*

2. *Run Steps 2-4 of Protocol 2, testing that all shared blocks encode ℓ-tuples of binary secrets.*

3. *The servers write ϕ_C as a conjunction of (at most $m\ell$) atomic equality conditions, represented by a set $A \subseteq ([m] \times [\ell])^2$. For each type of atomic condition $(j, j') \in [\ell]^2$, do the following (in parallel):*
   For every pair (i, i') such that $((i, j), (i', j')) \in A$, let each server concatenate its share of block i and its share of block i'. Run Steps 2-4 of Protocol 2 to

> test that all concatenated vectors (each of length $2n$) are in the linear space $L_{j,j'}$ corresponding to atomic conditions of type (j, j').
> 4. Run Step 5 of Protocol 2 to produce the outputs.

Protocol 3 may result in eliminating the dealer P, or producing a set B of at most t servers. The final protocol will use parallel invocations of the above protocol with each player acting as a dealer, in an analogous way to the amortized RandDLC protocol described above. As in RandDLC, these invocations produce a global set B of eliminated players. The final m-tuple of blocks is obtained by summing up the contributions of all non-eliminated dealers. Again, in the case of a constant number of clients, we let the clients serve as dealers and refrain from eliminating clients that were not disqualified as (certain) cheaters.

The combined cost of all the testing procedures is bounded by a fixed polynomial in n, independently of m. As a result, the amortized complexity of the share generation protocol is the same as in the semi-honest case. Thus, we have:

Theorem 2 (Scalable protocol in plain model). *Suppose an easy PRG exists. Then, in the plain model, there is a general constant-round MPC protocol tolerating $t = \Omega(n)$ malicious servers, in which the communication complexity for computing a circuit C involves $O(|C| \cdot n \cdot \mathrm{poly}(k))$ bits of communication, and each input client must broadcast its (masked) inputs to the n servers. If the number of clients is constant, the complexity is reduced to $O(|C| \cdot \mathrm{poly}(k))$.*

Proof sketch: The main protocol is obtained by combining Protocol 1 with the above share generation protocols, namely:

1. Use parallel invocations of Protocol 3 and server elimination (as described above) to generate shared blocks of random secrets, replicated according to ϕ_C, substituting the linear preprocessing of Step 1 in Protocol 1. The outputs of the servers are consistent with the required pattern except, perhaps, for a set B of $O(t)$ eliminated servers.
2. Run Steps 2-4 of Protocol 1 using the outputs from the previous step to substitute linear preprocessing.

A proof of security can proceed as follows. If Step 1 were a fully secure protocol for the corresponding randomized functionality (i.e., with $B = \emptyset$), then the security of the protocol would follow directly from that of Protocol 1 and a composition theorem of [10]. The $O(t)$ outputs of servers in B (resulting from Step 1) have no effect on the outputs of output clients in the end of Step 2 because of the robustness property of this step. Moreover, the effect B has on the adversary's view can be easily incorporated into the simulation.   □

References

[1] B. Applebaum, Y. Ishai, and E. Kushilevitz. Cryptography in NC^0. In *Proc. FOCS 2004*, pages 165-175.
[2] B. Applebaum, Y. Ishai, and E. Kushilevitz. Computationally private randomizing polynomials and their applications. In *Proc. CCC 2005*, pages 260-274.

[3] B. Applebaum, Y. Ishai, and E. Kushilevitz. On Pseudorandom Generators with Linear Stretch in NC^0. In *Proc. RANDOM 2006*.

[4] O. Barkol and Y. Ishai. Secure Computation of Constant-Depth Circuits with Applications to Database Search Problems. In *Proc. Crypto 2005*, pages 395-411.

[5] D. Beaver. Efficient Multiparty Protocols Using Circuit Randomization. In *Proc. Crypto 1991*, pages 420-432.

[6] D. Beaver, J. Feigenbaum, J. Kilian, and P. Rogaway. Security with low communication overhead. In *Proc. Crypto 1990*, pages 62-76.

[7] D. Beaver, S. Micali, and P. Rogaway. The round complexity of secure protocols (extended abstract). In *Proc. STOC 1990*, pages 503-513.

[8] Z. Beerliova and M. Hirt. Efficient Multi-Party Computation with Dispute Control. In *Proc. TCC 2006*, pages 305-328.

[9] M. Ben-Or, S. Goldwasser, and A. Wigderson. Completeness theorems for non-cryptographic fault-tolerant distributed computation. In *Proc. STOC 1988*, pages 1-10.

[10] R. Canetti. Security and composition of multiparty cryptographic protocols. In *J. of Cryptology*, 13(1):143-202, 2000.

[11] R. Canetti. Universally Composable Security: A New Paradigm for Cryptographic Protocols. In *Proc. FOCS 2001*, pages 136-145.

[12] D. Chaum, C. Crépeau, and I. Damgård. Multiparty unconditionally secure protocols (extended abstract). In *Proc. STOC 1988*, pages 11-19.

[13] Richard Cleve. Limits on the Security of Coin Flips when Half the Processors Are Faulty (Extended Abstract). In *Proc. STOC 1986*, pages 364-369.

[14] R. Cramer, I. Damgård, and Y. Ishai. Share conversion, pseudorandom secret-sharing and applications to secure computation. In *Proc. TCC 2005*, pages 342-362.

[15] R. Cramer, I. Damgård, and J. Nielsen. Multiparty computation from threshold homomorphic encryption. In *Proc. Eurocrypt 2001*, pages 280-299.

[16] I. Damgård and Y. Ishai. Constant-Round Multiparty Computation Using a Black-Box Pseudorandom Generator. In *Proc. Crypto 2005*, pages 378-394.

[17] I. Damgård, and J. Nielsen. Universally Composable Efficient Multiparty Computation from Threshold Homomorphic Encryption. In *Proc. Crypto 2003*, pages 247-264.

[18] S. Even, O. Goldreich and A. Lempel. A Randomized Protocol for Signing Contracts. In *Communications of the ACM*, 28(6):637–647, 1985.

[19] P. Feldman and S. Micali. An Optimal Algorithm for Synchronous Byzantine Agreement. *SIAM. J. Computing*, 26(2):873–933, 1997.

[20] M. K. Franklin and S. Haber. Joint Encryption and Message-Efficient Secure Computation. In *Proc. Crypto 1993*, pages 266-277. Full version in *Journal of Cyptoglogy* 9(4): 217-232 (1996).

[21] M. K. Franklin and M. Yung. Communication Complexity of Secure Computation. In *Proc. STOC 1992*, pages 699-710.

[22] R. Gennaro, Y. Ishai, E. Kushilevitz and T. Rabin. The Round Complexity of Verifiable Secret Sharing and Secure Multicast. In *Proc. STOC 2001*, pages 580-589.

[23] O. Goldreich, S. Micali, and A. Wigderson. How to play any mental game (extended abstract). In *Proc. STOC 1987*, pages 218-229.

[24] M. Hirt and U. M. Maurer. Robustness for Free in Unconditional Multi-party Computation. In *Proc. Crypto 2001*, pages 101-118.

[25] M. Hirt, U. M. Maurer, and B. Przydatek. Efficient Secure Multi-party Computation. In *Proc. Asiacrypt 2000*, pages 143-161.

[26] M. Hirt and J. B. Nielsen. Upper Bounds on the Communication Complexity of Optimally Resilient Cryptographic Multiparty Computation. In *Proc. Asiacrypt 2005*, pages 79-99.

[27] M. Hirt and J. B. Nielsen. Robust Multiparty Computation with Linear Communication Complexity. These proceedings.

[28] Y. Ishai and E. Kushilevitz. Randomizing polynomials: A new representation with applications to round-efficient secure computation. In *Proc. FOCS 2000*, pages 294-304.

[29] M. Jakobsson and A. Juels. Mix and Match: Secure Function Evaluation via Ciphertexts. *Proc. Asiacrypt 2000*, pages 162-177.

[30] J. Katz and C.-Y. Koo. On Expected Constant-Round Protocols for Byzantine Agreement. These proceedings.

[31] Y. Lindell, A. Lysyanskaya, and T. Rabin. Sequential composition of protocols without simultaneous termination. In *Proc. PODC 2002*, pages 203-212.

[32] E. Mossel, A. Shpilka, and L. Trevisan. On ϵ-biased generators in NC^0. In *Proc. FOCS 2003*, pages 136-145.

[33] J. Naor and M. Naor. Small-bias probability spaces: Efficient constructions and applications. *SIAM J. Comput.*, 22(4):838–856, 1993. Preliminary version in Proc. STOC '90.

[34] M. Naor and K. Nissim. Communication preserving protocols for secure function evaluation. In *Proc. STOC 2001*, pages 590-599.

[35] M. Naor, B. Pinkas, and R. Sumner. Privacy preserving auctions and mechanism design. In *Proc. 1st ACM Conference on Electronic Commerce*, pages 129-139, 1999.

[36] A. Shamir. How to share a secret. *Commun. ACM*, 22(6):612–613, June 1979.

[37] A. C. Yao. How to generate and exchange secrets. In *Proc. FOCS 1986*, pages 162-167.

[38] Z. Zhang, M. Liu, and L. Xiao. Parallel Multi-Party Computation from Linear Multi-Secret Sharing Schemes. In *Proc. Asiacrypt 2005*.

Algebraic Geometric Secret Sharing Schemes and Secure Multi-Party Computations over Small Fields[*]

Hao Chen[1] and Ronald Cramer[2,3]

[1] Department of Computing and Information Technology, School of Information Science and Engineering, Fudan University, Shanghai, China
chenhao@fudan.edu.cn
[2] CWI, Amsterdam, The Netherlands
cramer@cwi.nl
[3] Mathematical Institute, Leiden University, The Netherlands

Abstract. We introduce algebraic geometric techniques in secret sharing and in secure multi-party computation (MPC) in particular. The main result is a linear secret sharing scheme (LSSS) defined over a finite field $\mathbb{F}_q$, with the following properties.

1. It is *ideal*. The number of players n can be *as large as* $\#C(\mathbb{F}_q)$, where C is an algebraic curve C of genus g defined over $\mathbb{F}_q$.
2. It is *quasi-threshold*: it is t-rejecting and $t+1+2g$-accepting, but not necessarily $t + 1$-accepting. It is thus in particular a ramp scheme. High information rate can be achieved.
3. It has *strong multiplication* with respect to the t-threshold adversary structure, if $t < \frac{1}{3}n - \frac{4}{3}g$. This is a multi-linear algebraic property on an LSSS facilitating zero-error multi-party multiplication, unconditionally secure against corruption by an active t-adversary.
4. The finite field $\mathbb{F}_q$ can be *dramatically smaller than* n. This is by using algebraic curves with many $\mathbb{F}_q$-rational points. For example, for each small enough ϵ, there is a finite field $\mathbb{F}_q$ such that for infinitely many n there is an LSSS over $\mathbb{F}_q$ with strong multiplication satisfying $(\frac{1}{3} - \epsilon)n \leq t < \frac{1}{3}n$.
5. Shamir's scheme, which requires $n > q$ and which has strong multiplication for $t < \frac{1}{3}n$, is a special case by taking $g = 0$.

Now consider the classical ("BGW") scenario of MPC unconditionally secure (with zero error probability) against an active t-adversary with $t < \frac{1}{3}n$, in a synchronous n-player network with secure channels. By known results it now follows that there exist MPC protocols in this scenario, achieving the same communication complexities in terms of the number of field elements exchanged in the network compared with known Shamir-based solutions. However, in return for decreasing corruption tolerance by a small ϵ-fraction, q may be dramatically smaller than n. This tolerance decrease is unavoidable due to properties of MDS codes. The techniques extend to other models of MPC. Results on less specialized LSSS can be obtained from more general coding theory arguments.

---

[*] The authors independently submitted similar results to CRYPTO 2006 [5,8]. This paper is the result of merging those two papers.

C. Dwork (Ed.): CRYPTO 2006, LNCS 4117, pp. 521–536, 2006.
© International Association for Cryptologic Research 2006

1   Introduction

This paper introduces the use of algebraic geometric techniques in secret sharing and in secure multi-party computation (MPC) in particular. MPC concerns the problem of a network of players who wish to compute an agreed function on respective private inputs in a secure way, i.e., guaranteeing correctness of the result and privacy of their respective inputs, even if some players are corrupted by an adversary.

Let $\mathcal{A}_{t,n}$ denote the t-threshold adversary structure on the players $\{1, \ldots, n\}$, i.e., it consists of all subsets of size at most t. Similarly, let $\Gamma_{r,n}$ denote the threshold access structure consisting of all subsets of size at least r. For a linear secret sharing scheme (LSSS) $\mathcal{M}$ on n players, let $\Gamma(\mathcal{M})$ denote the sets accepted by $\mathcal{M}$, and let $\mathcal{A}(\mathcal{M})$ denote the sets rejected by $\mathcal{M}$. Let $\mathbb{F}_q$ be a finite field, and let C be a smooth projective absolutely irreducible curve defined over $\mathbb{F}_q$, and let g denote its genus. The number of $\mathbb{F}_q$-rational points on C is denoted $\#C(\mathbb{F}_q)$.

We show that for any integer n with $1 < n < \#C(\mathbb{F}_q)$ and any integer t with $1 \leq t < n - 2g$ there exists an LSSS $\mathcal{M}$ over $\mathbb{F}_q$ with the following properties.

1. It is an *ideal* scheme on n players; the secret as well as each share consists of a single field element.
2. It is *quasi-threshold* (with a $2g$ gap). This means that $\mathcal{A}_{t,n} \subset \mathcal{A}(\mathcal{M})$ and $\Gamma_{r,n} \subset \Gamma(\mathcal{M})$ where $r = t + 1 + 2g$. It it thus in particular a ramp scheme. We also show how high information rate can be achieved.
3. It has *strong multiplication* [15] with respect to the $\mathcal{A}_{t,n}$ adversary structure provided that $t < \frac{1}{3}n - \frac{4}{3}g$.
   This is a specialized multi-linear algebraic property known to facilitate zero-error multi-party multiplication, unconditionally secure against active corruptions (see [15,14]). See Section 2 for the definition.
   It is also known to linearize in general the problem of recovery of the secret in the presence of corrupted shares [10].
4. Shamir's scheme, which requires $n > q$ and which has strong multiplication for $t < \frac{n}{3}$, is a special case of our scheme by taking $g = 0$. However, by taking $g > 0$ and by selecting suitable curves, q can be *dramatically smaller than the number of players n*, as elaborated below.

Consider the model of a synchronous network with pair-wise secure channels. It is a classical result due Ben-Or, Goldwasser and Wigderson [1] and Chaum, Crépeau and Damgaard [4] that efficient MPC unconditionally secure (*with zero error probability* [1]) against an active adversary bound by the threshold condition that fewer than a $1/3$-fraction of the players are corrupted is possible in this model. An active adversary is one who may arbitrarily influence and coordinate the behavior of the corrupted parties. The actual protocols make intricate use of Shamir's scheme.

Cramer, Damgaard and Maurer [15] show how to "efficiently bootstrap" MPC protocols from general LSSS, thereby providing means for dealing with general (i.e., "non-threshold") adversaries. The adversary structure capturing the resulting MPC's resilience is related to the access structure of the LSSS. Interestingly, these techniques will be used here for achieving security against a threshold adversary. Indeed, in the model we consider here, the properties of the proposed LSSS are for instance sufficient for the construction of an efficient (additively homomorphic) verifiable secret sharing scheme (VSS). [1] This is a fundamental primitive in MPC, as general MPC secure against an active adversary is essentially about performing secure arithmetic on VSS-ed secret values. The scheme is unconditionally secure (with zero error probability) against an active t-adversary. This is by requiring $t < \frac{1}{3}n - \frac{2}{3}g$: this renders the scheme $n - 2t$-accepting, which is needed to enforce error-freeness of the VSS. The strong multiplication property is immaterial in this case. The number of field elements exchanged is the same as in the VSS from [1] or as in later variations, except that the field over which it is defined can be much smaller than the number of players n. Shamir-based solutions require $n > q$. The price to be paid is that that the corruption tolerance is decreased by an (arbitrarily) small (constant) fraction of n.

An LSSS per se is not known to be sufficient for efficient MPC, even though, as said, additively homomorphic VSS as such can be constructed from it. This does of course enable secure computation of addition, or more generally, linear functionals. However, secure multiplication is only known to be possible if the underlying LSSS in addition satisfies certain multi-linear algebraic properties, such as the multiplication property or the strong multiplication property. In the particular "perfect, zero-error" MPC scenario considered here, the strong multiplication property is essential. It will enable the claimed MPC over small fields, just as with the VSS discussed above.

Note that there exists an efficient transformation that maps relevant LSSS to equivalent LSSS that additionally satisfy the multiplication property [15], increasing the share size by only a multiplicative factor of two. However, it is important to note that no efficient transformation at all is known for the case of strong multiplication.

LSSS that satisfy the multiplication property rather than the strong multiplication property can in particular be used as basis for MPC secure against a *passive* adversary or against an active one but with non-zero yet negligible error probabilities. Moreover, if the model is augmented with a *broadcast primitive*, an *active adversary* can be tolerated who corrupts fewer than a *1/2-fraction* of the players, at the cost of introducing non-zero yet negligible error probabilities [32]. For VSS it is required that the underlying LSSS is $n - t$-accepting and for secure multiplication the LSSS is required to satisfy the multiplication property. Using the LSSS introduced here, both are satisfied if $t < \frac{1}{2}n - 2g$. More details and extended results can be found in [16,7].

---

[1] Briefly, this strengthens a secret sharing scheme so as to withstand an active adversary who may corrupt part of the network, possibly including the dealer: it is a scheme that uses an interactive protocol to force even a possibly corrupted dealer to distribute shares consistent with the secret sharing scheme, that offers privacy to an honest dealer, and that offers unambiguous reconstruction of the secret.

In general, LSSS with strong multiplication admit an algorithm to recover the secret in the presence of corrupted shares. This algorithm is efficient if the secret sharing scheme itself is efficient to begin with. This follows from the results in [10] where it is shown that strong multiplication linearizes this "decoding problem," by means of a generalization of ideas taken from the Berlekamp-Welch algorithm. For completeness we include in this paper a description of the general procedure from [10] as it applies to our algebraic geometric secret sharing schemes, even though in the present case it also follows by efficient decoding algorithms for algebraic geometry codes.

The quasi-threshold property of our scheme is sufficient for our purposes, but it is also unavoidable. Indeed, threshold linear secret sharing schemes are equivalent to maximum distance separable codes (MDS). For such codes it is well-known that $q \geq \max\{t, n-t+1\}$ if $0 < t < n-1$, which gives for example a $2q/3$ lower bound when t is approximately $n/3$. Worse, the Main Conjecture for MDS codes implies $q \geq n-2$. Since VSS in the scenario we consider here requires an LSSS that is t-rejecting and $n - 2t$-accepting, the setting $n = 3t + 1$ would mean that the LSSS is required to be t-rejecting and $t + 1$-accepting. In other words, it would be a threshold scheme, and the MDS argument above applies. Similar reasoning applies to different scenarios of MPC.

Some remarks on ramp schemes are in order. An (a, b)-ramp scheme is in some sense as a threshold secret sharing scheme. It is a-rejecting (no information) and b-accepting (full information). However, $b > a + 1$ is allowed. This means that for sets whose size is between a and b anything might happen, including partial information. Ramp schemes may have higher information rate than secret sharing schemes; the size of a secret may be larger than the size of a share. It seems that all known *linear* ramp schemes required relatively large fields of definition, just as ordinary threshold secret sharing schemes. So it seems that our results may also be viewed as bearing on the theory of ramp schemes per se. We also show how to achieve high information rate, see Section 4. Finally, we also indicate in this paper how general coding theory arguments combined with (extensions of) known relationships between coding and secret sharing allow for a discussion of the ramp schemes claimed above (neglecting strong multiplication) without explicit reference to the algebraic geometric framework.

Mathematically, our construction is inspired by Goppa's algebraic geometric error correcting codes [23]. In particular, shares are defined by evaluating a function on the points of a curve, where the function is chosen from an appropriate Riemann-Roch space. However, several issues that are immaterial for coding theory play a role in the present context and indeed influence the definition, the analysis and the choice of parameters of the new scheme presented here. Shamir's scheme may be viewed as the genus 0 case.

Earlier work on secret sharing and secure computation that relies on techniques from algebraic number theory and/or algebraic geometry includes [17,13,12,9]. Earlier applications of algebraic curves to other areas in information theoretically secure cryptography include those to authentication codes (see e.g. [2,38,26,39]), and cover-free families and broadcast exclusion [25]. In general, algebraic geometry plays an increasingly important role in combinatorics, theoretical computer science and applied mathematics.

Before quantifying the possible advantages and trade-offs of the new scheme, note that trade-offs between communication complexity and corruption tolerance have been studied before. Indeed, Franklin and Yung [19] have shown that with a lower but still linear (in n) corruption bound, the same computation can be performed on many different inputs for the price of one such computation in the standard, classical case. Recently [11] it has been shown that a single, stand-alone secure multiplication can be performed with linear instead of quadratic communication (in n). Nevertheless, these results rely in an essential way on secret sharing techniques over a field of size at least n.

Let $N_q(g)$ denote the maximal number of $\mathbb{F}_q$-rational points on a genus g curve C defined over $\mathbb{F}_q$. The Hasse-Weil bound states that $q + 1 - 2g\sqrt{q} \leq N_q(g) \leq q + 1 + 2g\sqrt{q}$. Note that if C is a plane curve then the bound becomes $q + 1 - (d - 1)(d - 2)\sqrt{q} \leq N_q(g) \leq q + 1 + (d - 1)(d - 2)\sqrt{q}$. However, curves in higher dimensional spaces can have many more points.

The theoretical upper bound is an additional $2g\sqrt{q}$ players compared to using Shamir's scheme in secure computation, while lowering the maximal tolerable number of corrupted players by an additive factor at most $4g/3$.

Viewed from a different angle, and using the García-Stichtenoth curves [21], a family of non-plane curves with many rational points and celebrated for their optimal ratio between genus and number of rational points, one can achieve the following asymptotic bound. For each ϵ with $0 < \epsilon < \frac{1}{6}$, there is a finite field $\mathbb{F}_q$ such that for infinitely many n there exists a scheme with strong multiplication and with $(\frac{1}{3} - \epsilon) \cdot n \leq t < \frac{1}{3} \cdot n$. In particular all sets of size at least $n - 2t$ are accepted. This example is detailed at the end of this paper, together with other examples.

Note that there exists theoretically efficient constructions of the curves of García and Stichtenoth. Efficient construction means here that one can also efficiently work with the relevant Riemann-Roch spaces. There are a host of classes of curves with many rational points known from coding theory that allow for efficient construction. We do not further address computational issues here. This is deferred to the full version.

The results in this paper focus on application to error-free unconditionally secure MPC in the secure channels model, with synchronous communication and in the presence of an active adversary. The results can be adapted to other models of MPC or just to plain secret sharing or ramp schemes, e.g., in the context of secure and private storage. In ongoing work generalizations to higher dimensional varieties are studied [6] as well as the case of MPC in the broadcast model [7].

2 Preliminaries

This section contains some basic definitions and conventions about linear secret sharing and about algebraic curves over finite fields, as well as some relevant facts. The definitions concerning linear secret sharing below are slight adaptations of definitions from [15].

An adversary structure $\mathcal{A}$ on a finite player set $\mathcal{U}$ is a collection of subsets of $\mathcal{U}$ with the property that $A' \in \mathcal{A}$ and $A \subset A'$ implies $A \in \mathcal{A}$. An adversary structure is $Q3$ if for all $A, A', A'' \in \mathcal{A}$ it holds that $A \cup A' \cup A''$ is a proper subset of $\mathcal{U}$. $\mathcal{A}_{t,n}$ is the adversary structure consisting of all sets $A \subset \mathcal{U}$ of size at most t. The access structure $\Gamma_{r,n}$ consists of all sets $B \subset \mathcal{U}$ of size at least r.

A linear secret sharing scheme (LSSS) $\mathcal{M}$ over $\mathbb{F}_q$ on the player set $\mathcal{U}$ is given by a positive integer e, a sequence $V_1, \ldots, V_n$ of subspaces of the e-dimensional $\mathbb{F}_q$-vector space $\mathbb{F}_q^e$, and a non-zero vector $u \in \mathbb{F}_q^e$. Let V_A denote $\sum_{i \in A} V_i$, the $\mathbb{F}_q$-subspace spanned by all the V_i with $i \in A$. The access structure $\Gamma(\mathcal{M})$ of $\mathcal{M}$ consists of all sets $B \subset \mathcal{U}$ with $u \in V_B$. We set $u = (1, 0, \ldots, 0) \in \mathbb{F}^e$, without loss of generality. The structure $\mathcal{A}(\mathcal{M})$ consists of the sets $A \subset \mathcal{U}$ with $A \notin \Gamma(\mathcal{M})$. An LSSS as defined here is essentially a monotone span program [24]. An LSSS $\mathcal{M}$ is said to reject a given adversary structure $\mathcal{A}$ defined on $\mathcal{U}$ if $\mathcal{A} \subset \mathcal{A}(\mathcal{M})$. If B is a non-empty set with $B \subset \mathcal{U}$, then $\mathcal{M}_B$ denotes the LSSS on the player set B given by restricting to those V_i with $i \in B$. An ideal LSSS is one in which all V_i have dimension 1 and where for each i there is B in $\Gamma(\mathcal{M})$ that is minimal with respect to inclusion and for which $i \in B$.

Secret sharing based on an LSSS works in essence as follows. Suppose that bases for the V_i's are fixed. Let $s \in \mathbb{F}_q$ be a "secret value." Choose a random linear map $\phi : \mathbb{F}_q^e \to \mathbb{F}_q$ subject to $\phi(u) = s$, and give $\phi_{|V_i}$ to player i, i.e., the action of ϕ on each of the chosen basis vectors of V_i. It holds that $\{\phi_{|V_i}\}_{i \in A}$ determines the secret s uniquely if and only if $A \in \Gamma(\mathcal{M})$, and $\{\phi_{|V_i}\}_{i \in A}$ gives no information about s in all other cases, i.e., when $A \in \mathcal{A}(\mathcal{M})$. Note that by basic linear algebra $A \in \mathcal{A}(\mathcal{M})$ if and only if there exists a linear map $\kappa : \mathbb{F}_q^e \to \mathbb{F}_q$ such that κ vanishes on V_A, i.e., $\kappa_{|V_A} \equiv 0$, but $\kappa(u) = 1$.

For elements $x, \in \mathbb{F}_q^e$, let $(x_1, \ldots, x_e)$, $(y_1, \ldots, y_e)$ denote their respective coordinate vectors with respect to the standard basis. $x \otimes y$ denotes the vector with coordinates $(\ldots, x_i \cdot y, \ldots) \in \mathbb{F}_q^{e^2}$. Let $\widehat{V_i}$ denote the subspace $V_i \otimes V_i \subset \mathbb{F}_q^{e^2}$, i.e., the $\mathbb{F}_q$-vector space spanned by all elements of the form $x \otimes y$ with $x, y \in V_i$. $\widehat{V_B}$ denotes $\mathbb{F}_q \langle \{\widehat{V_i}\}_{i \in B} \rangle$, and $\widehat{u}$ denotes $u \otimes u$. For given LSSS $\mathcal{M}$, the LSSS $\widehat{\mathcal{M}}$ be defined by the tuple $(\mathbb{F}_q, \widehat{V_1}, \ldots, \widehat{V_n}, \widehat{v})$. $\mathcal{M}$ is said to have the multiplication property if $\widehat{u} \in V_{\mathcal{U}}$. $\mathcal{M}$ has the strong multiplication property with respect to an adversary structure $\mathcal{A}$ (on $\mathcal{U}$) if the following holds.

1. $\mathcal{M}$ rejects the adversary structure $\mathcal{A}$.
2. For all $B \subset \mathcal{U}$ with $B = \mathcal{U} \setminus A$ for some $A \in \mathcal{A}$, $\widehat{\mathcal{M}}_B$ has multiplication.

It is not hard to see that strong multiplication implies that if $A, A' \in \mathcal{A}$, then $\mathcal{U} \setminus A \cup A' \in \Gamma(\mathcal{M})$. Thus, in particular, in order for an LSSS to have strong multiplication with respect to an adversary structure $\mathcal{A}$, it must be so that $\mathcal{A}$ is $Q3$.[2]

It can be shown that for all finite fields $\mathbb{F}_q$ and for all $Q3$ adversary structures $\mathcal{A}$ there is an LSSS with strong multiplication. In general, however, the dimension

---

[2] In [10] it is shown how strong multiplication enables efficient error correction.

may be very large. The standard example for an ideal LSSS is Shamir's scheme. If $t < n/3$ in this scheme, it has strong multiplication with respect to $\mathcal{A}_{t,n}$. Note that $\Gamma(\mathcal{M}) = \Gamma_{t+1,n}$ and that it requires $q > n$.

In analogy to Shamir's scheme, in the case of an ideal LSSS it is sufficient to prove that "for each set $A \in \mathcal{A}$, the pair-wise local products of two vectors of shares belonging to the set $B = \mathcal{U} \setminus A$ jointly uniquely determine the product of the secrets." This is then by linear combination as a consequence. These facts are used implicitly when arguing about strong multiplication in the sequel.

As an aside we mention that it is known [15] how to efficiently enforce the multiplication property on all relevant LSSS; the dimension only goes up by a multiplicative factor 2. In the much more demanding case of strong multiplication the general question whether it can always be efficiently enforced on all relevant LSSS is completely open.

As indicated earlier in this paper, using the techniques from [15] one can construct efficient MPC protocols for the MPC scenario we consider in this paper from an LSSS that satisfies strong multiplication with respect to a $Q3$ adversary structure $\mathcal{A}$. Note that in the present paper we are using a slightly generalized definition of the adversary structure of an LSSS and what it means to satisfy strong multiplication with respect to it: in [15] the adversary structure is always $\mathcal{A}(\mathcal{M})$, and strong multiplication is always defined with respect to that structure only. In the present paper we have refined these notions, and allow for the definition to apply to an adversary structure $\mathcal{A}$ contained in $\mathcal{A}(\mathcal{M})$. This does not make any essential difference. [3]

We now give a quick overview of basics on algebraic geometry. In Section 3 we briefly point out how part of our result (i.e, neglecting strong multiplication) can be appreciated if one accepts some general results about algebraic geometric error correcting codes and (an extension of) a known connection between codes and secret sharing.

Let C be a smooth, projective, absolutely irreducible curve defined over $\mathbb{F}_q$, and let g denote the genus of C. Let $\overline{\mathbb{F}}_q$ denote the algebraic closure of $\mathbb{F}_q$. A plane such curve can be represented by some polynomial $F[X,Y] \in \mathbb{F}_q[X,Y]$ that is irreducible in $\overline{\mathbb{F}}_q[X,Y]$. The affine part of the curve is defined as the set of points $P \in \overline{\mathbb{F}}_q^2$ such that $F(P) = 0$. By taking its projective closure, which amounts to introducing an extra variable, homogenizing the polynomial and considering the zeroes in the two-dimensional projective space $\mathbb{P}^2(\overline{\mathbb{F}}_q)$, one obtains the entire curve. More generally, curves defined over $\mathbb{F}_q$ is the "set of zeroes" in $\mathbb{P}^m(\overline{\mathbb{F}}_q)$ of a homogeneous ideal $I \subset \mathbb{F}_q[X_0, \ldots, X_m]$, where I is such that its function field has transcendence degree 1 over the ground field, i.e., it is a one dimensional variety. Smoothness concerns not simultaneously vanishing partial (formal) derivatives.

$\overline{\mathbb{F}}_q(C)$ denotes the function field of the curve. Very briefly, it consists of all fractions of polynomials $a, b \in \overline{\mathbb{F}}_q[X_0, \ldots, X_m]$, $b \notin I$, such that both are

---

[3] For zero-error VSS from LSSS the condition that $\mathcal{A}$ is $Q3$ and that for all $A, A' \in \mathcal{A}$, $\mathcal{U} \setminus A \cup A' \in \Gamma(\mathcal{M})$ must be explicitly made. In case of strong multiplication this condition is implied, as pointed out.

homogeneous of the same degree, under the equivalence relation that $a/b \equiv a'/b'$ if $ab' \equiv a'b \bmod I$. The elements can be viewed as maps from the curve to $\overline{\mathbb{F}}_q$, and they have at most a finite number of poles and zeroes, unless it is the zero function. Their "multiplicities add up to zero."

Since C is smooth at each point $P \in C$ by assumption, the local ring $\mathcal{O}_P(C)$ of functions $f \in \overline{\mathbb{F}}_q(C)$ that are well-defined at P (equivalently, the ones that do not have a pole at P) is a discrete valuation ring. Thus, at each $P \in C$, there exists $t \in \overline{\mathbb{F}}_q(C)$ (a uniformizing parameter) such that $t(P) = 0$ and each $f \in \mathcal{O}_P(C)$ can be uniquely written as $f = u \cdot t^{\nu_P(f)}$. Here, $u \in \mathcal{O}_P(C)$ is a unit (i.e., $u(P) \neq 0$), and $\nu_P(f)$ is a non-negative integer. This valuation ν_P extends to all of $\overline{\mathbb{F}}_q(C)$, by defining $\nu_P(f) = -\nu_P(1/f)$ if f has a pole at P.

A divisor is a formal sum $\sum_{P \in C} m_p \cdot (P)$ with integer coefficients m_p taken over all points P of the curve C. Divisors are required to have finite support, i.e., they are zero except possibly at finitely many points. The divisor of $f \in \mathbb{F}_q(C)$ is defined as $\operatorname{div}(f) = \sum_{P \in C} \nu_P(f) \cdot (P)$. It holds that $\deg \operatorname{div}(f) = 0$. The degree $\deg D$ of a divisor D is the sum $\sum_{P \in C} m_P \in \mathbb{Z}$ of its coefficients m_P.

The Riemann-Roch space associated with a divisor D is defined as $\mathcal{L}(D) = \{f \in \overline{\mathbb{F}}_q(C) | \operatorname{div}(f) + D \geq 0\} \cup \{0\}$. This is an $\overline{\mathbb{F}}_q$-vector space. The (partial) ordering "$\geq$" refers to the comparison of integer vectors and declaring one larger than the other if this holds coordinate-wise. Its dimension is denoted $\ell(D)$. This dimension is equal to 0 if $\deg D < 0$. The Riemann-Roch Theorem is concerned with the dimensions of those spaces. It says that $\ell(D) - \ell(K - D) = \deg D + 1 - g$. Here K is a canonical divisor. These are the divisors K of degree $2g - 2$ and $\ell(K) = g$. It follows immediately that $\ell(D) = \deg D + 1 - g$ if $\deg D$ is large enough, i.e., at least $2g - 1$. This consequence suffices for the purposes in this paper.

An $\mathbb{F}_q$-rational point on C is one whose projective coordinates can be chosen in $\mathbb{F}_q$. Rational point shall mean $\mathbb{F}_q$-rational point. Note that non-plane curves can in principle harbor many more rational points than plane curves.

The divisors on C defined over $\mathbb{F}_q$ (or $\mathbb{F}_q$-rational divisors) are those that are invariant under the Galois group $\operatorname{Gal}(\overline{\mathbb{F}}_q/\mathbb{F}_q)$. This includes the divisors whose support consists of rational points only. In this paper all divisors are rational.

If D is rational, then $\mathcal{L}(D)$ admits a basis defined over $\mathbb{F}_q$. *In this paper $\mathcal{L}(D)$ is tacitly restricted to the $\mathbb{F}_q$-part of $\mathcal{L}$, i.e., the $\mathbb{F}_q$-linear span of such basis, or equivalently, the subspace of $\mathcal{L}(D)$ fixed under $\operatorname{Gal}(\overline{\mathbb{F}}_q/\mathbb{F}_q)$. This has $q^{\ell(D)}$ elements. As a consequence of this convention, if $P \in C$ is rational and if $f \in \mathcal{L}(D)$, then $f(P) \in \mathbb{F}_q$.*

For introductions to algebraic geometry, see for instance [20,27], or or textbooks that place special emphasis on curves over $\mathbb{F}_q$ such as [37,36]. For an accessible, high level overview of the technicalities sketched above see for instance [28].

3   Main Result

As before, let C be a smooth projective absolutely irreducible curve defined over $\mathbb{F}_q$, and let g denote its genus.

Let

$$Q, P_0, P_1, \ldots, P_n$$

be any distinct rational points on C, possibly exhausting all rational points of C. Let t be any fixed integer with $1 \leq t < n - 2g$. The divisor D is defined as

$$D = (2g + t) \cdot (Q).$$

Thus, it has support Q and degree $2g + t$.[4]

The claimed LSSS $\mathcal{M}$ works as follows.

Let $s \in \mathbb{F}_q$ be a secret value.

Select

$$f \in \mathcal{L}(D)$$

at random, subject to the constraint

$$f(P_0) = s.$$

There is always at least one such f, since $\mathcal{L}(D)$ contains in particular the constant functions. By the convention made earlier on, the choice of f is restricted to the $\mathbb{F}_q$-part of $\mathcal{L}(D)$, which is a vector space over $\mathbb{F}_q$ of dimension $g + t + 1$. So the random choice of f conditioned on the secret s consumes $g + t$ random field elements from $\mathbb{F}_q$.

Now define

$$f(P_1) = s_1 \in \mathbb{F}_q, \ldots, f(P_n) = s_n \in \mathbb{F}_q$$

as the shares. By definition of the divisor D and by the definition of the space $\mathcal{L}(D)$, the functions $f \in \mathcal{L}(D)$ only have a pole in Q. Thus the values $f(P_i)$ are always well-defined.

The construction above may be viewed as a combination of Goppa's algebraic-geometry error correcting codes [23] and Massey's construction of linear secret sharing schemes from error-correcting codes [29,30]. The latter result allows to study privacy and reconstruction in terms of properties of the underlying linear codes and their duals. More precisely, one observes that their respective minimum distances imply bounds on the parameters of a ramp scheme. This can be combined with known properties of algebraic-geometric codes to obtain bounds on privacy and reconstruction as we give below. Nevertheless, a slight generalization of the results of Massey is needed to be able to analyze our high information rate ramp scheme from Section 4. We have, however, chosen a self-contained presentation whose details can all be directly understood from the corollary of Riemann-Roch we stated before. Moreover, in order to prove the strong multiplication property as we do it is essential to be able to address the explicit structure of our secret sharing scheme.

---

[4] Any other divisor of that degree would do as well. However, with a small support of D the maximum value of n is greater.

LEMMA 1. *Let E be a divisor on C that is defined over $\mathbb{F}_q$, and suppose that $\ell(E) > 0$. Then each $f \in \mathcal{L}(E)$ is uniquely determined by evaluations of f on any $degE + 1$ rational points on C outside the support of E.*

PROOF. This is a standard argument. First note that for each $f \in \mathcal{L}(E)$ and for each rational point P on the curve outside the support of E the value $f(P)$ is well-defined, as f certainly has no poles there. This follows from the definition of $\mathcal{L}(E)$.

Write d for the degree of E, and consider rational points $Q_1, \ldots, Q_{d+1}$ on the curve, outside the support of E. The map

$$\phi : \mathcal{L}(E) \longrightarrow \mathbb{F}_q^{d+1},$$

$$f \mapsto (f(Q_1), \ldots, f(Q_{d+1}))$$

is an injective linear map of $\mathbb{F}_q$-vector spaces. Indeed, if $f, h \in \mathcal{L}(E)$ and $\phi(f) = \phi(h)$, then $f - h \in \mathcal{L}(E - (Q_1 + \ldots + Q_{d+1})) \subset \mathcal{L}(E)$. Here it is used that the support of E is disjoint from the Q_i's. The degree of the divisor $E - (Q_1 + \ldots + Q_{d+1})$ is negative, so $f = h$.

Note that, just as with Lagrange interpolation by polynomials, this interpolation is linear in the following sense. If Q_0 is a rational point on the curve different from the Q_i's and outside the support of E, then there are coefficients $\lambda_i \in \mathbb{F}_q$ such that for all $f \in \mathcal{L}(E)$

$$f(Q_0) = \sum_{i=1}^{d+1} \lambda_i \cdot f(Q_i).$$

Concretely, since ϕ is injective, there exists a surjective linear map

$$\chi : \mathbb{F}_q^{d+1} \longrightarrow \mathcal{L}(E)$$

such that $\chi \circ \phi$ is the identity on $\mathcal{L}(E)$. So

$$f = \chi(f(Q_1), \ldots, f(Q_{d+1})),$$

and

$$f(Q_0) = \chi_0 \circ \chi(f(Q_1), \ldots, f(Q_{d+1})),$$

where the linear map χ_0 is defined as

$$\chi_0 : \mathcal{L}(E) \longrightarrow \mathbb{F}_q$$

$$f \mapsto f(Q_0). \hspace{3cm} \triangle$$

PROPOSITION 1. $\mathcal{A}(\mathcal{M})$ *consists of all sets $A \subset \{1, \ldots, n\}$ such that*

$$\ell(D - (P_0 + \sum_{i \in A} P_i)) < \ell(D - (\sum_{i \in A} P_i)).$$

Equivalently, $\Gamma(\mathcal{M})$ consists of all sets $B \subset \{1, \ldots, n\}$ such that

$$\ell(D - (P_0 + \sum_{i \in B} P_i)) = \ell(D - (\sum_{i \in B} P_i)).$$

PROOF. Clearly, $A \in \mathcal{A}$ if and only if there exists $k \in \mathcal{L}(D)$ such that $k(P_i) = 0$ for all $i \in A$ and $k(P_0) = 1$. This is a general fact about linear secret sharing schemes, which is easily proved by linear algebra.

It holds generally that $\mathcal{L}(F) \subset \mathcal{L}(F')$ if $F \leq F'$. This follows immediately from the definitions. Since the support of D is disjoint from the P_i's, it therefore holds that

$$\mathcal{L}(D - (P_0 + \sum_{i \in A} P_i)) \subset \mathcal{L}(D - (\sum_{i \in A} P_i)) \subset \mathcal{L}(D).$$

All functions k' in the difference

$$\mathcal{L}(D - (\sum_{i \in A} P_i)) \setminus \mathcal{L}(D - (P_0 + \sum_{i \in A} P_i)),$$

if any, satisfy $k' \in \mathcal{L}(D)$, $k'(P_0) \neq 0$, and $k'(P_i) = 0$ for all $i \in A$. By normalizing at P_0, the desired function k is obtained. Clearly, the difference between those spaces is non-empty if and only if their dimensions differ. $\triangle$

COROLLARY 1. $\mathcal{A}_{t,n} \subset \mathcal{A}(\mathcal{M})$.

PROOF. If $|A| = t$, then

$$\deg(D - (P_0 + \sum_{i \in A} P_i)) = 2g - 1,$$

$$\deg(D - (\sum_{i \in A} P_i)) = 2g.$$

Therefore,

$$g = \ell(D - (P_0 + \sum_{i \in A} P_i)) < g + 1 = \ell(D - (\sum_{i \in A} P_i)).$$ $\triangle$

COROLLARY 2. $\Gamma_{2g+t+1,n} \subset \Gamma(\mathcal{M})$.

PROOF. First note that by definition $n \geq 2g + t + 1$. If $B \subset \{1, \ldots, n\}$ is a set of size $2g + t + 1$, then $\ell(D - \sum_{i \in B} P_i) = 0$, since the argument is a divisor of negative degree. Thus, $0 = \ell(D - (P_0 + \sum_{i \in B} P_i)) \leq \ell(D - \sum_{i \in B} P_i) = 0$ $\triangle$

PROPOSITION 2. $\mathcal{M}$ has strong multiplication with respect to $\mathcal{A}_{t,n}$ if $3t < n - 4g$. $\mathcal{M}$ has multiplication if $2t < n - 4g$.

PROOF. We only treat the strong multiplication case. Let $f, h \in \mathcal{L}(D)$. Using the basic fact that $\mathrm{div}(fh) = \mathrm{div}f + \mathrm{div}h$, it follows that

$$0 \leq (\mathrm{div}f + D) + (\mathrm{div}h + D) = \mathrm{div}(fh) + 2D.$$

Hence

$$f \cdot h \in \mathcal{L}(2D).$$

Thus M has strong multiplication if

$$n - t > \deg(2D) = 4g + 2t,$$

as follows by application of Lemma 1. Indeed, let B be any set with $B \subset \{1, \ldots, n\}$ and $|B| = 4g + 2t + 1$. Define linear maps

$$\hat{\phi}_0 : \mathcal{L}(2D) \longrightarrow \mathbb{F}_q$$

$$\hat{f} \mapsto \hat{f}(P_0),$$

and

$$\hat{\phi} : \mathcal{L}(2D) \longrightarrow \mathbb{F}_q^{4g+2t+1}$$

$$\hat{f} \mapsto (\hat{f}(P_i))_{i \in B},$$

and

$$\hat{\chi} : \mathbb{F}_q^{4g+2t+1} \longrightarrow \mathcal{L}(2D)$$

such that $\hat{\chi} \circ \hat{\phi}$ is the identity on $\mathcal{L}(2D)$.

Then, for all $f, h \in \mathcal{L}(D)$, it holds that

$$s \cdot s' = \hat{\phi}_0 \circ \hat{\chi}((s_i \cdot s_i')_{i \in B}),$$

where

$$s = f(P_0), s' = h(P_0), \text{ and for all } i \in B, s_i = f(P_i), s_i' = h(P_i). \qquad \triangle$$

An alternative proof of the strong multiplication property can be based on the observation that this LSSS has strong multiplication with respect to an adversary structure $\mathcal{A} \subset \mathcal{A}(\mathcal{M})$ if for all $A \in \mathcal{A}$ it holds that

$$\ell(2D - (P_0 + \sum_{i \in B} P_i)) = \ell(2D - (\sum_{i \in B} P_i)),$$

where $B = \{1, \ldots, n\} \setminus A$.

For completeness we show how strong multiplication linearizes the problem of recovering the secret in the presence of corrupted shared. It is a special case of the more general technique given in [10]. But also note that known techniques for decoding algebraic-geometry codes apply here. In any case, assume $t < (n - 4g)/3$. Let $u = (f(P_1), \ldots, f(P_n))$ be a share vector for the secret $s = f(P_0)$, with $f \in \mathcal{L}(D)$. Let $e \in \mathbb{F}_q^n$ be a vector of Hamming-weight at most t, i.e., its number of nonzero coordinates is at most t. For any $P \in \{P_1, \ldots, P_n\}$ write $c = u + e \in \mathbb{F}_q^n$, and write $c(P)$ for the coordinate of c "that corresponds to P." Now solve the following linear equation system $\forall P \in \{P_1, \ldots, P_n\} \; : \; h(P) = c(P) \cdot k(P)$, $k(P_0) = 1$, where $h \in \mathcal{L}(2D)$, $k \in \mathcal{L}(D)$. These are $n + 1$ equations in $(3g + 2t + 1) + (g + t + 1) = 4g + 3t + 2$ variables. There always exists a solution, and each solution $(h, k) \in \mathcal{L}(2D) \times \mathcal{L}(D)$ satisfies $h(P_0) = s$. This system of linear equations can be efficiently set up if the underlying curve supports efficient algorithms.

Some final remarks about the basic construction are in order. Often one may re-define D so that one extra player is supported. Indeed, by using the Weak Approximation Theorem (see [36]) an equivalent D' can be found whose support really lies in an extension field, thereby making all rational points available for players. There is an alternative approach to "winning points" in which all of the points in the support of any (positive divisor) D can be used as extra players. This involves redefining the embedding of $\mathcal{L}(D)$ by scaling at each point in the support of D with an appropriate power of a uniformizing parameter at that point.

4   Construction of Ramp Schemes with Large Information Rate

Instead of taking a single point P_0, consider taking a sequence of distinct points $P_0^1, \ldots, P_0^\ell$, disjoint from Q and $P_1, \ldots, P_n$ and where $2g + t + \ell \leq n$. It is not hard to show, by arguments virtually identical to the ones used before, that if ones takes $2g + t + \ell - 1$ instead of $2g + t$ as the degree of D, then the secrets may in fact be chosen arbitrarily from $\mathbb{F}_q^\ell$ instead of $\mathbb{F}_q$. The share size doesn't change. Thus it is a $(t, 2g + t + \ell, n)$-ramp scheme where each share is in $\mathbb{F}_q$, but where the secret can be chosen in $\mathbb{F}_q^\ell$. Strong multiplication and efficient error recovery can also be appropriately carried over to this variation.

5   Achievable Parameters

Below concrete numerical examples are given, using well-known classes of curves. The genus 0 case of our construction collapses to Shamir's scheme. As a first example with an advantage compared to known technique, consider elliptic curves, i.e., $g = 1$. It is well-known that $N_q(1) = q + 1 + \lfloor 2\sqrt{q} \rfloor$, unless a certain condition on q and the characteristic of p of $\mathbb{F}_q$ holds,[5] in which case this maximum number is just one less. Compared to using Shamir's scheme in secure computation, our scheme supports, over the same finite field $\mathbb{F}_q$, an additional $2\sqrt{q} - 1$ players. The maximal level of corruption tolerance is decreased by at most an additive

---

[5] p divides $\lfloor 2\sqrt{q} \rfloor$ and $\mathbb{F}_q$ is an extension of degree at least 3 over $\mathbb{F}_p$.

factor of 2 (just 1 if the number of players n is such that $n - 1$ is not divisible by 3).

Here is one example based on higher genus curves. Consider the Hermitian curves $X^{\sqrt{q}+1} + Y^{\sqrt{q}+1} = Z^{\sqrt{q}+1}$ over $\mathbb{F}_q$, where q is a square. These well-known curves hit the Hasse-Weil upper bound. The genus of such curves is equal to $\frac{1}{2}(q - \sqrt{q})$, and their number of $\mathbb{F}_q$-rational points is $q\sqrt{q} + 1$.

For example, working over $\mathbb{F}_{64}$, more than 500 players are supported and more than 130 corruptions are tolerated. In comparison, in Shamir's scheme q would be greater than 500 (instead of 64), but almost 40 more corruptions could be tolerated.

Finally, the well-known (non-plane) curves of García and Stichtenoth [21] from coding theory, prove useful here as well. Let q be a square. Then there is a family of curves $\{C_m\}_{m \in \mathbb{Z}_{>0}}$ defined over $\mathbb{F}_q$ such that

$$\#C_m(\mathbb{F}_q) \geq (q - \sqrt{q})\sqrt{q}^{m-1} \text{ and } g(C_m) \leq \sqrt{q}^m.$$

So the ratio here is

$$\frac{g(C_m)}{\#C_m(\mathbb{F}_q))} \leq \frac{1}{\sqrt{q} - 1}.$$

Consider a finite field $\mathbb{F}_q$ with q a square. Let t be chosen maximal such that

$$t < \left(\frac{1}{3} - c\right) \cdot n,$$

where $c = \frac{4}{3(\sqrt{q}-1)} < \frac{1}{3}$. This means that $q \geq 49$. It follows that for each ϵ with $0 < \epsilon < \frac{1}{6}$, there is a finite field $\mathbb{F}_q$ such that for infinitely many n there exists a scheme with strong multiplication and with $\left(\frac{1}{3} - \epsilon\right) \cdot n \leq t < \frac{1}{3} \cdot n$. Note that in particular all sets of size at least $n - 2t$ are accepted.

Note that there exists theoretically efficient constructions of the curves of García and Stichtenoth. Efficient construction means here that one can also efficiently work with the relevant Riemann-Roch spaces. There are a host of classes of curves with many rational points known from coding theory that allow for efficient construction. We do not further address computational issues here. This is deferred to future work. See also a table with known values of $N_q(g)$, see [22].

Acknowledgements

Ronald Cramer makes the following acknowledgments. Thanks to Ivan Damgaard, Iwan Duursma, Bas Edixhoven, Serge Fehr, Gerard van der Geer, Maribel González Vasco, Robbert de Haan, Javier Herranz, Martin Hirt, Robin de Jong, Keith Martin, Hendrik Lenstra, Carles Padró, Victor Shoup, Douglas Stinson and Chaoping Xing for useful discussions, inspiration, and suggestions. Special thanks to Hendrik Lenstra for encouraging study of secret sharing from the point of view of algebraic geometry in the first place. Hao Chen's work was supported by National Natural Science Foundation Distinguished Young Scholar Grant 10225106 and by Grant 60542006 (NNSF, China).

References

1. M. Ben-Or, S. Goldwasser, and A. Wigderson. Completeness theorems for non-cryptographic fault-tolerant distributed computation. Proceedings of STOC 1988, ACM Press, pp. 1–10.
2. J. Bierbrauer. Universal Hashing and Geometric Codes. Designs, Codes and Cryptography, vol. 11, pp. 207–221, 1997.
3. G. R. Blakley. Safeguarding cryptographic keys. Proceedings of National Computer Conference '79, volume 48 of AFIPS Proceedings, pp. 313–317, 1979.
4. D. Chaum, C. Crépeau, and I. Damgaard. Multi-party unconditionally secure protocols. Proceedings STOC 1988, ACM Press, pp. 11–19.
5. H. Chen. Linear secret sharing from algebraic-geometric codes. Merged with [8].
6. H. Chen, R. Cramer, C. Ding, and C. Xing. Secret sharing and secure multi-party compuation from projective algebraic subsets. Work in progress.
7. H. Chen, R. Cramer, S. Goldwasser, V. Vaikuntanathan, R. de Haan. Threshold MPC in the Rabin-Ben Or broadcast model unconditionally secure against corrupt minorities based on general error correcting codes. Work in progress.
8. R. Cramer. Algebraic geometric secret sharing and secure computation over small fields. Merged with [5].
9. R. Cramer, S. Fehr, and M. Stam. Blackbox secret sharing from primitive sets in algebraic number fields. Proceedings of CRYPTO 2005, volume 3621 of LNCS, pp. 344–360, Springer-Verlag, 2005.
10. R. Cramer, V. Daza, I. Gracia, J. Jimenez Urroz, G. Leander, J. Martí-Farré, and C. Padró. On codes, matroids and secure multi-party computation from linear secret sharing schemes. Proceedings of CRYPTO 2005, volume 3621 of LNCS, pp. 327–343, Springer-Verlag, 2005.
11. R. Cramer and R. de Haan. Atomic Secure Multi-Party Multiplication with Low Communication. Manuscript, 2004.
12. R. Cramer, S. Fehr, Y. Ishai, and E. Kushilevitz. Efficient multi-party computation over rings. Proceedings of EUROCRYPT 2003, volume 2656 of LNCS, pp. 596–613, Springer-Verlag, 2003.
13. R. Cramer and S. Fehr. Optimal black-box secret sharing over arbitrary Abelian groups. Proceedings of CRYPTO 2002, volume 2442 of LNCS, pp. 272–287, Springer-Verlag, 2002.
14. R. Cramer, I. Damgaard, and S. Dziembowski. On the complexity of verifiable secret sharing and multi-party computation. Proceedings of STOC 2000, ACM Press, pp. 325–334, 2000.
15. R. Cramer, I. Damgaard, and U. Maurer. General secure multi-party computation from any linear secret sharing scheme. Proceedings of EUROCRYPT 2000, volume 1807 of LNCS, pp. 316–334, Springer-Verlag, 2000.
16. R. Cramer, I. Damgaard, S. Dziembowski, M. Hirt, and T. Rabin. Efficient multi-party computations secure against an adaptive adversary. In Proceedings of EUROCRYPT 1999, volume 1592, Springer-Verlag, 1999.
17. Y. Desmedt and Y. Frankel. Homomorphic zero-knowledge threshold schemes over any finite abelian group. SIAM Journal of Discrete Mathematics, 7 (1994), pp. 667-679.
18. R. J. McEliece and D. V. Sarvate. On sharing secrets and Reed-Solomon codes. Comm. of the ACM, 22(11), November 1979, pp. 612–613.
19. M. Franklin and M. Yung. Communication complexity of secure computation. Proceedings of STOC 1992, ACM Press.

20. W. Fulton. Algebraic Curves. Advanced Book Classics, Addission-Wesley.

21. A. García and H. Stichtenoth. On the asymptotic behavior of some towers of function fields over finite fields. J. Number Theory, vol. 61, pp. 248–273, 1996.

22. G. van der Geer and M. van der Vlugt. Tables of curves with many points Mathematics of Computation 69 (2000), 797-810. See also `www.science.uva.nl/~geer` for regular updates.

23. V.D. Goppa. Codes on algebraic curves. Soviet Math. Dokl. 24: 170–172, 1981.

24. M. Karchmer and A. Wigderson. On span programs. Proceedings of the Eigth Annual Structure in Complexity Theory Conference, pp. 102–111, IEEE, 1993.

25. R. Kumar, S. Rajagopalan, and A. Sahai. Coding constructions for blacklisting problems without computational assumptions. Proceedings of CRYPTO 1999, Springer-Verlag, pp. 609–623, 1999.

26. K. Y. Lam, H. X. Wang, and C. Xing. Constructions of authentication codes from algebraic curves over finite fields. IEEE Transactions in Information Theory, Vol. 46, 886-892, 2000.

27. S. Lang. Algebra. Addison-Wesley, 1997.

28. J.H. van Lint. Introduction to Coding Theory. GTM, Springer Verlag.

29. J. L. Massey. Minimal codewords and secret sharing. Proceedings of the 6-th Joint Swedish-Russian Workshop on Information Theory Molle, Sweden, August 1993, pp. 269–279.

30. J. L. Massey. Some applications of coding theory in cryptography. Codes and Ciphers: Cryptography and Coding IV, 1995, pp. 33–47.

31. H. Niederreiter, H. Wang, and C. Xing. Function fields over finite fields and their application to cryptography. In: A. García and H. Stichtenoth (Ed.). Topics in geometry, coding theory and cryptography, Springer Verlag, to appear (2006).

32. T. Rabin and M. Ben-Or. Verifiable secret sharing and multiparty protocols with honest majority. In Proc. ACM STOC 1989, pages 73–85, 1989.

33. A. Shamir. How to share a secret. Comm. of the ACM, 22(11):612–613, 1979.

34. J. Silverman. The Arithmetic of Elliptic Curves. GTM, Springer Verlag.

35. K.W. Shum, I. Aleshnikov, V.P. Kumar, H. Stichtenoth, V. Deolaikar. A low-complexity algorithm for the construction of algebraic-geometric codes better than the Gilbert-Varshamov bound. IEEE Trans. IT 47 (2001), no. 6, 2225–2241.

36. H. Stichtenoth. Algebraic function fields and codes. Springer Verlag, 1993.

37. M. Tsfasman and S. Vladuts. Algebraic-geometric codes. Kluwer, 1991.

38. S. Vladuts. A note on authentication codes from algebraic geometry. IEEE Transactions in Information Theory 44, no. 3, pp. 1342–1345, 1998.

39. C. Xing. Authentication codes and algebraic curves. Proceedings of the 3rd European Congress of Mathematics, Bikhauser, Vol.2, pp. 239-244, 2001.

Automated Security Proofs
with Sequences of Games

Bruno Blanchet and David Pointcheval

CNRS, École Normale Supérieure, Paris
{blanchet, pointche}@di.ens.fr

Abstract. This paper presents the first automatic technique for proving not only protocols but also primitives in the exact security computational model. Automatic proofs of cryptographic protocols were up to now reserved to the Dolev-Yao model, which however makes quite strong assumptions on the primitives. On the other hand, with the proofs by reductions, in the complexity theoretic framework, more subtle security assumptions can be considered, but security analyses are manual. A process calculus is thus defined in order to take into account the probabilistic semantics of the computational model. It is already rich enough to describe all the usual security notions of both symmetric and asymmetric cryptography, as well as the basic computational assumptions. As an example, we illustrate the use of the new tool with the proof of a quite famous asymmetric primitive: unforgeability under chosen-message attacks (UF-CMA) of the Full-Domain Hash signature scheme under the (trapdoor)-one-wayness of some permutations.

1 Introduction

There exist two main frameworks for analyzing the security of cryptographic protocols. The most famous one, among the cryptographic community, is the "provable security" in the reductionist sense [8]: adversaries are probabilistic polynomial-time Turing machines which try to win a game, specific to the cryptographic primitive/protocol and to the security notion to be satisfied. The "computational" security is achieved by contradiction: if an adversary can win such an attack game with non-negligible probability, then a well-defined computational assumption is invalid (e.g., one-wayness, intractability of integer factoring, etc.) As a consequence, the actual security relies on the sole validity of the computational assumption. On the other hand, people from formal methods defined formal and abstract models, the so-called Dolev-Yao [21] framework, in order to be able to prove the security of cryptographic protocols too. However, these "formal" security proofs use the cryptographic primitives as ideal blackboxes. The main advantage of such a formalism is the automatic verifiability, or even provability, of the security, but under strong (and unfortunately unrealistic) assumptions. Our goal is to take the best of each framework, without the drawbacks, that is, to achieve automatic provability under classical (and realistic) computational assumptions.

C. Dwork (Ed.): CRYPTO 2006, LNCS 4117, pp. 537–554, 2006.

The Computational Model. Since the seminal paper by Diffie and Hellman [20], complexity theory is tightly related to cryptography. Cryptographers indeed tried to use $\mathcal{NP}$-hard problems to build secure cryptosystems. Therefore, adversaries have been modeled by probabilistic polynomial-time Turing machines, and security notions have been defined by security games in which the adversary can interact with several oracles (which possibly embed some private information) and has to achieve a clear goal to win: for signature schemes, the adversary tries to forge a new valid message-signature pair, while it is able to ask for the signature of any message of its choice. Such an attack is called an existential forgery under chosen-message attacks [23]. Similarly, for encryption, the adversary chooses two messages, and one of them is encrypted. Then the goal of the adversary is to guess which one has been encrypted [22], with a probability significantly better than one half. Again, several oracles may be available to the adversary, according to the kind of attack (chosen-plaintext and/or chosen-ciphertext attacks [34,35]). One can see in these security notions that computation time and probabilities are of major importance: an unlimited adversary can always break them, with probability one; or in a shorter period of time, an adversary can guess the secret values, by chance, and thus win the attack game with possibly negligible but non-zero probability. Security proofs in this framework consist in showing that if such an adversary can win with significant probability, within reasonable time, then a well-defined problem can be broken with significant probability and within reasonable time too. Such an intractable problem and the reduction will quantify the security of the cryptographic protocol.

Indeed, in both symmetric and asymmetric scenarios, most security notions cannot be unconditionally guaranteed (*i.e.* whatever the computational power of the adversary). Therefore, security generally relies on a computational assumption: for instance, the existence of one-way functions, or permutations, possibly trapdoor. A one-way function is a function f which anyone can easily compute, but given $y = f(x)$ it is computationally intractable to recover x (or any pre-image of y). A one-way permutation is a bijective one-way function. For encryption, one would like the inversion to be possible for the recipient only: a trapdoor one-way permutation is a one-way permutation for which a secret information (the trapdoor) helps to invert the function on any point.

Given such objects, and thus computational assumptions about the intractability of the inversion (without trapdoors), we would like that security could be achieved without any additional assumptions. The only way to "formally" prove such a fact is by showing that an attacker against the cryptographic protocol can be used as a sub-part in an algorithm (the reduction) that can break the basic computational assumption.

Observational Equivalence and Sequence of Games. Initially, reductionist proofs consisted in presenting a reduction, and then proving that the view of the adversary provided by the reduction was (almost) indistinguishable to the view of the adversary during a real attack. Such an indistinguishability was quite technical and error-prone. Victor Shoup [37] suggested to prove it by small changes [11], using a "sequence of games" (a.k.a. the game hopping technique) that the adversary plays, starting from the real attack game. Two consecutive games look either identical, or very close to each other in the view of the adversary, and thus involve a statistical distance, or a computational one. In the final game, the adversary has clearly no chance to win at all. Actually, the modifications of games can be seen as "rewriting rules" of the probability distributions of the variables involved in the games. They may consist of a simple renaming of some variables, and thus to perfectly identical distributions. They may introduce unlikely differences, and then the distributions are "statistically" indistinguishable. Finally, the rewriting rule may be true under a computational assumption only: then appears the computational indistinguishability.

In formal methods, games are replaced with processes using perfect primitives modeled by function symbols in an algebra of terms. "Observational equivalence" is a notion similar to indistinguishability: it expresses that two processes are perfectly indistinguishable by any adversary. The proof technique typically used for observational equivalence is however quite different from the one used for computational proofs. Indeed, in formal models, one has to exploit the absence of algebraic relations between function symbols in order to prove equivalence; in contrast to the computational setting, one does not have observational equivalence hypotheses (*i.e.* indistinguishability hypotheses), which specify security properties of primitives, and which can be combined in order to obtain a proof of the protocol.

Related Work. Following the seminal paper by Abadi and Rogaway [1], recent results [32,18,25] show the soundness of the Dolev-Yao model with respect to the computational model, which makes it possible to use Dolev-Yao provers in order to prove protocols in the computational model. However, these results have limitations, in particular in terms of allowed cryptographic primitives (they must satisfy strong security properties so that they correspond to Dolev-Yao style primitives), and they require some restrictions on protocols (such as the absence of key cycles).

Several frameworks exist for formalizing proofs of protocols in the computational model. Backes, Pfitzmann, and Waidner [5,6,3] have designed an abstract cryptographic library and shown its soundness with respect to computational primitives, under arbitrary active attacks. Backes and Pfitzmann [4] relate the computational and formal notions of secrecy in the framework of this library. Recently, this framework has been used for a computationally-sound machine-checked proof of the Needham-Schroeder-Lowe protocol [38]. Canetti [16] introduced the notion of universal composability. With Herzog [17], they show how a Dolev-Yao-style symbolic analysis can be used to prove security properties of protocols within the framework of universal composability, for a restricted class of protocols using public-key encryption as only cryptographic primitive. Then, they use the automatic Dolev-Yao verification tool ProVerif [12] for verifying protocols in this framework. Lincoln, Mateus, Mitchell, Mitchell, Ramanathan, Scedrov, and Teague [29,30,31,36,33] developed a probabilistic polynomial-time calculus for the analysis of cryptographic protocols. Datta *et al* [19] have designed a computationally sound logic that enables them to prove computational security properties using a logical deduction system. These frameworks can be used to prove security properties of protocols in the computational sense, but except for [17] which relies on a Dolev-Yao prover, they have not been automated up to now, as far as we know.

Laud [26] designed an automatic analysis for proving secrecy for protocols using shared-key encryption, with passive adversaries. He extended it [27] to active adversaries, but with only one session of the protocol. This work is the closest to ours. We extend it considerably by handling more primitives, a variable number of sessions, and evaluating the probability of an attack. More recently, he [28] designed a type system for proving security protocols in the computational model. This type system handles shared- and public-key encryption, with an unbounded number of sessions. This system relies on the Backes-Pfitzmann-Waidner library. A type inference algorithm is sketched in [2].

Barthe, Cerderquist, and Tarento [7,39] have formalized the generic model and the random oracle model in the interactive theorem prover Coq, and proved signature schemes in this framework. In contrast to our specialized prover, proofs in generic interactive theorem provers require a lot of human effort, in order to build a detailed enough proof for the theorem prover to check it.

Halevi [24] explains that implementing an automatic prover based on sequences of games would be useful, and suggests ideas in this direction, but does not actually implement one.

Our prover, which we describe in this paper, was previously presented in [13,14], but in a more restricted way. It was indeed applied only to classical, Dolev-Yao-style protocols of the literature, such as the Needham-Schroeder public-key protocol. In this paper, we show that it can also be used for the proof of security of cryptographic primitives. [13,14] considered only asymptotic proofs. In this paper, we have extended the prover for providing exact security proofs. We also extend it to the proof of authentication properties, while [13,14] considered only secrecy properties. Finally, we also show how to model a random oracle.

Achievements. As in [13,14], our goal is to fill the gap between the two usual techniques (computational and formal methods), but with a direct approach, in order to get the best of each: a computationally sound technique, which an automatic prover can apply. More precisely, we adapt the notion of observational equivalence so that it corresponds to the indistinguishability of games. To this aim, we also adapt the notion of processes: our processes run in time t and work with bit-strings. Furthermore, the process calculus has a probabilistic semantics, so that a measure can be defined on the distinguishability notion, or the observational equivalence, which extends the "perfect indistinguishability": the distance between two views of an adversary. This distance is due to the application of a transformation, which is purely syntactic. The transformations are rewriting rules, which yield a game either equivalent or almost equivalent under a "computational assumption". For example, we define a rewriting rule, which is true under the one-wayness of a specific function. The automatic prover tries to apply the rewriting rules until the winning event, which is executed in the original attack game when the adversary breaks the cryptographic protocol, has totally disappeared: the adversary eventually has a success probability 0. We can then upper-bound the success probability of the adversary in the initial game by the sum of all gaps.

Our prover also provides a manual mode in which the user can specify the main rewriting steps that the prover has to perform. This allows the system to prove protocols in situations in which the automatic proof strategy does not find the proof, and to direct the prover towards a specific proof, for instance a proof that yields a better reduction, since exact security is now dealt with.

2   A Calculus for Games

2.1   Description of the Calculus

In this section, we review the process calculus defined in [13,14] in order to model games as done in computational security proofs. This calculus has been carefully designed to make the automatic proof of cryptographic protocols easier. One should note that the main addition from previous models [33,28] is the introduction of arrays, which allow us to formalize the random oracle model [9], but also the authenticity (unforgeability) in several cryptographic primitives, such as signatures, message authentication codes, but also encryption schemes. Arrays allow us to have full access to the whole memory state of the system, and replace lists often used in cryptographic proofs. For example, in the case of a random oracle, one generally stores the input and output of the random oracle in a list. In our calculus, they are stored in arrays.

Contrarily to [13,14], we adopt the exact security framework [10], instead of the asymptotic one. The cost of the reductions, and the probability loss will thus be precisely determined. We also adapt the syntax of our calculus, in order to be closer to the usual syntax of cryptographic games.

In this calculus, we denote by T types, which are subsets of $bitstring_\perp = bitstring \cup \{\perp\}$, where $bitstring$ is the set of all bit-strings and $\perp$ is a special symbol. A type is said to be *fixed-length* when it is the set of all bit-strings of a certain length. A type T is said to be *large* when its cardinal is large enough so that we can consider collisions between elements of T chosen randomly with uniform probability quite unlikely, but still keeping track of the small probability. Such an information is useful for the strategy of the prover. The boolean type is predefined: $bool = \{\text{true}, \text{false}\}$, where $\text{true} = 1$ and $\text{false} = 0$.

The calculus also assumes a finite set of function symbols f. Each function symbol f comes with a type declaration $f : T_1 \times \ldots \times T_m \to T$. Then, the function symbol f corresponds to a function, also denoted f, from $T_1 \times \ldots \times T_m$ to T, such that $f(x_1, \ldots, x_m)$ is computable in time t_f, which is bounded by a function of the length of the inputs $x_1, \ldots, x_m$. Some predefined functions use the infix notation: $M = N$ for the equality test (taking two values of the same type T and returning a value of type $bool$), $M \wedge N$ for the boolean and (taking and returning values of type $bool$).

Let us now illustrate on an example how we represent games in our process calculus. As we shall see in the next sections, this example comes from the definition of security of the Full-Domain Hash (FDH) signature scheme [9]. This example uses the function symbols hash, pkgen, skgen, f, and invf (such that $x \mapsto \mathsf{invf}(sk, x)$ is the inverse of the function $x \mapsto \mathsf{f}(pk, x)$), which will all be explained later in detail. We define an oracle $Ogen$ which chooses a random seed r, generates a key pair (pk, sk) from this seed, and returns the public key pk:

$$Ogen() := r \stackrel{R}{\leftarrow} seed; pk \leftarrow \mathsf{pkgen}(r); sk \leftarrow \mathsf{skgen}(r); \mathbf{return}(pk)$$

The seed r is chosen randomly with uniform probability in the type $seed$ by the construct $r \stackrel{R}{\leftarrow} seed$. (The type $seed$ must be a fixed-length type, because probabilistic bounded-time Turing machines can choose random numbers uniformly only in such types. The set of bit-strings $seed$ is associated to a fixed value of the security parameter.)

Next, we define a signature oracle OS which takes as argument a bit-string m and returns its FDH signature, computed as $\mathsf{invf}(sk, \mathsf{hash}(m))$, where sk is the secret key, so this oracle could be defined by

$$OS(m : bitstring) := \mathbf{return}(\mathsf{invf}(sk, \mathsf{hash}(m)))$$

where $m : bitstring$ means that m is of type $bitstring$, that is, it is any bit-string. However, this oracle can be called several times, say at most qS times. We express this repetition by $\mathbf{foreach}\ iS \leq qS\ \mathbf{do}\ OS$, meaning that we make available qS copies of OS, each with a different value of the index $iS \in [1, qS]$. Furthermore, in our calculus, variables defined in repeated oracles are arrays with a cell for each call to the oracle, so that we can remember the values used in all calls to the oracles. Here, m is then an array indexed by iS. Along similar lines, the copies of the oracle OS itself are indexed by iS, so that the caller can specify exactly which copy of OS he wants to call, by calling $OS[iS]$ for a specific value of iS. So we obtain the following formalization of this oracle:

$$\mathbf{foreach}\ iS \leq qS\ \mathbf{do}\ OS[iS](m[iS] : bitstring) := \mathbf{return}(\mathsf{invf}(sk, \mathsf{hash}(m[iS]))) \tag{1}$$

Note that sk has no array index, since it is defined in the oracle $Ogen$, which is executed only once.

We also define a test oracle OT which takes as arguments a bit-string m' and a candidate signature s of type D and executes the event forge when s is a forged signature of m', that is, s is a correct signature of m' and the signature oracle has not been called on m'. The test oracle can be defined as follows:

$$OT(m' : bitstring, s : D) := \mathbf{if}\ \mathsf{f}(pk, s) = \mathsf{hash}(m')\ \mathbf{then}$$
$$\mathbf{find}\ u \leq qS\ \mathbf{suchthat}\ (\mathbf{defined}(m[u]) \wedge m' = m[u])\ \mathbf{then\ end} \tag{2}$$
$$\mathbf{else\ event}\ \mathsf{forge}$$

It first tests whether $\mathsf{f}(pk, s) = \mathsf{hash}(m')$, as the verification algorithm of FDH would do. When the equality holds, it executes the $\mathbf{then}$ branch; otherwise, it executes the **else** branch which is here omitted. In this case, it ends the oracle, as if it executed **end**. When the test $f(pk, s) = \mathsf{hash}(m')$ succeeds, the process performs an array lookup: it looks for an index u in $[1, qS]$ such that $m[u]$ is defined and $m' = m[u]$. If such an u is found, that is, m' has already been received by the signing oracle, we simply end the oracle. Otherwise, we execute the event forge and implicitly end the oracle. Arrays and array lookups are crucial in this calculus, and will help to model many properties which were hard to capture.

Finally, we add a hash oracle, which is similar to the signing oracle OS but returns the hash of the message instead of its signature:

$$\textbf{foreach } iH \leq qH \textbf{ do } OH[iH](x[iH] : bitstring) := \textbf{return}(\mathsf{hash}(x[iH]))$$

To lighten the notation, some array indexes can be omitted in the input we give to our prover. Precisely, when x is defined under **foreach** $i_1 \leq n_1 \ldots$ **foreach** $i_m \leq n_m$, x is always an array with indexes $i_1, \ldots, i_m$, so we abbreviate all occurrences of $x[i_1, \ldots, i_m]$ by x. Here, all array indexes in OS and OH can then be omitted.

We can remark that the signature and test oracles only make sense after the generation oracle $Ogen$ has been called, since they make use of the keys pk and sk computed by $Ogen$. So we define OS and OT after $Ogen$ by a sequential composition. In contrast, OS and OT are simultaneously available, so we use a parallel composition $Q_S \mid Q_T$ where Q_S and Q_T are the processes (1) and (2) respectively. Similarly, OH is composed in parallel with the rest of the process. So we obtain the following game which models the security of the FDH signature scheme in the random oracle model:

$$G_0 = \textbf{foreach } iH \leq qH \textbf{ do } OH(x : bitstring) := \textbf{return}(\mathsf{hash}(x))$$

$$\mid Ogen() := r \overset{R}{\leftarrow} seed; pk \leftarrow \mathsf{pkgen}(r); sk \leftarrow \mathsf{skgen}(r); \textbf{return}(pk);$$

$$(\textbf{foreach } iS \leq qS \textbf{ do } OS(m : bitstring) := \textbf{return}(\mathsf{invf}(sk, \mathsf{hash}(m)))$$

$$\mid OT(m' : bitstring, s : D) := \textbf{if } f(pk, s) = \mathsf{hash}(m') \textbf{ then}$$

$$\textbf{find } u \leq qS \textbf{ suchthat } (\textbf{defined}(m[u]) \wedge m' = m[u]) \textbf{ then end}$$

$$\textbf{else event } \mathsf{forge})$$

Our calculus obviously also has a construct for calling oracles. However, we do not need it explicitly in this paper, because oracles are called by the adversary, not by processes we write ourselves.

As detailed in [13,14], we require some *well-formedness invariants* to guarantee that several definitions of the same oracle cannot be simultaneously available, that bit-strings are of their expected type, and that arrays are used properly (that each cell of an array is assigned at most once during execution, and that variables are accessed only after being initialized). The formal semantics of the calculus can be found in [13].

2.2 Observational Equivalence

We denote by $\Pr[Q \rightsquigarrow a]$ the probability that the answer of Q to the oracle call $Ostart()$ is a, where $Ostart$ is an oracle called to start the experiment. We denote by $\Pr[Q \leadsto \mathcal{E}]$ the probability that the process Q executes exactly the sequence of events $\mathcal{E}$, in the order of $\mathcal{E}$, when oracle $Ostart()$ is called.

In the next definition, we use a context C to represent an algorithm that tries to distinguish Q from Q'. A context C is put around a process Q by $C[Q]$. This construct means that Q is put in parallel with some other process Q' contained in C, possibly hiding some oracles defined in Q, so that, when considering $C'[C[Q]]$, C' cannot call these oracles. This will be detailed in the following of this section.

Definition 1 (Observational equivalence). *Let Q and Q' be two processes that satisfy the well-formedness invariants.*

A context C is said to be acceptable *for Q if and only if C does not contain events, C and Q have no common variables, and $C[Q]$ satisfies the well-formedness invariants.*

We say that Q and Q' are observationally equivalent *up to probability p, written $Q \approx_p Q'$, when for all t, for all contexts C acceptable for Q and Q' that run in time at most t, for all bit-strings a, $|\Pr[C[Q] \leadsto a] - \Pr[C[Q'] \leadsto a]| \leq p(t)$ and $\sum_{\mathcal{E}} |\Pr[C[Q] \leadsto \mathcal{E}] - \Pr[C[Q'] \leadsto \mathcal{E}]| \leq p(t)$.*

This definition formalizes that the probability that an algorithm C running in time t distinguishes the games Q and Q' is at most $p(t)$. The context C is not allowed to access directly the variables of Q (using **find**). We say that a context C runs in time t, when for all processes Q, the time spent in C in any trace of $C[Q]$ is at most t, ignoring the time spent in Q. (The runtime of a context is bounded. Indeed, we bound the length of messages in calls or returns to oracle O by a value $\mathrm{maxlen}(O, \mathrm{arg}_i)$ or $\mathrm{maxlen}(O, \mathrm{res}_i)$. Longer messages are truncated. The length of random numbers created by C is bounded; the number of instructions executed by C is bounded; and the time of a function evaluation is bounded by a function of the length of its arguments.)

Definition 2. *We say that Q executes event e with probability at most p if and only if for all t, for all contexts C acceptable for Q that run in time t, $\sum_{\mathcal{E}, e \in \mathcal{E}} \Pr[C[Q] \leadsto \mathcal{E}] \leq p(t)$.*

The above definitions allow us to perform proofs using sequences of indistinguishable games. The following lemma is straightforward:

Lemma 1. *1. $\approx_p$ is reflexive and symmetric.*
2. If $Q \approx_p Q'$ and $Q' \approx_{p'} Q''$, then $Q \approx_{p+p'} Q''$.
3. If Q executes event e with probability at most p and $Q \approx_{p'} Q'$, then Q' executes event e with probability at most $p + p'$.
4. If $Q \approx_p Q'$ and C is a context acceptable for Q and Q' that runs in time t_C, then $C[Q] \approx_{p'} C[Q']$ where $p'(t) = p(t + t_C)$.
5. If Q executes event e with probability at most p and C is a context acceptable for Q that runs in time t_C, then $C[Q]$ executes event e with probability at most p' where $p'(t) = p(t + t_C)$.

Properties 2 and 3 are key to computing probabilities coming from a sequence of games. Indeed, our prover will start from a game G_0 corresponding to the initial attack, and build a sequence of observationally equivalent games $G_0 \approx_{p_1} G_1 \approx_{p_2} \ldots \approx_{p_m} G_m$. By Property 2, we conclude that $G_0 \approx_{p_1 + \ldots + p_m} G_m$. By Property 3, we can bound the probability that G_0 executes an event from the probability that G_m executes this event.

The elementary transformations used to build each game from the previous one can in particular come from an algorithmic assumption on a cryptographic primitive. This assumption needs to be specified as an observational equivalence $L \approx_p R$. To use it to transform a game G, the prover finds a context C such that $G \approx_0 C[L]$ by purely syntactic transformations, and builds a game G' such that $G' \approx_0 C[R]$ by purely syntactic transformations. C is the simulator usually defined for reductions. By Property 4, we have $C[L] \approx_{p'} C[R]$, so $G \approx_{p'} G'$. The context C typically hides the oracles of L and R so that they are visible from C but not from the adversary C' against $G \approx_{p'} G'$. The context $C'[C[]]$ then defines the adversary against the algorithmic assumption $L \approx_p R$.

If the security assumptions are initially not in the form of an equivalence $L \approx_p R$, one needs to manually prove such an equivalence that formalizes the desired security assumption. The design of such equivalences can be delicate, but this is a one-time effort: the same equivalence can be reused for proofs that rely on the same assumption. For instance, we give below such an equivalence for one-wayness, and use it not only for the proof of the FDH signature scheme, but also for proofs of encryption schemes as mentioned in Section 4.2. Similarly, the definition of security of a signature (UF-CMA) says that some event is executed with negligible probability. When we want to prove the security of a protocol using a signature scheme, we use a manual proof of an equivalence that corresponds to that definition, done once for UF-CMA in the long version of this paper [15].

The prover automatically establishes certain equivalences $G_0 \approx_p G_m$ as mentioned above. However, the user can give only the left-hand side of the equivalence G_0; the right-hand side G_m is obtained by the prover. As a consequence, the prover is in general not appropriate for proving automatically properties $L \approx_p R$ in which L and R are both given a priori: the right-hand side found by the prover is unlikely to correspond exactly to the desired right-hand side. On the other hand, the prover can check security properties on the right-hand side G_m it finds, for example that the event forge cannot be executed by G_m. Using $G_0 \approx_p G_m$, it concludes that G_0 executes forge with probability at most p.

3 Characterization of One-Wayness and Unforgeability

In this section, we introduce the assumption (one-wayness) and the security notion (unforgeability) to achieve.

3.1 Trapdoor One-Way Permutations

Most cryptographic protocols rely on the existence of trapdoor one-way permutations. They are families of permutations, which are easy to compute, but hard to invert, unless one has a trapdoor.

The Computational Model. A family of permutations $\mathcal{P}$ onto a set D is defined by the three following algorithms:

- The *key generation algorithm* kgen (which can be split in two sub-algorithms pkgen and skgen). On input a seed r, the algorithm kgen produces a pair (pk, sk) of matching public and secret keys. The public key pk specifies the actual permutation f_{pk} onto the domain D.
- The *evaluation algorithm* f. Given a public key pk and a value $x \in D$, it outputs $y = f_{pk}(x)$.
- The *inversion algorithm* invf. Given an element y, and the trapdoor sk, invf outputs the unique pre-image x of y with respect to f_{pk}.

The above properties simply require the algorithms to be efficient. The "one-wayness" property is more intricate, since it claims the "non-existence" of some efficient algorithm: one wants that the success probability of any adversary $\mathcal{A}$ within a reasonable time is small, where this success is commonly defined by

$$\mathsf{Succ}_{\mathcal{P}}^{\mathsf{ow}}(\mathcal{A}) = \Pr\left[r \xleftarrow{R} seed, (pk, sk) \leftarrow \mathsf{kgen}(r), x \xleftarrow{R} D, y \leftarrow \mathsf{f}(pk, x), \\ x' \leftarrow \mathcal{A}(pk, y) : x = x' \right].$$

Eventually, we denote by $\mathsf{Succ}_{\mathcal{P}}^{\mathsf{ow}}(t)$ the maximal success probability an adversary can get within time t.

Syntactic Rules. Let *seed* be a large, fixed-length type, *pkey*, *skey*, and D the types of public keys, secret keys, and the domain of the permutations respectively. A family of trapdoor one-way permutations can then be defined as a set of four function symbols: skgen : *seed* $\rightarrow$ *skey* generates secret keys; pkgen : *seed* $\rightarrow$ *pkey* generates public keys; f : *pkey* $\times$ D $\rightarrow$ D and invf : *skey* $\times$ D $\rightarrow$ D, such that, for each pk, $x \mapsto \mathsf{f}(pk, x)$ is a permutation of D, whose inverse permutation is $x \mapsto \mathsf{invf}(sk, x)$ when $pk = \mathsf{pkgen}(r)$ and $sk = \mathsf{skgen}(r)$.

The one-wayness property can be formalized in our calculus by requiring that LR executes **event** invert with probability at most $\mathsf{Succ}_{\mathcal{P}}^{\mathsf{ow}}(t)$ in the presence of a context that runs in time t, where

$$LR = Ogen() := r_0 \xleftarrow{R} seed; x_0 \xleftarrow{R} D; \mathbf{return}(\mathsf{pkgen}(r_0), \mathsf{f}(\mathsf{pkgen}(r_0), x_0));$$
$$Oeq(x' : D) := \mathbf{if}\ x' = x_0\ \mathbf{then}\ \mathbf{event}\ \text{invert}$$

Indeed, the event invert is executed when the adversary, given the public key pkgen(r_0) and the image of some x_0 by f, manages to find x_0 (without having the trapdoor).

In order to use the one-wayness property in proofs of protocols, our prover needs a more general formulation of one-wayness, using "observationally equivalent" processes. We thus define two processes which are actually equivalent unless LR executes **event** invert. We prove in the long version of this paper [15] the equivalence of Figure 1 where $p^{\mathsf{ow}}(t) = n_{\mathsf{k}} \times n_{\mathsf{f}} \times \mathsf{Succ}_{\mathcal{P}}^{\mathsf{ow}}(t + (n_{\mathsf{k}}n_{\mathsf{f}} - 1)t_{\mathsf{f}} + (n_{\mathsf{k}} - 1)t_{\mathsf{pkgen}})$, t_{f} is the time of one evaluation of f, and t_{pkgen} is the time of one evaluation of pkgen. In this equivalence, the function symbols pkgen$'$: *seed* $\rightarrow$ *pkey* and

$$\textbf{foreach } i_{\text{k}} \leq n_{\text{k}} \textbf{ do } r \xleftarrow{R} seed; (Opk() := \textbf{return}(\textsf{pkgen}(r))$$

$$| \textbf{ foreach } i_{\text{f}} \leq n_{\text{f}} \textbf{ do } x \xleftarrow{R} D; (Oy() := \textbf{return}(\textsf{f}(\textsf{pkgen}(r), x))$$

$$| \textbf{ foreach } i_1 \leq n_1 \textbf{ do } Oeq(x' : D) := \textbf{return}(x' = x)$$

$$| Ox() := \textbf{return}(x)))$$

$$\approx_{p^{\text{ow}}} \textbf{foreach } i_{\text{k}} \leq n_{\text{k}} \textbf{ do } r \xleftarrow{R} seed; (Opk() := \textbf{return}(\textsf{pkgen}'(r)) \qquad (3)$$

$$| \textbf{ foreach } i_{\text{f}} \leq n_{\text{f}} \textbf{ do } x \xleftarrow{R} D; (Oy() := \textbf{return}(\textsf{f}'(\textsf{pkgen}'(r), x))$$

$$| \textbf{ foreach } i_1 \leq n_1 \textbf{ do } Oeq(x' : D) :=$$

$$\textbf{if defined}(k) \textbf{ then return}(x' = x) \textbf{ else return}(\textsf{false})$$

$$| Ox() := k \leftarrow \textbf{mark}; \textbf{return}(x)))$$

Fig. 1. Definition of one-wayness

$\textsf{f}' : pkey \times D \to D$ are such that the functions associated to the primed symbols $\textsf{pkgen}', \textsf{f}'$ are equal to the functions associated to their corresponding unprimed symbol $\textsf{pkgen}, \textsf{f}$, respectively. We replace $\textsf{pkgen}$ and $\textsf{f}$ with $\textsf{pkgen}'$ and $\textsf{f}'$ in the right-hand side just to prevent repeated applications of the transformation with the same keys, which would lead to an infinite loop.

In this equivalence, we consider n_{k} keys $\textsf{pkgen}(r[i_{\text{k}}])$ instead of a single one, and n_{f} antecedents of $\textsf{f}$ for each key, $x[i_{\text{k}}, i_{\text{f}}]$. The first oracle $Opk[i_{\text{k}}]$ publishes the public key $\textsf{pkgen}(r[i_{\text{k}}])$. The second group of oracles first picks a new $x[i_{\text{k}}, i_{\text{f}}]$, and then makes available three oracles: $Oy[i_{\text{k}}, i_{\text{f}}]$ returns the image of $x[i_{\text{k}}, i_{\text{f}}]$ by $\textsf{f}$, $Oeq[i_{\text{k}}, i_{\text{f}}, i_1]$ returns true when it receives $x[i_{\text{k}}, i_{\text{f}}]$ as argument, and $Ox[i_{\text{k}}, i_{\text{f}}]$ returns $x[i_{\text{k}}, i_{\text{f}}]$ itself. The one-wayness property guarantees that when $Ox[i_{\text{k}}, i_{\text{f}}]$ has not been called, the adversary has little chance of finding $x[i_{\text{k}}, i_{\text{f}}]$, so $Oeq[i_{\text{k}}, i_{\text{f}}, i_1]$ returns false. Therefore, we can replace the left-hand side of the equivalence with its right-hand side, in which $Ox[i_{\text{k}}, i_{\text{f}}]$ records that it has been called by defining $k[i_{\text{k}}, i_{\text{f}}]$, and $Oeq[i_{\text{k}}, i_{\text{f}}, i_1]$ always returns false when $k[i_{\text{k}}, i_{\text{f}}]$ is not defined, that is, when $Ox[i_{\text{k}}, i_{\text{f}}]$ has not been called.

In the left-hand side of the equivalences used to specify primitives, the oracles must consist of a single return instruction. This restriction allows us to model many equivalences that define cryptographic primitives, and it simplifies considerably the transformation of processes compared to using the general syntax of processes. (In order to use an equivalence $L \approx_p R$, we need to recognize processes that can easily be transformed into $C[L]$ for some context C, to transform them into $C[R]$. This is rather easy to do with such oracles: we just need to recognize terms that occur as a result of these oracles. That would be much more difficult with general processes.)

Since $x \mapsto \textsf{f}(\textsf{pkgen}(r), x)$ and $x \mapsto \textsf{invf}(\textsf{skgen}(r), x)$ are inverse permutations, we have:

$$\forall r : seed, \forall x : D, \textsf{invf}(\textsf{skgen}(r), \textsf{f}(\textsf{pkgen}(r), x)) = x \qquad (4)$$

Since $x \mapsto \textsf{f}(pk, x)$ is injective, $\textsf{f}(pk, x) = \textsf{f}(pk, x')$ if and only if $x = x'$:

$$\forall pk : pkey, \forall x : D, \forall x' : D, (\textsf{f}(pk, x) = \textsf{f}(pk, x')) = (x = x') \qquad (5)$$

Since $x \mapsto f(pk, x)$ is a permutation, when x is a uniformly distributed random number, we can replace x with $f(pk, x)$ everywhere, without changing the probability distribution. In order to enable automatic proof, we give a more restricted formulation of this result:

$$
\begin{aligned}
&\textbf{foreach } i_k \leq n_k \textbf{ do } r \xleftarrow{R} seed; (Opk() := \textbf{return}(\mathsf{pkgen}(r)) \\
&\quad | \textbf{ foreach } i_f \leq n_f \textbf{ do } x \xleftarrow{R} D; (Oant() := \textbf{return}(\mathsf{invf}(\mathsf{skgen}(r), x)) \\
&\qquad\qquad\qquad\qquad\qquad | Oim() := \textbf{return}(x))) \\
&\approx_0 \textbf{foreach } i_k \leq n_k \textbf{ do } r \xleftarrow{R} seed; (Opk() := \textbf{return}(\mathsf{pkgen}(r)) \\
&\quad | \textbf{ foreach } i_f \leq n_f \textbf{ do } x \xleftarrow{R} D; (Oant() := \textbf{return}(x) \\
&\qquad\qquad\qquad\qquad\qquad | Oim() := \textbf{return}(\mathsf{f}(\mathsf{pkgen}(r), x))))
\end{aligned}
$$

(6)

which allows to perform the previous replacement only when x is used in calls to $\mathsf{invf}(\mathsf{skgen}(r), x)$, where r is a random number such that r occurs only in $\mathsf{pkgen}(r)$ and $\mathsf{invf}(\mathsf{skgen}(r), x)$ for some random numbers x.

3.2   Signatures

The Computational Model. A signature scheme $\mathsf{S} = (\mathsf{kgen}, \mathsf{sign}, \mathsf{verify})$ is defined by:

- The *key generation algorithm* kgen (which can be split in two sub-algorithms pkgen and skgen). On input a random seed r, the algorithm kgen produces a pair (pk, sk) of matching keys.
- The *signing algorithm* sign. Given a message m and a secret key sk, sign produces a signature σ. For sake of clarity, we restrict ourselves to the deterministic case.
- The *verification algorithm* verify. Given a signature σ, a message m, and a public key pk, verify tests whether σ is a valid signature of m with respect to pk.

We consider here *(existential) unforgeability under adaptive chosen-message attack* (UF-CMA) [23], that is, the attacker can ask the signer to sign any message of its choice, in an adaptive way, and has to provide a signature on a new message. In its answer, there is indeed the natural restriction that the returned message has not been asked to the signing oracle.

When one designs a signature scheme, one wants to computationally rule out existential forgeries under adaptive chosen-message attacks. More formally, one wants that the success probability of any adversary $\mathcal{A}$ with a reasonable time is small, where

$$
\mathsf{Succ}_{\mathsf{S}}^{\mathsf{uf-cma}}(\mathcal{A}) = \Pr\left[\begin{aligned} & r \xleftarrow{R} seed, (pk, sk) \leftarrow \mathsf{kgen}(r), (m, \sigma) \leftarrow \mathcal{A}^{\mathsf{sign}(\cdot, sk)}(pk) : \\ & \mathsf{verify}(m, pk, \sigma) = 1 \end{aligned} \right].
$$

As above, we denote by $\mathsf{Succ}_\mathsf{S}^{\mathsf{uf-cma}}(n_\mathsf{s}, \ell, t)$ the maximal success probability an adversary can get within time t, after at most n_s queries to the signing oracle, where the maximum length of all messages in queries is ℓ.

Syntactic Rules. Let *seed* be a large, fixed-length type. Let *pkey*, *skey*, and *signature* the types of public keys, secret keys, and signatures respectively. A signature scheme is defined as a set of four function symbols: $\mathsf{skgen} : seed \rightarrow skey$ generates secret keys; $\mathsf{pkgen} : seed \rightarrow pkey$ generates public keys; $\mathsf{sign} : bitstring \times skey \rightarrow signature$ generates signatures; and $\mathsf{verify} : bitstring \times pkey \times signature \rightarrow bool$ verifies signatures.

The signature verification succeeds for signatures generated by sign, that is,

$$\forall m : bitstring, \forall r : seed, \mathsf{verify}(m, \mathsf{pkgen}(r), \mathsf{sign}(m, \mathsf{skgen}(r))) = \mathsf{true}$$

According to the previous definition of UF-CMA, the following process LR executes **event** forge with probability at most $\mathsf{Succ}_\mathsf{S}^{\mathsf{uf-cma}}(n_\mathsf{s}, \ell, t)$ in the presence of a context that runs in time t, where

$$
\begin{aligned}
LR = Ogen() := r &\xleftarrow{R} seed; pk \leftarrow \mathsf{pkgen}(r); sk \leftarrow \mathsf{skgen}(r); \mathbf{return}(pk); \\
&(\mathbf{foreach}\ i_\mathsf{s} \leq n_\mathsf{s}\ \mathbf{do}\ OS(m : bitstring) := \mathbf{return}(\mathsf{sign}(m, sk)) \\
&|\ OT(m' : bitstring, s : signature) := \mathbf{if}\ \mathsf{verify}(m', pk, s)\ \mathbf{then} \quad (7) \\
&\quad \mathbf{find}\ u_\mathsf{s} \leq n_\mathsf{s}\ \mathbf{suchthat}\ (\mathbf{defined}(m[u_\mathsf{s}]) \wedge m' = m[u_\mathsf{s}]) \\
&\quad \mathbf{then\ end\ else\ event}\ \mathsf{forge})
\end{aligned}
$$

and ℓ is the maximum length of m and m'. This is indeed clear since **event** forge is raised if a signature is accepted (by the verification algorithm), while the signing algorithm has not been called on the signed message.

4   Examples

4.1   FDH Signature

The Full-Domain Hash (FDH) signature scheme [9] is defined as follows: Let $\mathsf{pkgen}, \mathsf{skgen}, \mathsf{f}, \mathsf{invf}$ define a family of trapdoor one-way permutations. Let hash be a hash function, in the random oracle model. The FDH signature scheme uses the functions pkgen and skgen as key-generation functions, the signing algorithm is $\mathsf{sign}(m, sk) = \mathsf{invf}(sk, \mathsf{hash}(m))$, and the verification algorithm is $\mathsf{verify}(m', pk, s) = (\mathsf{f}(pk, s) = \mathsf{hash}(m'))$. In this section, we explain how our automatic prover finds the well-known bound for $\mathsf{Succ}_\mathsf{S}^{\mathsf{uf-cma}}$ for the FDH signature scheme.

The input given to the prover contains two parts. First, it contains the definition of security of primitives used to build the FDH scheme, that is, the definition of one-way trapdoor permutations (3), (4), (5), and (6) as detailed in Section 3.1 and the formalization of a hash function in the random oracle model:

foreach $i_h \leq n_h$ **do** $OH(x : bitstring) :=$ **return**(hash(x)) $[all]$

$\approx_0$ **foreach** $i_h \leq n_h$ **do** $OH(x : bitstring) :=$

    find $u \leq n_h$ **suchthat** (**defined**$(x[u], r[u]) \wedge x = x[u]$) **then return**$(r[u])$

    else $r \xleftarrow{R} D$; **return**(r)

$$(8)$$

This equivalence expresses that we can replace a call to a hash function with a random oracle, that is, an oracle that returns a fresh random number when it is called with a new argument, and the previously returned result when it is called with the same argument as in a previous call. Such a random oracle is implemented in our calculus by a lookup in the array x of the arguments of hash. When a u such that $x[u]$, $r[u]$ are defined and $x = x[u]$ is found, hash has already been called with x, at call number u, so we return the result of that call, $r[u]$. Otherwise, we create a fresh random number r. (The indication $[all]$ on the first line of (8) instructs the prover to replace all occurrences of hash in the game.)

Second, the input file contains as initial game the process G_0 of Section 2.1. As detailed in Section 3.2, this game corresponds to the definition of security of the FDH signature scheme (7). An important remark is that we need to add to the standard definition of security of a signature scheme the hash oracle. This is necessary so that, after transformation of hash into a random oracle, the adversary can still call the hash oracle. (The adversary does not have access to the arrays that encode the values of the random oracle.) Our goal is to bound the probability $p(t)$ that **event** forge is executed in this game in the presence of a context that runs in time t: $p(t) = \mathsf{Succ}_S^{\mathsf{uf-cma}}(qS, \ell, t+t_H) \geq \mathsf{Succ}_S^{\mathsf{uf-cma}}(qS, \ell, t)$ where t_H is the total time spent in the hash oracle and ℓ is the maximum length of m and m'.

Given this input, our prover automatically produces a proof that this game executes **event** forge with probability $p(t) \leq (qH + qS + 1)\mathsf{Succ}_P^{\mathsf{ow}}(t + (qH + qS)t_f + (3qS + 2qH + qS^2 + 2qSqH + qH^2)t_{eq}(\ell))$ where ℓ is the maximum length of a bit-string in m, m', or x and $t_{eq}(\ell)$ is the time of a comparison between bit-strings of length at most ℓ. (Evaluating a **find** implies evaluating the condition of the **find** for each value of the indexes, so here the lookup in an array of size n of bit-strings of length ℓ is considered as taking time $n \times t_{eq}(\ell)$, although there are in fact more efficient algorithms for this particular case of array lookup.) If we ignore the time of bit-string comparisons, we obtain the usual upper-bound [10] $(qH + qS + 1)\mathsf{Succ}_P^{\mathsf{ow}}(t + (qH + qS)t_f)$. The prover also outputs the sequence of games that leads to this proof, and a succinct explanation of the transformation performed between consecutive games of the sequence. The input and output of the prover, as well as the prover itself, are available at http://www.di.ens.fr/~blanchet/cryptoc/FDH/; the runtime of the prover on this example is 14 ms on a Pentium M 1.8 GHz. The prover has been implemented in Ocaml and contains 14800 lines of code.

We sketch here the main proof steps. Starting from the initial game G_0 given in Section 2.1, the prover tries to apply all observational equivalences it has as hypotheses, that is here, (3), (6), and (8). It succeeds applying the security of the hash function (8), so it transforms the game accordingly, by replacing the left-hand side with the right-hand side of the equivalence. Each call to hash is then replaced with a lookup in the arguments of all calls to hash. When the argument of hash is found in one of these arrays, the returned result is the same as the result previously returned by hash. Otherwise, we pick a fresh random number and return it.

The obtained game is then simplified. In particular, when the argument m' of OT is found in the arguments m of the call to hash in OS, the **find** in OT always succeeds, so its **else** branch can be removed (that is, when m' has already been passed to the signature oracle, it is not a forgery).

Then, the prover tries to apply an observational equivalence. All transformations fail, but when applying (6), the game contains $\mathsf{invf}(sk, y)$ while (6) expects $\mathsf{invf}(\mathsf{skgen}(r), y)$, which suggests to remove assignments to variable sk for it to succeed. So the prover performs this removal: it substitutes $\mathsf{skgen}(r)$ for sk and removes the assignment $sk \leftarrow \mathsf{skgen}(r)$. The transformation (6) is then retried. It now succeeds, which leads to replacing r_j with $\mathsf{f}(\mathsf{pkgen}(r), r_j)$ and $\mathsf{invf}(\mathsf{skgen}(r), r_j)$ with r_j, where r_j represents the random numbers that are the result of the random oracle. (The term $\mathsf{f}(\mathsf{pkgen}(r), r_j)$ can then be computed by oracle Oy of (3) and r_j can be computed by Ox.) More generally, in our prover, when a transformation $\mathcal{T}$ fails, it may return transformations $\mathcal{T}'$ to apply in order to enable $\mathcal{T}$ [14, Section 5]. In this case, the prover applies the suggested transformations $\mathcal{T}'$ and retries the transformation $\mathcal{T}$.

The obtained game is then simplified. In particular, by injectivity of f (5), the prover replaces terms of the form $\mathsf{f}(pk, s) = \mathsf{f}(\mathsf{pkgen}(r), r_j)$ with $s = r_j$, knowing $pk = \mathsf{pkgen}(r)$. (The test $s = r_j$ can then be computed by oracle Oeq of (3).)

The prover then tries to apply an observational equivalence. It succeeds using the definition of one-wayness (3). This transformation leads to replacing $\mathsf{f}(\mathsf{pkgen}(r), r_j)$ with $\mathsf{f}'(\mathsf{pkgen}'(r), r_j)$, r_j with $k_j \leftarrow \mathsf{mark}; r_j$, and $s = r_j$ with **find** $u_j \le N$ **suchthat** (**defined**$(k_j[u_j]) \wedge \mathsf{true}$) **then** $s = r_j$ **else** false. The difference of probability is $p^{\mathsf{ow}}(t + t') = n_k \times n_f \times \mathsf{Succ}_{\mathcal{P}}^{\mathsf{ow}}(t + t' + (n_k n_f - 1)t_f + (n_k - 1)t_{\mathsf{pkgen}}) = (qH + qS + 1)\mathsf{Succ}_{\mathcal{P}}^{\mathsf{ow}}(t + t' + (qH + qS)t_f)$ where $n_k = 1$ is the number of key pairs considered, $n_f = qH + qS + 1$ is the number of antecedents of f, and $t' = (3qS + 2qH + qS^2 + 2qSqH + qH^2)t_{\mathsf{eq}}(\ell)$ is the runtime of the context put around the equivalence (3).

Finally, the obtained game is simplified again. Thanks to some equational reasoning, the prover manages to show that the **find** in OT always succeeds, so its **else** branch can be removed. The prover then detects that the forge event cannot be executed in the resulting game, so the desired property is proved, and the probability that forge is executed in the initial game is the sum of the differences of probability between games of the sequence, which here comes only from the application of one-wayness (3).

4.2 Encryption Schemes

Besides proving the security of many protocols [14], we have also used our prover
for proving other cryptographic schemes. For example, our prover can show that
the basic Bellare-Rogaway construction [9] without redundancy (*i.e.* $\mathcal{E}(m, r) =$
$f(r) \| hash(r) \, xor \, m$) is IND-CPA, with the following manual proof:

`crypto hash`	apply the security of hash (8)
`remove_assign binder pk`	remove assignments to pk
`crypto f r`	apply the security of f (3) with random seed r
`crypto xor *`	apply the security of xor as many times as possible
`success`	check that the desired property is proved

These manual indications are necessary because (3) can also be applied without
removing the assignments to pk, but with different results: $f(pk, x)$ is computed
by applying f to the results of oracles Opk and Ox if assignments to pk are not
removed, and by oracle Oy if assignments to pk are removed.

With similar manual indications, it can show that the enhanced variant with
redundancy $\mathcal{E}(m, r) = f(r) \| hash(r) \, xor \, m \| hash'(hash(r) \, xor \, m, r)$ is IND-CCA2.
With an improved treatment of the equational theory of xor, we believe that it
could also show that $\mathcal{E}(m, r) = f(r) \| hash(r) \, xor \, m \| hash'(m, r)$ is IND-CCA2.

5   Conclusion

We have presented a new tool to automatically prove the security of both cryptographic primitives and cryptographic protocols. As usual, assumptions and
expected security notions have to be stated. For the latter, specifications are
quite similar to the usual definitions, where a "bad" event has to be shown to
be unlikely. However, the former may seem more intricate, since it has to be
specified as an observational equivalence. Anyway, this has to be done only once
for all proofs, and several specifications have already been given in [13,14,15]:
one-wayness, UF-CMA signatures, UF-CMA message authentication codes, IND-
CPA symmetric stream ciphers, IND-CPA and IND-CCA2 public-key encryption,
hash functions in the random oracle model, xor, with detailed proofs for the first
three. Thereafter, the protocol/scheme itself has to be specified, but the syntax is quite close to the notations classically used in cryptography. Eventually,
the prover provides the sequence of transformations, and thus of games, which
lead to a final experiment (indistinguishable from the initial one) in which the
"bad" event never appears. Since several paths may be used for such a sequence,
the user is allowed (but does not have) to interact with the prover, in order to
make it follow a specific sequence. Of course, the prover will accept only if the
sequence is valid. Contrary to most of the formal proof techniques, the failure
of the prover does not lead to an attack. It just means that the prover did not
find an appropriate sequence of games.

Acknowledgments. We thank Jacques Stern for initiating our collaboration
on this topic and the anonymous reviewers for their helpful comments. This work
was partly supported by ARA SSIA Formacrypt.

References

1. M. Abadi and P. Rogaway. Reconciling two views of cryptography (the computational soundness of formal encryption). *Journal of Cryptology*, 15(2):103–127, 2002.

2. M. Backes and P. Laud. A mechanized, cryptographically sound type inference checker. In *Workshop on Formal and Computational Cryptography (FCC'06)*, July 2006. To appear.

3. M. Backes and B. Pfitzmann. Symmetric encryption in a simulatable Dolev-Yao style cryptographic library. In *CSFW'04*. IEEE, June 2004.

4. M. Backes and B. Pfitzmann. Relating symbolic and cryptographic secrecy. In *26th IEEE Symposium on Security and Privacy*, pages 171–182. IEEE, May 2005.

5. M. Backes, B. Pfitzmann, and M. Waidner. A composable cryptographic library with nested operations. In *CCS'03*, pages 220–230. ACM, Oct. 2003.

6. M. Backes, B. Pfitzmann, and M. Waidner. Symmetric authentication within a simulatable cryptographic library. In *ESORICS'03*, LNCS 2808, pages 271–290. Springer, Oct. 2003.

7. G. Barthe, J. Cederquist, and S. Tarento. A machine-checked formalization of the generic model and the random oracle model. In *IJCAR'04*, LNCS 3097, pages 385–399. Springer, July 2004.

8. M. Bellare. Practice-Oriented Provable Security. In *ISW '97*, LNCS 1396. Springer, 1997.

9. M. Bellare and P. Rogaway. Random Oracles Are Practical: a Paradigm for Designing Efficient Protocols. In *CCS'93*, pages 62–73. ACM Press, 1993.

10. M. Bellare and P. Rogaway. The Exact Security of Digital Signatures – How to Sign with RSA and Rabin. In *Eurocrypt '96*, LNCS 1070, pages 399–416. Springer, 1996.

11. M. Bellare and P. Rogaway. The Game-Playing Technique and its Application to Triple Encryption, 2004. Cryptology ePrint Archive 2004/331.

12. B. Blanchet. Automatic proof of strong secrecy for security protocols. In *IEEE Symposium on Security and Privacy*, pages 86–100, May 2004.

13. B. Blanchet. A computationally sound mechanized prover for security protocols. Cryptology ePrint Archive, Report 2005/401, Nov. 2005. Available at http://eprint.iacr.org/2005/401.

14. B. Blanchet. A computationally sound mechanized prover for security protocols. In *IEEE Symposium on Security and Privacy*, pages 140–154, May 2006.

15. B. Blanchet and D. Pointcheval. Automated security proofs with sequences of games. Cryptology ePrint Archive, Report 2006/069, Feb. 2006. Available at http://eprint.iacr.org/2006/069.

16. R. Canetti. Universally composable security: A new paradigm for cryptographic protocols. In *FOCS'01*, pages 136–145. IEEE, Oct. 2001. An updated version is available at Cryptology ePrint Archive, http://eprint.iacr.org/2000/067.

17. R. Canetti and J. Herzog. Universally composable symbolic analysis of cryptographic protocols (the case of encryption-based mutual authentication and key exchange). Cryptology ePrint Archive, Report 2004/334, 2004. Available at http://eprint.iacr.org/2004/334.

18. V. Cortier and B. Warinschi. Computationally sound, automated proofs for security protocols. In *ESOP'05*, LNCS 3444, pages 157–171. Springer, Apr. 2005.

19. A. Datta, A. Derek, J. C. Mitchell, V. Shmatikov, and M. Turuani. Probabilistic polynomial-time semantics for a protocol security logic. In *ICALP'05*, LNCS 3580, pages 16–29. Springer, July 2005.

20. W. Diffie and M. E. Hellman. New Directions in Cryptography. *IEEE Transactions on Information Theory*, IT–22(6):644–654, November 1976.
21. D. Dolev and A. C. Yao. On the Security of Public-Key Protocols. *IEEE Transactions on Information Theory*, 29(2):198–208, 1983.
22. S. Goldwasser and S. Micali. Probabilistic Encryption. *Journal of Computer and System Sciences*, 28:270–299, 1984.
23. S. Goldwasser, S. Micali, and R. Rivest. A Digital Signature Scheme Secure Against Adaptive Chosen-Message Attacks. *SIAM Journal of Computing*, 17(2):281–308, April 1988.
24. S. Halevi. A plausible approach to computer-aided cryptographic proofs. Cryptology ePrint Archive, Report 2005/181, June 2005. Available at `http://eprint.iacr.org/2005/181`.
25. R. Janvier, Y. Lakhnech, and L. Mazaré. Completing the picture: Soundness of formal encryption in the presence of active adversaries. In *ESOP'05*, LNCS 3444, pages 172–185. Springer, Apr. 2005.
26. P. Laud. Handling encryption in an analysis for secure information flow. In *ESOP'03*, LNCS 2618, pages 159–173. Springer, Apr. 2003.
27. P. Laud. Symmetric encryption in automatic analyses for confidentiality against active adversaries. In *IEEE Symposium on Security and Privacy*, pages 71–85, May 2004.
28. P. Laud. Secrecy types for a simulatable cryptographic library. In *CCS'05*, pages 26–35. ACM, Nov. 2005.
29. P. D. Lincoln, J. C. Mitchell, M. Mitchell, and A. Scedrov. A probabilistic polytime framework for protocol analysis. In *CCS'98*, pages 112–121, Nov. 1998.
30. P. D. Lincoln, J. C. Mitchell, M. Mitchell, and A. Scedrov. Probabilistic polynomial-time equivalence and security protocols. In *FM'99*, LNCS 1708, pages 776–793. Springer, Sept. 1999.
31. P. Mateus, J. Mitchell, and A. Scedrov. Composition of cryptographic protocols in a probabilistic polynomial-time process calculus. In *CONCUR 2003*, LNCS 2761, pages 327–349. Springer, Sept. 2003.
32. D. Micciancio and B. Warinschi. Soundness of formal encryption in the presence of active adversaries. In *TCC'04*, LNCS 2951, pages 133–151. Springer, Feb. 2004.
33. J. C. Mitchell, A. Ramanathan, A. Scedrov, and V. Teague. A probabilistic polynomial-time calculus for the analysis of cryptographic protocols. *Theoretical Computer Science*, 353(1–3):118–164, Mar. 2006.
34. M. Naor and M. Yung. Universal One-Way Hash Functions and Their Cryptographic Applications. In *STOC'89*, pages 33–43. ACM Press, 1989.
35. C. Rackoff and D. R. Simon. Non-Interactive Zero-Knowledge Proof of Knowledge and Chosen Ciphertext Attack. In *Crypto '91*, LNCS 576, pages 433–444. Springer, 1992.
36. A. Ramanathan, J. Mitchell, A. Scedrov, and V. Teague. Probabilistic bisimulation and equivalence for security analysis of network protocols. In *FOSSACS'04*, LNCS 2987, pages 468–483. Springer, Mar. 2004.
37. V. Shoup. Sequences of games: a tool for taming complexity in security proofs, 2004. Cryptology ePrint Archive 2004/332.
38. C. Sprenger, M. Backes, D. Basin, B. Pfitzmann, and M. Waidner. Cryptographically sound theorem proving. In *CSFW'06*. IEEE, July 2006. To appear.
39. S. Tarento. Machine-checked security proofs of cryptographic signature schemes. In *ESORICS'05*, LNCS 3679, pages 140–158. Springer, Sept. 2005.

On Robust Combiners for Private Information Retrieval and Other Primitives

Remo Meier and Bartosz Przydatek

Department of Computer Science, ETH Zurich
8092 Zurich, Switzerland
remmeier@student.ethz.ch, przydatek@inf.ethz.ch

Abstract. Let $\mathcal{A}$ and $\mathcal{B}$ denote cryptographic primitives. A (k, m)-*robust $\mathcal{A}$-to-$\mathcal{B}$ combiner* is a construction, which takes m implementations of primitive $\mathcal{A}$ as input, and yields an implementation of primitive $\mathcal{B}$, which is guaranteed to be secure as long as at least k input implementations are secure. The main motivation for such constructions is the tolerance against *wrong assumptions* on which the security of implementations is based. For example, a $(1,2)$-robust $\mathcal{A}$-to-$\mathcal{B}$ combiner yields a secure implementation of $\mathcal{B}$ even if an assumption underlying *one* of the input implementations of $\mathcal{A}$ turns out to be wrong.

In this work we study robust combiners for private information retrieval (PIR), oblivious transfer (OT), and bit commitment (BC). We propose a $(1,2)$-robust PIR-to-PIR combiner, and describe various optimizations based on properties of existing PIR protocols. The existence of simple PIR-to-PIR combiners is somewhat surprising, since OT, a very closely related primitive, seems difficult to combine (Harnik *et al.*, Eurocrypt'05). Furthermore, we present $(1,2)$-robust PIR-to-OT and PIR-to-BC combiners. To the best of our knowledge these are the first constructions of $\mathcal{A}$-to-$\mathcal{B}$ combiners with $\mathcal{A} \neq \mathcal{B}$. Such combiners, in addition to being interesting in their own right, offer insights into relationships between cryptographic primitives. In particular, our PIR-to-OT combiner together with the impossibility result for OT-combiners of Harnik *et al.* rule out certain types of reductions of PIR to OT. Finally, we suggest a more fine-grained approach to construction of robust combiners, which may lead to more efficient and practical combiners in many scenarios.

Keywords: robust combiners, cryptographic primitives, reductions, private information retrieval, oblivious transfer, bit commitment.

1 Introduction

Consider a scenario when two implementations, I_1 and I_2, of some cryptographic primitive are given, e.g., two encryption schemes or two bit commitment schemes. Each implementation is based on some unproven computational assumption, α_1 resp. α_2, like for example the hardness of factoring integer numbers or the hardness of computing discrete logarithms. We would like to have an implementation I of the primitive, which is as secure as possible given the current state of knowledge. As it is often not clear, which of the assumptions α_1, α_2 is more likely to be

C. Dwork (Ed.): CRYPTO 2006, LNCS 4117, pp. 555–569, 2006.

correct, picking just one of the implementations does not work — we might bet on the wrong assumption! A better option would be to have an implementation which is guaranteed to be secure as long as *at least one* of the assumptions α_1, α_2 is correct. That is, given I_1 and I_2 we would like to construct an efficient implementation I, which is secure whenever at least one of the input implementations is. Such a construction is an example of a *(1,2)-robust combiner*, as it *combines* the input implementations and is *robust* against situations when one of the two inputs is insecure.

In general, robust combiners can use more than just two input schemes, and aim at providing a secure implementation of the output primitive assuming that sufficiently many of the candidates are secure. Moreover, the input candidates do not have to be necessarily implementing the same primitive, and the goal of a combiner may be a construction of a primitive different from the primitives given at the input. That is, a robust combiner can be viewed as a robust *reduction* of the output primitive to the input primitive(s).

The concept of robust combiners is actually not so new in cryptography and many techniques are known for combining cryptographic primitives to improve security guarantees, e.g., cascading of block ciphers. However, a more formal and rigorous study of combiners was initiated quite recently [Her05, HKN+05].

Robust combiners for some primitives, like one-way functions or pseudorandom generators, are rather simple, while for others, e.g., for oblivious transfer (OT), the construction of combiners seems considerably harder. In particular, in a recent work Harnik *et al.* [HKN+05] show that there exists no "transparent black-box" (1,2)-robust OT-combiner. Given the impossibility result for OT-combiners, it is interesting to investigate the existence of combiners for *single-database* private information retrieval (PIR), a primitive closely related, yet not known to be equivalent, to oblivious transfer. Potential PIR-combiners could lead to better understanding of relations between PIR, OT, and other primitives. Moreover, constructions of robust PIR-combiners are also of considerable practical interest, stemming from the fact that some of the most efficient PIR protocols are based on relatively new computational assumptions (e.g., [CMS99, KY01]), which are less studied and thus potentially more likely to be proved wrong.

Contributions. In this work we consider robust combiners for private information retrieval, bit commitment, and oblivious transfer. In particular, we present (1,2)-robust PIR-combiner, i.e. combiner which given two implementations of PIR yield an implementation of PIR which is secure if at least one of the input implementations is secure. We also describe various techniques and optimizations based on properties of existing PIR protocols, which yield PIR-combiners with better efficiency and applicability.

Furthermore, we construct $\mathcal{A}$-to-$\mathcal{B}$ combiners, i.e. "cross-primitive" combiners, which given multiple implementations of a primitive $\mathcal{A}$ yield an implementation of some other primitive $\mathcal{B}$, which is provably secure assuming that sufficiently many of the input implementations of $\mathcal{A}$ are secure. Specifically, we construct (1,2)-robust PIR-to-BC and PIR-to-OT combiners. To the best of our knowledge these are the first combiners of this type. While interesting in their own right, such combiners also offer insights into relationships and reductions between cryptographic primitives. In particular, our PIR-to-OT combiner together with the impossibility result of [HKN+05] rule out certain types of reductions of PIR to OT (cf. Corollary 4 in Section 4).

Finally, we suggest a more fine-grained approach to design of robust combiners. That is, we argue that in order to obtain combiners as efficient as possible, the constructions may take into account that some properties of the input candidates are proved to hold unconditionally, and hence cannot go wrong even if some computational assumption turns out to be wrong. Therefore, keeping in mind the original motivation for combiners, i.e. protection against wrong assumptions, a more fine-grained approach to design of robust combiners exploits the unconditionally secure properties and focuses on the properties which hold only under given assumptions. This change of focus yields sometimes immediately trivial constructions of combiners (as observed by Harnik et al. [HKN+05] for OT and BC), yet in many cases the resulting problems are still interesting and challenging (see Sections 2.3 and 5 for more details).

Related Work. As mentioned above, a more rigorous study of robust combiners was initiated only recently, by Herzberg [Her05] and by Harnik et al. [HKN+05]. On the other hand, there are numerous implicit uses and constructions of combiners in the literature (e.g., [AB81, EG85, DK05, HL05]).

Private information retrieval was introduced by Chor et al. [CKGS98] and has been intensively studied since then. The original setting of PIR consisted of multiple non-communicating copies of the database and guaranteed *information-theoretic* privacy for the user. Later, Kushilevitz and Ostrovsky [KO97] gave the first solution to *single-database* PIR, in which the privacy of the user is based on a computational assumption. The first PIR protocol with communication complexity polylogarithmic in the size of the database was proposed by Cachin et al. [CMS99], and in recent years more efficient constructions have been proposed (e.g. [Cha04, Lip05]). For more information about PIR we refer to the survey by Gasarch [Gas04].

The relationships between PIR and other primitives have been studied intensively in the recent years. In particular, Beimel et al. [BIKM99] proved that any non-trivial single-database PIR implies one-way functions, and Di Crescenzo et al. [DMO00] showed that such a PIR implies oblivious transfer. Kushilevitz and Ostrovsky [KO00] demonstrated that one-way trapdoor *permutations* are sufficient for non-trivial single-database PIR. On the negative side, Fischlin [Fis02] showed that there is no black-box construction of one-round (i.e., two-message) PIR from one-to-one trapdoor *functions*.

Techniques similar to the ones employed in the proposed PIR-combiners were previously used by Di Crescenzo et al. [DIO01] in constructions of universal service-providers for PIR.

Organization. Section 2 contains notation, definitions of cryptographic primitives and of robust combiners, and some general remarks about combining PIR protocols. In Section 3 we describe proposed constructions of (1,2)-robust

PIR-combiners. In Section 4 we turn to "cross-primitive" combiners and present PIR-to-BC and PIR-to-OT combiners. Finally, in Section 5 we discuss some general aspects of design of efficient combiners and point out some open problems.

2   Preliminaries

Notational Conventions. If x is a bit-string, $|x|$ denotes its length, and we write $x\|y$ to denote the concatenation of the bit-strings x, y. For an integer n we write $[n]$ to denote the set $\{1, \ldots, n\}$. The parties participating in the protocols and the adversary are assumed to be probabilistic polynomial time Turing machines, (PPTMs).

2.1   Primitives

We review shortly the primitives relevant in this work. For more formal definitions we refer to the literature.

Private Information Retrieval. is a protocol between two parties, a server holding an n-bit database $x = (x_1 \| \ldots \| x_n)$, and a user holding an index $i \in [n]$. The protocol allows the user to retrieve bit x_i without revealing i to the server, i.e. it protects user's privacy. In this work we consider only single-database PIR. Of interest are only *non-trivial* protocols, in which the total *server-side* communication (i.e. communication from the server to the user) is less than n bits. Moreover, of special interest are 2-message protocols, in which only two messages are sent: a *query* from the user to the server and a *response* from the server to the user.

Oblivious Transfer. [1] is a protocol between a sender holding two bits x_0 and x_1, and a receiver holding a choice-bit c. The protocol allows the receiver to get bit x_c so that the sender does not learn any information about receiver's choice c, and the receiver does not learn any information about bit x_{1-c}.

Bit Commitment. is a two-phase protocol between two parties Alice and Bob. In the *commit* phase Alice commits to bit b without revealing it, by sending to Bob an "encrypted" representation e of b. Later, in the *decommit* phase, Alice sends to Bob a decommitment string d, allowing Bob to "open" e and obtain b. In addition to correctness, a bit commitment scheme must satisfy two properties: *hiding*, i.e. Bob does not learn the bit b before the decommit phase, and *binding*, i.e. Alice cannot come up with decommitment strings d, d' which lead to opening the commitment as different bits. We consider also *weak* bit commitment, i.e. BC with *weak binding* property: Alice might be able to cheat, but Bob catches her cheating with noticeable probability [BIKM99].

---

[1] The version of oblivious transfer described here and used in this paper is more precisely denoted as *1-out-of-2 bit-OT* [EGL85]. There are several other versions of OT, e.g., *Rabin's OT*, *1-out-of-n bit-OT*, or *1-out-of-n string-OT*, but all are known to be equivalent [Rab81, Cré87, CK88].

2.2   Robust Combiners

The following definition is a generalization of the definition of combiners given in [HKN+05].

Definition 1 ((k,m)-robust $\mathcal{A}$-to-$\mathcal{B}$ combiner). *Let $\mathcal{A}$ and $\mathcal{B}$ be cryptographic primitives. A (k,m)-robust $\mathcal{A}$-to-$\mathcal{B}$ combiner is a PPTM which gets m candidate schemes implementing $\mathcal{A}$ as inputs and implements $\mathcal{B}$ while satisfying the following two properties:*

1. *If at least k candidates securely implement $\mathcal{A}$, then the combiner securely implements $\mathcal{B}$.*
2. *The running time of the combiner is polynomial in the security parameter κ, in m, and in the lengths of the inputs to $\mathcal{B}$.*

An $\mathcal{A}$-*to-*$\mathcal{A}$ *combiner* is called an $\mathcal{A}$-*combiner*. For completeness we recall three definitions from [HKN+05], which will be useful in our constructions.

Definition 2 (Black-box combiner [HKN+05]). *A $(1,2)$-robust combiner is called a* black-box combiner *if the following two conditions hold:*

BLACK-BOX IMPLEMENTATION: *The combiner is an oracle PPTM given access to the candidates via oracle calls to their implementation function.*

BLACK-BOX PROOF: *For every candidate there exists an oracle PPTM R^A (with access to A) such that if adversary A breaks the combiner, then R^A breaks the candidate.*

Definition 3 (Third-party black-box combiner [HKN+05]). *A third-party black-box combiner is a black-box combiner where the input candidates behave like trusted third parties. The candidates give no transcript to the players but rather take their inputs and return outputs.*

Definition 4 (Transparent black-box combiner [HKN+05]). *A transparent black-box combiner is a black-box combiner for an interactive primitive where every call to a candidate's next message function is followed by this message being sent to the other party.*

Note that the notion of *reduction* of primitive $\mathcal{B}$ to primitive $\mathcal{A}$, i.e. a construction of $\mathcal{B}$ from $\mathcal{A}$, can be viewed as a (1,1)-robust $\mathcal{A}$-to-$\mathcal{B}$ combiner. Therefore, the above definitions also include notions like a *transparent black-box reduction* or a *third-party black-box reduction*.

2.3   Remarks on Combiners for PIR Protocols

As pointed out by Harnik *et al.* [HKN+05], cryptographic primitives are mainly about security, while functionality issues are often straightforward. For example, a PIR protocol has to satisfy a security property, i.e. privacy of the user, and functionality properties: efficiency, completeness, and non-triviality. Usually[2] the

---

[2] We are not aware of any (single-database) PIR protocol not conforming to this characterization.

privacy of the user is based on some cryptographic assumption, and the remaining properties hold unconditionally. Moreover, in some cases a possible way of dealing with unknown implementations of primitives is to test them for the desired functionality, hence, even if the candidate input primitives are given as black-boxes, one can test them before applying a combiner (for a more detailed discussion of these issues see Section 3.1 in [HKN+05]).

For the above reasons we assume that the PIR candidates used as input by the combiners are guaranteed to have the desired functionality (i.e. efficiency, completeness, and non-triviality), and that explicit bounds on running time and on communication complexity are given as parts of the input to the combiner. Thus, the task of a combiner is to protect against wrong computational assumptions. This approach is especially relevant in the context of private information retrieval, since some of the most efficient PIR protocols are based on new computational assumptions (e.g., [CMS99, KY01]), which are less studied and so potentially more likely to be broken (cf. recent attack of Bleichenbacher *et al.* [BKY03] on [KY01]).

3   PIR-Combiners

In this section we assume that two (non-trivial) private information retrieval schemes are given, PIR_1 and PIR_2, where PIR_1 is a *two-message* PIR protocol with a query $q = Q_1(i)$ and a response $r = R_1(q, (x_1 \| \dots \| x_n))$, and where PIR_2 is an arbitrary (possibly multi-round) PIR protocol. We use $c_{s,1}(n)$ and $c_{s,2}(n)$ to denote the server-side communication complexities of the PIR-schemes, and $c_{u,1}(n)$ and $c_{u,2}(n)$ to denote the corresponding user-side complexities[3]. Without loss of generality we assume that these complexities give the exact number of communicated bits and are not just upper bounds.

First we describe a basic scheme for a (1,2)-robust PIR-combiner, and then present some variations of the scheme, resulting in better efficiency. Our constructions are black-box combiners, but not *transparent* black-box combiners because they require offline access to one of the candidates.

3.1   The Basic Scheme

Our basic PIR-combiner works as follows: to retrieve the i-th bit from a database $x = (x_1 \| \dots \| x_n)$, the database first defines n auxiliary databases $y_1, \dots, y_n$, where y_j is just a copy of x rotated by $(j-1)$ positions, i.e.

$$y_j = (x_j \| \dots \| x_{n-1} \| x_n \| x_1 \| \dots \| x_{j-1}).$$

The user picks a random $t \in [n]$ and sends to the server PIR_1-query $q = Q_1(t)$. For each database y_j, $j \in [n]$, the server computes the corresponding response $r_j = R_1(q, y_j)$, but instead of sending the responses back to the user, he stores them in a new database[4] $x' = (r_1 \| \dots \| r_n)$. Note that the new database x'

---

[3] In particular, $c_{s,1}(n) = |R_1(q, (x_1, \dots, x_n))|$ and $c_{u,1} = |Q_1(i)|$.

[4] A similar technique was used in [DIO01] for universal service-providers for PIR.

SERVER'S INPUT: n-bit string $x = (x_1 \| \ldots \| x_n)$
USER'S INPUT: $i \in [n]$
INPUT PIR PROTOCOLS:
    PIR$_1$: 2-message, with query $Q_1(j)$ and response $R_1(Q_1(j), x)$
    PIR$_2$: arbitrary

phase I:
1. server defines n databases y_j for each $j \in [n]$ as:
$$y_j = (x_j \| \ldots \| x_{n-1} \| x_n \| x_1 \| \ldots \| x_{j-1})$$
2. user picks a random $t \in [n]$ and sends PIR$_1$-query $q = Q_1(t)$ to the server
3. server computes n PIR$_1$-responses $r_j = R_1(q, y_j)$ for each $j \in [n]$

phase II:
1. user computes $k = ((i - t) \mod n) + 1$
2. user retrieves from server response r_k bit-by-bit, by running $|r_k|$ instances of PIR$_2$ with $x' = (r_1 \| r_2 \| \ldots \| r_n)$ as server's input
3. user computes x_i from the PIR$_1$-response r_k

Fig. 1. The basic (1,2)-robust PIR-combiner

contains a PIR$_1$-response for each bit x_j of the original database x, but with the positions rotated by $(t - 1)$. Finally the user retrieves bit-by-bit the response r_k for $k = ((i - t) \mod n) + 1$, by running $c_{s,1}(n)$ instances of PIR$_2$, and computes x_i from r_k. Figure 1 presents the combiner in more detail, and the following theorem summarizes the properties of the combiner.

Theorem 1. *There exists a black-box (1,2)-robust PIR-combiner for input candidates* PIR$_1$ *and* PIR$_2$*, where* PIR$_1$ *is a 2-message protocol, and where the candidates' server-side communication complexities satisfy*

$$c_{s,1}(n) \cdot c_{s,2}(n \cdot c_{s,1}(n)) < n \ . \tag{1}$$

The user-side communication of the resulting PIR scheme is equals

$$c_{u,1}(n) + c_{s,1}(n) \cdot c_{u,2}(n \cdot c_{s,1}(n)) \ .$$

Proof. *(sketch)* Consider the construction presented in Figure 1. It is clear that this construction is efficient. It remains to show that the resulting protocol is a secure, non-trivial PIR if at least one of the input candidates is a secure PIR. Let $\overline{\text{PIR}}$ denote the PIR protocol resulting from the combiner. If PIR$_1$ is insecure, then the server learns the random index t. However, as in this case PIR$_2$ remains secure, the server obtains no information about index k when the user retrieves the response r_k, and so does not gain information about the index i.

On the other hand, if PIR$_2$ is insecure, then the server learns index k, but as PIR$_1$ is now secure, the server gets no information about t. Since t is randomly chosen, knowledge of k does not give any information about index i.

Finally, we argue the non-triviality condition: it is easy to verify that the server-side communication of $\overline{\mathsf{PIR}}$ is

$$c_s(n) = c_{s,1}(n) \cdot c_{s,2}\left(n \cdot c_{s,1}(n)\right) ,$$

and the user-side communication is

$$c_u(n) = c_{u,1}(n) + c_{s,1}(n) \cdot c_{u,2}\left(n \cdot c_{s,1}(n)\right) .$$

Thus if $c_s(n) < n$ holds, i.e. if condition (1) is satisfied, then $\overline{\mathsf{PIR}}$ is non-trivial.   □

3.2   PIR-Combiners with Lower Communication

The basic combiner presented in the previous section is conceptually simple and works well for a wide range of candidate PIR-protocols, but leaves some space for improvements. In this section we describe some variations and optimizations of this basic combiner, which yield significant improvements in communication efficiency of the resulting PIR schemes. This results in combiners applicable to a wider range of input candidates.

First we describe how to reduce the cost of querying x' by using several databases in parallel. Then we discuss possible improvements in situations when the candidates return entire blocks of several bits instead of single bits.

Reducing Overall Communication. In the second phase of the basic scheme the user retrieves r_k bit-by-bit by running $|r_k| = c_{s,1}(n)$ instances of PIR_2 with server's input x' of length $n \cdot c_{s,1}(n)$. An alternative way of retrieving r_k is the following: we arrange all responses $r_1, \ldots, r_n$ into $l = |r_k|$ databases $x'_1, \ldots, x'_l$, each of length n, where x'_j contains the j-th bits of all responses $r_1, \ldots, r_n$. Then the user obtains r_k by retrieving the k-th bits from the databases $x'_1, \ldots, x'_l$. That is, user and server run $|r_k|$ instances of PIR_2, where in the j-th instance server's input is x'_j and user's input k. Thus we obtain the following corollary.

Corollary 1. *There exists a black-box (1,2)-robust PIR-combiner for input candidates PIR_1 and PIR_2, where PIR_1 is a 2-message protocol, and where the candidates' server-side communication complexities satisfy*

$$c_{s,1}(n) \cdot c_{s,2}(n) < n . \tag{2}$$

The user-side communication of the resulting PIR scheme is equals

$$c_{u,1}(n) + c_{s,1}(n) \cdot c_{u,2}(n) . \tag{3}$$

Note that if PIR_2 is also a 2-message PIR protocol, then only *one query* must be sent in the second phase of the combiner (for which $c_{s,1}(n)$ PIR_2-responses will be sent), thus reducing the user-side communication of the resulting PIR scheme even further, to merely

$$c_{u,1}(n) + c_{u,2}(n) .$$

Further Optimizations and Variations. If PIR_2 retrieves entire blocks rather than single bits (for example, the basic PIR protocol of [KO97] does exactly that), than the retrieval of r_k can be substantially sped-up, as it can proceed block-by-block rather than bit-by-bit. Moreover, if $|r_k|$ is not larger than the size of blocks retrieved by PIR_2, than just one execution of PIR_2 is sufficient.

Corollary 2. *There exists a black-box (1,2)-robust PIR-combiner for input candidates PIR_1 and PIR_2, where PIR_1 is a 2-message protocol, PIR_2 retrieves blocks of size at least $c_{s,1}(n)$, and where the candidates' server-side communication complexities satisfy*

$$c_{s,2}(n \cdot c_{s,1}(n)) < n \ .$$

The user-side communication of the resulting PIR scheme is equals

$$c_{u,1}(n) + c_{u,2}(n \cdot c_{s,1}(n)) \ .$$

Another simple optimization is possible when PIR_1 supports block-wise retrieval, i.e., when each PIR_1-response r_j allows retrieval of ℓ-bit blocks. In such a case it is sufficient to store in x' a subset of n/ℓ responses, so that the corresponding blocks cover the entire database — then in phase II user simply retrieves the block containing the desired bit x_i.

Finally, when the user-side communication of the candidate PIRs is much higher than the server-side communication, it is possible to balance the load better between the two parties with the so called *balancing technique*, which was introduced in the context of information-theoretic PIR [CKGS98], and which can be viewed as a simulation of block-wise retrieval: server partitions the database to u databases of size n/u. User provides then a single query for some index $j \in [n/u]$, which is answered for each of u databases, yielding a block of u bits.

Clearly, one can use multiple optimizations together (if applicable) to obtain the most efficient construction for the given candidate PIR protocols.

4   Combining PIR Protocols to Other Primitives

The research on robust combiners so far focused mainly on finding ways of combining candidate instances of a given primitive $\mathcal{A}$ to yield a secure instance of $\mathcal{A}$. In this section we describe robust combiners of a more general type, combining instances of primitive $\mathcal{A}$ to an instance of primitive $\mathcal{B}$. Such combiners can be viewed as a combination of robust combiners and reductions between primitives in one construction.

First we consider the problem of combining PIR protocols to obtain a bit commitment scheme, and present a third-party black-box (1,2)-robust PIR-to-BC combiner. Then we turn to combining PIR protocols to oblivious transfer, and present a black-box (1,2)-robust PIR-to-OT combiner. The existence of such a combiner is somewhat surprising, given the impossibility result of [HKN+05] and the fact that PIR and OT are very closely related.

ALICE'S INPUT: bit b
BOB'S INPUT: (none)
INPUT PIR PROTOCOLS: PIR_1, PIR_2

commit phase:

1. Alice picks two independent, uniformly random strings $x, y \in_R \{0,1\}^n$
   Bob picks two independent, uniformly random indices $i_1, i_2 \in_R [n]$
2. Alice and Bob execute PIR protocols $\mathsf{PIR}_1(x; i_1)$ and $\mathsf{PIR}_2(x; i_2)$, with Alice as server with database x, and Bob retrieving bit with index i_1 resp. i_2.
3. Alice sends to Bob y and $c = b \oplus \mathrm{IP}(x, y)$
   (where IP denotes inner product over $\mathrm{GF}(2)$)

decommit phase:

1. Alice sends to Bob x as decommitment string
2. Bob verifies that x is consistent with the bits retrieved during commit phase and computes $b = c \oplus \mathrm{IP}(x, y)$

Fig. 2. A (1,2)-robust PIR-to-(weak)BC combiner

4.1   PIR-to-BC Combiner

It is well-known that single-database PIR implies one-way functions [BIKM99], which in turn are sufficient to construct computationally hiding and statistically binding bit commitments schemes [Nao91]. It follows immediately that there exists a generic combiner going through these reductions and an OWF-combiner. However, such a combiner is quite inefficient, and it is not a third-party black-box combiner.

In this section we present a more efficient, third-party black-box PIR-to-BC combiner, which is basically a slight variation of the reduction of bit commitment to private information retrieval due to Beimel *et al.* [BIKM99]. In contrast to the generic combiner described above, the BC-scheme resulting from the proposed combiner is statistically hiding and computationally binding. We describe only a construction for *weak* bit commitment, which can then be strengthened by using multiple independent commitments to the same bit [BIKM99]. A detailed description of the combiner is presented in Figure 2.

Theorem 2. *There exists a third-party black-box (1,2)-robust PIR-to-BC combiner yielding a statistically hiding BC, for input candidates PIR_1 and PIR_2 with server-side communication complexities satisfying*

$$c_{s,1}(n) + c_{s,2}(n) \leq n/2 . \tag{4}$$

Proof. *(sketch)* As mentioned above, it is sufficient to show a combiner from PIR to *weak* bit commitment. Consider the construction presented in Fig. 2. The correctness and the efficiency of the scheme are straightforward. The hiding property follows from the bound on server-side communication complexities (4) and from high communication complexity of the inner product IP (see [BIKM99] for details). The weak binding property follows from the assumption that at least one of the PIR protocols is secure for the receiver, hence at least one of the indices i_1, i_2 remains unknown to Alice. Finally, it is straightforward to verify that this is a third-party black-box combiner.   □

Obviously, the bound $n/2$ in (4) is not tight. Since our focus in this work is on existence of efficient combiners, and since many practical PIR protocols have polylogarithmic communication bounds (which clearly satisfy (4)), we do not attempt to optimize this bound. Moreover, the PIR-to-OT combiner presented in the next section implies an alternative, efficient $(1, 2)$-robust PIR-to-BC combiner (cf. Corollary 3).

4.2   PIR-to-OT Combiner

The PIR-to-BC combiner presented in the previous section can be viewed as a variation of the general approach to construct (1,2)-robust PIR-to-BC combiners: first use a construction of unconditionally hiding BC from a single-database PIR to obtain BC_1 resp. BC_2, and then combine the two BC protocols using the fact that both are unconditionally secure for Alice and at most one not binding (if the corresponding PIR protocol is insecure). As we show in this section, a similar approach works for PIR-to-OT combiners.[5] That is, our proposed PIR-to-OT combiner first constructs OT protocols OT_1 and OT_2 based on candidates PIR_1 resp. PIR_2, and then combines OT_1 and OT_2 using the fact that both these protocols are unconditionally secure for the sender.

For completeness, Figure 4 in the appendix presents the construction of OT (unconditionally secure for the sender) based on single-database PIR [DMO00]. Using this construction, our proposed (1,2)-robust PIR-to-OT combiner works as follows: given two PIR protocols, PIR_1 and PIR_2, we use the reduction from Fig. 4 to obtain OT protocols OT_1 and OT_2, respectively. Now, as both resulting OT's are unconditionally secure for the sender, we can combine them by using a combiner which guarantees the privacy of the receiver as long as at least one of the two input OT's is secure. For this purpose we use the combiner[6] $R(\cdot, \cdot)$ from [HKN+05]. Figure 3 presents the proposed PIR-to-OT combiner in full detail, and we obtain the following theorem.

Theorem 3. *There exists a black-box (1,2)-robust PIR-to-OT combiner.*

Proof. *(sketch)* Consider the construction in Figure 3, and let $\overline{OT}$ denote the resulting OT protocol. Since *any* non-trivial single-database PIR implies OT with unconditional security for the sender [DMO00], OT_1 and OT_2 are well defined,

---

[5] Note that unlike in the case of PIR-to-BC combiners, it is unclear whether there exists (1,2)-robust PIR-to-OT combiner based on combiners for one-way functions: while it is known that non-trivial PIR implies one-way functions [BIKM99], it is unlikely that OT can be constructed from one-way functions only [IR89] (the most general assumptions known to be sufficient for OT are the existence of *enhanced* [EGL85, Gol04] or *dense* [Hai04] one-way trapdoor permutations).

[6] This combiner was used in [HKN+05] to construct a (2,3)-robust OT-combiner.

SENDER'S INPUT: two bits b_0, b_1
RECEIVER'S INPUT: choice bit c
INPUT PIR PROTOCOLS: PIR_1, PIR_2

1. parties construct OT protocols OT_1, OT_2 from PIR_1, PIR_2, respectively (using the construction from Fig. 4)
2. parties combine OT_1, OT_2, using combiner $R(\mathsf{OT}_1,\mathsf{OT}_2)(b_0,b_1;c)$ [HKN+05]:
    (i) sender picks a random bit r,
        receiver picks random bits c_1, c_2 s.t. $c_1 \oplus c_2 = c$
    (ii) parties run $\mathsf{OT}_1(r, r \oplus b_0 \oplus b_1; c_1)$ and $\mathsf{OT}_2(r \oplus b_0, r \oplus b_1; c_2)$
    (iii) receiver outputs the XOR of the two bits received in OT_1 and OT_2

Fig. 3. A (1,2)-robust PIR-to-OT combiner

and it is easy to verify the correctness of $\overline{\mathsf{OT}}$. Moreover, unconditional security of the sender in OT_1 and OT_2 means that the receiver obtains information about at most one of $(r, r \oplus b_0 \oplus b_1)$ and about at most one of $(r \oplus b_0, r \oplus b_1)$. This implies unconditional security of the sender in $\overline{\mathsf{OT}}$. The security of the receiver in $\overline{\mathsf{OT}}$ follows from the assumption that at least one of the input PIR protocols is secure. More precisely, security of at least one PIR implies security of at least one of OT_1, OT_2, hence, at least one of c_1, c_2 remains hidden from the sender, and consequently the sender obtains no information about c. Finally, it is easy to verify that this is a black-box combiner. □

Since the OT protocol resulting from the above combiner is unconditionally secure for the sender, we can use it to construct a statistically hiding BC scheme, hence we get the following corollary.

Corollary 3. *There exists a black-box (1,2)-robust PIR-to-BC combiner yielding a statistically hiding BC.*

Furthermore, recall that a reduction of a primitive $\mathcal{B}$ to a primitive $\mathcal{A}$ can be viewed as a (1,1)-robust $\mathcal{A}$-to-$\mathcal{B}$ combiner, hence a notion of *transparent black-box reduction* is well-defined. Note also that in the case of honest-but-curious parties or low-communication PIR-candidates, the PIR-to-OT combiner resulting from the proof of Theorem 3 is even a *third-party* black-box combiner (cf. Appendix and [DMO00]). Therefore, a combination of Theorem 3 with the impossibility result for (1,2)-robust transparent black-box OT-combiners [HKN+05] leads to the following corollary, which rules out certain types of reductions of PIR to OT.

Corollary 4. *There exists no transparent black-box reduction of single-database private information retrieval to oblivious transfer, even for honest-but-curious parties.*

5   Conclusions and Open Problems

We have presented constructions of (1,2)-robust PIR-combiners, and also "cross-primitive" combiners: PIR-to-BC and PIR-to-OT. The existence of simple and efficient PIR-combiners is somewhat surprising given the impossibility result for OT-combiners. Moreover, a closer look at the PIR-to-BC and PIR-to-OT combiners reveals a common theme — we use a reduction of the target primitive to the input primitive, and exploit additional security properties guaranteed by the reduction to obtain an efficient combiner. It seems that such a fine-grained approach to the design of combiners, i.e. taking explicitly into account that *some* properties of the candidates hold unconditionally can yield more efficient, practical combiners. Indeed, as pointed out by Harnik *et al.* [HKN+05], if the security of one of the parties is guaranteed, then constructing (1,2)-robust combiners for commitments is easy. The same observation holds for OT: it is easy to construct a (1,2)-robust OT-combiner if two candidate OTs are unconditionally secure for the sender (or receiver). Of course, while such combiners are very simple and efficient, they have somewhat limited applicability, as they require more knowledge about the input candidates. But given the apparent difficulty of efficient *general* (1,2)-robust combiners for primitives like OT or BC, a possible approach to obtain more practical combiners might be to consider constructions for "mixed" candidates, e.g., combiners that combine an unconditionally hiding bit commitment with an unconditionally binding one.

While the basic PIR-combiner we propose is applicable to many PIR protocols described in the literature, it is not universal in the sense that it does not work for *any non-trivial* PIR schemes — the combiners requires one *two-message* PIR and some bounds on communication complexities. It would be interesting to either find a universal combiner that does not need such assumptions or to further optimize the current combiner while maintaining its applicability.

An intermediate step towards universal (1,2)-robust PIR-combiners might be a construction of an universal (2,3)-robust PIR-combiner. Oblivious transfer and bit commitment, primitives considered to be hard to combine with (1,2)-robust combiners, do have very efficient universal (2,3)-robust combiners.

With regard to "cross-primitive" combiners, we have argued that there exists a PIR-to-BC combiner which yields statistically hiding bit commitment, and that there exists one yielding statistically binding bit commitment. However, for the later only an inefficient, generic construction via combiner for one-way functions is known. It would be interesting to find a more efficient, direct PIR-to-BC combiner yielding a statistically binding bit commitment scheme.

Acknowledgements. We would like to thank anonymous referees for useful comments and for pointing out the reference [DIO01].

References

[AB81]   C. A. Asmuth and G. R. Blakely. An efficient algorithm for constructing a cryptosystem which is harder to break than two other cryptosystems. *Computers and Mathematics with Applications*, 7:447–450, 1981.

[BIKM99] A. Beimel, Y. Ishai, E. Kushilevitz, and T. Malkin. One-way functions are essential for single-server private information retrieval. In *Proc. ACM STOC*, pages 89–98, 1999.

[BKY03] D. Bleichenbacher, A. Kiayias, and M. Yung. Decoding of interleaved Reed-Solomon codes over noisy data. In *Proc. ICALP 2003*, pages 97–108, 2003.

[Cha04] Y.-C. Chang. Single database private information retrieval with logarithmic communication. In *Proc. Information Security and Privacy: 9th Australasian Conference, ACISP 2004*, pages 50–61, 2004.

[CK88] C. Crépeau and J. Kilian. Achieving oblivious transfer using weakened security assumptions (extended abstract). In *Proc. IEEE FOCS '88*, pages 42–52, 1988.

[CKGS98] B. Chor, E. Kushilevitz, O. Goldreich, and M. Sudan. Private information retrieval. *J. ACM*, 45(6):965–981, 1998.

[CMS99] C. Cachin, S. Micali, and M. Stadler. Computationally private information retrieval with polylogarithmic communication. In *Proc. Eurocrypt '99*, pages 402–414, 1999.

[Cré87] C. Crépeau. Equivalence between two flavours of oblivious transfers. In *Proc. Crypto '87*, pages 350–354, 1987.

[DIO01] G. Di Crescenzo, Y. Ishai, and R. Ostrovsky. Universal service-providers for private information retrieval. *Journal of Cryptology*, 14(1):37–74, 2001.

[DK05] Y. Dodis and J. Katz. Chosen-ciphertext security of multiple encryption. In *Proc. TCC '05*, pages 188–209, 2005.

[DMO00] G. Di Crescenzo, T. Malkin, and R. Ostrovsky. Single database private information retrieval implies oblivious transfer. In *Proc. Eurocrypt '00*, pages 122–138, 2000.

[EG85] S. Even and O. Goldreich. On the power of cascade ciphers. *ACM Trans. Comput. Syst.*, 3(2):108–116, 1985.

[EGL85] S. Even, O. Goldreich, and A. Lempel. A randomized protocol for signing contracts. *Communications of the ACM*, 28(6):637–647, 1985.

[Fis02] M. Fischlin. On the impossibility of constructing non-interactive statistically-secret protocols from any trapdoor one-way function. In *Proc. CT-RSA*, pages 79–95, 2002.

[Gas04] W. I. Gasarch. A survey on private information retrieval (column: Computational complexity). *Bulletin of the EATCS*, 82:72–107, 2004.

[Gol04] O. Goldreich. *The Foundations of Cryptography*, volume II, Basic Applications. Cambridge University Press, 2004.

[Hai04] I. Haitner. Implementing oblivious transfer using collection of dense trapdoor permutations. In *Proc. TCC'04*, pages 394–409, 2004.

[Her05] A. Herzberg. On tolerant cryptographic constructions. In *CT-RSA*, pages 172–190, 2005. full version on Cryptology ePrint Archive, eprint.iacr.org/2002/135.

[HKN⁺05] D. Harnik, J. Kilian, M. Naor, O. Reingold, and A. Rosen. On robust combiners for oblivious transfer and other primitives. In *Proc. Eurocrypt '05*, pages 96–113, 2005.

[HL05] S. Hohenberger and A. Lysyanskaya. How to securely outsource cryptographic computations. In *Proc. TCC '05*, pages 264–282, 2005.

[IR89] R. Impagliazzo and S. Rudich. Limits on the provable consequences of one-way permutations. In *Proc. ACM STOC*, pages 44–61, 1989.

[KO97]   E. Kushilevitz and R. Ostrovsky. Replication is not needed: Single database, computationally-private information retrieval. In *Proc. IEEE FOCS '00*, pages 364–373, 1997.

[KO00]   E. Kushilevitz and R. Ostrovsky. One-way trapdoor permutations are sufficient for non-trivial single-server private information retrieval. In *Proc. Eurocrypt '00*, pages 104–121, 2000.

[KY01]   A. Kiayias and M. Yung. Secure games with polynomial expressions. In *Proc. ICALP 2001*, pages 939–950, 2001.

[Lip05]   H. Lipmaa. An oblivious transfer protocol with log-squared communication. In *Proc. Information Security, 8th International Conference, ISC 2005*, pages 314–328, 2005.

[Nao91]   M. Naor. Bit commitment using pseudorandomness. *J. Cryptology*, 4(2):151–158, 1991.

[Rab81]   M. O. Rabin. How to exchange secrets by oblivious transfer., 1981. Tech. Memo TR-81, Aiken Computation Laboratory, available at eprint.iacr.org/2005/187.

Appendix

For completeness, in Figure 4 we present the construction of OT (unconditionally secure for the sender) based on single-database PIR, due to Di Crescenzo *et al.* [DMO00]. This construction is used in our (1,2)-robust PIR-to-OT combiner (see Section 4.2).

Note that in this protocol the privacy of sender holds only against *honest-but-curious* receiver. It can however be transformed into a protocol resilient against arbitrary (possibly dishonest) parties [DMO00].

SENDER'S INPUT: two bits b_0, b_1
RECEIVER'S INPUT: choice bit c
COMMON INPUTS: PIR protocol, security param. κ, a param. m polynomial in κ

1. Sender and Receiver invoke m executions of PIR, with Sender as server and Receiver as user:
   for each execution $j \in [m]$ they pick independent, uniformly random inputs: Sender a string $x^j \in_R \{0,1\}^\kappa$, Receiver an index $i^j \in_R [\kappa]$
2. Receiver sets $(i_c^1, ..., i_c^m) := (i^1, ..., i^m)$, and picks random $(i_{1-c}^1, ..., i_{1-c}^m) \in [\kappa]^m$
3. Receiver sends $(i_0^1, ..., i_0^m)$ and $(i_1^1, ..., i_1^m)$ to Sender
4. Sender computes
$$z_0 := b_0 \oplus x^1(i_0^1) \oplus ... \oplus x^m(i_0^m)$$

   and

$$z_1 := b_1 \oplus x^1(i_1^1) \oplus ... \oplus x^m(i_1^m)$$

   (where $x^j(i)$ denotes the i-th bit of string x^j), and sends z_0, z_1 to Receiver
5. Receiver computes his output $b_c := z_c \oplus x^1(i^1) \oplus ... \oplus x^m(i^m)$

Fig. 4. Construction of (honest receiver) OT from single-database PIR [DMO00]

On the Impossibility of Efficiently Combining Collision Resistant Hash Functions

Dan Boneh[1,*] and Xavier Boyen[2]

[1] Computer Science Dept., Stanford University
dabo@cs.stanford.edu
[2] Voltage Inc., Palo Alto
xb@boyen.org

Abstract. Let H_1, H_2 be two hash functions. We wish to construct a new hash function H that is collision resistant if at least one of H_1 or H_2 is collision resistant. Concatenating the output of H_1 and H_2 clearly works, but at the cost of doubling the hash output size. We ask whether a better construction exists, namely, can we hedge our bets without doubling the size of the output? We take a step towards answering this question in the negative — we show that any secure construction that evaluates each hash function once cannot output fewer bits than simply concatenating the given functions.

1 Introduction

Let $H : \{0,1\}^* \to \{0,1\}^n$ be a function. A collision for H is a pair $M, M' \in \{0,1\}^*$ such that $H(M) = H(M')$ and $M \neq M'$. Roughly speaking, we say that H is collision resistant if no efficient algorithm can find collisions for H. Recent attacks [21,20,23,22] on functions previously believed to be collision resistant greatly stimulated the search for new constructions as well as methods to avoid collision resistance altogether [17]. One natural question is how to combine existing hash functions. Suppose we are given two hash functions

$$H_1, H_2 : \{0,1\}^* \to \{0,1\}^v$$

that are currently believed to be collision resistant. We worry that one of these functions will become insecure due to a future attack. Hence, we want to hedge our bet and construct a new hash function $H : \{0,1\}^* \to \{0,1\}^n$ that is collision resistant assuming at least one of H_1 or H_2 is collision resistant. This can be easily achieved by concatenation. Simply define:

$$H(x) := H_1(x) \parallel H_2(x)$$

in which case H outputs digests of length $n = 2v$. It is easy to see that if either H_1 or H_2 is collision resistant then so is H. More precisely, concatenation has the following property:

(∗)   Any collision on H leads to a collision on **both** H_1 and H_2

---

* Supported by NSF.

Therefore, if finding collisions for *either* H_1 or H_2 is difficult then finding collisions for H must also be difficult. Note that property $(*)$ holds no matter what hash functions H_1 and H_2 are used. We give precise definitions in the next section.

In this paper we ask whether there are more clever ways to construct H. Specifically, is there a construction that satisfies property $(*)$, but whose output size n is less than $2v$ bits? A positive answer will give a clean way to combine two hash functions, such as SHA-512 [18] and Whirlpool [2], without increasing the output size by too much. Here we only consider constructions for H that evaluate H_1 and H_2 at most once. Concatenation is such an example.

Unfortunately, we show that when $n < 2v$ it is not possible to satisfy property $(*)$. In other words, there is no generic construction that combines arbitrary collision resistant hash functions H_1, H_2 and whose output is any shorter than concatenation. We give a precise statement of the result in the next section. Roughly speaking, our proof shows that for any construction with $n < 2v$ we can construct functions H_1, H_2 for which property $(*)$ fails. It is worth noting that our counterexamples for H_1, H_2 are realistic functions. For instance, they can be very similar to SHA-512 and Whirlpool. Our results apply to an arbitrary number of underlying hash functions and show that we cannot do better than concatenation, as long as each function is only used once.

We begin by precisely defining what it means for a construction combining ℓ hash functions to be secure. Intuitively, the construction should satisfy property $(*)$ — a collision on H should lead to a collision on all ℓ given hash functions. Consequently, if one of the ℓ functions is collision resistant then so is H. We then state and prove our impossibility results.

1.1   Related Work

Robust combiners for various cryptographic primitives have received a steady amount of attention in the cryptographic community. Some examples include the early results of Asmuth and Blakely [1] on multiple encryption, Herzberg's work on combiners for commitment and one-way functions [9], Dodis and Katz's on encryption with chosen ciphertext security [7], and Harnik's *et al.* on key agreement [8]. Harnik *et al.* [8] provide lower bounds regarding the existence of "transparent" black-box combiners for oblivious transfer and secure computation [8]. We refer to Herzberg [9] for a survey of robust combiners for various cryptographic primitives as well as formal models. Hohenberger and Lysyanskaya recently investigated a related notion of "outsourcing" cryptographic operations by enlisting potentially malicious helpers to perform some of the calculations [11]. The very notion of increasing the security of encryption by using multiple independent keys dates back to the work of Shannon.

Our focus in this paper is on robust combiners for collision resistant hash functions. Many cryptographic hash functions today are based on the Merkle-Damgård paradigm [15,6], which repeatedly applies a given compression function on successive message blocks in order to hash messages of arbitrary size. Joux [12] showed that the concatenation of two Merkle-Damgård functions is not much more secure than the individual functions. In particular, let H_1, H_2 be Merkle-Damgård functions that output v bits each. Joux showed that a collision for $H = H_1 \parallel H_2$ can be found in expected time $O(v\, 2^{v/2})$. This is far less than the time for finding a collision on a random function outputting $2v$ bits. In other words, concatenation provides a good hedge, but does not increase security for Merkle-Damgård functions. Generalizations of Joux's attack were given by Nandi and Stinson [16] and Hoch and Shamir [10].

Kelsey [13] observed that truncating a collision resistant hash function need not be collision resistant. A special case of our results (the case $\ell = 1$ in Theorem 1) implies that *any* construction that evaluates a collision resistant function H and outputs fewer bits than the output of H need not be collision resistant.

For constructing hash functions, Lucks [14] studied the idea of increasing the internal state of the iterated hash beyond the size of the final output, specifically to defend it against Joux's attack. Along the same lines, Coron *et al.* [5] showed how to increase the security of an iterated hash function in which the compression function is viewed as a random oracle. They gave several techniques for building an economical Merkle-Damgård hash that becomes indistinguishable from a random oracle provided we stay below the birthday bound, by truncating the output or by encoding the input using a prefix-free code.

Generic constructions that attempt to build secure compression functions from ideal block ciphers have also been investigated. Preneel *et al.* [19] identified 12 constructions that were potentially secure. Black *et al.* [4] provided formal security proofs for all 12 constructions in the ideal cipher model. Some impossibility results have also been shown. Notably, Black *et al.* [3] proved that secure hash functions based on block ciphers cannot be "highly efficient", *i.e.*, require only one block cipher call per message block on a fixed set of keys.

2   Secure Combination of Collision Resistant Hashing

Suppose we wish to combine ℓ collision resistant hash functions $H_1, \ldots, H_\ell$ into a single function that is collision resistant if any of the H_i is collision resistant. We first define precisely what it means for such a construction to be secure. Throughout the section we let H_i be a function from $\{0,1\}^*$ to $\{0,1\}^{v_i}$ for $i = 1, \ldots, \ell$.

A **collision resistant combination** of $H_1, \ldots, H_\ell$ is a pair (C, P) where C is a boolean circuit and P is a probabilistic algorithm such that:

- The circuit C is a boolean circuit that includes special "oracle" gates for evaluating the hash functions $H_1, \ldots, H_\ell$. We refer to $C^{H_1, \ldots, H_\ell}$ as the hash function constructed from $H_1, \ldots, H_\ell$. The circuit takes as input a string in $\{0,1\}^{\leq m}$, namely a string of length at most m, and outputs a digest in $\{0,1\}^n$. The upper bound on the input size merely reflects the finiteness of the circuit. This upper bound only strengthens our results since it shows that efficient combiners are not possible even when the input space for C is finite. We assume C is compressing so that m is (much) larger than n. The output of C on input M and using hash functions $H_1, \ldots, H_\ell$ is denoted by $C^{H_1, \ldots, H_\ell}(M)$.

– Algorithm P is an oracle Turing machine. It provides the proof of security for C. The algorithm takes as input a pair of messages (M, M'). It then repeatedly queries the oracles $H_1, \ldots, H_\ell$ on inputs of its choice and finally outputs two vectors,

$$\boldsymbol{W} = (W_1, \ldots, W_\ell) \quad \text{and} \quad \boldsymbol{W}' = (W_1', \ldots, W_\ell')$$

The output of P on input (M, M') is denoted by $P^{H_1, \ldots, H_\ell}(M, M')$.

The purpose of algorithm P is to convert any collision on $C^{H_1, \ldots, H_\ell}$ into collisions on all $H_1, \ldots, H_\ell$. Thus, we define security of a pair (C, P) as follows.

– Let $H_1, \ldots, H_\ell$ be functions where $H_i : \{0,1\}^* \rightarrow \{0,1\}^{v_i}$.
– Let $M, M' \in \{0,1\}^{\leq m}$ be a collision for the resulting function $C^{H_1, \ldots, H_\ell}$.

We say that P **succeeds** on this $(H_1, \ldots, H_\ell)$ and (M, M') if the output of $P^{H_1, \ldots, H_\ell}(M, M')$ is two vectors $\boldsymbol{W} = (W_1, \ldots, W_\ell)$ and $\boldsymbol{W}' = (W_1', \ldots, W_\ell')$ such that

$$\text{For all} \quad i = 1, \ldots, \ell : \quad (W_i, W_i') \quad \text{is a collision for } H_i.$$

We define the "advantage" $\text{Adv}_P[(H_1, \ldots, H_\ell), (M, M')]$ as the probability that P succeeds on input (M, M') relative to the oracles $(H_1, \ldots, H_\ell)$. The probability is over the random bits used by algorithm P.

Definition 1. *We say that (C, P) is an ϵ-secure collision resistant combination if for all $H_1, \ldots, H_\ell$ and all collisions (M, M') on $C^{H_1, \ldots, H_\ell}$ we have that*

$$Adv_P[(H_1, \ldots, H_\ell), (M, M')] > 1 - \epsilon$$

For example, for the concatenation construction discussed in the introduction, the pair (C, P) is as follows:

– $C^{H_1, H_2}(M) = H_1(M) \parallel H_2(M)$,   and
– $P^{H_1, H_2}(M, M')$ simply outputs the two vectors (M, M) and (M', M').

This pair (C, P) is a secure collision resistant combination since for any H_1, H_2 and any collision (M, M') on C^{H_1, H_2} we have that

$$Adv_P[(H_1, H_2), (M, M')] = 1$$

Our goal is to show that for any construction (C, P) where C outputs fewer bits than concatenation, we have that $\text{Adv}_P[(H_1, \ldots, H_\ell), (M, M')]$ is negligible, for some $(H_1, \ldots, H_\ell)$ and some collision (M, M'). This will show that there is no provably secure way of combining generic hash functions that is better than concatenation. In this paper we focus on constructions (C, P) where C makes at most one call to each of $H_1, \ldots, H_\ell$. We do not discuss constructions where C uses some hash function multiple times.

3 Black-Box Impossibility Results

Our main result states that, given any pair (C, P) where C evaluates each function once, it will always be possible to find a collision for $C^{H_1,\ldots,H_\cdot}$ while preventing P from finding collisions for all the H_i. The only trivial exceptions are (1) when the output size of C is at least as large as the combined output sizes of all the H_i, which is to say that C does no better than concatenation, and (2) when the output of C is as large as its input, which is to say that C is non-compressing. Formally, we have the following theorem.

Theorem 1. *Let (C, P) be a collision resistant combination of oracles H_i : $\{0, 1\}^* \to \{0, 1\}^{v_\cdot}$ for $i = 1, \ldots, \ell$, where $C : \{0, 1\}^{\leq m} \to \{0, 1\}^n$. Let $w = v_1 + \ldots + v_\ell$. Suppose that $n < w \leq m - \log_2 \ell$. Suppose further that C calls each of the H_i at most once. Then, there exist inputs $M, M' \in \{0, 1\}^{\leq m}$ and functions $\hat{H}_i : \{0, 1\}^* \to \{0, 1\}^{v_\cdot}$ for $i = 1, \ldots, \ell$, relative to which:*

$$Adv_P[(\hat{H}_1, \ldots, \hat{H}_\ell), (M, M')] \leq q^2/2^{v+1},$$

where $v = \min_{i=1}^{\ell}\{v_i\}$ is the smallest oracle output size, and q is the maximum number of oracle queries made by P during each execution.

The theorem shows that if C outputs fewer than $w = v_1 + \cdots + v_\ell$ bits then the probability of defeating P is overwhelming, provided that $q \ll \sqrt{2^{v+1}}$, namely, $q \ll \sqrt{2^{v_\cdot + 1}}$ for all $i = 1, \ldots, \ell$. The condition $q \ll \sqrt{2^{v+1}}$ must hold since otherwise one of the H_i is trivially not collision resistant and should not be used to begin with.

We prove Theorem 1 in two steps. First, in Section 3.1 we assume the existence of $H_1, \ldots, H_\ell$ and a pair $M \neq M'$ that is a collision for $C^{H_1,\ldots,H_\cdot}$ satisfying a certain property. We use a randomization argument to build $\hat{H}_1, \ldots, \hat{H}_\ell$ that defeat P. Then, in Section 3.2 we show that for any C satisfying the conditions of Theorem 1 there must exist functions $H_1, \ldots, H_\ell$ and a collision $M \neq M'$ that satisfies the required property. This second step is the heart of the proof. These two results taken together will immediately prove Theorem 1.

Proof. Theorem 1 follows from Theorems 2 and 3.   □

3.1 Randomization

Suppose there exist functions $H_1, \ldots, H_\ell$ and two messages M, M' such that:

- M, M' are a collision for $C^{H_1,\ldots,H_\cdot}$, and
- there is some $j \in \{1, \ldots, \ell\}$ so that the process of evaluating $C^{H_1,\ldots,H_\cdot}(M)$ and $C^{H_1,\ldots,H_\cdot}(M')$ does not present a collision for H_j.

We show how to tweak the H_i into a new set of oracles, $\hat{H}_1, \ldots, \hat{H}_\ell$, such that P will fail to reduce a collision for $C^{\hat{H}_1,\ldots,\hat{H}_\cdot}$ to a collision on all the $\hat{H}_i$. In Section 3.2 we show that such $H_1, \ldots, H_\ell$ and M, M' must exist.

To state the result more precisely, we first introduce some notation. As before, we let $H_i : \{0,1\}^* \to \{0,1\}^{v_i}$ for $i = 1, \ldots, \ell$ be a set of functions, and denote by $C^{H_1, \ldots, H_\ell} : \{0,1\}^m \to \{0,1\}^n$ the function evaluated by C when it is given oracle access to the H_i. Consider an input message $M \in \{0,1\}^m$. For each $i = 1, \ldots, \ell$, we define:

- $\mathbf{W}_i(M)$ to be the set of oracle queries to H_i made by C while evaluating $C^{H_1, \ldots, H_\ell}(M)$,
- $\mathbf{V}_i(M) = \{H_i(W) : W \in \mathbf{W}_i(M)\}$ to be the set of corresponding values output by H_i.

These sets are taken without multiplicity, and hence, we have a collision for H_j upon evaluating $C(M)$ if and only if $|\mathbf{W}_j(M)| > |\mathbf{V}_j(M)|$. If $|\mathbf{W}_j(M)| = |\mathbf{V}_j(M)|$ then no collision occurred for H_j. We can now state the following theorem.

Theorem 2. *Let (C, P) be a collision resistant combination of ℓ oracles as previously defined. Assume that there exist oracles $H_i : \{0,1\}^* \to \{0,1\}^{v_i}$ for $i = 1, \ldots, \ell$ and a pair of messages M, M', such that:*

- *$M \neq M'$ and $C^{H_1, \ldots, H_\ell}(M) = C^{H_1, \ldots, H_\ell}(M')$, and*
- *$|\mathbf{V}_j(M) \cup \mathbf{V}_j(M')| = |\mathbf{W}_j(M) \cup \mathbf{W}_j(M')|$ for some $j \in \{1, \ldots, \ell\}$.*

Let $v = \min_{i=1}^{\ell}\{v_i\}$. Suppose that P never makes more than q oracle queries upon each execution. Then, there exist oracles $\hat{H}_i : \{0,1\}^* \to \{0,1\}^{v_i}$ for $i = 1, \ldots, \ell$, relative to which:

$$Adv_P[(\hat{H}_1, \ldots, \hat{H}_\ell), (M, M')] \leq q^2/2^{v+1}$$

Proof. Let thus M, M', and H_i for $i = 1, \ldots, \ell$, be as stated in the theorem. We construct the $\hat{H}_i$ by patching H_i in a way that breaks P without affecting the output of $C(M)$ and $C(M')$. Basically, we instruct $\hat{H}_i$ to emulate H_i on all queries that appear in the course of evaluating one or both of $C^{H_1, \ldots, H_\ell}(M)$ and $C^{H_1, \ldots, H_\ell}(M')$. On all other inputs, the output of $\hat{H}_i$ is set to a fresh random string, so that $\hat{H}_i$ behaves as a random function on these inputs.

We now give an explicit construction of the $\hat{H}_i$ based on the H_i and the circuit C. For each $i \in \{1, \ldots, \ell\}$ we pick an independent random function $R_i : \{0,1\}^* \to \{0,1\}^{v_i}$. Then, for $i \in \{1, \ldots, \ell\}$, we define $\hat{H}_i$ as follows:

$$\hat{H}_i(W) := \begin{cases} H_i(W) & \text{if } W \in \mathbf{W}_i(M) \cup \mathbf{W}_i(M') \\ R_i(W) & \text{otherwise.} \end{cases}$$

Notice that the random function R_i can be selected and evaluated efficiently using a lazy algorithm that would pick and store a randomly drawn $R_i(W) \in \{0,1\}^{v_i}$ upon each novel query for $R_i(W)$.

It is easy to see that the messages M and M' still collide under $C^{\hat{H}_1, \ldots, \hat{H}_\ell}$ for the new oracle functions, since C is a deterministic function, and its evaluation on inputs M and M' is unaffected by the change of oracles.

Now, recall that j is the index of (one of) the hash oracle for which no collision occurred during the evaluations of $C^{H_1,\dots,H.}(M)$ and $C^{H_1,\dots,H.}(M')$, or, equivalently, during the evaluations of $C^{\hat{H}_1,\dots,\hat{H}.}(M)$ and $C^{\hat{H}_1,\dots,\hat{H}.}(M')$. At least one such $j \in \{1,\dots,\ell\}$ exists by our assumptions.

Next, we consider the set of all distinct queries to $\hat{H}_j$ that are made during an execution of P on input (M, M'). Each output value $\hat{H}_j(x)$ returned by $\hat{H}_j$ on query x is either:

1. "unpatched", i.e., so that $\hat{H}_j(x) = H_j(x)$, corresponding to a query in $\mathbf{W}_j(M) \cup \mathbf{W}_j(M')$;
2. "patched", i.e., such that $\hat{H}_j(x) = R_j(x)$, in response to any other query.

We know that the few unpatched outputs of $\hat{H}_j$ are all distinct, otherwise the evaluations of $C(M)$ and $C(M')$ would have caused a collision on H_j and thus $\hat{H}_j$, contradicting our assumption. As for the patched outputs, by construction they are random binary strings of v_j bits.

Therefore, using the union bound we obtain the following bound on the probability that an arbitrary set of queries to $\hat{H}_j$ will result in a collision on $\hat{H}_j$:

$$\Pr\left[\begin{array}{c} q \text{ queries cause} \\ \text{a collision on } \hat{H}_j \end{array}\right] \leq \sum_{\substack{a \neq b \in \\ \{0,\dots,q-1\}}} \Pr\left[\begin{array}{c} \text{query } a \text{ collides} \\ \text{with query } b \end{array}\right] \leq$$

$$\leq \sum_{\substack{a \neq b \in \\ \{0,\dots,q-1\}}} \frac{1}{2^{v.}} \leq \frac{q^2}{2^{v.+1}} \leq \frac{q^2}{2^{v+1}}$$

We conclude that the probability that $P^{\hat{H}_1,\dots,\hat{H}.}(M, M')$ outputs a collision on $\hat{H}_j$ is at most $\frac{q^2}{2^{v.+1}}$. The probability is over the random choice of $R_1,\dots,R_\ell$ and the random bits used by P. But then by Markov there must exist some fixed setting of $R_1,\dots,R_\ell$ so that for the corresponding $\hat{H}_1,\dots,\hat{H}_\ell$, algorithm $P^{\hat{H}_1,\dots,\hat{H}.}(M, M')$ outputs a collision on $\hat{H}_j$ with probability at most $\frac{q^2}{2^{v.+1}}$. This time the probability is only over the random bits of P. Therefore, the advantage $\mathrm{Adv}_P[(\hat{H}_1,\dots,\hat{H}_\ell),(M, M')]$ is at most $q^2/2^{v+1}$ as required.   □

3.2   Existence Argument

We now turn to the second step of the proof of Theorem 1, which is the main part of the proof. Let C be a circuit that outputs elements in $\{0,1\}^n$ and uses hash functions (i.e. oracles) $H_1,\dots,H_\ell : \{0,1\}^* \to \{0,1\}^{v.}$. As in Section 3, let $w = v_1 + \dots + v_\ell$. We assume that $n < w$ so that the output of C is at least one bit less than the concatenation of all the outputs of $H_1,\dots,H_\ell$. In this section we will occasionally use the standard notation $A \to B$ to denote $(\neg A) \vee B$ where A and B are boolean expressions.

Suppose C uses each function H_i for $i = 1,\dots,\ell$ at most once. We show that there exists $H_1,\dots,H_\ell$ and $M, M' \in \{0,1\}^m$ such that: M, M' are a collision for $C^{H_1,\ldots,H_\ell}$, but for some H_j $(1 \leq j \leq \ell)$ evaluating $C(M)$ and $C(M')$ causes no internal collision on H_j. The argument of Section 3.1 then shows that this M, M' pair cannot be used to find collisions on H_j proving that C cannot be a secure combination of $H_1, \ldots, H_\ell$.

More precisely, we state the main result of this section in the following theorem. We use the notation introduced at the beginning of Section 3.1.

Theorem 3. *Let C be a circuit computing a function from $\{0,1\}^m$ to $\{0,1\}^n$ using a set of oracles $H_i : \{0,1\}^* \rightarrow \{0,1\}^{v_i}$ for $i = 1, \ldots, \ell$. Furthermore, we assume that each H_i is used at most once in C. Let $w = v_1 + \ldots + v_\ell$. Then, whenever $n < w \leq m - \log_2 \ell$, there exist functions $H_1, \ldots, H_\ell$ and messages M, M' such that:*

- $M \neq M'$ *and* $C^{H_1,\ldots,H_\ell}(M) = C^{H_1,\ldots,H_\ell}(M')$, *and*
- $|\mathbf{V}_j(M) \cup \mathbf{V}_j(M')| = |\mathbf{W}_j(M) \cup \mathbf{W}_j(M')|$ *for some $j \in \{1, \ldots, \ell\}$.*

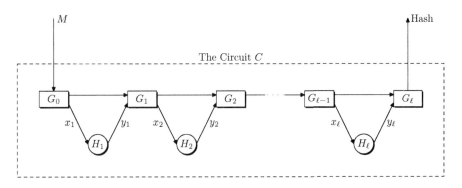

Fig. 1. Generic hash construction – the circuit C

Clearly the circuit C must use all hash functions $H_1, \ldots, H_\ell$. Otherwise, there is no hope of using a collision on C to get a collision on all H_i. Since we assumed that each H_i is used only once we know that C makes exactly ℓ oracle calls, one for each of $H_1, \ldots, H_\ell$. An inductive argument on the structure of C shows that, up to re-ordering the oracles, there exist circuits $G_0, G_1, \ldots, G_\ell$ so that C can be written as shown in Figure 1. Note that the wires connecting G_i and G_{i+1} can be quite thick, and in particular preserve all of the input and the state so far.

For an input message $M \in \{0,1\}^m$ let $\bar{y} = (y_1, \ldots, y_\ell)$ be the outputs of $H_1, \ldots, H_\ell$ while evaluating $C(M)$. Clearly the output of C is fully determined by M and $\bar{y}$. For $i = 1, \ldots, \ell$ we define the function:

$$X_i(M, y_1, \ldots, y_{i-1}) \quad : \quad \{0,1\}^{m+v_1+\ldots+v_{i-1}} \quad \rightarrow \quad \{0,1\}^*$$

to be the input x_i given to H_i when evaluating $C(M)$ and assuming that the output of the hash functions $H_1, \ldots, H_{i-1}$ is $y_1, \ldots, y_{i-1}$ respectively.

Theorem 3 can now be restated more directly. The theorem states that there exist functions $H_1, \ldots, H_\ell$ and inputs $M \neq M'$ such that $C(M) = C(M')$ and the following condition holds. Let $(x_1, y_1), \ldots, (x_\ell, y_\ell)$ be the input/output pairs for $H_1, \ldots, H_\ell$ while evaluating $C(M)$. Similarly let $(x'_1, y'_1), \ldots, (x'_\ell, y'_\ell)$ be the input/output pairs while evaluating $C(M')$. Then there exists a $j \in \{1, \ldots, \ell\}$ for which

$$(y_j = y'_j) \rightarrow (x_j = x'_j)$$

In other words, no collision occurred on H_j.

Proof (of Theorem 3). The plan is to consider the set S of *all* 2^{m+w} tuples of the form $(M, y_1, \ldots, y_\ell)$. We often write $\bar{y} = (y_1, \ldots, y_\ell)$. Observe that a tuple $(M, \bar{y})$ in S completely defines the output of $C(M)$: simply define the output of hash function i to be y_i for all $i = 1, \ldots, \ell$. We view a tuple $(M, \bar{y})$ as an abstract object that is not derived from a specific instantiation of the hash functions.

We first partition S into equivalence classes based on the output of C. That is, two tuples $(M, \bar{y})$ and $(M', \bar{y}')$ are in the same equivalence class if C evaluates to the same output on both. Note that this partition of S is dependent only on C. It is independent of the choice of $H_1, \ldots, H_\ell$. Since C outputs elements in $\{0,1\}^n$ there are at most 2^n equivalence classes. It follows that there exists an equivalence class $E \subseteq S$ of size at least 2^{m+w-n}. Since $w > n$ this equivalence class E must be of size strictly greater than 2^m. We will use this E extensively in the remainder of the proof. First, we argue in the following simple lemma that E must contain tuples with certain properties.

Lemma 1. *Under the conditions of Theorem 3 the set E must contain tuples satisfying the following properties:*

1. *There exist distinct $M^{(0)}, \ldots, M^{(\ell)} \in \{0,1\}^m$ and some $\bar{y} \in \{0,1\}^w$ such that $(M^{(i)}, \bar{y})$ is in E for all $i = 0, \ldots, \ell$. In other words, E contains $\ell + 1$ tuples that all share the same $\bar{y}$.*
2. *There exists a tuple $(M', \bar{y}')$ in E such that $\bar{y}' \neq \bar{y}$.*

Proof. The first property is easy to see. Let us partition E into equivalence classes based on $\bar{y}$. Two tuples are in the same equivalence class if they have identical $\bar{y}$ components. If every class had only ℓ tuples in it then the size of E would be at most $2^w \ell$. But $2^w \ell \leq 2^m$ which contradicts the fact that $|E| > 2^m$. Hence, there must exist $\ell + 1$ tuples in E with the same $\bar{y}$.

Similarly, we prove the second property by a counting argument. Since $|E| > 2^m$ there must exist in E two tuples $T_1 = (M, \bar{y}_1)$ and $T_2 = (M, \bar{y}_2)$ where $\bar{y}_1 \neq \bar{y}_2$. Then either $\bar{y} \neq \bar{y}_1$ or $\bar{y} \neq \bar{y}_2$. Thus, either T_1 or T_2 is the required tuple. This completes the proof of the lemma.   □

Note that M' may be contained in $\{M^{(0)}, \ldots, M^{(\ell)}\}$. This lemma relies on the fact that C outputs fewer than w bits. Part (2) of the lemma would not hold otherwise. For example, part (2) does not hold for the concatenation construction.

Back to the proof of Theorem 3, we will use the tuples that are guaranteed to exist by the lemma to complete the proof of the theorem. We need a little more notation. For $i = 0, 1, \ldots, \ell$ and $\bar{y} = (y_1, \ldots, y_\ell)$ we let:

$$x_1^{(i)} = X_1(M^{(i)}),$$
$$x_2^{(i)} = X_2(M^{(i)}, y_1),$$
$$\vdots$$
$$x_\ell^{(i)} = X_\ell(M^{(i)}, y_1, \ldots, y_{\ell-1})$$

and write $\bar{x}^{(i)} = (x_1^{(i)}, \ldots, x_\ell^{(\ell)})$. These are the inputs to the hash functions corresponding to the tuple $(M^{(i)}, \bar{y}^{(i)})$. Similarly, we write $\bar{y}' = (y_1', \ldots, y_\ell')$, and let:

$$x_1' = X_1(M'),$$
$$x_2' = X_2(M', y_1'),$$
$$\vdots$$
$$x_\ell' = X_\ell(M', y_1', \ldots, y_{\ell-1}')$$

Again, we write $\bar{x}' = (x_1', \ldots, x_\ell')$.

We now construct the functions $H_1, \ldots, H_\ell$ and the inputs M, M' needed to prove Theorem 3. For $i = 1, \ldots, \ell$ let H_i be some arbitrary function from $\{0,1\}^*$ to $\{0,1\}^v$. We show how to modify these functions in at most two points to obtain a collision on C without an internal collision on some H_j. The difficulty is in ensuring that our modifications to the H_i are legal. That is, we never map the same input to two different outputs, otherwise we will not be able to instantiate the H_i. We consider two cases.

Case 1: There exist $j \in \{1, \ldots, \ell\}$ and $a, b \in \{0, \ldots, \ell\}$ such that $x_j^{(a)} = x_j^{(b)}$ and $a \neq b$. In this case, for $i = 1, \ldots, \ell$ we define $\hat{H}_i$ as follows:

$$\hat{H}_i(x) = \begin{cases} y_i & \text{if } x = x_i^{(a)} \\ y_i & \text{if } x = x_i^{(b)} \\ H_i(x) & \text{otherwise} \end{cases}$$

Defining $\hat{H}_i(x)$ in this way is legal since we never map the same input to two different outputs.

Recall that both tuples $(M^{(a)}, \bar{y})$ and $(M^{(b)}, \bar{y})$ are in E. Moreover, we just forced each $\hat{H}_i$ to evaluate to y_i when C is given $M^{(a)}$ or $M^{(b)}$ as input. Therefore,

$$C^{\hat{H}_1, \ldots, \hat{H}_\cdot}(M^{(a)}) = C^{\hat{H}_1, \ldots, \hat{H}_\cdot}(M^{(b)})$$

Since $a \neq b$, we know by the lemma that $M^{(a)} \neq M^{(b)}$. But $x_j^{(a)} = x_j^{(b)}$ and hence there is no collision on $\hat{H}_j$. Thus, $M^{(a)}$ and $M^{(b)}$ cause a collision on C but not a collision on $\hat{H}_j$, as required.

Case 2: Suppose Case 1 did not happen. In this case we know that the following holds:

- Since $\bar{y} \neq \bar{y}'$, there exists an $r \in \{1, \ldots, \ell\}$ where: $y_r \neq y_r'$.
- Since Case 1 did not happen, then for all $j = 1, \ldots, \ell$ the set $\{x_j^{(i)}\}_{i=0}^{\ell}$ contains $\ell + 1$ distinct elements.

Observe that the second bullet implies that for any vector $\bar{x} = (x_1, \ldots, x_\ell)$ there exists some $k \in \{0, 1, \ldots, \ell\}$ such that $x_j \neq x_j^{(k)}$ for all $j = 1, \ldots, \ell$. The reason is that there are $\ell + 1$ possible values for k, but there are only ℓ coordinates in $\bar{x}$. Therefore, there must exist some $k \in \{0, 1, \ldots, \ell\}$ such that $\bar{x}$ differs from $\bar{x}^{(k)}$ in all the coordinates.

In particular, there exists $k \in \{0, 1, \ldots, \ell\}$ such that $\bar{x}'$ differs from $\bar{x}^{(k)}$ in all the coordinates. Then for $i = 1, \ldots, \ell$ we define $\hat{H}_i$ as follows:

$$
\hat{H}_i(x) = \begin{cases} y_i' & \text{if } x = x_i' \\ y_i & \text{if } x = x_i^{(k)} \\ H_i(x) & \text{otherwise} \end{cases}
$$

Since the vectors $\bar{x}'$ and $\bar{x}^{(k)}$ differ in all the coordinates, defining $\hat{H}_i(x)$ in this way is legal. We never map the same input to two different outputs.

Both tuples $(M^{(k)}, \bar{y})$ and $(M', \bar{y}')$ are in E and therefore

$$
C^{\hat{H}_1, \ldots, \hat{H}_\cdot}(M^{(k)}) = C^{\hat{H}_1, \ldots, \hat{H}_\cdot}(M')
$$

Furthermore, since $x_1^{(k)} \neq x_1'$ we know that $X_1(M^{(k)}) \neq X_1(M')$ and hence $M^{(k)} \neq M'$. In addition, since $y_r \neq y_r'$ there is no collision on $\hat{H}_r$. Thus, once again, M' and $M^{(k)}$ cause a collision on C but not a collision on $\hat{H}_r$, as required.

We see that in both cases we were able to produce the required functions $\hat{H}_1, \ldots, \hat{H}_\ell$ and pair of inputs M, M'. This completes the proof of Theorem 3. □

4   Discussion and Future Work

We discuss a few directions for future work in this area.

Stronger Impossibility Results: Our current impossibility results apply to any construction C that uses each of the underlying hash functions at most once. Is there a similar impossibility result for constructions that use each hash function multiple times? To extend our results all that is needed is a stronger existence argument, namely a version of Theorem 3 that applies to more general circuits. Ideally, Theorem 3 can be strengthened to handle adaptive constructions, namely constructions that evaluate the underlying hash functions a variable number of times. For example, the number of times that C evaluates the function H_1 may depend on the length of M. Nevertheless, even an existence argument for circuits C that evaluate the hash functions a *constant* number of times would be progress. Until a complete impossibility argument is obtained, we cannot rule out the existence of space-efficient combiners.

Non-Blackbox Results: Our impossibility results build *generic* hash functions $H_1, \ldots, H_\ell$ that cause a construction C and its proof of security P to fail. In practice, however, hash functions typically follow the Merkle-Damgård paradigm. Thus, a natural question is whether one can give an impossibility proof where the counter-example hash functions $H_1, \ldots, H_\ell$ are Merkle-Damgård functions. Alternatively, is there a space efficient combiner that is proven secure only when instantiated with Merkle-Damgård functions? (even though the combiner is necessarily insecure when instantiated with general hash functions).

We note that a stronger impossibility proof, as mentioned in the first paragraph, will also provide an impossibility proof for efficiently combining Merkle-Damgård functions. Simply view the ℓ Merkle-Damgård chains as part of the circuit C and then apply the impossibility argument to the ℓ compression functions. A complete impossibility proof will provide ℓ compression functions for which the proof of security P fails. Then the derived hash functions $H_1, \ldots, H_\ell$ cause the proof P to fail and are Merkle-Damgård functions as required.

Impossibility Results for Other Hashing Concepts: It seems natural that similar ideas can be applied to other hash function concepts such as second pre-image resistance. Can one prove that concatenation is the best secure combiner for second pre-image resistant functions?

5 Conclusion

We studied the problem of combining multiple hash functions into a single function that is collision resistant whenever at least one of the original functions is. The hope was that, given a number of plausibly secure hash constructions (*e.g.*, SHA-512 and Whirlpool), one might be able to hedge one's bet and build a new function that is at least as secure as the strongest of them. The combination should be space efficient in that the final output should be smaller than the concatenation of all hashes.

We showed that no such efficient black-box combination is possible assuming each hash function is evaluated once. We leave for future work the question of generalizing our proof to the case where the same hash can be evaluated more than once — or building a working construction that exploits this condition.

Acknowledgments

We thank Anupam Datta and the anonymous reviewers for their helpful comments.

References

1. Charles Asmuth and G. R. Blakely. An efficient algorithm for constructing a cryptosystem which is harder to break than two other cryptosystems. *Computers and Mathematics with Applications*, 7:447–450, 1981.
2. Paulo S. L. M. Barreto and Vincent Rijmen. The Whirlpool hashing function. In *First open NESSIE Workshop*, Leuven, Belgium, November 2000.
3. John Black, Martin Cochran, and Thomas Shrimpton. On the impossibility of highly efficient blockcipher-based hash functions. In *Proceedings of Eurocrypt '05*, volume 3494 of *LNCS*. Springer-Verlag, 2005.
4. John Black, Phillip Rogaway, and Thomas Shrimpton. Black-box analysis of the block-cipher-based hash-function constructions from PGV. In *Proceedings of Crypto '02*, LNCS. Springer-Verlag, 2002.
5. Jean-Sebastian Coron, Yevgeniy Dodis, Cecile Malinaud, and Prashant Puniya. Merkle-Damgård revisited: how to construct a hash function. In *Proceedings of Crypto '05*, volume 3621 of *LNCS*. Springer-Verlag, 2005.
6. Ivan B. Damgård. A design principle for hash functions. In *Proceedings of Crypto '89*, LNCS. Springer-Verlag, 1989.
7. Yevgeniy Dodis and Jonathan Katz. Chosen ciphertext security of multiple encryption. In *Proceedings of TCC '05*, pages 188–209, 2005.
8. Danny Harnik, Joe Kilian, Moni Naor, Omer Reingold, and Alon Rosen. On robust combiners for oblivious transfer and other primitives. In *Proceedings of Eurocrypt '05*, pages 96–113, 2005.
9. Amir Herzberg. Tolerant combiners: Resilient cryptographic design. Cryptology ePrint Archive, Report 2002/135, 2002. http://eprint.iacr.org/.
10. Jonathan Hoch and Adi Shamir. Breaking the ICE - finding multicollisions in iterated concatenated and expanded (ICE) hash functions. In *Proceedings of FSE '06*, 2006.
11. Susan Hohenberger and Anna Lysyanskaya. How to securely outsource cryptographic computations. In *Proceedings of TCC '05*, pages 264–282, 2005.
12. Antoine Joux. Multicollisions in iterated hash functions. application to cascaded constructions. In *Proceedings of Crypto '04*, volume 3152 of *LNCS*, pages 306–316. Springer-Verlag, 2005.
13. John Kelsey. Can we get a reduction proof for truncated hashes? Crypto '05 rump session, 2005.
14. Stefan Lucks. Design principles for iterated hash functions. Cryptology ePrint Archive, Report 2004/253, 2004. http://eprint.iacr.org/.
15. Ralph C. Merkle. One way hash functions and DES. In *Proceedings of Crypto '89*, LNCS. Springer-Verlag, 1989.
16. Mridul Nandi and Doug R. Stinson. Multicollision attacks on some generalized sequential hash functions. Cryptology ePrint Archive, Report 2006/055, 2006.
17. Moni Naor and Moti Yung. Universal one-way hash functions and their cryptographic applications. In *Proceedings of STOC'89*, pages 33–43, 1989.
18. NIST. Secure hash standard. Federal Information Processing Standards (FIPS) Publication 180-2, 2002.
19. Bart Preneel, René Govaerts, and Joos Vandewalle. Hash functions based on block ciphers: A synthetic approach. In *Proceedings of Crypto '93*, LNCS. Springer-Verlag, 1993.
20. Xiaoyun Wang, Xuejia Lai, Dengguo Feng, Hui Chen, and Xiuyuan Yu. Cryptanalysis of the hash functions MD4 and RIPEMD. In *Proceedings of Eurocrypt '05*, volume 3494 of *LNCS*, pages 1–18. Springer-Verlag, 2005.

21. Xiaoyun Wang, Andrew Yao, and Frances Yao. New collision search for SHA-1. Rump Session Crypto'04, 2004.
22. Xiaoyun Wang, Yiqun Lisa Yin, and Hongbo Yu. Finding collisions in the full SHA-1. In *Proceedings of Crypto '05*, volume 3621 of *LNCS*. Springer-Verlag, 2005.
23. Xiaoyun Wang and Hongbo Yu. How to break MD5 and other hash functions. In *Proceedings of Eurocrypt '05*, volume 3494 of *LNCS*, pages 19–35. Springer-Verlag, 2005.

On the Higher Order Nonlinearities of Algebraic Immune Functions

Claude Carlet[*]

INRIA Projet CODES, BP 105, 78153 Le Chesnay Cedex, France
claude.carlet@inria.fr

Abstract. One of the most basic requirements concerning Boolean functions used in cryptosystems is that they must have high algebraic degrees. This simple criterion is not always well adapted to the concrete situation in which Boolean functions are used in symmetric cryptography, since changing one or several output bits of a Boolean function considerably changes its algebraic degree while it may not change its robustness. The proper characteristic is the r-th order nonlinearity profile (which includes the first-order nonlinearity). However, studying it is difficult and almost no paper, in the literature, has ever been able to give general effective results on it. The values of the nonlinearity profile are known for very few functions and these functions have little cryptographic interest. A recent paper has given a lower bound on the nonlinearity profile of functions, given their algebraic immunity. We improve upon it, and we deduce that it is enough, for a Boolean function, to have high algebraic immunity, for having non-weak low order nonlinearity profile (even when it cannot be evaluated), except maybe for the first order.

Keywords: stream cipher, block cipher, algebraic attack, Boolean function, algebraic immunity, algebraic degree, higher order nonlinearity.

1 Introduction

Boolean functions, that is, F_2-valued functions defined over the vector space F_2^n of all binary vectors of a given length n, are used in the S-boxes of block ciphers and in the pseudo-random generators of stream ciphers. They play a central role in their security.

In stream ciphers, the main model for the generation of the keystream consists of a linear part, producing a sequence with a large period, usually composed of one or several LFSRs, and a nonlinear combining or filtering function f which produces the output, given the state of the linear part. In the nonlinear combiner sub-model, the outputs to several LFSRs are combined using a nonlinear Boolean function to produce the keystream. In the nonlinear filter sub-model, the content of some of the flip-flops in a single (longer) LFSR constitute the input to a nonlinear Boolean function which produces the keystream. These models which

---

[*] Also member of the University of Paris 8 (MAATICAH).

C. Dwork (Ed.): CRYPTO 2006, LNCS 4117, pp. 584–601, 2006.
© International Association for Cryptologic Research 2006

are very efficient, in particular in hardware, have undergone a lot of cryptanalysis and to resist those attacks, different design criteria have been proposed for
both the LFSRs and the combining Boolean function. The main classical cryptographic criteria for designing the function f are balancedness (f is balanced
if its Hamming weight equals 2^{n-1}) to prevent the system from leaking statistical information on the plaintext when the ciphertext is known, a high algebraic
degree (that is, a high degree of the algebraic normal form of the function) to
prevent the system from Massey's attack by the Berlekamp-Massey algorithm
(cf. [31,39], see also [34]), the non-existence of (non-zero) linear structure $a \in F_2^n$
(such that $f(x + a) + f(x)$ is constant) so that the function effectively depends
on all its variables, a high order of correlation immunity (and more precisely,
of resiliency, since the functions must be balanced - a function is t-resilient if
each of its restrictions obtained by keeping constant t input bits is balanced)
to counter correlation attacks (in the case of combining functions), and a high
nonlinearity (that is, a large Hamming distance to affine functions) to withstand
correlation attacks (again) and linear attacks.

The recent algebraic attacks [15] have led to further characteristics that a cryptographic Boolean function must have. These attacks cleverly use over-defined
systems of multivariate nonlinear equations to recover the secret key (the idea
of using such systems comes from C. Shannon [40], but the improvement in the
efficiency of the method is recent). The core of the analysis in the standard algebraic attack is to find out low degree functions $g \neq 0$ and h such that $fg = h$.
It has been shown in [37] that this is equivalent to the existence of a low degree
nonzero annihilator of f or of $1 + f$, that is, of a function g such that $fg = 0$ (i.e.
whose support is disjoint of that of f) or $(1 + f)g = 0$. The minimum degree of
such g is called the (basic) algebraic immunity of f and must be as high as possible (the maximum being $\lceil \frac{n}{2} \rceil$). This condition is not sufficient, since a function
can have sufficiently high algebraic immunity and be weak against fast algebraic
attacks [16]. If one can find g of low degree and $h \neq 0$ such that $fg = h$, then
a fast algebraic attack is feasible if the degree of h is not too high, see [16,1,24].
Since $fg = h$ implies $fh = ffg = fg = h$, we see that h is then an annihilator of
$f + 1$ and its degree is then at least equal to the algebraic immunity of f. This
means that having a high algebraic immunity is not only a necessary condition
for a resistance to standard algebraic attacks but also for a resistance to fast
algebraic attacks.

Some of the criteria above play also important roles for S-boxes in block
ciphers: the nonlinearity (cf. the linear attack by Matsui [32], see also [12]) and
the algebraic degree (the complexity of the "higher order differential attack" on
block ciphers due to Knudsen and Lai [26,28] depends on the algebraic degrees
of the Boolean functions involved in the system).

But these criteria must be considered in an extended way: suppose that, given
a function which does not satisfy some criterion, it is possible, by changing one
or a few bits in its output (that is, in its truth-table) to obtain a function which
satisfies the criterion; then this criterion cannot have a general relevance to
cryptography, since this change does not fundamentally modify the robustness

of the system using this function (however, this situation is not quite the same according to whether the function is used in a synchronous stream cipher, a self-synchronizing stream cipher or a block cipher). Some papers in the literature have already addressed this problem for some criteria: see [38] for the criterion of non-existence of nonzero linear structure (Meier and Staffelbach considered the so-called distance to linear structures) and [29,10] for the resiliency criterion. Some other criteria – for instance the nonlinearity – do not change much when a few bits of the output to the function are changed; hence, they do not need such extension. On the contrary, changing one single bit in the output to an n-variable function of algebraic degree at most $n - 1$ moves its degree to n, and if, starting from a balanced function we want to keep balancedness, changing two bits moves it almost surely to $n - 1$. A natural way of putting this right is to do as Meier and Staffelbach did for the linear structures, considering, for every $r < n$, the minimum Hamming distance to all functions of degrees at most r (whose set is the so-called r-th order Reed-Muller code and will be denoted by $RM(r)$). This distance is usually called the r-th order nonlinearity of the function (and more simply its nonlinearity in the first-order case).

We shall call the *nonlinearity profile* of a function the sequence whose r-th term, for $r = 1, \ldots, n - 1$, equals the r-th order nonlinearity of the function. Note that the nonlinearity profile is extended-affine-invariant, in the sense that if ϕ is an affine automorphism of F_2^n and if ℓ is an affine Boolean function on F_2^n, then for every n-variable Boolean function f, the nonlinearity profile of the function $f \circ \phi + \ell$ equals that of f. The best known upper bound on the r-th order nonlinearity of general functions has been given in [11]. Several papers have shown the role played by this parameter in relation to some cryptanalyses (note that, contrary to the first order nonlinearity, it must have low value for allowing the attacks to be realistic) and studied it from an algorithmic viewpoint [14,23,25,27,33,35]. However, very few have attempted to give constructions of functions with reasonably good nonlinearity profile or to show general properties of this parameter [25]. In fact, until recently, almost nothing relevant was known on this criterion (see Section 2). Fortunately, it has been shown lately that, if the algebraic immunity $AI(f)$ of a function f is known, then we can deduce a lower bound on its r-th order nonlinearity for every $r \leq AI(f) - 1$: for the first order, see [17] and the improvement of [30]; for any order, see [8] (see Section 2 for a recall of these bounds). This changes completely the situation with the nonlinearity profile. In this paper, we obtain a new bound which improves upon the bound of [8] for all values of $AI(f)$ when the number of variables is smaller than or equal to 12, and for most values of $AI(f)$ when the number of variables is smaller than or equal to 22 (which covers the practical situation of stream ciphers). It also improves asymptotically upon it.

The paper is organized as follows. In Section 2 are given the necessary definitions and properties of the main cryptographic criteria on Boolean functions. In Section 3, we study the dimension of the vector space of annihilators with prescribed algebraic degrees, of a Boolean function with given algebraic degree. The results of this section are used in Section 4 to obtain the lower bound on the r-th order nonlinearity of a function of given algebraic immunity. Finally, in Section 5, we show that the bound of Section 4 simplifies the question of designing cryptographic Boolean functions meeting all necessary criteria.

2   Preliminaries

Let n be any positive integer. Any Boolean function f on n variables admits a unique algebraic normal form (ANF):

$$f(x_1, \ldots, x_n) = \sum_{I \subseteq \{1, \ldots, n\}} a_I \prod_{i \in I} x_i,$$

where the a_I's are in F_2. The terms $\prod_{i \in I} x_i$ are called *monomials*. The *algebraic degree* $d°f$ of a Boolean function f equals the maximum degree of those monomials whose coefficients are nonzero in its algebraic normal form. *Affine functions* are those Boolean functions of degrees at most 1.

A slightly different form for the ANF is $f(x) = \sum_{u \in F_2^n} a_u x^u$, where $a_u \in F_2$ and where $x^u = \prod_{i=1}^n x_i^{u_i}$. Then $d°f$ equals $\max_{a_u \neq 0} wt(u)$, where $wt(u)$ denotes the Hamming weight $|\{i = 1, \ldots, n \,/\, u_i = 1\}|$ of u. Note that, for every $x \in F_2^n$, we have then $f(x) = \sum_{u \preceq x} a_u$, where $u \preceq x$ means that every coordinate of u is upper bounded by the corresponding coordinate of x, that is, that the support of u is included in the support of x.

The Hamming weight of a Boolean function is the Hamming weight of its list of values, that is, the size of its support $\{x \in F_2^n \,/\, f(x) = 1\}$. The Hamming distance between two Boolean functions is the Hamming weight of $f + g$, that is $d(f, g) = |\{x \in F_2^n \,/\, f(x) \neq g(x)\}|$.

Definition 1. *Let $f : F_2^n \to F_2$ be an n-variable Boolean function. Let r be a positive integer such that $r \leq n$. The r-th order nonlinearity of f is the minimum Hamming distance between f and all n-variable functions of algebraic degrees at most r.*

We shall denote the r-th order nonlinearity of f by $nl_r(f)$. The first-order nonlinearity of f is simply called the nonlinearity of f and denoted by $nl(f)$.

Clearly we have $nl_r(f) = 0$ if and only if f has degree at most r. So, the knowledge of the nonlinearity profile (i.e. of all the nonlinearities of orders $r \geq 1$) of a Boolean function includes the knowledge of its algebraic degree. It is in fact a much more complete cryptographic parameter than are the single (first-order) nonlinearity and the algebraic degree: the former is not sufficient for knowing the cryptographic behavior of a function (it does not allow to quantify for instance the resistance to Berlekamp-Massey attack) and the latter is still less sufficient, as explained in introduction.

As far as we know, the nonlinearity profile (or a great part of it) is known in general only for quadratic functions (the functions of algebraic degrees at most 2) and for their sums with functions of very small Hamming weights. Indeed, the first-order nonlinearities of quadratic functions are known (see [36]) and the r-th order nonlinearity of a quadratic function is obviously null for every $r \geq 2$. The first-order nonlinearities of the functions of degrees greater than or equal to 3 are unknown, except for some particular primary constructions of Boolean functions (such as the indicators of flats, or some concatenations of such indicators or of quadratic functions - including the Maiorana-McFarland functions) and for some secondary constructions; nothing is known on the second-order nonlinearities of functions of degrees at least 3 (except for functions of small weights). In the case of functions of small Hamming weights (e.g. the indicators of flats of small dimensions), the r-th order nonlinearity is equal, for sufficiently low values of r (namely, for $2^{n-r} > 2wt(f)$), to the weight $wt(f)$ itself.

The algebraic immunity [37] of a Boolean function f quantifies the resistance to the standard algebraic attack of the pseudo-random generators using it as a nonlinear function.

Definition 2. *Let* $f : F_2^n \rightarrow F_2$ *be an n-variable Boolean function. We call annihilator of f any n-variable function g whose product with f is null (i.e. whose support is included in the support of $f+1$, or in other words any function which is a multiple of $f + 1$). The algebraic immunity of f is the minimum algebraic degree of all the nonzero annihilators of f or of $f + 1$.*

We shall denote the algebraic immunity of f by $AI(f)$.

A very useful property is its affine invariance: for every affine automorphism ϕ of F_2^n, we have $AI(f \circ \phi) = AI(f)$. This comes from the affine invariance of the algebraic degree.

Clearly, since f is an annihilator of $f + 1$ (and $f + 1$ is an annihilator of f) we have $AI(f) \leq d^\circ f$.

As shown in [15], we always have $AI(f) \leq \lceil \frac{n}{2} \rceil$. This bound is tight (see below). Also, we know that almost all Boolean functions have algebraic immunities close to this optimum; more precisely, for all $a < 1$, AI(f) is almost surely greater than $\frac{n}{2} - \sqrt{\frac{n}{2} \ln \left(\frac{n}{a \ln 2} \right)}$ when n tends to infinity: see [21].

Even when restricting ourselves to functions with optimum algebraic immunity, the algebraic attacks oblige to use now functions on at least 13 variables, see [4,8] (this number is a strict minimum and is in fact risky; a safer number of variables would better be near 20).

Very few functions are known (up to affine equivalence) with provably optimum algebraic immunities: the functions whose construction is introduced in [18] (see in [8] their further properties) and some functions which are symmetric (that is, whose outputs depend only on the Hamming weights of their inputs) [19,3]. These functions have some drawbacks: all of them have insufficient nonlinearities and all but one are non-balanced. Moreover, the functions studied in [19,3], and to a slightly smaller extent the functions introduced in [18], have not a good behavior against fast algebraic attacks, see [2,20]. But the research in this domain is very active and it is probable that better examples of functions will be found in the future.

It was shown in [17] that the weight of a function f with given algebraic immunity satisfies: $\sum_{i=0}^{AI(f)-1} \binom{n}{i} \leq wt(f) \leq \sum_{i=0}^{n-AI(f)} \binom{n}{i}$. In particular, if n is odd and f has optimum algebraic immunity, then f is balanced.

The first lower bound on the (first-order) nonlinearity of functions with given algebraic immunity has been obtained in [17]: $nl(f) \geq \sum_{i=0}^{AI(f)-2} \binom{n}{i}$. In [30], M. Lobanov has improved upon this lower bound: $nl(f) \geq 2 \sum_{i=0}^{AI(f)-2} \binom{n-1}{i}$. In [8], an easy generalization to the r-th order nonlinearity of the bound obtained in [17] has been given: $nl_r(f) \geq \sum_{i=0}^{AI(f)-r-1} \binom{n}{i}$. In the present paper, we extend Lobanov's bound into a bound which improves, asymptotically and in most cases of practical situations, upon the bound obtained in [8]. This bound is related to the dimension of the annihilators with prescribed algebraic degrees of Boolean functions with given algebraic degrees.

3   The Dimension of the Vector Space of Prescribed Degree Annihilators of a Function

The number of linearly independent low degree annihilators of a given Boolean function f and of the function $f + 1$ is an important parameter for evaluating the complexity of algebraic attacks on the systems using this function. We shall see in the next section that it plays also an important role in relation to the r-th order nonlinearity.

Definition 3. *Let h be an n-variable Boolean function. We denote by $An_k(h)$ the vector space of those annihilators of degrees at most k of h and by $d_{k,h}$ the dimension of $An_k(h)$.*

Little is known on the behavior of $d_{k,h}$. For $k = n$, we have clearly $d_{n,h} = 2^n - wt(h)$ since $An_n(h)$ contains all functions whose supports are disjoint of that of h. It is also shown in [17,8] that:

- for $k = AI(h)$, we have $d_{k,h} \leq \binom{n}{k}$,
- if h is balanced and has algebraic immunity $\frac{n}{2}$ (n even), then $d_{\frac{n}{2},h} \geq \frac{1}{2} \cdot \binom{n}{\frac{n}{2}}$,
- if h has algebraic immunity $\frac{n+1}{2}$ (n odd), then $d_{\frac{n+1}{2},h} = \binom{n}{\frac{n+1}{2}}$.

Also, Lobanov [30] showed that for every non-constant affine function h and every k, we have $d_{k,h} = \sum_{i=0}^{k-1} \binom{n-1}{i}$.

Before introducing an upper bound on $d_{k,h}$ which is valid for all functions, we generalize Lobanov's result by determining the values of $d_{k,h}$ for several classes of functions. This will be useful in the sequel.

Proposition 1. *Let h be any n-variable function of degree r, such that $0 \leq r \leq n$, and of weight 2^{n-r}. Then for every $k \geq 0$ we have $d_{k,h} = \sum_{i=0}^{k} \binom{n}{i} - \sum_{i=0}^{k} \binom{n-r}{i}$.*

Proof. We know that h is the indicator of an $(n - r)$-dimensional flat (see e.g. [36]), and thanks to the affine invariance of the algebraic immunity, we may without loss of generality assume that it equals $(x_1+1)(x_2+1)\cdots(x_r+1)$. The system characterizing the elements of $An_k(h)$, that is, the system of all equations $\sum_{u \preceq x \mid wt(u) \leq k} a_u = 0$ where x ranges over the support $\{(0,\ldots,0)\} \times F_2^{n-r}$ of h, does not involve any unknown a_u such that $(u_1,\ldots,u_r) \neq (0,\ldots,0)$. And when considering it as a system with unknowns a_u such that $(u_1,\ldots,u_r) = (0,\ldots,0)$, it is the system obtained when characterizing the $(n-r)$-variable annihilators of degrees at most k of the constant function 1. This last system has rank $\sum_{i=0}^{k} \binom{n-r}{i}$, since the function 1 admits the null function as only annihilator, and this implies that $d_{k,h} = \sum_{i=0}^{k} \binom{n}{i} - \sum_{i=0}^{k} \binom{n-r}{i}$. Note that, in the case $r=1$, this is the value given by Lobanov for non-constant affine functions. $\qquad\square$

Proposition 2. *Let h be any n-variable function of degree r, such that $0 \leq r \leq n$, and of weight $2^n - 2^{n-r}$. Then for every $k \geq 0$ we have $d_{k,h} = \sum_{i=0}^{k-r} \binom{n-r}{i}$.*

Proof: $h+1$ is the indicator of an $(n-r)$-dimensional flat, and we may without loss of generality assume that it equals $x_1 x_2 \cdots x_r$. Then the elements of $An_k(h)$ are the products of $x_1 x_2 \cdots x_r$ with those functions in the variables $x_{r+1},\ldots,x_n$ whose degrees are at most $k-r$. Then $d_{k,h} = \sum_{i=0}^{k-r} \binom{n-r}{i}$. In the case $r=1$, this is also the value given by Lobanov. $\qquad\square$

Proposition 3. *Let t be an integer such that $t \leq n$ and let h be the symmetric function defined by $h(x) = 1$ if and only if $wt(x) < t$. Then, for every k, we have $d_{k,h} = \sum_{i=t}^{k} \binom{n}{i}$.*

Proof: The coefficients in the ANFs of the elements of $An_k(h)$ are the solutions of the system of equations $\sum\limits_{u \preceq x \mid wt(u) \leq k} a_u = 0$, where x ranges over the set of vectors of weights strictly smaller than t. If $k \geq t-1$, then these equations become $\sum_{u \preceq x} a_u = 0$ and these ANFs are the polynomials such that $a_u = 0$ if $wt(u) < t$ (and a_u is any element of F_2 if $t \leq wt(u) \leq k$). Otherwise, it is clear that $An_k(h) = \{0\}$. $\qquad\square$

If $t \leq \lceil \frac{n}{2} \rceil$, we have then $AI(h) = t$, since it is easy to show then that $An_k(h+1)$ does not contain nonzero functions of degrees strictly smaller than t.

Note that, denoting by f the majority function (i.e. the symmetric function of support $\{x \in F_2^n / wt(x) \geq \lceil \frac{n}{2} \rceil\}$), Proposition 3 with $t = \lceil \frac{n}{2} \rceil$ (resp. with $t = \lfloor \frac{n}{2} \rfloor + 1$) gives the value of $d_{k,f+1}$ (resp. of $d_{k,f}$, thanks to affine invariance).

The functions studied in Propositions 1 and 2 are balanced when $r=1$ only, and those studied in Proposition 3 are balanced when $t = \frac{n+1}{2}$ (n odd). We study in the next proposition a more general case of balanced functions.

Proposition 4. *Let h be an n-variable function of weight 2^{n-r} ($1 < r \leq n-1$) and ℓ a non-constant affine function such that $h + \ell$ is balanced. Then*

$$d_{k,h+\ell} = \sum_{i=0}^{k-1} \binom{n-1}{i} - \sum_{i=k-r+1}^{k-1} \binom{n-r-1}{i} + \sum_{i=0}^{k-r-1} \binom{n-r-1}{i}.$$

Proof. We may without loss of generality assume that $h(x)$ equals $(x_1 + 1)(x_2 + 1) \cdots (x_r + 1)$ and that $l(x)$ is the function x_{r+1}. The annihilators of $h + l$ are then the multiples of $(x_1 + 1)(x_2 + 1) \cdots (x_r + 1) + (x_{r+1} + 1)$.

Let $\left((x_1 + 1)(x_2 + 1) \cdots (x_r + 1) + (x_{r+1} + 1)\right) \left(\sum_{u \in F_2^n} a_u x^u\right)$ be such a multiple.

1. For every u such that $u_{r+1} = 1$ and for which there exists $i \leq r$ such that $u_i = 1$, we have $((x_1 + 1)(x_2 + 1) \cdots (x_r + 1) + (x_{r+1} + 1)) x^u = 0$. Hence, we must not take this case into account when quantifying the dimension.

2. For every u such that $u_{r+1} = 0$ and for which there exists $i \leq r$ such that $u_i = 1$, the corresponding multiple $((x_1 + 1)(x_2 + 1) \cdots (x_r + 1) + (x_{r+1} + 1)) x^u$ equals $(x_{r+1} + 1) \prod_{i=1}^n x_i^{u_i}$. Its degree equals $1 + wt(u_1, \dots, u_n)$.

3. For every u such that $u_1 = \cdots = u_r = 0$ and $u_{r+1} = 1$, the corresponding multiple equals $((x_1 + 1)(x_2 + 1) \cdots (x_r + 1)) \prod_{i=r+1}^n x_i^{u_i}$. Its degree equals $r + 1 + wt(u_{r+2}, \dots, u_n)$.

4. For every u such that $u_1 = \cdots = u_r = u_{r+1} = 0$, the corresponding multiple equals $((x_1 + 1)(x_2 + 1) \cdots (x_r + 1) + (x_{r+1} + 1)) \prod_{i=r+2}^n x_i^{u_i}$. Its degree equals $r + wt(u_{r+2}, \dots, u_n)$.

The functions of cases 2, 3 and 4 are linearly independent. Then $d_{k,h+l}$ equals

$$\sum_{i=0}^{k-1} \binom{n-1}{i} - \sum_{i=0}^{k-1} \binom{n-r-1}{i} + \sum_{i=0}^{k-r-1} \binom{n-r-1}{i} + \sum_{i=0}^{k-r} \binom{n-r-1}{i}. \qquad \square$$

Remark: The knowledge of $d_{k,h}$ for some function h gives information on $d_{k,h'}$ for some other functions h':

1. For every n-variable functions h and h' and every positive integer k, we have $|d_{k,h} - d_{k,h'}| \leq \max(wt(h(h' + 1)), wt((h + 1)h')) \leq wt(h + h')$, since the ranks of the systems characterizing $An_k(h)$ and $An_k(h')$ satisfy the same inequality (indeed, adding equations to a system increases its rank by at most the number of added equations, and the system characterizing the ANFs of the annihilators of h' can be obtained from the system characterizing the ANFs of those of h by adding $wt(h'(h + 1))$ equations and suppressing $wt(h(h' + 1))$ equations).

2. Let h be an n-variable function and let h' be the $(n + 1)$-variable function $h'(x_1, \dots, x_{n+1}) = h(x_1, \dots, x_n)$. Let k be an integer. The ANFs of the elements of $An_k(h)$ are the solutions of the system of equations $\sum_{u \in F_2^n \,|\, wt(u) \leq k} a_u x^u = 0$, where x ranges over $supp(h)$ and the ANFs of the elements of $An_k(h')$ are the solutions of the system of equations $\sum_{v \in F_2^{n+1} \,|\, wt(v) \leq k} b_v y^v = 0$, where y ranges over $supp(h') = supp(h) \times F_2$. This last equation is equal to $\sum_{u \in F_2^n \,|\, wt(u) \leq k} b_{u,0} x^u = 0$

if $y = (x, 0)$ and to $\sum_{u \in F_2^n \,|\, wt(u) \leq k} b_{u,0} x^u + \sum_{u \in F_2^n \,|\, wt(u) \leq k-1} b_{u,1} x^u = 0$ if $y = (x, 1)$.

Hence, $d_{k,h'} = d_{k,h} + d_{k-1,h}$.

In the next lemma, we extend to all Boolean functions the result from [30], recalled above, which dealt only with affine functions.

Lemma 1. *Let n be a positive integer. Let r and k be positive integers smaller than or equal to n. Let h be any n-variable Boolean function of algebraic degree r. Then*

$$d_{k,h} \leq \min \left(\sum_{i=AI(h)}^{k} \binom{n}{i}, \sum_{i=0}^{k} \binom{n}{i} - \sum_{i=0}^{k} \binom{n-r}{i} \right).$$

Proof. We first prove that $d_{k,h} \leq \sum_{i=AI(h)}^{k} \binom{n}{i}$. If two elements of $An_k(h)$ have the same degree k part $\sum_{u \in F_2^n \mid wt(u)=k} x^u$, then their sum belongs to $An_{k-1}(h)$. We deduce that $d_{k,h}$ is smaller than or equal to the sum of $d_{k-1,h}$ and of the dimension of $RM(k)/RM(k-1)$, where $RM(k)$ is the Reed-Muller code of order k. This proves $d_{k,h} \leq d_{k-1,h} + \binom{n}{k}$ and we deduce the relation $d_{k,h} \leq \sum_{i=AI(h)}^{k} \binom{n}{i}$ by induction on k, since, by definition, $d_{AI(h)-1,h} = 0$.

We prove now that $d_{k,h} \leq \sum_{i=0}^{k} \binom{n}{i} - \sum_{i=0}^{k} \binom{n-r}{i}$. Since h has degree r and since the dimension of $An_k(h)$ is invariant under affine equivalence, we can assume without loss of generality that $h(x) = x_1 x_2 \cdots x_r + h'(x)$, where h' has degree at most r and where the term $x_1 x_2 \cdots x_r$ has null coefficient in its ANF. For any choice of $n-r$ bits $u_{r+1}, \ldots, u_n$, the restriction $h_{u_{r+1},\ldots,u_n}$ of h obtained by fixing the variables $x_{r+1}, \ldots, x_n$ to the values $u_{r+1}, \ldots, u_n$ (respectively) has then degree r, and has therefore odd weight, since r is the number of its variables. Hence it has weight at least 1. For every $(u_{r+1}, \ldots, u_n) \in F_2^{n-r}$, let us denote by $x_{u_{r+1},\ldots,u_n}$ a vector x such that $(x_{r+1}, \ldots, x_n) = (u_{r+1}, \ldots, u_n)$ and $h(x) = 1$.

A function $g(x) = \sum_{u \in F_2^n \mid wt(u) \leq k} a_u x^u$ is an annihilator of h if and only if, for every $x \in F_2^n$ such that $h(x) = 1$, we have $g(x) = 0$. A necessary condition is that $g(x) = 0$ for every $x = x_{u_{r+1},\ldots,u_n}$. If, in each of the resulting equations, we transfer all unknowns a_u such that $(u_1, \ldots, u_r) \neq (0, \ldots, 0)$ to the right hand side, we obtain a system S' in the unknowns a_u such that $(u_1, \ldots, u_r) = (0, \ldots, 0)$. Replacing the right hand sides of the resulting equations by 0 (i.e. considering the corresponding homogeneous system S'_0) gives the system that we obtain when we characterize the $(n-r)$-variable annihilators of degrees at most k of the constant function 1, considered as a function in the variables $x_{r+1}, \ldots, x_n$. Since the constant function 1 admits only the null function as annihilator, this means that the matrix of S'_0 has full rank $\sum_{i=0}^{k} \binom{n-r}{i}$. Hence the rank of the whole system of equations $\sum_{u \in F_2^n \mid wt(u) \leq k} a_u x^u = 0$, where x ranges over the support of h, is at least $\sum_{i=0}^{k} \binom{n-r}{i}$ and the dimension of $A_k(h)$ is at most $\sum_{i=0}^{k} \binom{n}{i} - \sum_{i=0}^{k} \binom{n-r}{i}$. $\square$

Note that, for $k = n$, Lemma 1 gives $wt(h) \geq \max(\sum_{i=0}^{AI(h)-1} \binom{n}{i}, 2^{n-r})$ since $d_{n,h} = 2^n - wt(h)$, but we already knew that $wt(h) \geq 2^{n-r}$, see [36].

Remarks:

1. The bound of Lemma 1, which generalizes and improves upon the bound $d_{AI(h),h} \leq \binom{n}{AI(h)}$ of [8], is tight for every value of $AI(h)$ (upper bounded by $\lceil \frac{n}{2} \rceil$) and for every $k \leq n$. When $AI(h) = 1$, it is achieved at least by all functions of weight 2^{n-r}, according to Proposition 1 (all of these functions have algebraic immunity 1). When $AI(h) = t > 1$, it is achieved by the function of

Proposition 3. Note that in this case, the value of r is large. We do not assert that the bound is also tight for every r.

2. We show in Appendix that the bound of Lemma 1 can be improved in some cases. We do not know whether the stronger inequality $d_{k,h} \leq \sum_{i=AI(h)}^{k} \binom{n}{i} - \sum_{i=AI(h)}^{k} \binom{n-r}{i}$ can be true for every function of degree r and for some values of k (depending on the value of $AI(h)$). The functions of Proposition 1 are not counter-examples since they have algebraic immunity 1, and those of Proposition 3 neither since they have degree $r > n - t$. □

4   The Lower Bound on the r-th Order Nonlinearity

We need a preliminary result before stating our main result.

Proposition 5. *Let f be a Boolean function in n variables and let r, r' be non-negative integers such that $r' \leq r$ and $AI(f) - r - 1 \geq 0$. For every n-variable function h of degree r and of algebraic immunity r', we have*

$$wt(fh) \geq \max \left(\sum_{i=0}^{r'-1} \binom{n}{i}, \sum_{i=0}^{AI(f)-r-1} \binom{n-r}{i} \right) \quad \text{if } r' \leq AI(f) - r - 1,$$

$$\geq \sum_{i=0}^{AI(f)-r-1} \binom{n}{i} \quad \text{if } r' > AI(f) - r - 1.$$

In all cases, we have:

$$wt(fh) \geq \sum_{i=0}^{AI(f)-r-1} \binom{n-r}{i}.$$

Proof. Let k be any non-negative integer. A Boolean function of degree at most k belongs to $An_k(fh)$ if and only if the coefficients in its ANF satisfy a system of $wt(fh)$ equations in $\sum_{i=0}^{k} \binom{n}{i}$ variables. Hence we have: $\dim(An_k(fh)) \geq \sum_{i=0}^{k} \binom{n}{i} - wt(fh)$.

According to Lemma 1: $\dim(An_k(h)) \leq \min \left(\sum_{i=r'}^{k} \binom{n}{i}, \sum_{i=0}^{k} \binom{n}{i} - \sum_{i=0}^{k} \binom{n-r}{i} \right)$.

If $\dim(An_k(fh)) > \dim(An_k(h))$, then there exists an annihilator g of degree at most k of fh which is not an annihilator of h. Then, gh is a nonzero annihilator of f and has degree at most $k + r$. Thus, if $k = AI(f) - r - 1 \geq 0$, we arrive to a contradiction. We deduce that $\dim(An_{AI(f)-r-1}(fh)) \leq \dim(An_{AI(f)-r-1}(h))$. This implies that $\sum_{i=0}^{AI(f)-r-1} \binom{n}{i} - wt(fh)$ is upper bounded by:

$$\min \left(\sum_{i=r'}^{AI(f)-r-1} \binom{n}{i}, \sum_{i=0}^{AI(f)-r-1} \binom{n}{i} - \sum_{i=0}^{AI(f)-r-1} \binom{n-r}{i} \right),$$

that is:

$$wt(fh) \geq \max \left(\sum_{i=0}^{r'-1} \binom{n}{i}, \sum_{i=0}^{AI(f)-r-1} \binom{n-r}{i} \right) \quad \text{if } r' \leq AI(f) - r - 1,$$

and

$$wt(fh) \geq \sum_{i=0}^{AI(f)-r-1} \binom{n}{i} \quad \text{if } r' > AI(f) - r - 1. \qquad \square$$

Theorem 1. *Let f be any Boolean function in n variables and let r be any nonnegative integer such that $AI(f) - r - 1 \geq 0$. Then $nl_r(f) \geq 2 \sum_{i=0}^{AI(f)-r-1} \binom{n-r}{i}$. More precisely, we have $nl_r(f) \geq \max_{r' \leq n} (\min(\lambda_{r'}, \mu_{r'}))$, where:*

$$\lambda_{r'} = 2 \max \left(\sum_{i=0}^{r'-1} \binom{n}{i}, \sum_{i=0}^{AI(f)-r-1} \binom{n-r}{i} \right) \quad \text{if } r' \leq AI(f) - r - 1,$$

$$= 2 \sum_{i=0}^{AI(f)-r-1} \binom{n}{i} \quad \text{if } r' > AI(f) - r - 1,$$

$$\mu_{r'} = \sum_{i=0}^{AI(f)-r-1} \binom{n-r}{i} + \sum_{i=0}^{AI(f)-r'} \binom{n-r'+1}{i}.$$

Proof. Let h be a function of degree at most r such that $nl_r(f) = wt(f + h) = wt(f(h+1)) + wt((f+1)h)$. Proposition 5 implies $nl_r(f) \geq 2 \sum_{i=0}^{AI(f)-r-1} \binom{n-r}{i}$. Let r' be any nonnegative integer. If $AI(h) \geq r'$, then Proposition 5 applied to the functions f and $h+1$ and to the functions $f+1$ and h shows that $wt(f(h+1))$ and $wt((f+1)h)$ are lower bounded by:

$$\max \left(\sum_{i=0}^{r'-1} \binom{n}{i}, \sum_{i=0}^{AI(f)-r-1} \binom{n-r}{i} \right) \quad \text{if } r' \leq AI(f) - r - 1,$$

$$\sum_{i=0}^{AI(f)-r-1} \binom{n}{i} \quad \text{if } r' > AI(f) - r - 1.$$

If $AI(h) < r'$, then there exists $g \neq 0$ such that either $g \in An_{r'-1}(h + 1)$, and therefore $supp(g) \subseteq supp(h)$, or $g \in An_{r'-1}(h)$, and therefore $supp(g) \subseteq supp(h + 1)$. If $supp(g) \subseteq supp(h)$, then we apply Proposition 5 (last sentence of) to the functions f and $h + 1$ and to the functions $f + 1$ and g. We obtain: $wt(f(h + 1)) \geq \sum_{i=0}^{AI(f)-r-1} \binom{n-r}{i}$ and $wt((f + 1)h) \geq wt((f + 1)g) \geq \sum_{i=0}^{AI(f)-r'} \binom{n-r'+1}{i}$. The case where $supp(g) \subseteq supp(h + 1)$ is similar. $\qquad \square$

Remarks:

1. Let δ be the algebraic degree of f. We assume that $r < \delta$ since, otherwise, $nl_r(f)$ is null. According to McEliece's theorem (see [36] or [7]), $nl_r(f)$ is divisible by $2^{\lceil \frac{n}{\delta} \rceil - 1} = 2^{\lfloor \frac{n-1}{\delta} \rfloor}$.

2. The bound of Theorem 1 improves upon the bound $nl_r(f) \geq \sum_{i=0}^{AI(f)-r-1} \binom{n}{i}$ of [8] for $r = 1$ (this is Lobanov's result). For $r > 1$, it improves upon it for every $n \leq 12$ and if we go up to $n \leq 22$, it improves upon it for every $AI(f)$ and most of the possible values of r (all of them, unless $AI(f)$ is very large). We give in Appendix the table of the values of the bounds of Theorem 1 and of [8] when $AI(f) = \lceil \frac{n}{2} \rceil$ (i.e. when $AI(f)$ is optimal - this is the worst case for the bound of Theorem 1), for all values of n from 13 to 22 and for all values of r from 1 to $AI(f) - 1$.

3. Both bounds of Theorem 1 and [8] show that, for most values of r, functions with high algebraic immunities are (much better than) "r-th order bent", in the sense of the paper [25]. In particular, they give more robust functions than those constructed in this paper.

4. In the case of optimal $AI(f) = \lceil \frac{n}{2} \rceil$, the bound of Theorem 1 gives $nl_r(f) \geq 2 \left(2^{n-r-1} - \sum_{i=n/2-r}^{(n-r)/2-1} \binom{n-r}{i} - \frac{1}{2} \binom{n-r}{(n-r)/2} \right)$ when n and r are even; it gives $nl_r(f) \geq 2 \left(2^{n-r-1} - \sum_{i=n/2-r}^{(n-r-1)/2} \binom{n-r}{i} \right)$ when n is even and r is odd, $nl_r(f) \geq 2 \left(2^{n-r-1} - \sum_{i=(n+1)/2-r}^{(n-r)/2-1} \binom{n-r}{i} - \frac{1}{2} \binom{n-r}{(n-r)/2} \right)$ when n and r are odd and finally $nl_r(f) \geq 2 \left(2^{n-r-1} - \sum_{i=(n+1)/2-r}^{(n-r-1)/2} \binom{n-r}{i} \right)$ when n is odd and r is even. This is in all cases asymptotically greater than $2^{n-r} - (r+1)\binom{n-r}{\lfloor \frac{n-r}{2} \rfloor} \approx 2^{n-r}\left(1 - \frac{(r+1)\sqrt{2}}{\sqrt{\pi n}}\right)$ when n tends to ∞ and r is fixed.

5. For $r \geq 2$, the asymptotic lower bound $2^{n-r}\left(1 - \frac{(r+1)\sqrt{2}}{\sqrt{\pi n}}\right)$ given by Theorem 1 is far from the asymptotic lower bound $2^{n-1} - 2^{n/2} n^{r/2}\sqrt{\frac{\ln 2}{2r!}}$ given in [13]. But this last bound, which can be shown by a simple argument (counting the number of those functions whose r-th order nonlinearities are smaller than this number and showing that it is negligible compared to the number of all functions), is only a proof of existence and gives absolutely no idea of what can be explicitly a function with greater r-th order nonlinearity. Before our bound, as far as we know, obtaining for every low $r > 1$ an *effective* way of designing functions with provably non-weak r-th order nonlinearities was open. Moreover (and most importantly), the functions with high algebraic immunities satisfy our bound *for every r of reasonably low value*.

5   The Consequences on the Necessary Criteria for the Use of Boolean Functions in Symmetric Ciphers

The impact of our result may be greater for block ciphers and for self-synchronizing stream ciphers than for synchronous stream ciphers. Indeed, the role of the r-th order nonlinearity relatively to the currently known attacks has been more clearly shown for the former than for the latter (see [14,23,25,27,33,35]). As far as we know, and except in extreme situations, no explicit attack is known, that can use the approximation by a low degree function of the combining or filtering function in the pseudo-random generator of a stream cipher (except that an approximate pseudo-random sequence can be generated with a lower complexity, which may allow predicting bits after the observation of a part of the sequence). However, such attack may be found in the future and we know now that a high algebraic immunity will help in this matter. Note that a high algebraic immunity does not necessarily prevent the system from fast algebraic attacks, see e.g. [2], but these attacks were not the subject of the present paper.

The requirements concerning the Boolean functions used in symmetric ciphers are going towards a simplification. We know that the functions with optimum algebraic immunity are balanced, for n odd. A very high (first-order) nonlinearity takes care of the distance to linear structures, since we know that if a function has a nonzero linear structure then its nonlinearity is upper bounded by $2^{n-1} - 2^{\frac{n-1}{2}}$ and, therefore that, if its distance to linear structures is d then its nonlinearity is upper bounded by $2^{n-1} - 2^{\frac{n-1}{2}} + d$. Lobanov's bound does not guarantee a resistance to attacks using approximations by affine functions, since such resistance needs (see e.g. [22,5]) a high nonlinearity, but Theorem 1 (with $r \geq 2$) shows that having a high algebraic immunity helps protecting against attacks by approximations by non-affine functions of low degrees, since the complexity of such attacks increases fastly with the degree.

However, the problem of determining functions meeting all the criteria needed for the combining or the filtering model of stream ciphers is still wide open.

References

1. F. Armknecht. Improving Fast Algebraic Attacks. FSE 2004, number 3017 in Lecture Notes in Computer Science, pp. 65–82. Springer Verlag, 2004.
2. F. Armknecht, C. Carlet, P. Gaborit, S. Knzli, W. Meier and O. Ruatta. Efficient computation of algebraic immunity for algebraic and fast algebraic attacks. Advances in Cryptology, EUROCRYPT 2006, Lecture Notes in Computer Science 4004 , pp. 147-164, 2006.
3. A. Braeken and B. Preneel. On the Algebraic Immunity of Symmetric Boolean Functions. Indocrypt 2005, LNCS 3797, pp. 35–48, 2005. Some false results of this reference have been corrected in Braeken's PhD thesis entitled "Cryptographic properties of Boolean functions and S-boxes" and available at URL http://homes.esat.kuleuven.be/ abraeken/thesisAn.pdf.
4. A. Canteaut. Open problems related to algebraic attacks on stream ciphers. Proceedings of WCC 2005, pp. 1-10, 2005.
5. A. Canteaut and M. Trabbia. Improved fast correlation attacks using parity-check equations of weight 4 and 5. EUROCRYPT 2000, number 1807 in Lecture Notes in Computer Science, pp. 573–588. Springer Verlag, 2000.
6. C. Carlet. On bent and highly nonlinear balanced/resilient functions and their algebraic immunities. Proceedings of AAECC 16, LNCS 3857, pp. 1-28, 2006.

7. C. Carlet. Boolean Functions for Cryptography and Error Correcting Codes. Chapter of the monography *Boolean Methods and Models*, Y. Crama and P. Hammer eds, Cambridge University Press, to appear in 2006. Preliminary version available at http://www-rocq.inria.fr/codes/Claude.Carlet/pubs.html

8. C. Carlet, D. Dalai, K. Gupta and S. Maitra. Algebraic Immunity for Cryptographically Significant Boolean Functions: Analysis and Construction. To appear in IEEE Transactions on Information Theory, vol. 52, no. 7, July 2006.

9. C. Carlet and P. Gaborit. On the construction of balanced Boolean functions with a good algebraic immunity. Proceedings of BFCA (First Workshop on Boolean Functions: Cryptography and Applications), Rouen, France, March 2005, pp. 1-14.

10. C. Carlet, P Guillot and S. Mesnager. On immunity profile of Boolean functions. Proceedings of SETA'06 (International Conference on Sequences and their Applications). To appear in Lecture Notes in Computer Science.

11. C. Carlet and S. Mesnager. Improving the upper bounds on the covering radii of binary Reed-Muller codes. C. Carlet et S. Mesnager. To appear in IEEE Transactions on Information Theory, 2006.

12. F. Chabaud and S. Vaudenay. Links between Differential and Linear Cryptanalysis. EUROCRYPT'94, Advances in Cryptology, Lecture Notes in Computer Science 950, Springer Verlag, pp. 356-365, 1995.

13. G. Cohen, I. Honkala, S. Litsyn and A. Lobstein. *Covering codes*. North-Holland, 1997.

14. N. Courtois. Higher order correlation attacks, XL algorithm and cryptanalysis of Toyocrypt. Proceedings of ICISC 2002, LNCS 2587, pp. 182-199.

15. N. Courtois and W. Meier. Algebraic attacks on stream ciphers with linear feedback. Advances in Cryptology - EUROCRYPT 2003, number 2656 in Lecture Notes in Computer Science, pp. 345–359. Springer Verlag, 2003.

16. N. Courtois. Fast algebraic attacks on stream ciphers with linear feedback. CRYPTO 2003, number 2729 in Lecture Notes in Computer Science, pp. 176–194. Springer Verlag, 2003.

17. D. K. Dalai, K. C. Gupta and S. Maitra. Results on Algebraic Immunity for Cryptographically Significant Boolean Functions. Indocrypt 2004, Chennai, India, December 20–22, pp. 92–106, number 3348 in Lecture Notes in Computer Science, Springer Verlag, 2004

18. D. K. Dalai, K. C. Gupta and S. Maitra. Cryptographically Significant Boolean functions: Construction and Analysis in terms of Algebraic Immunity. Workshop on Fast Software Encryption, FSE 2005, pages 98–111, number 3557, Lecture Notes in Computer Science, Springer-Verlag.

19. D. K. Dalai, S. Maitra and S. Sarkar. Basic Theory in Construction of Boolean Functions with Maximum Possible Annihilator Immunity. Cryptology ePrint Archive, http://eprint.iacr.org/, No. 2005/229, 15 July, 2005. To be published in Designs, Codes and Cryptography.

20. D. K. Dalai, K. C. Gupta and S. Maitra. Notion of algebraic immunity and its evaluation related to fast algebraic attacks. Paper 2006/018 in http://eprint.iacr.org/

21. F. Didier. A new upper bound on the block error probability after decoding over the erasure channel. Preprint available at http://www-rocq.inria.fr/codes/Frederic.Didier/
A revised version will appear in IEEE Transactions on Information Theory, 2006.

22. R. Forré. A fast correlation attack on nonlinearly feedforward filtered shift register sequences. EUROCRYPT '89, Lecture Notes in Comput. Sci. 434, pp. 586-595, Springer, 1990.

23. J. Golic. Fast low order approximation of cryptographic functions. Proceedings of EUROCRYPT'96, LNCS 1070, pp. 268-282, 1996.

24. P. Hawkes and G. G. Rose. Rewriting Variables: The Complexity of Fast Algebraic Attacks on Stream Ciphers. CRYPTO 2004, LNCS 3152, pp. 390–406. Springer Verlag, 2004.

25. T. Iwata and K. Kurosawa. Probabilistic higher order differential attack and higher order bent functions. Proceedings of ASIACRYPT'99, LNCS 1716, pp. 62-74, 1999.

26. L.R. Knudsen. Truncated and higher order differentials. Fast Software Encryption, Second International Workshop, Lecture Notes in Computer Science, n 1008. pp. 196–211. – Springer-Verlag, 1995.

27. L.R. Knudsen and M. J. B. Robshaw. Non-linear approximations in linear crypt-analysis. Proceedings of EUROCRYPT'96, LNCS 1070, pp. 224-236, 1996.

28. X. Lai. Higher order derivatives and differential cryptanalysis. Proc. "Symposium on Communication, Coding and Cryptography", in honor of J. L. Massey on the occasion of his 60'th birthday. 1994.

29. K. Kurosawa, T. Johansson and D. Stinson. Almost k-wise independent sample spaces and their applications. J. of Cryptology, vol. 14, no. 4, pp. 231-253, 2001.

30. M. Lobanov. Tight bound between nonlinearity and algebraic immunity. Paper 2005/441 in http://eprint.iacr.org/

31. J.L. Massey. Shift-register synthesis and BCH decoding. IEEE Transactions on Information Theory, vol. 15, pp. 122–127, 1969.

32. M. Matsui. Linear cryptanalysis method for DES cipher. Advances in Cryptology - EUROCRYPT'93, number 765 in Lecture Notes in Computer Science. Springer-Verlag, pp. 386-397, 1994.

33. U. M. Maurer. New approaches to the design of self-synchronizing stream ciphers. Proceedings of EUROCRYPT'91. LNCS 547, pp. 458-471, 1991.

34. A. Menezes, P. van Oorschot, and S. Vanstone. *Handbook of Applied Cryptography.* CRC Press Series on Discrete Mathematics and Its Applications, 1996.

35. W. Millan. Low order approximation of cipher functions. Cryptographic Policy and Algorithms. LNCS 1029, pp. 144-155, 1996.

36. F. J. MacWilliams and N. J. Sloane. *The theory of error-correcting codes*, Amsterdam, North Holland. 1977.

37. W. Meier, E. Pasalic and C. Carlet. Algebraic attacks and decomposition of Boolean functions. EUROCRYPT 2004, number 3027 in Lecture Notes in Computer Science, pp. 474–491. Springer Verlag, 2004.

38. W. Meier and O. Staffelbach. Nonlinearity Criteria for Cryptographic Functions. EUROCRYPT' 89, Lecture Notes in Computer Science 434, Springer Verlag, pp. 549-562, 1990.

39. R. A. Rueppel *Analysis and design of stream ciphers* Com. and Contr. Eng. Series, Berlin, Heidelberg, NY, London, Paris, Tokyo, 1986

40. C.E. Shannon. Communication theory of secrecy systems. Bell system technical journal, 28, pp. 656-715, 1949.

Appendix

Remark on Lemma 1

For every choice of an element x in the support of h, let us denote by E_x the corresponding equation $\sum_{u \in F_2^n \mid wt(u) \leq k} a_u x^u = 0$. We obtain an equivalent system by replacing every equation E_x such that $x \notin \{x_{v_{r+1},\ldots,v_n}; (v_{r+1}, \ldots, v_n) \in F_2^{n-r}\}$ (see the proof of Lemma 1) by the equation $E_x + E_{x_{v_{r+1},\ldots,v_n}}$, where $v_{r+1}, \ldots, v_n$ are the $n - r$ last coordinates of x. Note that, in this equation, every a_u such that $(u_1, \ldots, u_r) = (0, \ldots, 0)$ vanishes. The resulting system contains the 2^{n-r} equations $E_{x_{v_{r+1},\ldots,v_n}}$ (in the $\sum_{i=0}^{k} \binom{n}{i}$ unknowns a_u such that $wt(u) \le k$) and $wt(h) - 2^{n-r}$ equations in the $\sum_{i=0}^{k} \binom{n}{i} - \sum_{i=0}^{k} \binom{n-r}{i}$ unknowns a_u such that $wt(u) \le k$ and $(u_1, \ldots, u_r) \ne (0, \ldots, 0)$. Let us denote these two sub-systems by S_k^1 and S_k^2, respectively. Moving in each of the equations $E_{x_{v_{r+1},\ldots,v_n}}$ of S_k^1 all unknowns such that $(u_1, \ldots, u_r) \ne (0, \ldots, 0)$ to the right hand side, we obtain a sub-system expressing the a_u's such that $(u_1, \ldots, u_r) = (0, \ldots, 0)$ by means of the a_u's such that $(u_1, \ldots, u_r) \ne (0, \ldots, 0)$, if such a_u's can exist (note that the number of equations in S_k^1 is greater than the number of the a_u's such that $(u_1, \ldots, u_r) = (0, \ldots, 0)$). Indeed, we have seen that this system has full rank $\sum_{i=0}^{k} \binom{n-r}{i}$. Hence, a solution of S_k^2 may allow zero or one solution of S_k^1. Let $d'_{k,h}$ be the dimension of the vector space of solutions of S_k^2. Then we have $d_{k,h} \le d'_{k,h}$ and $d'_{k,h}$ equals $\sum_{i=0}^{k} \binom{n}{i} - \sum_{i=0}^{k} \binom{n-r}{i}$ minus the rank of the system S_k^2. Similarly, $d'_{k-1,h}$ equals $\sum_{i=0}^{k-1} \binom{n}{i} - \sum_{i=0}^{k-1} \binom{n-r}{i}$ minus the rank of the system S_{k-1}^2. We deduce that $d'_{k,h} \le d'_{k-1,h} + \binom{n}{k} - \binom{n-r}{k}$, since the rank of S_{k-1}^2 is upper bounded by the rank of S_k^2 (the system S_{k-1}^2 can be obtained from S_k^2 by erasing the unknowns a_u such that $wt(u) = k$). We deduce by induction on k that $d_{k,h} \le \sum_{i=AI'(h)}^{k} \binom{n}{i} - \sum_{i=AI'(h)}^{k} \binom{n-r}{i}$, where $AI'(h)$ equals the minimum value of k such that S_k^2 has non-trivial solutions. Note that $AI'(h) \le AI(h)$ and that $AI'(h)$ may be strictly smaller than $AI(h)$, since S_k^2 may have non-trivial solutions when $S_k^1 \cup S_k^2$ has none. Hence, we cannot deduce that $d_{k,h} \le \sum_{i=AI(h)}^{k} \binom{n}{i} - \sum_{i=AI(h)}^{k} \binom{n-r}{i}$. But the bound $d_{k,h} \le \sum_{i=AI'(h)}^{k} \binom{n}{i} - \sum_{i=AI'(h)}^{k} \binom{n-r}{i}$ may be better, in many concrete situations, than the bound of Lemma 1.

In particular, for $k = n$, it implies $wt(h) \ge \sum_{i=0}^{AI'(h)-1} \binom{n}{i} + \sum_{i=AI'(h)}^{n} \binom{n-r}{i}$, and therefore (applying it also to $h + 1$): $\sum_{i=0}^{AI'(h)-1} \binom{n}{i} + \sum_{i=AI'(h)}^{n} \binom{n-r}{i} \le wt(h) \le \sum_{i=0}^{n-AI'(h)} \binom{n}{i} + \sum_{i=AI'(h)}^{n} \binom{n-r}{i}$ and, when $AI(h)$ is not large and $AI'(h)$ is not too small, this is better than the double inequality $\max(\sum_{i=0}^{AI(h)-1} \binom{n}{i}, 2^{n-r}) \le wt(h) \le \min(\sum_{i=0}^{n-AI(h)} \binom{n}{i}, 2^n - 2^{n-r})$ implied by Lemma 1 for $k = n$.

Table

We give in the next table the values of the lower bounds of Theorem 1 and of [8], for n ranging from 13 to 22 (this covers all practical cases, currently for stream ciphers), for optimum algebraic immunity $\lceil \frac{n}{2} \rceil$ (note that this is the most unfavorable case for the bound of Theorem 1) and for r ranging from 1 to $AI(f) - 1$.

Table 1. THE LOWER BOUNDS ON $nl_r(f)$ GIVEN BY THEOREM 1 AND BY [8], FOR $13 \leq n \leq 22$, $r \leq AI(f) - 1$ AND $AI(f)$ OPTIMUM

n	$AI(f)$	r	The bound of Th. 1	The bound of [8]
13	7	1	3172	2380
		2	1124	1093
		3	352	378
		4	184	92
		5	28	14
		6	2	1
14	7	1	4760	3473
		2	1588	1471
		3	464	470
		4	212	106
		5	30	15
		6	2	1
15	8	1	12952	9949
		2	4760	4944
		3	1588	1941
		4	1026	576
		5	242	121
		6	32	16
		7	2	1
16	8	1	19898	14893
		2	6946	6885
		3	2186	2517
		4	1392	697
		5	274	137
		6	34	17
		7	2	1
17	9	1	52666	41226
		2	19898	21778
		3	6946	9402
		4	2186	3214
		5	1668	834
		6	308	154
		7	36	18
		8	2	1
18	9	1	82452	63004
		2	29786	31180
		3	9888	12616
		4	2942	4048
		5	1976	988
		6	344	172
		7	38	19
		8	2	1

Table 1. (*Continued*)

n	$AI(f)$	r	The bound of Th. 1	The bound of [8]
19	10	1	213524	169766
		2	82452	94184
		3	29786	43796
		4	9888	16664
		5	6415	5036
		6	2320	1160
		7	382	191
		8	40	20
		9	2	1
20	10	1	339532	263950
		2	126008	137980
		3	43556	60460
		4	13770	21700
		5	8826	6196
		6	2702	1351
		7	422	211
		8	42	21
		9	2	1
21	11	1	863820	695860
		2	339532	401930
		3	126008	198440
		4	43556	82160
		5	15094	27896
		6	15094	7547
		7	3124	1562
		8	464	232
		9	44	22
		10	2	1
22	11	1	1391720	1097790
		2	527900	600370
		3	188368	280600
		4	62360	110056
		5	18804	35443
		6	18218	9109
		7	3588	1794
		8	508	254
		9	46	23
		10	2	1

New Proofs for NMAC and HMAC: Security Without Collision-Resistance

Mihir Bellare

Dept. of Computer Science & Engineering 0404, University of California San Diego
9500 Gilman Drive, La Jolla, CA 92093-0404, USA
mihir@cs.ucsd.edu
http://www-cse.ucsd.edu/users/mihir

Abstract. HMAC was proved in [3] to be a PRF assuming that (1) the underlying compression function is a PRF, and (2) the iterated hash function is weakly collision-resistant. However, recent attacks show that assumption (2) is false for MD5 and SHA-1, removing the proof-based support for HMAC in these cases. This paper proves that HMAC is a PRF under the sole assumption that the compression function is a PRF. This recovers a proof based guarantee since no known attacks compromise the pseudorandomness of the compression function, and it also helps explain the resistance-to-attack that HMAC has shown even when implemented with hash functions whose (weak) collision resistance is compromised. We also show that an even weaker-than-PRF condition on the compression function, namely that it is a privacy-preserving MAC, suffices to establish HMAC is a secure MAC as long as the hash function meets the very weak requirement of being computationally almost universal, where again the value lies in the fact that known attacks do not invalidate the assumptions made.

1 Introduction

HMAC [3] is a popular cryptographic-hash-function-based MAC. The basic construct is actually NMAC, of which HMAC can be viewed as a derivative.

THE CONSTRUCTIONS. Succinctly:

$$\text{NMAC}(K_{\text{out}} \| K_{\text{in}}, M) = H^*(K_{\text{out}}, H^*(K_{\text{in}}, M))$$
$$\text{HMAC}(K_{\text{out}} \| K_{\text{in}}, M) = H(K_{\text{out}} \| H(K_{\text{in}} \| M)) \, .$$

Here H is a cryptographic hash function, eg. MD5 [28], SHA-1 [26], or RIPEMD-160 [17]. Let $h \colon \{0,1\}^c \times \{0,1\}^b \to \{0,1\}^c$ denote the underlying compression function. (Here $b = 512$ while c is 128 or 160.) Let h^* be the iterated compression function which on input $K \in \{0,1\}^c$ and a message $x = x[1] \ldots x[n]$ consisting of b-bit blocks, lets $a[0] = K$ and $a[i] = h(a[i-1], x[i])$ for $i = 1, \ldots, n$, and finally returns $a[n]$. Then $H^*(K, M) = h^*(K, M^*)$ and $H(M) = H^*(\text{IV}, M)$, where M^* denotes M padded appropriately to a length that is a positive multiple of b and IV is a public c-bit initial vector that is fixed as part of the description of H. Both NMAC and HMAC use two keys, which in the case of NMAC are of length

C. Dwork (Ed.): CRYPTO 2006, LNCS 4117, pp. 602–619, 2006.

c bits each, and in the case of HMAC of length b bits each and derived from a single b-bit key. HMAC is a non-intrusive version of NMAC in the sense that it uses the cryptographic hash function only as a black box, making it easier to implement.

USAGE. HMAC is standardized via an IETF RFC [22], a NIST FIPS [25] and ANSI X9.71 [1], and implemented in SSL, SSH, IPsec and TLS amongst other places. It is often used as a PRF (pseudorandom function [19]) rather than merely as a MAC. In particular this is the case when it is used for key-derivation, as in TLS [16] and IKE (the Internet Key Exchange protocol of IPsec) [20]. HMAC is also used as a PRF in a proposed standard for one-time passwords [24].

WHAT'S KNOWN. The results are for NMAC but can be lifted to HMAC. It is shown in [3] that NMAC is a secure PRF if (1) the underlying compression function h is a secure PRF, and also (2) that the hash function H is *weakly collision resistant* (WCR). The latter, introduced in [3], is a relaxation of collision resistance (CR) that asks that it be computationally infeasible for an adversary, given an oracle for $H^*(K, \cdot)$ under a hidden key K, to find a collision, meaning distinct inputs M_1, M_2 such that $H^*(K, M_1) = H^*(K, M_2)$.

THE PROBLEM. HMAC is usually implemented with MD5 or SHA-1. But, due to recent attacks [32,31], these functions are *not* WCR. Thus the assumption on which the proof of [3] is based is not true. This does not reflect any actual weaknesses in the NMAC or HMAC constructs, on which no attacks are known. (Being iterated MACs, the generic birthday based forgery attacks of [27] always break NMAC and HMAC in time $2^{c/2}$, but we mean no better-than-birthday attacks are known.) But it means that we have lost the proof-based guarantees. We are interested in recovering them.

LOSS OF WCR. First we pause to expand on the claim above that our main hash functions are not WCR. Although WCR appears to be a weaker requirement than CR due to the hidden key, in fact, for iterated hash functions, it ends up not usually being so. The reason is that collision-finding attacks such as those on MD5 [32] and SHA-1 [31] extend to find collisions in $H^*(\mathrm{IV}, \cdot)$ for an arbitrary but given IV, and, any such attack, via a further extension attack, can be used to compromise WCR, meaning to find a collision in $H^*(K, \cdot)$, given an oracle for this function, even with K hidden. This was pointed out in [3,21], and, for the curious, we recall the attack in [2].

MAIN RESULT. We show (Theorem 1) that NMAC is a PRF under the *sole* assumption that the underlying compression function h is itself a PRF. In other words, the additional assumption that the hash function is WCR is dropped. (And, in particular, as long as h is a PRF, the conclusion is true even if H is *not* WCR, let alone CR.)

The main advantage of our result is that it is based on an assumption that is not refuted by any known attacks. (There are to date no attacks that compromise the pseudorandomness of the compression functions of MD5 or SHA-1.) Another feature of our result is that it is the first proof for NMAC that is based solely on an assumption about the compression function rather than also assuming something about the entire iterated hash function.

TECHNIQUES. We show (Lemma 1) that if a compression function h is a PRF then the iterated compression function h^* is *computationally almost universal* (cAU), a computational relaxation of the standard information-theoretic notion of almost-universality (AU) of [13,33,30]. We then conclude with Lemma 2 which says that the composition of a PRF and a cAU function is a PRF. (This can be viewed as a computational relaxation of the Carter-Wegman paradigm [13,33].)

RELATED WORK. If h^* were a PRF, it would imply it is cAU, but h^* is not a PRF due to the extension attack. It is however shown by [4] that if h is a PRF then h^* (which they call the cascade) is a "pf-PRF" (prefix-free PRF), meaning a PRF as long as no query of the adversary is a prefix of another query. It was pointed out to us by Shoup after seeing an early draft of our paper that it is possible to apply this in a black-box way to show that h^* is cAU. However Lemma 1 is a somewhat stronger result and bound whose proof distills and strengthens the ideas of [4] and also involves some new ones. For comparison, we do present the indirect proof in [2].

Dodis, Gennaro, Håstad, Krawczyk and Rabin show [18, Lemma 4] that the cascade over a family of *random* functions is AU as long as the two messages whose collision probability one considers have the same length. (In this model, $h(K, \cdot)$ is a random function for each $K \in \{0,1\}^c$. That is, it is like Shannon's ideal cipher model, except the component maps are functions not permutations.) This does not imply Lemma 1 (showing the cascade h^* is cAU if h is a PRF), because we need to allow the two messages to have different lengths, and also because it is not clear what implication their result has for the case when h is a PRF. (A PRF does *not* permit one to securely instantiate a *family* of random functions.) A second result [18, Lemma 5] in the same paper says that if $h^*(K, M)$ is close to uniformly distributed then so is $h^*(K, M\|X)$. (Here M is chosen from some distribution, K is a random but known key, and X is a fixed block.) This result only assumes h is a PRF, but again we are not able to discern any implications for the problems we consider, because in our case the last block of the input is not fixed, we are interested in the cAU property rather than randomness, and our inputs are not drawn from a distribution.

ANOTHER RESULT. The fact that compression functions are underlain by block ciphers, together with the fact that no known attacks compromise the pseudo-randomness of the compression functions of MD5, SHA-1, may give us some confidence that it is ok to assume these are PRFs, but it still behooves us to be cautious. What can we prove about NMAC without assuming the compression function is a PRF? We would not expect to be able to prove it is a PRF, but what about just a secure MAC? (Any PRF is a secure MAC [6,9], so our main result implies NMAC is a secure MAC, but we are interested in seeing whether this can be proved under weaker assumptions.) We show (Theorem 2) that NMAC is a secure MAC if h is a *privacy-preserving MAC* (PP-MAC) [8] and h^* (equivalently, H^*) is cAU. A PP-MAC (the definition is provided in Section 4) is stronger than a MAC but weaker than a PRF. This result reverts to the paradigm of [3] of making assumptions both about the compression function and its iteration, but the point is that cAU is a very weak assumption compared to WCR and PP-MAC is a weaker assumption than PRF.

FROM NMAC TO HMAC. The formal results (both previous and new) we have discussed so far pertain to NMAC. However, discussions (above and in the literature) tend to identify NMAC and HMAC security-wise. This is explained by an observation of [3] which says that HMAC inherits the security of NMAC as long as the compression function is a PRF when keyed via the data input. (So far when we have talked of it being a PRF, it is keyed via the chaining variable.) In our case this means that HMAC is a PRF if the compression function is a "dual-PRF," meaning a PRF when keyed by either of its two inputs.

However, the analysis above assumes that the two keys K_{out}, K_{in} of HMAC are chosen *independently* at random, while in truth they are equal to $K \oplus \mathsf{opad}$ and $K \oplus \mathsf{ipad}$ respectively, where K is a random b-bit key and $\mathsf{opad}, \mathsf{ipad}$ are fixed, distinct constants. We apply the theory of PRFs under related-key attacks [7] to extend the observation of [3] to this single-key version of HMAC, showing it inherits the security of NMAC as long as the data-input-keyed compression function is a PRF under an appropriate (and small) class of related key attacks. Assuming additionally that the compression function is a PRF in the usual sense, we obtain a (in fact, the first) security proof of the single-key version of HMAC. These results are in Section 5.

2   Definitions

NOTATION. We denote by $s_1\|s_2$ the concatenation of strings s_1, s_2, and by $|s|$ the length of string s. Let b be a positive integer representing a block length, and let $B = \{0,1\}^b$. Let B^+ denote the set of all strings of length a positive multiple of b bits. Whenever we speak of blocks we mean b-bit ones. If $M \in B^+$ then $\|M\|_b = |M|/b$ is the number of blocks in M, and $M[i]$ denotes its i-th b-bit block, meaning $M = M[1]\ldots M[n]$ where $n = \|M\|_b$. If $M_1, M_2 \in B^+$, then M_1 is a prefix of M_2, written $M_1 \subseteq M_2$, if $M_2 = M_1\|A$ for some $A \in B^*$. If S is a set then $s \xleftarrow{\$} S$ denotes the operation of selecting s uniformly at random from S. An adversary is a (possibly randomized) algorithm that may have access to one or more oracles. We let

$$A^{\mathcal{O}_1,\cdots}(x_1, \ldots) \Rightarrow 1 \quad \text{and} \quad y \xleftarrow{\$} A^{\mathcal{O}_1,\cdots}(x_1, \ldots)$$

denote, respectively, the event that A with the indicated oracles and inputs outputs 1, and the experiment of running A with the indicated oracles and inputs and letting y be the value returned. (This value is a random variable depending on the random choices made by A and its oracles.) Either the oracles or the inputs (or both) may be absent, and often will be.

A family of functions is a two-argument map $f\colon Keys \times Dom \to Rng$ whose first argument is regarded as a key. We fix one such family $h\colon \{0,1\}^c \times B \to \{0,1\}^c$ to model a compression function that we regard as being keyed via its c-bit chaining variable. Typical values are $b = 512$ and $c = 128$ or 160. The *iteration* of family h: $\{0,1\}^c \times B \rightarrow \{0,1\}^c$ is the family of functions h^*: $\{0,1\}^c \times B^+ \rightarrow \{0,1\}^c$ where $h^*(K, M)$ (for $K \in \{0,1\}^c$ and $M \in B^+$) is defined by the following code: $n \leftarrow \|M\|_b$; $a[0] \leftarrow K$; For $i = 1, \ldots, n$ do $a[i] \leftarrow h(a[i-1], M[i])$; Return $a[n]$. This represents the Merkle-Damgård [23,14] iteration method used in all the popular hash functions but without the "strengthening," meaning that there is no $|M|$-based message padding.

PRFs. A prf-adversary A against a family of functions f: $Keys \times Dom \rightarrow Rng$ takes as oracle a function g: $Dom \rightarrow Rng$ and returns a bit. The prf-advantage of A against f is the difference between the probability that it outputs 1 when its oracle is $g = f(K, \cdot)$ for a random key $K \xleftarrow{\$} Keys$, and the probability that it outputs 1 when its oracle g is chosen at random from the set $\mathsf{Maps}(Dom, Rng)$ of all functions mapping Dom to Rng, succinctly written as

$$\mathbf{Adv}_f^{\mathrm{prf}}(A) \;=\; \Pr\left[A^{f(K,\cdot)} \Rightarrow 1\right] - \Pr\left[A^{\$} \Rightarrow 1\right]. \tag{1}$$

In both cases the probability is over the choice of oracle and the coins of A.

cAU AND COLLISION-PROBABILITY. Let F: $\{0,1\}^k \times Dom \rightarrow Rng$ be a family of functions. cAU is measured by considering an almost-universal (au) adversary A against F. It (takes no inputs and) returns a pair of messages in Dom. Its au-advantage, denoted $\mathbf{Adv}_F^{\mathrm{au}}(A)$, is

$$\Pr\left[F(K, M_1) = F(K, M_2) \;\wedge\; M_1 \neq M_2 \;:\; (M_1, M_2) \xleftarrow{\$} A\,;\; K \xleftarrow{\$} Keys\right].$$

This represents a very weak form of collision-resistance since A must produce M_1, M_2 without being given any information about K. WCR [3] is a stronger notion because here A gets an oracle for $F(K, \cdot)$ and can query this in its search for M_1, M_2.

For $M_1, M_2 \in Dom$ it is useful to let $\mathsf{Coll}_F(M_1, M_2) = \Pr[F(K, M_1) = F(K, M_2)]$, the probability being over $K \xleftarrow{\$} \{0,1\}^k$.

3   Security of NMAC

Let h: $\{0,1\}^c \times \{0,1\}^b \rightarrow \{0,1\}^c$ be a family of functions that represents the compression function, here assumed to be a PRF. Let pad denote a padding function such that $s^* = s\|\mathsf{pad}(|s|) \in B^+$ for any string s. (Such padding functions are part of the description of current hash functions. Note the pad depends only on the length of s.) Then the family NMAC: $\{0,1\}^{2c} \times D \rightarrow \{0,1\}^c$ is defined by $\mathsf{NMAC}(K_{\mathrm{out}}\|K_{\mathrm{in}}, M) = h(K_{\mathrm{out}}, h^*(K_{\mathrm{in}}, M^*)\|\mathsf{fpad})$ where $\mathsf{fpad} = \mathsf{pad}(c) \in \{0,1\}^{b-c}$ and h^* is the iterated compression function as defined in Section 2. The domain D is the set of all strings up to some maximum length, which is 2^{64} for current hash functions.

It turns out that our security proof for NMAC does not rely on any properties of pad beyond the fact that $M^* = M\|\mathsf{pad}(|M|) \in B^+$. (In particular, the Merkle-Damgård strengthening, namely inclusion of the message length in the padding, that is used in current hash functions and is crucial to collision resistance of the hash function, is not important to the security of NMAC.) Accordingly, we will actually prove the security of a more general construct that we call generalized NMAC. The family GNMAC: $\{0,1\}^{2c} \times B^+ \to \{0,1\}^c$ is defined by GNMAC$(K_{\text{out}} \| K_{\text{in}}, M) = h(K_{\text{out}}, h^*(K_{\text{in}}, M) \| \text{fpad})$ where fpad is any (fixed) $b-c$ bit string. Note the domain is B^+, meaning inputs have to have a length that is a positive multiple of b bits. (But can be of any length.) NMAC is nonetheless a special case of GNMAC via NMAC$(K_{\text{out}} \| K_{\text{in}}, M) = $ GNMAC$(K_{\text{out}} \| K_{\text{in}}, M^*)$ and thus the security of NMAC is implied by that of GNMAC. (Security as a PRF or a MAC, respectively, for both.)

3.1   The Results

MAIN LEMMA. The following says that if h is a PRF then its iteration h^* is cAU.

Lemma 1. *Let $B = \{0,1\}^b$. Let h: $\{0,1\}^c \times B \to \{0,1\}^c$ be a family of functions, and let A^* be an au-adversary against h^*. Assume that the two messages output by A^* are at most $n_1, n_2 \geq 1$ blocks long, respectively. Then there exists a prf-adversary A against h such that*

$$\mathbf{Adv}_{h^*}^{\text{au}}(A^*) \leq (n_1 + n_2 - 1) \cdot \mathbf{Adv}_h^{\text{prf}}(A) + \frac{1}{2^c} \ . \tag{2}$$

Furthermore, A has time-complexity at most $O((n_1 + n_2)T_h)$, where T_h is the time for one evaluation of h, and makes at most 2 oracle queries. ∎

The proof is in Section 3.3. The quality of the reduction is good because the time-complexity of the constructed adversary A is small and in particular independent of the time-complexity of A^* (the proof shows how this is possible) and furthermore A makes only two oracle queries.

One might ask whether stronger results hold. For example, assuming h is a PRF, (1) Is h^* a PRF? (2) Is h^* WCR? (Either would imply that h^* is cAU.) But the answer is NO to both questions. The function h^* is *never* a PRF due to the extension attack. On the other hand it is easy to give an example of a PRF h such that h^* is not WCR. Also MD5 and SHA-1 are candidate counter-examples, since their compression functions appear to be PRFs but their iterations are not WCR.

PRF(CAU)=PRF LEMMA. The *composition* of families h: $\{0,1\}^c \times \{0,1\}^b \to \{0,1\}^c$ and F: $\{0,1\}^k \times D \to \{0,1\}^b$ is the family hF: $\{0,1\}^{c+k} \times D \to \{0,1\}^c$ defined by $hF(K_{\text{out}} \| K_{\text{in}}, M) = h(K_{\text{out}}, F(K_{\text{in}}, M))$. The following lemma says that if h is a PRF and F is cAU then hF is a PRF.

Lemma 2. *Let $B = \{0,1\}^b$. Let h: $\{0,1\}^c \times B \to \{0,1\}^c$ and F: $\{0,1\}^k \times D \to B$ be families of functions, and let hF: $\{0,1\}^{c+k} \times D \to \{0,1\}^c$ be defined by*

$$hF(K_{\text{out}} \| K_{\text{in}}, M) = h(K_{\text{out}}, F(K_{\text{in}}, M))$$

for all $K_{\text{out}} \in \{0,1\}^c, K_{\text{in}} \in \{0,1\}^k$ and $M \in D$. Let A_{hF} be a prf-adversary against hF that makes at most $q \geq 2$ oracle queries, each of length at most n, *and has time-complexity at most t. Then there exists a prf-adversary A_h against h and an au-adversary A_F against F such that*

$$\mathbf{Adv}_{hF}^{\mathrm{prf}}(A_{hF}) \leq \mathbf{Adv}_{h}^{\mathrm{prf}}(A_h) + \binom{q}{2} \cdot \mathbf{Adv}_{F}^{\mathrm{au}}(A_F) \,. \tag{3}$$

Furthermore, A_h has time-complexity at most t and makes at most q oracle queries, while A_F has time-complexity $O(T_F(n))$ and the two messages it outputs have length at most n, where $T_F(n)$ is the time to compute F on an n-bit input. ∎

This extends the analogous PRF(AU)=PRF lemma by relaxing the condition on F from AU to cAU. (The PRF(AU)=PRF lemma is alluded to in [4,11], and variants are in [11,12].) A simple proof of Lemma 2, using games [10,29], is in [2].

GNMAC IS A PRF. We now combine the two lemmas above to conclude that if h is a PRF then so is GNMAC.

Theorem 1. *Assume $b \geq c$ and let $B = \{0,1\}^b$. Let $h: \{0,1\}^c \times B \to \{0,1\}^c$ be a family of functions and let $\mathsf{fpad} \in \{0,1\}^{b-c}$ be a fixed padding string. Let GNMAC: $\{0,1\}^{2c} \times B^+ \to \{0,1\}^c$ be defined by*

$$\mathsf{GNMAC}(K_{\mathrm{out}} \| K_{\mathrm{in}}, M) = h(K_{\mathrm{out}}, h^*(K_{\mathrm{in}}, M) \| \mathsf{fpad})$$

for all $K_{\mathrm{out}}, K_{\mathrm{in}} \in \{0,1\}^c$ and $M \in B^+$. Let A_{GNMAC} be a prf-adversary against GNMAC that makes at most q oracle queries, each of at most m blocks, and has time-complexity at most t. Then there exist prf-adversaries A_1, A_2 against h such that

$$\mathbf{Adv}_{\mathsf{GNMAC}}^{\mathrm{prf}}(A_{\mathsf{GNMAC}}) \leq \mathbf{Adv}_{h}^{\mathrm{prf}}(A_1) + \binom{q}{2}\left[2m \cdot \mathbf{Adv}_{h}^{\mathrm{prf}}(A_2) + \frac{1}{2^c}\right] \,. \tag{4}$$

Furthermore, A_1 has time-complexity at most t and makes at most q oracle queries, while A_2 has time-complexity at most $O(mT_h)$ and makes at most 2 oracle queries, where T_h is the time for one computation of h. ∎

Proof (Theorem 1). Let $F: \{0,1\}^c \times B^+ \to \{0,1\}^b$ be defined by $F(K_{\mathrm{in}}, M) = h^*(K_{\mathrm{in}}, M) \| \mathsf{fpad}$. Then GNMAC $= hF$. Apply Lemma 2 (with $k = c$, $D = B^+$ and $A_{hF} = A_{\mathsf{GNMAC}}$) to get prf-adversary A_1 and au-adversary A_F with the properties stated in the lemma. Note that $\mathbf{Adv}_{F}^{\mathrm{au}}(A_F) = \mathbf{Adv}_{h^*}^{\mathrm{au}}(A_F)$. (Because a pair of messages is a collision for $h^*(K_{\mathrm{in}}, \cdot) \| \mathsf{fpad}$ iff it is a collision for $h^*(K_{\mathrm{in}}, \cdot)$.) Now apply Lemma 1 to $A^* = A_F$ to get prf-adversary A_2. ∎

3.2   Tightness of Bound

The best known attack on GNMAC is the birthday one of [27]. They show that it is possible to break NMAC with about $2^{c/2}/\sqrt{m}$ queries of at most m blocks each. We now want to assess how close to this is the guarantee we can get from Theorem 1 gets. If t is a time-complexity then let $\bar{t} = t/T_h$. Assume that the best attack against h as a PRF is exhaustive key search. (Birthday attacks do not apply since h is not a family of permutations.) This means that $\mathbf{Adv}_{h}^{\mathrm{prf}}(A) \leq \bar{t} \cdot 2^{-c}$ for any prf-adversary A of time complexity t making $q \leq \bar{t}$ queries. Plugging this into (4) and simplifying, the upper bound on the prf-advantage of any adversary against GNMAC who has time-complexity t and makes at most q queries is $O(\bar{t} + m^2 q^2 T_h) \cdot 2^{-c}$. If we ignore the T_h term, then this hits 1 when $q \approx 2^{c/2}/m$. This means that the bound justifies NMAC up to roughly $2^{c/2}/m$ queries, off from the number in the above-mentioned attack by a factor of $\sqrt{m}$. It is an interesting open problem to improve our analysis and fill the gap.

3.3   Proof of Lemma 1

Proofs of claims below are omitted and may be found in [2].

SOME DEFINITIONS. In this proof it will be convenient to consider prf-adversaries that take inputs. The advantage of A against h on inputs $x_1, \ldots$ is defined as

$$\mathbf{Adv}_h^{\mathrm{prf}}(A(x_1, \ldots)) = \Pr\left[A^{h(K, \cdot)}(x_1, \ldots) \Rightarrow 1\right] - \Pr\left[A^{\$}(x_1, \ldots) \Rightarrow 1\right],$$

where in the first case $K \xleftarrow{\$} \{0,1\}^c$ and in the second case the notation means that A is given as oracle a map chosen at random from $\mathsf{Maps}(\{0,1\}^b, \{0,1\}^c)$.

OVERVIEW. To start with, we ignore A_{hF} and upper bound $\mathsf{Coll}_F(M_1, M_2)$ as some appropriate function of the prf-advantage of a prf-adversary against h that takes M_1, M_2 as input. We consider first the case that $M_1 \subseteq M_2$ (M_1 is a prefix of M_2) and then the case that $M_1 \not\subseteq M_2$, building in each case a different adversary.

THE CASE $M_1 \subseteq M_2$. We begin with some high-level intuition. Suppose $M_1 \subseteq M_2$ with $m_2 = \|M_2\|_b \geq 1 + m_1$, where $m_1 = \|M_1\|_b$. The argument to upper

---

Game $G_1(M_1, M_2, l)$

$\quad m_1 \leftarrow \|M_1\|_b$; $m_2 \leftarrow \|M_2\|_b$
$\quad a[l] \xleftarrow{\$} \{0,1\}^c$
$\quad$ For $i = l + 1$ to m_2 do
$\quad\quad a[i] \leftarrow h(a[i-1], M_2[i])$
$\quad$ If $a[m_1] = a[m_2]$ then return 1
$\quad\quad$ else return 0

Adversary $A_3^g(M_1, M_2)$

$\quad m_1 \leftarrow \|M_1\|_b$; $m_2 \leftarrow \|M_2\|_b$
$\quad l \xleftarrow{\$} \{1, \ldots, m_1 + 1\}$
$\quad$ If $l = m_1 + 1$ then return $A_2^g(M_1, M_2)$
$\quad$ Else return $A_1^g(M_1, M_2, l)$

Adversary $A_1^g(M_1, M_2, l)$

$\quad m_1 \leftarrow \|M_1\|_b$; $m_2 \leftarrow \|M_2\|_b$
$\quad a[l] \leftarrow g(M_2[l])$
$\quad$ For $i = l + 1$ to m_2 do
$\quad\quad a[i] \leftarrow h(a[i-1], M_2[i])$
$\quad$ If $a[m_1] = a[m_2]$ then return 1
$\quad\quad$ else return 0

Adversary $A_2^g(M_1, M_2)$

$\quad m_1 \leftarrow \|M_1\|_b$; $m_2 \leftarrow \|M_2\|_b$
$\quad a[m_1 + 1] \leftarrow g(M_2[m_1 + 1])$
$\quad$ For $i = m_1 + 2$ to m_2 do
$\quad\quad a[i] \leftarrow h(a[i-1], M_2[i])$
$\quad y \xleftarrow{\$} B \setminus \{M_2[m_1 + 1]\}$
$\quad$ If $h(a[m_2], y) = g(y)$ then return 1
$\quad\quad$ else return 0

Fig. 1. Games and adversaries taking input distinct messages M_1, M_2 such that $M_1 \subseteq M_2$. The adversaries take an oracle $g \colon \{0,1\}^b \to \{0,1\}^c$. For Game $G_1(M_1, M_2, l)$, the input l is the range $0 \leq l \leq \|M_1\|_b$ while for adversary A_1 it is in the range $1 \leq l \leq \|M_1\|_b$.

bound $\mathsf{Coll}_{h^*}(M_1, M_2)$ has two parts. First, a hybrid argument is used to show that $a[m_1] = h^*(K, M_1)$ is computationally close to random when K is drawn at random. Next, we imagine a game in which $a[m_1]$ functions as a key to h. Let $a[m_1 + 1] = h(a[m_1], M_2[m_1 + 1])$ and $a[m_2] = h^*(a[m_1 + 1], M_2[m_1 + 2] \ldots M_2[m_2])$. Now, if $a[m_2] = a[m_1]$ then we effectively have a way to recover the "key" $a[m_1]$ given $a[m_1+1]$, amounting to a key-recovery attack on $h(a[m_1], \cdot)$ based on one input-output example of this function. But being a PRF, h is also secure against key-recovery.

In the full proof that follows, we use the games and adversaries specified in Figure 1. Adversaries A_1, A_2 represent, respectively, the first and second parts of the argument outlined above, while A_3 integrates the two.

Claim 1. Let $M_1, M_2 \in B^+$ with $M_1 \subseteq M_2$ and $1 + \|M_1\|_b \le \|M_2\|_b$. Suppose $1 \le l \le \|M_1\|_b$. Then

$$\Pr\left[A_1^{\$}(M_1, M_2, l) \Rightarrow 1\right] = \Pr\left[G_1(M_1, M_2, l) \Rightarrow 1\right]$$

$$\Pr\left[A_1^{h(K, \cdot)}(M_1, M_2, l) \Rightarrow 1\right] = \Pr\left[G_1(M_1, M_2, l - 1) \Rightarrow 1\right]. \qquad \blacksquare$$

Recall the notation means that in the first case A_1 gets as oracle $g \xleftarrow{\$} \mathsf{Maps}$ $(\{0,1\}^b, \{0,1\}^c)$ and in the second case $K \xleftarrow{\$} \{0,1\}^c$.

Claim 2. Let $M_1, M_2 \in B^+$ with $M_1 \subseteq M_2$ and $1 + \|M_1\|_b \le \|M_2\|_b$. Then

$$\Pr\left[A_2^{\$}(M_1, M_2) \Rightarrow 1\right] = 2^{-c}$$

$$\Pr\left[A_2^{h(K, \cdot)}(M_1, M_2) \Rightarrow 1\right] \ge \Pr\left[G_1(M_1, M_2, m_1) \Rightarrow 1\right]. \qquad \blacksquare$$

Claim 3. Let $M_1, M_2 \in B^+$ with $M_1 \subseteq M_2$ and $1 + \|M_1\|_b \le \|M_2\|_b$. Let $m_1 = \|M_1\|_b$. Then

$$\mathbf{Adv}_h^{\mathrm{prf}}(A_3(M_1, M_2)) \ge \frac{1}{m_1 + 1}\left(\mathsf{Coll}_{h^*}(M_1, M_2) - 2^{-c}\right). \qquad \blacksquare$$

THE CASE $M_1 \not\subseteq M_2$. For $M_1, M_2 \in B^+$ with $\|M_1\|_b \le \|M_2\|_b$ and $M_1 \not\subseteq M_2$, we let $\mathrm{LCP}(M_1, M_2)$ denote the length of the longest common blockwise prefix of M_1, M_2, meaning the largest integer p such that $M_1[1] \ldots M_1[p] = M_2[1] \ldots M_2[p]$ but $M_1[p+1] \ne M_2[p+1]$. We consider the games and adversaries of Figure 2.

Claim 4. Let $M_1, M_2 \in B^+$ with $M_1 \not\subseteq M_2$, and $\|M_1\|_b \le \|M_2\|_b$. Suppose $1 \le l \le \|M_1\|_b + \|M_2\|_b - \mathrm{LCP}(M_1, M_2) - 1$. Then

$$\Pr\left[A_4^{\$}(M_1, M_2, l) \Rightarrow 1\right] = \Pr\left[G_2(M_1, M_2, l) \Rightarrow 1\right]$$

$$\Pr\left[A_4^{h(K, \cdot)}(M_1, M_2, l) \Rightarrow 1\right] = \Pr\left[G_2(M_1, M_2, l - 1) \Rightarrow 1\right]. \qquad \blacksquare$$

Game $G_2(M_1, M_2, l)$ // $0 \leq l \leq \|M_1\|_b + \|M_2\|_b - \mathrm{LCP}(M_1, M_2) - 1$

200 $m_1 \leftarrow \|M_1\|_b$; $m_2 \leftarrow \|M_2\|_b$; $p \leftarrow \mathrm{LCP}(M_1, M_2)$

210 If $0 \leq l \leq m_1$ then

220    $a_1[l] \xleftarrow{\$} \{0,1\}^c$; For $i = l+1$ to m_1 do $a_1[i] \leftarrow h(a_1[i-1], M_1[i])$

230 If $m_1 + 1 \leq l \leq m_1 + m_2 - p$ then $a_1[m_1] \xleftarrow{\$} \{0,1\}^c$

240 If $0 \leq l \leq p$ then $n \leftarrow l$; $a_2[n] \leftarrow a_1[n]$

250 If $p + 1 \leq l \leq m_1$ then $n \leftarrow p + 1$; $a_2[n] \xleftarrow{\$} \{0,1\}^c$

260 If $m_1 + 1 \leq l \leq m_1 + m_2 - p$ then $n \leftarrow l - m_1 + p + 1$; $a_2[n] \xleftarrow{\$} \{0,1\}^c$

270 For $i = n+1$ to m_2 do $a_2[i] \leftarrow h(a_2[i-1], M_2[i])$

280 If $a_1[m_1] = a_2[m_2]$ then return 1 else return 0

Adversary $A_4^g(M_1, M_2, l)$ // $1 \leq l \leq \|M_1\|_b + \|M_2\|_b - \mathrm{LCP}(M_1, M_2) - 1$

a00 $m_1 \leftarrow \|M_1\|_b$; $m_2 \leftarrow \|M_2\|_b$; $p \leftarrow \mathrm{LCP}(M_1, M_2)$

a10 If $1 \leq l \leq m_1$ then

a20 $a_1[l] \leftarrow g(M_1[l])$; For $i = l+1$ to m_1 do $a_1[i] \leftarrow h(a_1[i-1], M_1[i])$

a30 If $m_1 + 1 \leq l \leq m_1 + m_2 - p$ then $a_1[m_1] \xleftarrow{\$} \{0,1\}^c$

a40 If $1 \leq l \leq p$ then $n \leftarrow l$; $a_2[n] \leftarrow a_1[n]$

a50 If $l = p + 1$ then $n \leftarrow p + 1$; $a_2[n] \leftarrow g(M_2[n])$

a51 If $p + 2 \leq l \leq m_1$ then $n \leftarrow p + 1$; $a_2[n] \xleftarrow{\$} \{0,1\}^c$

a60 If $m_1 + 1 \leq l \leq m_1 + m_2 - p$ then $n \leftarrow l - m_1 + p + 1$; $a_2[n] \leftarrow g(M_2[n])$

a70 For $i = n+1$ to m_2 do $a_2[i] \leftarrow h(a_2[i-1], M_2[i])$

a80 If $a_1[m_1] = a_2[m_2]$ then return 1 else return 0

Fig. 2. Games and adversaries taking input distinct messages $M_1, M_2 \in B^+$ such that $M_1 \not\sqsubseteq M_2$ and $\|M_1\|_b \leq \|M_2\|_b$. For Game $G_2(M_1, M_2, l)$, the input l is the range $0 \leq l \leq \|M_1\|_b + \|M_2\|_b - \mathrm{LCP}(M_1, M_2) - 1$ while for adversary A_4 it is in the range $1 \leq l \leq \|M_1\|_b + \|M_2\|_b - \mathrm{LCP}(M_1, M_2) - 1$.

We now define prf adversary $A_5^g(M_1, M_2)$ against h as follows. It picks $l \xleftarrow{\$} \{1, \ldots, \|M_1\|_b + \|M_2\|_b - \mathrm{LCP}(M_1, M_2) - 1\}$ and returns $A_4^g(M_1, M_2, l)$.

Claim 5. Let $M_1, M_2 \in B^+$ with $M_1 \not\sqsubseteq M_2$ and $\|M_1\|_b \leq \|M_2\|_b$. Let $m = \|M_1\|_b + \|M_2\|_b - \mathrm{LCP}(M_1, M_2) - 1$. Then

$$\mathbf{Adv}_h^{\mathrm{prf}}(A_5) \geq \frac{1}{m} \cdot \left(\mathsf{Coll}_{h^*}(M_1, M_2) - 2^{-c} \right) . \qquad \blacksquare$$

PUTTING IT TOGETHER. The final step to construct the prf-adversary A against h, claimed in the lemma, is to appropriately combine A_3, A_5. We assume wlog that the two messages M_1, M_2 output by A^* are always distinct, in B^+, and satisfy $\|M_1\|_b \leq \|M_2\|_b$. We first define

Adversary $A_6^g(M_1, M_2)$	**Adversary** A_7^g
If $M_1 \sqsubseteq M_2$ then return $A_3^g(M_1, M_2)$	$(M_1, M_2) \xleftarrow{\$} A^*$
Else return $A_5^g(M_1, M_2)$	Return $A_6^g(M_1, M_2)$

Claim 6. $\mathbf{Adv}_{h^*}^{\mathrm{au}}(A^*) \leq 2^{-c} + (n_1 + n_2 - 1) \cdot \mathbf{Adv}_h^{\mathrm{prf}}(A_7)$ $\qquad \blacksquare$

Prf-Adversary A_7 achieves the prf-advantage we seek, but has time-complexity that of A^* since it runs the latter. We now use a standard "coin-fixing" argument to reduce this time-complexity. Note that

$$\mathbf{Adv}_h^{\mathrm{prf}}(A_7) \;=\; \mathbf{E}_{M_1,M_2}\left[\mathbf{Adv}_h^{\mathrm{prf}}(A_6(M_1,M_2))\right]$$

where the expectation is over $(M_1, M_2) \xleftarrow{\$} A^*$. Thus there must exist distinct $M_1, M_2 \in B^+$ ($\|M_1\|_b \leq \|M_2\|_b$) such that $\mathbf{Adv}_h^{\mathrm{prf}}(A_6(M_1, M_2)) \geq \mathbf{Adv}_h^{\mathrm{prf}}(A_7)$. Let A be the prf-adversary that has M_1, M_2 hardwired as part of its code and, given oracle g, runs $A_6^g(M_1, M_2)$. Since the latter has time complexity $O(mT_h)$, the proof of Lemma 1 is complete.

4   MAC-Security of NMAC Under Weaker Assumptions

Since any PRF is a secure MAC [6,9], Theorem 1 implies that NMAC is a secure MAC if the compression function is a PRF. Here we show that NMAC is a secure MAC under a weaker-than-PRF assumption on the compression function —namely that it is a privacy-preserving MAC— coupled with the assumption that the hash function is cAU. This is of interest given the still numerous usages of HMAC as a MAC (rather than as a PRF). This result can be viewed as attempting to formalize the intuition given in [3, Remark 4.4].

MAC FORGERY. Recall that the mac-advantage of mac-adversary A against a family of functions $f\colon Keys \times Dom \to Rng$ is

$$\mathbf{Adv}_f^{\mathrm{mac}}(A) \;=\; \Pr\left[A^{f(K,\cdot),\mathrm{VF}_f(K,\cdot,\cdot)} \text{ forges } : K \xleftarrow{\$} Keys\right]\;.$$

The verification oracle $\mathrm{VF}_f(K, \cdot, \cdot)$ associated to f takes input M, T, returning 1 if $f(K, M) = T$ and 0 otherwise. Queries to the first oracle are called mac queries, and ones to the second are called verification queries. A is said to forge if it makes a verification query M, T the response to which is 1 but M was not previously a mac query. Note we allow multiple verification queries [9].

PRIVACY-PRESERVING MACs. The privacy notion for MACs that we use adapts the notion of left-or-right indistinguishability of encryption [5] to functions that are deterministic, and was first introduced by [8] who called it indistinguishability under distinct chosen-plaintext attack. An oracle query of an ind-adversary A against family $f\colon \{0,1\}^k \times \{0,1\}^l \to \{0,1\}^L$ is a pair of l-bit strings. The reply is provided by one or the other of the following games:

Game Left	Game Right
$K \xleftarrow{\$} \{0,1\}^c$	$K \xleftarrow{\$} \{0,1\}^c$
On query (x_0, x_1):	**On query** (x_0, x_1):
Reply $f(K, x_0)$	Reply $f(K, x_1)$

Each game has an initialization step in which it picks a key; it then uses this key in computing replies to all the queries made by A. The ind-advantage of A is

$$\mathbf{Adv}_f^{\mathrm{ind}}(A) \;=\; \Pr\left[A^{\mathrm{Right}} \Rightarrow 1\right] - \Pr\left[A^{\mathrm{Left}} \Rightarrow 1\right]\;.$$

However, unlike for encryption, the oracles here are deterministic. So A can easily win (meaning, obtain a high advantage), for example by making a pair of queries of the form $(x, z), (y, z)$, where x, y, z are distinct, and then returning 1 iff the replies returned are the same. (We expect that $h(K, x) \neq h(K, y)$ with high probability over K for functions h of interest, for example compression functions.) We fix this by simply outlawing such behavior. To be precise, let us say that A is *legitimate* if for any sequence $(x_0^1, x_1^1), \ldots, (x_0^q, x_1^q)$ of oracle queries that it makes, $x_0^1, \ldots, x_0^q$ are all distinct l-bit strings, and also $x_1^1, \ldots, x_1^q$ are all distinct l-bit strings. (As a test, notice that the adversary who queried $(x, z), (y, z)$ was not legitimate.) It is to be understood henceforth that an ind-adversary means a legitimate one. When we say that f is privacy-preserving, we mean that the ind-advantage of any (legitimate) practical ind-adversary is low.

Privacy-preservation is not, by itself, a demanding property. For example, it is achieved by a constant family such as the one defined by $f(K, x) = 0^L$ for all K, x. We will however want the property for families that are also secure MACs.

PP-MAC $<$ PRF. We claim that a privacy-preserving MAC (PP-MAC) is strictly weaker than a PRF, in the sense that any PRF is (a secure MAC [6,9] and) privacy-preserving, but not vice-versa. This means that when (below) we assume that a compression function h is a PP-MAC, we are indeed assuming less of it than that it is a PRF. Let us now provide some details about the claims made above. First, the following is the formal statement corresponding to the claim that any PRF is privacy-preserving:

Proposition 1. *Let* $f \colon \{0,1\}^k \times \{0,1\}^l \to \{0,1\}^L$ *be a family of functions, and* A_{ind} *an ind-adversary against it that makes at most q oracle queries and has time-complexity at most t. Then there is a prf-adversary A_{prf} against f such that* $\mathbf{Adv}_f^{\mathrm{ind}}(A_{\mathrm{ind}}) \leq 2 \cdot \mathbf{Adv}_f^{\mathrm{prf}}(A_{\mathrm{prf}})$. *Furthermore, A_{prf} makes at most q oracle queries and has time-complexity at most t.* ∎

The proof is a simple exercise and is omitted. Next we explain why a PP-MAC need not be a PRF. The reason (or one reason) is that if the output of a family of functions has some structure, for example always ending in a 0 bit, it would disqualify the family as a PRF but need not preclude its being a PP-MAC. To make this more precise, let $f \colon \{0,1\}^k \times \{0,1\}^l \to \{0,1\}^L$ be a PP-MAC. Define $g \colon \{0,1\}^k \times \{0,1\}^l \to \{0,1\}^{L+1}$ by $g(K, x) = f(K, x) \| 0$ for all $K \in \{0,1\}^k$ and $x \in \{0,1\}^l$. Then g is also a PP-MAC, but is clearly not a PRF.

RESULTS. The following implies that if h is a PP-MAC and F is cAU then their composition hF is a secure MAC.

Lemma 3. *Let* $B = \{0,1\}^b$. *Let* $h \colon \{0,1\}^c \times B \to \{0,1\}^c$ *and* $F \colon \{0,1\}^k \times D \to B$ *be families of functions, and let* $hF \colon \{0,1\}^{c+k} \times D \to \{0,1\}^c$ *be defined by*

$$hF(K_{\mathrm{out}} \| K_{\mathrm{in}}, M) = h(K_{\mathrm{out}}, F(K_{\mathrm{in}}, M))$$

for all $K_{\mathrm{out}} \in \{0,1\}^c, K_{\mathrm{in}} \in \{0,1\}^k$ *and* $M \in D$. *Let* A_{hF} *be a mac-adversary against hF that makes at most q_{mac} mac queries and at most q_{vf} verification queries, with the messages in each of these queries being of length at most n.*

Suppose A_{hF} has time-complexity at most t. Let $q = q_{mac} + q_{vf}$ and assume $2 \leq q < 2^b$. Then there exists a mac-adversary A_1 against h, an ind-adversary A_2 against h, and an au-adversary A_F against F such that

$$\mathbf{Adv}_{hF}^{mac}(A_{hF}) \leq \mathbf{Adv}_h^{mac}(A_1) + \mathbf{Adv}_h^{ind}(A_2) + \binom{q}{2} \cdot \mathbf{Adv}_F^{au}(A_F) . \quad (5)$$

Furthermore, A_1 makes at most q_{mac} mac queries and at most q_{vf} verification queries and has time-complexity at most t; A_2 makes at most q oracle queries and has time-complexity at most t; and A_F outputs messages of length at most n, makes 2 oracle queries, and has time-complexity $O(T_F(n))$, where $T_F(n)$ is the time to compute F on an n-bit input. ∎

The proof is in [2]. As a corollary we have the following, which says that if h is a PP-MAC and h^* is cAU then GNMAC is a secure MAC.

Theorem 2. *Assume $b \geq c$ and let $B = \{0,1\}^b$. Let $h\colon \{0,1\}^c \times B \to \{0,1\}^c$ be a family of functions and let $\mathsf{fpad} \in \{0,1\}^{b-c}$ be a fixed padding string. Let $\mathsf{GNMAC}\colon \{0,1\}^{2c} \times B^+ \to \{0,1\}^c$ be defined by*

$$\mathsf{GNMAC}(K_{out}\|K_{in}, M) = h(K_{out}, h^*(K_{in}, M)\|\mathsf{fpad})$$

for all $K_{out}, K_{in} \in \{0,1\}^c$ and $M \in B^+$. Let A_{GNMAC} be a mac-adversary against GNMAC that makes at most q_{mac} mac queries and at most q_{vf} verification queries, with the messages in each of these queries being of at most m blocks. Suppose A_{GNMAC} has time-complexity at most t. Let $q = q_{mac} + q_{vf}$ and assume $2 \leq q < 2^b$. Then there exists a mac-adversary A_1 against h, an ind-adversary A_2 against h, and an au-adversary A^* against h^* such that

$$\mathbf{Adv}_{\mathsf{GNMAC}}^{mac}(A_{\mathsf{GNMAC}}) \leq \mathbf{Adv}_h^{mac}(A_1) + \mathbf{Adv}_h^{ind}(A_2) + \binom{q}{2} \cdot \mathbf{Adv}_{h^*}^{au}(A^*) . \quad (6)$$

Furthermore, A_1 makes at most q_{mac} mac queries and at most q_{vf} verification queries and has time-complexity at most t; A_2 makes at most q oracle queries and has time-complexity at most t; and A^* outputs messages of at most m blocks, makes 2 oracle queries, and has time-complexity $O(mT_h)$, where T_h is the time for one computation of h. ∎

We remark that Lemma 3 can be extended to show that hF is not only a MAC but itself privacy-preserving. (This assumes h is privacy-preserving and F is cAU. We do not prove this here.) This implies that GNMAC is privacy-preserving as long as h is privacy-preserving and h^* is cAU. This is potentially useful because it may be possible to show that a PP-MAC is sufficient to ensure security in some applications where HMAC is currently assumed to be a PRF.

5 Security of HMAC

In this section we show how our security results about NMAC lift to corresponding ones about HMAC. We begin by recalling the observation of [3] as to how this works for HMAC with two independent keys, and then discuss how to extend this to the single-keyed version of HMAC.

THE CONSTRUCTS. Let $h\colon \{0,1\}^c \times \{0,1\}^b \to \{0,1\}^c$ as usual denote the compression function. Let pad be the padding function as described in Section 3, so that $s^* = s\|\mathsf{pad}(|s|) \in B^+$ for any string s. Recall that the cryptographic hash function H associated to h is defined by $H(M) = h^*(\mathrm{IV}, M^*)$, where IV is a c-bit initial vector that is fixed as part of the description of H and M is a string of any length up to some maximum length that is related to pad. (This maximum length is 2^{64} for current hash functions.) Then $\mathsf{HMAC}(K_{\mathrm{out}}\|K_{\mathrm{in}}, M) = H(K_{\mathrm{out}}\|H(K_{\mathrm{in}}\|M))$, where $K_{\mathrm{out}}, K_{\mathrm{in}} \in \{0,1\}^b$. If we write this out in terms of h^* alone we get

$$\mathsf{HMAC}(K_{\mathrm{out}}\|K_{\mathrm{in}}, M) = h^*(\mathrm{IV},\ K_{\mathrm{out}}\|h^*(\mathrm{IV}, K_{\mathrm{in}}\|M\|\mathsf{pad}(b+|M|))\|\mathsf{pad}(b+c)) .$$

As with NMAC, the details of the padding conventions are not important to the security of HMAC as a PRF, and we will consider the more general construct $\mathsf{GHMAC}\colon \{0,1\}^{2b} \times B^+ \to \{0,1\}^c$ defined by

$$\mathsf{GHMAC}(K_{\mathrm{out}}\|K_{\mathrm{in}}, M) = h^*(\mathrm{IV},\ K_{\mathrm{out}} \,\|\, h^*(\mathrm{IV}, K_{\mathrm{in}}\|M) \,\|\, \mathsf{fpad}) \qquad (7)$$

for all $K_{\mathrm{out}}, K_{\mathrm{in}} \in \{0,1\}^b$ and all $M \in B^+$. Here $\mathrm{IV} \in \{0,1\}^c$ and $\mathsf{fpad} \in \{0,1\}^{b-c}$ are fixed strings. HMAC is a special case, via $\mathsf{HMAC}(K_{\mathrm{out}}\|K_{\mathrm{in}}, M) = \mathsf{GHMAC}(M\|\mathsf{pad}(b+|M|))$ with $\mathsf{fpad} = \mathsf{pad}(b+c)$, and thus security properties of GHMAC (as a PRF or MAC) are inherited by HMAC, allowing us to focus on the former.

THE DUAL FAMILY. To state the results, it is useful to define $\overline{h}\colon \{0,1\}^b \times \{0,1\}^c \to \{0,1\}^c$, the *dual* of family h, by $\overline{h}(x,y) = h(y,x)$. The assumption that h is a PRF when keyed by its data input is formally captured by the assumption that $\overline{h}$ is a PRF.

SECURITY OF GHMAC. Let $K'_{\mathrm{out}} = h(\mathrm{IV}, K_{\mathrm{out}})$ and $K'_{\mathrm{in}} = h(\mathrm{IV}, K_{\mathrm{in}})$. The observation of [3] is that

$$\mathsf{GHMAC}(K_{\mathrm{out}}\|K_{\mathrm{in}}, M) = h(K'_{\mathrm{out}}, h^*(K'_{\mathrm{in}}, M)\|\mathsf{fpad})$$
$$= \mathsf{GNMAC}(K'_{\mathrm{out}}\|K'_{\mathrm{in}}, M) . \qquad (8)$$

This effectively reduces the security of GHMAC to GNMAC. Namely, if $\overline{h}$ is a PRF and $K_{\mathrm{out}}, K_{\mathrm{in}}$ are chosen at random, then $K'_{\mathrm{out}}, K'_{\mathrm{in}}$ will be computationally close to random. Now (8) implies that if GNMAC is a PRF then so is GHMAC. The formal statement follows.

Lemma 4. *Assume $b \geq c$ and let $B = \{0,1\}^b$. Let $h\colon \{0,1\}^c \times B \to \{0,1\}^c$ be a family of functions. Let $\mathsf{fpad} \in \{0,1\}^{b-c}$ be a fixed padding string and $\mathrm{IV} \in \{0,1\}^c$ a fixed initial vector. Let $\mathsf{GHMAC}\colon \{0,1\}^{2b} \times B^+ \to \{0,1\}^c$ be defined by (7) above. Let A be a prf-adversary against GHMAC that has time-complexity at most t. Then there exists a prf-adversary $A_{\overline{h}}$ against $\overline{h}$ such that*

$$\mathbf{Adv}^{\mathrm{prf}}_{\mathsf{GHMAC}}(A) \leq 2 \cdot \mathbf{Adv}^{\mathrm{prf}}_{\overline{h}}(A_{\overline{h}}) + \mathbf{Adv}^{\mathrm{prf}}_{\mathsf{GNMAC}}(A) .$$

Furthermore, $A_{\overline{h}}$ makes only 1 oracle query, this being IV, and has time-complexity at most t. ∎

The proof is simple and can be found in [2]. Combining this with Theorem 1 yields the result that GHMAC is a PRF assuming $h, \overline{h}$ are both PRFs. Note that the PRF assumption on $\overline{h}$ is mild because $A_{\overline{h}}$ makes only one oracle query.

SINGLE-KEYED HMAC. HMAC, GHMAC as described and analyzed above above use two keys that are assumed to be chosen independently at random. However, HMAC is in fact usually implemented with these keys derived from a single b-bit key. Here we provide the first security proofs for the actually-implemented single-key version of HMAC.

Specifically, let opad, ipad $\in \{0,1\}^b$ be distinct, fixed and known constants. (Their particular values can be found in [3] and are not important here.) Then the single-key version of HMAC is defined by

$$\mathsf{HMAC\text{-}1}(K, M) = \mathsf{HMAC}(K \oplus \mathsf{opad} \| K \oplus \mathsf{ipad}, M) .$$

As before, we look at this as a special case of a more general construct, namely GHMAC-1: $\{0,1\}^b \times B^+ \to \{0,1\}^c$, defined by

$$\mathsf{GHMAC\text{-}1}(K, M) = \mathsf{GHMAC}(K \oplus \mathsf{opad} \| K \oplus \mathsf{ipad}, M) \tag{9}$$

for all $K \in \{0,1\}^b$ and all $M \in B^+$. We now focus on GHMAC-1. We will show that GHMAC-1 inherits the security of GNMAC as long as $\overline{h}$ is a PRF against an appropriate class of related key attacks. In such an attack, the adversary can obtain input-output examples of h under keys related to the target key. Let us recall the formal definitions following [7].

A related-key attack on a family of functions $\overline{h} \colon \{0,1\}^b \times \{0,1\}^c \to \{0,1\}^c$ is parameterized by a set $\Phi \subseteq \mathsf{Maps}(\{0,1\}^b, \{0,1\}^b)$ of *key-derivation* functions. We define the function RK: $\Phi \times \{0,1\}^b \to \{0,1\}^b$ by $\mathrm{RK}(\phi, K) = \phi(K)$ for all $\phi \in \Phi$ and $K \in \{0,1\}^b$. A rka-adversary $A_{\overline{h}}$ may make an oracle query of the form ϕ, x where $\phi \in \Phi$ and $x \in \{0,1\}^c$. Its rka-advantage is defined by

$$\mathbf{Adv}^{\mathrm{rka}}_{\overline{h}, \Phi}(A_{\overline{h}}) = \Pr\left[A_{\overline{h}}^{\overline{h}(\mathrm{RK}(\cdot, K), \cdot)} \Rightarrow 1 \right] - \Pr\left[A_{\overline{h}}^{G(\mathrm{RK}(\cdot, K), \cdot)} \Rightarrow 1 \right] .$$

In the first case, K is chosen at random from $\{0,1\}^b$ and the reply to query ϕ, x of $A_{\overline{h}}$ is $\overline{h}(\phi(K), x)$. In the second case, $G \xleftarrow{\$} \mathsf{Maps}(\{0,1\}^b \times \{0,1\}^c, \{0,1\}^c)$ and $K \xleftarrow{\$} \{0,1\}^b$, and the reply to query ϕ, x of $A_{\overline{h}}$ is $G(\phi(K), x)$. For any string $s \in \{0,1\}^b$ let $\Delta_s \colon \{0,1\}^b \to \{0,1\}^b$ be defined by $\Delta_s(K) = K \oplus s$.

Lemma 5. *Assume $b \geq c$ and let $B = \{0,1\}^b$. Let $h \colon \{0,1\}^c \times B \to \{0,1\}^c$ be a family of functions. Let $\mathsf{fpad} \in \{0,1\}^{b-c}$ be a fixed padding string, IV $\in \{0,1\}^c$ a fixed initial vector, and opad, ipad $\in \{0,1\}^b$ fixed, distinct strings. Let GHMAC-1: $\{0,1\}^b \times B^+ \to \{0,1\}^c$ be defined by (9) above. Let $\Phi = \{\Delta_{\mathsf{opad}}, \Delta_{\mathsf{ipad}}\}$. Let A be a prf-adversary against GHMAC-1 that has time-complexity at most t. Then there exists a rka-adversary $A_{\overline{h}}$ against $\overline{h}$ such that*

$$\mathbf{Adv}^{\mathrm{prf}}_{\mathsf{GHMAC\text{-}1}}(A) \leq \mathbf{Adv}^{\mathrm{rka}}_{\overline{h}, \Phi}(A_{\overline{h}}) + \mathbf{Adv}^{\mathrm{prf}}_{\mathsf{GNMAC}}(A) .$$

Furthermore, $A_{\overline{h}}$ makes 2 oracle queries, these being $\Delta_{\mathsf{opad}}, \mathrm{IV}$ and $\Delta_{\mathsf{ipad}}, \mathrm{IV}$, and has time-complexity at most t. ∎

The proof is simple and can be found in [2]. Combining this with Theorem 1 yields the result that GHMAC-1 is a PRF assuming h is a PRF and $\overline{h}$ is a PRF under Φ-restricted related-key attacks, where Φ is as in Lemma 5. We remark that Φ is a small set of simple functions, which is important because it is shown in [7] that if Φ is too rich then no family can be a PRF under Φ-restricted related-key attacks. Furthermore, the assumption on $\overline{h}$ is rendered milder by the fact that $A_{\overline{h}}$ makes only two oracle queries, in both of which the message is the same, namely is the initial vector.

LIFTING THE RESULTS OF SECTION 4. The procedure above to lift the NMAC results of Section 3 to HMAC applies also to lift the results of Section 4 to HMAC. Specifically, if h is a PP-MAC, h^* is AU and $\overline{h}$ is a PRF then GHMAC is a (privacy-preserving) MAC. Also if h is a PP-MAC, h^* is AU and $\overline{h}$ is a PRF under Φ-restricted related-key attacks, with Φ as in Lemma 5, then GHMAC-1 is a (privacy-preserving) MAC. Note that the assumption on $\overline{h}$ continues to be that it is a PRF or PRF against Φ-restricted related-key attacks. (Namely, this has not been reduced to its being a PP-MAC.) This assumption is however mild in this context since (as indicated by Lemmas 5 and 4) it need only hold with respect to adversaries that make very few queries and these of a very specific type.

REMARKS. Let h: $\{0,1\}^{128} \times \{0,1\}^{512} \to \{0,1\}^{128}$ denote the compression function of MD5 [28]. An attack by den Boer and Bosselaers [15] finds values x_0, x_1, K such that $h(x_0, K) = h(x_1, K)$ but $x_0 \neq x_1$. In a personal communication, Rijmen has said that it seems possible to extend this to an attack that finds such x_0, x_1 even when K is unknown. If so, this might translate into the following attack showing h is not a PRF when keyed by its data input. (That is, $\overline{h}$ is not a PRF.) Given an oracle g: $\{0,1\}^{128} \to \{0,1\}^{128}$, the attacker would find x_0, x_1 and obtain $y_0 = g(x_0)$ and $y_1 = g(x_1)$ from the oracle. It would return 1 if $y_0 = y_1$ and 0 otherwise. How does this impact the above, where we are assuming $\overline{h}$ is a PRF? Interestingly, the actual PRF assumptions we need on $\overline{h}$ are so weak that even such an attack does not break them. In Lemma 4, we need $\overline{h}$ to be a PRF only against adversaries that make just one oracle query. (Because $A_{\overline{h}}$ makes only one query.) But the attack above makes two queries. On the other hand, in Lemma 5, we need $\overline{h}$ to be a related-key PRF only against adversaries that make two related-key queries in both of which the 128-bit message for $\overline{h}$ is the same, this value being the initial vector used by the hash function. Furthermore, the related key functions must be $\Delta_{\mathsf{opad}}, \Delta_{\mathsf{ipad}}$. The above-mentioned attack, however, uses two different messages x_0, x_1 and calls the oracle under the original key rather than the related keys. In summary, the attack does not violate the assumptions made in either of the lemmas.

Acknowledgments

Thanks to Ran Canetti, Hugo Krawczyk, Mridul Nandi, Vincent Rijmen, Phillip Rogaway, Victor Shoup, Paul Van Oorschot and the Crypto 2006 PC for comments and references. Author supported in part by NSF grants ANR-0129617, CCR-0208842 and CNS-0524765 and by an IBM Faculty Partnership Development Award.

References

1. American National Standards Institution. ANSI X9.71, Keyed hash message authentication code, 2000.
2. M. Bellare. New Proofs for NMAC and HMAC: Security without Collision-Resistance. Full version of this paper. *Cryptology ePrint Archive*: Report 2006/043, 2006.
3. M. Bellare, R. Canetti and H. Krawczyk. Keying hash functions for message authentication. *Advances in Cryptology – CRYPTO '96*, Lecture Notes in Computer Science Vol. 1109, N. Koblitz ed., Springer-Verlag, 1996.
4. M. Bellare, R. Canetti and H. Krawczyk. Pseudorandom functions revisited: The cascade construction and its concrete security. `http://www-cse.ucsd.edu/users/mihir`. (Preliminary version in *Proceedings of the 37th Symposium on Foundations of Computer Science*, IEEE, 1996.)
5. M. Bellare, A. Desai, E. Jokipii and P. Rogaway. A concrete security treatment of symmetric encryption. *Proceedings of the 38th Symposium on Foundations of Computer Science*, IEEE, 1997.
6. M. Bellare, J. Kilian and P. Rogaway. The security of the cipher block chaining message authentication code. *Journal of Computer and System Sciences*, Vol. 61, No. 3, Dec 2000, pp. 362–399.
7. M. Bellare and T. Kohno. A theoretical treatment of related-key attacks: RKA-PRPs, RKA-PRFs, and applications. *Advances in Cryptology – EUROCRYPT '03*, Lecture Notes in Computer Science Vol. 2656, E. Biham ed., Springer-Verlag, 2003.
8. M. Bellare, C. Namprempre and T. Kohno. Authenticated Encryption in SSH: Provably Fixing the SSH Binary Packet Protocol. *ACM Transactions on Information and System Security (TISSEC)*, Vol. 7, Iss. 2, May 2004, pp. 206–241.
9. M. Bellare, O. Goldreich and A. Mityagin. The power of verification queries in message authentication and authenticated encryption. *Cryptology ePrint Archive*: Report 2004/309, 2004.
10. M. Bellare and P. Rogaway. The game-playing technique and its application to triple encryption. *Cryptology ePrint Archive*: Report 2004/331, 2004.
11. J. BLACK, S. HALEVI, H. KRAWCZYK, T. KROVETZ AND P. ROGAWAY. UMAC: Fast and Secure Message Authentication. *Advances in Cryptology – CRYPTO '99*, Lecture Notes in Computer Science Vol. 1666, M. Wiener ed., Springer-Verlag, 1999.
12. J. Black and P. Rogaway. CBC MACs for arbitrary-length messages: The three-key constructions. *Advances in Cryptology – CRYPTO '00*, Lecture Notes in Computer Science Vol. 1880, M. Bellare ed., Springer-Verlag, 2000.
13. L. CARTER AND M. WEGMAN. Universal classes of hash functions. *Journal of Computer and System Sciences*, Vol. 18, No. 2, 1979, pp. 143–154.
14. I. Damgård. A design principle for hash functions. *Advances in Cryptology – CRYPTO '89*, Lecture Notes in Computer Science Vol. 435, G. Brassard ed., Springer-Verlag, 1989.
15. B. den Boer and A. Bosselaers. Collisions for the compression function of MD5. *Advances in Cryptology – EUROCRYPT '93*, Lecture Notes in Computer Science Vol. 765, T. Helleseth ed., Springer-Verlag, 1993.
16. T. Dierks and C. Allen. The TLS protocol. Internet RFC 2246, 1999.

17. H. Dobbertin, A. Bosselaers and B. Preneel. RIPEMD-160: A strengthened version of RIPEMD. *Fast Software Encryption '96*, Lecture Notes in Computer Science Vol. 1039, D. Gollmann ed., Springer-Verlag, 1996.
18. Y. Dodis, R. Gennaro, J. Håstad, H. Krawczyk, and T. Rabin. Randomness extraction and key derivation using the CBC, Cascade, and HMAC modes. *Advances in Cryptology – CRYPTO '04*, Lecture Notes in Computer Science Vol. 3152, M. Franklin ed., Springer-Verlag, 2004.
19. O. GOLDREICH, S. GOLDWASSER AND S. MICALI. How to construct random functions. *Journal of the ACM*, Vol. 33, No. 4, 1986, pp. 210–217.
20. D. Harkins and D. Carrel. The Internet Key Exchange (IKE). Internet RFC 2409, 1998.
21. S. Hirose. A note on the strength of weak collision resistance. *IEICE Transactions on Fundamentals*, Vol. E87-A, No. 5, May 2004, pp. 1092–1097.
22. H. Krawczyk, M. Bellare and R. Canetti. HMAC: Keyed-hashing for message authentication. Internet RFC 2104, 1997.
23. R. Merkle. One-way hash functions and DES. *Advances in Cryptology – CRYPTO '89*, Lecture Notes in Computer Science Vol. 435, G. Brassard ed., Springer-Verlag, 1989.
24. D. M'Raihi, M. Bellare, F. Hoornaert, D. Naccache, O. Ranen. HOTP: An HMAC-based one time password algorithm. Internet RFC 4226, December 2005.
25. National Institute of Standards and Technology. The keyed-hash message authentication code (HMAC). FIPS PUB 198, March 2002.
26. National Institute of Standards and Technology. Secure hash standard. FIPS PUB 180-2, August 2000.
27. B. Preneel and P. van Oorschot. On the security of iterated message authentication codes. *IEEE Transactions on Information Theory*, Vol. 45, No. 1, January 1999, pp. 188–199. (Preliminary version, entitled "MD-x MAC and building fast MACs from hash functions," in CRYPTO 95.)
28. R. RIVEST. The MD5 message-digest algorithm. Internet RFC 1321, April 1992.
29. V. Shoup. Sequences of games: A tool for taming complexity in security proofs. *Cryptology ePrint Archive*: Report 2004/332, 2004.
30. D. Stinson. Universal hashing and authentication codes. *Designs, Codes and Cryptography*, Vol. 4, 1994, 369–380.
31. X. Wang, Y. L. Yin and H. Yu. Finding collisions in the full SHA-1. *Advances in Cryptology – CRYPTO '05*, Lecture Notes in Computer Science Vol. 3621 , V. Shoup ed., Springer-Verlag, 2005.
32. X. Wang and H. Yu. How to break MD5 and other hash functions. *Advances in Cryptology – EUROCRYPT '05*, Lecture Notes in Computer Science Vol. 3494, R. Cramer ed., Springer-Verlag, 2005.
33. M. WEGMAN AND L. CARTER. New hash functions and their use in authentication and set equality. *Journal of Computer and System Sciences*, Vol. 22, No. 3, 1981, pp. 265–279.

Author Index

Lecture Notes in Computer Science

For information about Vols. 1–4024

please contact your bookseller or Springer

Vol. 4068: H. Schärfe, P. Hitzler, P. Øhrstrøm (Eds.), Conceptual Structures: Inspiration and Application. XI, 455 pages. 2006. (Sublibrary LNAI).

Vol. 4067: D. Thomas (Ed.), ECOOP 2006 – Object-Oriented Programming. XIV, 527 pages. 2006.

Vol. 4066: A. Rensink, J. Warmer (Eds.), Model Driven Architecture – Foundations and Applications. XII, 392 pages. 2006.

Vol. 4065: P. Perner (Ed.), Advances in Data Mining. XI, 592 pages. 2006. (Sublibrary LNAI).

Vol. 4064: R. Büschkes, P. Laskov (Eds.), Detection of Intrusions and Malware & Vulnerability Assessment. X, 195 pages. 2006.

Vol. 4063: I. Gorton, G.T. Heineman, I. Crnkovic, H.W. Schmidt, J.A. Stafford, C.A. Szyperski, K. Wallnau (Eds.), Component-Based Software Engineering. XI, 394 pages. 2006.

Vol. 4062: G. Wang, J.F. Peters, A. Skowron, Y. Yao (Eds.), Rough Sets and Knowledge Technology. XX, 810 pages. 2006. (Sublibrary LNAI).

Vol. 4061: K. Miesenberger, J. Klaus, W. Zagler, A. Karshmer (Eds.), Computers Helping People with Special Needs. XXIX, 1356 pages. 2006.

Vol. 4060: K. Futatsugi, J.-P. Jouannaud, J. Meseguer (Eds.), Algebra, Meaning, and Computation. XXXVIII, 643 pages. 2006.

Vol. 4059: L. Arge, R. Freivalds (Eds.), Algorithm Theory – SWAT 2006. XII, 436 pages. 2006.

Vol. 4058: L.M. Batten, R. Safavi-Naini (Eds.), Information Security and Privacy. XII, 446 pages. 2006.

Vol. 4057: J.P.W. Pluim, B. Likar, F.A. Gerritsen (Eds.), Biomedical Image Registration. XII, 324 pages. 2006.

Vol. 4056: P. Flocchini, L. Gąsieniec (Eds.), Structural Information and Communication Complexity. X, 357 pages. 2006.

Vol. 4055: J. Lee, J. Shim, S.-g. Lee, C. Bussler, S. Shim (Eds.), Data Engineering Issues in E-Commerce and Services. IX, 290 pages. 2006.

Vol. 4054: A. Horváth, M. Telek (Eds.), Formal Methods and Stochastic Models for Performance Evaluation. VIII, 239 pages. 2006.

Vol. 4053: M. Ikeda, K.D. Ashley, T.-W. Chan (Eds.), Intelligent Tutoring Systems. XXVI, 821 pages. 2006.

Vol. 4052: M. Bugliesi, B. Preneel, V. Sassone, I. Wegener (Eds.), Automata, Languages and Programming, Part II. XXIV, 603 pages. 2006.

Vol. 4051: M. Bugliesi, B. Preneel, V. Sassone, I. Wegener (Eds.), Automata, Languages and Programming, Part I. XXIII, 729 pages. 2006.

Vol. 4049: S. Parsons, N. Maudet, P. Moraitis, I. Rahwan (Eds.), Argumentation in Multi-Agent Systems. XIV, 313 pages. 2006. (Sublibrary LNAI).

Vol. 4048: L. Goble, J.-J.C.. Meyer (Eds.), Deontic Logic and Artificial Normative Systems. X, 273 pages. 2006. (Sublibrary LNAI).

Vol. 4047: M. Robshaw (Ed.), Fast Software Encryption. XI, 434 pages. 2006.

Vol. 4046: S.M. Astley, M. Brady, C. Rose, R. Zwiggelaar (Eds.), Digital Mammography. XVI, 654 pages. 2006.

Vol. 4045: D. Barker-Plummer, R. Cox, N. Swoboda (Eds.), Diagrammatic Representation and Inference. XII, 301 pages. 2006. (Sublibrary LNAI).

Vol. 4044: P. Abrahamsson, M. Marchesi, G. Succi (Eds.), Extreme Programming and Agile Processes in Software Engineering. XII, 230 pages. 2006.

Vol. 4043: A.S. Atzeni, A. Lioy (Eds.), Public Key Infrastructure. XI, 261 pages. 2006.

Vol. 4042: D. Bell, J. Hong (Eds.), Flexible and Efficient Information Handling. XVI, 296 pages. 2006.

Vol. 4041: S.-W. Cheng, C.K. Poon (Eds.), Algorithmic Aspects in Information and Management. XI, 395 pages. 2006.

Vol. 4040: R. Reulke, U. Eckardt, B. Flach, U. Knauer, K. Polthier (Eds.), Combinatorial Image Analysis. XII, 482 pages. 2006.

Vol. 4039: M. Morisio (Ed.), Reuse of Off-the-Shelf Components. XIII, 444 pages. 2006.

Vol. 4038: P. Ciancarini, H. Wiklicky (Eds.), Coordination Models and Languages. VIII, 299 pages. 2006.

Vol. 4037: R. Gorrieri, H. Wehrheim (Eds.), Formal Methods for Open Object-Based Distributed Systems. XVII, 474 pages. 2006.

Vol. 4036: O. H. Ibarra, Z. Dang (Eds.), Developments in Language Theory. XII, 456 pages. 2006.

Vol. 4035: T. Nishita, Q. Peng, H.-P. Seidel (Eds.), Advances in Computer Graphics. XX, 771 pages. 2006.

Vol. 4034: J. Münch, M. Vierimaa (Eds.), Product-Focused Software Process Improvement. XVII, 474 pages. 2006.

Vol. 4033: B. Stiller, P. Reichl, B. Tuffin (Eds.), Performability Has its Price. X, 103 pages. 2006.

Vol. 4032: O. Etzion, T. Kuflik, A. Motro (Eds.), Next Generation Information Technologies and Systems. XIII, 365 pages. 2006.

Vol. 4031: M. Ali, R. Dapoigny (Eds.), Advances in Applied Artificial Intelligence. XXIII, 1353 pages. 2006. (Sublibrary LNAI).

Vol. 4029: L. Rutkowski, R. Tadeusiewicz, L.A. Zadeh, J.M. Zurada (Eds.), Artificial Intelligence and Soft Computing – ICAISC 2006. XXI, 1235 pages. 2006. (Sublibrary LNAI).

Vol. 4028: J. Kohlas, B. Meyer, A. Schiper (Eds.), Dependable Systems: Software, Computing, Networks. XII, 295 pages. 2006.

Vol. 4027: H.L. Larsen, G. Pasi, D. Ortiz-Arroyo, T. Andreasen, H. Christiansen (Eds.), Flexible Query Answering Systems. XVIII, 714 pages. 2006. (Sublibrary LNAI).

Vol. 4026: P.B. Gibbons, T. Abdelzaher, J. Aspnes, R. Rao (Eds.), Distributed Computing in Sensor Systems. XIV, 566 pages. 2006.

Vol. 4025: F. Eliassen, A. Montresor (Eds.), Distributed Applications and Interoperable Systems. XI, 355 pages. 2006.

Lecture Notes in Computer Science

For information about Vols. 1–4024

please contact your bookseller or Springer